ECONOMIC AND SOCIAL COMMISSION
FOR ASIA AND THE PACIFIC

ASIAN DEVELOPMENT BANK

State of the Environment in Asia and the Pacific

2000

UNITED NATIONS
New York, 2000

ST/ESCAP/2087

UNITED NATIONS PUBLICATION
Sales No. E.01.II.F.8
Copyright © United Nations 2000
ISBN: 92-1-120019-9

The designations employed and the presentation of the material in this publication do not imply the expression of any opinion whatsoever on the part of the Secretariat of the United Nations concerning the legal status of any country, territory, city or area, or of its authorities, or concerning the delimitation of its frontiers or boundaries.

With regard to figures 22.3 and 22.5, the dotted line represents approximately the Line of Control in Jammu and Kashmir agreed upon by India and Pakistan. The final status of Jammu and Kashmir has not been agreed upon by the parties.

This document has been issued without formal editing.

PREFACE

The State of the Environment in Asia and the Pacific 2000 is the fourth in a series of reports published every five years on environmental trends in the region. It reviews the dynamics of the region's environmental conditions and the status of national and regional response to the changing environmental situation. Its objective is to provide a general evaluation of the quality of environment and emerging sustainable development trends, so that major issues can be brought to the attention of governments, intergovernmental organizations, the private sector, non-governmental and community-based organizations and other public interest groups in the region.

While providing a series of statistics on environmental conditions and trends and emerging issues, the present report critically analyses the aggregate effect of policy response on environment and sustainable development across the region. It also illustrates successful local, national, regional and international initiatives and best practices which may be usefully replicated both inside and outside the region. Finally, the report identifies key challenges and areas in which urgent action is needed, particularly in regard to regional and international collaboration efforts.

The report strikes a sombre note on environmental degradation in eroded croplands, falling water tables, declining forests and biodiversity, and increased pollution. Apart from this gloomy scenario, the positive trends are the decline in birth rates, increased life expectancy, improved nutritional and poverty levels, growing awareness and public participation and the enhanced role of governments and the private sector in the promotion of sustainable development. The principal environmental challenges in the twenty-first century are promoting growth while safeguarding natural resources, promoting eco-efficiency, countering the negative effects of globalization and enhancing public participation while empowering communities to become custodians of the environment.

An important lesson that can be discerned from the report is that environmental initiatives in the Asian and Pacific region cannot be effective unless they are implemented in the context of development policies. Failure to consider the environmental implications of developments in agriculture, urbanization and industrialization is having serious environmental consequences. Overpopulation, poverty and difficulties in meeting basic needs are also aggravating factors. Lack of appropriate planning and foresight in the development process will further exacerbate the degradation of the environment.

These interdependent and interconnected environmental challenges for the twenty-first century contain an important inherent message for the countries of the Asian and Pacific region. That message is to broaden the vision of economic growth to the holistic perspective of sustainable development in the new century.

This report is the result of close collaboration between ESCAP and the Asian Development Bank. We wish to acknowledge the cooperation extended by the member governments of ESCAP, United Nations agencies and other international organizations in the preparation of the report, in particular the valuable support of the United Nations Environment Programme, the World Health Organization and the United Nations Children's Fund, which provided data, information, comments and research assistance. The generous financial support of the Government of Japan towards the preparation of the report is deeply appreciated.

Tadao Chino
Tadao Chino
President
Asian Development Bank

Kim Hak-Su
Executive Secretary
Economic and Social Commission
for Asia and the Pacific

CONTENTS

PREFACE — iii
LIST OF BOXES — vi
LIST OF FIGURES — ix
LIST OF TABLES — xii
ABBREVIATIONS — xvii
INTRODUCTION — xxvii

PART I — ENVIRONMENTAL CONDITIONS

CHAPTER 1	LAND	2
CHAPTER 2	FORESTS	24
CHAPTER 3	BIODIVERSITY	42
CHAPTER 4	INLAND WATER	74
CHAPTER 5	COASTAL AND MARINE	98
CHAPTER 6	ATMOSPHERE AND CLIMATE	122

PART II — HUMAN ECOSYSTEMS-EMERGING ISSUES

CHAPTER 7	URBAN ENVIRONMENT	146
CHAPTER 8	WASTE	168
CHAPTER 9	POVERTY AND ENVIRONMENT	196
CHAPTER 10	FOOD SECURITY	214

PART III — NATIONAL AND REGIONAL RESPONSES

CHAPTER 11	INSTITUTIONS AND LEGISLATION	238
CHAPTER 12	MECHANISMS AND METHODS	258
CHAPTER 13	PRIVATE SECTOR	280
CHAPTER 14	MAJOR GROUPS	300
CHAPTER 15	EDUCATION, INFORMATION AND AWARENESS	316

PART IV — SUBREGIONAL OUTLOOK

CHAPTER 16	SOUTH ASIA	342
CHAPTER 17	SOUTHEAST ASIA	352
CHAPTER 18	SOUTH PACIFIC	362
CHAPTER 19	NORTHEAST ASIA	372
CHAPTER 20	CENTRAL ASIA	380

PART V — FUTURE OUTLOOK

| CHAPTER 21 | GLOBAL AND REGIONAL ISSUES | 390 |
| CHAPTER 22 | ASIA AND THE PACIFIC INTO THE 21st CENTURY | 424 |

REFERENCES — 443

LIST OF BOXES

PART I ENVIRONMENTAL CONDITIONS

		Page
Box 1.1	The Aral Sea Basin	7
Box 1.2	Land Degradation in Kazakhstan	8
Box 1.3	Technique for Establishing a Shelterbelt System Around an Oasis and its Benefit in Xinjiang, People's Republic of China	9
Box 1.4	Mining Land Rehabilitation in Australia	13
Box 1.5	Traditional Rotational Use of Pastureland in Mongolia	15
Box 1.6	Cost of Australian Dryland Salinity	17
Box 1.7	Implementation of Convention to Combat Desertification (CCD) in Asia and the Pacific	20
Box 2.1	The Loss of Large Frontier Forests	28
Box 2.2	Forest Plantations in Turkmenistan	29
Box 2.3	Adapting Tenure to Encourage Forestry	36
Box 2.4	Urban Forestry	38
Box 2.5	Forest Certification	40
Box 3.1	Biodiversity of Papua New Guinea	44
Box 3.2	Wetlands on Fire	48
Box 3.3	Risk to Food Security as a Result of Monoculture and Biodiversity Loss	55
Box 3.4	Public Acceptance of Transgenic Crops in Asia and the Pacific: A Case of Transgenic Rice in the Philippines	57
Box 3.5	Patenting of Genetic Materials and Plight of Developing Countries in Asia and the Pacific	58
Box 3.6	Conserving Biodiversity Through Eco-development	61
Box 3.7	The Biotrade Initiative: A New Integrated Approach to Biodiversity Conservation	68
Box 3.8	Conservation of Migratory Species in Mongolia	70
Box 4.1	Water Quality in People's Republic of China	84
Box 4.2	Millions in Bangladesh Face Slow Arsenic Poisoning	86
Box 4.3	Pasig River Rehabilitation Programme in the Philippines	90
Box 4.4	Prevention of Water Pollution Accidents in the Republic of Korea	91
Box 4.5	From Effluent to Affluent: A Case in Thailand	92
Box 4.6	Ganges Water Sharing Treaty	94
Box 5.1	Australia's National Plan to Combat Pollution of the Sea by Oil	101
Box 5.2	REEF CHECK-A Global Coral Reef Monitoring System	104
Box 5.3	The Evolution of Fisheries Management in New Zealand	111
Box 5.4	Coastal Zone Management in Sri Lanka	113
Box 5.5	The Bantay Puerto Programme in the Philippines	114
Box 5.6	The Funafuti Conservation Area Project in Tuvalu	116
Box 5.7	Aquaculture and the Environment in the Republic of Korea	117
Box 6.1	The Montreal Protocol: A Successful Example of International Cooperation	129
Box 6.2	Anthropogenic Emission and Climate Change	130
Box 6.3	El Nino and Climate Change: Extreme Natural Events	131
Box 6.4	Global Warming: Threat to Coral Reefs	132
Box 6.5	Illustrating the Impacts of Sea-Level Rise: Bangladesh	133
Box 6.6	Measures by Local Government in Japan to Combat Climate Change	141

LIST OF BOXES (continued)

Page

PART II HUMAN ECOSYSTEMS-EMERGING ISSUES

Box 7.1	Costs of Urban Air Pollution	153
Box 7.2	Urban Environmental Protection: The Case of Hong Kong, China	155
Box 7.3	Policy Objectives of the Habitat II Agenda	160
Box 7.4	Environmental Responsibilities in Local Government: The Case of Kitakyushu	161
Box 8.1	Recycling: Fortunes and Costs	179
Box 8.2	PVC Waste Export to Asia Despite International Agreement	184
Box 8.3	Waste to Profits: Some Success Stories of Waste Minimization through Cleaner Production (CP) Programmes	187
Box 8.4	Private Sector Initiative towards Urban Waste Management in Pakistan	188
Box 8.5	Volume-based Collection Fee System for Municipal Waste in the Republic of Korea	190
Box 8.6	Integrated Solid Waste Management – Key Strategic Considerations	194
Box 9.1	A Project Addressing Desertification through Local Solutions – the "Barefoot Approach" in Rajasthan, India	203
Box 9.2	Community-based Approaches to Improve the Living Environment and Address Poverty in Poor Urban Settlements – The Orangi Pilot Project in Pakistan	206
Box 9.3	The Interrelationship: Environmental Deterioration, Poverty and Natural Disaster	207
Box 9.4	Fighting Poverty – The Grameen Bank in Bangladesh	212
Box 10.1	The 1990s Agricultural Crisis in Democratic People's Republic of Korea	222
Box 10.2	Population Growth, Declining Arable Land, and Environmental Degradation in Nepal	228
Box 10.3	Policy Reform and Sustainable Agriculture in Viet Nam	229
Box 10.4	Agricultural Water Pricing Policies as Market-based Instruments in People's Republic of China	230
Box 10.5	Food Security in Cities: Urban Agriculture in Istanbul	231

PART III NATIONAL AND REGIONAL RESPONSES

Box 11.1	Philippine Council for Sustainable Development (PCSD): A Mechanism for Integrating Environmental Considerations into Overall Economic Policies in the Philippines	243
Box 11.2	Synchronization of Environmental Policies, Laws and Administration: A Case Study of Japan	247
Box 11.3	Impact of International Events and Conventions on Institutional and Legislative Development: The Case of Fiji	252
Box 11.4	Plant Varieties Act of India	253
Box 12.1	Devolution of Power and Environmental Regulation in India	262
Box 12.2	Control of Air Pollution in Bhutan	265
Box 12.3	Greening of Government Operations in Japan	267
Box 12.4	Environmental Tax Reforms in People's Republic of China	270
Box 12.5	Indicators for Sustainable Development in Asia and the Pacific	277
Box 12.6	The Introduction of Economic Instruments is Working Better than Prosecutions in India	278
Box 13.1	Private Investment in Infrastructure	285
Box 13.2	Construction Companies and the WASTE-WISE Programme in Australia and New Zealand	287
Box 13.3	Sustainable Corporate Management and ISO 14000 in People's Republic of China	291
Box 13.4	The Global Reporting Initiative: An Emerging Tool for Corporate Accountability	293

LIST OF BOXES (continued)

		Page
Box 13.5	Business and Environment Programme in Thailand	294
Box 13.6	International Hotels Environmental Action Plan for the Asian and Pacific Region	297
Box 13.7	The Kandalama Hotel: An Eco Experience	298
Box 14.1	Harnessing the Power of the World Wide Web	303
Box 14.2	NGOs Working to Improve Media Coverage of Environmental Issues	305
Box 14.3	Using Education to Encourage Participation: Success Stories	309
Box 14.4	Community Based Projects	310
Box 14.5	Empowering Women	312
Box 14.6	NGO Contribution Toward Making Cities Child-friendly	313
Box 15.1	PICCAP: A Training Success in the South Pacific	325
Box 15.2	NETTLAP: Building Regional Capacity through Environmental Research, Training and Education	328
Box 15.3	Awareness Raising Campaigns	330
Box 15.4	New Environmentalism in People's Republic of China	333
Box 15.5	Moving Pictures to Save the Environment	334
Box 15.6	Community Radio: A Growing Cacophony of Local Voices	335

PART IV   SUBREGIONAL OUTLOOK

Box 16.1	Sri Lanka and its Biodiversity	346
Box 16.2	Coral Reef Degradation in the Indian Ocean (CORDIO)	347
Box 16.3	Air Pollution In India	348
Box 17.1	ASEAN Regional Centre for Biodiversity Conservation (ARCBC)	359
Box 17.2	ASEAN Combat Against Forest Fires in Indonesia	359
Box 18.1	The Vatthe Conservation Area – Big Bay, Espirito Santo, Vanuatu	369
Box 18.2	The Samoa Fisheries Extension and Training Project	370
Box 19.1	Air Quality Improvement in Japan	375
Box 19.2	Improvement of Urban Environment in Northeast Asia through Public-Private Partnerships: Delivering Urban Environmental Services in People's Republic of China	377
Box 20.1	Human Induced Land Degradation in Turkmenistan	383
Box 20.2	Radiation and Human Health	384
Box 20.3	Joint Declaration of the Environmental Protection Ministers of the Central Asia	386

PART V   FUTURE OUTLOOK

Box 21.1	The Social Impact of the Asian Economic Crisis	416
Box 22.1	Community Action for the Environment: Bhaonta-Kolyala Village in Rajasthan	430
Box 22.2	Private Sector Initiatives for Environmental Rehabilitation – the Use of Indigenous Techniques	431
Box 22.3	Integrated Environment and Business Planning – The Case of Two Shrimp Farms in Indonesia	437
Box 22.4	ECO ASIA and the Promotion of Regional Cooperation for the Environment	440

LIST OF FIGURES

Page

PART I ENVIRONMENTAL CONDITIONS

Figure 1.1	Land Use Trends in Asia and the Pacific 1850-1980	4
Figure 1.2	Land Use Trends in Asia and the Pacific 1980 and 1995	4
Figure 1.3	Land Use in Asia and the Pacific and the World 1998	4
Figure 1.4	Land Use in Subregions (by Selected countries) of the Asian and Pacific Region, 1996	5
Figure 1.5	Population Density per Hectare of Arable Land in Selected Countries and Subregions of the Asian and Pacific Region, 1996	11
Figure 1.6	Percentage of Land Degradation in Asia and the Pacific and the World, by Land Use Patterns	11
Figure 1.7	Distribution of Main Degradation Types in Asia as Percentage of the Total Degraded Area	12
Figure 1.8	Pattern of Dryland Desertification in the Asian and Pacific Region Compared with the World	14
Figure 2.1	Distribution of Forests by Major Geographic Regions and Subregions in Asia and the Pacific	26
Figure 2.2	Per Capita Forest Areas in Selected Countries of the Asian and Pacific Region, 1992 and 1995	26
Figure 2.3	Forest Product Trade in Asia and the Pacific, 1995	33
Figure 3.1	Wetland Types in Asia and the Pacific	46
Figure 3.2	Biodiversity "Hot Spots" in Asia and the Pacific	48
Figure 3.3	Total Number of Plant Species in Asia and the Pacific by Subregion and the Percentage of Threatened Plant Species	49
Figure 3.4	Number of Threatened Animal Species in Asia and the Pacific by Subregion	51
Figure 3.5	Percentage Share in Number of Protected Areas by Subregions in Asia and the Pacific	70
Figure 3.6	Percentage Share in Number of Protected Areas by Countries in Asia and the Pacific	70
Figure 3.7	Percentage Share in Areal Extent of Protected Areas by Subregion	70
Figure 3.8	Percentage Share in Areal Extent of Protected Areas by Countries in Asia and the Pacific	70
Figure 4.1	Variability of Precipitation in Selected Countries of the Asian and Pacific Region	77
Figure 4.2	Annual Water Resources Per Capita in Selected Countries of the Asian and Pacific Region	77
Figure 4.3	Major River Systems of Selected countries in the Asian and Pacific Region	77
Figure 4.4	Estimates of Annual Water Resources Per Capita in the Asian and Pacific Region	78
Figure 4.5	Decline in Water Resources Per Capita in the Asian and Pacific Region	78
Figure 4.6	Annual Renewable Freshwater Per Capita Under Three Long-Range United Nations Population Projections: India and People's Republic of China	80
Figure 4.7	Water Withdrawals Against Water Resources (1900-2000)	80
Figure 4.8	Demand and Uses of Water in a River Basin	81
Figure 4.9	Domestic Water Use in Selected Cities in the Asian and Pacific Region as a Percentage of the Overall Water Use	82
Figure 4.10	Selected Cases of Water Pollution in the Asian and Pacific Region	83
Figure 5.1	Distribution of Pollution "Hot Spot" In South China Sea	102
Figure 5.2	Estimated Loss of Original Mangrove Areas in Asia and the Pacific	103
Figure 5.3	Distribution of Coral Reefs in the Asian and Pacific Region	103
Figure 5.4	High Risk Areas for Oil Pollution in the South China Sea	110

LIST OF FIGURES (continued)

		Page
Figure 6.1	Ambient Levels of Air Pollutants in Selected Large Asian Cities	125
Figure 6.2	Excess Levels of Acid Deposition Projected by the RAIN-Asia Model	127
Figure 6.3	Decline in the Global Production of Ozone Depleting CFCs	128
Figure 6.4	Correlation between Levels of Carbon Dioxide, Methane and Surface Temperature	128
Figure 6.5	Regional Shares of Carbon Dioxide Emissions from the Use of Fossil Fuels	134
Figure 6.6	Total and Per Capita Emissions of Carbon Dioxide from Fossil Fuels in the Largest Emitting Countries, 1996	134
Figure 6.7	Share of Energy Supplied by Traditional Fuels in Selected Asian Countries	135

PART II HUMAN ECOSYSTEMS-EMERGING ISSUES

Figure 7.1	Rate of Urbanization in Asia and the Pacific, 1995-2030	150
Figure 7.2	Components of Urban Growth in Asia and the Pacific, 1990-2005	150
Figure 7.3a	Number of Cities by City Size in Asia and the Pacific in 2000 and 2015	151
Figure 7.3b	Percentage of Cities by City Size in Asia and the Pacific in 2000 and 2015	151
Figure 7.4	Annual Growth Rates of the Present or Likely to Become Mega-cities During 1975-1995 and 1995-2015 in the Asia and the Pacific	151
Figure 7.5	Urban Population without Access to Safe Water Supply in Selected Countries of Asia and the Pacific	156
Figure 7.6	Water Use in Selected Urban Areas in Asia and the Pacific	157
Figure 7.7	Urban Population Without Access to Sanitation in Selected Countries of Asia and the Pacific	157
Figure 7.8	Annual Cost of Time Delay (in US$ million) in some Asian Cities Due to Traffic Jams	158
Figure 8.1	Municipal Solid Waste Generation in Different Groups of Countries in the Region	171
Figure 8.2	Estimated Generation of Municipal Solid Waste in Different Subregions	171
Figure 8.3	Approximate Composition of Municipal Solid Waste in Selected Cities of ESCAP Member Countries	171
Figure 8.4	Waste Intensity of Industrial Production in Selected Countries in the Region	172
Figure 8.5	Proportionate Annual Production of Agricultural Waste in People's Republic of China	173
Figure 8.6	Proportionate Annual Production of Agricultural Waste in Malaysia	173
Figure 9.1	Poverty by Developing Regions of the World	199
Figure 9.2	Vicious Cycle of Poverty and Environment Degradation in Developing Countries	201
Figure 9.3	Environmental Entitlements	204
Figure 9.4	Large Outbreaks	208
Figure 9.5	Unexpected Outbreaks	208
Figure 9.6	Leading Infectious Killers	209
Figure 10.1	Food Security Indices for Selected Countries (1986-1997)	216
Figure 10.2	Annual Growth Rates of Production of Major Food Crops by Selected Region/Area	218
Figure 10.3	Per Capita Production of Major Food Crops in 18 Asian Countries	219
Figure 10.4	Per Capita Production of Total Cereals by Region	219
Figure 10.5	Estimated Total Cereal Stock Carryover in Selected Asian Countries	226

PART III NATIONAL AND REGIONAL RESPONSES

Figure 11.1	Role of Planning Commission on Environment in India	244
Figure 12.1	Step by step approach adopted for the development of Indicators for Sustainable Development (ISD)	276

LIST OF FIGURES (continued)

		Page
Figure 13.1	Private and Public Investment in East Asia and the Pacific	284
Figure 13.2	Private and Public Investment in South Asia	284
Figure 13.3	Total Investment in Energy Projects With Private Participation in Developing Countries by Region, 1990-99	284
Figure 13.4	Private Participation in Water and Sewerage in Developing Countries 1990-97	286
Figure 13.5	Status of Eco-business in Japan 1996	287
Figure 13.6	Share of Asia and the Pacific in Worldwide ISO 14000 Certification	290
Figure 13.7	ISO 14000 Certification Growth in the Region from 1995 to end of 1998	290
Figure 13.8	ISO 14000 Certification Growth in Selected Countries of the Region from 1995 to end of 1998	290

PART IV   SUBREGIONAL OUTLOOK

Figure 16.1	Access to Safe Water and Sanitation (% of Urban Population) in South Asia	346
Figure 17.1	Share of Selected Southeast Asia Urban Household's Connection to Basic Services	354
Figure 18.1	Percentage of Population in Selected Pacific Island Countries with Access to Safe Water	366
Figure 18.2	Estimated Levels of Vulnerability to Specific Natural Hazards in Selected South Pacific Islands	366
Figure 19.1	Energy Efficiency (1996 GDP $ per kg oil equivalent) in Northeast Asia	374
Figure 19.2	CO_2 Emissions of Northeast Asian Countries for the Year 1996	375

PART V   FUTURE OUTLOOK

Figure 21.1	Trends in ODA and Private Direct Investments	393
Figure 21.2	Percentage of Countries in Asia and the Pacific that are Parties to Conventions	399
Figure 21.3	Ozone Protocol and Amendments Ratification Status	400
Figure 21.4	Multilateral Ozone Cumulative Funds Approved and CFC Tonnes Phase Out in the World	402
Figure 21.5	Impact of the Montreal Protocol on Ozone Depletion	402
Figure 21.6	Regional Distribution of Development Finance, 1989	412
Figure 21.7	GDP Growth Rates in Developing Asian and Pacific Economies	418
Figure 21.8	GDP Growth Rates in Newly Industrializing Economies of Asia and the Pacific	418
Figure 21.9	Causal Links between Environmental Scarcity and Violence	419
Figure 22.1	Environment and Development Trends in Asia and the Pacific 1995-2005	427
Figure 22.2	Fertilizer Consumption Trends in Asia and the Pacific 1980-1997	429
Figure 22.3	Projected Water Resources per Capita in 2050	433
Figure 22.4	Current and Projected Total Primary Energy Supply in Asia and the Pacific 1997 and 2010	434
Figure 22.5	Suspended Particulate Matters in Asia and the Pacific	435
Figure 22.6	Projected Municipal Solid Wastes Generation in Asia and the Pacific 2030	436

LIST OF TABLES

Page

PART I ENVIRONMENTAL CONDITIONS

Table 1.1	Changes in Agricultural Land Use and Associated Degradation Problems in Asia and the Pacific	6
Table 1.2	Current Assessment of the Extent of Various Types of Land Degradation in India	8
Table 1.3	Land Degradation/Desertification "Hot Spots" in Asia and the Pacific	9
Table 1.4	Trend in Irrigated Area Expansion in Selected Developing Countries and the World	17
Table 2.1	Distribution of Forest Ecosystems in Asia and the Pacific by Subregion	27
Table 2.2	Deforestation in Selected Countries in Asia and the Pacific	27
Table 2.3	Fuelwood and Charcoal Production (in 1 000 cu m) in Asia and the Pacific	31
Table 2.4	Production (1995) and Average Growth (1985-1995) of Industrial, Sawlogs, Sawnwood and Wood-based Panels	31
Table 2.5	Categorization of Non-Wood Forest Products by End-use	34
Table 2.6	Value of Exports for the Major Asian and Pacific Non-Wood Forest Products, 1996	34
Table 2.7	Historical Perspective on Forests and Forestry	37
Table 3.1	Agents of Wetland Change	47
Table 3.2	Higher Plant Diversity and Endemism in Megadiversity Countries of Asia and the Pacific	49
Table 3.3	Terrestrial Animal Species Diversity and Endemism in Megadiversity Countries of Asia and the Pacific	50
Table 3.4	Top Five Countries in Asia and the Pacific by Each Group of Threatened Species and Percentage of the Regional Total	50
Table 3.5	Total Numbers of Threatened Freshwater Fishes in Selected Countries of Asia and the Pacific	54
Table 3.6a	National Strategies for Conservation and Sustainable Use of Biodiversity in Asia and the Pacific	64
Table 3.6b	National Strategies for Conservation and Sustainable Use of Biodiversity in Asia and the Pacific	65
Table 3.7	Component of Biotrade Initiatives	68
Table 4.1	Water Resources and Use in Selected Countries of the Asian and Pacific Region	76
Table 4.2	Freshwater Withdrawal by Sector in the Region and the World	79
Table 4.3	Relative Severity of Water Pollution in the Asian and Pacific Region	83
Table 4.4	Rivers/Canals Clean-up Programmes in Selected Countries of the Asian and Pacific Region	89
Table 4.5	Water Vision of the Subregions in Asia and the Pacific	95
Table 5.1	Economic Losses from Red Tides in Fisheries and Aquaculture Facilities in Selected Countries of the Region	102
Table 5.2	Status of the Coral Reef in the Asian and Pacific Region	104
Table 5.3	Comparison Between Estimated Fisheries Potentials and Average Landings of the last 5 years (1990-1994) in Million Tonnes in Various Parts of the Marine Environment in Asia and the Pacific	105
Table 5.4	Small Pelagic Fisheries in the South China Sea, 1978-1993	106
Table 5.5	Fisheries Potential of the South China Sea	106
Table 5.6	Pollutant Fluxes from Rivers of Selected Countries to the South China Sea	109
Table 5.7	Marine Protected Areas in the Asian and Pacific Region	116
Table 5.8	Regional Organizations for the Sustainability of Coastal and Marine Resources in South Pacific	120

LIST OF TABLES (continued)

		Page
Table 6.1	Indoor Concentrations of Particulate Matter due to Biomass Combustion	124
Table 6.2	Estimated Health Benefits of Reducing Air Pollution in Jakarta	126
Table 6.3	Health Effects Associated with Common Air Pollutants	126
Table 6.4	The World's Major Vector-borne Diseases Ranked by Population Currently at Risk	133
Table 6.5	Emissions of Carbon Dioxide from Energy Use in 11 Asian Countries (in Million Tonnes of Carbon Dioxide)	134
Table 6.6	Carbon Dioxide Emissions from Land Use Changes in the ALGAS Participating Countries	135
Table 6.7	Emissions of Methane and Nitrous Oxide in ALGAS Participating Countries	135
Table 6.8	Ambient Air Quality Guidelines Recommended by the World Health Organization	136
Table 6.9	Ambient Air Quality Standards in People's Republic of China and India (micrograms per cubic metre)	137
Table 6.10	Consumption of Ozone Depleting Substances in Azerbaijan, Turkmenistan and Uzbekistan	139
Table 6.11	Options for Reducing Emissions from the Energy Sector in 11 Asian countries	140
Table 6.12	Installed Capacity of Wind Power in Indian States, 1997-1998	141
Table 6.13	Options for Reducing Future Emissions of Methane in the Agricultural Sector in Selected Countries	142

PART II   HUMAN ECOSYSTEMS-EMERGING ISSUES

Table 7.1	Urbanization Trends in the Asian and Pacific Region, 1999-2030	149
Table 7.2	Degree of Urbanization in the Asian and Pacific Region, 1999	150
Table 7.3	Annual Urban Growth Rate in the Asian and Pacific Region, 1999	150
Table 8.1	Sources and Types of Solid Wastes	170
Table 8.2	Quantities and Types of MSW Generated in Selected South Pacific Countries	172
Table 8.3	Approximate Estimate of Annual Production of Agricultural Waste and Residues in Selected Countries in the Region	173
Table 8.4	Conservative Estimate of Annual Production of Hazardous Waste in Selected Countries and Territories in the Asian and Pacific Region	174
Table 8.5	Impacts of Various Categories of Wastes on Water, Soil and Air in Selected Countries of Different Subregions	175
Table 8.6	Comparison of Typical Solid Waste Management Practices	176
Table 8.7	Various Categories of Materials Recycled from MSW in Singapore in 1997	178
Table 8.8	Disposal Methods for Municipal Solid Waste in Selected Countries of the Region	180
Table 8.9	Disposal Methods of Agricultural Waste and Residues in Selected Countries in the Region/Area	181
Table 8.10	Current Status of Overall Waste Management in Selected Countries of the Region	186
Table 8.11	Fees Payable for Disposal of Solid Waste in Singapore	191
Table 9.1	Poverty Measurement Indices in Selected Countries of the Asian and Pacific Region	200
Table 9.2	Demographic, Poverty and Economic Indicators for Asia and the Pacific	200
Table 9.3	Balance of Payments and International Reserves for Selected Countries of Asia and the Pacific (millions of dollars) (1996)	202
Table 9.4	Aid and Financial Flows Relating to Selected Countries in Asia and the Pacific (millions of dollars) (1996)	202
Table 9.5	Health Effects of Environmental Degradation in Selected Asian Countries	210
Table 10.1	Cereals Production (Including Rice in Milled Form) and Growth by Selected Region/Area	217
Table 10.2	Recent Trends in Production of Major Cereals in Selected Asian Countries	219
Table 10.3	Sources of Growth	220

LIST OF TABLES (continued)

		Page
Table 10.4	Growth of Production of Meat, Milk and Fish (Per Cent Per Annum) in Selected Countries	221
Table 10.5	Estimated Levels of Undernourishment in Selected Countries	223
Table 10.6	Average Annual Net Trade in Food for Selected Countries	224
Table 10.7	Recent Trends in Cereal Trade in Selected Asian Countries (Million Tonnes)	225
Table 10.8	Mechanization of Agriculture in the Selected Asian and Pacific Countries (Number of Tractors)	227
Table 10.9	Selected Indicators of Pesticide Consumption for the Asian and Pacific Region	228
Table 10.10	Lending for Agriculture by the World Bank, ADB and IFAD (Current US$ Billion)	234

PART III NATIONAL AND REGIONAL RESPONSES

Table 11.1	Forms of Governmental Environment Protection Institutions in Asia and the Pacific	241
Table 11.2	Some Framework Laws for Environmental Management in Selected Countries in Asia and the Pacific	251
Table 11.3	Some EIA Laws and Regulations for Selected Countries in Asia and the Pacific	255
Table 12.1	Governmental Environment Protection Vision in Selected Countries of Asia and the Pacific	261
Table 12.2	Market Based Instruments and Their Sample Applications in Asia and the Pacific	271
Table 13.1	Companies/Agencies Involved in Environment-Business in Asia and the Pacific	282
Table 13.2	World Rank of Revenues from Selected Japanese Environment-related Companies, 1995	283
Table 13.3	Green Financing in Japan	285
Table 13.4	Prominent Eco-Labels instigated in Asia and the Pacific	292
Table 13.5	A Review of Various Initiatives on Measuring and Reporting Aspects of Corporate Sustainability	293

PART IV SUBREGIONAL OUTLOOK

Table 16.1	Key Environmental Issues and Causes in South Asia	345
Table 16.2	Organic Water Pollution Resulting from Industrial Activities in South Asia	346
Table 16.3	Major Socio-Economic Indicators for South Asian Countries	348
Table 17.1	Key Environmental Issues and Causes in Southeast Asia	355
Table 17.2	Projected Coastal Populations	356
Table 17.3	Major Socio-Economic Indicators for Southeast Asian Countries	358
Table 18.1	Key Environmental Issues and Causes in the South Pacific Region	364
Table 18.2	Indicative List of Potential Impacts of Climate Change and Sea-level Rise Requiring Adaptive Responses in the South Pacific Subregion	365
Table 18.3	Major Socio-Economic indicators for South Pacific Countries	367
Table 19.1	Key Environmental Issues and Causes in Northeast Asia	374
Table 19.2	Major Socio-Economic Indicators for Northeast Asian Countries	376
Table 20.1	Key Environmental Issues and Causes in Central Asia	382
Table 20.2	Major Socio-Economic Indicators for Central Asian Countries	385

LIST OF TABLES (continued)

PART V FUTURE OUTLOOK

		Page
Table 21.1	Programme Areas of the Regional Action Programme and the Core Themes of Agenda 21	397
Table 21.2	Basic Information on the Different Conventions on the Environment	401
Table 21.3	Percentage of Parties to UNFCC in Asia and the Pacific (As of 1 March 1999)	404
Table 21.4	Percentage of Parties in CBD in Asia and the Pacific (As of 1 March 1999)	406
Table 21.5	Percentage of Parties in CITES in Asia and the Pacific (As of 1 March 1999)	406
Table 21.6	Percentage of Parties to RAMSAR in Asia and the Pacific (As of 1 March 1999)	407
Table 21.7	Parties to the Basel Convention and Ratification (As of 1 March 1999)	407
Table 21.8	Percentage of Parties to CCD in the Subregion of Asia and the Pacific (As of 1 March 1999)	408
Table 21.9	Aggregate Net Long-term Resource Flows to Developing Countries, 1990-96 (US$ billion)	410
Table 21.10	Aggregate Net Private Capital Flows to Developing Countries, 1990-96 (US$ billion)	410
Table 21.11	Net Private Capital Flows to Developing Countries by Country Group, 1990-96	411
Table 21.12	Official Net flows of Development Finance, 1990-96	411

ABBREVIATIONS

ACAP	Asian Conservation Awareness Programme
ACEID	Asia Pacific Centre for Educational Innovation for Development
ACFOD	Asian Cultural Forum on Development
ACT	Advanced Control Technologies
ADB	Asian Development Bank
AEEN	Australian Environmental Education Network
AEWA	African-Eurasian Migratory Waterbird Agreement
AFD	Agriculture and Fisheries Department
AFDEC	Andaman Sea Fisheries Development Centre Organization
AFEJ	Asia-Pacific Forum of Environmental Journalists
AGDP	Agricultural Gross Domestic Product
AGRIS	Agricultural Information System
AIBD	Asia-Pacific Institute for the Development of Broadcasting
AIJ	Activities Implemented Jointly
AKRSP	Agha Khan Rural Support Programme
AMCS	ASEAN Member Countries
AMIC	Asian Media Information and Communication Centre
AMSA	Australian Maritime Safety Authority
ANGOC	Asian NGO Coalition for Agrarian Reform and Rural Development
AP2000	Asia Pacific 2000
APC	Association of Progressive Communications
APCTT	Asian and Pacific Centre for the Transfer of Technology
APEC	Asia Pacific Economic Cooperation
APPEN	Asia-Pacific People's Environmental Network
ASIA	Asian Alliance of Appropriate Technology Practitioners
ARC	Alliance of Religions and Conservation
ARCBC	ASEAN Regional Centre for Biodiversity Conservation
ARCTIS	African Regional Centre for Technology Information System
ASEAN	Association of South-East Asian Nations
ASEP I	1st ASEAN Sub-Regional Environment Programme for the Human Tropics
ASOCON	Asia Soil Conservation Network
ASOEN	ASEAN Senior Officials on the Environment
ASSOD	Assessment of Human-Induced Soil Degradation in South and South-East Asia
AusAID	Australian Aid
BAT	Best Available Technology
BCRMF	Batangas Bay Coastal Resources Management Foundation
BCSD	Business Councils for Sustainable Development
BCT	Basic Control Technologies
BELA	Bangladesh Environmental Law Association
BGCI	Botanic Gardens Conservation International
BIMP-EAGA	Brunei Darussalam/Indonesia/Malaysia/Philippines East Asian Growth Area
BIOTHAI	Thai Network on Community Rights and Biodiversity
BLP	Best Practices and Local Leadership Programme
BMA	Bangkok Metropolitan Administration

BOD	Biological Oxygen Demand
BOOT	Build-Own-Operate-Transfer
BOT	Build-Operate-Transfer
BRAC	Bangladesh Rural Advancement Committee
Bt	Bacillus thuringiensis
CARIS	Current Agricultural Information System
CAST	China Association for Science and Technology
CBD	Convention on Biological Diversity
CBFM	Community-based Forest Management
CCD	Convention to Combat Desertification
CCD	Coast Conservation Department
CCEE	Certificate Course in Environment Education
CCOL	Coordinating Committee on the Ozone Layer
C & D	Construction and Development
CDM	Clean Development Mechanism
CDS	Community Development Society
CEC	Centre for Environmental Concerns
CEE	Centre for Environment Education
CEEC	Centre for Environmental Education and Communication
CEIT	Countries with Economies in Transition
CEP	Caspian Environment Programme
CEPOM	Committee on Environmental Policy and Management (Sri Lanka)
CERES	Coalition for Responsible Economies
CERs	Certified Emission Reductions
CEST	Centre for Environmentally Sustainable Technology
CETV	China Educational Television
CFC	Chlorofluorocarbon
CFCI	International Child-Friendly Cities Initiative
CFS	Cubic Feet per Second
CGIAR	Consultative Group on International Agricultural Research
CIDA	Canadian International Development Agency
CIFOR	Centre for International Forestry Research
CIIF	Coconut Industry Investment Fund
CIMMYT	International Centre for Maize and Wheat Improvement
CITES	Convention on International Trade in Endangered Species of Fauna and Flora
CITYNET	Regional Network of Local Authorities for the Management of Human Settlements
CLASP	Scheme for the Conservation of Lands in Asia and the Pacific
CLEAN	Community Led Environmental Action Network
CMS	Convention on the Conservation of Migratory Species
CNPPA	Commission on National Parks and Protected Areas
COP	Conference of Parties
CORDIO	Coral Reef Degradation in the Indian Ocean
CP	Cleaner Production
CPOL	Convention for the Protection of the Ozone Layer
CRC	Convention on the Rights of the Child
CRIMP	Centre for Research on Introduced Marine Pests
CRMP	Coastal Resources Management Program

CROP	Council for Regional Organizations of the Pacific
CSD	Commission on Sustainable Development
CSE	Centre for Science and Environment
CSIR	Council of Scientific and Industrial Research (India)
CSIRO	Commonwealth Scientific and Industrial Research Organization
CSR	Corporate Social Responsibility
CST	Committee on Science and Technology
CWTC	Chemical Waste Treatment Centre
CZM	Coastal Zone Management
DA	Development Alternatives
DAC	Development Assistance Committee
DANIDA	Danish International Development Agency
DBO	Design Build and Operate
DCC	Development Coordination Committee
DDP	Desert Development Programme
DEQP	Department of Environmental Quality Promotion
DESA	Department of Social Affairs and Development
DESIRE	Demonstration in Small Industries for Reducing Waste
DoE	Department of Environment
DRS	Deposit-Refund System
EANET	East Asia Acid Rain Monitoring Network
EC	European Commission
ECANET	Environmental Communication Asia Network
ECC	Environmental Clearance Certificate
ECCA	Environmental Camps for Conservation Awareness
ECOSOC	Economic and Social Council of the United Nations
EdNA	Education Network Australia
EEPSEA	Economy and Environment Programme for South-East Asia
EEZ	Exclusive Economic Zone
EGF	Environmental Guarantee Fund
EIA	Environmental Impact Assessment
EICnet	Environmental Information & Communication Network
EIS	Environmental Impact Statement
EMAS	European Eco-Management & Audit Scheme
EMINWA	Environmentally Sound Management of Inland Water
ENV	Ministry of the Environment
ENVIS	Environment Information System
EPA	US Environmental Protection Agency
EPD	Environmental Protection Department
EPL	Environment Protection Licensing
EQA	Environmental Quality Act
ERA	Environmental Risk Assessment
ERIN	Environmental Resources Information Network
ESCAP	Economic and Social Commission for Asia and the Pacific
ESDO	Environment and Social Development Organisation
ESTs	Environmentally Sound Technologies
EU	European Union

EU	Environmental Unit
FADINAP	Fertilizer Advisory Development and Information for Asia and the Pacific
FAI	Fixed-Asset Investment
FALCAP	Framework for Action on Land Conservation in Asia and the Pacific
FAO	Food and Agriculture Organization
FCA	Funafuti Conservation Area
FDI	Foreign Direct Investments
FEJB	Forum of Environmental Journalists of Bangladesh
FFA	Forum Fisheries Agency
FGD	Flue Gas Desulphurisation
FKI	Federation of Korean Industries
FLM	Flexible Lending Mechanism
FMA	Forest Management Agreement
FPC	Fish Protein Concentrate
FRA	Forest Resource Assessment
FREEP	Forestry Research Education and Extension Project (Indian)
FSC	Forest Stewardship Council
FSM	Federated States of Micronesia
FUGs	Forest-User Groups
GAA	Global Aquaculture Alliance
GATT	General Agreement on Trade and Tariff
GBMPA	Great Barrier Reef Marine Park Authority
GC-MS	Gas Chromatography and Mass Spectrophotometry
GCMs	Global Climate Models
GCRMN	Global Coral Reef Monitoring Network
GDP	Gross Domestic Product
GECIB	Global Village Environmental Culture Institute of Beijing
GEF	Global Environment Facility
GEO	Global Environmental Outlook
GHG	Greenhouse Gas
GIES	Global Information and Early Warning System
GIS	Geographic Information System
GLASOD	Global Assessment of Soil Degradation
GLOBE	Global Learning and Observations to Benefit the Environment
GM	Genetically Modified
GMO	Genetically Modified Organism
GMS	Greater Mekong Sub-region
GOOS	Global Ocean Observing System
GNP	Gross National Product
GRI	Global Reporting Initiative
GtC	Giga tonnes of Carbon
GWP	Global Water Partnership
Ha	Hectare
HIPC	Highly Indebted Poor Countries
HOMS	Hydrological Operational Multipurpose System
IAEA	International Atomic Energy Authority
ICAO	International Civil Aeronautics Organization

ICARDA	International Centre for Agricultural Research in the Dry Areas
ICAS	Interstate Council on the Aral Sea
ICC	International Chamber of Commerce
ICESD	Inter-Agency Sub-committee on the Environment and Sustainable Development
ICFTU	International Confederation of Free Trade Unions
ICIMOD	International Centre for Integrated Mountain Development
ICLARM	International Centre for Living Aquatic Resources Management
ICZM	Integrated Coastal Zone Management
ICRAF	International Centre for Research in Agroforestry
ICRI	International Coral Reef Initiative
ICRISAT	International Crops Research Institute for the Semi-Arid Tropics
ICWC	Interstate Commission for Water Coordination
IDA	International Development Association
IEEA	Integrated Environmental and Economic Accounting
IEFR	International Emergency Food Reserve
IETC	International Environmental Technology Centre
IFAD	International Fund for Agricultural Development
IFAS	Interstate Fund for the Aral Sea
IFCS	Inter-governmental Forum on Chemical Safety
IFF	Inter-governmental Forum on Forests
IFPRI	International Food Policy Research Institute
IFMA	Industrial Forest Management Agreement
IFPRI	International Food Policy Research Institute's
IHA	International Hotel Association
IHEI	International Hotels Environment Initiative
IIASA	International Institute for Applied Systems Analysis
IKLAS	National Institute for Environmental Skill and Training
ILO	International Labour Organization
IMF	International Monetary Fund
IMO	International Maritime Organization
IMPECT	Inter Mountain Peoples Education and Culture in Thailand Association
INC	Inter-governmental Negotiating Committee
INCD	International Negotiating Committee for Desertification
INR	Institute of Natural Resources
IOC	International Oceanographic Commission
IOCU	International Organization of Consumer Unions
IPCC	Inter-governmental Panel on Climate Change
IPF	Inter-governmental Panel on Forests
IPM	Integrated Pest Management
IPNS	Integrated Plant Nutrition Systems
IPR	Intellectual Property Rights
IBRD	International Bank for Reconstruction and Development
IRRI	International Rice Research Institute
ISD	Indicators for Sustainable Development
ISPs	Internet Service Providers
ISRIC	International Soil Reference and Information Centre
ISSS	International Society of Soil Science

ITQ	Individual Transferable Quota
ITU	International Technology Union
IUCN	International Union for the Conservation of Nature
IWEP	Industrial Waste Exchange of the Philippines
IWICM	International Workshop on Integrated Coastal Management
IWMI	International Water Management Institute
JFEJ	Japanese Forum of Environmental Journalists
LA21	Local Agenda 21
LIFE	Lanka International Forum on Environmental and Sustainable Development
LMP	Land Management Programme
LOGOTRI	Local Government Training and Research Institutes
MAI	Multilateral Agreement on Investment
MATREM	Malaysia Training and Research in Environmental Management
MCCW	Malaysian Council for Child Welfare
MCR	Mahaweli Community Radio
MDF	Medium-Density Fibreboard
MEA	Multilateral Environment Agreement
MEIP	Metropolitan Environment Improvement Programme
MINAS	Minimum National Environmental Standards
MINSOC	Management Institute for Social Change
MOE	Ministry of Environment
MOSTE	Ministry of Science, Technology and Environment
MOTIE	Ministry of Trade, Industry and Energy
MPN	Most Probable Number
MRC	Mekong River Commission
MSS	Myanmar Selection System
MSW	Municipal Solid Waste
MT	Million Tonnes
NACOM	Nature Conservation Movement
NBA	National Biodiversity Authority
NBRUs	National Biodiversity Reference Units
NCC	National Coordination Committee
NCPCs	National Cleaner Production Centres
NCWO	National Council for Women's Organizations
NEAEC	North-East Asian Conference on Environmental Cooperation
NEAP	National Environmental Action Plan
NEAP	National Ecotourism Accreditation Programme
NEASPEC	North-East Asia Sub-Regional Programme of Environmental Cooperation
NEB	National Environment Board
NEC	Nature Education Centre
NEDA	National Economic Development Agency
NEF	National Environment Forum
NEFEJ	Nepal Forum of Environmental Journalists
NEMPs	National Environmental Management Plans
NEPA	National Environmental Protection Agency
NESDB	National Economic and Social Development Board
NETTLAP	Network for Environmental Training at Tertiary Level in Asia and the Pacific

NFAP	National Forestry Action Programmes
NFYOB	National Federation of Youth Organizations in Bangladesh
NGO	Non-Governmental Organization
NORAD	Norwegian Agency for Development
NOWPAP	North-West Pacific Action Plan
NPAs	National Plan of Action
NPC	National Planning Commission
NRIC	National Resource Information Centre
NRPTC	National Register of Potentially Toxic Chemicals
NSC	National Steel Corporation
NTFPs	Non-Timber Forest Products
NWDPRA	National Watershed Development Project for Rainfed Areas
NWFP	North West Frontier Province
NWFPs	Non-wood Forest Product
NZ	New Zealand
OAD	Development Assistance
ODA	Official Development Assistance
ODS	Ozone Depleting Substances
OECD	Organization for Economic Cooperation and Development
OECF	Overseas Economic Cooperation Fund
OLS	Ordinary Least Squares
O/M	Operation and Maintenance
OPP2	Second Outline Perspective Plan (Malaysia)
OSB	Oriented-Strandboard
PAF	Penang Organic Farm
PATA	Pacific Asia Travel Association
PATLEPAM	Philippine Association of Tertiary Level Educational Institutions in Environmental Protection and Management
PBR	Plant Breeders Right
PCAF	Pollution Control and Abatement Fund
PCD	Pollution Control Department
PCSC	Pollution Control Service Corporation
PCSD	Philippine Council for Sustainable Development
PDS	Public Distribution System
PE	Pollution Equivalent
PEEC	Provincial Environmental Education Centres
PFEC	Philippine Federation for Environmental Concern
PICCAP	Pacific Islands Climate Change Assistance Programme
PINA	Pacific Islands News Association
PLS	Pollution Levy System
PM_{10}	Particulate Matter (10 micron)
PNG	Papua New Guinea
POA	Programme of Action
POP	Persistent Organic Pollutant
POs	People's Organizations
PPBV	Parts Per Billion by Volume
PPP	Public-Private Partnerships

PPP	Polluter-Pays Principle
PPTV	Parts Per Trillion by Volume
PRIA	Participatory Research in Asia
PRRC	Pasig River Rehabilitation Commission
PSC	Philippines Sinter Corporation
PSI	Pollution Standard Index
PTDP	Preliminary Testing and Development Phase
PTO	Patent and Trademark Office
PVA	Plant Varieties Act
PVC	Polyvinylchloride
PVs	Photovoltaics
PWBLF	Prince of Wales Business Leaders Forum
RAINs	Regional Air Pollution Information and Simulation
RAP	Rapid Assessment Programme
RCFTC	Regional Community Forestry Training Centre
R & D	Research and Development
REOF	Regional Community Forestry Training Centre
RHAP	Regional Haze Action Plan
RICAP	Regional Inter-agency Committee for Asia and the Pacific
ROAP	Regional Office for Asia and the Pacific (UNEP)
RWMC	Radioactive Waste Management Centre
SA 8000	Social Accountability 8000
SAARC	South Asian Association for Regional Cooperation
SACEP	South Asia Cooperative Environment Programme
SALT	Sloping Agricultural Land Technology
SAM	Special Area Management (Plan)
SARM	Sustainable Agricultural Resource Management
SASEANEE	South and South-East Asia Network for Environmental Education
SCOPE	Society for the Conservation and Protection of the Environment
SDB	Sustainable Development Bill
SEC	Singapore Environment Council
SEEA	United Nations System for Integrated Environmental and Economic Accounting
SEHD	Society for Environment and Human Development
SELF	Solar Electric Light Fund
SEP	South-East Asian Countries Environment Programme
SEPA	State Environmental Protection Agency
SFM	Sustainable Forest Management
SGP	Small Grants Program
SIDA	Swedish International Development Agency
SIDS	Small Island Developing States
SLEJF	Sri Lanka Environmental Journalists Forum
SLR	Sea Level Rise
SMEs	Small and Medium Enterprises
SMIs	Small and Medium Industries
SNA	System of National Accounts
SoE	State of the Environment
SPBCP	South Pacific Biodiversity Conservation Programme

SPC	Secretariat for the Pacific Community
SPM	Suspended Particulate Matter
SPREP	South Pacific Regional Environment Programme
SRI	Socially Responsible Investment
SST	Sea Surface Temperature
START	Global Change System for Analysis, Research, and Training
SUSTRANS	Sustainable Transport Action Network for Asia and the Pacific
TACIS	Technical Assistance for the Commonwealth of Independent States
TACs	Total Allowable Catches
TBCSD	Thailand Business Council for Sustainable Development
TCBS	Training and Capacity-Building Section
TCMBCP	Tumen Coastal and Marine Biodiversity Conservation Programme
TCSP	Tourism Council for the South Pacific
TEI	Thailand Environment Institute
TEMM	Tripartite Environment Ministers Meeting
TERI	Tata Energy Research Institute
THAITREM	Thailand Training and Research in Environmental Management
TICE	Tertiary Institutions Council for the Environment
TPNs	Thematic Programme Networks
TRIPs	Trade Related Aspects of Intellectual Property Rights
TSP	Total Suspended Particulates
TTE	Television Trust for the Environment
TVE	Television Trust for the Environment
TWN	Third World Network
UK	United Kingdom
UMP	Urban Management Programme
UMPAP	Urban Management Programme for Asia and the Pacific
UNCCD	United Nations Convention to Combat Desertification
UNCED	United Nations Conference on Environment and Development
UNCHS	United Nations Centre for Human Settlements
UNCLOS	The United Nations Convention on the Law of the Sea
UNCTAD	United Nations Conference on Trade and Development
UNDP	United Nations Development Programme
UN ECE	United Nations Economics Commission for Europe
UNEP	United Nations Environment Programme
UNESCO	United Nations Educational, Scientific and Cultural Organization
UNFCCC	United Nations Framework Convention on Climate Change
UNFPA	United Nations Population Fund
UNICEF	United Nations International Children's Emergency Fund
UNIDO	United Nations Industrial Development Organization
UPLB	University of the Philippines Los Baños
UPOV	International Convention for the Protection of New Varieties of Plants
URBNET-ASIA	Regional Network of Urban Experts in Asia and the Pacific
USA	United State of America
US-AEP	United States-Asia Environmental Partnership
USAID	United States Agency for International Development
USDA	United States Department of Agriculture

UV-B	Ultraviolet B
UWP	Urban Water Programme
UWSEA	United World College of South East Asia
V & A	Vulnerability and Adaptation Assessment
VBA	Viet Nam Bank for Agriculture
VFC	Village Forest Committee
VOC	Volatile Organic Compound
VRC	Video Resource Centre
VWU	Viet Nam Women's Union
VYU	Viet Nam Youth Union
WAP	Waste Analysis Protocol
WBCSD	World Business Council for Sustainable Development
WCMC	World Conservation Monitoring Centre
WE	Women in Environment
WFP	World Food Programme
WHO	World Health Organization
WICE	World Industry Council for the Environment
WMO	World Meteorological Organization
WRI	World Resources Institute
WTO	World Trade Organization
WWC	World Water Council
WWF	Worldwide Fund for Nature

INTRODUCTION

The ESCAP region encompasses a vast territory extending from the Russian Federation in the north, to New Zealand in the south; from the Cook Islands in the east, to the Republic of Armenia and Georgia in the west. The environmental diversity of Asia and the Pacific is therefore vast, and is contrasted by the region's coldest and hottest deserts, verdant tropical rainforests, extensive grasslands, and rich alluvial plains. The region also embraces two of the world's three largest oceans (the Pacific and the Indian), together with its highest mountains (the Himalayas and Karakoram). It is this great variation in geography, topography and climate that provides the unique diversity found in the region's ecosystems. For example, Asia and the Pacific houses about thirty per cent of the world's tropical forests, and its marine and coastal environments are amongst the most productive in the world, supporting over two-thirds of the world's coral reefs and two fifths of its mangrove habitats.

The pressures on these rich natural resources and environmental systems have, however, been continuously increasing over the last decade. Rapid population growth, urbanisation, rising economic output and consumptive lifestyles, coupled with an increasing incidence of poverty, have all contributed to the region's shortfall in meeting the sustainable development goals which it set for itself at Rio. Environmental disruptions can be seen in the form of increasing atmospheric pollution, destruction of biodiversity, depletion of aquifers, and the pollution of aquatic and marine ecosystems, as well as increasing loads of municipal, industrial and hazardous wastes.

The population of the region more than doubled in the latter half of 20th Century, from 1.7 billion in 1960, to 3.6 billion in the year 2000; according to the United Nations it will reach about 5 billion by 2025. To date, the accompanying demands of this burgeoning population have largely been satisfied through increased economic output, which has quadrupled in the last twenty years. While economic growth has no doubt assisted in reducing poverty in the region, the adoption of unsustainable consumption patterns is now becoming a severe problem. Traditional concerns in relation to population and the environment have largely focused on aggregate population levels. However, it is currently accepted that the impact of humanity on the world environment is as much a function of per capita consumption as overall population size. For example, to replicate the pattern of grain consumption as evidenced in the United States today, by 2025 the regional requirement would be 4.5 billion tonnes of grain, or the harvest of more than two planets at earth's current output levels. Moreover, while the affluent consume, the poor are compelled to enhance processes of environmental erosion through cutting trees, growing crops on steep slopes or marginal land, and exploiting fragile resources, simply to meet their basic needs for survival. It is therefore the enlarging polarisation of lifestyles, both globally and intra-regionally, that is presenting the overwhelming challenge for sustainable development in Asia and the Pacific.

An additional cause that has struck the environment harshly in recent years has been the negative effect of globalization, involving trade, investments and debts. A large proportion of the region's exports, which have in turn contributed to the depletion of its natural resources, have been to meet the industrial and other needs of developed countries of the world. Globalization has contributed to the region's rapid economic growth, although investments have declined following the economic crisis in the latter half of 1997. Ironically, the environment has again suffered because of this decline, as the economic downturn inevitably led to sustainable development retreating down the list of national priorities. In most nations, budgetary allocations for the environment have been reduced, leading to fewer investments in conservation and protection activities, and delays in investments in capital renewal and cleaner technologies. As the region continues to undergo spontaneous transformation and economic development, the impacts of associated activities on the terrestrial, aquatic, and atmospheric ecosystems are anticipated to become more acute with time. In addition, whilst globalization has strengthened the economic linkages of the world, some countries could find themselves short of valuable financing overnight, as capital and financial flows change direction.

In the forthcoming Rio+10, in 2002, the greatest unexpressed fear of the diplomatic community is that it will be a reprise of Rio+5. The promotion of effective regional and sub-regional cooperation is therefore even more vital for providing opportunities for a coordinated response to global initiatives. The dwindling availability of international financial resources, lack of technological transfer between developed and developing nations, and unfavourable trade regimes, demand enhanced regional unity in a proactive response to offset the negative trends identified throughout this report.

In order to review the dynamics of the region's environmental conditions and the status of national, regional and international responses to the changing environmental situation, the following State of the Environment Report, 2000 has been divided into five related parts, as illustrated in Figure I. The content of each of these parts is summarized as follows.

- *Part One* presents the prevailing environmental conditions in different components of terrestrial, aquatic and marine ecosystems. It has six chapters, on land, forest, biodiversity, inland water, marine and coastal environment, and the atmosphere.
- *Part Two* presents the key emerging issues in human ecosystems, which are described in terms of urban environment, wastes, poverty, and food security.
- *Part Three* highlights the responses to the problem of environment. It covers the national and regional response by various actors including the governments, industry and private sector as well as NGOs and other major public interest groups. Topics are presented as institutions and legislation, mechanisms and methods, private sector, major groups, and environmental education and awareness.
- *Part Four* provides an overview of the key environmental issues, together with the causes and trends in environmental management and cooperation in the five sub-regions of Asia and the Pacific, i.e. South Asia, South-East Asia, South Pacific, North-East Asia, and Central Asia.
- *Part Five* discusses the prevailing conditions and trends in the region's physical environment, their impacts on the health and well-being of the population, and the management and policy responses which have been adopted at all levels (international, regional and sub-regional) to address them. It goes on to examine projected trends and future scenarios for the environment, and concludes with a discussion of future prospects for the region.

Figure I Framework of the State of the Environment Report in Asia and the Pacific

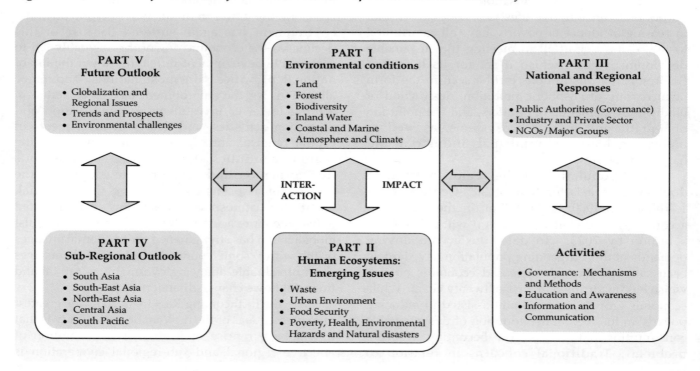

PART ONE

CHAPTER 1-6

PART I ENVIRONMENTAL CONDITIONS

CHAPTER 1 LAND
CHAPTER 2 FORESTS
CHAPTER 3 BIODIVERSITY
CHAPTER 4 INLAND WATER
CHAPTER 5 COASTAL AND MARINE
CHAPTER 6 ATMOSPHERE AND CLIMATE

PART II

CHAPTER 7 URBAN ENVIRONMENT
CHAPTER 8 WASTE
CHAPTER 9 POVERTY AND ENVIRONMENT
CHAPTER 10 FOOD SECURITY

PART III

CHAPTER 11 INSTITUTIONS AND LEGISLATION
CHAPTER 12 MECHANISMS AND METHODS
CHAPTER 13 PRIVATE SECTOR
CHAPTER 14 MAJOR GROUPS
CHAPTER 15 EDUCATION, INFORMATION AND AWARENESS

PART IV

CHAPTER 16 SOUTH ASIA
CHAPTER 17 SOUTHEAST ASIA
CHAPTER 18 SOUTH PACIFIC
CHAPTER 19 NORTHEAST ASIA
CHAPTER 20 CENTRAL ASIA

PART V

CHAPTER 21 GLOBAL AND REGIONAL ISSUES
CHAPTER 22 ASIA AND THE PACIFIC INTO THE 21st CENTURY

Chapter One

Fighting land degradation: spraying of bitumen mulch in Islamic Republic of Iran to control sand dune migration.

1

Land

INTRODUCTION
LAND USE
LAND DEGRADATION
CAUSES OF LAND DEGRADATION
PROCESSES AND MEASUREMENTS
DESERTIFICATION
CONSEQUENCES OF LAND DEGRADATION
KEY ISSUES AND TRENDS
POLICIES AND PROGRAMMES
CONCLUSION

CHAPTER ONE

INTRODUCTION

Humankind throughout history has derived sustenance from land. Demand for the resources which it provides increases with population growth. If these demands are managed inappropriately the result can be a reduction or altogether loss in productivity, or degradation of the land. This chapter provides an overview of condition of the land in Asia and the Pacific, including its use, trends, productivity patterns and policies undertaken for its management.

LAND USE

There are three important uses of land for sustenance: arable or croplands; permanent pastures and grazing land; and forest and woodlands. The land use pattern in the region has undergone a major change over the years with a sharp increase in cropland but a marked decline in forest and (in recent years) pastures, as shown in Figures 1.1 and 1.2. An FAO study (Dent 1990) estimated that only 14 per cent of the region's land was free from soil related constraints to agricultural production. In 1999, arable land accounted for 16 per cent of the total land area of the region (Figure 1.3). This may imply that lands with low production potential are increasingly being used. Such areas (where environmental degradation is most severe) are generally used by landless people and their livestock who migrate from overpopulated or inequitably distributed high potential land.

A more positive development in land use during the last decade was an expansion in plantation of trees and wood lots (Chapter 2). However, plantations are no substitute for natural forest and increasing deforestation needs to be addressed in the region. Land use by subregion is given in Figure 1.4. Southeast Asia has substantial land (49 per cent) under forests. All other subregions have depleted natural forest cover due to excessive exploitation. Northeast Asia also has substantial land under forests and woodlands but this is partly due to plantations. Deforestation and over exploitation of the natural vegetative cover are the major causative factors of human-induced soil degradation in the forest and woodlands of the region.

Out of the five subregions of Asia and the Pacific, South Asia has the biggest area under crops followed by Northeast Asia. Land use details for Central Asia are not available, but the subregion has about 24 200 ha of croplands and 37 100 ha under rangelands (Kruzhilin 1995 and Glazovsky 1997).

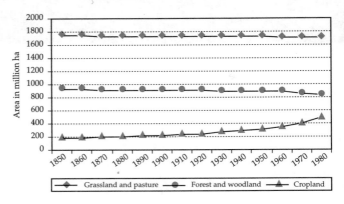

Figure 1.1 Land Use Trends in Asia and the Pacific 1850-1980

Source:
1. Rapetto R. and Gillis M eds 1988
2. FAO Statistical Data 1997

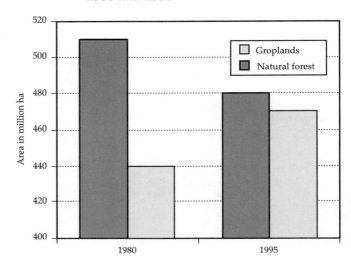

Figure 1.2 Land Use Trends in Asia and the Pacific 1980 and 1995

Source: 1. FAO Statistical Data 1997

Figure 1.3 Land Use in Asia and the Pacific and the World 1998

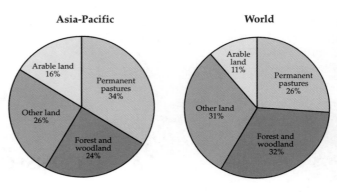

Source: FAO 1999

Figure 1.4 Land Use in Subregions (by Selected countries) of the Asian and Pacific Region, 1996

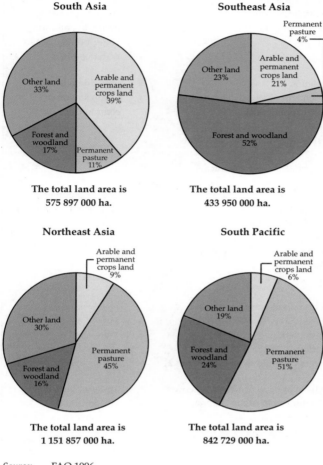

Source: FAO 1996

LAND DEGRADATION

Land use changes in Asia and the Pacific, particularly those related to agricultural expansion (area and yield) (Chapter 10) were associated with considerable degradation of land resources (Table 1.1). Land degradation has been defined as, *'the reduction in the capability of the land to produce benefits from a particular land use under a specified form of land management'* (FAO 1999). Degradation processes include erosion, compaction and hard setting, acidification, declining soil organic matter, soil fertility depletion, biological degradation and soil pollution (Lal and Stewart 1990).

A. Extent of land degradation

Land degradation in Asia and the Pacific is affecting areas of both low and high production potential. The areas of low production potential include marginal lands that have lower quality or degradation prone soils and are subject to more harsh or variable climatic conditions. The most serious form of land degradation affecting low production potential areas is water and wind erosion. Under accelerated erosion, soil loss takes place faster than the formation of new soil. The two major human activities contributing to the removal of vegetative cover and resultant accelerated erosion are shifting cultivation and deforestation.

High agricultural production potential areas generally have high (or potentially high) soil fertility and can sustain intensive cropping using existing technologies, with either irrigation or reliable and adequate rainfall. In Asia, at least 82 per cent of cereals are grown on high-potential, naturally flooded or irrigated land. The "population carrying capacity" of these lands can be raised using existing agricultural technology. The Green Revolution success story of the 1970s and early 1980s was developed for, and took place in, these high potential areas. However, many of the development projects undertaken have not been environmentally sound. Some intensive irrigation programmes have neglected drainage and proper water management practices, leading to waterlogging and salinization. Soil fertility decline associated with fertilizer application (directly or indirectly) is also a cause for concern. Few countries have made estimates of the extent of the problem; but reports of organic matter depletion, negative soil nutrient balance, unbalanced fertilizer application, and the onset of secondary and micronutrient deficiencies are beginning to emerge. In India, analysis of long-term fertilizer experiments carried out on a wide variety of soils over the period 1885 to 1985 (Nambiar and Abrol 1989) showed a clear declining trend in productivity, even with the application of mineral fertilizers under modern intensive farming conditions. This failure of increasing yields from fertilizer use is particularly worrying, as with declining land availability, yield growth is increasingly important. Concern is also being expressed that all the important rice-wheat systems of Southeast and South Asia may be over-exploitative of the natural resource base and that there will be consequent degradations in soil health and fertility, and would also have impact on water quality.

1. Land Degradation in Central Asia

In terms of land degradation, one of the worst affected subregions is Central Asia, where the most common forms of land degradation are from water and wind erosion, salinization, water logging, compaction and land pollution. Seven per cent of land is severely affected by desertification, 34 per cent is moderately affected and the rest is slightly degraded (Kharin 1996), however the situation is demonstrably worse in the Aral Sea basin (Box 1.1).

CHAPTER ONE

Table 1.1 Changes in Agricultural Land Use and Associated Degradation Problems in Asia and the Pacific

Land type	Main changes	On-site soil degradation	Other kinds of degradation
Irrigated lands	Increase in irrigated area, increased multiple cropping	• Salinization and waterlogging • Nutrient constraints under multiple cropping • Biological degradation (agro-chemicals)	• Nutrient pollution in ground/surface water • Pesticide pollution • Water-borne diseas • Water conflicts
High-quality rainfed lands	Transition from short fallow to continuous cropping, HYVs, mechanization	• Nutrient depletion • Soil compaction and physical degradation from over cultivation, machinery • Acidification • Removal of natural vegetation, perennials • Soil erosion • Biological degradation (agrochemicals)	• Pesticide pollution • Deforestation of commons
Densely populated marginal lands	Transition from long to short fallows or continuous cropping; cropping in new landscape niches	• Soil erosion • Soil fertility depletion • Removal of natural vegetation, perennials from landscape • Soil compaction, physical degradation from overcultivation • Acidification	• Loss of biodiversity • Watershed degradation
Extensively managed marginal lands	Immigration and land-clearing for low input agriculture	• Soil erosion from land-clearing • Soil erosion from crop/livestock production • Soil nutrient depletion • Weed infestation • Biological degradation from top soil removal	• Deforestation • Loss of biodiversity • Watershed degradation
Urban and peri-urban agricultural lands	Rapid urbanization; diversification of urban food markets; rise in urban poverty	• Soil erosion from poor agricultural practices • Soil contamination from urban pollutants • Overgrazing and compaction	• Water pollution • Air pollution • Human disease vectors

Source: Scherr, S.J. 1999

Salinization affects 87 per cent of the irrigated land in Turkmenistan, 60 per cent in Uzbekistan, and 60-70 per cent in Kazakhstan (Khakimov 1989, Glazovsky 1995, Mainguet and Letolle 1998). In Armenia, soil erosion is taking place over an area of 12 000 km^2; in Azerbaijan, 13 700 km^2 and in Kyrgyzstan, 5 500 km^2. In Tajikistan, eroded land has reached 97.9 per cent of all agricultural area and has stimulated migratory programmes. Degradation of the plant cover of rangelands can be observed in over 600 000 km^2 in Kazakhstan (Box 1.2).

The deflation and formation of dunes also contributed to land degradation in Central Asia. During the seasonal periods for dust storms, deflation reaches disturbing proportions (Plit *et al* 1995). The areas most affected are, the Aralo-Caspian depression, northern Kazakhstan, and south-western Siberia, a total area of over one million km^2.

2.   Land Degradation in Other Subregions

Land degradation is also a severe problem in South Asia where problems vary according to land-use practices. For example, in India the move towards intensification of land use by expanding cultivation into pasture/grazing lands, cultivable waste, reducing fallow, unsustainable irrigation, and other unsustainable activities have contributed substantially to land degradation (Kaul et al 1999; Table 1.2). In Northeast Asia, the worst sufferers from land degradation are People's Republic of China, the Russian Federation and Mongolia. In People's Republic of China, many arable areas suffer serious salinization, desertification and soil erosion despite continuous reclamation and protection efforts (Box 1.3). Desertification affected 262.2 million ha in 1996, accounting for 27.3 per cent of total national territory. Of this, the area of salinized land was 44 million ha of which over 10 per cent was arable land.

Deflation and sand dune formation is also prevalent in the Russian Federation where the upper Volga and south-western Siberia are the most severely affected. In the southern Urals and western Siberia, at least 25 per cent of arable land is subject to erosion. Several million ha of irrigated land have been

Box 1.1 The Aral Sea Basin

The Aral Sea basin is the largest contiguous area on our planet affected by human-induced desertification extending to a vast area of 1.8 million km². The population affected by desertification is about 400 million, which is more than twice the number of people that lived in the area in the early 1960s. Five republics in Central Asia (in Kazakhstan, Uzbekistan, Turkmenistan, Kyrgystan, and Tajikistan) share the Aral Sea Basin. In an ambitious effort to increase productivity and control desertification, the waters of two large rivers flowing into the Aral Sea – the Amudarya and the Syrdarya – were redirected for irrigation purposes in the river deltas. This mega-project caused profound and seriously negative environmental effects.

Due to the large scale irrigation programmes, the total inflow of water from the two main rivers into the Aral Sea declined from 40 km³ in 1960 to zero in the 1980s, and only slightly increased between 1987 and 1991. As a result, the Aral Sea water level declined between 1961 and 1991 from 53 metres to 30 metres. The surface area of the sea diminished from 68 000 km² to 37 000 km², and the volume declined by 70 per cent from 1 090 to 340 km³. The total area of the dry seabed is more than 30 000 km. The lowered water table in the wide surroundings of lakes, rivers, and pastures has had disastrous effects on soil quality and productivity. The land emerging after the sea dried up is not fertile, and mainly consists of sand dunes or salted soil. The Aral Sea has also become severely polluted with pesticide residue. This has had an enormous impact on the environment, extending up to 400 kilometres from the Sea, as well as on the health and livelihood of the local communities.

According to research assessments, 8.5 per cent of the Basin (121 700 km²) is severely affected by desertification, 33 per cent (474 356 km²) is moderately affected, and the remaining land (837 714 km²) is slightly degraded. The main aspects of desertification are degradation of the vegetation cover, salinization of irrigated farmlands and of the borders of the Aral Sea, and water erosion. Land degradation is particularly serious in Kazakhstan, Uzbekistan, and Turkmenistan, with large areas of irrigated farmlands (mainly cotton) being affected in the areas bordering the Aral Sea.

Because of the contraction of the Aral Sea, the irrigated delta lands in the surrounding areas suffer from serious salinization, the loss of soil structure and nutrients, and the increase of salty dust storms. Soil salinity has quadrupled from 10 g/l to 40 g/l, adversely affecting the productivity of vast tracts of lands. The groundwater is polluted with salt, pesticides, herbicides and defoliants, affecting food production and public health. According to a report by the Desert Research Institute in Turkmenistan, the toxicity of the ground water, which is used as drinking water, is 5 to 10 times the acceptable level. The drying and salinization of the delta lands has also led to a diminishing resistance of the ecosystems to human activities such as woodcutting and grazing. As a consequence, intensified soil degradation and desertification are continuing, and the capacity for regeneration is diminishing. During the last 25 years, the desert growth around the Aral Sea is estimated at about 100 000 ha per year, which makes an average annual rate of approximately 4 per cent. The degradation processes (pollution, salinization, decreasing sea volume) are considered irreversible.

Sources: 1. UNCCD 1998
2. Kharin, N. 1994
3. Mainguet, M. and Letolle, R. 19981

---

salinized and waterlogged. Millions of ha of other agricultural land, outside irrigated areas, have also been lost as a result of the discharge of irrigation water, flooding and salinization of the soil, and the adverse effects of construction work (Glazovsky 1997).

In the South Pacific, Australia is one of the countries that is facing massive challenges due to widespread land degradation, particularly in regard to dryland salinity (AFFA Joint Release 1999). Currently, 25 million km² of land is affected by salt, and this is likely to increase six-fold in the coming decades. Western Australia, one of the worst affected states, has 18 million km² of salt affected land and this is increasing at a rate equal to one football field an hour.

3. *Critically Degraded Areas ("Hot Spots")*

Hot spots or land degradation/desertification 'tension zones' around the world have been defined and located by Eswaran et al (1999). Hot spots are where the potential decline in land quality is so severe as to trigger a whole range of negative socio-economic conditions that could threaten political stability, sustainability, and the general quality of life. 'Tension zones' are created by factors such as, excessive and continuous soil erosion resulting from over and improper use of lands especially marginal and sloping land; nutrient depletion and/or soil acidification; waterlogging and salinization; or soil pollution from excessive use of organic and inorganic agro-chemicals (WRI 1999). The probability of hot spot occurrence is highest in areas with the following:

- *systematic reduction* in crop performance leading to failure in rainfed and irrigated systems;
- *reduction* in land cover and biomass production in rangeland;
- *removal* of available biomass for fuel and increased distances to harvest them;

CHAPTER ONE

Box 1.2 Land Degradation in Kazakhstan

Kazakhstan has a total area of 272.5 million ha including 182.3 million ha of pastures, 31.9 million ha of ploughed fields, which includes 1.8 million ha of irrigated land, 5.1 million ha of hayfields, 2.8 million ha of fallow land, and 10.4 million ha of forest. At present, 179.9 million ha, or 60 per cent, of the country's territory suffers from desertification, cutting across all kinds of land uses. Wind erosion affects over 45 million ha of agricultural and grazing lands. Water erosion extends to more than 30 million ha. Soil dehumification is observed on 11 million ha of uncultivated lands in the steppe zone. Salinization due to poor irrigation practices affects 20 per cent of all irrigated lands. In addition, soil salinization related to receding lakes has affected a large portion of the country.

Desertification in Kazakhstan has been caused both by natural and man-made factors. Amongst natural causes, climate aridization and an increase in average annual temperatures (by 0.2°C every 10 years over the last hundred years) are important factors, as are the recurrence of droughts and dust storms, weather extremes, and natural disasters (mountain torrents, floods, etc.). Human-induced factors are the same as in the majority of arid countries. These include excessive grazing, poor agricultural systems, mining, regulation of river outlets and the construction of unsustainable reservoirs and canals, for example, those which caused the Aral Sea crisis, and the drying of the Balkhash lake. Unplanned and unregistered forest cuttings, hay cuttings, fuel and forage collection and industrial pollution of soils and underground water, as well as urbanization, all have contributed to land degradation.

The causes of desertification of Kazakhstan's territory are not only national but also international. For example, the Aral Sea crisis (Box 1.1) is a regional problem as is the rising of Caspian Sea Level.

Overall damages from desertification in Kazakhstan are estimated at thousands of millions of US dollars. It is predicted that if present trends continue, desertification will inevitably lead to irreversible loss of biodiversity, a reduction of land fertility, and a corresponding deterioration in living standards. Already, this has not only caused a reduction in crop productivity, but also in animal (cow milk) productivity as a result of the decline in forage resources.

Combating desertification and preserving the land have to be tackled both at the national and subregional level. At the national level, the government of Kazakhstan has already adopted a National Action Plan to Combat Desertification. The plan proposes to tackle the problem through actions on both economic and ecological fronts, and recommends numerous measures. The plan, however, needs urgent implementation with the active participation of all stakeholders, including the State's administrative, legislative and executive organs, as well as non-government organizations and the local population. Moreover, support is also needed from neighbouring states to control the transboundary factors which are contributing to desertification in Kazakhstan.

Source: Baitulin. I. and Beiturova, G. 1997

Table 1.2 Current Assessment of the Extent of Various Types of Land Degradation in India

Type of land degradation	Area (million ha)	Per cent of total geographical area
Erosion by water	57.15	17.42
Erosion by wind	10.46	3.18
Ravine formation	2.67	0.81
Salt affliction	6.32	1.92
Waterlogging	3.19	0.97
Mining and industrial wastes	0.25	0.08
Shifting cultivation	2.37	0.72
Degraded forest	24.89	7.58
Special problems	0.11	0.30

Source: Kaul et al 1999

- *significant reduction* in water from overland flows or aquifers and a reduction in water quality;
- *encroachment* of sand and crop damage by sand-blasting and wind erosion; and
- *increased gully* and *sheet* erosion by torrential rain.

"Hot spots"(or Tension zones) exhibiting these effects in the Asian and Pacific Region are given in Table 1.3. An expert consultation on land degradation, convened in 1995 as part of International Food Policy Research Institute's (IFPRI) 2020 Vision initiative, identified several "hot spots" for soil degradation in irrigated agriculture. Salinization was considered a potential problem in the Indus river basin, northeastern Thailand and People's Republic of China. Soil quality may also limit yields in the rice-wheat system of South and Central Asia, and in irrigated rice production under intensive management in the island of Java in Indonesia, People's Republic of China, the Philippines, and Viet Nam (Scherr and Yadav 1996).

Box 1.3 Technique for Establishing a Shelterbelt System Around an Oasis and its Benefit in Xinjiang, People's Republic of China

Among provinces of People's Republic of China, Xinjiang houses the largest arid desert. It has a fragile ecosystem which suffers from heat, insufficient water resources, scanty vegetation, population pressure, erosion and land degradation. In order to protect the oasis in Xinjiang from land degradation, a shelterbelt system was designed to fit the special characteristics of the area. Shrub-grass belts, wind- and sand-breaks were built. In the periphery of the oasis, forest belts were grown at the border, and forest networks were also established in the inner area of the oasis. The key to the establishment of a shelterbelt system in the area was the efficient utilization of limited available water resources. The proportion of the farmland had to be arranged in relation to the forest area and the crop structure had to be adjusted accordingly. Part of the spring floods were allocated/used for afforestation instead of agriculture. Additionally canals were also built for irrigation.

The ecological and economic benefits of a shelterbelt system were remarkable. The shrub-grass belt in the periphery, of the oasis, for example, prevented the fringe area from sand erosion and limited the accumulation of sand around the oasis. Larger-scale stem wind- and sand-breaks not only prevented erosion and sand accumulation, but also controlled the place and amount of accumulated sand, depending on their structure and the tree species used. The main part of the shelterbelt system was farmland protective shelterbelts inside the oasis. Aside from the positive effects on crop production by improving the soil, these shelterbelts have also supplied timber and other forestry products.

The shelterbelt system in the Xinjiang province thus played an important role in dealing with wind erosion and desertification, and also enhanced oasis stability and promoted the development of sustainable agriculture and animal husbandry. Development of such shelterbelts in areas with similar ecological conditions in Asia and the Pacific could bring similar benefits.

Source: UNDP Office to Combat Desertification and Drought

Table 1.3 **Land Degradation/Desertification "Hot Spots" in Asia and the Pacific**

Nutrient depletion	Salinization	Erosion	Agrochemical pollution
Mid-altitude hills of Nepal	Indus river basin	Foothills of the Himalayas	Heavy use of pesticides on cotton in Pakistan
Poor soil quality in areas of northern India in transition to permanent agriculture	Northern Thailand and PR China (constraints to yield increases of rice, wheat etc.)	Sloping areas in Southern and Southeast Asia	Water pollution in high-density areas and coastal areas in the Asian and Pacific region
Nutrient mining of sandy soils of northern Thailand and remote upland areas in the region	Stagnant yields of intensive irrigated rice in dense areas of Java, PR China, the Philippines and Viet Nam (waterlogging, nutrient imbalance)	–	Pollution from peri-urban agriculture
Poor quality soil in Myanmar degrading in transition to permanent agriculture	–	–	Coastal and delta degradation due to sedimentation in Southeast Asia

Source: Sara, J. and Satya Yadav 1996

CAUSES OF LAND DEGRADATION

The causes of land degradation can be divided into *natural hazards*, *direct causes* and *underlying causes* (FAO 1994). *Natural hazards* relate to factors of the biophysical environment that increase the risk of land degradation for example landslides or water erosion on steep slope. *Direct causes* are unsuitable land use and inappropriate land management practices. *Underlying causes* are the reasons for which inappropriate types of land use or land management are practiced. These may relate to socio-economic circumstances of the land users and/or the social, cultural, economic and policy environment in which they operate.

A. Natural Hazards

There are many natural causes of land degradation, for example a considerable proportion of the eroded sediment found in river systems can be attributed to natural erosion and various on-going geomorphologic processes associated with the shaping of upland landscapes. Even with excellent forest cover, the soil can become totally saturated during periods of heavy and prolonged rainfall resulting in high levels of natural runoff that can often concentrate into a single channel, causing natural erosion by gullying. The major natural hazards in the Asian and Pacific region include:

CHAPTER ONE

- monsoon rains of high intensity, particularly on steep slopes of the mountain and hill lands;
- soils with low resistance to water erosion (e.g., silty soils, and topsoils low in organic matter);
- semi-arid to arid climates with high rainfall variability and with liability to drought spells and low but torrential rains; and
- soils with low resistance to wind erosion (e.g., sandy soils).

Floods are also a major natural hazard that occur in many countries every year. For example, in India, floods affect an average of 8 million ha, of which 3.7 million ha are cropped. Annual estimated damages due to floods has been put at Rs. 6 268.5 million (approximately US$157 million). In some cases, these natural hazards are of sufficient intensity to give rise to unproductive land without human interference. Examples include the naturally saline soils, which occur in some interior basins of dry regions such as Australia and People's Republic of China; and areas of natural gullying (badlands) which are common in Potowar plateau of Pakistan. In many cases, however, land shortage within the region has led to the widespread use of land prone to natural hazards for agricultural purposes, enhancing considerably the process of land degradation.

B. Direct Causes

A number of human-induced factors such as unsuitable agricultural practices, deforestation, poor quality irrigation water, absence or bad maintenance of erosion control measures, untimely or too frequent use of heavy machinery or improper crop rotations, directly contribute to land degradation. Overgrazing and trampling by livestock causes erosion and soil compaction, and enhanced water and/or wind erosion. The extension of cultivation or grazing onto lands of lower potential and/or high natural hazards also contributes to land degradation.

In areas of non-saline groundwater, the technology of tubewells has led to abstraction of water in excess of natural recharge by rainfall and river seepage and a progressive lowering of the water table promoting aridity/desertification. Over extraction of water (for irrigation, urban and industrial use) from rivers and other surface water sources have reduced downstream availability and in certain cases, incursion of sea water resulting in salinity (see Chapter 4). Further, in some cases, used water may have a higher salt content and/or be polluted from agro-industrial chemicals and human wastes causing salinity or pollution of the soil. Industrial activities including power generation, infrastructure and urbanization, waste handling, traffic, etc. may also result in land pollution. The most significant example of desertification by human-induced factor is seen in Central Asia where the effects of overgrazing, over-irrigation, and pollution (Kharin 1996) etc. are clearly visible.

C. Underlying Causes

Within the Asian and Pacific Region the most important single underlying cause of land degradation is poverty. Lack of alternative income generating activities (off- and on-farm) means that the majority of the region's rural households remain dependent on land-based small-scale farming and/or forestry activities for their livelihood. In particular the indigenous and migrant population in the more remote upland and mountain areas, are generally very poor and often struggle to meet their basic survival needs. The poor are often unable to apply long term sustainable practices and remain reliant on short-term production goals, such as food production for survival.

In Papua New Guinea, in the highlands where population densities are relatively high (13.5 per cent of the total land with 36 per cent of the population), the practice of subsistence agriculture causes land degradation (Darkoh 1996). Most at risk is land that is cleared and subsequently utilized continuously or where the period of fallow is inadequate for the recovery of its previous condition. Shortening of fallow cycles and frequent burn-offs have led to the conversion of primeval forests to secondary forests and ultimately to bush and grasslands in the upland areas. Shifting agriculture is principally responsible for the occurrence of anthropogenic grasslands, over an area of 3.5 to 4.0 million ha in Papua New Guinea.

The promotion of commercial agriculture in an effort to improve exports and national income has also resulted in land degradation. For example, foreign exchange earning crops have been expanding into forest lands and have displaced subsistence farmers onto alternative forested areas or to marginal agricultural lands, therein, contributing to land degradation. Land degradation may be further enhanced where commercial farming adopts exhaustive and input-intensive monoculture.

Another major underlying cause of land degradation is the pressure on arable land, from population growth demanding food and competing for land. The population density per hectare of arable land in different subregions and countries of Asia and the Pacific is presented in Figure 1.5. Both South and Northeast Asia bearing a population density of about 15 per ha of land are under heavy human pressure compared with mean values for Asia and the Pacific (7 people per ha) and the World (4 people per ha).

Figure 1.5 Population Density per Hectare of Arable Land in Selected Countries and Subregions of the Asian and Pacific Region, 1996

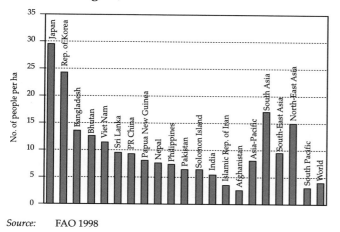

Source: FAO 1998

Figure 1.6 Percentage of Land Degradation in Asia and the Pacific and the World, by Land Use Patterns

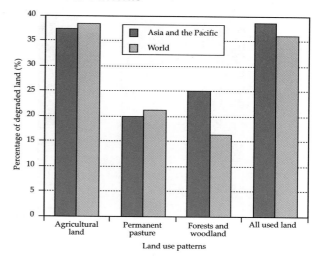

Source: Oldeman et al 1991

Notes: Figures for all used land are combined between degraded and seriously degraded land. In the GLASOD study, a lightly degraded soil is defined as having somewhat reduced agricultural suitability, but is suitable in local farming systems. Original biotic functions are still largely intact, and restoration to full productivity is possible through modifications in farm management. A moderately degraded soil is soil that offers greatly reduced productivity, but is still suitable for use in local farming systems. Major improvements are needed that are typically beyond the means of local farmers; the original biotic functions are partially destroyed. An extremely degraded soil is defined as a human-induced wasteland, unreclaimable, beyond restoration, and with biotic functions that are fully destroyed. Data for permanent pasture and forests and woodland include arable and non-arable land.

PROCESSES AND MEASUREMENTS

A. Measurement

1. GLASOD and ASSOD

The Global Assessment of Soil Degradation (GLASOD), based on the formal survey of regional experts, was the first worldwide comparative analysis that focused specifically on soil degradation (Oldeman 1994). GLASOD was designed to provide continental estimates of the extent and severity of degradation from World War II to 1990[1].

The extent of soil degradation in Asia and the Pacific was evaluated in five major studies in the 1980s and 1990s. GLASOD estimates of soil degradation for the region compared to the world by land uses is presented in Figure 1.6. Overall, approximately 25 per cent of all land used in the region is degraded, with 13 per cent classified as seriously degraded. Out of the 851 million ha degraded land of the region, 747 million ha (88 per cent) is in Asia and the remainder 104 million ha (12 per cent) is in the South Pacific subregion.

The Assessment of Human-Induced Soil Degradation in South and Southeast Asia (ASSOD) provided a more detailed and nationally representative GLASOD-type study (Van Lynden and Oldeman 1997). The study was undertaken several years after GLASOD and found the decline in soil fertility and organic matter to be 20 times greater, with triple the extent of salinization, and nearly 100 times the extent of waterlogging found than in the GLASOD study. Agricultural activity was found to have led to degradation of 27 per cent of all land and deforestation, 11 per cent.

B. Processes of Land Degradation

There is considerable variation between estimates of GLASOD's and ASSOD's studies related to the distribution of land degradation processes in the region, due to slightly different methodologies used and the years in which the studies were undertaken. Figure 1.7 presents a comparison between the two studies and also between the region and the world average.

---

[1] The objective of GLASOD (discussed later in this Chapter) was to create awareness about the status of soil degradation. Over 250 soil and environmental scientists cooperated in preparing 21 regional maps of human-induced soil degradation, using a common methodology. Following delineation of physiographic units with homogeneous topography, climate, soils, vegetation and land use, each unit was evaluated for its degree, relative extent, and recent past rate of degradation, as well as for the forms of human intervention causing degradation. Types of degradation were ranked in importance. Map segments were compiled and reduced to the final 1:10 million scale of the GLASOD map. The map units were digitized and linked to a GLASOD database to calculate the real extent of degradation. Since the maps rely on expert evaluation they may be subjective.

Figure 1.7 Distribution of Main Degradation Types in Asia as Percentage of the Total Degraded Area

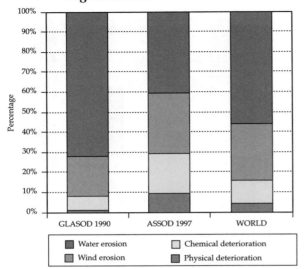

Source: FAO 1998 and United Nations 1995

1. *Erosion*

(a) Water Erosion

Water erosion is the most pervasive cause of land degradation and constitutes over 70 per cent of all the degraded land of the region under GLASOD assessment, and almost 50 per cent under the ASSOD study. Although common to all subregions, the most severe water erosion occurs throughout the Himalayas, Central Asia, People's Republic of China and the South Pacific subregion, especially Australia. Also particularly affected are Islamic Republic of Iran, India, Afghanistan and Pakistan.

(b) Wind Erosion

Wind erosion in both the GLASOD and ASSOD studies has been estimated to affect some 20 per cent of the total degraded area. Wind erosion occurs in three forms: loss of topsoil, terrain deformation and over-blowing. The destruction of natural vegetation cover resulting from excessive grazing and the extension of agriculture into marginal areas are the main causes through human intervention (Noor 1993). Wind erosion is a widespread phenomenon in the arid and semi-arid climates, such as South and Central Asia, and People's Republic of China and Australia. In the South Asia subregion, Islamic Republic of Iran, India, Pakistan and Afghanistan are worst affected. In Pakistan, wind erosion occurs in low rainfall areas in Thar, Thal, the sand desert of Cholistan and Balochistan, affecting a total of 4.8 million ha, of which 35 per cent is severely eroded. Erosion by wind is the dominant factor in western India, affecting some 13 million ha. In addition, wind erosion creates a transboundary problem. For example, "yellow dust" from northern China has been observed to affect the Republic of Korea and Japan.

2. *Degradation*

(a) Physical Degradation and Loss

The main processes of physical degradation or land productivity loss are waterlogging and aridification, along with sub-forms such as compaction, crusting and sealing and subsidence of organic soils. According to GLASOD (1990), physical soil degradation accounts for 1 per cent of the total degraded area in the region, whereas in ASSOD (1997) it was 9.1 per cent. Waterlogging has affected India, Pakistan and Bangladesh. Subsidence of organic soils is mainly found in the coastal swamps of Southeast Asia. Soil compaction is primarily caused by sealing and crusting, and has relatively small impact in the region. Agricultural mis-management and overgrazing are the two major causative factors of human-induced physical soil degradation (FAO 1999). Loss of productive land has been particularly high in a number of countries, such as in People's Republic of China, Philippines, Thailand, India and Pakistan, due to urbanization, industrialization and infrastructure development (see Chapter 16).

(b) Chemical Soil Degradation

Chemical soil degradation occurs due to the loss of nutrients and/or organic matter, salinization, acidification, or pollution from industrial activities, such as mining. In ASSOD (1997) almost 25 per cent of degraded land is estimated to be from chemical degradation, as compared with only 7 per cent in GLASOD (1990).

Agricultural mismanagement and deforestation are major causes of chemical soil degradation, whilst industrial and bio-industrial activities are the main sources of pollution. Soil toxicity can also result from the presence of municipal, industrial, radioactive or oily wastes, which occur mainly around towns, industrial areas and mines. Such toxicity, although usually local in nature, may lead to significant problems in some countries. For instance tailings from former tin mining operations have affected extensive areas in Malaysia. However, best practices in mining reveal that related problems can be mitigated with careful planning and implementation of rehabilitation activities (Box 1.4). In Australia for example, the New South Wales government provides an annual grant of $500 000 for cleaning and rehabilitation of the derelict sites. It has cleaned up around 90-110 ha a year during the past 10 years, at a cost of around $120 000 a year. During 1996-97,

Box 1.4 Mining Land Rehabilitation in Australia

The Australian government has taken a lot of interest in rehabilitating mined lands and has encouraged such initiatives by the private sector. Nabarlek Uranium mining project is a case in point. Uranium was mined from the Nabarlek ore body in a single 143-day campaign during the dry season of 1979. It was stockpiled on a specially prepared site while the mill was constructed. The ore was processed in the mill over the subsequent 10-year period. The rehabilitation aspect of the area after mining and processing of Uranium was an important component of the project from its very inception. It was for the same reason that topsoil from the mine and mill construction was placed in a stockpile and allowed to stand until required in the final rehabilitation. Tailings from the milling operation were returned directly to the mined out pit. The waste rock was placed to the south of the site and planted with an exotic grass species to provide erosion control. During the mine planning process, the final decommissioning and rehabilitation programme was developed as a series of specific component plans including an earthmoving and revegetation document. Throughout the life of the mine, these components were reviewed at intervals and updated to take account of changes in mine development as well as incorporating the results of site-specific research and new technology.

During preparation for final decommissioning, the site topsoil dump was investigated. It was found that, due to its 10 years in store, the material was of little value to the rehabilitation process. The soil had lost much of its micro flora and faunal populations, it had been leached of nutrients and had become a source of weed seeds. Few viable propagules of potentially "useful" plants had survived. Therefore the topsoil was used in the rehabilitation work but not as a final cover as this would have spread undesirable weeds across the site. The waste rock dump had been untended during the life of the mine and had become well vegetated with a wide range of native species of trees and shrubs. This material was selected for the final cover for reshaped and rehabilitated landforms. The rehabilitation objective, as agreed with the traditional owners and the supervising authorities, was to establish a landscape that matched the surrounding areas as closely as possible and would permit traditional hunting and gathering activities to be pursued.

The earthmoving plan placed all mine wastes in the mined out pit together with scrap metal etc. This was then covered with a layer of waste rock up to 15 metres thick and the final landform left as a mound over the pit to allow for subsidence and to still provide a water shedding cover. The original cover design was of great importance as it was required to act as a barrier to radon gas and to contain the tailings and radioactive waste for thousands of years.

Earthmoving for the final landform shaping was carried out during the dry season of 1995. Apart from demolishing earthworks, including substantial pond walls, the work also required the land surface over most of the site to be returned to approximately its original contours. The ponds were filled in and the waste rock was spread and incorporated the degraded topsoil lower down the soil profile.

One concern while completing the rehabilitation earthworks was the amount of compaction caused over the site as a result of the constant passage of trucks and other mobile plant. At the end of earthmoving, therefore, a large bulldozer fitted with a winged deep ripping tyne was used to rip the whole site to loosen the surface and provide improved conditions for seed germination. During this operation some oversize rocks were brought to the surface, which were collected into piles and spread randomly across the site to provide refuges for small animals and reptiles that were anticipated would re-colonize the site.

The final domed cover over the pit was designed following research and shaped to provide shorter runoff paths and so reduce runoff water velocities. A single, low, central ridge was established to facilitate these shorter flow paths. Seeding was carried out at the end of earthmoving, immediately before the onset of the monsoon rains of the 1995-96 wet season as previous work on site had shown that this was likely to be the most successful revegetation approach. The rehabilitation of the site is progressing well and continued monitoring is in place to establish when the site can be returned to the traditional owners.

Nabarlek story is unique and offers practical approaches towards planning and executing the rehabilitation process of mined lands.

Source: Government of Australia 1999

---

23 sites and 72 ha of derelict mining land were rehabilitated at a total cost of $501 800.

(c) Biological Degradation

Biological soil degradation is associated with lowering or depletion of soil organic matter, continuing negative soil nutrient balance, imbalance in fertilizer application, and secondary and micronutrient deficiencies. Negative soil nutrient balances have been reported in South Asia (Tandon 1992) for all three major nutrients in Bangladesh and Nepal; for phosphorus and potassium in Sri Lanka and a large deficit of potassium in Pakistan. In India, it has been estimated that the nutrient deficit is 60 Kg/ha per year, or nine million tonnes for the whole country. Sulphur deficiencies have also been reported in Bangladesh, India, Pakistan and Sri Lanka, and zinc deficiencies in Bangladesh, India and Pakistan. Micronutrient deficiencies are being increasingly reported in Pakistan.

CHAPTER ONE

DESERTIFICATION

Desertification has been defined as *"land degradation in arid, semi-arid and dry sub-humid areas resulting from various factors, including climate variation and human activities"* (UNCED 1992). A recent study conducted by UNCCD (1998) indicated that Asia has 1 977 million ha of drylands, which is 46 per cent of the continent's and 32 per cent of the world's total surface area, over half of which is suffering from desertification. The worst affected area is Central Asia (with a total of nearly 60 per cent of land affected in some form) followed by South Asia (nearly 50 per cent) and Northeast Asia (nearly 30 per cent). In Central Asia, the worst affected country is Turkmenistan, where desertification affects two-thirds of the territory. In South Asia, Afghanistan is the most severely affected with over four-fifths of its land affected, followed by India, Pakistan and Islamic Republic of Iran. The estimates for People's Republic of China vary from 8 per cent to 27 per cent of the total area. In Mongolia, about 41 per cent of the total area has been affected by desertification in some form with more than 90 per cent of the cropland eroded by wind and water.

Drylands in the region have three major land use systems: irrigated croplands, rainfed croplands and rangelands. In particular, the region has the biggest portion of irrigated land in the world (8 per cent), with over 164 million ha. In five countries of the region, People's Republic of China, Islamic Republic of Iran, Democratic Republic of Korea, Pakistan and Republic of Korea, more than 40 per cent of all arable land was under irrigation in 1996. Figure 1.8 shows the pattern of dryland desertification in the region, as compared to the world.

A. Processes of Desertification

1. *Irrigated Croplands*

The major processes of desertification in dry irrigated lands are waterlogging and salinity. For example, in Central Asia, a rise of the groundwater table around irrigated areas and canals has resulted in salinization and waterlogging of the soil, spoiling vast area of grazing lands, and cotton plantations, occupying 60 per cent of the irrigated land, are seriously affected by an almost irreversible salinization; 15 to 90 tonnes per year of salt can accumulate in one hectare of irrigated soil (Kharin 1994).

Human-induced soil salinization is also a major threat to the sustainability of irrigated agriculture in People's Republic of China, India and Pakistan and is instrumented by faulty irrigation management, such as the poorly drained irrigation lands in hot climates with high evaporation. In addition, inadequate drainage causes the water table to rise, bringing saline groundwater into contact with plant roots. People's Republic of China is increasingly suffering from salinity throughout its arid regions as new irrigation oases are being established and old ones expanded. Natural and human-induced salinity affects more than a fifth of land in People's Republic of China in particular in Ningxia and Hetao irrigated plains along the Yellow River, northern (North China Plain), central, and western areas. Naturally saline areas are especially extensive in the gravelly (gobi) soils of Xinjing and the Tibetan Plateau in western China.

Waterlogging and salinization also affects between 2 and 3 million ha in India and Pakistan respectively (FAO 1994). In coastal zones, such as the South Pacific Islands, over abstraction of groundwater and sea level rise, has resulted in salinization from salt water encroachment. One very serious effect has been its impact on freshwater sources and affect the roots of pit-grown taro, coconut palms and other tree crops (ADB/SPREP 1992).

2. *Rainfed Croplands*

The principal causes of desertification in rainfed lands are erosion by wind and water. In particular, India, Pakistan, Indonesia and Thailand are suffering from desertification of rainfed areas.

In the prime rainfed lands farmers have greatly increased cropping intensities, in response to high and rapidly growing rural populations and the

Figure 1.8 Pattern of Dryland Desertification in the Asian and Pacific Region Compared with the World

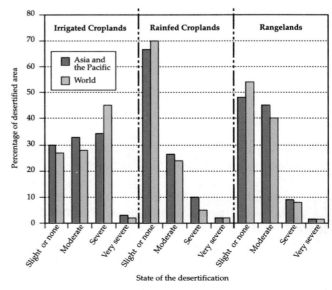

Source: FAO 1997

> **Box 1.5 Traditional Rotational Use of Pastureland in Mongolia**
>
> Animal husbandry history in Mongolia dates back thousands of years. Over time, Mongolians have developed a nomadic lifestyle, in accordance with their environment, which included creation of a unique way of utilizing pastureland and developing appropriate herding technologies.
>
> Traditionally, Mongolians used pastureland rotationally by frequently moving from sites between seasons. Winter and Spring pasture was divided into three to four different classes, such as the main pasture, reserved pasture, and pastures for hobbled horses and milking cows. The classes were then divided into smaller camps to increase utilization. The pasture directly around the camp was used in rotation on a daily basis during the growing season as well as during cold days in Winter and hot days in Summer. Pasture within a distance of four to five kilometres from the camp was used during warm days in winter and cold days in summer. Throughout the grazing activity, the herders were very sensitive to degradation of pastureland, moving from one camp to the next every 15-25 days, depending on pasture condition.
>
> This traditional rotational use of pastureland had economic and environmental as well as social benefits. Economically, it enabled Mongolians to utilize pastureland as efficiently as possible, which in turn was environmentally sound. Socially, the rotational use demanded a steady movement by the herders, having a positive impact in terms of information exchange and social contacts, etc. on society and culture. During the period of central planning, however, these technologies and practices were given up and large numbers of livestock were made to concentrate around dairy farms, resulting in excessive pressure on carrying capacities of these areas, and their subsequent degradation.
>
> With the dawn of new era, it is important that some of the traditional methods and technologies used in past are promoted on a wider scale to prevent further degradation of pastureland in Mongolia. This is extremely important for a country of about 30 million livestock, and where more than 90 per cent of land is under natural pastures. These traditions and technologies could also be replicated to promote sustainable grazing in pasturelands elsewhere in the region.
>
> *Sources:* FAO Regional Office for Asia and the Pacific, Bangkok

development of agricultural markets. In India, 70 per cent of the wet cropped area is rainfed, producing 40 per cent of the country's food grain (Kaul et al 1999). The Green Revolution, which brought increased use of hybrids, chemicals, mechanization and a trend towards monocropping, also played a pivotal role in these areas. In some cases, inappropriate use of machinery has led to soil compaction; poor vegetation management has exposed soils to erosion; and substitution of organic inputs with chemical fertilizers has led to declining organic matter and acidification of vulnerable soils. In addition, cultivation has spread into more marginal land, such as steep slopes, with poorer and more vulnerable soils, and where human settlements compete for use of agricultural lands. Often the intensive farming practices of the high potential areas are inappropriately applied to the marginal lands, where over exploitation for subsistence and commercial uses often leads to loss of vegetation for soil cover and fallow periods being reduced. This results in soil erosion and nutrient depletion, although there is some evidence that intensification has led to greater use of soil-protecting practices (Scherr 1997).

3. *Rangelands*

Rangelands, claimed from permanent pasture and grazing land for cropping purposes, are also highly prone to desertification. The Asian and Pacific Region has the largest area of degraded rangelands, at about 1.56 billion ha, which amounts to 34 per cent of the total rangeland area in the world. These frontier areas have lower intrinsic soil quality or pose higher production risks due to factors such as steep slopes and extremes of rainfall. The soils are degraded by the land-clearing process, decreasing fallow periods that deplete nutrients, and by widespread burning to control weeds and pests. Large areas have been abandoned due to nutrient and organic matter depletion and weeds. Intensive extraction of groundwater and over grazing also cause desertification in rangeland. For example, in the Islamic Republic of Iran, nomad herds have acted in the short term by converting submarginal sandy rangelands to melon farms, using water from wells, unaware that over-extraction from the well depletes the aquifer source in the long term (Kowsar 1998). Sustainable management of grazing, as in Mongolia (Box 1.5), will reduce impacts on the land, however grazing beyond the carrying capacity of land may result in desertification.

CONSEQUENCES OF LAND DEGRADATION

A. **Productivity and Economic Loss**

The most important consequence of land degradation is lost productivity. According to one estimate, erosion, conversion to nonagricultural uses, salinization, inundation and toxification decreases the cropland area of the world by about 12 million ha a year (Kowsar 1998). Monetary valuation of this loss is very difficult to calculate. Early crude estimates of the annual cost of soil erosion in the world hovered around US$26 billion and according to UNEP, about half of this cost was borne by developing countries (UNEP 1980). A decade later, Dregne and Chou (1992) proposed US$28 billion per year as the cost of dryland degradation. Using GLASOD data, Crosson (1995b) estimated an aggregate global loss of 12-13 per cent of agricultural supply.

A UNDP/UNEP/FAO study on land degradation in South Asia attempted to assess the economic cost of land degradation for the subregion utilizing a resource accounting approach with cautious assumptions. On this basis, the best estimate that could be obtained is that land degradation is costing the countries of the subregion more than US$10 billion per year, equivalent to seven per cent of their combined agricultural gross domestic product (AGDP) (Dent 1990). In Pakistan, the value of reduced wheat production (due to waterlogging and salinization) in 1993 approximated to 5 per cent of AGDP, while in India, annual cereal production loss amounted to about 5 per cent of AGDP. Desertification damage includes lost income and the cost to rehabilitate degraded land (Kharin 1998). In Central Asia, desertification costs amount to about 3 per cent of the subregion total national income (NAP 1996, Kharin 1996 and CCD Interim Secretariat Report 1998).

Severe desertification of the Aral Sea Basin (Box 1.1) led to the decline of fish catches, an increase in the cost of supplying water to villages, sand encroachment, massive salt accumulation (30 to 60 tonnes per hectare at northern and eastern sides) and dust storms with inter-regional implications. In Kazakhstan and Uzbekistan, this has resulted in the subsequent loss of employment in agriculture, fishery and industry and has led to mass migration. It is estimated that about 460 000 people are severely affected by the ecological crisis and two million more live in vulnerable conditions. About 65 000 people a year are migrating in Uzbekistan and Kazakhstan. Reportedly, almost 300 000 people left their homes (called environmental refugees) in 1995 due to the severity of desertification in Kazakhstan. The inhabitants of the Aral Sea region also suffer from poor health as a result of dust and salt in the air. This has led to a high number of contagious diseases (especially typhus and gastro-intestinal illnesses), eye diseases and a very high infant mortality rate of 10 per cent (Plit *et al* 1995). Calculations of direct losses (loss of production from natural resources) and indirect losses (cost of regeneration and desertification control) in Turkmenistan amount up to US$347 million a year (Kharin 1998).

In Southeast Asia, the densely populated and intensively cultivated island of Java, Indonesia appears to have experienced high soil degradation (De Graaff and Wiersum 1992 and Diemont et al 1991). Magrath and Arens (1989) calculated that agricultural productivity was declining at the rate of 2-5 per cent a year due to soil erosion, creating annual economic losses of nearly one per cent of the gross national product (GNP) or approximately 3 per cent of GDP.

Degradation has reduced People's Republic of China's production of grain yields by as much as 5.6 million metric tonnes of grain per year, roughly equivalent to 30 per cent of yearly grain imports in the early 1990s. It is possible that without environmental degradation, rice yields would have grown 12 per cent faster in the 1980s (Huang and Rozelle 1994 and 1996; Huang et al 1996). The same authors calculated that the economic loss from soil degradation in People's Republic of China in the late 1980s reached $700 million (at 1990 prices), an amount equal to People's Republic of China's budget for rural infrastructure investment.

In Australia, lost agricultural production exceeds AUS$130 million annually, all of which is attributed to salinity Box 1.6 (AFFA Joint Release 1999).

Irreparable damage is also being done to the topsoil of the Australian countryside by cattle farming (Asia-Pacific Agribusiness Report 1996), and as topsoil is lost, desertification will occur. For example, in the State of Queensland, where the sheep industry yearly brings in profits of AUS$160 million estimated costs from resulting land degradation totals half that amount.

KEY ISSUES AND TRENDS

The current trends in Asia and the Pacific indicate that soil degradation problems especially related to irrigation, could intensify in the future. The greatest policy challenges in coming decades will be in densely populated areas, with soils of lower resistance and higher sensitivity to degradation, and where degradation is increasingly limiting agricultural production, economic growth, and rural welfare. For countries with limited high-quality rainfed and irrigated land, the impacts may be especially acute.

A. Irrigated Agriculture

It is projected that the expansion of irrigation worldwide will slow down significantly between now

Box 1.6 Cost of Australian Dryland Salinity

"Australia currently has 2.5 million ha of salt affected land and this is likely to increase six fold in the coming decades". Costs associated with dryland salinity include (in Australian dollars) $130 million in lost agricultural production, $100 million in infrastructure costs, $100 million in local supply catchments. As well, the salinity affects the biodiversity of wetlands. The salt was deposited in some areas of the country by mildly saline rainfall falling onto a semi-arid landscape and evaporating over billions of years. In other parts, notably New South Wales, the salt originated in ancient seabeds. "In South Australia, at least 20 per cent of surface water resources are sufficiently saline to be above desirable limits for human consumption". These limits will be pushed further and for longer periods in the present millennium.

Australia's original ecosystems coped with salinity, but European settlement during the last 200 years has dramatically changed those ecosystems, and the massive conversion from native bush to agriculture in the last 50 years has created a major problem. Officials say that up to 30 per cent of regional roads are being affected with major highway reconstruction costing up to $1 million per kilometre. Australia's National Dryland Salinity Programme, has been extended to more than 80 regional towns and cities recently on costs related to salinity. The costs include damage to building foundations, bridges, pipelines and roads.

To improve this situation, the scientists are calling for a major intervention. For example, tree planting is needed across most of the landscape, including 50-70 per cent of catchment areas, to achieve significant reductions in the ultimate extent of the salinity. It is also opportunity to explore the broader options of saline aquaculture and growing various salt tolerant pastures, crops and woody perennials.

However, according to experts the response times of salinity control will be long and it is unlikely that these systems will be completely restored within normal human time scales but looking at the enormous environmental and economic cost in the long term, it appears to be a worthwhile effort.

Source: Environmental News Network published on 25 June 1999

---

and 2020. In developing countries, irrigated area is expected to increase by only about 40 million ha (to 227 million ha), at an annual growth rate of only 0.7 per cent, compared with 1.7 per cent during 1982-93. Of all the irrigated area in developing countries in 2020, it is thought that 80 per cent will be located in India, People's Republic of China, West Asia, North Africa, and Pakistan (Table 1.4).

This has two major implications for future soil degradation. First, problems of salinization and waterlogging are likely to increase, as recently developed systems with inadequate drainage infrastructure or water management get older. Whether governments and local people will be willing to divert infrastructure investment capital to provide proper drainage in new systems and prevent degradation or rehabilitate older systems will depend on the general profitability of irrigated agriculture. Systems that depend on flushing large amounts of water to manage salinization may become much more vulnerable to degradation as water pricing is introduced. Second, without proactive efforts, a considerable amount of irrigated land will go out of production. Indeed, where irrigation systems were built under unsustainable conditions, this will be inevitable. In some countries, this loss of irrigated land will affect aggregate agricultural supply (see Chapter 10). In far more cases, especially in South Asia, however, it will have serious local repercussions on economic growth and poverty alleviation.

Table 1.4 Trend in Irrigated Area Expansion in Selected Developing Countries and the World

Subregion	1993 (ha)	2020 (ha)	Rate of increase (%)
Southeast Asia	14 316	16 195	13.13
PR China	49 872	53 075	6.42
Other East Asia	2 877	2 878	0.03
India	50 101	68 619	36.96
Pakistan	17 120	20 538	19.96
Other South Asia	7 526	8 719	15.85
West Asia and North Africa	23 819	31 186	30.93
Latin America	17 147	18 748	9.34
Sub-saharan Africa	4 850	7 375	52.06
All developing countries	187 628	227 332	21.16
World	253 003	295 964	16.98

Sources: FAO 1994 and Pinstrup-Andersen et al 1997

B. High-Quality Rainfed Lands

Few hot spots are identified in the high-quality rainfed lands. However, erosion and compaction problems caused by mismanaged mechanization are considered important in these areas along with agrochemical pollution due to poor nutrient management practices, for example, in high density

CHAPTER ONE

and coastal farming in Southeast Asia (Scherr and Yadav 1996).

C. Densely Populated Marginal Lands

Several hot spots for soil degradation in densely populated marginal lands have been identified. Nutrient depletion was considered critical in the mid-altitude hills of Nepal and the sandy soils of northeastern Thailand; technological constraints to yield increases were perceived to be a major threat in the marginal arable lands in Islamic Republic of Iran; and erosion was cited as a particular problem in the Himalayan foothills, the Southeast Asian hill country, and the Asian rangelands that have been converted to grain production.

In situations where current land pressure is moderate, technology is available for sustainable intensification, and the economic incentives for its use are favourable, some types of soil degradation (water erosion, for example) can be expected to decline in future. However, grave economic effects from further soil degradation can be expected in areas with high population growth rates, where technologies for more intensive, sustainable soil management practices are still unobserved, and where unfavorable economic policies and incentives undermine agricultural investment.

D. Extensive Agriculture in Marginal Lands

It is likely that in the future, much of the land which has already been cleared and extensively cultivated will be under semi-permanent cultivation or else abandoned due to degradation. Currently identified hot spots include areas with nutrient depletion in remote upland areas in Northeast and Southeast Asia, as well as poor quality soils in north-eastern India in transition to permanent agriculture. Erosion will also continue to be a major problem in sloping areas in southern China and Southeast Asia (Scherr and Yadav 1996).

There will be new opportunities for rehabilitation of degraded lands through sustainable pasture management systems, and improved fallows using agroforestry. Development programmes are likely to promote "mosaic" landscapes (Forman 1995), with areas maintained under natural vegetation and crops and management systems adapted to various production niches. Production systems which are economically appropriate for low land-use intensity will also be used.

E. Urban and Peri-Urban Lands

With the rapid growth of urbanization, urban and peri-urban agricultural land will expand. Urban agriculture may play a growing environmental role in the recycling of urban solid waste and wastewater.

It may, however, also contribute to health problems from contaminated food, pollution caused by insecticides, wastes and agrochemicals, and downstream flooding due to poor farming practices on slopes and streambanks. Contamination of soils with heavy metals, chemicals, waste, and other urban pollutants may pose a health hazard to consumers and also hinder production.

POLICIES AND PROGRAMMES

A. National Initiatives

In order to combat land degradation, significant efforts have been made at the national level throughout the region for formulating and implementing appropriate policies, plans and programmes, often with the assistance of UNDP/ESCAP. The most significant steps, have been taken under the Convention to Combat Desertification (CCD) in the Asian and Pacific region (further discussed in the next Section), in the development of National Plans for Action (NPAs) for the control of desertification. Various countries, including Mongolia, Pakistan, the Islamic Republic of Iran, Kazakhstan and Turkmenistan have already developed such plans, and a number of other countries are in the process formulating them.

Generally, the country NPAs for desertification include issues such as the provision of monitoring, coordination, educational, legal, and other institutional support. Specific projects have been identified to address the critical issues in each country. For example in Mongolia, the NPA includes the promotion of sustainable pastoral land use systems, integrated management and rehabilitation of crop lands and the sustainable management of forest resources. In Central Asia, NPAs have been developed for Turkmenistan and Kazakhstan with external funding assistance. For Kazakhstan the NPA highlights the prevention of soil degradation from salinization, wind erosion and other types of degradation, restoration of fertility of arable lands and rangelands and recultivation of industrial lands. Additionally, the importance of realization of concrete regional projects to overcome the cross-boundary consequences of desertification is stressed. The NPA in Turkmenistan highlights issues such as the development of a strategy of utilization of water resources, conservation of biodiversity, melioration of irrigated farmlands, range management and improvement, stabilization of moving sand dunes, improvement of nature conservation legislation, and international cooperation (Kharin 1996).

Since the early 1990s, People's Republic of China's nationwide projects and programmes to combat desertification have been strongly intensified.

Since 1991, more than 4 million ha of degraded lands have been rehabilitated by reforestation or sand dune fixation, and 3.76 million ha of desertified land was treated. Additionally, around 5.5 million ha of salinized farmland was treated, accounting for 12 per cent of total area suffering from salinization. The second phase of the programme, covering the period 1993-2002, aims to introduce higher efficiency and quality in crop production (NEPA 1993 and UNCCD 1998). In 1994 combating desertification was included in the Chinese Agenda 21 and the Chinese government signed the CCD. People's Republic of China's activities include the construction of sand-break forests, dune fixation afforestation, the installation of drainage networks in salinized areas, and a variety of erosion control measures.

In Australia, land administrators, managers and scientists are responding to past mistakes and land-use changes in a variety of ways. Restrictions have been placed on land clearing in most areas and much of the native fauna is now protected. A taxation system is being imposed to promote better land management, and soil conservation works are considered as tax deductible. It is not only the government that is making changes; community-based action programmes are now seen as crucial in combating land degradation and they make up a major part of the National Landcare Programme. A joint effort of the National Farmers' Federation and the Australian Conservation Foundation resulted in a national programme with 2 200 Landcare groups nationwide, involving one-third of all farm families.

National land use policies in India, take into account the environmental, social, demographic, economic and legal issues (Kaul *et al* 1999). From 1985-1997, 23 million ha were afforested under such policies. With the realization of the fragility of the arid land ecosystem, the Desert Development Programme (DDP) was initiated in 1977. One component is the Indira Gandhi Canal, in northwestern Rajasthan, which covers part of the Thar Desert, providing irrigation facilities to over 2.5 million ha of the Desert, of which 1.2 million ha is cultivable command area. Under an externally aided Overseas Economic Cooperation Fund (OECF) project, a total area of 33 725 ha has been protected against sand deposition.

Integrated watershed management programmes in India have also been implemented extensively. These programmes planned to cover 86 million ha, of which 26 million ha (27 river valley catchments and eight in flood-prone rivers) in highly critical areas have been assigned priority under 35 centrally sponsored projects. In addition, over 30 000 ha of shifting and semi-stable sand dunes have been treated with shelterbelts and strip cropping as a conservation measure. The National Watershed Development Project for Rainfed Areas (NWDPRA), initiated in 1990, was redesigned in 1993 with a focus on development of micro-watersheds as models of comprehensive and integrated development in different agro-climatic regions of the country.

The Department of Soil and Water Conservation in Nepal, is conducting watershed management programmes in critically affected or degraded areas, such as the Kulekhani Watershed Management Project and the Phewa Tal Watershed, and has had considerable success in reducing the extent of land degradation. Involvement of the local communities at every stage in the implementation of the projects has also ensured the sustainability (Sharma and Wagley 1995). Bhutan has also been carrying out integrated watershed management projects, particularly in the highlands, to improve the condition of lightly degraded land and avoid future degradation.

Countries in the Asian and Pacific region have also developed and followed appropriate technologies for reducing land degradation. For example, minimum tillage techniques (resulting in better yields, lower input costs, higher profits and the reduction of erosion) have been applied in countries such as Australia. Other techniques include Sloping Agricultural Land Technology (SALT), which has been developed in the Philippines. Here, the technology is based on a system of agroforestry and contour cultivation, allied with a number of practices which control erosion and lead to enhanced production and farm incomes.

B.   International and Regional Initiatives

1.   Convention to Combat Desertification (CCD)

A global and regional pursuit to saving the land from degradation and desertification centres on the implementation of the CCD (Box 1.7), which reached a significant phase following the convening of the First Conference of Parties (COP) in Rome in October 1997. The COP clearly manifested the commitment of governments to move the CCD to the next levels of implementation. Since the Convention's adoption in 1994, the International Negotiating Committee for Desertification (INCD) has taken steps in laying the foundation for a unified action to address desertification issues. Supervised by the Interim Secretariat, human and financial resources, however limited, were effectively mobilized to lay the cornerstones for a concerted effort to arrest the devastating effects of desertification. The CCD is implemented through regional and national action programmes which begin with long-term strategies and priorities (UNCCD 1999). National governments commit to providing

CHAPTER ONE

> **Box 1.7 Implementation of Convention to Combat Desertification (CCD) in Asia and the Pacific**
>
> The Convention on Desertification adopted in June 1994, as the first multilateral legal instrument adopted after Rio, has successfully integrated, innovative aspects to address a major threat to sustainability in the world's drylands. In addition to stressing the need for national ownership of the whole process of implementation, it places the principal focus on activities at local or community levels to improve livelihood security, using a genuine participatory approach with all the concerned stakeholders. Enabling measures would also be taken at national level to support and sustain these activities. As the very first step, the Convention calls for the building of partnerships within the country and with external partners as a pre-requisite for ensuring the fullest possible coordination and for sustained support to its implementation. Finally, it calls for the development of innovative resource mobilization strategies to ensure effective implementation.
>
> Many Asian and the Pacific countries are affected by desertification and drought in one way or another. As of 5 May 1997, 18 Asian and Pacific countries had already ratified or acceded to the Convention (as of March 1999, this number had increased to 22). National, subregional and regional consultations took place throughout the region, from 1995 onward, and are continuing, with participation from all stakeholders. The UNCCD Secretariat and ESCAP have worked together to strengthen international cooperation both among countries and within the Asian and Pacific region. The Convention requires member parties to prepare national action programmes (NAP) with the consultation of affected populations in the drylands. In addition, the CCD also calls for preparation of regional action programmes (RAP).
>
> The Regional Action Programme (RAP) for the CCD-Asia region was conceived and developed at various meetings held in Beijing, New Delhi and Bangkok. The most important amongst these meetings was the ministerial conference on the implementation of the CCD in Asia, held in Beijing, People's Republic of China from 13-15 May 1997. This conference further elaborated the Regional Action Framework for Asia on the basis of NAPs. Participants included ministers in charge of desertification control as well as representatives from associated donors, international organizations and NGOs. The RAP for Asia is launching six thematic programme networks (TPNs) related to: desertification monitoring; soil conservation and forestry; rangeland management and sand dune fixation; water management in the drylands; drought preparedness and; local area development initiatives. Each of these TPNs are being hosted by a particular country in the Asian region.
>
> At the national level, in response to the recommendation of CCD, many countries of the Asian and Pacific region have prepared or are in the process of preparing National Action Plans to Combat Desertification. These plans provide a review of programmes adopted to respond to combat desertification and drought as well as the cluster of priority programme areas to be developed and implemented under the NAP process. They also include some concrete urgent actions in response to the immediate needs of local populations and measures to be taken for their quick implementation. The plans have promoted awareness campaigns and decentralized consultations. The campaigns (general and targeted) undertaken at the appropriate levels were basically intended to identify main stakeholders; build long-term commitment; mobilize support; create consensus for action and clearly identify responsibilities.
>
> It may be noted that the Regional Action programme and National Action programmes to Combat Desertification are means and not end in themselves. In fact they initiate beginning of a continuous process going beyond the production of a document to a genuine, flexible, participatory approach aimed at integrating dryland issues in the nation's overall economic and social development.
>
> *Sources:* 1. UNCCD 1999
> 2. UNSO 1999

an "enabling environment", involving local communities in formulating action programmes specifying the practical steps and measures to be taken. Among the important initiatives supported by the INCD and the Secretariat in Asia and the Pacific were activities that dealt with the promotion of regional and subregional collaboration and linkages. These activities were pursued along with promotion, advocacy and individual effort of countries in preparing their respective plans.

The CCD also established a Committee on Science and Technology which is composed of government representatives to advise the COP on scientific and technological matters relevant to desertification and drought (UNCCD 1999). The Convention promoted international cooperation in scientific research and observation as well as technological cooperation, whilst protecting traditional and local technologies. Finally, the Convention encouraged developed countries to support capacity-building efforts that will enable developing countries to combat desertification more effectively through the exchange of information.

2. Other International & Regional Actions

The FAO (1999) has made policy recommendations aimed to protect and improve agricultural lands as part of its Sustainable Agricultural Resource Management (SARM) strategies. In addition, the approach of the International Scheme for the Conservation of Lands in Asia and the Pacific (CLASP), provides means by which the countries of Asia and the Pacific can develop their own programmes to fight land degradation (FAO 1999). The approach of CLASP intends action on different levels: national for improving land use; and regional and international for information exchange and advance training.

FAO also established an Asian Network on Problem Soils in 1989 involving 13 countries. The network is mainly concerned with the rational use, management and conservation of problem soils within the Asian and Pacific Region in a sustainable and environmentally sound manner. In cooperation with the Asia Soil Conservation Network for the Humid Tropics (ASOCON), FAO is developing a Framework for Action on Land Conservation in Asia and the Pacific (FALCAP). The aim of ASOCON, (with funding from UNDP and technical assistance from FAO), has been to assist its member countries (People's Republic of China, Indonesia, Malaysia, Papua New Guinea, the Philippines, Thailand and Viet Nam) to enhance the skills and expertise of personnel, who are supposed to assist small-scale farmers in the development and dissemination of soil and water conservation practices. Its ultimate objective has been to enable small farmers to use their land in a sustainable and productive manner and to enable member countries to develop their own programmes to combat land degradation. However, despite its strong co-operative arrangement, long collective experience, proven track record and great potential for work, the activities of ASOCON are greatly hampered by limited internal funding.

The World Bank's 'Dryland Programme' is targeted at fighting desertification in the marginal areas of developing countries. Under this programme, the World Bank mobilizes financial resources and links national and international programmes. The main goal of the Dryland Programme is to improve the management of natural resources in dry-land ecosystems. From 1990 to 1995, the World Bank financed 108 projects worth US$6.8 billion, which worked on the improvement of natural resource management in dry areas. Currently World Bank has 57 projects (totalling US$2.3 billion) in the pipeline targeting desertification over the next three years.

International centres such as the International Soil Reference and Information Centre (ISRIC) and International Centre for Integrated Mountain Development (ICIMOD) are also contributing towards the control of land degradation, including: compilation and development of a soil and terrain information system; assessment of soil and terrain resources for sustainable utilization; strengthening the capacities of soil institutions in the developing countries; and enhancement of accessibility of soil and terrain information. Finally the conference on Land Degradation at Adana in 1996 recommended the establishment of an International Task Force on Land Degradation, under the auspices of the International Society of Soil Science (ISSS) (Land Degradation Newsletter 1998). The International Task Force on Land Degradation has a responsibility to assess the severity and cost of land degradation worldwide and to stimulate cooperation among scientists in understanding and solving the degradation problem (Dregne 1998).

CONCLUSION

There has been an increasing pressure on land to meet the basic needs of the growing population in the Asian and Pacific region. In the past, expansion in cropland has been the major means of enhancing agricultural production in the countries of the region. However, since 1976 the expansion in cropland has reduced substantially. Moreover, land degradation is increasing, through fertility loss by erosion, salinization and pollution, amongst other causes.

In the most severe or extreme cases, land degradation has been irreversible, either placing the land beyond restoration or in a position requiring major and costly engineering work to restore productivity. About 460 million ha, or 13 per cent of all used land in the region, falls into this category. Data for the arable land lost is not available, but assuming even a representative 13 per cent loss, it amounts to some 46 million ha. Such an area could feed about 650 million people.

A significant proportion of degraded land is still in production, though much of it is less fertile than before. Productivity losses to degradation have been estimated, for each degradation category indicating that highly and moderately degraded lands yielded about 10 per cent less in 1990 than they would have without degradation. When strongly and extremely degraded (i.e. non-productive) lands are factored in, the estimated production loss from degradation rises to more than 18 per cent.

Erosion is the most pervasive form of soil degradation, accounting for 90 per cent of degraded areas in the region. Salinization affects a much less extensive area (about 9 per cent) but because this damage is common to irrigated land, its impact is intensified. Land degradation in drylands or sub-humid lands is mostly caused by the process of desertification, which severely affects the drylands of Central, South and Northeast Asia.

The cost of every hectare of cropland lost or degraded is increasingly being realized by the planners and decision makers, as a result of which numerous policies and programmes have been formulated and are being implemented in the region. Specific measures adopted have included watershed management, soil and water conservation, sand-dune stabilization, reclamation of waterlogged and saline land, forest and rangeland management, and the replenishment of soil fertility in arable lands by use of green manures and cultivation of appropriate crops. The fight against land degradation however continues apace as pressures on the land increase in response to the region's growing population.

Chapter Two

Clearing of tropical forest for cultivation results in a major environmental loss.

2

Forests

INTRODUCTION

DISTRIBUTION AND DIVERSITY OF FORESTRY RESOURCES

DEFORESTATION AND AFFORESTATION RATES

CAUSES AND CONSEQUENCES OF FOREST LOSS AND DEGRADATION

PRODUCTION, CONSUMPTION AND TRADE OF TIMBER AND NON-WOOD RODUCTS

FOREST POLICIES AND STRATEGIES

CONCLUSION

CHAPTER TWO

INTRODUCTION

For centuries, communities in Asia and the Pacific have established an intricate and dependent relationship with forests. They serve as a repository for natural wealth and play a significant role in social and economic development. Today, the Asian and Pacific Region's forests are under pressure from a rapidly expanding population, the need for development and the pressure for land from agriculture. In recent years however, there has been an emerging recognition of the wider values (direct and indirect) of forests. Such values include the appreciation of traditional and customary practices for forest management, the link between forests and climate change, the critical role of forests in providing environmental stability and quality, the conservation of biological diversity, the protection of groundwater resources and the multiple uses of forest resources.

This Chapter assesses the current state of the Region's forest resources in terms of rates of deforestation and afforestation, the impact of demand for forest production and trade and management and policy issues at the national, regional and international level.

DISTRIBUTION AND DIVERSITY OF FORESTRY RESOURCES

In 1995, world forest resources, including natural forests and plantations, were estimated at 3 454 million hectares (ha) or 26.6 per cent of the total land area of the world (FAO 1997). Figure 2.1 presents the distribution of forests by major geographic regions and at the subregional level. The Asian and Pacific Region accounts for 16 per cent of the global forests. Of the seven countries which contain more than 60 per cent of the world's forests, two, namely People's Republic of China and Indonesia, are in the Asian and Pacific Region. Within the region, Southeast Asia has the largest forest area followed by Northeast Asia, South Pacific and South Asia respectively. A comparison of forest area per capita in 1995, shows that at least 15 countries within the region exceeded the world average of 0.6 ha per capita, while 6 countries fell within the regional average of 0.17 ha per capita. Figure 2.2 provides a comparative review of forest area per capita for a sample of countries in the region.

Temperate and tropical forests are two of the major forest types in the Asian and Pacific Region. The 1995 Forest Resource Assessment (FRA) estimated that 40 to 45 per cent of the forests in the region are located in temperate areas of Australia, New Zealand, Japan, Democratic People's Republic of Korea, Republic of Korea, Mongolia and in some areas of People's Republic of China. In addition, the temperate countries have at least 60 per cent of the region's plantation resources which account for more than 16 per cent of the forest area in the Temperate Zone, but only 7 per cent in tropical countries. With respect to natural forests, Southeast Asia is particularly important, as it has 40 per cent of the region's total forest resources.

Based on the data classification used by the World Conservation Monitoring Centre (WCMC), the major forest ecosystems in the region are mangroves,

Figure 2.1 Distribution of Forests by Major Geographic Regions and Subregions in Asia and the Pacific

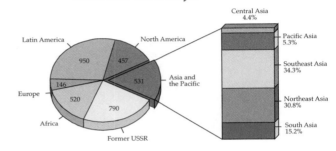

Source: FAO 1998

Figure 2.2 Per Capita Forest Areas in Selected Countries of the Asian and Pacific Region, 1992 and 1995

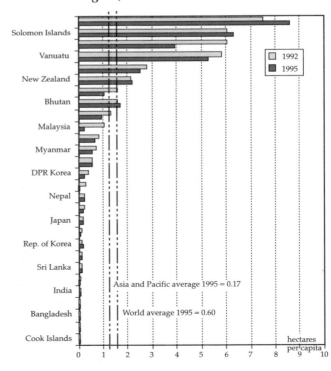

Source: FAO 1998

Table 2.1 Distribution of Forest Ecosystems in Asia and the Pacific by Subregion

Subregions	Forest Ecosystems (000 ha)			
	Mangroves	Tropical	Non-Tropical	Sparse Trees and Parkland
South Asia	496	49 828	27 936	466
Southeast Asia (Continental)	629	56 271	11 507	0
Southeast Asia (Insular)	2 558	104 153	0	0
Northeast Asia		109	96 416	41 918
South Pacific	5 413	53 219	27 089	101 485
Central Asia		0	5 558	0

Source: WRI 1998

Table 2.2 Deforestation in Selected Countries in Asia and the Pacific

Country	1990-1995[2]			1981-1985[1] Ranking
	Annual rate of deforestation	Area Deforested Annually (thousand ha)	Current Ranking	
Group I (High rate of deforestation (>1) and big losses of area)				
Indonesia	1	1 084	1	6
Malaysia*	2.4	400	2	1
Myanmar*	1.4	387	3	8
Thailand	2.7	330	4	2
Philippines	3.5	262	5	11
Uzbekistan	2.7	226	6	
Afghanistan	6.8	188	7	
Cambodia	1.6	164	8	
Lao, People's Dem. Republic	1.2	148	9	3
Iran, Islamic Republic	1.7	142	10	
Viet Nam	1.4	135	11	12
Pakistan	3	55	12	15
Nepal	1.1	55	13	4
Sri Lanka	1.1	21	14	5
Group II (Low rate of deforestation (<1) but big losses of area)				
Papua New Guinea	0.4	133	15	10
PR China	0.1	87	16	
Japan	0.1	13	17	
Korea, Republic	0.2	13	18	
Group III (High rate of deforestation with relatively small losses)				
Armenia	2.7	8	19	
Samoa	1.1	2	20	
Group IV (Low rate of deforestation with relatively small losses of area)				
Bhutan	0.3	9.4	21	16
Bangladesh	0.8	8.8	22	
Vanuatu	0.2	8	23	
Solomon	0.2	5	24	
Fiji	0.4	4	25	
Brunei Darussalam	0.6	3	26	13
New Caledonia	0.1	1	27	

Source(s): 1. United Nations 1990
2. WRI 1998

Note: * According to Government of Malaysia the forest area since 1990 decreased by 500 000 ha only. According to Government of Myanmar annual rate of deforestation is 0.68% and the loss of forest area is only 220 000 ha per annum.

tropical forests, non-tropical, and sparse trees and parkland (Iremonger 1997). Table 2.1 summarizes the distribution of forest ecosystems in the region. In addition to these classifications, frontier forest can be identified which are forest areas large enough to support their biodiversity and ensure the long-term survival of their plant and animal species. Using the World Resources Institute (WRI) (WRI 1997) set of criteria for frontier forest, Indonesia and Papua New Guinea have the majority of the world's remaining frontier forest.

DEFORESTATION AND AFFORESTATION RATES

A. Deforestation

Between 1980 and 1990, the Asian and Pacific Region annually lost approximately 4.4 million ha of natural forest. Roughly 1.6 million ha per annum of these forests were converted to non-forest land uses, and 2.2 million ha to other wooded land. Although the establishment of forest plantations, at 2.7 million ha per annum, partially offset the loss of natural forests, the net loss of forests is still approximately 1.7 million ha per annum. Table 2.2 shows that annual areas lost from deforestation (1990 to 1995) are particularly high for many Southeast Asian countries, namely Indonesia, Malaysia, Myanmar, Thailand, Philippines, Cambodia, Lao People's Democratic Republic and Viet Nam.

The region has lost almost 95 per cent of its frontier forests (WRI 1999c) and the situation in the South Pacific will be similar if strict control is not exercised (Box 2.1). The Islamic Republic of Iran and Afghanistan have lost all such forests, and People's Republic of China and India today have just 20 per cent of their original forest cover. Of these remaining forests, less than 10 per cent can be classified as frontier.

However, the deforestation trend has not been uniform across all countries in the region. The forest and wooded areas are increasing in countries such as India, People's Republic of China, Australia, New Zealand, Armenia, Kazakhstan and Uzbekistan, mostly due to large forest plantations. Forest plantations, though, are no substitute for natural forest which serves as a reservoir of biodiversity that once gone, cannot be recovered.

CHAPTER TWO

Box 2.1 The Loss of Large Frontier Forests

A significant number of frontier forests in the Asian and Pacific Region that have survived into the 20th Century are threatened today. The region has lost almost 95 per cent of its frontier forests. Apart from the Mediterranean and Middle East – where all such forests have disappeared – this represents the world's greatest loss of frontier forest outside of Europe. People's Republic of China and India today have just 20 per cent of their original forest cover. Of these remaining forests, less than 10 per cent can be classified as frontier. In the 20 years between 1960 and 1980 alone, the region lost almost a third of its tropical forest cover, the highest rate of forest conversion in the world.

Most of the region's remaining frontier forest is confined to the islands of Borneo, Sumatra, Sulawesi and Irian Jaya in Indonesia. Even here, however, loggers have exploited most accessible forests along coasts and major rivers. Agriculture and poorly planned resettlement programmes however also take their toll. Between 1969 and 1994, Indonesia's transmigration programme moved 8 million people to the nation's forested islands where 1.7 million ha of tropical forest were soon stripped.

On mainland Southeast Asia, most frontier forests are gone. The isolated pockets which were left are confined primarily to Myanmar, Lao People's Democratic Republic and Cambodia, where war and civil unrest until recently inhibited development. With peace have come new threats from commercial loggers who have already exhausted forests in Thailand and peninsular Malaysia – where harvest and import restrictions now encourage logging companies to move on to neighbouring nations.

A great long-term worry is the burgeoning population of the region and its ever-increasing demand for food and agricultural land. Between 1990 and 1995 alone, the region's largely rural population grew by more than 270 million people. The world's most densely populated region, Asia had more than 1 person for every hectare of land in 1995. Most of the forests that are left have been heavily altered by humans and often rendered into a patchwork of smaller fragmented forest areas.

The fragmentation process is one of the most serious consequences of the current deforestation and degradation of the world's forests. Frontier forests differ significantly from the dissected, human-modified forests that dominate the region today. For one thing, frontier forests are large and natural enough to ensure the long-term survival of their plant and animal species, including the biggest mammals with the most extensive home ranges. As secure habitats for native species, frontier forests are invaluable refuges for regional biodiversity. Forests are home to between 50 and 90 per cent of the world's terrestrial species – plants and animals that have provided much of the food and other basics that humans need to survive.

Frontier forests also contribute a large portion of the ecological services that make the planet habitable. They take up tremendous amounts of carbon dioxide (CO_2) – more carbon than is likely to be released by fossil fuel burning and cement manufacture over the next 70 years or so.

Keeping Earth's last frontier forests will require a fundamental shift in how we view them. From the American Wild West of the 1800s to Russian Federation's Far East and the South American Amazon today, frontier forests have been seen as limitless providers of land, timber, wildlife, and other sources of wealth. Careless and wasteful, a typical frontier economy mines the forest for a quick profit and moves on. It is time to replace this outdated thinking with a concept of frontier that is based on stewardship – taking responsibility for the forest and ensuring that its riches will be available for future generations. Good stewardship may mean complete protection of some frontier forests combined with careful management of portions of others for both timber and non-timber products. The change must happen soon: over the coming years, citizens, policy-makers, industry leaders and others have a chance to decide the fate of the world's last frontier forests. The key decisions before us are windows of opportunity that may never open again.

Source: WRI 1998

B. Afforestation and Plantations

The Asian and Pacific Region maintains its distinction of having the largest area of forest plantation, accounting for 83 per cent (66.9 million ha) of the world's total planted area. Of the seven countries with forest plantation estates greater than 1 million ha in 1990, five are in the region: People's Republic of China, India, Indonesia, Republic of Korea and Viet Nam. This shift to plantation forests relieves pressure on natural forests whilst developing reliable sources of industrial raw material for the future. In Turkmenistan, for example, a programme has been developed with the aim of protecting natural forests in parallel to the creation of a new plantation resource (Box 2.2). Several countries are also pursuing plantation development in tandem with policies of withdrawing natural forests from production.

People's Republic of China has one of the most extensive plantation programmes in the region, oriented towards wasteland reclamation and land stabilization. The country plans to plant an additional 26 million ha of forests in the Yangtze and Yellow River basins by the year 2030. The high rate of plantation establishment in the region reflects, in general, the commitment of governments to plantations as the principal source of raw materials for the future. In some countries, a wide array of incentives is offered to promote plantation

> **Box 2.2 Forest Plantations in Turkmenistan**
>
> Turkmenistan possesses small areas covered with forests. The primary natural forests are the existing mountain walnut forests that have survived from the Cretaceous period, and the riparian tugai forests which formed at the same time as the river valleys. Now, when the original forest areas are considerably reduced, and the areas with the extreme soil and climatic conditions are expanding, the importance of planted forest shelter belts has increased, since they have become one of the basic factors ensuring the protection of the environment.
>
> Significant progress in reforestation and forest growing has been achieved since adoption of the Decree of the President of Turkmenistan – "On development of horticulture and planting of trees in Turkmenistan" (09.11.1992). Soon after adopted in 1993, 3 680 000 young plants were planted in cities and populated areas, along highways, along the edges of irrigated fields, and on slopes of hills for laying Juniper and Pistachio plantations. From 1994-1997 an average of 4 million saplings were planted annually. Furthermore, in these years, an area of 100 ha of Juniper, 300 ha of Pistachio and 30 000 ha of pasture protective forest belts were planted.
>
> The process continues to date. For example, in Autumn 1998, a large park zone of 3 000 ha, with 2.5 million young plants was planted in the foothills of the Kopetdag mountains, south of Ashgabat. In the Autumn-Winter period of 1999, along the Berzenghee-Khindivar highway, 1 million conifers (pine-tree, cypress, thuya, juniper, etc.) were planted. Also in 1999, along other highways and in the parks of Ashgabat, 2 100 000 saplings were planted over an area of 3 600 ha, and works begin on establishing 6-row forest belts, fringing Ashgabat, Velayat and Etrap.
>
> The marine plantation efforts undertaken in Turkmenistan will hopefully lead to considerable environmental improvements in the country.
>
> *Source:* The State of Environment of Turkmenistan

establishment. For example, in the Philippines, both large- and small-scale industrial plantations are provided with financial and non-fiscal incentives; India provides free seedlings and other extension services; and Indonesia offers tax concessions, and promotes joint ventures between state forest enterprises and private-sector investors.

However, there has been a declining trend in the region for establishing large industrial tree plantations, with a shift towards community woodlots, farm forestry and agro-forestry plantations. For example, in India, farmers in Uttar Pradesh, Punjab and Haryana States established 26 000 ha of poplar (*Papulus deltoides*) in 1990, which is now being sold for the manufacture of matches, plywood and other wood products.

CAUSES AND CONSEQUENCES OF FOREST LOSS AND DEGRADATION

A. Causes

Deforestation and forest degradation is caused by a number of factors. These include natural factors, such as fire, disease and weather-induced stress, but more often than not, are factors exacerbated by human activities, such as land clearing for agriculture, overgrazing, over-extraction of timber and harmful logging practices. The direct causes of deforestation and degradation, however, are obscured by the underlying causes of poverty, inequitable resource tenure, population pressures, corruption, misguided policies, and institutional failures.

A major direct cause of deforestation in the Asian and Pacific Region is the clearing of forests, both planned and unplanned, for permanent cropland and pasture. Poor forest harvesting practices, followed by encroachments, once concession operations have ceased, contribute significantly to the degradation of the remaining forests. In many cases, shifting cultivation and overgrazing have caused widespread degradation of forests in the region, and even complete deforestation in extreme instances. The estimated number of people involved in shifting cultivation in the region varies between 25 million and 40 million.

Although the destruction of forests directly attributed to felling of industrial timber is extensive, it is difficult to accurately estimate the actual extent. The harvesting operations for logging concessionaires are usually defined by government regulations and codes of practice, which in some areas, particularly in Southeast Asia will include a reforestation programme. However, the limited resources of governments often prohibit adequate monitoring of the logging practices of concessionaires.

In recent years, fires have caused serious damage to forests in the region. It is reported that in India, fires affect about 53 per cent of the forest area, or about 35 million ha, each year. In People's Republic of China, 43 690 ha were burnt in 1999; in Mongolia more than 3 million ha were burnt in 1996;

CHAPTER TWO

and in Indonesia, 3.2 million ha were burnt in 1982-1983, and 160 000 in 1994. The period 1997-1998 was the worst for forest fires in the region. The wild fires that raged in Indonesia, Papua New Guinea, Australia and Mongolia prompted their respective governments to declare national disasters and seek international support to fight the fires. In Indonesia alone, the 1997-1998 fires burned an estimated 2 million ha in Sumatra and Kalimantan. Large quantities of smoke generated by slow burning ground fires affected human health both *in situ* and in the neighbouring countries. The smoke also interfered with transportation systems and disrupted the multi-million dollar tourism industry (see Box 17.3). The fire was exacerbated by the drought associated with the El Nino weather pattern, which turned most forests into drier habitats, and increased the flammability of forest vegetation.

Severe outbreaks of insects, pest and diseases can potentially be environmentally devastating with costly economic repercussions, in particular for plantation forestry (with a reduced diversification of species). The causes of pests and diseases vary, depending largely on a host of factors that include climate, human interference, destruction of habitats of the pest's natural predators, or even the introduction of exotic plant species that are vulnerable to insect pest and disease attacks.

B. Consequences

The costs and negative effects of deforestation are well documented. Where deforestation occurs in an unplanned and wasteful manner, the economic losses can be substantial, particularly from the loss of timber and other commercial resources. At community forest level, deforestation causes severe hardships and social disruptions for forest-dwelling and forest-dependent people. While the effects of deforestation are relatively simple to identify, the effects of forest degradation are subtler. Degradation can lead to a host of problems including loss of soil fertility and nutrient recycling capability, reduction in productivity and growth, decline in species richness, erosion of genetic diversity, decline in stock density and crown cover, reduction in the economic value of timber crops, and decreases in wildlife populations.

Another serious negative impact of deforestation and degradation is the loss of wildlife habitat. Forests, particularly natural forest areas, are the single most important repositories of terrestrial biological diversity.

PRODUCTION, CONSUMPTION & TRADE OF TIMBER & NON-WOOD PRODUCTS

Production and consumption are major indicators of the pressure on forest resources. This section outlines the major trends in production, consumption and trade of timber and non-wood products.

A. Trends in Production and Consumption

Consumption and production of forest products is dominated by: wood fuels; industrial wood products; and paper and paperboard. For the past 25 years, the Asian and Pacific Region exhibited impressive growth in production and consumption of forest products and commodities. The increase in forest products has been demand driven by the increasing population and wealth in a number of countries in the region. Between 1980 and 1998, population in the Asian and Pacific region increased by 29 per cent. Although growth was not evenly distributed, the overall wealth in the region (GNP per capita) also increased by 40 per cent. Such advances in population and demand for forest products will put continued pressure on forestry resources.

1. Wood products

The Asian and Pacific Region produced approximately 1.2 billion m^3 of roundwood (including industrial roundwood) in 1994, or about one-third of the world's total. The region currently produces approximately 19 per cent of the world's industrial roundwood, with production rising steadily in the past decade, from 250 million m^3 in 1983 to 294 million m^3 in 1995 (FAO 1997). Most industrial roundwood is used locally in value added processes.

(a) Wood Fuels

Fuelwood and charcoal remain the most commonly used sources of energy for a substantial proportion of rural and urban dwellers, in developing countries in the region. Table 2.3 summarizes the region's fuelwood and charcoal production from 1985 to 1995. Production and consumption of fuelwood generates substantial employment among the rural households, particularly among women. Demand for fuelwood is dominated by domestic uses, such as cooking. In Nepal and Bangladesh, for example, cooking consumes 75 per cent and 66 per cent respectively of all biomass energy. Even in urban

Table 2.3   Fuelwood and Charcoal Production (in 1 000 cu m) in Asia and the Pacific

Region/Subregion	1985	1992	1993	1994	1995
Developing Asia and the Pacific countries in:					
South Asia	297 307	343 754	350 225	356 493	362 875
Southeast Asia	245 351	280 234	285 534	290 140	295 012
Northeast Asia	9 623	8 864	8 916	8 955	8 994
South Pacific	5 806	5 802	5 802	5 802	5 802
Developed ESCAP Countries	3 465	3 326	3 308	3 308	3 308
Asia and the Pacific	561 532	641 980	653 785	664 700	675 991
World Total	1 643 161	1 789 164	1 794 950	1 816 953	1 839 040

Source: FAO 1998

Table 2.4   Production (1995) and Average Growth (1985-1995) of Industrial, Sawlogs, Sawnwood and Wood-based Panels

Region/Subregion	Industrial Roundwood		Sawlogs/Veneer		Sawnwoods/Sleepers		Wood-based Panels	
	Production (000 m³)	Growth (%) (1985-95)	Production (000 m³)	Growth (%) (1985-95)	Production (000 m³)	Growth (%) (1985-95)	Production (000 m³)	Growth (%) (1985-95)
Developing ESCAP Countries	127 017	0.8	98 516	0.7	41 758	0.8	20 377	8.6
Developed ESCAP Countries	59 318	0.2	37 212	1.7	31 134	-1.1	9 295	-1.2
Asia and the Pacific	186 335	0.6	135 728	0.9	72 892	-0.1	29 672	4.5
World	1 492 043	-1.0	919 469	-0.8	427 477	-1.6	146 021	1.7

Source: FAO 1998

areas, fuelwood demand is significant. Potential fuelwood supplies exceed consumption, both at present and projected needs. Fuelwoods can be produced from forests, wooded land, trees on agricultural land, and from land clearing to other land use (FAO 1997). This is demonstrated in Bangladesh, Pakistan, Sri Lanka, India, Thailand, People's Republic of China and the Philippines, which have less than 10 per cent forest cover but are still able to supply large volumes of locally produced fuelwood.

(b)   Industrial Wood Products

Over the past 25 years, production and consumption of industrial wood products has expanded significantly making the region a dominant consumer, as well as producer of wood products. The region now surpasses Europe in the consumption of all major wood products, and is second only to North America in the consumption of sawn timber, paper and paperboard. The growth in production of wood products is given in Table 2.4. An increase in the efficiency of processing (through technological advances) of wood products, with more products being produced from fewer raw materials has occurred. In the Asian and Pacific Region, industrial roundwood is predominantly produced in People's Republic of China, Indonesia, Malaysia, Japan and Australia, whilst consumption is highest in People's Republic of China, Japan, Republic of Korea, Indonesia and Malaysia.

Since 1987, the total production of sawnwood has declined by approximately six per cent, while consumption has slightly increased. People's Republic of China, Japan and India have remained the lead producers of sawnwood in the region. Consumption of sawnwood has increased in Southeast Asia and in the newly industrializing economies. South-Asia and the South Pacific have maintained a balance in production and consumption. During 1990-1995, the use of sawn timber for domestic consumption shifted to remanufacturing processed products for export.

The growth in the consumption and production of panel products has been the most spectacular trend in the Asian and Pacific Region over the past decade. People's Republic of China, Indonesia and Japan are the largest individual producers of panel products. Rapid growth in Chinese production of plywood accounts for the majority of the increased production (and consumption) in Northeast Asia since 1990. During the same period, Japan's production declined significantly, and was replaced primarily by imports of plywood from Indonesia. Similarly, production declined in the Republic of Korea, although imports (mostly from Indonesia) allowed the country to increase its consumption of wood-based panels nearly three-fold between 1991 and 1995. Southeast Asia consumes only a quarter of its own wood-based panel output. The bulk of the remainder is exported to Northeast Asia and Japan. Region-wide, the plywood industry is no longer expanding. This reflects the diminishing supply of large hardwood peeler logs and that medium-density fibreboard (MDF) and oriented-strandboard (OSB) are replacing plywood for many uses.

(c) Paper and Paperboard

The rapid economic expansion in the Asian and Pacific Region has been accompanied by major expansions in paper and paperboard production and consumption. The installation of significant new productive capacity since 1987, resulted in an increase of 83 per cent in the production of paper and paperboard. The most rapid expansion occurred in Northeast Asia and Southeast Asia, which increased production by 230 per cent and 400 per cent respectively. However, Northeast Asia still dominates the total production and consumption of paper and paperboard products in the region, owing to China's enormous population and the high levels of economic development in Japan and the Republic of Korea.

The enormous growth in utilization of recovered paper in Northeast Asia, is largely a result of a concerted effort in Japan to develop its paper-recycling industry (paper recovery and utilization in Japan now exceeds 50 per cent) and has obvious potential for reducing pressure on forests.

2. *Non-Wood Forest Products (NWFPs)*

Non-wood forest products (NWFPs) have attracted considerable interest in recent years due to an increasing recognition of their contribution to household economies, and to environmental objectives, including the conservation of biological diversity. It is estimated that 80 per cent of the population of the developing world use NWFPs to meet some of their health and nutritional needs (FAO 1998). NWFPs are also used in many village level artisanal and craft activities in developing countries in the region. Large-scale industrial processing of some NWFPs for products such as foods and beverages, confectionery, flavourings, perfumes, medicines, paints or polishes is undertaken.

NWFPs may represent the major actual or potential source of income from forests with low timber production potential, such as those in arid and semi-arid zones. It is unlikely, however, for them to compete on financial terms with production forests. NWFPs should simply be recognized as a supplement to the returns from timber, rather than to replace it as a source of revenue.

B. **Forest Industries**

A major transformation has taken place in forest industry in recent years, which has implications for the promotion of sustainable forestry. The Asian and Pacific Region has some of the worlds most competitive forest industries, with market-leading companies in both the developing and the developed countries. Developing countries have generally capitalized on abundant forest resources and low labour costs, while the developed countries have remained competitive by emphasizing processing efficiency and marketing acumen. A move to developing value-adding processing and exporting is evident with countries such as Malaysia and Thailand moving progressively into exports of furniture and joinery. Similarly, considerable attention is now being devoted to developing processing capacity for new forest products in the region. Most notable is the rapid expansion of medium-density fibreboard (MDF) and pulp and paper, particularly in Indonesia, Malaysia, People's Republic of China and Viet Nam. A major new product development has been the recent rapid increase in rubberwood processing, especially in Malaysia and Thailand. More than 80 per cent of all furniture produced in Malaysia now uses rubberwood as the base raw material. This is a good example of a move towards sustainable development objectives. Some countries, such as the Philippines, are undergoing major restructuring of the industrial processing sector to adapt to the realities of reduced raw material supplies, smaller logs, and increased dependency on plantation-grown wood. Other companies are internationalizing their operations as domestic forest resources decline. While Japanese and Korean firms have long practiced this adaptive strategy, Indonesian, Malaysian and Thai companies have only recently adopted it. Many of these companies have recently expanded operations into Cambodia, Myanmar, Papua New Guinea, Solomon Islands and Vanuatu, and also beyond Asia into Latin America and Africa. This could have implications for the rate of future deforestation.

C. **Trade in Forest Products**

The Asian and Pacific Region is a major net importer of forest products, as illustrated for 1996 in Figure 2.3. Much of the imbalance is attributable to Northeast Asia, especially Japan, which imported over US$19 billion of forest products in 1996. Other

Figure 2.3 Forest Product Trade in Asia and the Pacific, 1995

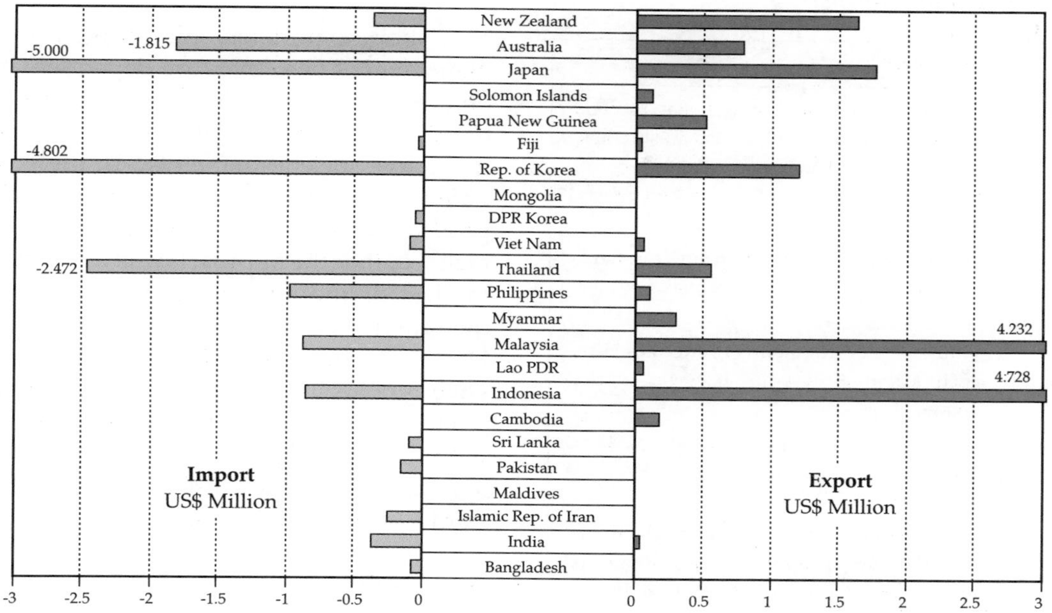

[Bar chart showing Import (US$ Million) and Export (US$ Million) for countries in Asia and the Pacific. Selected values: Japan Import -5,000; Rep. of Korea Import -4,802; Thailand Import -2,472; Australia Import -1,815; Malaysia Export 4,232; Indonesia Export 4,728. Countries listed: New Zealand, Australia, Japan, Solomon Islands, Papua New Guinea, Fiji, Rep. of Korea, Mongolia, DPR Korea, Viet Nam, Thailand, Philippines, Myanmar, Malaysia, Lao PDR, Indonesia, Cambodia, Sri Lanka, Pakistan, Maldives, Islamic Rep. of Iran, India, Bangladesh.]

Source: FAO 1998

major importers are Republic of Korea, Thailand, and Australia. Indonesia, Malaysia, New Zealand and Papua New Guinea are significant net exporters, with net exports of over US$500 million from each country. Within the last decade, exports have also increased rapidly in Myanmar, Cambodia, Lao People's Democratic Republic, Fiji, and Solomon Islands, indicating an expansion of logging operations in these countries.

(1)   Wood Products

In the Asia-Pacific region, since 1970, export volumes of sawnwood more than doubled to 7.5 million m^3; wood pulp exports increased eight-fold to 1.0 million metric tonnes; wood-based panel exports increased fivefold to 15.2 million m^3; and paper and paperboard exports increased sevenfold to 6.3 million metric tonnes. Industrial roundwood exports, however, have decreased by 3.2 per cent to 31.2 million m^3, following world trends.

The regional trade patterns have also changed substantially over the past decade, particularly through increasing trade between developing countries. People's Republic of China, Republic of Korea, Thailand and the Philippines have not only increased their imports from outside the region, but also from other countries within the region. Japan has changed some of its sources, and some of the products it imports. It is importing from new suppliers outside the region such as Africa and Scandinavia, as well as from new countries within the region. These changes reflect a wide range of factors including changes in resource availability from natural forests, investment decisions, market preferences and recent attitudes to environmental issues. However they also again indicate encroachment of companies and logging practices into neighbouring countries to satisfy growing internal pressures of demand.

(2)   Non-Wood Forest Products (NWFPs)

NWFPs are generally traded by countries in the region, in raw or semi-processed form and, while most NWFPs are traded in small quantities, exports of some products, such as honey, gum Arabic, rattan, cork, forest nuts and mushrooms, essential oils, and pharmaceutical products reach substantial levels. Table 2.5 categorizes NWFPs by end-use.

The present export value of NWFPs is US$5.4 billion in 1996, of which natural rubber is the most significant with a value of US$3.9 billion (Table 2.6). People's Republic of China is the largest NWFP exporter (excluding rubber), having exported US$900 million in 1996. In addition to the well-known NWFP exports, People's Republic of China exports commercial volumes of Chinese dates, gingko, tea oil, Chinese tallow, and tung oil. It's forest chemicals industry has also progressed significantly in recent years, earning US$180 million in foreign exchange in 1992.

NWFPs tend to face low trade barriers, however they are likely to be affected by measures such as health standards and phytosanitary rules, as many enter food industries.

Table 2.5 Categorization of Non-Wood Forest Products by End-use

End-use	Examples
Food products and additives	Wild meat, edible nuts, fruits, honey, bamboo shoots, birds nests, oil seeds, mushrooms, palm sugar and starch, spices, culinary herbs, food colorants, gums, caterpillars and insects, fungi
Ornamental plants and parts of plants	Wild orchids, bulbs, cycads, palms, tree ferns, succulent plants, carnivorous plants
Animals and animal product	Plumes, pelts, cage birds, butterflies, Lac, cochineal dye, cocoons, beeswax, snake venom
Non-wood construction materials	Bamboo, rattan, grass, palm, leaves, bark fibres
Bio-organic chemicals	Phytopharmaceuticals, aromatic, chemicals and flavours, fragrances, agrochemicals insecticides, bio-diesel, tans, colours, dyes

Source: FAO 1999

Table 2.6 Value of Exports for the Major Asian and Pacific Non-Wood Forest Products, 1996

Non-wood forest product	1996 Value of exports (US$1 000)	Major exporters	(US$1 000)
Lac, natural gums, resins	46 428	Indonesia	23 272
Essential oils	164 506	PR China	88 687
Natural rubber, balata, gutta-percha, guayule, chicle and similar natural gums	3 927 053	Indonesia	1 920 055
		Malaysia	1 395 147
Plaits and similar products of plaiting materials	265 476	PR China	224 132
Honey	143 667	PR China	110 665
Edible products of animal origin	66 613	Singapore	28 458
Ambergris, castoreum, civet and musk, etc.	8 086	PR China	4 172
Mushrooms (fresh or chilled)	151 860	PR China	118 276
Mushrooms and truffles (dried)	181 514	PR China	147 827
Walnuts	40 876	PR China	40 297
Chestnuts	178 741	Republic of Korea	109 022
Ginseng roots	125 183	Republic of Korea	69 576
Bamboos	30 719	PR China	26 414
Rattans	57 552	Singapore	35 388
TOTAL	**5 388 274**		

Source: FAO 1999

FOREST POLICIES AND STRATEGIES

Forest policies, planning, approaches and management objectives are dependent on the nature of forest resources with which countries are endowed, the development priorities of governments and the respective societal demands. In the past, a primary objective of forest management had been the production of wood products. However, it was realized, that the approach was not sustainable, given the fast dwindling natural resources. The 1992 Rio de Janeiro UN Conference on Environment and Development (UNCED) influenced many governments in the region to focus their management strategy on overall sustainable forest management with multiple objectives. Since then, there has been a notable shift in the forest management practices of most countries in the region.

A. National

1. *Management Approach*

(a) Natural Forests

Most natural forests in the region are generally publicly owned, with the exception of some countries in the South Pacific subregion, where indigenous people and local communities own most forests under respective customary laws. As natural forest areas are state-owned, the responsibility for their management and protection rests with governments. The approach to managing these areas is usually based on the principles of timber production, where forest resources are harvested under a selective felling system based on the sustained yield principle. Well-defined forest management systems include, for example, Indonesia where in 1989, the Indonesian Selective Cutting and Planting System or TPTI was introduced, a system which placed greater importance on natural regeneration and enrichment planting.

By comparison, the Malaysian Selection System is one of the oldest forest management systems used in the tropics, and has served as a model for many countries. However, with the increasing utilization of less known species, greater mechanization, and poor harvesting techniques, degradation of natural forests started to occur. This resulted in the introduction in 1997 of harvesting codes of practices and low impact management techniques such as helicopter logging. Since 1989, Sri Lanka has enforced a moratorium on logging of all natural forests in the country. Relaxation of the moratorium is unlikely as most of the important forest areas are badly degraded and need respiration for another 20-25 years.

In the South Pacific subregion the structure of the New Zealand forest estate, with a large exotic plantation resource in addition to indigenous virgin or regenerated forests, has enabled the country to implement a distinctive approach to Sustainable Forest Management (SFM). The system involves a significant separation of functions and specialization in management roles where the management of indigenous forests is oriented toward conservation functions (such as biodiversity, wildlife habitat and ecotourism), while management of plantations is focused on economic roles, such as the provision of industrial raw material. To put this complementary system into effect, New Zealand has also separated its legal and institutional management.

In summary, the theoretical underpinning of the systems adopted by countries is sound. In many countries in the region, however, the application of these systems has been inconsistent and has resulted in the failure and perceived ineffectiveness of the management practices, thus contributing to the diminution of forest resources.

(b) Protected Areas

The management objectives for protected areas and biodiversity (Chapter 3) clearly differ from those for natural forests, and present special challenges. Protected areas management places emphasis on maintaining environmental and ecosystem integrity, minimizing human impacts, conserving biological diversity, and enhancing wildlife habitats. The management record of protected areas in the region is variable, with generally good management in the developed countries and weak management in the developing countries. New and innovative ways are needed to combine conservation objectives (especially in high population areas) with people's livelihoods, such as in India, where compatible co-existence of people and conservation are being tried (Ahmed 1997). Several innovative funding initiatives have been undertaken in the region. Cook Islands, for example, imposes a tax on tourists to generate conservation revenue; Bhutan has a conservation fund to which donors are encouraged to contribute; and New Zealand has an elaborate set of grants, covenants and other ways to finance conservation on private lands.

(c) Plantation Forests

The objectives of plantation management are different from managing natural forests as they are usually focused on intensive production of wood, using only a few selected tree species. This makes plantations relatively easy to manage, although they require large investments at the time of establishment and have inherent weakness to cope with environmental problems. The growth and yield of plantations in the developing countries of the region generally leaves much to be desired due to the use of inferior planting materials that result in poor survival rates and slow growth rates, and the lack of silvicultural tending, impeding the full development of plantation operations. The situation is compounded by the difficult situation faced by the forest and wood industry, largely due to the declining profitability of forest products, particularly timber.

Government incentive packages to assist private plantations vary from country to country and are dictated largely by the market and the purpose for which the plantations are established. In Australia and New Zealand plantations will continue to expand given their competitive edge in processing forest products. However in developing countries the provisions of incentives to expand forest plantations are driven by different reasons. For example in 1999, the Philippines prohibited logging within old-growth forests, canceling many commercial-logging concessions. In order to meet the future wood requirements of the country, emphasis was given to plantation development and an array of incentives (including export tax exemptions and duty free imports) was offered to the private sector. Industrial Forest Management Agreements (IFMAs) were promoted, which fundamentally combined natural forest management objectives with industrial tree plantations. A socialized industrial forest management programme has also been launched which caters to individuals, families, cooperatives and small-scale corporations intending to engage in plantation management.

The motivation for plantations, however, is not confined to production alone but is also aesthetic. In Japan, multi-storied plantations, which closely resemble natural forests, are developed using low impact harvesting techniques that reduce the denudation of mountain sides and at the same time provide various aesthetic services for people to enjoy.

(d) Management of Non-Wood Forest Products (NWFPs)

Only recently have government forestry agencies in the Asian and Pacific Regions recognized the economic potential of NWFPS and the need for appropriate management. Forestry development officials and planners have tended to overlook the effective traditional management systems, adopted by forest boundary communities, and have been biased towards conventional and often incompatible timber production practices.

Aside from some traditional and localized management systems, the extraction of NWFPs is generally carried out in a haphazard, opportunistic and inefficient manner. The absence of preliminary inventories and the dispersed nature of collectors of NWFP resources are major impediments for its effective management. However, selling for a fixed rate the rights to collect NWFPs from large units of forests over a specified period (often one to five years) controls commercial harvesting, although there is rarely close supervision of the collection after the rights have been assigned.

(e) Social Forestry Management

There is a growing trend in the region for greater involvement of NGOs, community organizations, and local people in managing public forests. Under a range of collaborative forest management mechanisms, such as the Joint Forest Management in India and management by forest user groups of Nepal, local people are now increasingly being given full or partial forest management responsibilities. Local people generally have the right to collect certain forest products for their own use at no cost. Some communities are promised a share of the proceeds from future harvests of forests that regenerate as a result of protection provided by local people (See Box 2.3). In Nepal, for example, the government hands over forests to forest-user groups (FUGs). Negotiated management agreements include provision for managed utilization of forest products, including grass, fodder, fuelwood and NWFPs. Recently, there have been moves to initiate FUG-managed sawmills and the harvesting of timber from community forests, although these proposals are reportedly meeting resistance from within the Forest Department.

Collaborative management in most countries is still in the early stages of development. Only two to three per cent of India's public forest estate is estimated to be under community protection recognized by forest departments.

Box 2.3 Adapting Tenure to Encourage Forestry

Papua New Guinea: With nearly 80 per cent of the country's land under customary ownership, the people own most forests. The state is challenged to find ways of awarding rights to forest use that protect the people's interests, ensure responsible management or exploitation, and allow the investors adequate returns and security of access to resources. A number of instruments have been tried. The latest approach, intended for application on a large scale, is the Forest Management Agreement (FMA). Under the FMA approach, the PNG Forest Authority secures a commitment from the resource owners to follow recommended forest management practices, while simultaneously offering investors access to the forest for a minimum of 35 years. Implementation may involve the state issuing a Timber Permit under which it manages the forest on behalf of the customary owners for the duration of the FMA. State management roles can be implemented through a developer.

On a much smaller scale, provincial governments, upon the recommendation of provincial forest management committees and the consent of the National Forest Board, issue "Timber Authorities". Such agreements allow the execution of harvest agreements concluded directly with the landowners. It is reported that many foreign companies have abused this facility. As such, new policies on Timber Authorities are now in place to protect forest resources and prevent unnecessary forest clearance on customarily owned land.

Maldives: In the Maldives, the community forest lands on inhabited and uninhabited islands are leased to individual developers of agriculture and tourism. However, uncertainty of land tenure acts as a major disincentive to developing and protecting forests and agroforests and is a significant obstacle to the promotion of growing trees. Very recently, a more secure system of leasing land has been adopted (for example, renting of uninhabited islands for a fixed term of not more than 20 years). This will encourage tree planting and will develop sustainable land use systems throughout the country.

Philippines: Community-based Forest Management (CBFM) has been adopted as the national strategy for management and sustainable development of forest resources in the Philippines, pursuant to a 1995 Presidential Executive Order. To date, more than 500 000 ha of national forests have been turned over to communities, mostly of indigenous peoples. Unlike previous programmes that granted tenure over denuded and/or degraded forest lands (e.g. Agroforestry initiatives), the CBFM approach extends tenure and use rights to well-stocked forests.

Organized communities operate within allowable cut limits set by the government. They harvest timber and other forest products to sell, to use for their own needs, or to process. Timber harvesting by communities typically follows a labour-intensive, low-impact approach. Felling uses small chainsaws, flitching and/or quarter sawing is done at the stump, and animals are used to skid logs to roadside landings. Income from the sale of timber, rattan, bamboo and other forest products has created new income opportunities in upland communities where poverty is severe. Slash-and-burn forest destruction and illegal logging have declined dramatically in all areas where the CBFM concept has been introduced. In the words of one community leader, "Why should we burn or overcut the forest now that it belongs to us and not to some rich man from Manila?" Results thus far augur well for an expanded programme. The Philippines Master Plan for Forest Development envisions CBFM coverage of 2.0 million ha within the next decade.

Source(s): 1. Papua New Guinea 1996
2. P. Dugan (Personal communication) 1997

(f) Urban Forestry Management

There has been a notable surge of interest in the roles that forests can play in meeting the needs of urban dwellers and improving urban environments. The focus of urban forestry management has recently broadened beyond landscape architecture and horticulture for aesthetic purposes, to include concerns related to air quality, cooling of cities, protection of water supplies and nature conservation.

The urban landscape is rapidly changing. Presently only 34 per cent of population live in cities, but it is projected that by 2025 it will jump to 55 per cent. The status of urban forestry development varies greatly throughout the Asian and Pacific region (Box 2.4), particularly in the developing countries with some cities, such as Delhi, with a negligible per cent of city area as a green area, as compared with over 25 per cent Seoul (1996 figures). In Australia and New Zealand, there is a considerable area under urban reserves, unlike very densely populated (and less developed) cities such as Jakarta, Colombo and Dhaka. Green space per city dweller in the poor cities of the region are generally far below the international minimum standard of nine square metres, set by the World Health Organization (WHO).

It is also important to recognize the importance of managing trees in peri-urban and even more distant locations to meet the needs of cities. In Japan, for example, improvement of forests located in suburbs and villages is a priority concern of policy for the future intention of bringing nature closer to people.

2. *New Trends in Forestry Management*

(a) *Forestry Planning*

Over the past decade, countries in Asian and the Pacific region have significantly progressed in reorienting their forest policies and strategies to lay the foundation for sustainable forest management consistent with UNCED and Agenda 21. The realities of inadequate government resources, a long history of government forest management failures, a desire to reform economies towards stronger market orientation, and a commitment to more socially oriented forest management, have all been factors in recent moves by governments towards decentralization and devolution of forest management responsibilities to local governments, user groups, and local communities. The shift had far reaching implications in the conventional approach to forest management and administration. Many institutions are struggling to identify their roles and adapt to the rapidly changing resource management environments.

Table 2.7 Historical Perspective on Forests and Forestry

19th Century Viewpoint	Viewpoint Today
Forest outputs are timber, game, fuelwood, and water	Outputs are many, including various goods but also complex ecosystem services and social values
The natural world can be managed and controlled	The effects of human interventions on the natural world can be difficult to predict and control
Forest-dependent communities are local villages and farms	Local, national, regional and global "communities" demand goods and services from forests
Management aims to produce commodities through sustained harvest yields	Management treats the forest as a complex ecosystem and seeks to maintain its productive, protective and social values, and to preserve future options
The forester is an expert and a decision-maker	The "public" is the decision-maker through democratic processes; the forester is a technical advisor and facilitator

Source: FAO 1997

Conventional forestry sector planning placed emphasis on assessing forest timber resources and formulating strategies, primarily in relation to forest industries. With emerging issues that underscore the multiple value of forests, its scope has been expanded to address the causes of deforestation, needs for reforestation, contribution of forests to food security and rural energy, and building capacity of national forestry administrations. Table 2.7 presents the evolving perspective in forestry. Countries in the region are now recognizing the critical importance of redefining their planning approaches by emphasizing on the iterative and participative process instead of preparing blueprints and static plans.

Many countries in the region have not conducted a comprehensive and statistically sound forest inventory since the 1970s or early 1980s. New methods, such as remote sensing, are expanding the ability to observe large changes in land cover. However, the fact remains that, without reasonably updated forest inventories, it is increasingly difficult to assess change in forest quality and function, and to draw useful conclusions about the sustainability of their use. This limitation remains a critical gap in forest planning in the region if forest management is to improve.

(b) Forestry Policy Formulation

National Forestry Action Programmes (NFAP), which have been endorsed by the Intergovernmental

CHAPTER TWO

Box 2.4 Urban Forestry

Evidence indicates that considerable damage has been done to forests in the areas surrounding cities. The country with perhaps the best data on this process in the region is India. Between 1972-75 and 1980-82, the forested area within 100 kilometres of India's nine largest cities collectively diminished by one third. In well under a decade, the loss of forested area ranged from a comparatively moderate 15 per cent around Coimbatore in the south to a staggering 60 per cent around Delhi.

In recent years a growing consciousness has emerged, not only to save the urban forest but also to enhance the forest areas in cities. Good examples of effective urban forest management are in Singapore, Kuala Lumpur and Hong Kong, China.

In Singapore, in the late 1880s, forest reserves were established and catchment areas of new reservoirs were put under protection. The Centre of the island remains forested today, protected as nature reserves and a catchment area managed by public utilities. Although planting of ornamental trees in Singapore dates form the middle of the last century (with the active involvement of the Singapore Botanic Garden), the most active programmes in street tree planting and urban greening have taken place since the 1970s. Rapid population growth between the end of World War II and the mid-1960s, led to urban congestion and housing shortages. The low rate of population increase after the mid-1960s, and rising affluence, were important factors underlying the successful planning and revitalization of the urban environment in Singapore. Beautification of the city through tree planting was part of an overall strategy to create a more pleasant and healthy living environment and to stimulate economic growth by attracting foreign investment and business development in the city.

In Hong Kong, China major reforestation programmes were carried out from the late 1800s to the early 1900s, and again starting in the early 1950s, following extensive deforestation during World War II. These reforestation programmes had a dual purpose: watershed protection and wood production. Government plantations and village woodlots were established to provide fuelwood, poles and timber for the rapidly growing population. In the late 1960s, these programmes were scaled down due to a decrease in the demand for fuelwood and increased availability of alternative fuels. Greening efforts picked up again in the 1970s with establishment of county parks and, later, with various tree planting initiatives. The extremely rapid growth of Hong Kong, China's population and dense infrastructure development without planning for green space earlier in the century now limits the scope for street tree planting and urban green space in the older parts of urban areas. Nevertheless, particularly outside the urban core, various public, private and citizen-led efforts are being made in tree planting along streets and roads and in parks and other public areas.

In Kuala Lumpur there are still significant areas of forest within the city limits and extensive forest reserves in the peri-urban area, providing for watershed protection, recreation and nature conservation functions. Pressure on the urban and peri-urban forest for provision of wood products is relatively low. Demand for fuelwood is small because the average income level is relatively high and people have access to alternative household fuels. There are also ample supplies of timber for construction and other needs from other sources in the country. Similar to Singapore, Kuala Lumpur has put great emphasis on beautification of the city, both for the benefit of urban dwellers and to attract businesses.

This comparative study of the urban forestry in Hong Kong, China, Singapore and Kuala Lumpur provides important lessons for cities that are at earlier stages in development of their urban forestry programmes. Afforestation of water catchment areas and the protection of remnant forests were the earliest and most important urban forestry activities in both Singapore and Hong Kong, China. Wood production was also a driving force behind afforestation efforts in Hong Kong, China. As the economies of all three cities have developed, street tree planting and urban green space for recreation have increased in importance.

Source: FAO 1998

Panel on Forests (IPF), provide the unifying framework for re-orienting the forestry policy formulation process. The role of FAO in this process has been lauded, as it played a catalytic role in linking donor countries with recipient developing countries in the region.

Many countries in the region, such as Bangladesh, Bhutan, People's Republic of China, Fiji, Indonesia, Lao People's Democratic Republic, Malaysia, Nepal, Pakistan, Papua New Guinea, Philippines, Sri Lanka, Thailand and Viet Nam have adopted the NFAP framework. These countries are presently at various stages of programme execution often with the help of the Asian Development Bank (ADB), World Bank, and various bilateral organizations. Indonesia, Nepal, Papua New Guinea, Philippines and Sri Lanka developed NFAPs several years ago, and are now in the process of reviewing and revising them. It is encouraging to note that planning activities are underway, or soon to be initiated, in India, Myanmar, Samoa, Solomon Islands, Tonga and Vanuatu. Experience shows that countries who adopted NFAPs, have significantly benefited from the process through improvements in overall forest management. Other improvements include the adoption of new forestry legislation; institutional reorganization; redefinition of the role of the state; decentralization of forest management responsibilities; transfer of responsibility to communities and local groups; participation in the decision-making process; and coordination and harmonization of actions within coherent, holistic and intersectoral strategic frameworks.

(c) Supporting Forestry Research and Educational Institutions

Worldwide forestry research has suffered from a lack of resources, and it has not been sufficiently interdisciplinary to provide an integrated view of forestry. This situation is compounded by the on-going structural adjustments in many developing countries, where forestry research institutions have always been the first targets for downsizing or decentralization. To cope with this situation, some institutions have taken measures to provide continuing support to research activities, such as including research components in development projects; using universities to undertake research; and privatizing research works.

The experience from these measures has been mixed and it is clear that governments need to provide continuing overall support for research activities. Forestry educational institutions are also facing the pressure to respond to the new challenges of producing graduates with more rounded skills capable of meeting the demands of sustainable forest management. Curriculum reforms, expanding continuing education opportunities and redefining the learning environments for forestry education are some of the actions already taken. The Asian and Pacific region has a large number of forestry research and education institutions. These include the Centre for International Forestry Research (CIFOR), the Southeast Asian Regional Research Programme of the International Centre for Research in Agroforestry (ICRAF), the Regional Community Forestry Training Centre (REOF), the ASEAN Tree Seed Centre, and the International Centre for Integrated Mountain Development (ICIMOD).

(d) Decentralizing Forest Management

Local governments, community organizations, and the private sector, are being given increasing levels of responsibility for forest management and protection. The movement is most pronounced in South Asia, where social forestry programmes have evolved considerably since the 1970s. In the South Pacific, tribal and clan ownership and management of forest resources has been a long-standing tradition. In Fiji, the Native Land Trust Board assists in the management of forestland.

In some countries in the region, centralized management and protection of forests has been ineffective due to its inability to manage forest resource conflict at the local level. This view provided the impetus for decentralizing forest management functions. The idea has gained adherence in the region, with the Philippines leading the process and gaining considerable experience for other countries to learn from. Such lessons include that transfer of responsibility should be accompanied by: sufficient authority with transparent accountabilities; mechanisms for generating local revenues or provision of financial resources to sustainably support the devolved responsibilities; strengthened technical capacity and administrative capability; and clearly defined functions to be transferred.

Advantages of the devolution process are that it brings closer the objective of "empowerment" where local governments are allowed to become more responsive and accountable to the needs of the local stakeholders, and local communities are given the stewardship of forest resources.

Local, as well as international, NGOs are rapidly gaining power and influence in many countries. In Thailand, for example, advocacy groups have strongly influenced the pace of development, the orientation of the Master Plan for Forestry, and the development of a community forestry policy. In Indonesia, a consortium of environmental groups recently succeeded in elevating the debate over reallocation of government reforestation funds to international levels by challenging the government in court. International NGOs have been particularly helpful in developing local capacities in countries such as Bhutan, Cambodia, People's Republic of China, Lao People's Democratic Republic and Viet Nam, which previously had limited links to international organizations.

Generally, NGOs enjoy a considerable degree of acceptance and credibility among local communities, and play a significant role in sustainable forest management in the region. Their ability to mobilize funds from international donors (with private and multi-lateral agencies), and to influence policies and national programme directives has now been widely recognized. This has been demonstrated in a number of instances, where environmental NGOs have advocated for better deals for indigenous communities and forest dependent people.

Although the private sector is anticipated to continue wielding considerable influence on forestry in the region, there may be notable shifts in their involvement. The private sector is increasingly becoming involved in initiatives that promote environmental sensitivity in their operations. The industry is making voluntary efforts to be more responsive to the demands of its customers and to the public for improved management of forests under its control, and for the use of environmentally friendly production technologies. To elicit support for these initiatives, the private sector is also reporting its efforts to its stockholders and the public. Other private sector initiatives that are being pursued, include the industry supported forest management programmes; support to forest certification efforts; and active participation in global discussions such as the World Business Council for Sustainable Development (WBCSD), the FAO Advisory Committee on Paper and Wood products and other forums.

B. Regional and Global Initiatives

Throughout the Asian and Pacific Region, attention is now being focused on bringing forests under sustainable management following the UNCED process. National initiatives, tailored to specific local needs and conditions, exist in nearly every country of the region. These country-level initiatives are complemented and supported by a wide array of regional and international initiatives. Relevant examples include India, who (with Britain) sponsored a meeting to agree on UNCED reporting arrangements on forestry; Malaysia, who (with Canada) sponsored a review process three years after Rio; Indonesia, who sponsored the Bandung Global Forestry Forum to promote global consensus; Japan, who (with Canada) sponsored work on criteria and indicators for sustainable forest management; and Australia, who supported a post-UNCED review of forest products certification and trade.

Several countries in the Asian and Pacific Region are working to mainstream the principles of UNCED by adopting voluntary codes of responsible practice. The South Pacific code on logging and trade in products from natural forests pioneered this approach, and was followed by the Code of Practice for Forest Harvesting in Asian and Pacific Region and various national codes. Indonesia and Malaysia are voluntarily introducing sustainability produced timber products. Similarly, policies are in place or being adopted in accordance with the principles laid out by the UNCED process. The challenge, however, lies in translating these into operational terms, using the ITTO Guidelines for Sustainable Management of Natural Tropical Forests, UNCED's Forest Principles, and other broad forest management concepts. In this respect, the ITTO, "Year 2000 Objective" is to make all producer countries enter a commitment to produce their tropical product exports from sustainably managed sources by the year 2000. Most tropical timber producing countries in Asia and the Pacific Region, and most of the region's major consumers subscribe fully to ITTO's Year 2000 Objective. While most producer countries in the region are working towards meeting the objective, only Indonesia and Malaysia are judged to have a reasonable prospect of bringing their entire production forests under sustainable management by year 2000.

Australia, People's Republic of China, Japan, New Zealand and the Republic of Korea are active members of the Montreal Process, working to develop criteria and indicators for sustainable forest management in temperate and boreal forests. Indonesia and Malaysia recently proposed an initiative to develop regional criteria and indicators for Asia and the Pacific Region.

Various donor organizations are also helping the developing countries in the region to translate the concept of sustainable forest management into

---

Box 2.5 Forest Certification

Certification systems are based on evaluating the standard of forest management being practised. The two main approaches, whose relative merits is a subject of considerable debate, are those of Forest Stewardship Council (FSC) and International Standard Organization (ISO). FSC favours a performance-based approach, i.e. stipulating that a specified level of forest management, covering all aspects – including social aspects – must be achieved. ISO's approach is based on the management system, i.e. stipulating that specific management systems and commitment to specified actions and procedures must be in place. The FSC approach also permits chain-of-custody monitoring, while other certification approaches do not, as the chain-of-custody monitoring is considered too difficult and expensive to contemplate.

In the region, only two countries are known to be in the advance stages of formulating their respective national certification system – Malaysia and Indonesia. The Lembaga Indonesia was initiated in late 1993 through the Ecological Working Group, which has been developing a certification system for Indonesian forest products. The LEI's scheme is premised on sustaining values along three dimensions: production functions; ecological functions; and socio-cultural functions of forest areas. The criteria of sustainable forest management at the management unit level are: clear and secure land tenure; timber production and regeneration; financial feasibility of the management unit; efficiency of forest resource utilization; and evidence of professional management. The system has been tested in three concessions in 1994, and the Centre for International Forestry Research (CIFOR) and Forest Stewardship Council (FSC) have been consulted to develop and further refine the criteria and indicators to be more widely used.

Evolution of such forest certification systems are likely to assist both in sustainable development of forests as well as in promoting forestry related trade.

Source: FAO 1998 and Evans, B. 1996

action at the field level, through initiatives in reduced-impact logging, harvest planning and monitoring, "Model Forest" management, application of codes of forest practice, community-based forest management and resource inventory. The World Bank and the ADB, two multilateral funding organizations supporting forestry in the region, continue to play key roles in providing financial assistance. The World Bank's total loans to the region, for the forestry sector alone, was approximately US$365 million in 1998. Under the ADB's loan programme, forestry loans to the region, cumulatively totaled to US$890 million.

Several governments and NGOs in the region are currently exploring the feasibility of forest product certification and the ecolabelling of forest products. The Indonesian Ecolabelling Institute working with CIFOR and the Forest Stewardship Council to develop certification procedures is one among several site-level certifications that have already been made. (See Box 2.5)

CONCLUSION

The 21st Century has been called the Asian and Pacific Century, given the tremendous dynamism the region has shown over the past 50 years. Despite the crises it has faced, countries in the region have manifested their ability for rapid recovery, and it is anticipated that forestry sector development will remain a key component in the continued growth of many countries. The dominant forestry issues in the past twenty years remain key concerns. With the increasing population, and economic development competing demands for the region's forest resources are correspondingly increasing, thereby adding to the already formidable requirements the region places on its forests. Some countries are coping with these pressures and are moving rapidly towards more sustainable forest management, others, however, are likely to experience continued, or even accelerated, forest degradation in the short term.

On a more positive note, the region is paving the way for some innovative approaches to forest management. It leads the world in having more areas reforested, and has demonstrated that it can mobilize civil societies and local governments to become effective stewards of forest resources. The initiatives for increasing efficiency and becoming environmentally friendly in producing and utilizing forest resources, the improvements in forest planning capability, and the revision of forestry legislation are some of the key achievements that the region has shown in response to the new realities of forest management.

The move towards a more market-based economy is having a profound impact in the region. With emphasis on market competition, it is likely that the Asian and Pacific Region will face increased competition from other areas which produce forestry products. This may force the less competitive companies and producing areas to go out of business, but others are likely to adapt by improving efficiency, diversifying product lines, developing or acquiring new raw material sources, and targeting niche markets.

In summary, the major trends in forestry in recent years the Asian and Pacific Region are the re-orientation of forestry towards local people, a more participatory approach to forest management, and the development of strategic alliances to meet, simultaneously, the needs of local communities, industries, and national and global environmental interests. Forestry organizations and institutions are undergoing major restructuring to accommodate and facilitate this movement. Similarly, policies are being redefined to suit the needs of the people. It is likely that such restructuring and re-orientation will continue into the next millennium, for the betterment of the forestry sector in the Asian and Pacific Region.

Chapter Three

**The rich biodiversity of the region is under serious threat. The endangered Bali Starling.
Courtesy: Government of Indonesia (Photo Alan Compost)**

3

Biodiversity

INTRODUCTION

BIODIVERSITY: STATUS AND TRENDS

CAUSES AND CONSEQUENCES OF
BIODIVERSITY LOSS

POLICIES AND PROGRAMMES FOR
BIODIVERSITY CONSERVATION

CONCLUSION

CHAPTER THREE

INTRODUCTION

Biodiversity constitutes the most important working component of a natural ecosystem. It helps maintain ecological processes, creates soils, recycles nutrients, has a moderating effect on the climate, degrades waste, controls diseases and above all, provides an index of health of an ecosystem. Providing food, medicines and a wide range of useful products, it is the natural wealth that exists on land, in freshwater and in the marine environment. Plant diversity alone offers more than just food security and healthcare for the one-quarter of humanity who live their lives at or near subsistence levels; it provides them with a roof over their heads and fuel to cook, and, on average, meets 90 per cent of their material needs (Tuxill 1999).

BIODIVERSITY: STATUS AND TRENDS

The three bio-geographic realms of the Asian and Pacific Region are of immense significance to the world, and include all of the major ecosystems to be found, including mountains, forests, grasslands, desert, wetlands, and seas. The rich biodiversity of the region, however, has been under serious threat from a variety of human induced factors, which can be measured by a reduction in the natural habitat, loss of species and depletion of genetic diversity. Indicators of this change, however, are not easy to define, since only a limited number of species have been identified and catalogued to date (Box 3.1); nevertheless, the impact is serious.

A. Habitat Diversity: Patterns and Trends

The area of natural habitat in the Asian and Pacific Region is rapidly shrinking. The major ecosystems in the Indo-Malayan realm are estimated to have lost almost 70 per cent of their original vegetation, with habitat losses being most acute in the Indian subcontinent and the Peoples Republic of China. Thailand, the island of Java in Indonesia, and the central islands of the Philippines have also experienced an extensive reduction in natural habitat. Habitat losses have been comparatively less severe in the South Pacific, with the exception of some of the small island ecosystems and coral systems, which have been lost or degraded.

1. Terrestrial Habitat

(a) Forests

The forests of Asia and the Pacific are the habitat of countless numbers of plants, mammals, birds and insects and are home to between 50 to

---

Box 3.1 Biodiversity of Papua New Guinea

The island nation of Papua New Guinea is approximately the size of California and contains some of the largest and most important remaining blocks of tropical forest wilderness. It also claims some of the least disturbed coral reef systems left on Earth. Papua New Guinea has an extremely diverse culture – an estimated 875 distinct languages are still spoken there. The combination of its rich biodiversity and cultural heritage makes Papua New Guinea a high conservation priority.

The Lakekamu Basin is one of the largest remaining pristine lowland rain forests in Papua New Guinea, covering an area approximately 975 square miles in the Gulf Province. Virtually uninhabited, the Basin has until now been spared from human destruction, offering excellent opportunities for conservation. In October and November of 1996, Conservation International sent an expedition to this area of Papua New Guinea, under its Rapid Assessment Programme (RAP). Over a four-week period, the RAP team, comprising world-renowned experts and host country scientists, surveyed the Lakekamu Basin to create a first-cut assessment of the biological resources in this poorly known area. A research station was also established in the region, which will serve as a base for further research and field training of in-country scientists.

The expedition discovered nearly 44 new species of frogs, fish, ants, bees, wasps, reptiles, and dragonflies. Species new to science included 22 species of ants, bees and wasps, 11 species of frogs, 7 species of reptiles, 3 species of fish and 1-3 new species of dragonflies and damselflies. More importantly, over 250 species of ants were found in a one square-kilometre area, making the Basin a record-setting site for the greatest animal diversity outside South America. The expedition's data, together with previous work in the Basin, will provide essential data for guiding Papua New Guinea's development. The working paper of the expedition makes recommendations for conservation measures in the Basin that incorporate the economic interests of the local landowners. Like most of Papua New Guinea, indigenous people own much, if not all of the land in the basin. It is clear from the large number of new species discovered during just one month's work of the expedition that there is an urgent need for more biological inventories and taxonomic studies in this area. Ironically, however, while the expedition continued to identify the species collected in the Basin, forests were already being logged without collecting biological information that is critical to ensuring that conservation efforts precede logging and development. Undoubtedly, with some of the largest and biologically richest tracts of tropical forest remaining on the planet, Papua New Guinea and other mega diversity countries of the region are today at a critical juncture in their history, as pressure mounts on the developing nations to exploit their natural treasures.

Source: Conservation International 1998a

90 per cent of the world's terrestrial species (WRI 1999b). However, they have been under serious human assault during this century, from activities such as agricultural development, construction and urban development, with an average annual loss of over 4 million ha per year during the latter part of the century (see Chapter 2). According to FAO (1998), the rate of deforestation dropped slightly during its last survey period, 1990-1995, and was offset to some extent by the enhanced plantation of trees and wood lots.

However, from an ecological point of view, the state of forests is not simply a matter of their extent; more important are their health, genetic diversity, and age profile. In addition, much of the remaining natural forest has been reduced to a patchwork of small-forested areas. This process of fragmentation leaves very little natural or frontier forest of a size and extent which can support species such as the large mammals; the tiger being a prime example.

(b) Grasslands

Grasslands constitute about 24 per cent of land cover in Asia and about 19 per cent in the South Pacific. The natural Asiatic steppe originally extended from Manchuria westwards to Europe, as far as the land which now forms the countries of Bulgaria and Hungary, occupying the broad zone between the taiga (coniferous boreal forest) in the north, and the deserts or mountains to the south. The continental climate of this vast area, with its hot, dry summers and very cold winters, is inimical to the growth of trees, and the area has not supported forest since a more favourable climate prevailed in one of the earlier interglacial period. This belt holds the grasslands of both Northeast and Central Asia. The grasslands of the South Pacific are primarily concentrated in Australia.

A high proportion of the natural Asiatic steppe has now been lost, particularly in Central Asia, where extensive tracts of land in countries such as Kazakhstan and Uzbekistan were turned over to irrigation during the Soviet Era. The grasslands habitats in Northeast Asia have also been disturbed by extensive agricultural, industrial and transport development, and in South Asia, a widespread cycle of vegetation clearance, fire, overgrazing, erosion and abandonment has taken place for many centuries, leaving countries like India with apparently no surviving primary grasslands at all (although there is continuing dispute over the origin of hill grasslands in the southwest). The overall effect is a prevalence of vegetation which is relatively poorly endowed with perennial herbaceous plants, has low floristic diversity, and which in general supports a poor representation of mammals and other wildlife.

Although much of the grassland habitats of Australia in the South Pacific have also been subject to traditional patterns of burning and clearance for many hundreds of years by the Aboriginal people, this has generally been carried out in a rotational system, with the burns being carefully timed in relation to season and weather. This has kept vast areas of the Australian hinterland in a broadly open condition, and has increased productivity for grazing animals. However, during the last one hundred years or so, as more of Australia has become settled, the traditional burning patterns have been disrupted. Many areas are burned annually, allowing few of the native plants and animals to survive, and creating an environment in which introduced European plants are increasingly dominating pastures. Conversely, large areas of land which were previously grazed by the Aboriginals are now burnt much less frequently, with two important consequences: firstly, species of grassland and other open habitat decline because the vegetation becomes too thick, woody and tall; and secondly, when fires do happen, they burn at a higher temperature and are more destructive (WCMC 1992).

2. *Marine Habitat*

The marine environment in Asia and the Pacific is extremely rich in biodiversity; its mangroves, coral reefs and sea grass beds are some of the most productive and diverse ecosystems in the world. In general, the coastal waters of the region support far more life than the open ocean or the deep sea because they contain the most abundant food sources. Approximately 20 per cent of marine plant production occurs in the 10 per cent area of the sea that occupies continental shelves, an area which extends on average to about 70 km from the shore. Here, microscopic phytoplankton and bottom dwelling plants thrive on the nutrients delivered from the land by rivers.

Coral reefs are regarded as the marine equivalent of the tropical rainforests, as they provide a wide variety of habitats to a large number of species. Unfortunately, these ecosystems are being degraded throughout Asia and the Pacific by the consequences of a wide range of human activities, including pollution from sewage, agricultural runoff and industrial waste; disturbance and pollution from aquaculture; sedimentation as a result of inland soil erosion; dynamite fishing and commercial collection of coral; and mineral prospecting and ocean mining (see Chapter 5).

3. *Freshwater Habitat*

Freshwater habitats are home to a wide variety of fauna and flora, including fish, amphibians, invertebrates, plants and microorganisms. However, compared to terrestrial and marine habitats, freshwater systems offer fewer chances for biodiversity to adjust to environmental change, since they are relatively discontinuous and offer less opportunity for species to disperse when conditions become unfavourable. Consequently, freshwater biodiversity is extremely sensitive to environmental disturbance. Being highly localized, however, lakes, rivers or streams can often harbour unique, locally evolved forms of life. In particular, some of the ancient lakes can have extremely high levels of endemism and spectacular species diversity; for example it is said that 90 per cent of the species in Lake Baikal in the Russian Federation are unique to the lake.

Freshwater habitats in Asia and the Pacific have been significantly reduced and degraded over the last century by a combination of factors, relating to water use, pollution and physical disturbance. These include abstraction for both irrigation, which has occurred on a vast scale in many countries of South and Central Asia, and water supply, which has risen in combination with rising population and growing industrial demand. In addition to over abstraction, natural freshwater ecosystems have also been degraded by a range of activities (see Chapter 4), such as the construction of dams and reservoirs, the drainage of wetlands, and the biological, chemical and thermal pollution of water bodies by industry (WRI 1999a). A classic example of the impact on biodiversity from the degradation of freshwater habitats relates to the Aral Sea in Central Asia, where receding water levels in the latter half of this century have reduced the number of nesting bird species in the delta of the Amudarya River from 319 to 168, and the number of mammal species from 70 to 30 (Mainguet and Letolle 1998).

4. *Wetlands*

Wetlands are ubiquitous to Asia and the Pacific, and provide an important and unique habitat for the region's biodiversity. They extend from the low-lying Pacific islands to the mountain lakes with fringing marshes which are abundant in the Himalayas. They exist in the mangrove forests of India and Bangladesh, and also in the extensive floodplain systems of the Ganges and Brahmaputra rivers. It is estimated that there are some 120 million ha of wetlands of international importance in the region (Scott and Poole, 1989), over 80 per cent of which occurs in just seven countries: Indonesia, China, India, Papua New Guinea, Bangladesh, Myanmar and Viet Nam. Although wetland types are extremely diverse, of the 40 types of wetlands of international importance identified by the Ramsar Convention on Wetlands, the three most commonly recorded in the Asian and Pacific countries are permanent rivers, permanent freshwater lakes, and permanent freshwater marshes (Figure 3.1).

Figure 3.1 Wetland Types in Asia and the Pacific

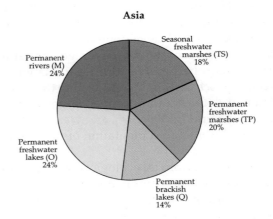

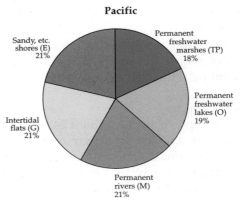

Remark: These wetland types are defined under Ramsar Pacific represents Micronesia, Samoa, New Caledonia, Vanuatu/New-Hebrides, French Polynesia, Tahiti, Australia, New Zealand, Pacific Islands, Guam, Saipan, Papua New Guinea, Fiji, Solomon Islands

Source: Frazier 1999

Over the centuries, vast wetland habitats in the region have been lost due to a variety of human induced and natural causes (Table 3.1). The most serious cause of wetland loss has been conversion (and usually drainage) for alternative uses, such as agricultural or urban development (Frazier 1999; Moser et al 1996). Other degrading influences include inundation from impoundment schemes, changes in water quality through pollution, unsustainable extraction of wetland products, the introduction of new (exotic) species, and, in recent years, fire (Box 3.2). The apparent extent of wetland loss throughout the region is significant. For example, no trace remains of the natural floodplain wetlands of the Red River Delta in Viet Nam, which originally covered almost two million hectares. Likewise, there is virtually no trace left of the one million hectares of natural floodplain vegetation that once covered the

Table 3.1   Agents of Wetland Change

ITEM	Estuaries*	Open Coasts	Flood-plains	Freshwater Marshes	Lakes	Peatlands**	Swamp Forest
Human actions: direct							
Drainage for agriculture and forestry; mosquito control	●	●	●	●	•	●	●
Dredging and stream channelization for navigation; flood protection	●	○	○	•	○	○	○
Filing for solid waste disposal; roads; commercial, residential and industrial developments	●	●	●	●	•	○	○
Conversion for aquaculture/Mari culture	●	○	○	○	○	○	○
Construction of dikes, dams and levees; seawalls for flood control, water supply, irrigation and storm protection	●	•	●	•	•	○	○
Discharges of pesticides, herbicides and nutrients from domestic sewage; agricultural runoff; sediments	●	●	●	●	●	○	○
Mining of wetland soils for peat, coal, gravel, phosphate and other minerals	•	•	•	○	●	●	●
Groundwater abstraction	○	○	•	●	•	○	○
Human actions: indirect							
Sediment diversion by dams, deep channels and other structures	●	●	●	●	○	○	○
Hydrological alterations by canals, roads, and other structures	●	●	●	●	●	○	○
Subsidence due to extraction of groundwater, oil, gas and other minerals	○	○	•	●	○	○	○
Natural causes							
Subsidence	•	•	○	○	•	•	•
Sea-level rise	●	●	○	○	○	○	●
Drought	●	●	●	●	•	•	•
Hurricanes and other storms	●	●	●	●	•	•	•
Erosion	●	●	•	○	○	•	○
Biotic effects	○	○	●	●	●	○	○

Source: UNEP and Wetland International 1997

● common and important;   • present;   ○ absent or exceptional
* without mangroves
** including peat swamp forest

Sylhet Basin in Bangladesh, or the six million hectares of floodplain wetlands in the lowlands of central Myanmar. Recent studies (Moser et al 1996) have tried to quantify the loss of certain types of wetland. Figures produced demonstrate the significance of the loss of mangrove systems in certain countries, such as Singapore (97 per cent loss), Philippines (78 per cent loss) and Thailand (22 per cent loss), and also the loss of peatlands in countries such as Thailand (82 per cent loss), Malaysia (42 per cent loss), Indonesia (18 per cent loss) and China (13 per cent loss).

In the South Pacific, little published quantitative information exists for the extent of wetland loss in the small island developing states, despite a recent survey in the subregion. Cromarty (1996) estimates a loss of 90 per cent of the original wetland area in New Zealand. For Australia, the recently published national wetland directory estimates losses of 27 per cent for Victoria (freshwater and marine), 89 per cent for the south eastern part of South Australia and the Swan Plain Coast of South Australia. The most detailed study is for Victoria, which shows losses of freshwater marshes exceeding

Box 3.2 Wetlands on Fire

Peat swamp forests are waterlogged forests growing on a layer of dead leaves and plant material, up to 20 metres thick. The continued survival of these wetlands depends on a naturally high water level, which prevents the soil from drying out to expose combustible peat matter. Peat swamp forests provide a variety of goods and services, both directly and indirectly, in the form of forestry and fisheries products, energy, flood mitigation, water supply and groundwater recharge.

The countries of Southeast Asia, in particular Malaysia and Indonesia, have more than 20 million ha or 60 per cent of the global resource of tropical peatlands. Fires have become a major threat to these peatlands. Initial estimates indicate that the fires have spread to forests covering 800 000 ha of peatlands. Fires in these peatlands are unique in that they create many times more smoke per hectare than other forest types, and they are almost impossible to extinguish without restoring the naturally high water levels in these swamps. The fires go deep underground and can burn uncontrolled and unseen in the peat deposits for several months.

In the past 10 years, there has been an increasing incidence of major fires in the peat swamp forests of the Southeast Asian region. In East Kalimantan, Indonesia, one fire which started in September 1982 lasted for 10 months and effected more than 35 000 ha. This fire followed an almost unprecedented period of drought in the region associated with "El Nino", the same climatic event which is being blamed for the severity of the 1998 forest fire in Indonesia.

The contribution of tropical peatlands to the global carbon cycle is higher than those of most of the temperate zones as 15 per cent of the global peatland carbon may reside in tropical peatlands. These prolonged peat fires are releasing a massive amount of carbon dioxide and particulate matter which has very serious implication for global warming and long-term climate disruption. It is therefore extremely important to safeguard the peat swamps from forest fires through national and cooperative action.

Source: Standing Committee of the Ramsar Convention 1997

---

70 per cent, although there have been some gains through the creation of some artificial wetlands (such as reed-bed wastewater treatment systems).

B. Species Diversity: Patterns and Trends

Of the seventeen-megadiversity nations in the world, which collectively claim more than two-thirds of the Earth's biological resources, seven – Australia, People's Republic of China, India, Indonesia, Malaysia, Philippines and Papua New Guinea – are in Asia and the Pacific (CI 1998b). These countries are also home to a significant number of threatened endemic species, in relation to which a number of biodiversity "hot spots" have been identified in the region (Figure 3.2). These include the Indian Ocean Islands, Indo-Burma, the Philippines, Eastern Himalayas, southwestern Australia, Polynesia and Micronesia Island Complex, New Caledonia, Western Ghats and Sri Lanka, and New Zealand.

1. Terrestrial Diversity

(a) Plant Diversity

Asia and the Pacific has a wealth of over 165 000 vascular plant species. Forests in the tropical part of the region account for a major share of these species, although the Hindu Kush Himalayan belt also has as many as 25 000 plant species, comprising 10 per cent of the world's flora (Shengji 1998).

Amongst subregions, Southeast Asia has the highest plant diversity (Figure 3.3), primarily because of its tropical forests. Amongst nations, Indonesia has the single highest plant diversity in the region (Table 3.2), estimated at 37 000, and is also in the top five countries worldwide. The country is also first in the world in terms of palm diversity, with 477 species (of which 47 per cent are endemic), and one of its provinces, Irian Jaya, is home to the very rare "birds-of-paradise" plant, which is only found elsewhere in the Moluccas and Australia (CI 1998c). Australia has far and away the highest rate of endemism in the region, at 92 per cent (Table 3.2), although high rates of endemism are also reported for the Hindu Kush Himalayan area (Shengji 1998), where for instance, of the 9 000 plant species found in the Eastern Himalayas, 3 500 species are endemic to the region.

Figure 3.2 Biodiversity "Hot Spots" in Asia and the Pacific

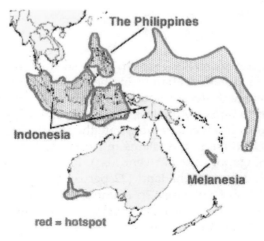

Source: Conservation International 1998g

Table 3.2 Higher Plant Diversity and Endemism in Megadiversity Countries of Asia and the Pacific

Country	Total higher plant diversity	Number of endemic species	Ranks by Endemism	Rate of Endemism (%)	Endemics as % of global diversity of higher plants
Indonesia	37 000	14 800 - 18 500	1	39 - 49	5.9 - 7.4
PR China	27 100 - 30 000	10 000	4	33	4.0
Papua New Guinea	15 000 - 21 000	10 500 - 16 000	2	50 - 76	4.2 - 6.4
India	17 000 +	7 025 - 7 875	5	41 - 46	2.8 - 3.2
Australia	15 638	14 458	3	92	5.8
Malaysia	15 000	6 500 - 8 000	6	43 - 53	2.6 - 3.2
The Philippines	8 000 - 12 000	3 800 - 6 000	7	48 - 50	1.5 - 2.4

Source: Conservation International 1998b

Likewise, the Indian Himalayas contain more than 50 per cent of India's endemic flora.

It is not known how many species of plants in Asia and the Pacific have already become extinct as a consequence of human activities, however, according to the IUCN (1998) Red List, more than 10 000 existing plant species in the region are threatened. Amongst subregions, Southeast Asia has the largest proportion of threatened plants (Figure 3.3). Amongst countries of the region, Turkey has the highest percentage of threatened plant species estimated at 21 per cent, closely followed by French Polynesia, at approximately 20 per cent. In terms of actual numbers of threatened species, Australia and Turkey have the highest estimates, at 2 245 and 1 876, respectively. More than 1 200 species of vascular plants in India, 707 in Japan, and about 500 each in Malaysia and New Caledonia are also threatened.

(b)   Animal Diversity

The mega diversity countries identified earlier also possess a large proportion of the animal diversity of Asia and the Pacific, a large number of which are endemic species (Table 3.3). Vertebrates provide a good indicator of species diversity, since they generally occupy the top rungs in food chains, i.e. habitats healthy enough to maintain a full complement of native vertebrates will have a good chance of retaining the invertebrates, plants, fungi, and other small or more obscure organisms to be found there. Conversely, ecological degradation can often be read most clearly in native vertebrate population trends (Baille and Groombridge 1996). As with plant diversity, terrestrial vertebrates are most abundant in the tropical forests of the region (WRI 1999d).

(i)   Birds

Birds were the first animals that IUCN comprehensively surveyed, in 1992, followed by full re-assessments in 1994 and 1996. They are recognized as good indicators of biodiversity because their

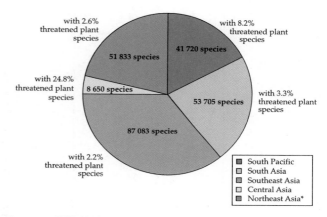

Figure 3.3 Total Number of Plant Species in Asia and the Pacific by Subregion and the Percentage of Threatened Plant Species

Source: WCMC 1997

* excluding the Russian Federation

distribution is well known and they are sensitive to environmental change. Amongst the countries of the region, Indonesia has both the largest number of bird species (Table 3.3), equating to about 17 per cent of the world's total, and also one of the highest rates of endemism in the region (only Australia and the Philippines are higher). Endemic bird areas that are of highest priority for conservation in the region are located in the Lesser Sundas in Indonesia, Eastern Himalayas, and Luzon (especially Mindoro) in the Philippines.

Close to 1 100 of the region's bird species are threatened, of which 40 per cent are in the Southeast Asian subregion (Figure 3.4), and 104 are in Indonesia alone (Table 3.4). The most threatened major groups include rails and cranes (both specialized wading birds), parrots, terrestrial game birds (pheasants, partridges, grouse and gitans), and pelagic seabirds (albatrosses, petrels and shearwaters). About one quarter of the species in each of these groups is currently threatened. Only 9 per cent of songbirds

CHAPTER THREE

Table 3.3 Terrestrial Animal Species Diversity and Endemism in Megadiversity Countries of Asia and the Pacific

Country	Mammals			Birds			Reptiles			Amphibians			Total number of non-fish vertebrate species	
	No. of Species	No. of Endemism	% of Endemism	No. of Species	No. of Endemism	% of Endemism	No. of Species	No. of Endemism	% of Endemism	No. of Species	No. of Endemism	% of Endemism	No. of Species	No. of Endemism	% of Endemism
Indonesia	515	201	39.0	1 531	397	25.9	511	150	29.4	270	100	37.0	2 827	848	30.0
Australia	282	210	74.5	751	355	47.3	755	616+	81.6+	196	169	86.2	1 984	1 350	68.0
PR China	499	77	15.4	1 244	99	8.0	387	133	34.4	274	175	63.9	2 404	484	20.1
Philippines	201	116	57.7	556	183	32.9	193	131	67.9	63	44	69.8	1 013	474	46.8
India	350	44	12.6	1 258	52	4.1	408	187	45.8	206	110	53.4	2 222	393	17.7
Papua New Guinea	242	57	23.6	762	85	11.2	305	79	25.9	200	134	67.0	1 509	355	23.5
Malaysia	286	27	9.4	738	11	1.5	268	68	25.4	158	57	36.1	1 450	163	11.2

Source: Conservation International 1998b

Table 3.4 Top Five Countries in Asia and the Pacific by Each Group of Threatened Species and Percentage of the Regional Total

Rank	Mammals			Birds			Reptiles			Amphibians			Fish			Invertebrates		
	Country	No. of Species	% total	Country	No. of Species	% total	Country	No. of Species	% total	Country	No. of Species	% total	Country	No. of Species	% total	Country	No. of Species	% total
1	Indonesia	128	13.2	Indonesia	104	9.6	Australia	37	11.5	Australia	25	50	Indonesia	60	20.2	Australia	281	48.5
2	India	75	7.7	PR China	90	8.3	Myanmar	20	6.2	Japan	10	20	Australia	37	12.5	Japan	45	7.8
	PR China	75	7.7															
3	Australia	58	6.0	Philippines	86	8.0	Indonesia	19	5.9	India	3	6	PR China	28	9.4	Indonesia	29	5.0
																French Polynesia	29	5.0
4	Papua New Guinea	57	5.9	India	73	6.8	India	16	5.0	Turkey	2	4	Philippines	26	8.8	Russian Federation	26	4.5
							Thailand	16	5.0	Islamic Rep. of Iran	2							
										Philippines	2							
5	Philippines	49	5.1	Viet Nam	47	4.4	PR China	15	4.7	PR China	1	2	Turkey	18	6.1	India	22	3.8
										Viet Nam	1							
										New Zealand	1							
										Fiji	1							

Source: WCMC 1997

Figure 3.4 Number of Threatened Animal Species in Asia and the Pacific by Subregion

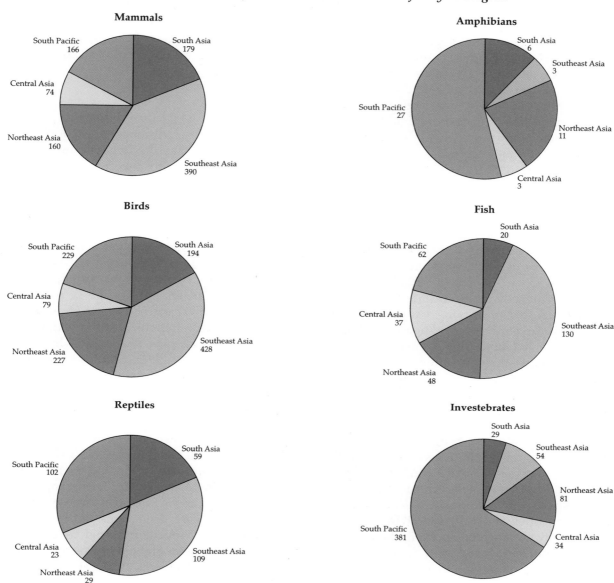

Source: WCMC 1997

are threatened, but they still contribute the single largest group of threatened species because they are far and away the most species-rich (Baille and Groombridge 1996). Other threatened bird species of the region include the crested ibis, a wading bird that has been eliminated from its former range in Japan, the Korean peninsula, and the Russian Federation, and is now down to one small population in the remote Qinling mountains of China, the Philippine monkey-eating eagle and the blue-winged Racquet-tail parrot.

The relatively high numbers of threatened species in the island environments are in part a reflection of the fact that island birds tend naturally to have smaller ranges and numbers, making them more susceptible to habitat disturbance. In the Philippines a number of bird species are believed to have disappeared due to habitat loss as a result of deforestation from the islands of Cebu, Negros, Panay and Mindoro. In Indonesia, the endemic bird habitats of the islands of Sumbu, Banggai and Sula are also under pressure. Another threat to bird species in the region, and in particular to the small island states of the Pacific, is due to predators and competition from introduced domestic animals.

(ii) Mammals

Asia and the Pacific has the largest number of mammal species in the world. Amongst the countries of the region, Indonesia has the largest number of mammal species (Table 3.3), and also one of the highest rates of endemism in the region (only Australia and the Philippines are higher). Almost 1 000 of the region's mammal species are currently threatened, of which 40 per cent are in Southeast Asia (Figure 3.4, and 128 are in Indonesia alone (Table 3.4). Moreover, Indonesia is also home to the orang-utan, whose populations decreased to such low levels in the early 1990s, when more than half were reportedly lost, that they were listed by the Convention on International Trade in Endangered Species of Fauna and Flora (CITES) as the most endangered species in the world (Conservation International 1998b). Current estimates suggest that less than 30 000 may still be surviving in Borneo (Van Schaik 1999), however, this figure could even be lower since these estimates were made before fires devastated millions of hectares of forests in Borneo in the late 1990s.

Other threatened species include the tiger, the largest of all cats that once ranged from Turkey to Bali and the Russian Far East. Wild tigers now barely total 3 000 to 5 000 individuals, many in small, isolated populations that are under severe threat (Tuxill and Bright 1998). Species such as Caspian tiger (*Panthera tigris virgata*) and Near Eastern leopard (*Panthera pardus ciscaucasica*) have already disappeared this century from Uzbekistan (Sievers 1999). The tiger is also regarded as extinct in the Republic of Korea, where the Siberian leopard, the fox wolf and sika are also no longer observed (National Biodiversity Report of Korea 1998). Other endangered species in the Republic of Korea include the musk deer, otter and Eurasian flying squirrel. Another critically endangered specie in the region is the Bulmer's fruit bat, found only in Papua New Guinea (Conservation International 1998b).

(iii) Amphibians

Asia and the Pacific is home to a large number of amphibian species. Amongst the countries of the region, People's Republic of China (closely followed by Indonesia) has the largest number of amphibian species (Table 3.3), and also the highest number of endemic species in the region. The Philippines, although comparatively low in amphibian species has high rate of endemism. The three introduced amphibian species are *Bufo marinus*, a marine toad introduced in the 1930s to control beetle infestation of sugarcane; *Rana catesbeiana*, a bullfrog introduced in the 1970s for breeding and export as food; and *Rana rugulosa*, introduced in the 1990s for breeding and export as food (PAWB-DENR 1998).

Of the 50 species of amphibian which are currently threatened in Asia and the Pacific, over 50 per cent are in South Pacific (Figure 3.4), and almost all of these (25) are in Australia alone (Table 3.4). There are also a high number of threatened amphibian species in Japan, which accounts for some 20 per cent of the overall total for the region.

(iv) Reptiles

Asia and the Pacific is home to a large number of reptile species. Amongst the countries of the region, Australia has the largest number of reptile species (Table 3.3), and also has by far the highest rate of endemism in the region (over 80 per cent).

Over 300 of the region's mammal species are currently threatened, of which over 65 per cent are in the Southeast Asian and South Pacific subregions (Figure 3.4); Australia, Myanmar and Indonesia having the highest individual totals (Table 3.4). Among reptiles, the status of turtles, crocodilians, and tuataras (an ancient lineage of two lizard-like species living on scattered islands off New Zealand) has been comprehensively surveyed (Tuxill and Bright 1998), but most snakes and lizards remain unassessed (Cogger 1992). Studies aside, however, a glaring case of reptile extinction is the sea turtle whose population has declined by more than 50 per cent worldwide, largely as a result of the over-exploitation of their eggs (Abas 1999). In some localities of Southeast Asia, such as Khram Island in Thailand, only 50 green turtles and 10 hawksbills are now found compared to 158 and 45, respectively, in the 1950s. In Malaysia, the leatherback turtle has been identified as the most endangered turtle specie, with a 99 per cent drop in its numbers since man began hunting it several hundred years ago. Barely 10 leatherbacks were sighted between 1993 and 1999, compared to 2 000 in the 1950s, and only 400 green turtles were spotted in Sarawak over the same period, compared to 4 500 in 1930s. Less than 100 green turtles can now be found in Ogasawara, Japan, as a result of an estimated harvesting of 1 800 adult green turtles each year.

Although less well known than their seagoing relatives, tortoise and river turtle species also are exploited intensively in certain areas of the region, to the point where populations have been greatly depleted (Tuxill and Bright 1998). In addition, certain species of crocodilians suffer from overhunting and also from pollution of their environment (such as the Indian gharial and the Chinese alligator), but this is one of the few taxonomic groups of animals whose overall fate has actually improved over the past two decades. Since 1971, seven alligator and crocodile species have been taken off IUCN's *Red Data* list, including Australia's huge estuarine crocodile.

(v) Invertebrates

Substantial uncertainty exists over the relative abundance of invertebrate species in the tropical forests of Asia and the Pacific (WRI 1999d), although they probably represent the richest group of biodiversity in the region. Until recently, the relative diversity of arthropods in the tropics, as compared to the temperate zone, was expected to be similar to that of better known groups, such as vascular plants or birds. However, the relatively recent discovery of a tremendous richness of invertebrate species in the canopy of tropical forest of Sulawesi, Indonesia, suggests that the richness of arthropods in the tropics is much greater. It has been estimated that as many as 30 million arthropod species-up to 96 per cent of the world's total for all species – may exist in tropical forests alone.

Almost 600 of the region's known invertebrate species are currently threatened, of which the vast proportion (almost 70 per cent) are in the South Pacific (Figure 3.4). Australia alone contains almost 50 per cent of the subregion's threatened species (Table 3.4).

2. *Marine Diversity*

Marine ecosystems harbour a myriad of life forms. The deep sea floor may contain about a million undiscovered species, and out of the 1.7 million species catalogued to date, around 0.25 million belong to the marine environment (WRI 1999f). However, the state of knowledge on marine species distribution and "hot spots" is currently poor because only about 7 per cent of the oceans have been sampled (WRI 1999e). Entirely new communities of organisms-hydrothermal vent communities – were found less than two decades ago in the marine ecosystems, and more than 200 new families or sub-families, 50 new genera, and 100 new species from these vents have already been identified. The highest overall diversity takes place in the tropical Indo-Western Pacific, a region that includes waters off the coasts of Asia, northern Australia, and the Pacific Islands. Within this region, some of the highest levels of marine species richness are found off the coasts of the Philippines, Indonesia, and Papua New Guinea. It is possible that, as a result of its marine diversity, Indonesia rivals Brazil for the title of the most biodiverse country on Earth.

Endemism in marine communities appears to be proportionately more dominant in areas surrounding isolated islands and thermal vents (WRI 1999e). Broad distributions of marine species seemingly indicate that they are less vulnerable to extinction than their terrestrial kin. However, not all marine species may be as wide-ranging as is currently believed. For example, a recent effort to map the distribution of coral reef fish revealed that, of the 950 species whose ranges were mapped (about 23 per cent of the total), one-third were limited to areas of less than 2 220 km^2. Although coral reefs share numerous attributes with tropical forests, their level of local species endemism is much lower (WRI 1999g). Within the Indo-Pacific, for example, the vast majority of coral species are found throughout the region. Because coral reef species disperse readily, locally endemic species occur only on isolated oceanic islands.

About 40 per cent of the world's mapped reefs are found in the Pacific (WRI 1999l), including the most extensive coral reef system in the world, the Great Barrier Reef of Australia, and many small islands in the South Pacific with extensive coral formations, such as Fiji, Guam, New Caledonia, Tonga, the French territories of the Pacific Islands and Tuvalu. Coral reefs and associated species diversity, however, is greatest in Southeast Asia, which has a cluster of coral hotspots within the Indo-West Pacific region (WRI 1999g). Containing one-quarter of the world's mapped reefs (WRI 1999i), this subregion support more than 16 per cent of the world's estimated 19 000 species of freshwater and marine fish (WRI 1999g). Indonesia and the Philippines account for a major portion of these reefs, each are containing at least 2 500 species of fish (WRI 1999i). Waters surrounding Polynesia and portions of Indian Ocean also contain areas with high levels of reef fish diversity.

Amongst marine fish, the coral reef variety makes up one-quarter of all known species (WRI 1999e). Some 4 000 species of fish and 800 species of reef-building coral have been described to date, but the total number of species associated with reefs is estimated to be more than one million (WRI 1999j). Levels of coral reef species diversity vary within the marine ecosystems depending on location. The most species-rich reefs are found in a swath extending through Southeast Asia to the Great Barrier Reef, off northeastern Australia. More than 700 species of corals alone are found in this area. The Great Barrier Reef, the world's largest system of coral reefs (covering 349 000 km^2) supports approximately 400 species of coral, 1 500 fishes, over 4 000 species of molluscs and populations of Indo-Pacific invertebrates, birds, turtles, dugong, whales, and dolphins (WRI 1999h). In addition, 252 species of bird's nest and breed on coral cays, five species of turtles live on the reef, and several species of whales and dolphins are associated with the habitat (WRI 1999j).

Like their terrestrial counterpart, marine species have also been subject to pressures. About 68 per cent of all threatened marine species suffer from over-exploitation (WCMC 1998). By 1994, 90 marine mammal species were listed as threatened or endangered (WRI 1999a). More than a quarter of the world's reefs are at high risk from human disturbance (WRI 1999j). In Asia and the Pacific, coral reefs are most threatened in the Southeast Asian subregion (see Chapter 5).

3.   Freshwater Diversity

The overall number of species in freshwater is low compared with marine and terrestrial groups. However, species richness in relation to habitat extent is extremely high, since the area occupied by freshwater is so much lower than land or sea. As a consequence, the importance of protecting freshwater habitats is correspondingly high.

Although most freshwater plant species are relatively cosmopolitan, the tropical regions of Asia appear to be one of the most rich in freshwater plant species which are restricted to a single continent, country or area (WCMC 1998). Sri Lanka, India, Myanmar and Indonesia hold localized species such as the Podostemaceae and a large number of narrowly endemic species in some cases, with several forms restricted to different stretches of a single river.

Animal species are considerably more diverse and numerous in inland waters than plants (WCMC 1998). More than half of all freshwater vertebrate species are fish, and more than 8 500 species (40 per cent) of the 25 000 known fish species exist in freshwater. Unfortunately, their distribution and systematic are inadequately known (WCMC 1998). Among the countries of Asia and the Pacific, IUCN has listed the highest number of freshwater species in Australia (which is also ranked third in the world), followed by Papua New Guinea and Turkey (Table 3.5). Scientists believe that Thailand may have as many as 1 000 species of freshwater fish but only some 475 have actually been recorded (WRI 1999a). Apart from fish, other important groups of inland water species in the region include crustacea, molluscs and insects.

Table 3.5   Total Numbers of Threatened Freshwater Fishes in Selected Countries of Asia and the Pacific

Country	Total species	Threatened species	Percentage of threatened species
Australia	216	27	13
Papua New Guinea	195	12	6
Turkey	174	18	11
Japan	150	7	4
Sri Lanka	90	8	9

Source: WCMC 1998

Accurate data required to evaluate the extinction of aquatic species are generally not available, although some 81 fish species are recorded as having become extinct during the past century, and a further 11 are extinct in the wild, but remain as captive populations (WCMC 1998). A recent survey in Malaysia found fewer than half of the 266 fish species previously known to exist in the country (WRI 1999a). Similarly, in Singapore, 18 out of 53 species of freshwater fish collected in 1934 could not be located some 30 years later, despite an exhaustive search. In the freshwater and marine ecosystems, over-fishing has also resulted in shifts in fish size, abundance and species composition. For example, some 20 species of edible fish were thriving in the Aral Sea during the 1940s, but in the early 1990s, only five species remained (Mainguet and Letolle 1998). At present, only one is thought to survive.

C.   Genetic Diversity: Patterns and Trends

The term "genetic resources" describes a category of biodiversity encompassing the diversity of genes in crop seeds, genetic materials such as germplasms or naturally occurring chemicals found in plant and animal species (Putterman 1999). The contribution of genetic resources to the global economy ranges from the use of genes in modern agriculture to enzymes used in industrial manufacturing, and from organic molecules used to design new pharmaceutical drugs to extracts of medicinal plants that are used to prepare herbal products.

Genetic resources yielding potentially valuable chemicals, enzymes or genes come from terrestrial and marine microbes, plants, insects, venomous animals and other marine organisms. Tropical rainforests recognized for their high biological diversity, are the major source of genetic wealth, but dryland ecosystems including dry forests, savannah, deserts and marine ecosystems such as coral reefs are also know to yield high levels of bioactive compounds (Putterman 1999). The wealth of genetic material which are available for medicine is amply demonstrated by World Health Organization statistics, which estimate that 3.5 billion people in developing countries (most of which are in Asia and the Pacific) rely on plant-based medicine for their primary health care (Tuxill 1999). Ayurvedic and other traditional healers in South Asia use at least 1 800 different plant species in treatments and are regularly consulted by millions of people. In People's Republic of China, where medicinal plant use goes back at least four millennia, healers employ more than 5 000 plant species (Baille and Groombridge 1996). Likewise, varied traditional communities in

Indonesia possess great knowledge on the use of biodiversity as a source of food and medicine etc. and Indonesian communities in their daily life use some 6 000 plants, 1 000 animals and 100 microbe species.

In addition to naturally occurring genetic resources, over the millennia, farmers have developed a wealth of distinctive genetic varieties within crops by selecting and replanting seeds and cuttings from uniquely favourable individual plants (Tuxill 1999), thus creating a series of folk varieties, or *landraces*. Many of these were perhaps ones that matured slightly sooner than others, were unusually resistant to pests, or possessed a distinctive colour or taste. Subsistence farmers have always been conscious of such varietal diversity, because it helped them cope with variability in their environment. India alone, for instance, probably had at least 30 000 rice landraces earlier this century (Swanson et al 1994).

These domesticated landraces have been in substantial decline in recent years. In most developing countries of the region, losses were minimal until the 1960s, when the Green Revolution introduced high yielding varieties of wheat, rice, corn, and other major crops (Tuxill 1999). Developed to boost food self-sufficiency in famine-prone countries, the Green Revolution varieties were widely distributed, often accompanied by government subsidies, and thereby displaced landraces from many prime farmland areas (NRC 1993). There is now a growing concern that these crops may have lost their genetic variability to such an extent that they may not be resistant to any future environmental pressures from, for example, global climate change (SOEAP 1995), or from attack by pesticide – resistant pests and diseases (Box 3.3).

An example of increasing genetic uniformity is provided by rice, the most important crop in the region. In India, it is estimated that by 2005, 75 per cent of rice production will come from less than ten varieties (Ryan 1992), as opposed to the thousands of varieties grown in 1950s. Similarly, in Indonesia, 1 500 local varieties of rice disappeared between 1975 and 1990, and nearly three quarters of the rice planted today descends from a single maternal plant (WRI/UNEP/IUCN 1990).

A reduction of genetic diversity is also becoming evident in animals, as new breeds selected for high output. A clear example is that of the domestic chicken, where, to improve production output, many traditional breeds have been abandoned in commercial terms in favour of US imports such as the white leghorn for egg production, and the Rhode Island Red for meat production.

D. Biotechnology, Biosafety and Bioprospecting: Prospects and Trends

The scale of manipulation of genetic diversity in recent years has been unprecedented. In particular, manipulation of pollens, seeds and other propagules through biotechnology and/or genetic engineering has contributed to changes in the patterns of occurrence of some genes. It has even resulted in transfer of genetic material from one species to another. This has brought revolutionary changes in agriculture, medicine and other fields. However, it

---

Box 3.3 Risk to Food Security as a Result of Monoculture and Biodiversity Loss

In the mid-1970s, a mysterious disease started devastating rice crops in Indonesia. The Green Revolution had encouraged the introduction of new, high-yielding but genetically uniform varieties of rice, highly vulnerable to attack by pathogens. Within a couple of years the virus was threatening more than a million ha, putting hundreds of millions of people throughout Southeast Asia at risk.

The International Rice Research Institute (IRRI) rapidly screened all 6 273 varieties in its collections for resistance to the virus. Only one possessed it – a low yielding, spindly species, collected from the wild in Southern India but thought to have no commercial value. The resulting new variety, IR36, is now planted as yet another genetic monoculture over millions of Asian hectares – vulnerable to the next pathogen whose natural selection may outpace the plant breeders.

Will there still be an uncultivated wild variety of rice possessing the genes needed to save people from starving when the next pathogen strikes? Or will it already be extinct? If climate change radically alters the patterns of agriculture throughout the world, as inevitably it will, where will the genetic material come from to produce the new crop varieties on which human survival will depend? Gene banks, like the IRRIs, may provide a partial answer, but the greatest gene bank of all is nature, and this is being destroyed at an increasing rate. The myriad of genes, species and ecosystems that collectively make up what we call Nature – may have taken four billion years to evolve, but it seems destined to be largely destroyed in just four human generations. Rates of species extinction are estimated to be 50-100 times the natural background rate and this could increase to 1 000 to 10 000 times with the forest loss projected for the nest 25 years. Unless direct action is taken now to protect biodiversity, the humanity will lose forever the opportunity of reaping its full potential benefits.

Source: Watson, T.R. et al, eds. 1996

CHAPTER THREE

has also brought new challenges for the protection and conservation of biodiversity in the region.

1. Agricultural Biotechnology

(a) Environmental and Health Related Risks

The most important outcome of agricultural biotechnology to date is the production of genetically engineered plant seeds, more commonly known as genetically modified (GM) seeds. Genetic engineers claim that the new seeds are to all intents and purposes the same as the original ones, except that they have been improved to produce a higher crop yield, for example through increased resistance to pests, and enhanced response to particular fertilizers. Critics of the new technology, however, assert that there are too many risks associated with genetic engineering, particularly in relation to environmental and human health. More specifically, they are concerned about biosafety of genetically engineered organisms (such as plants and micro-organisms), that is, their safe transfer, handling, use and disposal in the environment.

In terms of environmental risks, the consequences of releasing genetically engineered seeds into the environment have not been adequately assessed. Wind blown pollen from the modified seeds may, for example, fertilize related plants outside the growth area, and if it contains a gene for herbicide resistance, a new strain of "superweeds" could result. Similarly, it is difficult to predict with any certainty what effects the modified plants may have on other organisms. For example, genetic engineers have developed new strains of many plants that contain a gene from a bacterium *Bacillus thuringiensis* (Bt), which makes them resistant to insect pests. However, there is some evidence that the crops may have an impact not just on the pests, but also on beneficial insects such as bees, lacewing, and ladybirds and the birds that eat them.

In terms of health risks, it has been found that certain transgenic crop varieties (e.g. Bt-maize) contain a marker gene that codes for antibiotic resistance in the bacteria *Erwinia coli*. There is a risk that if animals or humans consume such products, for example in the form of cattle feed or starch, some antibiotics would be rendered useless. Another concern is that when crops such as rice or rape-seed with high Vitamin A concentrations are planted, there will be no way to distinguish them from normal crops, with the contingent risk of liver damage if too much Vitamin A is unwittingly consumed. Laboratory tests have also shown that certain genetically engineered materials are also associated with reproductive problems in rabbits.

In some countries, such as England and France, major supermarket chains have already banned genetically engineered products from their shelves, and in some Asian and Pacific Region countries, including Japan and Hong Kong, China, supermarkets have begun to introduce GM – free products to test consumers reaction (see Chapter 13). Reaction amongst some other countries is the biodiversity rich Asian and Pacific region, has also been hostile, for example, students from the Republic of Korea have blockaded government funded biotechnology greenhouses, and in India, farmers have made strong protests to the government. However, the reaction has been varied amongst different stakeholder groups throughout the region (Box 3.4).

(b) Intellectual Property Rights and Dependency

Intellectual property rights (IPR) is another area of major concern to some of the less industrialized countries of Asia and the Pacific which are rich in biodiversity; the worry being that they will be forced to accept or adopt some form of IPR, even before a full and frank international debate has taken place about the suitability of such systems to protect resources that are collectively owned. Moreover, if foreign researchers and transnational corporations can patent indigenous plants (e.g. varieties of rice) without adequately compensating the communities who provided them, there are fears that farmers will end up paying surcharges on products formed from their own knowledge and experience (Box 3.5). In response, some experts believe that IPR should also be applied in favour of the farming communities, to patent and protect their own varieties in order to retain any benefits, which are derived from them. However, this in turn gives rise to the problem of benefit sharing, how to define "community" and, in the case of plant breeding, whether it is possible to determine the provenance of a plant that is the product of generations of farmers' inputs.

Probably the most important and challenging risk to biodiversity rich countries of the region in relation to genetic engineering is the balance of power over food supply, which is fast becoming concentrated in the hands of a few major companies who dominate the world's seed market, such as Monsanto, Novartis, Calgene, Du Pont and AstraZenica. "Control of the plant gene" has aroused particular concern in this respect; a genetic technique which enables seed companies to ensure that farmers are not able to raise second generation (F2) populations from first generation (F1) seeds using mechanisms such as embryo abortion in the first generation seeds. Currently, 80 per cent of crops in the biodiversity rich world are grown each year from saved seed. The cost of buying expensive new seeds every year will clearly place a significant economic burden on many small farmers.

> **Box 3.4 Public Acceptance of Transgenic Crops in Asia and the Pacific:
> A Case of Transgenic Rice in the Philippines**
>
> With the advent of genetic engineering, *Bacillus thuringiensis* (Bt), a common natural resource potentially used as biopesticides, have been inserted directly in to several crop and have become widely commercialized. To understand the issue of public acceptance of this genetically engineered crop, a survey was conducted by the *Department of Agricultural Economics* at the *Swiss Federal Institute of Technology* (ETH) in cooperation with the *University of the Philippines Los Baños* (UPLB). The study aimed to examine the perception of risk and benefits of transgenic rice among the main political actors in the Philippines indicated that there were three major groups of perception with corresponding political weight in the debate:
>
> The *first group* dominated by NGOs, some large NGO networks, people's organizations (POs) and other public interest groups opposed biotechnology and did not see any potential for genetic engineering in agriculture. They anticipated that this technology had high risks and low benefits. In the survey, respondents perceived this group to have a major influence on public opinion since they seemed to be the most active in protest activities and other campaigns and were very effective in gathering and disseminating information. However, this group was not considered very important with regard to direct political decision-making processes in the Philippines, since they were not members of any legislative body.
>
> The *second group* included the majority of government officials and politicians with considerable influence on political decision-making processes and, to a certain extent, on public opinion. They had an important role in issuing directives and granting financial support. Respondents in this group had high expectations of the potential of genetic engineering for solving the problems confronting the Philippine rice economy. There was a contradiction in their expectation. On the one hand, they agreed that genetic engineering could only address agronomic problems, yet many of them also expected the technology to solve structural problems. Further, this group had a rather ambivalent attitude towards risks and benefits of genetic engineering. Half of them perceived the benefits while the other half perceived the risks. This may be explained by their perception that biotechnology was a tool that enabled plant breeders to solve those problems that could not be addressed through conventional technology. At the same time, this group doubted the sustainability of biotechnology since according to them insects would eventually develop resistance to Bt rice.
>
> The *third group* consisted of scientists of private companies, and national and international research centres. While scientists from the University of the Philippines were to be found in all three perception groups, in general the third group's view of the potential of genetic engineering in agriculture was more modest, although their attitude was definitely positive. Although they did not expect biotechnology to solve structural problems yet, they certainly saw the potential of genetic engineering for solving agronomic problems, including those caused by natural calamities. This group which received financial support from both national and international donors, was central in the debate on genetic engineering and genetically engineered rice and according to the survey respondents, represented the most important suppliers of information. Their influence on political decisions was felt to be relatively high, whereas their influence on public opinion was considered to be low. This group did not have direct access to the public; instead, those who had better access to the public, such as the NGOs and the mass media, gathered information from this group.
>
> A majority of the surveyed population indicated that labelling of transgenic food products and allowing farmers free choice of seeds was important for gaining public confidence. While real consumer behaviour could not be anticipated by the study, the survey indicated that consumer organizations had only a marginal stake in the debate and that health risks among the respondents were not perceived as very serious. Therefore, it is not anticipated that the average urban consumer in the Philippines would reject transgenic rice for fear of serious health risks. This can be considered as a major difference to opposition in industrialized countries.
>
> Given the NGOs' lack of direct influence in the political decision-making process, their opposition to genetic engineering will most likely not lead to restrictive legislation against genetic engineering in agriculture. This is probably because modern biotechnology is considered the 'flagship' of the government's 'Vision Philippines 2000' for national economic growth. However, it might have consequences on the future strategies in development cooperation in the Philippines since doubts about this technology, or lack of confidence in the responsible institutions, could lead to an increased polarization in the debate and may hinder future cooperation among all the actors.
>
> *Source:* Aerni, P. et al 1999

2. *Bio-prospecting*

Biodiversity prospectors search for genetic and biochemical resources that have a commercial value, particularly to the pharmaceutical, biotechnological, and agricultural industries. Biodiversity prospecting is not new – plant collectors from industrialized countries have ventured southward in search of valuable genetic material for agricultural plant breeding for many years – however, recent advances in molecular biology, and the increasing availability of sophisticated diagnostic tools for screening, have made it an increasingly cost-effective operation for pharmaceutical corporations and others to perform. As a result, the market for buying and selling exotic biological specimens is expanding rapidly (it is conservatively estimated that the market for

CHAPTER THREE

> **Box 3.5 Patenting of Genetic Materials and Plight of Developing Countries in Asia and the Pacific**
>
> In September 1997, the US company Ricetec, Inc., was granted a patent on Basmati rice. The patent is for a variety achieved by the crossing of Indian Basmati with semi-dwarf varieties, and it covers Basmati grown anywhere in the Western hemisphere. Ricetec can also put its brand on any breeding crosses involving 22 farmer-bred Basmati varieties from Pakistan and, on any blending of Pakistani or Indian Basmati strains with the company's other proprietary seeds. Ricetec also claims the right to use the Basmati name. The Indian government has challenged Ricetec's claim, arguing that the patent jeopardizes India's annual Basmati export market of around US$277 million, and threatens the livelihood of thousands of Punjabi farmers.
>
> The company is also marketing another proprietary rice called Jasmati, which is derived from a variety called Della, developed in the US. BIOTHAI (the Thai Network on Community Rights and Biodiversity) is concerned that by giving this variety a name that implies a cross between Jasmine rice and Basmati rice, the company is threatening the livelihoods of five million poor farmers in the northeast of Thailand too.
>
> A similar case is W.G. Grace and Co.'s series of patents on extracts from the neem tree, whose seeds and bark have been used for centuries in India and Pakistan for natural pesticides. Grace has estimated that the global market for the pesticide could reach US$50 million a year by 2000, and while there are no hard statistics on the impact on Indian farmers to date, Vandana Shiva says that farmers in the south of India, where neem is harvested, are already losing out because the processing and exporting of the seeds and oil is no longer available to them.
>
> Another example is the case of two Mississippi doctors who were granted a patent in 1995, for the traditional use of turmeric as a healing powder. India's Council of Scientific and Industrial Research (CSIR) petitioned the US Patent and Trademark Office (PTO) on the grounds that the 'discovery' was not original, but had been chronicled in traditional Indian texts. In August 1997, the PTO rejected the patent holders' claims. "What is being patented is not one invention of one individual or corporation, but the collective creativity and inventiveness of millions of people over millennia, a creativity...that is necessary for meeting the needs of our people in the future."
>
> Among developing countries of Asia and the Pacific, India is particularly vocal on farmers' rights and in order to cope with such patenting problems the Agriculture Ministry has drafted new legislation, which allows farmers to use, and exchange traditional varieties. In contrast to International Convention for the Protection of New Varieties of Plants (UPOV), the legislation will recognize the farmer as breeder first and foremost, before foreign commercial or research interests. India also commissioned a US$20 million national genebank in 1997, the third largest in the world, and the government is in the process of drafting legislation, in accordance with the CBD, which will regulate the export of germplasm and prohibit open access to the Indian genebank by American seed companies, access that was allowed under a 1988 agreement with the US. A National Biodiversity Authority (NBA) will be set up to enforce the law.
>
> *Source:* Spinney, L. 1998

natural product research specimens within the pharmaceutical industry alone is US$30-60 million per annum). The renaissance of bio-prospecting is also fuelled, in part, by the realization that species, their genetic material, and the ecosystems of which they are a part are rapidly disappearing from the face of the earth. In the mid-1980s, pharmaceutical industry analysts warned that each medicinal plant lost in the tropical rainforests could lose drug firms possible sales of more than US$200 million.

Valuable chemical compounds derived from plants, animals and micro-organisms are generally more easily identified and of greatest commercial value when collected with the assistance of, or in combination with, the knowledge of indigenous peoples. For example, scientists have found that 86 per cent of the plants used by Samoan healers displayed significant biological activity when tested in the laboratory. However, the contribution of indigenous peoples has rarely been acknowledged or rewarded in the past, and no matter how convincing the rhetoric today, conservation and equity are likely to remain secondary issues for the various corporations involved in the practice of bio-prospecting. Moreover, once indigenous peoples share information or genetic material, mechanisms such as monopoly patents can mean that they effectively lose control over those resources forever, regardless of whether or not they are compensated.

CAUSES AND CONSEQUENCES OF BIODIVERSITY LOSS

A. Causes of Biodiversity Loss

The Global Biodiversity Strategy (WRI/IUCN/UNEP 1992) identifies six fundamental causes of biodiversity loss (UNEP 1995):

- unsustainably high rates of natural resource consumption and human population growth;

- steadily narrowing spectrum of traded products from agriculture and forestry, and introduction of exotic species associated with agriculture, forestry and fisheries (including bio-engineered species);
- economic systems and policies that fail to value the environment and its resources;
- inequity in ownership and access to natural resources, including the benefits from use and conservation of biodiversity;
- inadequate knowledge and inefficient use of information;
- legal and institutional systems that fail to protect against unsustainable exploitation.

The above are a mixture of both direct and indirect (or underlying) causes, and are discussed in more detail in the following sections.

1.   Direct Causes of Biodiversity Loss

(a)   Habitat Loss and Degradation

Habitat loss and degradation is by far the leading cause of biodiversity loss in the region. In some cases, habitat alteration is intensive and large-scale, for example, when native forest is converted to plantation, or when a large dam inundates an area. In other instances, habitats are gradually eroded over time, as when a native forest or grassland is fragmented by expanding agricultural practice and/or population pressures (Baille and Groombridge 1996). In either case, plant species are lost and many animal species are forced from their natural habitats.

It is estimated that at least three quarters of all threatened bird species are in danger because of the pressures on their natural habitats, and that habitat loss is also the principal factor for the decline of at least three-quarters of all mammal species. In areas where forest degradation and conversion have been most intense, such as South and Southeast Asia, a significant proportion of the endemic primate species face extinction (Baille and Groombridge, 1996). Species of reptiles and amphibians are also declining for similar reasons; habitat loss accounts for the decline of 68 per cent of all threatened reptile species and 58 per cent of threatened amphibian species (Baille and Groombridge 1996).

Although degradation of terrestrial habitats such as forests may gain the most attention, both marine and freshwater habitats are also under serious pressure from human activities such as urbanization and agricultural expansion (Tuxill and Bright 1998). For example, river engineering works such as the construction of dams and/or levees have the effect of either inundating or drying up wetlands and backwaters which are important fish spawning grounds, and changing flow regimes to downstream lakes and estuaries, thereby altering their ecologies. A substantial proportion of marine ecosystems are also at risk, primarily due to the direct activities associated with coastal development, including dredging, filling, breakwater construction, mining and drilling, but also from secondary effects such as pollution and increased marine traffic (WRI 1999m).

The impacts of habitat pollution on biodiversity can be instantaneous or cumulative. For instance, oil wastes can asphyxiate and/or poison a wide range of marine life – from algae to seabirds – whereas other contaminants, such as radioactive wastes, pesticides, and other toxic chemicals, can take a while to build up and cause harm within individual organisms, especially within species high on the food chain. In addition, the impact of pollution can be secondary, for example, certain species of algae will capitalize on high nutrient conditions introduced by some forms of pollution, undergoing massive population explosions (known as blooms) which can in turn introduce toxins which are harmful to marine life such as fish and shellfish, or to consumers of those produce (WRI 1999m), or which can lower water clarity and oxygen content to the extent that marine life is depleted. Coral bleaching is another frequent outcome of pollution-induced stress (WRI 1999p).

(b)   Over-exploitation of Biological Resources

Throughout South and East Asia, a major factor in the excessive exploitation of wildlife has been over-hunting in response to the unsustainable demands of both national and international trade (see next section) in goods such as animal parts for traditional medicine or curios. Species which have been particularly affected include the tiger, which has been hunted almost to extinction in South Asia (Matthiesen 1997), and the seahorse, which is harvested in numbers approaching 20 million per year – a rate which is unlikely to be sustainable because of their low reproductive rate, complex social behaviour (they are monogamous, with males rearing the young) and low mobility. Already, it is thought that some 36-seahorse species are threatened by this growing, unregulated harvest (Vincent 1996). Tortoise and river turtle species also are exploited intensively in certain areas of the region; in Southeast Asia they have long been an important source of meat and eggs, but there is also now a burgeoning international trade for their use in traditional medicine, in countries such as People's Republic of China. At least five turtle species involved in this trade are now candidates for the most stringent listing available under the Convention on International Trade in Endangered

CHAPTER THREE

Species of Wild Flora and Fauna (CITES) (Baden-Daintree 1996).

Uncontrolled harvesting of fish is playing a major role in jeopardizing freshwater and marine ecosystems and their native biota (WRI 1999a, g), resulting in a situation whereby about 68 per cent of all threatened marine fish species in the region suffer from over-exploitation. A prime example is the sturgeon in Central Asia, where poaching is now so widespread that the few remaining stocks of this fish are nearly gone (Amstilavskii 1991; Birstein 1993; Baille and Groombridge 1996). Aside from overfishing of commercial stocks, poor management practices are also to blame for the decline in marine resources, which include blast fishing, fishing with cyanide and other poisonous chemicals, *muro-ami* netting (pounding reefs with weighted bags to scare fish out of crevices) and coral harvesting (WRI 1999m, n).

2. Underlying Causes of Biodiversity Loss

(a) International Trade

As discussed above, the growing international demand for traditional medicines, rare foods, curios and aquarium specimens etc. is leading to increasing over-exploitation of certain species within the region. The pressures on some individual species are illustrated by the value of their markets, for example, the market price for the body parts of a single tiger can be as much as US$5 million, and top-quality dried seahorses can sell for as much as US$1 200 per kilogram in parts of People's Republic of China. In addition, according to a recent report by TRAFFIC, a group that monitors the international wildlife trade, the annual Northeast Asian trade in tortoises and river turtles involves some 300 000 kilograms of live animals, with a value of at least US$1 million.

Traditional herbal therapies are also growing rapidly in popularity amongst industrialized countries of the world. The FAO estimates that between 4 000 and 6 000 species of medicinal plants are traded internationally, with People's Republic of China accounting for about 30 per cent of all such exports. In the early 1990s, the booming US retail market for herbal medicines reached nearly US$1.5 billion, and the current European market is thought to be even larger (Iqbal 1995; Shelton et al 1997).

The Asian rattan palm is another species of plant that has been in decline in recent years due to high international demand to supply an international furniture making industry worth US$3.5 to 6.5 billion annually. Rattan stocks are being depleted at an unsustainable rate throughout tropical Asia due to a combination of the loss of native rainforest and over-harvesting. As a result, in the past few years some Asian furniture makers have even begun importing rattan supplies from Nigeria and other central African countries (Network for Bamboo and Rattan 1997; Sunderland 1998).

(b) Population Growth and Poverty

Continued population growth and urbanization is exerting a constant and degrading pressure on biodiversity throughout the region, primarily due to encroachment into natural habitats and their conversion for human settlement, or for the expansion of agricultural production to meet increased demand. In addition, the intensification of agriculture production is also encouraging the use of hybrid seeds and agricultural chemicals, thereby further degrading natural habitats and biodiversity.

Poverty is another significant underlying factor in regard to the loss of biodiversity in the region. The poor are frequently forced to occupy and/or subsist on marginal lands, and thereby often encroach upon fragile ecosystems, such as wetlands in Bangladesh, hill forests in India and Nepal and mangroves in Thailand. Urgent, but pragmatic, responses are therefore needed to address this problem, for example focusing on developing alternative livelihood strategies for those who currently rely on protected natural habitats for their living (Box 3.6). It is worth mentioning, however, that it is by no means clear whether poverty, with its pressure to survive, or affluence, with its pressure to consume, ultimately leads to greater degradation of resources and the environment.

(c) Bioinvasion

Bioinvasions are the processes by which new (or "exotic") species are introduced into a habitat. They are rated second only to habitat loss as the major cause of biodiversity loss in the region, where they are deeply enmeshed in basic economic processes such as trade and travel. Such processes have created hundreds of exotic "pathways" for invasion (Bright 1996), for example, container shipments of used tires from Japan are believed to be responsible for the arrival of Asian tiger mosquito in New Zealand and Australia, and the ballast water of foreign ships was thought to contain the poisonous plankton which burst into the dinoflagellate "red tides" which devastated Australian fisheries in the 1980s (Hallegraeff and Bolch 1991; Carlton 1992b). Other examples include the introduction of the domestic cat to New Zealand, which is thought to have been the main factor in the loss of many bird populations in that country, and the invasion of "exotic" rats, particularly the black and the brown rat, which have also taken a similar toll on many island birds across the region. Another spectacular example involves the brown tree snake, which

Box 3.6 Conserving Biodiversity Through Eco-development

Using a strategy known as eco-development, India's Forestry Research Education and Extension Project (FREEP) is enlisting local communities in the effort to preserve precious biodiversity, and to move away from a more traditional, and confrontational, approach to safeguarding areas. The strategy involves developing alternative resources and sources of income for the many thousands of poor people who depend on protected natural habitat for their livelihood.

FREEP is conducting pilot eco-development programmes in two protected areas in the states of Tamil Nadu and Himachal Pradesh. The Kalakad-Mundanthurai Tiger Reserve (KMTR) in the southern state of Tamil Nadu is one of these areas; located in the southern part of the Western Ghats, the reserve contains a unique and varied array of flora ranging from thorn and dry teak to tropical evergreens, and it supports a rich variety of birds and mammals, including tigers, leopards, and elephants. The last tiger refuge in Tamil Nadu, the KMTR is one of 23 sites covered under the Indian government's Project Tiger, a programme receiving international assistance to enhance tiger habitat. The reserve consists of a core area of 536 square kilometres with large sections of undisturbed forest. It is bounded on the west, northwest, and south by protected forests. To the east is a belt about 5 kilometres wide (324 square kilometres in area) containing 145 villages, the inhabitants of which, depend heavily on KMTR resources for fuel, timber, and fodder. About 25 000 families, totaling almost 100 000 persons, live in these villages owning about 130 000 cattle and buffaloes, many of which routinely graze in the reserve.

The project was initiated in 1995, with the formation of a technical team, and training of staff in eco-development techniques. It also involved seven locally operating non-governmental organizations (NGOs), to implement its operational plan. A village's support was usually obtained through a visit by an eco-development team, comprising forestry department staff, NGO personnel, and a village representative, which met with villagers to discuss the need for biodiversity conservation and to help them identify what actions they might take to reduce their village's pressure on the reserve.

A village forest committee (VFC) was then formed with six members (at least three of whom were to be women), who pledged to collaborate to reduce their village's pressure on the forest. The VFC worked with a smaller group from the eco-development team to produce an eco-development microplan for the reserve. Under this plan, the village inhabitants were to agree to take measurable steps toward conservation in return for eco-development investments. For example, a village population could agree that it would reduce the number of headloads of fuelwood removed from the reserve by an agreed percentage in return for investments in fuel-saving devices or fuelwood plantations. A capital of Rs 250 000 (approximately US$7 100) per group of 200 families was usually set, and villagers had to agree to contribute at least 25 per cent toward the cost of microplan investments. To ensure transparency, funds could be released from the VFC account only with the approval of a two-thirds majority of the villagers participating in the microplan.

Over 100 villages are now participating in the project. These villages have only recently begun making investments, but the results are already significant. Communities and individual farmers have planted fuelwood and fodder plantations. Some villagers have installed cow dung-based gas plants for home fuel needs and are using fuel-saving pressure cookers and more efficient wood-burning stoves (smokeless *chulas*). VFCs have also made loans for a wide array of alternative income-generating activities such as dairy and poultry farming, tailoring, coconut leaf weaving, and setting up tea and dry goods shops. Because the people who encroach on reserves are typically the poorest in the community, the VFC loan programme is targeted at them. Many are receiving loans for the first time, but VFCs report practically 100 per cent repayment, reflecting borrowers' genuine sense of responsibility to their communities.

It has been reported that because of the project, forest encroachment has decreased dramatically. One woman from Mudaliarpatti village, along the reserve boundary, reports, "Now that I've a milk cow through a VFC loan, I no longer have to travel long distances to the forest in the hot sun to find wood to sell. I can stay home, take care of my cow, and earn income". More importantly, the project has transformed the relationship between villages and the state forestry department. Former adversaries have become collaborators in conserving biodiversity. Communities have become "social fences" reporting encroachments to forestry authorities. Some VFCs have even levied fines against members discovered entering the reserve. Thus the eco-development programme at the KMTR is rapidly coming to be seen as a model for conserving biodiversity through local participation.

Source: World Bank 1999

---

invaded Guam from Papua New Guinea around 1950 and has driven nine of the island's 18 native birds into extinction, along with several lizards and probably three species of bats (Savidge 1987; Fritts and Rodda 1995).

Although mammals have in general been less susceptible than birds to invasive species, a significant exception has been the unique marsupial and rodent fauna of Australia (Tuxill and Bright 1998). The introduction of non-native rabbits, foxes, cats, rats, and other animals in this country has combined with changing land-use patterns during the past two centuries to give Australia the world's highest rate of mammalian extinction. In total, 19 species of mammal have become extinct since European settlement in the Eighteenth Century, and at least one quarter of the remaining native mammalian fauna remains threatened.

As yet there are few data to analyse their impact, but the rapidly increasing use of genetically engineered micro-organisms, and their establishment (either directly or indirectly) within natural habitats, could increase the potential risks to existing natural biota (see section below). This could occur, for example, through engineered genetic traits which are harmful to non-target species upon which other native biota depend, through a mixing (and subsequent loss) of genetic stocks, or through their general competitive superiority and subsequent intrusiveness.

In agriculture, species which are considered favourable in some areas can become serious "biological polluters" in other some areas (Bright 1996). For example, the spice cardamom is a problem in lowland moist forests of Sri Lanka and southern India (Heywood 1989), whilst black pepper has invaded forest edges in Malaysia (Whitmore 1989). In addition, *Chromolaena odorata*, the shrub valued by small farmers in Indonesia as a fallow crop, has become a serious pest for many major tropical crops, such as rubber, oil palm, and coconut, and arguably the single most invasive plant in tropical nature reserves (Usher 1989; Baxter 1995).

B. Consequences of Biodiversity Loss

1. *Impacts on the Ecosystems*

One major consequence of biodiversity loss is the alteration and decline in species compositions, which may then trigger local and global extinctions (WRI 1999m). Evidence also suggests that removal of key herbivore and predator species may ultimately produce large-scale ecosystem changes. For instance, removal of triggerfish has been linked with explosions in burrowing urchin populations (their prey), which subsequently accelerate reef erosion through feeding activities (WRI 1999n). Moreover, the loss of a region's top predators or dominant herbivores is particularly damaging because it can trigger a cascade of disruptions in the ecological relationships among species that maintain an ecosystem's diversity and function. Large mammals tend to exert inordinate influence within their ecological communities by consuming and dispersing seeds, creating unique micro-habitats, and regulating populations of prey species. Similarly, decades of excessive whaling reduced the number of whales that die natural deaths in the open oceans. This may have adversely affected unique deep-sea communities of worms and other invertebrates that decompose the remains of dead whales after they have sunk to the ocean floor (Butman et al 1995). Loss of biodiversity through bioinvasion of exotics also has a serious repercussion as it could result in loss or alteration of genetic purity or genetic uniformity. Exotics can pose a kind of internal threat to natives as they may cause the mixing of genetic stocks. An exotic closely related to a native may interbreed with it, releasing its genes into the native gene pool (Bright 1996; WRI 1999m). Such genetic invasions can undermine the distinctiveness or stability of a native population by swamping it in foreign genes. Among plant species, many crops interbreed with wild relatives, and it is possible that these "exotic genes" could escape into wild plant populations – or that the crops themselves could escape. The appearance of herbicide tolerant or disease-resistant wild plants could obviously lead to major ecological- and agricultural-impacts (Rissler and Mellon 1995).

Mixing of genetic stocks could also occur in freshwater and marine species. In marine ecosystems, biodiversity loss has two important genetic consequences. Firstly, it affects the genetic variability within a species. Species exhibiting broad genetic diversity (the range of genetic variability found within different organisms in a population and between populations of a single species) are more likely to adapt to changing conditions than species with narrow genetic diversity. Population declines, by reducing genetic diversity, also reduce the ability of a species to adapt to changing conditions. Secondly, these losses can have cascading, unanticipated effects on other species within an ecosystem (WRI 1999a).

2. *Impacts on Humankind*

With the continued loss of terrestrial, freshwater and marine biodiversity in the region, fish, grains and other food and medicinal products which are derived from these ecosystems are also under increasing pressure. In most cases, these food and medicinal products are integral to the lives of poor and indigenous communities, and so it is they who are being forced to find alternative livelihoods, and who in many cases are suffering as a result. Moreover, although people who are more integrated into regional and national economies tend to use fewer natural resources, they still may depend on plant and animal diversity to generate cash income, for example in India, nearly six million people make a living by harvesting non-timber forest products, a trade that accounts for nearly half the revenues earned by Indian state forests.

Loss of biodiversity is also frequently associated with a decline in the quality of diet and/or intake of food for the poor, which can exacerbate the incidence of malnutrition and sickness, especially amongst children. Moreover, humanity derives many of its medicines and industrial products from the region's wealth of biodiversity, and as plant and animal species with medicinal properties are lost, primary healthcare for millions of people across the region, and in particular the poor, is at risk (Campbell and Schlarbaum 1994). In response, professional ethno botanists surveying medicinal plants used by different cultures are racing against time to document traditional knowledge before it vanishes with its last elderly practitioners (Gubler 1993).

POLICIES AND PROGRAMMES FOR BIODIVERSITY CONSERVATION

A. Implementation of International Conventions

International agreements have undoubtedly made a significant contribution towards to the preservation of biodiversity across the region. The most important of these is the Convention on Biological Diversity, which was signed by the majority of nations attending the Earth Summit in 1992. These, and others, are discussed below.

1. Convention on Biological Diversity (CBD)

The Convention on Biological Diversity (CBD) calls for increased international cooperation in the fight to conserve biodiversity across the globe, and provides a framework for individual countries to develop national strategies and programmes to this effect. The convention encompasses all ecosystems, species and genetic resources, and addresses both traditional conservation efforts and the sustainable exploitation of resources for commercial and economic benefit (including the relatively new field of bio-technology). Given its comprehensive scope, the CBD has led to action at both the national and international level.

(a) National Action

Recognizing that the most effective action will generally take place at the national level, the CBD requires governments to develop and implement national strategies and programmes to conserve biodiversity, which would include efforts to: monitor biodiversity levels (thus ensuring that action is based on sound scientific knowledge); integrate biodiversity concerns into national decision-making; adopt economic and social measures that encourage conservation and sustainable natural resources use; and support efforts by local populations to adopt more sustainable practices. In response, as of March 1999, some 36 countries of the region had ratified the Convention (see Chapter 21, Table 21.5) and a total of 21 countries have now prepared, or are close to preparing, national strategies and action plans.

Table 3.6a and b lists these countries, and broadly outlines the main components of their plans and strategies.

(b) International Action

At the international level, the CBD's main objective is to ensure that all benefits arising from the use of genetic resources are both fair and equitable. To achieve this, it seeks to promote cooperation to strengthen human resources and institutional capabilities worldwide, particularly in developing countries, through joint programmes for research and technology development, as well as exchange of information and expertise between participating countries. It also seeks to identify and provide funding to developing countries to achieve these aims, primarily by a financing mechanism currently being operated under GEF.

In order to promote international cooperation, the third Conference of the Parties (COP) of the CBD in November 1996, launched a major programme called the "Biotrade Initiative" (Box 3.7). The initiative consists of three complementary components:

- a country programme under which opportunities and constraints for the development of a sustainable bio-resource industry will be assessed in countries, and 'bio-partnership' will be facilitated.
- market research and policy analysis to include issues such as Intellectual Property Right (IPR), technology transfer, and benefit sharing mechanisms; and
- dissemination and exchange of country programmes' experiences through the establishment of Internet and communication services (Table 3.7).

The Biotrade country programme represents the most comprehensive part of the initiative. It analyses opportunities and constraints for the development of a sustainable bio-resource industry. To capture opportunities and solve identified problems, the country programmes develop proactive strategies, focusing on bio-business development, bio-partnerships, sustainable use, conservation, and benefit sharing incentives. The other two components are designed to systematically compile and analyse market data and policy issues. Information thus gathered will be disseminated through a web site, publications and briefings, which it is hoped will form the basis for a more transparent understanding of market dynamics and trends, market barriers, trade and investment flows, property right regimes, and bio-business development, as well as conservation,

CHAPTER THREE

Table 3.6a National Strategies for Conservation and Sustainable Use of Biodiversity in Asia and the Pacific

Strategy	National Strategy and Action Plan	Policy/ Legislation	Trust Fund	Conservation Programmes/ Plans	Human Resource Development	Community Participation	Public Education and Awareness	Incentive Measures	Protected Area System	Conservation and Access to Genetic Resources	Institutional Capacity Building and Skill Enhancement	Research	Survey and Monitoring	Land Use Planning	Database Establishment and Exchange of Information	Economic Valuation of Biodiversity Resources
Australia	✓	✓	✓	✓	✓	✓	✓	✓	✓	✓	✓	✓	✓			✓
Bhutan	✓	✓	✓	✓	✓	✓	✓		✓	✓	✓	✓	✓			✓
PR China	✓	✓	✓	✓	✓	✓	✓		✓	✓	✓	✓	✓			✓
Fiji	[1]	[1]		✓	✓	✓	✓		✓	✓	✓					✓
Indonesia	✓	✓		✓	✓	✓	✓		✓	✓	✓		✓			✓
Japan	✓	✓		✓	✓	✓	✓	✓	✓	✓	✓	✓	✓	✓		✓
Rep. of Korea	✓	✓		✓	✓	✓	✓	[2]	✓	✓	✓		✓			✓
Malaysia	✓	✓		✓	✓	✓	✓		✓	✓	✓	✓				✓
Maldives	✓	✓		✓	✓		✓		✓		✓		✓			✓
Marshall Islands	✓	✓	✓	✓	✓	✓	✓		✓		✓	✓	✓			
Mongolia	✓	✓	✓	✓	✓	✓	✓		✓	✓	✓	✓				✓
Nepal	✓	✓		✓	✓	✓	✓		✓	✓	✓		✓			
New Zealand	[1]	✓	✓	✓	✓	✓	✓		✓	✓	✓		✓			
Philippines	✓	✓	✓	✓	✓	✓	✓		✓	✓	✓	✓	✓			✓
Russian Federation	✓	✓		✓	✓	✓	✓		✓	✓	✓	✓	✓		✓	✓
Singapore	✓	✓		✓	✓	✓	✓		✓		✓	✓	✓		✓	
Sri Lanka	✓	✓		✓	✓	✓	✓	✓	✓	✓	✓	✓	✓		✓	✓
Thailand	✓	✓		✓	✓	✓	✓	✓	✓	✓	✓	✓	✓	✓		✓
Turkey	✓	✓		✓	✓	✓	✓		✓	✓	✓	✓	✓			
Uzbekistan	✓	✓		✓	✓	✓	✓		✓		✓		✓			✓
Viet Nam	✓	✓		✓	✓	✓	✓		✓	✓	✓	✓	✓		✓	

[1] In the process of formulating their national strategy
[2] Currently insufficient

Table 3.6b National Strategies for Conservation and Sustainable Use of Biodiversity in Asia and the Pacific

Strategy	Comments
Australia	• Incorporates measures for integrating conservation and sustainable use of biodiversity into sectoral strategies plans and programmes (Commonwealth of Australia 1998). • In-situ conservation is an important component of the Strategy. • Also established and maintained a wide range of measures and facilities for ex-situ conservation, through Commonwealth, States and Territory agencies as well as tertiary institutions and scientific organizations.
Bhutan	• Action Plan highlights alternative actions that can be taken to realize benefits from Bhutan's rich biodiversity (Government of Bhutan 1997). • Includes establishment and management of protected area system as well as development of management strategies for the buffer and enclave zones around and in protected areas. • Envisages both in-situ and ex-situ conservation of wild and domestic biodiversity resources. • Strategy includes supporting measures such as scientific research, surveys and monitoring, database establishment, land use planning, economic valuation of biodiversity resources, integrating of biodiversity in related sectorsû strategy and planning, etc.
PR China	• Action Plan has identified priority projects according to the urgency of conservation and their feasibility. • Country Study elaborates the strategic goal for national capacity building, human resources strengthening, conservation facility construction, development of science and technology, promotion of education and awareness, information management and international cooperation. • Other State Council departments have incorporated conservation concerns into their own departmental action plans.
Fiji	• Strategy and Action Plan is the cornerstone of the Sustainable Development Bill, which embodies together environmental protection and resource management as well as biodiversity conservation and national parks management (Department of Environment, Ministry of Local Government, Housing and Environment 1997). • Government has declared several areas as national parks, nature and forest reserves, and conservation and protected areas. • Present system is weak due to overlapping jurisdictions between the Departments of Environment, Fisheries and Forestry and the National Trust for Fiji.
Indonesia	• National Strategy provides guidance to all stakeholders, especially those involved in the management of biodiversity (Government of Indonesia 1997). • Action Plan sets out an action strategy for in-situ conservation in terrestrial parks and protected areas as well as outside the protected area network (i.e. in production forests, wetlands, agricultural areas, and coastal and marine environment) and also provides for ex-situ conservation. • Important component of the Strategy is prioritization of approaches which fulfil basic human needs and generate income for the poor.
Japan	• National Strategy describes the major legislation, and provides guidelines/administrative framework for conservation. • Major components of Strategy are purpose and objectives, basic directions for each sector, guidelines for cooperation between agencies, and its in-built monitoring and review process.
Rep. of Korea	• National Strategy identifies priority concerns, such as biodiversity survey, in-situ and ex-situ conservation, control of threatening activities, ecosystem rehabilitation, and follow-up monitoring (Government of Korea 1998). • Also covers measures for sustainable use of biodiversity resources in agriculture, fisheries, forestry, tourism and recreation, and genetic resources. • Advocates upgrading of capacity for biodiversity management through improved systems, incentives, strengthening education and research, raising awareness, exchange of information and technology
Malaysia	• National Policy developed to address biological diversity issues across various sectors of the economy (Ministry of Science, Technology and the Environment 1998). • 15 Strategies developed covering areas such as: improving scientific knowledge base; enhancing sustainable utilization of biodiversity; integrating conservation programmes into sectoral planning; enhancing staff skills, capabilities and competence; and, promoting institutional and public awareness.

Table 3.6b (continued)

Strategy	Comments
Maldives	• Environmental Protection and Preservation Act, a basis for a conservation strategy, is in place and marine protected areas and protected species have been declared (Ministry of Planning, Human Resources and the Environment 1997). • Two different governmental bodies deal with the issues of biodiversity: the Ministry of Environment and the Ministry of Fisheries and Agriculture. • Future conservation activities are planned in the 2nd National Environment Action Plan which includes development of a National Strategy, formulation of an Action Plan, drafting of First Report to the CDP, establishment of protected areas, conservation of coral reefs, and strengthening biodiversity awareness.
Marshall Islands	• In process of formulating, through participatory and analytical processes, a Strategy/Action Plan (Government of Marshall Islands 1997). • Focuses on assessment of the status and trends in biodiversity, and the collection and provision of a local information resource on biodiversity.
Mongolia	• Action Plan provides for the sustainable use of biological resources and their natural restoration (Ministry for Nature and the Environment/UNDP-GEF 1998). • Plan gives emphasis on institutional capacity building, policy development and planning, renewal and strengthening of legislation, survey, improved management of protected areas network, public education and awareness. • Large-scale investments to protect biological resources are also in place through the Mongolian Science and Technology Fund, Environmental Protection Fund, Endangered Species Fund and the Mongolian Environmental Trust Fund.
Nepal	• Conservation underpinned through establishment of protected areas in representative ecological zones, as well as adoption of policy and legal measures which focus on benefit sharing, and empowerment of the local communities (e.g. through community and leasehold forestry). • Environmental trust fund has been established • Preparing Action Plan, which will address cross-sectoral issues, refine priorities and develop investment proposal for implementing effective conservation programmes.
New Zealand	• National strategy not yet complete, but country has a range of institutional and legal arrangements in place to address biodiversity management issues (Government of New Zealand 1997).
Philippines	• Followed CBD guidelines in preparing national policies on bioprospecting, biosafety, biotechnology, marine conservation, indigenous knowledge, as well as their integration into sectoral plans and decision making (Protected Areas and Wildlife Bureau-Department of Natural Resources and Environment 1998). • Strategy/Action Plan formulated as blueprint for biodiversity conservation agenda.
Russian Federation	• GEF Biodiversity Conservation Programme launched in 1996 includes preparation of Strategy/Action Plan (State Committee of Russian Federation for Environmental Protection 1997). • Rapid expansion of the federal system of protected areas, creation of regional networks of protected areas and expansion of the network of organizations involved in ex-situ conservation of rare animal and plant species. • Enhanced role of Russian and international non-governmental ecological organizations in the conservation of biodiversity.
Singapore	• Green Plan describes the broad policy direction towards attaining a model green city (The National Parks Board and the Report Drafting Committee 1997). • Broad goals, objectives and approaches for biodiversity conservation have already been formulated, and several action programmes are also being implemented.
Sri Lanka	• Action Plan prepared through a participatory approach involving a large body of stakeholders, including state agencies, over 100 non-governmental organizations, local communities etc. (Government of Sri Lanka 1997). • Broad objectives include capacity building, developing programmes to enhance public awareness and encourage public participation in conservation. • One notable step taken by the government to fulfil its obligations to the CBD is the establishment of the National Experts Group on Biodiversity to advise/steer government policy on the subject.

Table 3.6b (continued)

Strategy	Comments
Thailand	• Strategy prepared through participatory approach (The Office of the Environmental Policy and Planning 1997). • Actions prioritized for: building institutional capacity; enhancing protected areas management; conserving species, population and ecosystems; monitoring and controlling activities which threaten biodiversity; and, promoting cooperation between international and national agencies/institutions.
Turkey	• Strategy/Action Plan outlines current status of biodiversity, with specific actions/recommendations (Government of Turkey 1997). • Strategy provides framework for action at all levels, and recognizes need for international cooperation.
Uzbekistan	• Strategy emphasizes three objectives: the conservation of biodiversity, the sustainable use of its components, and the fair and equitable sharing of benefits (Government of Uzbekistan 1997). • Includes plan for reorganization and expansion of protected areas system, enhancement of public awareness, education and participation in conservation. • Plan elaborates specific actions, and identifies responsible agencies and organizations, and timetable for implementation.
Viet Nam	• Action Plan for Viet Nam is being implemented through various activities with the objectives of conservation of terrestrial, marine and wetlands biodiversity as well as genetic resources. • Since the 1994 Law on Environment Protection, efforts towards public education and awareness building for biodiversity conservation have been intensified. • Several laws and statutes have been adopted to control over-exploitation of biological resources and the illegal trade in endangered species of wild fauna and flora.

sustainable use and benefit sharing. Guidance about the Biotrade Initiative is generally provided by an national advisory panel, which will bring together representatives of the private sector, local and indigenous communities, NGO's, academic institutes, governments and intergovernmental organizations.

Another major initiative under CBD is the ongoing negotiations concerning Biosafety protocol. These negotiations reflect the growing concerns about the potential risks posed by living modified organisms produced by modern biotechnology. Many countries with modern biotechnology industries do have domestic legislation, however, there are no binding international agreements addressing situations where living modified organisms cross national borders. The negotiations, therefore, focus on transboundary movements, and include consideration of informed agreement procedures that will enable governments to control imports and refuse those that are unwanted. Accidental releases of organisms are also being addressed in the protocol. A major concern is that many developing countries lack the technical, financial and institutional capacity to address biosafety. They need greater capacity for assessing and managing risks, establishing adequate information systems, and developing expert human resources in biotechnology.

Initiatives are also being promoted at subregional level in South Asia, Southeast Asia and South Pacific to assist countries in fulfilling the objectives of CBD as well as help conserve biodiversity. For example, with the assistance of European Union, ASEAN has established the ASEAN Regional Centre for Biodiversity Conservation (ARCBC) which serves as the main focal point for networking and institutional linkage among ASEAN Member Countries (AMCs) and between ASEAN and European Union (EU) partner organizations, to enhance the capacity of ASEAN in promoting biodiversity conservation (see Chapter 17). The EU provides the means for networking, applied research, training and technical assistance, whilst ASEAN provides office space, facilities and support personnel. The Department of Environment and Natural Resources (DENR) in the Philippines is the project's executing agency.

2. *Convention on International Trade in Endangered Species of Wild Fauna and Flora (CITES)*

CITES, which regulates the import and export of endangered species of wild flora and fauna, has become a very effective instrument for countering the loss of biodiversity caused by international trade, although there is still work to be done in Asia and the Pacific to improve its implementation and enforcement. The treaty operates through the issuance and control of export and import permits for species that can withstand current rates of exploitation, but prohibits trade in those facing extinction (CITES 1999a).

CHAPTER THREE

> **Box 3.7 The Biotrade Initiative: A New Integrated Approach to Biodiversity Conservation**
>
> The BIOTRADE initiative was launched at the third Conference of the Parties (COP3) of the *Convention on Biological Diversity (CBD)*, Buenos Aires, November 1996, to stimulate investment and trade in biological resources as a means of furthering the three objectives of the CBD, i.e. to promote: (1) conservation of biodiversity; (2) the sustainable use of its components; and (3) a fair and equitable sharing of benefits arising out of the utilization of biological resources. The initiative's objectives will be pursued by enhancing the capability of developing countries for sustainable use of biodiversity to produce new value-added products and services for both international and domestic markets.
>
> The initiative envisages to promote the profitable use of biodiversity by enhancing collaboration among different actors that are very often perceived a potential rivals, both in industrialized and developing countries, these include: the private sector, including both multinational corporations and local companies; governments; universities; financial institutions; and local and indigenous communities. If successfully implemented, cooperation could take place both at the local community and national levels. The core activity areas of the initiative cover Bioprospecting, harvest of non-timber forest products (NTFPs) and promotion of eco-tourism.
>
> The Biotrade Initiative stresses a so-called 'rights and benefits' approach, arguing that both should be addressed at the same time. It advocates that merely concentrating on fighting for rights, and waiting for the proper definition of all rights of local and indigenous communities would mean the loss of many years in the struggle for biodiversity conservation. In the same line, the initiative pursues a 'protect and promote' strategy, encouraging instruments and mechanisms that could protect and promote intellectual property, traditional knowledge, and biodiversity, for national or local producers as well as local and Indigenous communities. As part of the Biotrade country programmes, the Spanish patent office provides technical assistance for developing countries in these legal issues. However, it remains to be seen if this assistance will be of use for issues that go beyond the drafting of patent legislation.
>
> Though the Biotrade Initiative is still a new endeavour, it now constitutes a global network for the use of biological diversity to support social and economic improvements in developing countries. A number of projects are linking biodiversity conservation with remuneration for its sustainable use. However, the implementation of the initiative will be judged according to its ability to build entrepreneurial and institutional capacities to take advantage of the abundant genetic resources in the South; target multiple and profitable sectors and to handle the issues of access to genetic resources and traditional knowledge and the equitable sharing of benefits.
>
> *Sources:* CBD Secretariat 1999 and Rojas, M. 1999

Table 3.7 Component of Biotrade Initiatives

Market Research & Policy Analysis	Biotrade Country Programmes	Web Services & Communication
• Market Research • Economic Analysis • Technology Transfer • Property Right Regimes • Benefit Sharing • Conservation • Incentives • Bio-industry Development	• Training & Capacity Building • Country Assessment • Identification & Development of Market opportunities • Biopartnerships	• Research Up-dates • Country Experiences • Discussion Forums • Country Level Websites

Source: Biotrade 1998

As of March 1999, 25 countries from Asia and the Pacific had ratified the Convention (see Chapter 21, Table 21.6). Under this agreement, each member country is responsible for the implementation of CITES within its own jurisdiction, including the appointment of at least one Management Authority, and one Scientific Authority. Several countries have already taken moves to effectively implement the treaty. For example: in India, the Ministry of Environment and Forest established the National Coordination Committee (NCC) in 1995 which has been promoting effective inter-departmental coordination for the control of illegal trade in wildlife and related products (CITES 1999c); in Hong Kong, China, the Agriculture and Fisheries Department (AFD) launched a public awareness campaign about trade in endangered species (CITES 1999d); and in Thailand, the Thai Management Authority, in cooperation with the CITES Secretariat, has developed an Orchid Identification Guide to combat the problem of trade in rare orchid specimens (CITES 1999e).

3. Convention on Wetlands

The Ramsar Convention is aimed at the protection and conservation of internationally significant wetlands (Ramsar Convention Bureau 1999). Signatories are required to adopt the Ramsar Convention Strategic Plan 1997-2002, and to thereafter work with Wetlands International, at both the national and regional level, to implement the Convention. The main challenge for Ramsar signatories is to maintain the ecological character of their listed sites and all their wetlands through conservation and wise use (Frazier 1999). As of March 1999, 18 countries in the region had ratified the Convention (see Chapter 21, Table 21.7).

At the national level, collaboration is achieved through a combination of: development of national wetland strategies and policies (which have already been developed for Indonesia, Malaysia and Russian Federation); training programmes; wetland surveys; public education and awareness building; and demonstration management of selected wetland sites (Moser 1999). At the regional level, collaboration is achieved through the implementation of multi-country projects and regional wetland inventories.

In 1996, an international non-binding framework, the Asia-Pacific Migratory Waterbird Conservation Strategy: 1996-2000, was endorsed by contracting parties to the Ramsar convention to promote the conservation of migratory waterbirds and their wetland habitat in the Asian and Pacific Region. One of the Strategy's priorities is the establishment of networks of sites of internationally important wetland habitats for three groups of waterbird species: shorebirds (sandpipers, plovers and related species); cranes; and anatidae (ducks, geese and swans). There is already an East Asian Australasian Shorebird Reserve Network, launched in 1996, under which an Action Plan for the Conservation of Migratory Shorebirds in Asia and the Pacific 1998-2000 guides the work until the end of 2000 (Watkins 1999). A Shorebird Working Group of Experts from Australia, People's Republic of China, Japan, the Philippines and the Russian Federation coordinates the Action Plan. In the South Pacific, a number of tangible achievements in the implementation of the Convention have been noted. Potential Ramsar sites have been documented in Solomon Islands and Vanuatu, and a second Ramsar site has been designated for Papua New Guinea. An inventory of freshwater fish has also been conducted, revealing that far greater species richness is present on certain islands than was thought only 10 years ago.

4. Convention on the Conservation of Migratory Species of Wild Animals (CMS)

The Convention on the Conservation of Migratory Species of Wild Animals (also known as the Bonn Convention) aims to conserve terrestrial, marine and avian migratory species throughout their route or range. It is one of a small number of inter-governmental treaties concerned with the conservation of migratory wildlife and their habitats on a global scale. Seven countries from Asia and the Pacific (Australia, India, Mongolia, Pakistan, Philippines, Sri Lanka and Uzbekistan) are parties to the convention, all of whom work together to conserve migratory species and their habitat by providing strict protection for the endangered migratory species (see Box 3.8); by concluding multilateral agreements for the conservation and management of migratory species; and, by undertaking cooperative research activities (CMS 1999a).

Amongst the multilateral agreements formed under CMS, the African-Eurasian Migratory Waterbird Agreement (AEWA) is the largest. Worldwide, it covers 172 species of birds that are ecologically dependent on wetlands for at least part of their annual cycle, including many species of pelicans, storks, flamingos, swans, geese, ducks and waders (CMS 1999b). Other important agreements under CMS that concerns parts of the Asia are the Memorandum of Understanding Concerning Conservation Measures for the Slender-billed Curlew (CMS 1999c), and the Memorandum of Understanding Concerning Conservation Measures for the Siberian Crane (CMS 1999d).

B. Conservation of Ecosystems, Species and Genes

1. Protected Areas System

At present, about 2.4 million square kilometres of the region's surface area is officially designated as protected, or about five per cent of the total area. In all, there are over 4 000 protected areas in the region, more than two-thirds of which are in Northeast and Southeast Asia (Figure 3.5); the majority of these are in Indonesia and People's Republic of China (Figure 3.6). In terms of areal extent, People's Republic of China (in particular, Hong Kong, China) and the eastern regions of the Russian Federation (e.g. Lake Baikal) contain almost half of the officially designated protected area in the region (Figures 3.7 and 3.8). As a rule of thumb, it has been recommended that country's should aim to designate at least 10 per cent of their territory as protected

CHAPTER THREE

Box 3.8 Conservation of Migratory Species in Mongolia

Eastern Mongolia is home to vast herds of migratory Mongolian gazelles that were once widespread in Mongolia and neighbouring areas of Russian Federation and People's Republic of China, but are now limited largely to the Eastern Steppes of Mongolia due to habitat destruction and hunting in People's Republic of China and Russian Federation, and disruption of migration routes. There are estimated to be over two million gazelles, but migrating species always pose a formidable conservation challenge as they cannot be confined to protected areas and they readily cross international borders. The Mongolian gazelle has suffered a massive reduction in range and population size over the last few decades, and its continued survival is threatened if present trends continue.

In order to assist Mongolia with conservation of this globally important ecosystem supporting migratory species, the Global Environment Facility have provided funds through the United Nations Development Programme to implement a conservation project. Executed by the Ministry for Nature and the Environment and the United Nations Office of Project Services, the project, referred to as the Eastern Steppe Biodiversity Project, started in 1998 and is expected to run for 7 years. The project's vision is, in effect, of the whole Eastern Steppes being managed for economic development without depleting natural resources or adversely affecting ecological processes. It sees vast herds of Mongolian gazelle continuing to roam over the grasslands, and it sees their survival safeguarded by a well enforced system of protected areas and legislation, by economic incentives for sustainable harvests or tourism operations, and by effective land-use planning to reconcile the needs of economic development and infrastructure with the needs of the gazelles to follow their migration routes. It sees an integrated approach to rangeland management leading to the coexistence of traditional herding practices and wildlife, and it sees many local residents finding alternative livelihoods and small business opportunities. Most importantly it sees the participation of all stakeholders – the herders and other local citizens and government officials – as the key to successful management of the grasslands.

Source: UNDP 1999

Figure 3.5 **Percentage Share in Number of Protected Areas by Subregions in Asia and the Pacific**

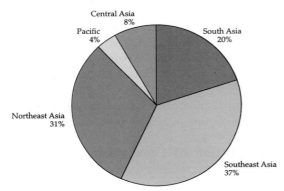

Source: WCMC 1997

Figure 3.7 **Percentage Share in Areal Extent of Protected Areas by Subregion**

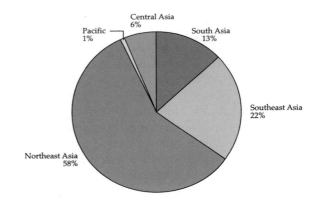

Source: WCMC 1997

Figure 3.6 **Percentage Share in Number of Protected Areas by Countries in Asia and the Pacific**

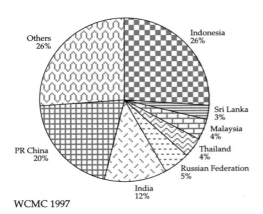

Source: WCMC 1997

Figure 3.8 **Percentage Share in Areal Extent of Protected Areas by Countries in Asia and the Pacific**

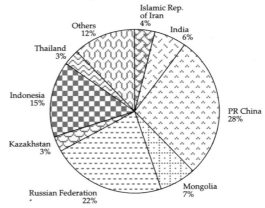

Source: WCMC 1997

(McNeely and Miller 1984; ADB and IUCN 1995). Few countries of the region have been able to reach this benchmark (Figure 3.8). However, it should be emphasized that percentage share of protected areas is only a very crude statistic of the effectiveness of ecosystem coverage in any one country, and one which does not reflect the relative diversity of the areas protected.

No doubt considerable achievements have been made in the region through designation and management of protected areas. However, there are some serious shortcomings with current efforts. For example, protected areas do not always target sites of high biological diversity (Tuxill 1999). Moreover, many protected areas, which are officially decreed as such on paper, are simply not protected in practice, due to reasons such as under-funding and/or understaffing of natural resource agencies. The lack of commitment by governments to implement protected area legislation is another shortcoming (Tuxill and Bright 1998), as a result of which, officially designated reserves are often subjected to agricultural development, mining, extensive poaching, and other forms of degradation. An example is the Narayan Sarovar Sanctuary in India, which is home to a rich assembly of wildlife, including wolves, desert cats, and the largest known population of Indian gazelle. In 1995, this reserve system was turned over to mining companies eager to harvest the coal, bauxite, and limestone deposits found in the area. Problems also arise when strictly defined borders conflict with the cultural and economic interests of local communities. An example is the current conflict between the Government of the Philippines and inhabitants in and around Mt. Kitanglad National Park, a biodiversity rich area in the southern part of the country. The Government's plans for the area have been strongly opposed by mountain farmers, who object to being forced to switch from traditional monocropping practices (e.g. potatoes) into agroforestry, which is perceived as less profitable.

(a) Botanical Gardens and Zoological Parks

Botanical gardens and zoological parks play an important role in both *in situ* and *ex situ* conservation of biological diversity. Since botanical gardens and zoological parks store myriads of plant and animal accessions, these represent a vast conservation resource of stored and managed biodiversity. In addition, botanical gardens have also played a crucial role in the development, designation, care and management of conservation areas through activities such as: habitat restoration; wild plant population research, recovery or management; individual species recovery programmes; development and maintenance of databanks; identification and development of economically important species in commercial horticulture, forestry and agriculture and bioprospecting, etc.

Five countries-Australia, People's Republic of China, India, Japan and the Russian Federation-house the majority of botanical gardens in the region. A recent initiative to link botanical gardens within a coordinated global network is the founding of Botanic Gardens Conservation International (BGCI), whereby member-countries are working towards the implementation of a worldwide Botanic Gardens Conservation Strategy and Action Plan for plant conservation (BGCI no date), providing technical guidance, data and supporting for botanical gardens, and helping create or strengthen national and regional networks of gardens in many countries, including Australia, People's Republic of China, India and Indonesia.

(b) Gene Banks

The conventional solution to the conservation of plant germplasm has been the establishment of genebanks, which serve as storehouses for germplasm collections, including seed banks, field genebanks, and tissue collections in culture (Jackson 1999). In Asia and the Pacific, they are commonly integrated with botanical gardens and have been an important source of material for plant breeding programmes and other research activities, thereby forming a basic element in biodiversity conservation programmes in the region (Tripp and van der Heide 1996).

Maintaining botanical gardens and gene banks, however, is costly, time-consuming and labour-intensive (Tuxill 1999). For example, the current estimated total annual cost of maintaining samples in gene banks is US$50 per accession. Many countries in the region do not have the facilities and are hard-pressed for operating funds; therefore, they cannot maintain samples of a number of species under optimal physical conditions.

C. **Bioregional Management**

Policymakers and environmental managers in several countries of the region have begun to adopt an integrated, "total ecosystem" strategy for regulating activities which adversely affect biodiversity (WRI 1999). Such an approach, whether on land or at sea, can be used to balance conservation needs with the economic and social demands of the people living within or close to the habitats which are being protected. Integrated coastal zone management programmes provide a good example of the concept of bioregional management.

By far the largest application of bioregional management within the region is the Regional Seas Programme initiated by UNEP, an action-oriented strategy focusing on the mitigation or elimination of both the causes and consequences of environmental degradation (UNEP 1999b). It has a comprehensive, integrated, and results-oriented approach towards combating environmental problems through rational management and development of marine and coastal areas. Each subregion is required to formulate their own action plans (see Chapter 5) to promote the parallel development of intra-regional legal agreements.

Another significant example of bioregional management is the International Coral Reef Initiative (ICRI), created to protect coral reefs and the associated ecosystems such as sea grass beds and mangroves. ICRI currently involves Australia, Japan, and the Philippines (UNEP 1999c) and seeks to implement those components of Agenda 21 which call on states to take a special care of marine ecosystems exhibiting high levels of biodiversity and productivity.

Establishment of transboundary-protected areas is also taking place in many parts of the region in order to promote bioregional management. For example, the Mekong River Commission has been tasked to coordinate activities related to the use and development of water and associated resources of the Lower Mekong Basin (United Nations 1995), and the concept of transboundary conservation within the Hindu-Kush Himalayan area is being facilitated by international organizations such as ICIMOD and WWF (Shengji 1998). Another good example of a transboundary reserve is the Turtle Islands group, a well-defined rookery of green turtles shared by nine islands, of which six are within Philippine territory (United Nations 1995). Within this group, the Philippines, Malaysia and Indonesia are all cooperating with one another in conducting scientific and sociological studies to promote turtle conservation. For example, the Philippine-Sabah Turtle Islands and Berau, Indonesia support the only major green sea turtle breeding populations in the world. These countries also collaborate in tagging and monitoring activities, hatcheries operations, and monitoring of traffic of turtle eggs between islands. Finally, in the Brunei Darussalam/Indonesia/ Malaysia/Philippines East Asian Growth Area (BIMP-EAGA) subregion, the Working Group on Environmental Management are also implementing various projects with the involvement of the private sector (UNEP 1999a).

Several inter-governmental regional agreements such as the SACEP in South Asia, ASEAN in Southeast Asia and SPREP in the South Pacific also deal with aspects of biodiversity conservation. For example, SACEP's Strategy and Programme includes regional cooperation in wildlife conservation and genetic resources and conservation of corals, mangroves, deltas, and coastal areas (see also Chapter 16). In addition, ASEAN established a regional framework for the promotion of conservation and sustainable use of heritage areas and endangered species (see Chapter 17). In the South Pacific biodiversity-related projects are included (Chapter 18) under the South Pacific Biodiversity Conservation Programme (SPBCP).

CONCLUSION

Asia and the Pacific is extremely rich in biodiversity, and possesses an immense variety of ecological habitats and climatic conditions. Seven out of seventeen mega diversity countries of the world are located in the region, which not only have a wealth of biodiversity, but are also noted for their high rates of species endemism. The region's biodiversity, however, has been under serious threat as a result of habitat loss, over-exploitation of resources, and the introduction of exotic species. A range of economic and social pressures on the environment exacerbates these problems.

According to the IUCN (1998) Red List, more than 10 000 higher plant species and over 3 000 vertebrate animal species in Asia and the Pacific are threatened. As natural habitats rapidly shrink, over a thousand birds, about a thousand mammals, and several hundred reptiles and fish species are at risk. Genetic diversity is also diminishing, particularly in relation to domesticated crops and livestock. In the short-term, this means that existing genetic resources are less able to adapt to conditions of stress, for example, in relation to pathogens or pests, drought or temperature extreme. In the long-term, it means the loss of well-adapted genetic varieties.

In an effort to preserve biodiversity in the region, policies and programmes have been promoted at the national level, and coordination of these programmes has been promoted at the regional and international level. Many countries in the region have prepared national strategies and action plans in response to international conventions such as the Convention on Biological Diversity. These strategies and plans emphasize actions in areas such as: the enhancement of institutional capacity; improved management systems for *in situ* and *ex situ* conservation; sustainable utilization of biodiversity by mainstreaming within other sectoral strategies; enhancing public awareness; research, surveys, and monitoring; partnerships with stakeholders, including increasing community participation in the management of biodiversity; and the promotion of national and international cooperation. All these components, if properly implemented, could have significant bearing on the region's biodiversity.

However, inadequate human and financial resources, particularly in the developing countries of the region, are hampering implementation of these strategies and action plans.

Protected area systems in the region are still limited in extent, as they constitute only five per cent of the total area, against an IUCN guideline of 10 per cent. Moreover, many important habitats are either un-represented or under-represented in the present system of protected areas. The coverage of wetlands and marine ecosystems, for example, is still extremely limited. Clearly, there remains considerable scope for expanding the network further, through conservation and restoration of natural areas or traditionally maintained land and/or seascapes. Another deficiency includes a preponderance of small and fragmented areas, which jeopardizes their integrity and long-term viability.

In terms of genetic resources, the concept of intellectual property rights and the lack of a clear, multilateral system for these rights distract stakeholders in countries of the region from the task of conservation of these resources. Governments in Asia and the Pacific should support (and legislate for) the right of subsistence farmers to save and adopt the seeds they plant, which is arguably the most important mechanism for sustaining agro-biodiversity. It is important to build institutional capacity in this regard, and to promote measures such as bioprospecting that could contribute both to conservation of biodiversity and enhancement of resources for conservation. Research, training and information to help expand the capacity for conservation of genes, species and ecosystems also remains vital.

Biotechnology and genetic engineering also appear to have significant potential for generating income amongst the region's poor, however, these are new fields, and much remains to be found out about the potential environmental and health risks from the interaction of genetically engineered or modified organisms with natural ecosystems, and strong regulatory and legal frameworks need to be introduced to protect against these impacts.

Finally, there is a need to provide appropriate incentives to individuals, institutions and industries that depend directly on biodiversity for their well-being; to invest in strategies that will help conserve ecosystems whilst promoting and sustaining developing country economies. There is a simultaneous need to involve these various stakeholders in policy and decision-making through multi-stakeholder programmes, environmental steering committees, public consultations and public awareness programmes.

Chapter Four

**Freshwater is a scarce commodity in the region: Water vendors in the Philippines.
(Photo credit: Roges Brown)**

4

Inland Water

INTRODUCTION
WATER RESOURCES: STATUS AND AVAILABILITY
WATER QUALITY: GROWING POLLUTION AND ITS IMPACTS
ENVIRONMENTAL IMPACTS OF UNSUSTAINABLE WATER USE
POLICIES AND STRATEGIES FOR SUSTAINABLE WATER MANAGEMENT
CONCLUSIONS

CHAPTER FOUR

INTRODUCTION

Freshwater is a critical resource for the people of the Asian and Pacific Region. The abstraction of freshwater from rivers, lakes and underground reservoirs is increasing in line with population growth, urbanization and economic expansion. The increasing abstraction is causing a growing imbalance between supply and demand that has already led to shortages and depletion of reserves. Moreover, the scarcity of water is being accompanied by deterioration in the quality of available water due to pollution and environmental degradation.

This chapter discusses the status and supply of inland water resources, highlights the environmental impacts of unsustainable water withdrawal and the degradation of water quality in the region. The policies and programmes undertaken promoting sustainable development of water resources are also reviewed.

WATER RESOURCES: STATUS AND AVAILABILITY

A. Sources and Status of Supply

Table 4.1 presents the distribution and utilization of freshwater resources across the region. The variation in the availability and consumption of water is determined by the individual country's physical topography, climate and catchment size as well as the accessibility of water resources and the level of socio-economic development (ESCAP 1997a). For example, the distribution of precipitation in the region is extremely varied both geographically and seasonally. Precipitation is abundant on the southern slopes of the Himalayas, western slopes of the mountains of India and Indo-China and the islands of Indonesia, all of which receive from 1 500 mm to in excess of 3 000 mm of rain annually. By contrast, almost the entire northwestern part of the region is extremely dry, with an annual precipitation of less than 200 mm (Figure 4.1). In the countries bordering the Indian and Pacific Oceans, cyclical monsoon rainfall is the dominant pattern, with distinctive dry and wet seasons; during the long dry season, temporary water shortages are experienced in many river basins, whilst the wet season is often accompanied by flooding (ESCAP 1997). The total run-off per year in the Asian and the Pacific region is approximately 13 260 km^3, a third of the global total (ESCAP 1997). In absolute terms, the annual renewable water resources are considerable in many developing countries of the region, although not all of this is available for exploitation. The highest absolute quantities of water resources are in People's Republic of China, Indonesia and Pakistan.

However, on a per capita basis, a different picture emerges; the variation in the annual per capita internal (within the territory of a country) renewable resources of the region is shown in Figure 4.2. In Southeast Asia, annual per capita internal renewable water resources range from about 172 m^3 a year in Singapore to more than 21 000 m^3 in Malaysia. Singapore is currently meeting its freshwater demands by importing some of its supply from Malaysia and People's Republic of China (SEPA 1997). In South Asia, India, the Islamic Republic of Iran, the Republic of Korea and Pakistan, freshwater supplies are between 1 400 and 1 900 m^3 per capita per year, which is considerably below the supply in Malaysia

Table 4.1 Water Resources and Use in Selected Countries of the Asian and Pacific Region

Country	Water resources (m^3/yr)	Water resources use (m^3/yr)	Water use as a percentage of overall water resources
Afghanistan	60	26	43
Australia	398	24	6
Bangladesh	115	23	20
Bhutan	95	<1	1
Cambodia	88	1	1
PR China	2 812	500	18
Dem. People's Rep. of Korea	67	14	21
Fiji	29	<1	3
India	1 142	552	48
Indonesia	2 986	49	2
Islamic Republic of Iran	130	75	58
Japan	435	90	21
Lao People's Dem. Republic	270	1	<1
Malaysia	556	12	2
Mongolia	25	<1	4
Myanmar	606	4	<1
Nepal	207	12	6
New Zealand	397	2	<1
Pakistan	247	180	73
Papua New Guinea	801	<1	<1
Philippines	356	105	30
Republic of Korea	70	30	42
Solomon Island	45	<1	2
Sri Lanka	47	10	21
Thailand	210	33	16
Viet Nam	318	65	20

Source: ESCAP 1996a and ESCAP 1999

and some other Southeast Asian countries. By contrast, Bhutan and Lao People's Democratic Republic have around 50 000 m³ per capita and Papua New Guinea an enormous 174 000 m³ per capita a year (WRI, UNEP, UNDP and World Bank 1998).

Figure 4.1 Variability of Precipitation in Selected Countries of the Asian and Pacific Region

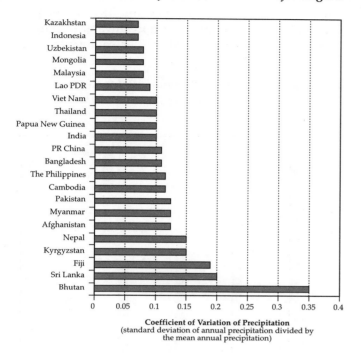

Source: Raskin, et al 1997

Figure 4.2 Annual Water Resources Per Capita in Selected Countries of the Asian and Pacific Region

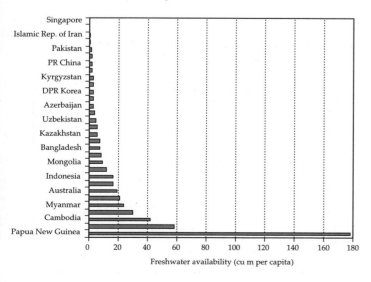

Source: Raskin, et al 1997

1. *Surface Water*

Rivers, lakes and man-made reservoirs are the main sources of surface water abstraction. The Asian and Pacific Region has several important river systems (Figure 4.3) with 400 major rivers in India, 200 in Indonesia, 108 in Japan, 50 in Bangladesh and 20 in Thailand. International rivers in the region include the Mekong which flows through Viet Nam, Lao People's Democratic Republic, Cambodia, Myanmar and Thailand; and the Ganges, Brahmaputra and Meghna River Systems which are shared by India, People's Republic of China, Nepal, Bangladesh and Bhutan. The region is also endowed with a substantial number of lakes; amongst the largest and most utilized are Dongting – hu in People's Republic of China, Tonle Sap in Cambodia, Lake Toba in Indonesia, Kasumigaura in Japan, Laguna de Bay in the Philippines, and Lake Songkhla in Thailand and Lake Issy Kul in Kyrgyzstan.

The development of reservoirs has usually been for irrigation, flood control and hydropower. There are over 800 major reservoirs in the region with a total capacity of about 2 000 km³, with thousands of small reservoirs particularly in People's Republic of China, India, Sri Lanka and the Russian Federation (Avakyan and Iakovleva 1998).

Figure 4.3 Major River Systems of Selected Countries in the Asian and Pacific Region

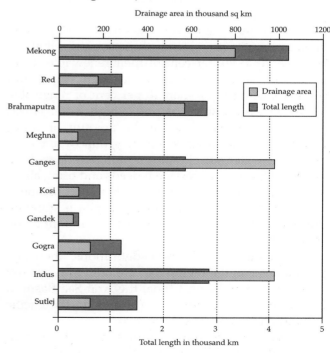

Source: United Nations 1995

CHAPTER FOUR

2. Groundwater

Groundwater in the region occurs in many different rock types, ranging from ancient crystalline basement rocks (which store relatively small quantities of water in their shallow weathered and jointed layers) to alluvial plain sediments (which may extend to depths of several hundred metres and contain substantial volumes of water) (BGS, ODA, UNEP, and WHO 1996). Certain parts of the region contain vast groundwater reservoirs that receive extensive amounts of water from the abundant rainy season recharge. Bangladesh, India, Indonesia, Nepal and Myanmar have particularly large and deep aquifers. Many countries in the region are dependent on groundwater exploitation to supplement scarce surface water resources; this dependency reaches 30 to 35 per cent of the total supply in Bangladesh, India and Pakistan (ADB 1998).

In small island nations of the South Pacific, such as Cook Islands, Fiji, Maldives, Niue, Papua New Guinea, Samoa, Solomon Islands and Tonga, groundwater exists in freshwater lenses above saline water tables. The formation of these lenses on low-lying islands, particularly the coral atolls and the small limestone islands, is influenced by the amount and distribution of the recharge to groundwater. The size and shape of the island, the permeability of the sediments and reef rock, and the magnitude of the tides also constitute important factors in this respect (UNESCO 1992).

B. Water Scarcity and Stress

Countries that use 10 to 20 per cent of the available water resources are classified as being under moderate water stress; while those countries that use between 20 to 40 per cent of their renewable water resources are under medium to high stress, and countries using more than 40 per cent of renewable resources are classified as under high water stress. Water scarcity occurs when the amount of water withdrawn from lakes, rivers or groundwater is so great that water supplies are no longer adequate to satisfy all human or ecosystem needs and bring about increased competition among potential consumers. Scarcities increase if demand per capita is growing due to changes in consumption pattern or population (ADB 1998).

On the basis of these definitions, there are several areas in the region that are under high water stress, including north China, the Aral Sea Basin in Central Asia and a number of islands in the Pacific and Indian Oceans (ESCAP 1998a).

With the current population of the region estimated at 3.7 billion, overall average water per capita has been estimated at about 3 700 m^3 per year. The per capita water resource is considered as being low and has been declining at the rate of 1.6 per cent per annum as a result of expanding demand within the region. The estimates for changes in the amount of water resources per capita by subregion over the period 1950-2000 are presented in Figure 4.4 (ESCAP 1997); due to the increasing population, the amount of per capita water resources available is considerably less in 2000 compared to that in 1950 (Figure 4.5).

A widely accepted threshold for water adequacy is 1 600 m^3 of renewable freshwater per capita per year. Countries with freshwater resources in the range of 1 000-1 600 m^3 per capita per year

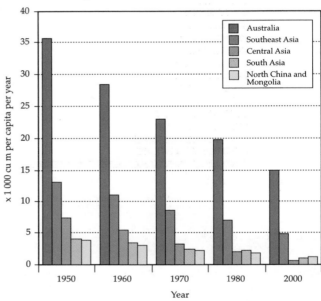

Figure 4.4 Estimates of Annual Water Resources Per Capita in the Asian and Pacific Region

Source: Shiklomanov 1997

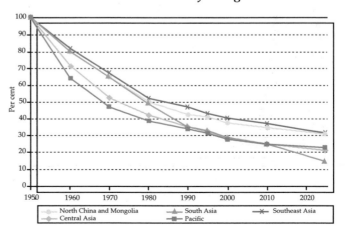

Figure 4.5 Decline in Water Resources Per Capita in the Asian and Pacific Region

Source: Shiklomanov 1997

face water stress, with major problems occurring in drought years. When annual renewable water resources are less than 1 000 m³ per capita, countries are considered water scarce. When these criteria are applied to the countries of the region, it is apparent that the Republic of Korea is currently approaching water stress, Singapore is already water scarce and the Maldives has chronic water scarcity, with the figure of 114 m³ per capita per year (FAO 1999). In India, water scarcity is expected to intensify as the country's population is predicted to exceed 1.4 billion by 2025 (United Nation's medium projection). People's Republic of China, the most populous nation (1990 annual per capita water resources: 2 427 m³), will only narrowly miss the water stress benchmark in 2025, according to the United Nations' projections (Das Gupta 1996). In that year, according to the medium scenario, People's Republic of China's projected 1.5 billion citizens will have water resources amounting at 1 818 m³ per person (Figure 4.6). It should also be noted that not all the renewable resources are available for use due to various constraints to the exploitation of the water resources.

Current freshwater use is increasing rapidly in almost all countries of the region and is expected to continue to rise in the future, leading to critical shortages in some areas. Agricultural, municipal/domestic and industrial sectors are the main consumers, although the annual water withdrawal for agriculture (84 per cent) is far more than that for industrial (10 per cent) and domestic (6 per cent) purposes (Table 4.2). As illustrated in Figure 4.7, amongst the subregions, Central Asia will push the upper limits of withdrawals as a proportion of its total available water by the year 2020, whereas South Asia's projected withdrawal will be about 70 per cent of the total available.

C. Sectoral Use and Conflicts

With the continuing economic expansion, the competition for water and the potential for conflicting demands between various sectors are increasing (see Figure 4.8). The control of water competition and the resolution of potential conflicts are increasingly being subjected to the assessment of consumption trade-offs on the basis of utilization efficiencies. For example, a thousand tonnes of water used in agriculture produces about a tonne of wheat worth US$200, whilst the same amount used in industry could expand its output by US$14 000, thereby producing financial gains that are 70 times greater. Similarly, if the goal is to produce jobs, using scarce water in industry is far more productive than using it for irrigation. Since the economics of water use do not favour agriculture, that sector is generally assigned lower priority (Brown, et al 1999). However, financial gain is seldom the primary consideration in water resource management and a range of political, cultural and social considerations provide sufficient weight to ensure that agriculture or "food security" are given greater priority in the water resource planning of many of the region's water resources (Chapter 10).

1. Agriculture/Food Production

Irrigation remains the largest consumptive use of water in the region and accounts for between 60 and 90 per cent of annual water withdrawals in most countries. The Indian sub-continent in South Asia and islands of the South Pacific have the highest level of water withdrawals for agriculture accounting for 92 and 90 per cent of the total consumption, respectively. These two areas together account for 82 per cent of the total irrigated land in Asia (FAO 1999). Of the countries within these geographic areas,

Table 4.2   Freshwater Withdrawal by Sector in the Region and the World

Subregion	Annual Water Withdrawal by Sector								
	Agricultural		Domestic		Industrial		Total Withdrawal		
	m³	% of total	m³	% of total	m³	% of total	m³	% of Asia	m³ per inhab.
Indian Sub-continent	510.7	92	27.2	5	15.5	3	553.4	38	500
East Asia	418.3	77	26.8	5	95.5	18	540.1	37	428
Far East	73.5	64	23.2	20	18.4	16	115.1	8	674
Southeast	82.1	88	3.9	4	7.0	8	93.0	6	476
Pacific Islands	127.9	90	10.4	7	4.3	3	142.6	10	483
Asia and the Pacific	1 212.5	84	91.5	6	140.2	10	1 444.2	100	476
World	2 310.5	71	290.6	9	652.2	20	3 253.3	100	564
Asia and the Pacific as % of the World	52.2		31.5		21.5		44.4		

Source: FAO 1999

Figure 4.6 Annual Renewable Freshwater Per Capita Under Three Long-Range United Nations Population Projections: India and People's Republic of China

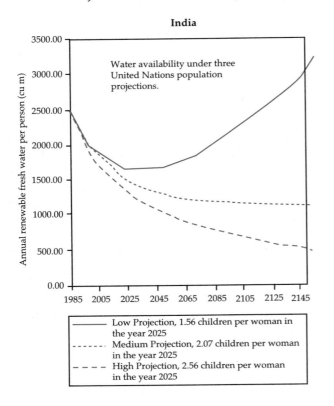
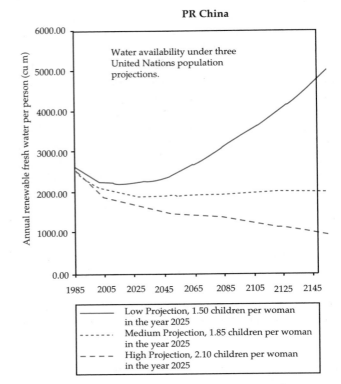

Source: Adapted from Engelman and Le Roy 1993

only in Bhutan and Malaysia was the share taken by agriculture below 60 per cent.

The main sources of irrigation water are rivers and lakes, although in some countries, such as India and People's Republic of China, groundwater is a significant source (BGS, ODA, UNEP and WHO 1996). The spread of privately owned wells and the increased water requirements of new high yielding crop varieties have led to an increase in demand for irrigation, particularly during the dry season. For example, until recently farmers in India and Pakistan required irrigation only during drought periods. However, the new improved varieties of seed have now resulted in increased irrigation throughout the year. Whilst significantly increasing food output, it has resulted in exhaustion of cheap supplies of water and the resultant water scarcity is likely to become a major constraint on increased food production in these countries (Asian Media Information and Communication Centre 1997).

2. *Household/Domestic/Municipal*

Household water consumption is typically for drinking, cooking and personal hygiene supplemented by a range of additional uses including flushing toilets, laundering of clothes and general

Figure 4.7 Water Withdrawals Against Water Resources (1900-2000)

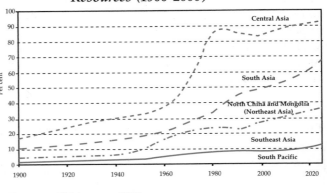

Source: Shiklomanov 1997

domestic usage. Being relatively clean, groundwater is the most popular source for domestic supplies and in most countries of the region supplies more than 50 per cent of requirements. In the arid and semi-arid zones of the region and small islands where surface supply is deficient or unsuitable, groundwater is the only source of water (BGS, ODA, UNEP and WHO 1996). Water consumption for domestic purposes ranges from 20 to 200 litres per person per day, depending on the level of affluence of users and the availability of water. The significant variation in

Figure 4.8 Demand and Uses of Water in a River Basin

Source: Newson 1992

domestic water consumption in selected cities of the region is shown in Figure 4.9. Increases in domestic demand are expected to be in the range of between 70 and 345 per cent between 1995 & 2025 (ADB 1998). In the urban areas of the region, domestic water supply and sanitation investments are unable to keep pace with population growth (see Chapter 7).

3. *Industrial*

Many industrial processes require significant quantities of water. This water is eventually released back to the environment, but is often contaminated in the process and can be recycled only after treatment. Industrial water supply in the region is provided by both surface water and groundwater sources. However, groundwater is the major source as it is usually cheaper and more reliable (BGS, ODA, UNEP and WHO 1996). Normally, water-consuming industries are located near water-bodies in order to have access to an inexpensive and reliable water supply. If water is readily available and tariffs are low or non-existent, industries tend to use

CHAPTER FOUR

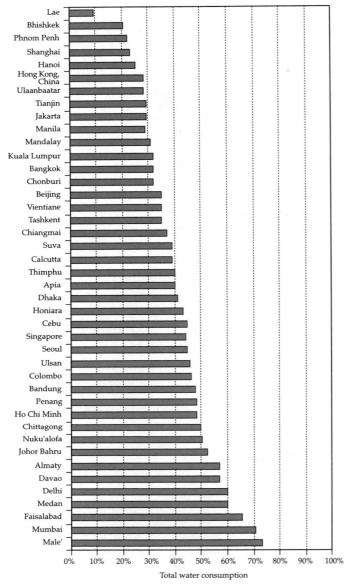

Figure 4.9 Domestic Water Use in Selected Cities in the Asian and Pacific Region as a Percentage of the Overall Water Use

Source: ADB 1997

water-wasting "once-through" technologies and simple cooling ponds instead of more water efficient but expensive technologies. Often, small and medium sized industries located in metropolitan areas utilize high quality drinking water from public supply systems and, in the older industrial city areas, the proportion of potable water used for industrial purposes is estimated to be as high as 40 per cent of the total water consumption. Elsewhere, industries are concentrated in special zones that are supplied with low quality water; in Thailand raw water is supplied to fast growing industrial zones in the country's eastern seaboard by the East Water Company, which was specifically set up for this purpose (ESCAP 1998a).

With continued industrialization, the industrial water requirement is certain to increase over the next decade. However the rate of water consumption per unit of industrial output is expected to decline as technological improvements are introduced with the expanding application of regulatory and economic measures (ESCAP 1998a).

4. Other Uses

Freshwater is also an important non-consumptive resource for nature conservation, hydropower developments, inland fisheries, tourism and recreation, transportation and flood control (UNEP 1994). Hydropower accounts for over 50 per cent of the electricity generated in Bhutan, Nepal, Afghanistan, Democratic People's Republic of Korea, Lao People's Democratic Republic, New Zealand and Sri Lanka, and is also a key source of energy in Central Asia. Fish and other aquatic macrophytes from the region's rivers, natural lakes and reservoirs provide the main sources of dietary proteins in countries such as India, People's Republic of China, Sri Lanka, Viet Nam, Cambodia and Lao People's Democratic Republic (Avakyan and Iakovleva 1998) and the total inland water fishery catch in the region amounted to 15.6 million tonnes in 1996 (ESCAP 1998c).

The use of inland water for the transportation of people and goods is widespread within the region and Bangladesh, India, Myanmar, Thailand and Viet Nam rely upon their rivers as major transportation arteries. In the Philippines, Laguna de Bay Lake has become an important transportation alternative to Manila's worsening road traffic congestion.

Lakes and reservoirs are also used extensively in the regulation of river flows and the management of floodwaters. For example, in the rainy season the Cambodian lake of Tonle Sap holds and stores excess water and sediment from the Mekong River and thereby maintains a constant flow regime in downstream Cambodia and Viet Nam, enhancing the biological and economic productivity of the river system.

WATER QUALITY: GROWING POLLUTION AND ITS IMPACTS

The threats to freshwater quality in the region come from a range of sources and the type and extent of water pollution varies by location, ecosystem characteristics, land-use and the degree and type of development. The relative severity of water quality problems in the subregions is summarized in

INLAND WATER

Table 4.3   Relative Severity of Water Pollution in the Asian and Pacific Region

Pollutant	Northeast Asia	Southeast Asia	South Asia	South Pacific	Central Asia
Suspended Solids		**	**		*
Fecal Coliforms		***	**	**	**
BOD		***	**		***
Nitrates	**	*	***		**
Lead	**	***	*		***

Source: ADB 1997 and 1999
* = moderate   ** = severe   *** = very severe

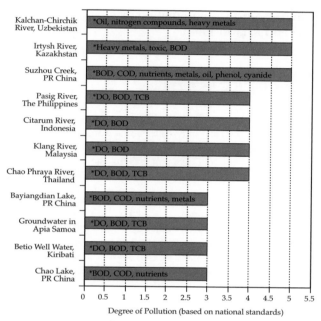

Figure 4.10   Selected Cases of Water Pollution in the Asian and Pacific Region

Source: ADB 1998
* significant water quality parameters

Table 4.3 while selected cases of water pollution have been highlighted in Figure 4.10. Among the rivers of the region, the Yellow River (People's Republic of China), Ganges (India), Amu and Syr Darya (Central Asia) top the list of the worlds most polluted rivers according to a report of the World Commission on Water (The Independent 1999).

A.   Pathogens, Organic Matters and Nutrients

Pathogenic bacteria and viruses are found in areas where untreated sewage and effluents from intensive animal husbandry operations drain into waterways. They also enter water supplies from stormwater run-off, or as a result of leaching from open solid waste dumpsites or agricultural areas where untreated wastewater is used on crops (UNEP 1991). The level of pathogens is usually in direct proportion to the density of population and level of socio-economic development in proximity to the water. The sewerage systems in many of the developing countries of the region are poorly developed and only 10 per cent of wastewater undergoes any form of treatment (Sweden's Ministry of Foreign Affairs 1999).

Many of the region's rivers contain up to three times as much bacteria from human waste (fecal coliform) as the world average and more than ten times the standards set out in the OECD guidelines. The reported median fecal coliform count in the rivers of the Asian landmass, for instance, is 50 times higher than the WHO guidelines and is even more serious in the Southeast Asian subregion (ADB 1997). Drinking or bathing in water contaminated by animal or human excreta facilitates transmission and proliferation of disease vectors. The most common water-borne infectious and parasitic diseases include hepatitis A, diarrhoea diseases, typhoid, roundworm, guinea worm, leptospirosis, and schistomiasis (WRI 1999).

Organic matter also constitutes a significant pollutant in the water bodies of the region, with industries such as pulp and paper, textile, tanneries and food processing contributing substantially. The geographic distribution of organic matter pollution largely coincides with that of pathogenic contamination (Goluveb 1993). Growth in biological pollution in total water pollution loads in high-growth areas of the region has been projected to increase 18 times between 1995 and 2005 (UNIDO 1996).

Many rivers and lakes in South Asia, Southeast Asia, and People's Republic of China are severely polluted by organic matter from sewage and industrial processes (Box 4.1). In India, for example, 114 towns and cities dump raw sewage into the Ganges, whilst the Vrishabavathi River near Bangalore receives substantial amounts of human and industrial waste, which eventually flows into Byramangala Lake, a traditional feeding ground for thousands of waterfowl. In People's Republic of China, during the mid-1990s, the Huaihe River received 7 million tonnes of untreated domestic and industrial waste each year, rendering the river water unsuitable for both domestic consumption and agricultural purposes (Asian Media Information and Communication Centre 1997).

Organic matter has also been the cause of groundwater pollution in many cases. Pollution sources include leachates from unsanitary dumping of refuse and other solid waste, and from the excessive use of fertilizers. In India, groundwater pollution of the Chennai urban aquifer is caused by a range of pollutants including nitrates, heavy metals, and micro-organisms (Somasundaram *et al* 1993), whilst in the Jaffna Peninsula of Sri Lanka, latrines contribute to the increasing nitrate problem of the shallow groundwater aquifer (Hiscock 1997).

Agricultural inputs, including fertilizers, pesticides and animal wastes, are another growing source of freshwater organic pollution in the region, particularly in People's Republic of China and the countries of South and Southeast Asia. In New Zealand, the increase in dairy farm and fertilizer use is intensifying pollution in groundwater as well as shallow lakes and streams (Smith, et al 1993). In New South Wales, it is estimated that around 90 per cent of rivers currently experience water quality problems due to excessive nutrients.

Among the rivers of the region, approximately 50 per cent have exceedingly high levels of nutrients while another 25 per cent have a moderate problem where nutrient levels occasionally exceed desirable levels (ESCAP 1998). In Central Asia, nutrients from the excessive use of fertilizers, herbicides, pesticides and defoliants is leading to health hazards due to water resource contamination (Mainguet and Letolle 1998 and Kharin 1996).

Box 4.1 Water Quality in People's Republic of China

Seven large river systems, lakes, reservoirs, and underground water in some regions of People's Republic of China were polluted to varying degrees in 1997. The arid and semi-arid regions in the north as well as many cities suffered from serious water shortage. Seventy eight per cent of rivers which flowed through cities could not be used for potable supplies, and 50 per cent of the groundwater in cities was polluted (1996 State of Environment Report, People's Republic of China). The deficiency of water resources and the pollution of water systems according to the report had become one of the main factors affecting China's social and economic development.

In terms of effluents the total amount of discharged wastewater was 41.6 billion tonnes, of which 22.7 billion tonnes was industrial wastewater and 18.9 billion tonnes was municipal wastewater. With regard to industrial wastewater, industries at the county level and above discharged 18.8 billion tonnes, while TVIE (Township and Villages Industrial Enterprises) discharged 3.9 billion tonnes. In discharged wastewater, the total amount of Chemical Oxygen Demand (COD) was 17.6 million tonnes, including 10.7 million tonnes of COD discharged with industrial wastewater and 6.8 million tonnes of COD discharged with municipal wastewater.

In order to cope with situation, People's Republic of China has enhanced the dynamic of pollution control measures along its rivers and it has listed, (in its environmental plan within its ninth Five-year plan) the water control project of the "Three Rivers and Three Lakes" as the most important undertaking. In addition, the following steps have been taken to solve the problem.

- Deadlines have been set for enterprises to comply with the state standards industrial pollution control and treatment has been strengthened.

- An investment loan of 250 million yuan has been arranged to support 28 key pollution control projects. With the support of the central government and joint efforts of the governments at all levels in the river basin, the problems of drinking water supply in heavily polluted areas are being gradually addressed.

- The treatment rate of industrial wastewater was 79.1 per cent in 1997. Moreover, during the implementation of China's Trans-century Green Project Plan, 99 projects for water pollution treatment were completed and 325 projects for water pollution prevention and treatment are under construction.

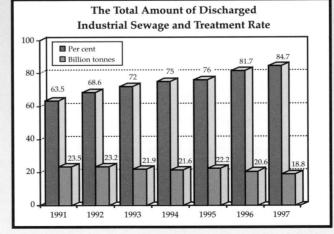

These measures are gradually improving the quality of water and it is expected that their strengthening will lead to further improvements in future.

Source: SEPA 1997

B. Heavy Metals and Toxic Chemicals

The measured concentrations of heavy metals (such as arsenic, cadmium, mercury and lead) exceed basic water quality standards in many of the region's water bodies. The concentrations of DDT, PCB's, industrial solvents and other toxic chemicals, which originate primarily from mining, oil refineries, chemical works and in textile, wood pulp, and pesticide factories, are also rising.

Within the region, the water bodies of the Southeast Asian sub-region are the most heavily polluted with heavy metals and toxic chemicals (ADB 1997). For example, sixteen rivers in the Johor State in Malaysia were found to have mercury levels exceeding the national standard, whilst in other rivers, the mercury, lead, cadmium, zinc and copper levels exceed the national standards. Heavy metal pollution in Malaysian rivers is caused mainly by industrial discharge or mining (Encyclopaedia of Malaysia 1997). An extensive survey of river water quality in Thailand revealed heavy lead contamination (in a number of major rivers including the Pattani and Colok in the south, the Moon river in the Northeast, the Pa Sak river in the north and the Mae Klong river in the central region), mercury contamination (in the lower central region's Pranburi River, the Mae Long, Chao Phraya and Petchburi rivers of the central region and the Wang River in the northern region) and high levels of arsenic poisoning in the Tambon Ron Phibun (Asian Media Information and Communication Centre 1997). Arsenic was also found to be the cause of groundwater pollution in several villages in southern Thailand situated near tin mines.

Cyanide and mercury used in gold mining are severely polluting the rivers and waterways of the Philippines' mountain provinces and islands and other areas including Malaysia and Papua New Guinea. One of the major lakes on the island of Mindanao had been declared biologically dead, since its basin was converted into a tailings pond by a copper and gold mining company in 1979. In Bangladesh, the recent case of slow poisoning from arsenic-contaminated water supply has gained worldwide attention (Box 4.2).

Rivers in People's Republic of China have also been affected by industrial pollution. For example, every day seven million tonnes of untreated industrial and domestic waste are dumped into the Huaihe River and similar practices have been reported from the Yangtze River delta in Jiangsu Province. In Japan, a study of 15 cities showed that 30 per cent of all groundwater supplies are contaminated by chlorinated solvents from industry; in some cases the solvents from spills travelled as far as 10 km from the source of pollution (UNEP 1996).

C. Sediments and Salts

With increasing deforestation and land conversion, soil erosion is exacerbating the natural process of siltation of water bodies and greater quantities of sediment are accumulating in the rivers, dams and reservoirs of the region. For example, in the Ganges, Brahmaputra and Yellow River basins, erosion is responsible for an annual yield of over 1 000 tonnes of sediment per square kilometre of land. The siltation in Pakistan's Tarbela Dam on the Indus River accumulates 200 million cubic metre of silt each year filling the reservoir at a rate of two per cent per year (Asian Media Information and Communication Centre 1997). In Cambodia, heavy siltation of Lake Tonle Sap, resulting from deforestation in the upper catchment, is significantly reducing the lake's depth and this has effected the yield of the lake's fisheries. In Malaysia, the Dungun River in Kuala Terrengganu has been polluted by sandy sediments, exacerbated by dredging activities along the river (Encyclopaedia of Malaysia 1997).

Increasing salinity of ground and surface waters renders land unsuitable for agriculture and leaves water unsuitable for domestic and industrial use (ESCAP 1998). Increasing salinity in the region is occurring not only as a consequence of excessive irrigation but also due to rapid deforestation and mining and other industrial activities that mobilize salts, naturally present in groundwater, rocks and soil, and contributes to their concentration in surface waters and groundwater aquifers. Moreover, the effluents from power stations and industrial cooling systems, paper mills and other industrial processes are also sources of salt accumulation in receiving water bodies. Other causes of increasing salinity are rising water tables due to excessive seepage of water from canal systems and capillary rise and evaporation of saline groundwater. Inadequate availability of water from rains or excessive irrigation without adequate drainage is also one of the important causes of salinity. According to FAO (1990), the Asian and Pacific Region has the world's greatest concentration of salt affected soils (see Chapter 1).

CHAPTER FOUR

Box 4.2 Millions in Bangladesh Face Slow Arsenic Poisoning

Millions of people in rural areas of Bangladesh are being slowly poisoned as they drink water contaminated with small but potentially fatal quantities of arsenic. Estimates by World Bank and other experts claim that from 18 to 50 million people out of a total population of about 120 million in the country are at risk. Thousands are already showing symptoms of poisoning. Nineteen rural districts covering an area of 500 km² near the border of Bangladesh and India have arsenic-contaminated wells. Many villages adjacent to the capital city, Dhaka, are also affected. In the neighbouring Indian state of West Bengal, an estimated 6 million Indians are drinking contaminated water and 300 000 are showing signs of poisoning. Many victims are children (45 per cent) who have been consuming the poisoned water since birth. The contaminated water comes from underground tube wells introduced widely over the last 20 years as a cheap alternative water supply to prevent outbreaks of diseases such as diarrhoea and cholera. Tube wells are steel cylinders sunk into the ground to varying depths to provide underground water for irrigation and drinking.

The United Nations Children's Fund (UNICEF) initiated well-drilling as a means of providing clean water in rural areas in Bangladesh. When the programme began, no water or soil tests were carried out. It was estimated that there are now 5 million tube wells, providing 95 per cent of all water to over 120 million people. Testing is meant to be carried out on new installations but mainly takes place at government installed wells. Of the 20 000 tube wells tested so far, 25 per cent have dangerous levels of arsenic, 40 per cent have unsafe levels and only 35 per cent were safer or below 0.01 milligrams/litre of arsenic. The World Health Organization recommends a level of 0.01 mg/L of arsenic but the governments of Bangladesh and India regard 0.05 mg/L – a level five times higher – as acceptable.

Various theories for the contamination have been advanced. According to one theory, overuse of the water supply has increased oxygen levels in underground waterways, resulting in higher rates of leaching of minerals containing arsenic. Other scientists say that biological processes may be involved. In 1997, the Bangladesh Centre for Advanced Studies hypothesized that only the upper 150 metres of sediment contained high levels of arsenic. Some experts consider long term and sustained use of pesticides and the waste products from industry as the main contributing factors.

Concerns over arsenic-contaminated water first began to emerge in 1980s in West Bengal but efforts on testing were so little that most of the danger warnings were dismissed even until early 1990s. Despite the mounting evidence of widespread water contamination, little has been done to identify the extent of the problem let alone provide any solutions. Facing a growing health crisis, a conference to bring together international specialists and medical experts were organized by Dhaka Community Hospital to find solution in 1994. A regional consultation was also held in New Delhi in April 1997. In August 1998, WHO, UNICEF and other international agencies, including the World Bank, agreed to provide funds to conduct more research and attempt to find alternative supply to safe drinking water. Some cheap solutions and interim measures are now under consideration.

There is no lack of water as Bangladesh forms a huge river delta system, though flooding afflicts the country, yet the most obvious solution – a long term plan to control the flow of river and water treatment plants to provide clean drinking water – remains far off reach in the wake of deficient financial resources.

Source: Government of Bangladesh

ENVIRONMENTAL IMPACTS OF UNSUSTAINABLE WATER USE

A. Surface water

1. *Diminishing Supply/Over-abstraction*

Over-abstraction can cause serious environmental degradation. Water abstraction from the Aral Sea in Central Asia (Box 1.1) has reduced the lake's surface area by 40 per cent and its volume by as much as 60 per cent, resulting in the loss of almost all of the native fish species. Similar symptoms have been reported from other arid zone lakes, including Lake Balkhash in Kazakhstan and Lake Chinghai in People's Republic of China.

The adverse effect of surface water mismanagement is also apparent in the hill districts of Uttar Pradesh in India where 2 300 of the 2 700 drinking water projects have failed as surface water sources have dried up. In the southern state of Tamil Nadu, the custom of celebrating the mid-August arrival of freshwater floods on the Amaravathi River is no longer practised as the flow is now reduced to a mere trickle of water. This is a result of the construction of a major dam 50 kilometres upstream and the development of a sugar mill in the vicinity that utilizes substantial amounts of water (Asian Media Information and Communication Centre 1997).

2. *Declining Productivity of Land*

Irrigation has increased food productivity by many folds but often it has resulted in the degradation of croplands and water quality. In particular, the neglect of the need to remove water and salt from confined or slowly draining alluvial basins has led to a loss in land productivity and to environmental degradation (see Chapter 1 and Chapter 10).

3. Negative Effects of Dams and Reservoirs

Dams and impounded reservoirs form an essential part of many water management systems, enabling hydropower, irrigation, flood control and providing water resources regardless of the seasons. However, the construction of many dams has resulted in high environmental, economic and social costs.

In recent years, the construction of dams and reservoirs has resulted in the loss of increasingly shrinking wilderness areas with high levels of biodiversity. In Malaysia, the Temenggor Dam flooded valuable forests and threatened the survival of 100 species of mammals and 300 species of birds. In India, a network of canals, proposed as part of the Sandar Sarovar Project, disrupted much of the wild ass sanctuary by destroying the vegetation of the species' primary grazing and breeding areas and the changes in irrigation, land-use patterns and soil moisture regimes threaten the survival of other species such as blackbuck, desert fox and indigenous plant species which are xerophytic and salt tolerant (Asian Media Information and Communication Centre 1997).

In the Russian Federation, eight million hectares of highly productive land has been lost to inundation by man-made lakes. These have resulted in a deterioration in water quality, the disruption in the hydrological regimes of downstream rivers, the interception of nutrients by dams, a rise in groundwater levels, the translocation of people and the disruption of traditional economic activities and fisheries (Goluveb 1993).

B. Groundwater

1. Land Subsidence

Since the early 1980s evidence has been accumulating of the substantial and widespread breakdown of the piezometric surface in many Asian cities, as a result of heavy exploitation of aquifers (Ramnarong and Buapeng 1991 and Sharma 1986). In many areas falling water tables signify that groundwater withdrawals have exceeded the rate of renewal. For instance, in some parts of India extraction exceeds recharge by a factor of at least two (Seckler, et al 1999). Water tables are falling throughout much of the Punjab and Haryana States, whilst in Gujarat groundwater levels declined by 90 per cent during the 1980s. In People's Republic of China, the water table in parts of Beijing has dropped 37 metres over the last four decades (Asian Media Information and Communication Centre 1997). In Japan, an estimated 12 per cent of habitable land, mainly in the industrial regions along the coast, is affected by subsidence.

Significant subsidence has also occurred in and around Manila and Jakarta (Postel 1996). Similarly, in Bangkok thousands of wells pump out more than a million cubic metres of water daily, thereby slowly draining the aquifers. Some central areas of the city have sunk by as much as 1.6 metres in the past half-century. Although the rate of pumping has decreased in the last decade, the city, which is only 0.4 metres above sea level, faces the perennial threat of inundation during the floods and seasurges of the rainy season.

2. Saline Intrusion

Pakistan is probably the country most severely affected by saline water because of the high salinity of much of its soils. With limited natural drainage from the primary agricultural areas, levels of salinity in the major rivers are progressively concentrated such that the water is rendered unusable for downstream users (Seckler, et al 1999).

In some coastal cities of the region, over-pumping has resulted in the movement of salty seawater inland. Known as "saline intrusion", this occurs when water levels in freshwater aquifers are lowered to a point where salt-water can invade through the water-bearing beds in the direction of the wells. For example, in Dhaka and Metro Manila seawater intrusion into aquifers presents a major problem, whilst in the major river basins and coastal plains of Viet Nam, the average salinity of groundwater is approximately 3 000-4 000 ppm, a level unsuitable for drinking (Asian Media Information and Communication Centre 1997). Saline intrusion has also occurred in Indian state of Gujarat, where irrigators have heavily overpumped local aquifers near the coast (Postel 1996).

POLICIES AND STRATEGIES FOR SUSTAINABLE WATER MANAGEMENT

A. National Level

Traditionally, governments' policies and strategies on water management have been aimed at the expansion of supply in order to meet the ever-increasing water demands of the domestic, agriculture and industrial sectors. However, increasingly policy frameworks are focused on an integrated approach to water resources management by placing emphasis on demand management, water-use efficiency, conservation and protection, institutional arrangements, legal regulatory and economic instruments, public information and interagency cooperation. An integrated approach, as articulated in Agenda 21, implies the use of a

CHAPTER FOUR

dynamic, interactive, iterative and multi-sectoral approach to water resources management and advocates the integration of sectoral water plans and programmes within the framework of national economic and social policies.

Many countries of the region have reviewed or revised their national policy on water resources development and management. Common elements in the national policies and strategies adopted include the integration of water resources development and management into national socio-economic development; the assessment and monitoring of water resources; the protection of water and associated resources; the provision of safe drinking water supply and sanitation; the conservation and sustainable use of water for food production and other economic activities; institutional and legislative development; and public participation. In 1997, ESCAP conducted a survey to report on the progress achieved in the implementation of Agenda 21 on freshwater resources management. The results of the survey showed that although the implementation levels of the elements outlined above were not fully realized, some encouraging progress had been made in a number of areas including an increase in the potential for regional cooperation.

1. *Integration of Water Resources Management into National Development Plans*

In a number of countries in the region, national water resources action programmes or water master plans have been prepared with a scope that ranges from single-purpose to more comprehensive development strategies. In Bangladesh, the National Water Plan II for the period of 1990-2010 has been prepared as an updated continuation of the National Water Plan, which covered the period between 1985 and 2005. In the Islamic Republic of Iran, the Second Five-Year Socio-economic and Cultural Development Plan (1995-1999) included a number of objectives related to the environment and water resources. In the Maldives, an action plan has been developed which gives priority to the wise use of groundwater resources.

In several countries, the preparatory work for the formulation or revision of national action plans has also been initiated, often with the assistance of international organizations. For instance, Mongolia is currently planning an update to the national Water Master Plan, first prepared in the early 1970's, to reflect the social and economic changes associated with the country's transition from a centrally controlled to a market economy. Sri Lanka is to formulate a national action plan for water resources management that will synthesize and use the results of a series of comprehensive sub-sectoral plans. The action plan is expected to have a positive effect on the quality of investments in irrigation, water supply, power generation and environmental protection and to strengthen their linkage with national development goals (ESCAP 1997).

The concept of water resources management within a river basin or water catchment area, with a focus on the integration of land and water related issues, has also been applied in some countries including Australia, People's Republic of China and Japan. In India, the national water policy asserts that water resources planning be undertaken for a hydrological unit, such as drainage basin or sub-basin. In Indonesia, institutions for water resources management have been established for some river basins, although these are yet to become fully functioning. The current challenge to many countries of the region is to overcome fragmented sub-sectoral approaches and to design and implement integrated mechanisms, particularly for the implementation of projects that transcend sub-sectors. The largely fragmented approach that has traditionally been applied has allowed conflicts and competition, and has led to the over-exploitation of scarce water resources.

2. *Assessment and Monitoring of Water Resources*

An essential pre-requisite to the integrated management and sustainable development of water resources in the region is the assessment of the quantity and quality of available surface and groundwater resources. The inadequacy of quantitative and qualitative information for planning and decision-making has often resulted in over-estimation of water resources potential and the development of agricultural and industrial projects that were constrained by water scarcity. It has also allowed complacency, and consequential wastage, in the use of water resources. In many small island countries, such as the Maldives, the lack of information on the assessment of groundwater reserves has also been identified as a major constraint for future planning and decision-making.

The main challenges to water resources assessment in the region are coordination in the collection and processing of data on water resources, standardization of assessment and monitoring procedures, and integration of information on current and future uses of water and related resources, including catchment protection under development scenarios. Comprehensive water resource assessment also necessitates local and community participation in the assessment to contribute and validate information, verify assumptions and develop effective resource management modalities (ADB 1998).

The problem of assessment and monitoring of water resources is aggravated by extremely uneven density of hydrometeorological observation networks, varying from the comprehensive network established in Japan, to the scattered gauges available over the vast areas of eastern China and the Himalayas and to occasional solitary gauges on a Pacific atoll. In view of the lack of information on water resources, a number of governments have devised strategies to develop comprehensive data and information system on water resources to assist in policy formulation and decision-making. Sri Lanka, for example, proposes to undertake a complete reappraisal of the available water resources and devised a strategy for their future development and conservation in a national water master plan. In India, a project is being implemented for undertaking systematic collection and analysis of hydrological data. Mongolia is also undertaking action to develop an adequate information system for both surface and groundwater, in terms of both water quantity and quality. An integrated, nation-wide initiative called the Urban Water Programme (UWP) in Australia highlights the importance of water assessment, flow mapping and modelling (Speers 1999) as a major step towards the country's drive for efficiency in water service provision and water allocation.

3. *Protection of Water Quality and Associated Resources*

Several countries of the region are implementing large-scale and ambitious programmes and action plans aimed at rehabilitating degraded and depleted water sources. These programmes and plans are typically given legislative or statutory authority such as that provided by Thailand's National Water Quality Act, the Philippines' Water Quality Code, India's Environment Protection Act, China's Water Law or the Republic of Korea's Water Quality Preservation Act (ESCAP 1999a).

Success stories with respect to the rehabilitation and protection of water quality of rivers come from those countries where water policies promote a multi-sectoral and multi-disciplinary approach to the management of water resources (ESCAP 1999a). Clean up campaigns for rivers, canals, lakes and other water bodies have become widespread throughout the region (see Table 4.4). Initiated by a range of organizations, including government bodies, business associations, NGO's, national or international development agencies and community groups, the programmes have often been successful in achieving water quality improvement and occasionally have led to the adoption of new water quality standards, water-use regulation, creation of governing bodies, delineation of responsibilities, allocation of resources, and promotion of sectoral and public awareness and participation (see Box 4.3). There has also been an increase in awareness regarding the reduction of pollutant loads through proper wastewater treatment, reuse and recycling of domestic sewage and industrial wastewater, introduction of appropriate low-waste technologies and strict control on industrial and municipal effluent. Programmes on protection and rehabilitation of lakes have been initiated through coordinated efforts of the government, industries, NGO's and the general public in a range of countries including People's Republic of China, Japan and the Republic of Korea.

A number of success stories have been documented regarding water reuse and recycling in the industrialized countries of the region attributed to the emergence of various corporate environmental management approaches such as cleaner production,

Table 4.4   Rivers/Canals Clean-up Programmes in Selected Countries of the Asian and Pacific Region

Country	Programme	River
Australia	Murray-Darling Basin Agreement	Murray- Darling River
Bangladesh	Save Buriganga Programme	Buriganga River
PR China	Pollution control plan on three rivers	Huihe, Haihe, Liaohe rivers
India	Ganges Action Plan	Ganges River
Indonesia	Prokasih	Various rivers in Indonesia
Malaysia	Operation Clogged Drains Love Our Rivers Campaign	Sungai Merliwan, Malacca Klang River, Kuala Lumpur
The Philippines	Ilog Ko, Irog Ko (My River, My Love) Sagip Pasig Movement (Save Pasig Movement) Piso Para sa Pasig (A Peso for Pasig)	Pasig River, Metro Manila
Thailand	Clean Up Chao Phraya River and Bangkok Canals	Chao Phraya River and Bangkok Waterways
Singapore	Clean River Programme	Singapore River and Kallang River

Source: ESCAP 1999a

environmental management systems, ISO 14000, environmental auditing and reporting and industrial eco-zoning (see Chapter 13).

Despite improving corporate performance and stricter enforcement of environmental regulations, a major challenge to the ongoing protection of water quality remains the management of catastrophic pollution events (see Box 4.4).

4. *Provision of Safe Drinking Water Supply*

Considerable progress was made in improving and extending the provision of water supply and sanitation during the International Water Supply and Sanitation Decade (1981-1990). Millions of people gained access to safe water and adequate sanitation for the first time (ESCAP 1998a). A regional consultation organized by World Health Organization (WHO) in 1996 also provided an impetus to countries in the region to draw up their own Action Plan for the Development of National Drinking Water Quality Surveillance and Control Programmes. The plan was intended to provide guidelines, within a regional framework, for implementing national drinking water quality surveillance programmes (WHO 1996).

5. *Conservation and Sustainable Water Use for Food Production and other Economic Activities*

Demand side management policies have led to the promotion of efficiency, conservation, rationalization of prices and the involvement of

---

Box 4.3 Pasig River Rehabilitation Programme in the Philippines

The 27-kilometre Pasig River in Metro Manila, Philippines is directly connected to the Laguna Lake and Manila Bay, and cuts through the heart of heavily urbanized Metro Manila. It is considered as one of the most polluted rivers in the region. As only seven per cent of Metro Manila's households are connected to a piped sewerage system, the river receives the direct discharge of untreated wastewater from the 11 municipalities located within the river basin, raising the biochemical oxygen demand (BOD) far above the absorptive capacity of the river. Likewise, the indiscriminate discharge from industrial sources and the dumping of solid wastes contribute to the severe level of pollution. The areas along the riverbanks and adjacent to the river are lined with squatter settlements, estimated to be around 12 000 families. These people are exposed to flooding, and major public health risks such as cholera, gastroenteritis, dysentery and intestinal parasites.

The President of the Philippines, on January 6, 1999, signed Executive Order No. 54 creating the Pasig River Rehabilitation Commission (PRRC) to oversee and coordinate all activities related to the rehabilitation and development of the river. Its overall objective is to restore the river water quality to Class C standard, which is capable of sustaining aquatic life, supporting secondary contact sports like boating, and, after treatment, may be used for industrial processing as a potable source. Among the major activities of PRRC are: massive dredging and de-silting; removal of derelicts clogging the waterways; construction of river walls and flood control structures; and construction of waterfront facilities, roads and walkways along the riverbanks in order to provide recreational areas and contribute to the environmental enhancement of these areas. The government also plans to build larger and better ferry service and terminals not only at several points along the river system but also along the Laguna Bay, thus providing an alternative transport system for Metro Manila. Resettling riverside squatter families is among the key environmental targets of the PRRC.

In terms of coordination and participation, the Metro Manila Development Authority Chairman was delegated to coordinate all government and private efforts at reviving the Pasig River and its major tributaries. Lead government agencies supporting PRRC include: Departments of Budget and Management, Environment and Natural Resources, Tourism, the Public Works and Highways; Housing and Urban Development Coordinating Council; and the Philippine Ports Authority. Partnerships with private organizations were also developed to support the PRRC. For example, the Department of Environment and Natural Resources (DENR) has forged ties with Coconut Industry Investment Fund (CIIF) Group of Companies for an effective waste management and pollution reduction systems. General public participation is also being promoted through awareness and education programmes and projects like the "Lakbay Ilog" (River Travel) and "Task Force Sagip-Ilog" (Save the River Task Force).

The project is estimated to cost some 12 billion Pesos (US$468 million) and will take about six to seven years to complete. Technical and financial assistance were also solicited from Danish International Development Assistance. The ADB is also currently studying a US$350 to US$400 million loan for a five-year development plan of Pasig River.

There has been some observed improvement in the quality of water in the river since the project started. The volume of floating debris has been reduced, sunken derelicts have been removed, clearing of illegal structures is ongoing, works on riverwalks are also on-going and 14 riverside parks have been developed. Some 4 104 families have already been resettled to safer grounds in anticipation of the effects of heavy monsoon rains. These relocated families have also been given livelihood support. With the PRRC's optimistic beginning, the programme shows that a decisive political will, effective focalized coordination among all agencies involved, private and public participation as well as international support are critical success factors towards rehabilitating a river like Pasig.

Source: Compiled from news at http://home.ease.lsoft.com/BALITA News

stakeholders in water conservation and management. Such policies include the promotion of efficiency in the use of irrigation water and in food production, consumption and leakage reduction programmes and encouragement of industries to reduce and recycle process water.

In view of the current low efficiency of irrigation water use, many countries are adopting policy measures for improving irrigation and taking steps towards the modernization of technologies and the improvement of existing methods of irrigation. For instance, in India the National Water Board of the Ministry of Water Resources has recently finalized the draft of an irrigation management policy for consideration and adoption by the National Water Resource Council. The policy aims at improving water application efficiency through the use of modern technologies such as drip/sprinkler irrigation and better on-farm irrigation methods. It also envisages providing adequate finances by linking provision of such finances to irrigation revenues earned. In Pakistan, where about 10 per cent of the best agricultural land is already adversely affected by salinity, Phase I of the National Drainage Programme (1996-2002) was launched in January 1996 in order to combat the problems of waterlogging and salinization more effectively. In the Republic of Korea, where agriculture utilizes 50 per cent of total available water resources, the government has drawn the agricultural water resources development plan for the 21st century highlighting seven point measures that relate to increased food production with efficient water use (Kwun 1999).

Demand management through pricing has been particularly effective in industrial and domestic sectors leading to the promotion of conservation, reuse and recycling measures. In Japan, industries are promoting recycling of water and sludge in many industrial plants, whilst in People's Republic of China wastewater recycling is being effectively incorporated into the country's water management strategy. In Thailand, several industries have begun to adopt a strategy of water recycling and factories have installed water-treatment plants and using the treated water to irrigate lawns and gardens. Many proactive companies are also promoting innovative wastewater recycling initiatives (see Box 4.5).

Box 4.4 Prevention of Water Pollution Accidents in the Republic of Korea

With the rise in industrial activity, more and more oil and hazardous substances are being handled and transported in the Republic of Korea resulting in more frequent water pollution accidents. A total of 75 water pollution accidents occurred in 1996 alone. An analysis of these accidents revealed that 52 of these were the result of traffic collisions and carelessness in handling and transporting oil and hazardous substances, or water pollution due to illegal dumping or emission of polluted wastes or toxic substances by industrial stes, while eight per cent involved negligence of persons in charge of handling toxic substances, including the collision of vehicles.

In response to increasing accidents, the government formulated "Integrated Guidelines on Water Quality Control" on 30 May 1994 by Prime Ministerial Directive No. 196. In addition, a system of cooperation between the related ministries responsible for water quality control and construction of dams and other water-resource supply related works was established. A Water Quality Control Consultation Committee for each water system around the four major rivers (Han, Nakdong, Keum, and Youngsan) was also constituted, and responsibilities for taking preventive measures against water pollution accidents were assigned to each of the appropriate authorities.

For effective surveillance under the new system, 1 932 government officials were assigned to 1 559 locations where the danger of water pollution accidents was greatest. To further strengthen the public water surveillance system, the Military Manpower Management Office provided 1 712 public service personnel to be assigned the duties of surveillance and protection of water supply sources in 1996, which were further increased in 1997.

Under the strengthened water quality analysis system, 43 locations were identified where water pollution accidents were most likely to occur and water quality analysis was done one to three times a day to detect water pollution accidents as early as possible and to take prompt response to such accidents.

During the drought season when even the flow of small amounts of pollutants into rivers may cause water pollution, administration was strengthened to control effluent discharges. The Water Quality Preservation Act was revised in December 1995 and put into effect on 1 July 1996. In an accident case involving water pollution, it required the polluter to file a report and take preventive measures. In addition, legal grounds for punishment of polluters were provided. The law aims at inducing operators of businesses to make voluntary efforts to prevent water pollution accidents.

The comprehensive measures adopted in the Republic of Korea to prevent water pollution accidentally has enabled the country to avoid a major disaster in terms of water pollution and maintain the quality of its valuable and limited water resources.

Source: Government of Republic of Korea 1997

CHAPTER FOUR

Water pricing has been a very successful instrument in controlling wastage of water in the domestic sector. For example, in Bangkok, introduction of appropriate water charges led to reduced use and a lowering in pumping rate of groundwater by one per cent per year. Differential prices have also been used as a mechanism in People's Republic of China to save water from fragile sources such as groundwater.

A major problem in water conservation is unaccounted water or distribution loss, which is high in urban areas averaging about a third of the total volume. However, Singapore and Bangkok have excellent records for reduction in transmission loses. Leak detection programmes, metering of production and consumption, regular maintenance and repair and replacement of defective metres have been keys to success.

The main challenges confronting many countries with respect to conservation and efficient water use are: how to increase investments in new water delivery systems that will efficiently meet customer demands in different uses and also include cost recovery for water services provided; how to upgrade and manage existing systems to reduce demand and run more efficiently including the maintenance of large common assets with long economic life expectancies; and how to safeguard social equity by ensuring the provision of basic affordable water services to all consumers.

6. Institutional Development, Legislation and Public Participation

Some countries in the region have created major national institutions with comprehensive responsibilities for water resources assessment, planning and development and other related functions (ESCAP 1997a). However, most countries retain institutional responsibilities for water resources development, management and conservation amongst a number of fragmented, sectoral agencies and central and provincial authorities; this fragmentation has been a major obstacle to the introduction of integrated water resources management.

In some countries, such as People's Republic of China, India, the Philippines and Thailand, the integration of these separate administrative and functional authorities has been addressed through the creation of inter-agency coordination committees and groups composed of high-ranking representatives of various agencies and ministries dealing. For example, in the Philippines the National Water Resources Board coordinates the activities of eleven government agencies with functions that directly relate to water resources management (Baltazar 1996). A similar approach is taken in India where the role of coordinating surface water management resides with the Central Water Commission, which is within the Central Ministry of Water Resources (Rao 1996).

Box 4.5 From Effluent to Affluent: A Case in Thailand

The Ban Pong Tapioca Starch Flour Company Ltd., in Ratchaburi Province, Thailand, is showing ways to turn waste effluent into marketable products. It has done so in harmony with its environment and at a profit.

The first innovation taken by the company was to introduce a filtration system that separates the cellulosic pulp from the pentose sugar-rich wastewater generated by the cassava starch plants. The pulp is then dried on concrete slabs and sold to local feedmillers as cheap raw materials with good nutritive value for ruminant feed pellets. The second development was a collaboration with King Mongkut Institute of Technology, Thonburi (KMITT) to build an anaerobic digester to convert the organic-rich wastewater into biogas as fuel for the plant's boiler. After successfully testing a smaller, upflow anaerobic sludge blanket digester, the plant installed a scaled-up system capable of handling the entire effluent load from the plant. When operating at full capacity, the digester produces enough fuel gas to meet the plant's energy requirements. The third innovation was to use the nutrient-rich effluent from the digester to grow algae in oval raceways. *Spirulina* is in demand as a health food and feed ingredient because of its high protein (60-65 per cent) high fat (12-25 per cent), and rich content of the orange pigment, *Beta carotene*. Effluent discharged after the algae harvesting, has a BOD of 15-20 mg per litre, well within the EPA standards. The effluent canal supports a rich population of flora and fauna and is linked with the irrigation system serving the area.

The innovations of this company provide a lesson that protecting the environment makes business sense. The proactive stance of the company shows that the wastewater can be another resource for another system to gain value added effects. The waste that could have been dumped into the rivers or elsewhere is transformed through technological creativity into useful by – products which are safe, cost effective, and environmental friendly.

Source: Asian Media Information and Communication Centre 1997

Decentralized water management is also being encouraged particularly in large countries such as People's Republic of China and India. In People's Republic of China, the regional rights in the development and management of water resources have been reinforced and the city/provincial authorities have been assigned the appropriate power to enable direct management of water resources. In India, multi-disciplinary units in charge of preparing comprehensive master water plans have been established in some states.

Water pollution control legislation has set standards for water usage and effluents and has established regulatory units to oversee the implementation and enforcement of standards. For example, implementing rules and regulations made under the Water Code of the Philippines impose permits for surface and groundwater use, pollution control measures and restrictions on disposal of wastewater. Although most countries of the region have adopted water legislation, there is a need in some countries, especially those with the economies in transition, to review the existing water legislation in order to incorporate relevant provisions associated with the economic value of water and its rational use as well as protection of the aquatic environment, etc.

Stakeholder participation is also being increasingly promoted in the management of water resources. Experience with small scale locally managed irrigation projects in countries such as People's Republic of China, Indonesia, the Philippines, Sri Lanka and Thailand has been extremely positive especially where water users associations have taken the managerial responsibility. Similarly, by involving the benefiting communities in the development and maintenance of water supply, sanitation and pollution control facilities the cost of supply have been greatly reduced. The involvement of the private sector has also reduced the financial burden on the government in implementing water development programmes, particularly in People's Republic of China, Malaysia and Viet Nam, where French and British companies have been actively seeking to invest in water supply projects (ESCAP 1997).

Public awareness and participation is also receiving greater priority in water resources management. For example, the National Land Agency in Japan organizes various activities and initiates public campaigns aimed at enhancing the public understanding of the limited availability of water resources and the importance of sustainable water resources development.

Each year, since its inauguration in 1993, World Water Day (22nd March) has been increasingly observed throughout the region and awareness building activities carried out in connection with this Day are targeted at both urban and rural population. Non-governmental organizations are also taking initiatives on public awareness. Popular in the city of Bangkok is one such initiative called the "Magic Eyes Campaign" which has brought about a higher degree of social consciousness not only towards cleaner water but overall environmental protection.

B. Regional/International Level

1. Transboundary Water Cooperation

Although managing or resolving transboundary water resources problems and conflicts in the region is a slow and cumbersome process, some headway has been made particularly in Southeast and Northeast Asian subregions where integrated basin-wide management approaches have been promoted. Such an approach is beneficial to the management of a large number of transboundary water systems through cooperation in the formulation and implementation of development plans. For example, the Mekong River Commission is identifying and implementing various projects from the Indicative Plan for the development of land, water, and related resources in the Lower Mekong. A similar approach is being adopted in Northeast Asia, where the Tumen River Area Development Programme has fostered participation and cooperation between People's Republic of China, the Democratic People's Republic of Korea, and Russian Federation.

The Indus Basin water sharing accord between India and Pakistan, the "Water Sharing Treaty" between India and Bangladesh (Box 4.6), the India-Bhutan cooperation in hydropower development and India-Nepal cooperation in harnessing transboundary rivers are further positive examples of transboundary cooperation on water management in the region (South Asia Technical Advisory Committee 1999).

In Central Asia, the agreed management of transboundary waters in the Aral Sea Basin by Kazakhstan, Kyrgyzstan, Tajikistan, Turkmenistan and Uzbekistan is also being strengthened and enhanced through consultation and coordination. The 1994 Aral Sea Basin Programme is founded on an agreed Regional Water Resources Management Strategy, aimed at achieving sustainability of water resources in the Aral Sea Basin through increased water use efficiency and improved water quality.

2. Roles and Programmes of International Organizations

International organizations have played a range of roles in water resources management within the region through undertaking sustainable water resource development and management programmes as well as investment projects with related activities for irrigation and drainage, flood control, fisheries, hydropower, water supply, sanitation, urban drainage, inland navigation and port development.

The United Nations Development Programme (UNDP) promotes capacity building for sustainable water resources development. The World Meteorological Organization (WMO) supports assessments and forecasting of water resources potential, and helps the development of information systems for better planning and decision-making. The Economic and Social Commission for Asia and the Pacific (ESCAP) has carried out studies on water resources assessment, management, water pricing, investment promotion, dissemination of water-saving technologies and the involvement of women in water conservation and development. Moreover, ESCAP has been serving as the secretariat of the Inter-agency Sub-committee of Water for Asia and the Pacific, which was established in 1978. The United Nations Educational, Scientific and Cultural Organization (UNESCO) funds the International Hydrological Programme to monitor hydrological data at regional and national levels, whilst the Food and Agriculture Organization (FAO) focuses on irrigation and water for food production and has developed guidelines for water policy review and strategy formulation as well as pioneering water legislation in many countries (ADB 1998).

The United Nations Environment Programme (UNEP) focuses on environmental legislation, legal training and data collection on water quality. One notable programme is the Environmentally Sound Management of Inland Water (EMINWA), a comprehensive approach to management and development of freshwater resources on a basin-wide scale (UNEP Freshwater Programme 1995). The programme is designed to assist governments to integrate environmental considerations into the management and development of inland water resources with a view to reconciling conflicting interests and ensuring the regional development of water resources in harmony with the environment (David et al 1988; and Tolba 1988).

The ADB is assisting countries to meet the challenges of water problems through investment and assistance in effective policy formulation and implementation. The recent important milestone of ADB is a series of Regional Water Policy Consultations, which serve as the venue for in-house analysis and dialogue among policy stakeholders which include governments, private sector and non-government organizations (NGOs), external support agencies like United Nations and World Bank, international private sectors, international research institutes, and experts from the region and beyond. The consultation defined ADB's role in redirecting its water operations and targeting water sector in long-term investment, as well as catalyzing partnership and cooperation in the region (ADB 1998). On the financing side, the World Bank has played an important role by primarily investing in many countries, with recent focus on the operation and maintenance of irrigation systems ensuring water supplies and improving water quality.

The international professional organizations that are active in the water sector include the

---

Box 4.6 Ganges Water Sharing Treaty

The Farakka transboundary water dispute between India and Bangladesh has come to an end with the signing of the long-awaited Ganges Water Treaty Between India and Bangladesh. The treaty has recently been signed by the Prime Ministers of the two countries in New Delhi. Putting aside the legal parlance of the treaty, the agreement in general has enkindled a hope that the problems and sufferings that ensued from the upstream withdrawal will end soon.

The treaty provides a formula for sharing of the Ganges water from January 1 to May 31 in 10-day periods. According to the formula, when the flow is more than 75 000 cu sec (CFS), India's share will stand at 40 000 cu sec, Bangladesh receiving the rest. When the flow is between 70 000 and 75 000 cu sec, Bangladesh will receive 35 000 cu sec and India will receive the rest. When the water available at Farakka is 70 000 cu sec or less, India and Bangladesh will share equally. If the water availability at Farakka falls below 5 000 cu sec, the two countries will meet immediately to decide the shares. The sharing is based on the recorded average water availability in these 10-day periods from 1949-1988. The focal point of this treaty is to ensure 35 000 cu sec of water for Bangladesh in alternate 10-day periods during the least water available period (March 1 to May 10). However, the "guarantee clause" of the previous 1977 agreement in Bangladesh's favour has not been included in the treaty.

The Ganges water-sharing treaty has finally broken the deadlock in the water sharing negotiations between the two countries and created a favourable atmosphere for agreement on sharing of the water of 53 other common rivers. The treaty is a manifestation that cooperation for mutual benefit can be attained through negotiation and strong political will of the involved parties.

Source: IFCDR 1996

Table 4.5   Water Vision of the Subregions in Asia and the Pacific

Subregion	Water Vision	Main Principle
Aral Sea	Formulated	Improvement of health; sufficiency in food supply; guaranteed security against floods and droughts; secured shelter in winter (energy from hydropower for heating); safeguard of environment; and increase of wealth from efficient water use in industries, services and agriculture.
South Asia	Formulated (July 1999)	Poverty eradication and upliftment of living conditions of people to sustainable level of comforts, health and well-being through coordinated and integrated development and management of water resources.
The Russian Federation	Formulated (April, 1999)	All encompassing approaches to address the aspects of water availability and demand in Russian Federation focusing on the conventional water world, water crisis, and sustainable water world.
PR China	Formulated (May 1999)	Conventional way with appropriate technology and institution catching up with fast economic growth of the country. Control of catastrophies and water-related multiple disasters affecting food, health and society, and sustainable way with tapping alternative water resources and self-reliant water environment.
Australia	Formulated	Equitable sharing of the water resources; efficiency in water use and re-use; sustainable water resource management; governance and integrated national policy.
Southeast Asia	Formulated (November 1999)	Safe water for life: political commitment and institutional strength to sustainably manage its water resources for efficient use and equitable access and distribution.

Source: World Water Vision 1999

International Water Resources Association, the International Commission on Irrigation and Drainage, the International Commission on Large Dams, the International Water Users Association, and the International Water Supply Association. The International Water Management Institute (IWMI) in Colombo has recently broadened its mandate from irrigation to water management. Two new initiatives recently undertaken are the Global Water Partnership (GWP) and the World Water Council (WWC). Launched with the support of the World Bank, UNDP, and the Swedish International Development Agency (SIDA), GWP aims to build a network for sustainable water management that brings together developing countries and external support agencies, to direct programmes in the water sector more effectively and efficiently.

The World Water Council (WWC) founded in 1996 is a water policy think tank that draws on international agencies, governments, and private groups from both developed and developing countries, to raise awareness on the need for reforms at governmental level. An important undertaking of the WWC that would spell the future of water resources development, management and policy is the establishment of the World Commission on Water for the 21st Century with the support of the FAO, UNEP, UNDP, UNESCO, WHO, WMO and the World Bank.

The main task of the Commission is to guide the development of the "World Water Vision" – the long-term vision on water, life, and environment for the 21st Century that WWC is preparing. The primary objective of the vision is to develop a widely shared concept on the future actions required for tackling water issues globally and regionally. The visioning exercise, which was started in September 1998, is ongoing with a series of studies, regional consultations, and promotions. The first interim results were discussed at the 1999 World Water Day meeting in Cairo and the August 1999 Stockholm Water Symposium. The final results were presented at the 2nd World Water Forum and Ministerial Conference which was held from March 17-22, in the Hague, The Netherlands (World Water Council 1998). To date, various visions have been formulated at country and subregional levels in Asia and the Pacific; a selection of these is provided by Table 4.5.

CONCLUSIONS

Growing population, urbanization and economic development are putting great pressure on the quantity and quality of the region's freshwater, whilst massive withdrawals from rivers, lakes and underground reservoirs have led to an imbalance between supply and demand. Sectoral competition and conflicts have become critical. Due to excessive abstractions, the volume of water in some rivers and lakes has depleted while water tables in underground aquifers have sunk leading to land subsidence and salt water intrusion.

Unfortunately, the growing scarcity of water is accompanied by deteriorating water quality due to pollution and environmental degradation. Discharges of waste, sewage and effluents from domestic, industrial and agricultural sources have rendered water from many rivers, lakes and some aquifers unsuitable for human consumption. Environmental damage to aquatic ecosystem through loss of biodiversity, sedimentation and siltation have also resulted in big economic losses through loss of production and increased costs of control or remedial measures.

The main policy and management issues in the water sector in the region in recent years have been: the lack of adequate legislation including water rights or entitlements; fragmented and overlapping responsibilities in water management projects; lack of coordination in implementing water-related projects; ineffective water resource planning and management; insufficient political and public awareness; lack of public and stakeholders participation, and a general shortage of institutional capacity to meet the increasing needs in service delivery and resource management. Responses to these issues have been through both national and international actions and water policies and management measures undertaken in many countries of the region have focused on integrated and holistic approaches. Such integrated approaches have led to review or refocus of national water policies with a view to integrating the concepts and principles articulated in Agenda 21.

Common elements in the revised or new national water policies adopted include the integration of water resources development and management into national socio-economic development; assessment and monitoring of water resources; protection of water and associated resources; provision of safe drinking water supply and sanitation; conservation and sustainable use of water for food production and other economic activities; institutional and legislative development; and, undertaking measures to promote public participation. Underpinning each element are actions and programmes that have been implemented, to various degrees, by countries of the region.

The role of regional and international organizations in support of these efforts has been extremely important, particularly by providing technical and financial assistance in the conservation and efficient utilization of water, mitigating scarcity, restoring water quality and adopting appropriate policies and management approaches.

Despite the remarkable achievements of a number of transboundary groupings and individual member countries in recent years, significant challenges remain in the improvement of water resources management in the region.

Chapter Five

Destruction of mangroves for aquaculture is seriously damaging the marine environment in the region.

5

Coastal and Marine

INTRODUCTION
STATE OF THE COASTAL AND MARINE ENVIRONMENT
CAUSES OF COASTAL AND MARINE ENVIRONMENTAL PROBLEMS
POLICIES AND PROGRAMMES
CONCLUSION

CHAPTER FIVE

INTRODUCTION

The oceans and coastal resources of the Asian and Pacific Region are currently facing three separate, but interactive threats. The first is pollution from both land and sea based sources, which causes direct damage to specialized ecosystems (such as mangroves, coral reefs and seagrasses), whilst weakening the ability of marine plants and animals to survive fluctuations in their environmental conditions. The second is a direct threat to the biomass and ecological balance of the marine environment through overfishing and unsustainable extraction of resources and, the third, direct physical damage to coastal and marine ecosystems from urban and tourist related development.

This chapter examines the major environmental issues facing the coastal and marine environment of the region and highlights the current and planned response to these challenges.

STATE OF THE COASTAL AND MARINE ENVIRONMENT

A. Damage to Coastal Zones and Habitat

Direct physical loss of, or damage to, coastal habitats arise from a range of development activities. Dredging of harbours and shipping channels, the construction of embayment (harbours) and marinas, and the reclamation of coastal wetlands for development purposes have each had a profound effect upon the ecological resources of the estuarine and coastal systems of the region. For example, New Zealand has reclaimed more than 85 per cent of its wetlands, mostly to create new pastureland, but also for housing and industrial sites. Many of the remaining wetlands have been degraded by drainage, pollution, animal grazing and the introduction of new plant species (Government of New Zealand 1996).

In addition to the direct physical damage to a range of in-shore, littoral and shoreline habitats, development within the coastal zone has also affected the geomorphological processes associated with accretion and deposition of coastal sediments. In order to guard against local coastal erosion, high cost sea walls and groynes have been constructed with significant impacts on the wider patterns of erosion and deposition. For example, a government quarantine facility, built in the 1940s on Makaluva Island in Fiji, now lies beneath the sea, some 100 metres offshore. The construction of the coastal defences around the quarantine facility accelerated the erosional processes such that the island has shrunk in size (United Nations 1995).

Coastal erosion triggered by human activity is also evident in many other countries of the region. In Malaysia, for example, coastal erosion has affected every state of the country and by 1998, of a total shoreline of 4 809 km^2, some 1 400 km^2 or 29 per cent was eroded.

B. Pollution of Coastal and Marine Environment

1. Marine Pollution: Status and Trends

Coastal and marine water pollution has increased throughout the region, mainly due to domestic and industrial effluent discharges, atmospheric deposition, oil spills and other wastes and contaminants from shipping. Most of the pollutants entering the marine environment come from land based sources and comprise sand/silt, nutrients, toxic chemicals and oil. The suspended load (primarily silt) per km^2 of drainage basin in the region is three to eight times higher than the world average and contributes to the high turbidity of coastal waters. Two thirds of the world's total sediment transport to oceans occurs in South and East-Asia, due to a combination of active tectonics, heavy rainfall, steep slopes and erodible soils, disturbed by unsound agricultural and logging practices (UNEP 1999). These sediments impact upon not only the shallow inshore water habitats, but also on the wider oceanic ecosystem.

The urban and agricultural areas of the region produce significant quantities of organic wastes in such concentrations that the nutrient filtering mechanisms of the coastal zone are unable to neutralize their effects. Rivers running through Cambodia, People's Republic of China, Malaysia, Thailand and Viet Nam deliver at least 636 840 tonnes of nitrogen to coastal waters overlying the Sunda Shelf. Of these, People's Republic of China contributes at least 55 per cent, Viet Nam and Thailand, contribute 21 and 20 per cent, respectively (Talaue-McManus, L. 2000). The issues associated with the discharge of pollutants to inland waters are discussed in greater detail in Chapter 4.

Industry, and commercial agriculture, contaminates the flow of natural wastes with a wide range of materials that include persistent organic pesticides, heavy metals like mercury and lead, plastics of all kinds, and a cocktail of hazardous industrial chemicals. The relative contribution of various sources of oil pollution varies and comprehensive and strategic management initiatives are necessary in its monitoring and control (Box 5.1). The factors affecting oil pollution include population density, extent of shipping and mineral exploration, and the degree of industrialization of littoral

COASTAL AND MARINE

> **Box 5.1 Australia's National Plan to Combat Pollution of the Sea by Oil**
>
> Australia's National Plan to Combat Pollution of the Sea by Oil is managed by the Australian Maritime Safety Authority (AMSA) and funded by a levy on the shipping industry. The National Plan is a collaborative arrangement between AMSA, the States and Northern Territory governments, the shipping, oil and exploration industries and the Australian Marine Oil Spill Centre, at Corio Quay, Victoria. The Centre was established by the oil industry to assist in responding to major oil spills around the Australian coast and in adjacent areas where Australian-based companies operate.
>
> Under the Plan, pollution-response equipment is stockpiled at strategic ports and oil terminals, with a response capability for an oil spill of up to 10 000 tonnes. Whilst fully laden tankers typically carry 60 000 tonnes of oil, the result of most collisions is the rupturing of only one or two internal tanks, such that any oil spill is typically much less than the tankers' fully laden capacity. Furthermore, in most cases oil is lost progressively such that the amount of oil that needs to be managed in a spill increases over time. The Kirki oil spill, for example, happened over a two-week period. If a spill larger than 10 000 tonnes occurs, Australia may need to seek international assistance through arrangements under the international Oil Pollution Response and Cooperation Convention. Australia has concluded a memorandum of understanding with New Zealand under this Convention, by which assistance will be provided to each other in cases of pollution incidents in either country. Similar agreements are currently being negotiated with Papua New Guinea and Indonesia.
>
> Australia has been at the forefront of regional initiatives to protect the marine environment through the regulation of international navigation. In 1990, the Great Barrier Reef was the first area in the world designated as a 'Particularly Sensitive Area' by the International Maritime Organization requiring all vessels of more than 70 metres in length or those carrying oil, chemicals or liquefied gas to carry Australian-licensed pilots when using the designated routes within the Great Barrier Reef Marine Park (Zann 1995). In addition, Australia and Papua New Guinea are co-operating in the development and provision of specific preventive and response measures to protect the Torres Strait area from oil spills. In support of these two specific initiatives, Australia is currently upgrading existing navigational aids and charts and establishing protocols for the management of ship passages through the Torres Strait and the Great Barrier Reef areas.
>
> *Source:* CSIRO 1996 and Zann, L. 1995

countries. In the South China Sea, with the intensification of these factors absolute oil inputs are likely to increase. A study of the South China Sea (Talaue-McManus 2000) identified pollution hotspots; the locations of these are shown in Figure 5.1.

The quality of marine water is not monitored in many countries of the Asian and Pacific Region. Where monitored, the results are far from satisfactory. For example in People's Republic of China, a 1997 study found that only 19 per cent of China's coastal waters met Grade I water quality standards, and by 1999, this had reduced to 15 per cent, with the most severe sewage and agricultural pollution in coastal areas of the Pearl River Estuary. Of China's four major sea areas, the East China Sea was the most polluted, followed by the Bohai Sea, the Yellow Sea and the South China Sea. (Government of China 1997 and 1999).

As pollution intensifies, the destabilized coastal and estuarine systems undergo wild gyrations in population densities – some species dying off while others bloom in huge numbers.

2. *Blooms and Diebacks*

Over the last twenty years, toxic blooms (or "red tides") have become increasingly common, with major outbreaks in Australia, People's Republic of China, Japan, the Philippines, New Zealand and the Republic of Korea. Since 1986, for example,

China's State Oceanographic Administration reported five major episodes (with two in 1998), each affecting more than 500 square kilometres of coastal waters.

In 1992, red tides, caused by outbreaks of massive numbers of toxic dinoflagellates, occurred in New Zealand and caused massive contamination of seafood resources. Over 200 cases of food poisoning were reported with symptoms ranging from diarrhoea, nausea, vomiting, muscular aches and weakness to dizziness, loss of memory, tingling, numbness and respiratory problems. (Government of New Zealand 1996). In addition to the human health impacts, toxic blooms have caused tremendous economic losses to countries in the region. For example, in 1997, a red tide outbreak in Kerala, India forced the closure of shellfish beds, leaving nearly 1 000 families without work. In Hong Kong, China, toxic blooms wiped out US$10 million worth of fish in 1997 and another US$32 million worth of high value fish in 1998 from its mariculture industry. In the Republic of Korea, 126 incidences of red tides were reported in 1996 alone, with losses to aquaculture estimated at US$ 10 million (Table 5.1).

It may also be noted that when the blooms of algae use all available nutrients and die, the organic enrichment of sediments cause long-term changes in benthic habitats, populations, and community structure. Increased sedimentation and nutrients from catchments have been linked with massive dieback of seagrasses in many areas. New South Wales, for example, lost half of the *Zostera* seagrass in its estuaries and the seagrass die-off in Queensland resulted in mortalities of endangered dugongs (Zann 1995).

Figure 5.1 Distribution of Pollution "Hot Spots" In South China Sea

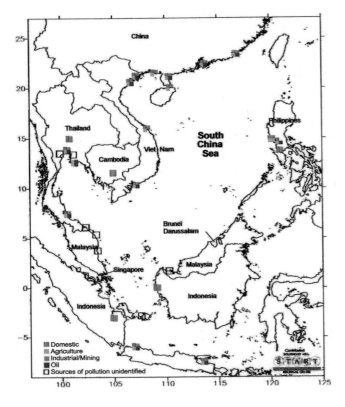

Source: Talaue-McManus 2000

Table 5.1 Economic Losses from Red Tides in Fisheries and Aquaculture Facilities in Selected Countries of the Region

Country	Year	Species	Loss (million US dollars)
Japan	1972	Yellowtail	47
	1977	Yellowtail	20
	1978	Yellowtail	22
	1987	Yellowtail	15
Rep. of Korea	1978	Oyster	4.6
	1981	Oyster	>60
	1991-92		133
	1996		10
Hong Kong, China	1998	Farmed fish	32

Source: Brown, L. R. et al eds. 1999

3. *Increased Dissolved Carbon Dioxide*

Observations showed that levels of dissolved carbon dioxide were abnormally elevated in seawater in the early 1970's, although no ill effects were detected. However, in 1998, researchers found that elevated dissolved CO_2 was slowing down calcium carbonate deposition in coral communities and possibly, by inference, in other marine organisms that secrete calcium carbonate in the formation of shells and exoskeletons. Elevated dissolved CO_2 was raising the acidity of seawater and thereby making carbonate formation more difficult.

Currently, the oceans absorb about two billion tonnes of carbon dioxide a year, about one third of the carbon dioxide produced by the burning of fossil fuels. If the normal process of binding carbon into calcium carbonate is impaired, less carbon dioxide will be removed from the atmosphere causing an even faster rise in both atmospheric and dissolved CO_2. (Langdon et al 1999).

C. **Specialized Ecosystem and Resources Status and Trends**

1. *Mangroves*

More than 40 per cent of the world's currently estimated 18 million hectares of mangrove forest occur in South and Southeast Asia. These subregions also have the highest diversity of mangrove species in the world. The South Pacific subregion provide a further 2 million ha of mangroves or some 10 per cent of the total mangrove areas in the world.

The resource needs of the region's growing population have exerted considerable pressure on the mangrove ecosystems. Large areas of mangrove have been removed for industrial, residential and leisure developments and, in particular, for the establishment of ponds for fish and prawns aquaculture. It is estimated that over 60 per cent of Asia's mangrove forests have already been converted to aquaculture ponds (Figure 5.2) equivalent to losses of 11 million ha. More than 3 million ha of mangroves have been converted to aquaculture ponds in Southeast Asia alone.

Agriculture has also affected mangroves both directly through landtake and indirectly through freshwater diversion for irrigation and through the addition of agricultural residues in the run-off. Freshwater interceptions for agricultural schemes have severely affected mangroves in the Indus delta of Pakistan and Ganges delta in the western part of Sunderbans.

2. *Coral Reefs*

Although coral reefs occupy less than one quarter of one per cent of the marine environment, they are home to more than a quarter of all known

Figure 5.2 Estimated Loss of Original Mangrove Areas in Asia and the Pacific

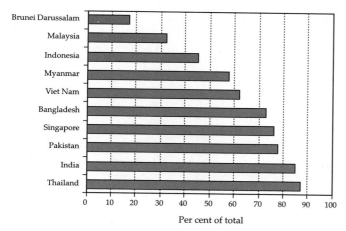

Source: Global Aquaculture Alliance 1998

Figure 5.3 Distribution of Coral Reefs in the Asian and Pacific Region

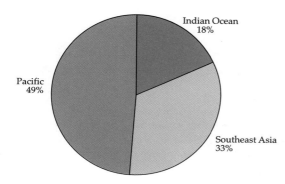

Source: Bryant, D. et al 1998

marine fish species and have been termed the "rainforests of the marine world" (Bryant, D. et al 1998). Coral reef systems also provide a useful indicator of marine biodiversity and ecological health (Box 5.2). About four fifths of the world coral reefs are in the Asian and Pacific Region, approximately half of these are in the Pacific, one third in Southeast Asia and the remainder in South Asia (Figure 5.3). Coral reefs and their associated plants and animals provide human populations with seafood, medicine and other products. In developing countries, coral reefs contribute about one quarter of the total fish catch and provide food, according to one estimate, for one billion people in Asia alone (Jameson S.C. et al 1995 and Hinrichsen D. 1997). Coral reefs also act as buffer zone to break the intensity of wave action and impact of storms and provide recreational resources to the tourism industry.

However, in the Asian and Pacific Region, coral reefs are threatened by a range of human activities including coastal development, over exploitation and destructive fishing practices, as well as land and marine based pollution. An extensive study by the WRI has categorized the threat to the coral reefs of the region (Table 5.2). In 1998, coral bleaching, which is a phenomenon associated with global warming killed large areas of coral reef within the region (see Chapter 16, Box 16.2).

The coral reefs of Indonesia and the Philippines are noted for their extraordinarily high levels of diversity, each containing at least 2 500 species of fish, although, at present, only 30 per cent of these reefs are in good or excellent condition. Due to the richness of these reef areas, the coastal zone policy and management decisions made by these two countries will have a major impact on the global heritage of coral reef diversity (Bryant, D. et al 1998).

In comparative terms, the reefs of the South Pacific subregion are under less immediate threat than those of Southeast Asia (WRI 1999j). Forty-one per cent of the Pacific reefs are classified as threatened, and just 10 per cent face a high risk. Those near population centres face significant human pressures including the reef communities off southeastern Papua New Guinea, the Solomon Islands, Vanuatu, and Fiji. Fiji's reefs are an important tourist draw and, according to a 1992 estimate, a major source of food for local people, generating close to $200 million annually in fisheries and tourism revenues alone (Bryant, D. et al 1998).

3. *Seagrasses*

Seagrasses are common throughout the tropical and temperate coastal waters of the Asian and the Pacific Region. Often associated with mangrove habitats in coastal waters and with coral reefs in deeper waters, seagrasses perform the crucial ecological function of trapping fine sediments that remain suspended after passing through estuarine and mangrove areas. Typically, therefore, seagrass beds are depositional areas, providing a mechanism for clearing waters of sediment – particularly important to the survival of sun-loving corals.

Seagrasses provide important habitats and food sources for a range of marine fauna including commercially important species, such as the tiger prawns of Northeast Australia. The generic richness of seagrass beds is centred in the Indo-West Pacific Region (Heck and McCoy 1978), whilst species diversity is highest in the area defined by Indonesia, Borneo, Papua New Guinea and northern Australia. Northeast and Southeast Asia harbours the second highest number of sea grass species at 20 of the 50-recorded species worldwide (Fortes 1994 and 1995; Sudara et al 1994).

Table 5.2 Status of the Coral Reef in the Asian and Pacific Region

Region	Reef Area in Km² Total	By Threat Category[a] Low	Medium	High	Percentages (%) Low	Medium	High	Coastal Population Density[b] (pp/km²)	Marine Protected Areas[c] Number	Area (km²)
Indian Ocean	36 100	16 600	10 500	9 000	46	29	25	135	66	15 100
Southeast Asia	68 100	12 300	18 000	37 800	18	26	56	128	57	36 263
Pacific	108 000	63 500	33 900	10 600	59	31	10	98	92	372 809
Global Total	255 300	108 400	79 000	67 900	42	31	27	101	367	475 298
Selected Country and Geographic Grouping Statistics										
Region	Total	Low	Medium	High	Low	Medium	High	(pp/km²)	Number	Area (km²)
Australia	48 000	33 700	13 700	600	70	29	1	12	12	374 967
Fiji	10 000	3 300	4 800	1 900	33	48	19	91	1	1
French Polynesia	6 000	4 900	1 100	0	82	18	0	38	1	124
India	6 000	1 400	500	4 100	23	8	68	412	2	288
Indonesia	42 000	7 000	14 000	21 000	17	33	50	93	26	30 405
Lesser Antilles	1 500	0	300	1 200	0	20	80	159	2	253
Maldives	9 000	7 900	1 100	0	88	12	0	n.a.	n.a.	n.a.
Marshall Islands	6 000	5 800	200	0	97	3	0	n.a.	2	163
New Caledonia	6 000	5 000	800	200	83	13	3	6	5	530
Papua New Guinea	12 000	6 000	4 500	1 500	50	38	13	7	8	2 149
Philippines	13 000	50	1 900	11 050	0	15	85	174	12	458
Solomon Islands	6 000	3 000	2 500	500	50	42	8	8	0	0

Source: Bryant, D. et al 1998

[a] Reef Area Estimates by Region and Threat Category (sq km) and percentages
Reef area estimates are based on WCMC's dataset Shallow Coral Reefs of the World and Spaiding and Grenfell (1997). Estimates of shallow reef area for Australia, Indonesia and the Philippines are significantly smaller than other published estimates.

[b] Average Coastal Population Density (pp/sq km), Statistics are for populated areas within 60 kilometres of the coastline. Population data come from Gridded Population of the World data set from the National Centre for Geographic Information and Analysis-Global Demography Project.
Data are unavailable for some small island areas.

[c] Marine Protected Areas (Number and Area Estimates)
Marine protected area counts and area estimates are summaries of the WCMC dataset Marine Protected Areas of the World, and are incomplete for some countries. Area statistics for protected sites are for the entire protected area, which include non-reef.

Box 5.2 REEF CHECK – A Global Coral Reef Monitoring System

In 1997-1998, the Institute for Environment and Sustainable Development of the Hong Kong University of Science and Technology, designed and implemented the REEF CHECK programme as a means of gaining a global assessment of the health of surviving coral reef systems.

Co-ordinating the efforts of marine scientists from more than 40 countries, REEF CHECK was able to derive snapshot assessments of the biodiversity and state of health of over 400 reef systems. The results also provided a basis for tracking changes to coral reef systems (at the individual reef level or at national, regional and global scales) in response to specific events, such as the 1998 coral bleaching (see Chapter 16, Box 16.2), or processes, such as overfishing or coral mining. REEF CHECK was very effective in building up community support for the conservation and management of reefs and many user groups, including SCUBA diving clubs and dedicated REEF CHECK associates, are requesting advice on how to use the method for repeated monitoring of reef health.

The methods employed proved to be flexible and robust, allowing modification for specific local circumstances, such as the addition of specific indicator organisms of local economic or social significance. For many areas of the world, the ideal monitoring programme will include a large number of broad-brush surveys carried out by the local community and using standardized methods, such as those developed for REEF CHECK, as well as a smaller number of more focused, specialized surveys using methods such as those in GCRMN's Survey Manual for Tropical Marine Resources.

Participants in the REEF CHECK/GCRMN network can compare their results locally, regionally and globally via the Internet. By increasing the number of sites surveyed using a standard methodology, there is a proportional increase in the chance of detecting subtle changes in the health of reef systems at the local, regional and international levels.

Source: Hodgson, G. 1999

It has been estimated that between 20 and 25 per cent of seagrass areas in Indonesia, Malaysia, Philippines and Thailand have been damaged by a combination of coastal development, elevated sedimentation, destructive fishing methods and land based pollution, thermal discharge, petroleum product spills, dredge and fill operations.

D. Food Resources

1. Fisheries

Fish catches in the Asian and Pacific Region continue to increase. The scale and range of the region's fisheries vary significantly with countries, such as People's Republic of China, Japan, Thailand, Indonesia, Democratic People's Republic of Korea and the Republic of Korea, engaged in large-scale industrial fishing, often venturing to distant waters to catch fish, whilst the small island states and the highly populated, least developed Asian states, concentrate on local food production from small (but numerous) coastal fisheries.

The comparison between estimated potential and average landing from various fishing grounds in the region by FAO is given in Table 5.3. Of all the fisheries in the world, the Indian Ocean fishery is believed to have the most potential for future development, but the data are unreliable and may not have taken into account natural population fluctuations (FAO 1997).

(a) The Eastern Indian Ocean

The Eastern Indian Ocean extends from the Bay of Bengal in the north, the Andaman Sea and northern part of the Malacca Straits in the east, to the waters around the west and south of Australia. The main fishing areas are on the continental shelves of the Bays of Bengal and Martaban and the narrower shelf areas on the western and southern sides of Indonesia and Australia. Knowledge of fish stocks is generally poor and management actions taken have usually been on an *ad hoc* basis, in most cases with little or no scientific rationale. India, Indonesia, Malaysia, Myanmar and Thailand accounted for some 90 per cent of the total catch in 1994. Whilst Australia caught only 3 per cent of the total by weight, the economic value of the Australian catch represented a much higher proportion.

Most of the catch from the coastal fisheries was used for local consumption, whilst shrimp and tuna were the main export commodities. Over exploitation of shrimp resources in coastal waters reduced the amount of exports from capture fisheries, although there is a continuing trend towards the exporting of shrimp from the aquaculture sector in almost all countries in the region. While the majority of tuna catches were from coastal fisheries, these were supplemented by offshore catches of skipjack and yellowfin tuna.

(b) The Western Indian Ocean

The Western Indian Ocean offers considerable potential for fisheries development, although lack of data on fish stocks and on the current levels of fishery activity are hampering the management of the sector. Rapid shifts in productivity (in part associated with the cycles of the monsoon seasons) and fluctuations in phytoplankton productivity are other factors that hinder a clear understanding of the long term potential of the fishery industry.

However, given the scarcity of alternative employment, fishing intensity is expected to remain high, increasing whenever the catch rates and economic conditions will allow.

(c) Northwest Pacific Fisheries

The Northwest Pacific is the second most productive fishery area in the world and is endowed with a broad continental shelf and high natural levels

Table 5.3 Comparison Between Estimated Fisheries Potentials and Average Landings of the last 5 years (1990-1994) in Million Tonnes in Various Parts of the Marine Environment in Asia and the Pacific

Marine Areas	Estimated Potential (A)	Year Potential Reached	Subjective Degree of Reliability[1]	Landings 1990-94 (B)	Difference (A-B)	Status[2]
East Indian	10	2037	Unreliable	3	7	I
West Indian	13	2051	Unreliable	4	9	I
Northwest Pacific	26	1998	**	24	2	I
Southwest Pacific	1	1991	**	1	0	O
Central-Western Pacific	11	2003	**	8	3	I
WORLD	82	1999	**	83	-1	

Source: FAO 1997a

[1] **Reasonably reliable regression
[2] Overfished, Increasing (when rate of increase = zero)

of nutrients. Although the sector has been heavily fished for many centuries, modern industrial fishing has had a significant impact on the marine resources of the area.

Currently, the largest fish catches in the Northwest Pacific are Alaskan pollock, with the largest portion being taken by the Russian fishery, which landed 1.75 million tonnes in 1994. All the major stocks are believed to be currently at substantially lower biomass levels than existed in the 1980s and there are forecasts of catch trends continuing downwards for several years into the future.

The fishing capacity of the Northwest Pacific fleets continued to rise, despite falling catches. China's fishery capacity grew rapidly between 1980 and 1997, with their fleet of decked fishing vessels increasing from about 60 000 to 460 000 vessels and the numbers of full time fishermen increasing from 2.9 million to 5.3 million. The amount of fish caught per unit of effort declined over the same period by a factor of 3. Other indications of overfishing, especially in coastal areas of the East China and Yellow Seas, was a shift in catches from large high-valued fish to lower-valued smaller fishes, from demersal and pelagic predator fishes to pelagic plankton-feeding fishes and from mature fish to immature fish (FAO 1997a).

(d) Central Western Pacific

The Central Western Pacific extends from the coast of Southeast Asia down to north and east Australia and further eastwards to the smaller island countries of the South Pacific. Much of the continental shelf in this area lies within the Exclusive Economic Zones of Southeast Asian countries and is rich in demersal resources, including penaeid shrimps, and small pelagic resources, while the oceanic waters have rich tuna resources.

Total catches in the area have increased almost continuously since 1950, although the rate of increase slowed in the 1990s. Unlike temperate fisheries where single species dominated the catch, tropical fisheries were a composite of many species. In 1994 seven species groups accounted for 87 per cent of the total: miscellaneous fishes, tunas, jacks, herrings, redfishes, mackerels, shrimps. The multispecies nature of tropical coastal resources is reflected by the high proportion of the miscellaneous group and the relatively even spread of catches between the other groups.

Two studies in the South China Sea indicate that most of the conventional small pelagic species are already fully exploited. The first study (Table 5.4) covered the period from 1978 to 1993, during which peak years were identified. It noted that after 1987, most of the 12 small pelagic fisheries had reached full levels of exploitation. The second study (Table 5.5), on a habitat division basis, showed that only a few sections of the shelf can sustain further expansion.

Countries in the region have introduced various conventional management measures such as closed seasons, closed areas or zoning, mesh-size

Table 5.4   Small Pelagic Fisheries in the South China Sea, 1978-1993

Group	Peak landings (mt)	Peak year
Round scads	596 000	1991
Selar scads	229 000	1990
Jacks, cavalla and trevallies	147 000	1993
Indian mackerel	357 000	1992
Indo-Pacific mackerel	212 000	1993
Spanish mackerel	114 000	1993
Kawakawa	283 000	1992
Frigate and bullet tunas	128 000	1992
Sardines	716 000	1993
Anchovies	419 000	1993

Source: Yanagawa, H. 1997

Table 5.5   Fisheries Potential of the South China Sea

Subdivision	Area (10^3/km^2)	Primary Production (tonnes/km^2/year)	Potential catch 10^3/tonne/year	Actual catch 10^3/tonne/year
Shallow areas to 10 m	172	3 650	No estimate but fully exploited	1 046
Reef flats and seagrasses to 10 m	21	4 023	No estimate but fully exploited	275
Gulf of Thailand to 50 m	133	3 650	No estimate but fully exploited	1 242
Viet Nam & PR China shelf to 50 m	280	3 003	1 860	453
Northwest Philippines. to 10 m	28	913	No estimate	315
Bornean shelf to 10 m	144	913	257	105
Southwest shelf to 10 m	112	2 433	No estimate but fully exploited	962
Coral reefs, 10-50 m	77	2 766	295	291
Deep shelf 50-200 m	928	730	1 688	176
Open ocean 200-400 m	1 605	400	1 686	80
Total South China Sea	3 500	Mean = 1 143	5 786	4 945

Source: Pauly, D. and V. Christensen 1993

regulations etc. to control fish decline. However, fishing pressure kept increasing despite attempts by some countries to reduce fishing pressure in coastal areas. The use of explosives to capture fish caused widespread damage to coral reefs. Fishing with cyanide for live food fish from the coral reefs resulted in severe reductions of juvenile and prey fish on coral reefs as well as over exploitation of target species.

(e) The Pacific Islands

There are three main types of fisheries in the Pacific islands: industrial fisheries (mainly tuna); coastal fisheries for export, mother-of-pearl shells; and coastal fisheries for domestic consumption.

Tuna is the target of the only significant industrial fisheries (purse seine, longline, pole and line, and troll) off small islands in the South Pacific. They are especially abundant in the Exclusive Economic Zones (EEZs) of Papua New Guinea, the Solomon Islands and Kiribati, but are also taken off the other Pacific island nations. Large-scale fishing is carried out by distant water fishing nations like People's Republic of China, Indonesia, Japan, Republic of Korea, Philippines, and the USA, which pay fees to gain access to South Pacific islands' EEZs. In the 1970s and 1980s, few Pacific island nations were fishing for cannery-quality skipjack and albacore. Recently, their participation in tuna fishing has increased with the advent of small-scale longline fisheries for sashimi-quality yellowfin and bigeye tuna. These fisheries operate mainly off Federated States of Micronesia, Fiji, French Polynesia, Guam, Marshall Islands, New Caledonia, Palau, Samoa and Tonga, although these island fleets still take only 6.5 per cent of the weight of tuna caught in the sector.

About 80 per cent of fish captured in the Pacific islands, estimated at around 100 000 tonnes annually, is consumed or bartered locally. Some remote atolls have a per-capita fishery product consumption of over 250 kg annually. Most Pacific islanders live within the coastal zone and in rural areas, and many people fish, mostly for subsistence purposes. A great variety of marine organisms are consumed. For example, over 100 species of finfish and 50 species of invertebrates are included in the fish market statistics in Fiji, and the number of species consumed in the subsistence fishery is nearly twice this.

(f) Southwest Pacific

The fisheries area of the Southwest Pacific, including New Zealand and Southern Australian waters, has been heavily fished since the 1970s and the stocks declined fairly rapidly in response. Thirty-seven species were under quota management in the 1995/96 fishery season, which has been success in curbing some of the overfishing of the past.

The orange roughly fishery provides a typical example: in 1979, the first catches of orange roughly (5 000 tonnes) were reported in New Zealand and by the early 1980s catches had climbed to around 40 000 tonnes before peaking, in 1990, at around 90 000 tonnes. Since this peak, catches have declined by over 50 per cent and catches in 1993/94 were back at around the 40 000 tonnes level. Stocks of orange roughly continue to decline, but this decline has been offset to some extent by the continuing discovery of new stocks.

Other species have shown similar drastic declines. For example, landings of greenback horse mackerel dropped by 90 per cent between 1991 and 1992 and catches of snoek declined by 40 per cent from their 1993 peak and by 1996 had declined to their lowest level since 1979. Harvests of squid started in 1972, and catches of Wellington flying squid collapsed in 1981 (from 63 000 to 1 000 tonnes) before picking up again and increasing, albeit with large variability, reaching a new peak of 58 000 tonnes in 1994.

E. Aquaculture

The Asian and Pacific Region accounts for 87 per cent of the total world production of marine aquaculture. The top 10 Asian aquaculture producers are People's Republic of China, India, Japan, Republic of Korea, the Philippines, Indonesia, Thailand (also the top seven producers in the world), Bangladesh, Viet Nam and Democratic Peoples Republic of Korea. In 1997, the production of 22 Asian countries and territories alone was 30.7 million metric tonnes (mt) valued at US$ 37.7 billion, an increase of 11.2 per cent and 10.6 per cent in weight and value respectively compared to 1995 when Asia accounted for 90 per cent of world aquaculture production. By comparison, figures for the whole of Asia show that aquaculture production in the region has been growing more than 4 times faster than landing from capture fisheries with aquaculture's share of total fisheries landing in the region increasing by nearly two fold from 20.7 per cent in 1984 to 38 per cent in 1995 (FAO 1997).

In terms of species, freshwater finfish, in particular Chinese and Indian carp, account for the greatest share (42 per cent) of total aquaculture production. Aquatic plants production is valued at nearly US$ 5 billion, 70 per cent of which comes from People's Republic of China. Successful hatchery operations is key factor in the rapid production growth of aquaculture.

While finfish make up almost the total volume of freshwater aquaculture, they represent less than 10 per cent of salt-water aquaculture. Seaweed, especially kelp, was the most commonly grown marine organism, followed by oysters, carp, and scallops. Oysters were grown on wooden or net structures in shallow, estuarine tidal flats and were most valuable in temperate climates. Australia and New Zealand have extensive oyster farms and have also been increasingly successful in farming finfish, especially salmon.

In terms of value, the most important aquaculture species was the giant tiger prawn, worth nearly US$ 4 billion in 1996, followed by oysters at slightly over US$ 3 billion. Prawns and shrimps are especially important to Asian countries, as they are tropical species and are grown in large tidal ponds, generally excavated in mangrove areas. With a 1997 output of some 175 000 tonnes, Thailand continued to be the world's main supplier of cultured shrimp.

Mangrove clearance for shrimp culture development has been discussed earlier. The additional environmental concern associated with aquaculture is the potential hazard posed by the accidental release of exotic species (particularly predator species) or the spread of diseases from the aquaculture facility to the surrounding natural environment.

F. Other Marine Resources

1. Mineral

More than 4 billion tonnes of oil and 5.8 trillion cubic metres of natural gas reserves have been found on the continental shelves of the Asian and Pacific Region. Offshore oil and gas production is developing rapidly on the west coast of India, the Gulf of Tongking, the Gulf of Thailand, East of Malaysia, West of Borneo and Palawan, West of Japan, in the Celebes Sea, off the North West and the South coast of Australia, the West coast of New Zealand and Papua New Guinea. Mining activities in the coastal zone include the extraction of sands, gravel and rock, whilst each year more than 6 millions tonnes of salt is extracted from seawater in the region.

The vast quantities of sand, gravel and rock quarried from the region's coastal areas are used for infilling, road building and as aggregate in construction concrete. Coral mining represents one of the few sources of building materials in the smaller islands of the region. In Tahiti (French Polynesia), for example, 3.2 million tonnes of coral were taken from the fringing reef during dredging and filling operations for the airport and the port (Gabriel et al, 1995). Elsewhere, corals are mined for the manufacture of agricultural and construction lime.

2. Tourism

The region's coastal areas are one of the major attractions of the tourism industry, which is the fastest growing sector in the regional economy. Many countries in the region have sought to exploit the tourist potential of their coastal areas through the development of seaside resorts, sport fishing and scuba diving capacity, sailing, whale watching and other ocean recreational activities. Tourism developers and governments have been quick to recognize the importance of preserving the major attractions of the coastal areas, including beautiful beaches, vibrant coral reefs, lively fisheries and natural scenic splendour. Concepts of "eco-tourism" and "environmentally sustainable tourism" have been the focus of much debate within the international tourism industry and, in a number of cases, have led to formal partnerships between tourism and conservation interests to protect ecologically valuable and interesting sites. Tourism, for example, is often cited as a justification for the creation of parks and reserves, compensating local people and fishing communities for the loss of potential earnings and promoting sustainable management practices.

However, "eco-tourism" remains a relatively small, niche market (albeit one with significant growth potential) and the demands of the mass tourism market within the region (which primarily comprises intra-regional holidaymakers) remains driven by cost and by a demand for comfort levels that can only be provided through significant resource consumption and infrastructure provision.

CAUSES OF COASTAL AND MARINE ENVIRONMENTAL PROBLEMS

A. Disruption and Modification of Habitats

1. Physical Habitat

Physical modifications to habitats by natural forces or by human influence threaten the physical integrity of many coastal ecosystems in the region. Modification may result from: major storms or earthquakes; filling of intertidal or subtidal habitats during coastal development; loss of tidal wetlands, submerged aquatic vegetation or coral reefs due to a decline in water quality or changes in sedimentation; or changes in the hydrodynamics of coastal systems. The human activities that contribute to the loss and degradation of mangrove, coral reefs and seagrass habitats were discussed earlier.

More subtle changes, such as the increasing plastic burden on the ocean floor, can also damage coastal habitats. While some of these modifications are reversible over time (by clean-ups, revegetation of submerged aquatic vegetation or restoration of salinity conditions), the likelihood of recovery for many modified habitats is uncertain. Modification of shallow water habitats, including coral and other reefs, wetlands, and seagrass beds, pose perhaps the greatest threat to the biological diversity of marine and other aquatic organisms and can have significant consequences on the production of species that depend on these habitats for shelter or food at critical life stages.

2. Hydrologic and Hydrodynamic Disruption

The hydrology of watersheds draining to the coast has been significantly altered as a result of landscape changes, dredging and damming, consumptive water uses and diversion to other drainage basins. Such hydrological changes can alter salinity patterns and circulation within coastal systems and the delivery of nutrients, toxicants, and sediment to the coast by enhancing their concentrations.

The consequences of such changes may be profound. For example, an increase in the supply of fluvial sediments as a result of land clearing, mining or agricultural practices caused decreased light availability and the smothering or shoaling of benthic habitats in Papua New Guinea, Fiji, New Caledonia, New Zealand, Australia, and the Philippines.

B. Land and Sea Based Pollutants

1. Land Based Sources

Much of the pollution affecting the marine environment derives from land-based human activities and enters the oceans and coastal zones of the region as either direct discharge, via the outflow of the region's rivers or through atmospheric deposition. Table 5.6 shows fluxes from rivers in seven selected countries bordering the South China Sea. Fluxes, obtained by multiplying average concentrations with annual discharge rates, indicate the amount of material conveyed by river systems to the sea, as a combination of load from all sources (agricultural, domestic and industrial).

Land based sources also contribute litter, plastics pathogens and hazardous waters including pesticides to the coastal and marine water. For example, the increased use of chemicals in agriculture is leading to the transport of an estimated 1 800 tonnes of pesticides into the Bay of Bengal where they reappear as toxic residues in finfish and shellfish (Holmgren 1994). The liberal use of agricultural fertilizers, on the other hand, enhances the productivity of coastal waters, favouring nuisance organisms such as phytoplankton species that cause red tides and other similar problems.

Litter, especially plastics, is also a major problem in the coastal and marine environment and may endure for decades once submerged. Pathogens enter the natural environment, both from inappropriately managed hospital wastes and from aquaculture. Surface pollutants from atmospheric deposits are also impacting fish eggs and plankton species which occupy surface waters, in addition to increasing the level of UV-B penetration. Aquaculture has also contributed to the discharge of nutrients and other wastes, introduction of exotic species, the use of chemicals such as pesticides, antibiotics and

Table 5.6 Pollutant Fluxes from Rivers of Selected Countries to the South China Sea

Country/River	Catchment Area (km^2)	Annual discharge (km^3)	Biological Oxygen Demand (BOD) (t/y)	Total Nitrogen (IP) (t/y)	Total Phosphorus (t/y)	Total Suspended Solid (t/y)	Oil (t/y)
Cambodia							
Tonle Sap Lake-River System	69 355	36.45	6 022	1 084	303	13 250	No data
Coastal rivers	13 406	21.79	No data	No data	No data	No data	No data
Mekong River, Cambodia Section	72 060	128.38	4 964	894	255	10 950	No data
PR China							
Guangdong:							
Han, Rong, Pearl, Moyang, Jian	488 802	422.2	566 385	(340 050)	(3 768)	(58 531 000)	9 698
Quanqxi:							
Nanliu, Qing, Maoling	14 051	24.9	57 668	(8 602)	(507)	No data	823
Hainan:							
Nandu, Changhua, Wanquanhe	15 865	31	140	No data	No data	No data	368
Thailand							
Central, Eastern, Southern rivers	320 553	144.2	299 224	130 044	7 137	12 587	No data
Total South China Sea for continental countries			1 015 936	636 840	58 202	58 642 827	

Source: Talaue-McManus, L. 2000

hormones. In addition, there has been an overall deterioration of water quality and hindrance to access posed by extensive pond systems (Barg 1992).

2. *Pollution from sea based activities*

The sources of marine pollution from sea-based activities are related to fishing, recreational boating, marine transportation and offshore mineral exploration and production activities. Accidental oil spills have been frequently reported along oil transport routes or at the points of discharge and loading for oil carriers. In the Straits of Malacca alone, 490 shipping accidents were reported during the period from 1988 to 1992, resulting in a considerable amount of oil spillage (*Straits Times* 1993). Although accurate data for the total amount of oil spills are not available, their frequency and distribution has led to the development of strict control regulations in many countries of the region (see Figure 5.4)

Organotin antifouling, which is used for the keels of ocean-going vessels, is especially toxic to marine molluscs and their larvae. Unlike copper antifouling, organotin toxins become trapped in biological food chains and cause cumulative pollution of harbours and marinas. In consequence, organotin was banned by the IMO in 1998. Vessels also require regular painting and utilize a wide range of highly toxic paints, paint removers, solvents, degreasers, and other compounds.

C. Global Climate Change

A rise in sea temperature has already been linked to the extensive bleaching of coral reefs in the Pacific and Indian Oceans. A more detailed discussion on global climate change, and its implications for the ecosystems of the region, is provided in Chapter 6.

D. Unsustainable Exploitation of Resources

1. *Overfishing:*

Overfishing can cause serious, long-term damage to fish resources and the targeting of particular species can disrupt the ecological balance, depleting the prey of other species and reducing populations of top predators.

Attempts to control overfishing result in considerable economic conflict as livelihoods are at stake. In New Zealand, for example, government attempts to reduce inshore commercial snapper fisheries were immediately challenged in court-leading to prolonged discussions and high legal costs for the government (Box 5.3). An FAO investigation (FAO 1998) revealed that in spite of the overexploitation of fisheries resources, most marine capture fisheries remain economically viable, generating sufficient income to cover costs-including allowances for depreciation as well as the opportunity cost of capital, with adequate levels of remuneration to the owners and crews and a surplus remaining for reinvestment.

Subsidies provided by governments have sometimes contributed to overfishing. Diesel fuel tax exemptions are common fishing subsidies in the Region. In Australia, and Japan, diesel fuel used in fishing and shipping is exempt from standard fuel taxes. However, a FAO study (1998) found that the number of subsidies in developing countries has been greatly reduced in recent years and the remaining subsidies are for offshore fishing, artisanal fisheries and fisheries co-operatives as well as for fishing operations in remote and underdeveloped areas.

2. *Poor Integration and Coordination of Stakeholders*

Government sectors that interact with fisheries were (and remain) largely excluded from the fisheries development process. There is, for example little, if any, communication between fisheries agencies and tourism (sport fishing, diving, resorts, parks), environment (parks and habitat protection), planning (macroeconomics and finance), agriculture and forestry (responsible for water siltation and

Figure 5.4 High Risk Areas for Oil Pollution in the South China Sea

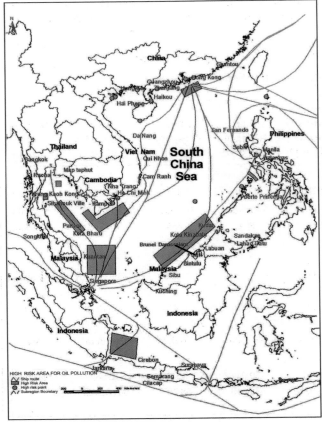

Source: Talaue-McManus 2000

> **Box 5.3 The Evolution of Fisheries Management in New Zealand**
>
> Capacity problems in New Zealand's inshore fisheries began to manifest themselves in the 1960s. Local fishermen complained that the government was licensing foreign offshore fishing vessels, whilst restricting access to domestic vessels. Consequently, in 1963 the government removed the restrictions on fishing effort applied to local fishermen and, in 1965, provided guarantees on loans for fishing vessel purchases. Through these measures the government established both an open access policy with regard to fishery resources and provided mechanisms to aid a marked increase in the capacity of the country's fishing fleet.
>
> While the intention was to base fisheries development on the offshore resources, the fishing effort in the prime inshore fisheries also expanded rapidly. By the early 1980s, overfishing of species in these zones and overcapitalization within the inshore fleets were rapidly depleting fish stocks. In response, the government (i) set up controlled inshore fishing zones, a new licensing regime that limited vessel numbers and prohibited new entrants to the inshore fisheries; (ii) removed "part-time" fishermen from the inshore fisheries; (iii) enabled regulation of fisheries using management plans formulated after extensive public consultation on the resources to be managed and the regulatory controls (on fisheries inputs) to be applied.
>
> During the lengthy consultation and planning process, overfishing and overcapitalization intensified. By 1984, the inshore harvesting sector was over capitalized by an estimated NZ$28 million and correction would mean the retirement of 44 per cent of the existing fishing capacity. Ultimately, Government and industry agreed to introduce total allowable catches (TACs) to ensure stock conservation, and individual transferable quotas (ITQs) to facilitate industry restructuring. Both parties agreed the initial TACs and ITQs would be set so as to effect a reduction in fishing activity. The main elements of the scheme were: (i) the allocation of a case history to each fisherman, on a national basis (with case history defined as the fisherman's catch in two of the three years between 1981 and 1983); and (ii) the buy-back of case histories to a level that is equivalent to the TAC for each fishery.
>
> The government ultimately spent over NZ$45 million in legal fees and in buying out 15 800 tonnes of fishermen's case histories. The important outcome was that a viable and more sustainable future was secured for the affected fisheries and the industry in general. The remaining fishermen could buy, sell or lease their entitlements without undue government restrictions or the requirement of consent. In addition, they could shift their vessels throughout the year between different fisheries for which they had quotas. The government benefited by being able to purchase case histories at prices that did not reflect their full value, owing to the absence of an established ITQ market at the time.
>
> The extensive consultation with fishing industry representatives in the planning, development and implementation of the quota management system was an important element in the successful introduction of ITQs. ITQ management was established for 29 species, including 21 inshore and eight deep-water species. By 1996, 33 species were managed under ITQs, representing some 80 per cent of the total commercial catch from New Zealand's EEZ.
>
> There are approximately 117 species currently outside the quota management system and these are being managed by a system of permits and regulations. The government intends to bring additional species into the quota system and, at present, a moratorium has been placed on the issuance of new permits for non-ITQ species as a means of controlling the fishing effort prior to these species' inclusion in the quota management system.
>
> The introduction of ITQs, together with the financial assistance in restructuring, retired 15 800 tonnes of catch from New Zealand fisheries. The reduction in the size of the fleet, whether it was due to this assistance scheme or to the subsequent introduction of ITQs, was dramatic. The number of vessels dropped by 22 per cent between 1983/84 and 1986/87 and there was a further 53 per cent reduction resulting from the use of ITQs between 1986/87 and 1994/95. However, as this rationalization primarily occurred in the country's inshore fisheries, it helped their conservation and redirected investment to harvest deep-water fisheries.
>
> *Source:* Ministry of Fisheries, New Zealand

subsequent damage to inshore fishery habitats), lands and survey (responsible for filling of mangrove and coastal areas) or public works (responsible for building waste dumps, coastal roads and causeways that diminish fish nursery areas, cut off fish migration routes, and disturb water flows in bays and estuaries).

3. *Damaging and Destructive Fishing Techniques*

Commercial fishing gear and practices are not always selective for the species and sizes being targeted and fishermen discard a wide range of species with little or no commercial value. Annual discards from the world's fisheries were estimated to be about 20 million tonnes, or about 25 per cent of the annual production from marine fisheries.

Catch quotas increase the incentive to discard, especially in mixed species fisheries where several of the species are subject to a quota. Fishermen, required to discard that proportion of the catch taken in excess of the quota, sort the most valuable specimens and discard the smaller or damaged specimens of the quota species as well as lower value species. Discarding by-catch has long been recognized as a wasteful, but inevitable feature of commercial fishing as it constitutes a loss of valuable food, has negative consequences for the environment and biodiversity and can be aesthetically offensive. By-catch attracted considerable public and political attention in reaction to the incidental capture of dolphins in tuna purse seine nets, turtles in shrimp trawls and marine mammals, birds, turtles and fish in high seas squid driftnets. However, most measures aimed at reducing the quantities being discarded carry substantial implementation costs.

In the subsistence fisheries that dominate much of Asia and the Pacific islands, especially in highly populated, undeveloped and rural areas, almost everything caught is either consumed or used as feed or fertilizer and little is discarded.

Physical damage to marine ecosystems from fishing derive from three activities: (i) damage from fish capture gear, such as trawls; (ii) damage by fishers during capture of sea foods such as walking on corals, breaking coral or rocks using iron bars or explosives; (iii) damage by anchors, ship groundings, propeller washes, and dredging and filling associated with vessel movements and construction of port facilities. Some forms of fishing also disrupt the physical habitat. For example, bottom trawling can change the physical habitat and biological structure of ecosystems and accidentally capture and destroy a large number of non-target species. Trawl fishing inshore of the Great Barrier Reef in Australia is reported to have changed the physical character of the seafloor, increased turbidity, and removed key microhabitats for coral reef juvenile fish (Zann 1995).

In the last decade, the live fish industry has expanded throughout Asia and the Pacific. Wide use of chemical toxins, such as cyanide, has had serious impacts on juvenile fish and other reef creatures. Collection of tropical fish and other marine creatures for the aquarium trade is common in the tropical areas of the region and the recent phenomenon of the "live rock" trade involves the taking of coral rock from reefs to be used as decorative and biologically active components of salt water aquaria.

POLICIES AND PROGRAMMES

A. National Initiatives

1. Country Experiences in Coastal Zone Management

While some countries, such as Sri Lanka (Box 5.4) began active coastal zone management in the late seventies, most started in the mid to late eighties with the formation of committees and the preparation of reports, recommendations and legislation. In recent years, significant progress has been made in the development of coastal zone management plans for Bangladesh, Sri Lanka, Pakistan, Philippines, Tonga, Indonesia, Singapore, Thailand and Malaysia.

A key factor in the successful implementation of such management plans has been the extent of community participation as demonstrated by the experience of the Philippines, which was one of the first countries to experiment with community partnerships (Box 5.5).

In Malaysia, the most recent initiative towards integrated coastal zone management is the pilot project being undertaken in Sabah, Sarawak and Pulau Pinang to formulate integrated coastal zone management plans at the State level.

The core of an Australian initiative is a programme to develop integrated coastal area management strategies and programmes (referred to as Coastcare) based on partnerships (and shared funding) between the Commonwealth, State and Local governments, the community and industry. Coastcare provides opportunities and resources (including federal grants of about US$ 82 million per year) for community, business and interest groups to become actively involved in coastal management and decision-making (Australia Department of Environment 1997).

New Zealand completely revolutionized the process of decentralization of government with its Resource Management Act of 1991, which required the formation of a New Zealand Coastal Policy Statement 1994 to provide a policy framework for the sustainable management of the natural and physical resources of the coastal environment (NZMFE 1996).

Remote sensing and GIS databases for marine resource and coastal zone management are becoming more important in the region. India and Japan both have ocean satellites that provide data to fishers showing likely spots for fishing. Satellites provide enforcement of fishing agreements in offshore waters. Remote sensing techniques provided important base-line information of ecosystem conditions in the coastal zone and marine environment. In New Caledonia, for example, satellite images were used to detect changes in coral reefs over a ten-year interval, proving the effectiveness of a marine park in the territory and the vulnerability of coral reefs to local development activities (Bour 1990). In 1998, satellite images tracked the movement of hot surface water in the Indian Ocean to predict where coral reefs would dieback and thus provide conclusive proof that coral bleaching was associated with localized increases in sea level temperatures.

Box 5.4 Coastal Zone Management in Sri Lanka

Sri Lanka's 18 million people share 1 562 km of coastline and increased migration to coastal areas since Sri Lanka's independence in 1948 has created a range of consequences including coastal erosion, degradation of valuable coastal habitats and resource use conflicts. A realization of the conflicts and challenges associated with the management of coastal resources in the late 1970s led to the establishment of the Coast Conservation Department (CCD) within the Ministry of Fisheries and the enactment of the 1981 Coast Conservation Act.

In 1986, the CCD began a programme focused on the management of four key issues in the narrowly defined coastal strip: shorefront development, coastal erosion, habitat loss and the decline of recreational and cultural sites. The first outcome was a regulatory programme designed to reduce coastal erosion through a coastal permit applications system (primarily for house construction and sand mining), an extensive programme of public education and the construction of some specific coastal protection works. The second outcome was the development of provincial-level Coastal Zone Management (CZM) Implementation Plans and, in 1995, a series of local-level Special Area Management Plans. Local communities were encouraged to become actively involved in the formulation and implementation of the coastal zone management programme and this bottom-up approach enabled the local community to be "fully aware of and integrated into the planning effort so that it is truly participatory."

The strategic Coastal 2000 Plan recommended a second-generation coastal resources management programme with a "twin-track" approach, in which plans are implemented simultaneously at both the national and local levels. One of the initiatives of Coastal 2000 was the Special Area Management (SAM) Plan; in the early 1990s, two locations were chosen for the development of SAM Plans: Hikkaduwa, a small town on the west coast known for its coastal tourism and marine sanctuary; and Rekawa Lagoon, important for its local fisheries, mangroves, beaches and agriculture.

In 1992, CCD staff and representatives from the Coastal Resources Management Programme (CRMP) began the process of SAM planning at both locations. Government officials in selected agencies at the national level were contacted, and their interest and support was solicited. At the same time, CCD and CRMP staff began to work with community organizations to identify appropriate groups to be consulted in identifying community perceptions of resource management problems and priorities. Over the next three years, government officials, community groups and interest group representatives identified priority resource management issues and technical questions. Special Area Co-ordinating Committees, comprising both community representatives and government officials, were established and technical studies were commissioned, including environmental profiles for each Plan area. Resource management issues and strategies were identified and compiled. The SAM Plans for Hikkaduwa and Rekawa Lagoon were adopted by their respective co-ordinating committees in 1996.

The SAM planning process at Hikkaduwa facilitated the effective management of the Hikkaduwa Marine Sanctuary, heightened awareness amongst tourists and residents of the need to protect and manage the coastal environment, initiated a waste management strategy and encouraged a glass-bottom boat owner association. In management of Rekawa Lagoon, habitat, fishery and livelihood issues have taken highest priority.

In late 1996, the SAM planning and management processes were evaluated to determine the degree to which coastal management efforts integrated multiple agencies and programmes, levels of government and technical analysis. The evaluation indicated that as the two plans were developed by multidisciplinary teams working with community groups and national, provincial and local government officials, overall integration was excellent. The plans are based on regulatory activities, coastal development projects, research, monitoring and organizational efforts undertaken by both government agencies and community groups. Coordinating committees at both sites are working to maintain a comprehensive approach to improving resource conditions.

SAM plans are a bottom-up strategy for managing coastal resources that complements the existing top-down regulatory approach in Sri Lanka. They allow for intensive, comprehensive management of coastal resources in a well defined geographic setting (as contrasted with a use-by-use regulation-by-permit approach). Participation by community residents or stakeholders in planning and management is central to the SAM concept. Government agencies serve as catalysts or facilitators to help organize communities for resource management. Government provides technical support, and acts as mediators to help balance competing demands in resource management and as partners of communities engaging in co-management with community groups.

Source: Dr Ampai Harakunarak, Centre for the Study of Marine Policy

2. Pollution Abatement and Control

(a) Land Based Sources

Pollution abatement and control varies considerably within the region. A number of developed countries have enacted and enforced regulations for point source discharges into rivers and harbours, particularly from large industrial sources. These regulations have been effective in reducing industrial wastes in Australia, People's Republic of China, Republic of Korea, Japan, and New Zealand. Pollution from smaller, more scattered industries has proven more difficult to control as such industries often lack the capital or expertise to achieve pollution abatement and under such circumstances, enforcement of regulations is often arbitrary and ineffective.

CHAPTER FIVE

> **Box 5.5 The Bantay Puerto Programme in the Philippines**
>
> Community participation has played an important part in coastal zone management in the Philippines. The City of Puerto Princesa on the island of Palawan provides a representative example of the Philippines' approach to resource management. The forests and coastal resources of Palawan had become severely degraded by both commercial activities and the rapid increase in both population and poverty. The City government developed and implemented the *Bantay Puerto Programme* in an effort to protect, conserve and rehabilitate the city's forest and marine resources so as to improve the residents' quality of life and to increase the city's economic contribution to the country by utilising its resources in ways that are ecologically sustainable, socially equitable and economically viable.
>
> Through the mobilization of the local community action groups, the *Bantay Dagat* (dubbed "Baywatch") and *Bantay Gubat* ("Forest Watch") programmes proved extremely successful in curbing forest destruction and coastal resource degradation.
>
> The Programme's key management concept is simple: protect what is there, rehabilitate what has been destroyed or damaged and plan for the management of resource utilization that is environmentally sustainable. Rehabilitation included replanting mangrove trees and setting aside marine conservation areas.
>
> The success of the project was noted by the national government and the 1987-1992 National Development Plan adopted a community-based approach to resource management. Under this approach, local communities were empowered to manage the resources, whilst government provided the necessary enabling conditions, including the provision of appropriate incentives and expertise to properly manage resources. The efficacy of the community-based approach to resource management may be seen from the success of a number of subsequent initiatives anchored to such an approach.
>
> *Source:* Piedad S. Geron 1998

A number of countries have implemented comprehensive programmes and subregional policies aimed at reducing pollution entering inland waterways in addition to addressing the impacts to inland and marine habitats caused by transboundary pollution (see Chapter 4).

(b) Sea Based Sources

Australia and New Zealand have developed impressive oil spill response capabilities. For example, oil from a break in a coastal oil transfer site in South Australia in June 1999 was sprayed with oil surfactants by air within hours and oil protection booms set in place to protect nearby beaches well before any oil washed ashore.

As with control of industries, pollution control agreements and legislation is effective for large ships, but in many Asia and Pacific countries, smaller fishing vessels and houseboats are difficult to control or monitor. In Australia and New Zealand, legislation controlling recreational and fishing boats is more effective. For example, organotin antifouling is prohibited and most marine repair yards now incorporate catch basins to prevent paint scrapings washing into the coastal water. Sewage pump-out stations have been set up in many marinas and in some ports, such as Sydney, it is illegal for people to live aboard their boats for more than three days.

Within the region, a number of steps have also been taken to prevent the introduction of marine exotic species. These include increased training and regulations for quarantine officers, research into potential routes of entry, voluntary controls on shipping and promotion of international action through the International Maritime Organization. Two major strategies being used are: to encourage ships to exchange ballast water in the ocean or flush it en route; and to set up quarantine inspection of ballast water prior to discharge. Scientists at CSIRO's recently established Centre for Research on Introduced Marine Pests (CRIMP) are investigating various options to reduce and manage marine pests. The recently formed Australian Ballast Water Management Council will coordinate these activities and implement principles to ensure adequate quarantine and to reduce the risk of the accidental displacement of species.

3. *National Experiences with Restricting Fishing Capacity*

Fisheries governance has traditionally been based on command and control, with a variety of regulations based on catch limits, seasons, closed areas and size limits for individual species. While this is marginally effective for commercial fisheries and in countries with strong enforcement capability, it is ineffective for subsistence fishing, particularly in countries with poor enforcement capabilities. All too often, command and control of ocean fishing generated increasing levels of conflict and non-cooperation. Starting in the early 1980's some governments began a new approach to resource management that incorporated the views of the stakeholders in management decisions and conferred rights to the resource users. In New Zealand, control of industrial fisheries at the national level was based on conferring rights to trade (buy, sell or lease) the entitlement to fish in a particular managed fishery. The rights were generally provided in the form of individual transferable quotas (ITQs) or as a limited number of licences to fish. Internationally, institutions such as the World Bank are also encouraging the adoption of rights-based management.

In subsistence, recreational, and small-scale artisanal fisheries, rights are based on recognition of community control over particular territories. For example, in the Pacific Islands, governments formed partnerships between national fisheries institutions and community stakeholders. Traditional marine tenure and resource allocation mechanisms have been legally and politically recognized in Fiji, Samoa, Vanuatu, the Solomon Islands, and other Pacific island nations. Participatory fisheries governance has proven successful, especially in the small states where community ties remain strong (see Chapter 18). There is an emerging consensus that establishment of specific use or property rights will improve community interest in sustainable management. This has proved difficult after long standing open access approach to sea resources, but early indications from Sri Lanka were that the formal allocation of user rights that give communities greater control over the factors affecting their well-being is a successful solution (UNDP/Government of Sri Lanka 1991).

4.   *Conservation of Coastal and Marine Resources*

Throughout the region, scientists and resource managers have recognized the importance of including key coastal and marine habitats within designated conservation areas. Ideally, the protected zones within a particular area should enable representative habitats to be preserved and should focus, in particular, on habitats that play a key role in the early life stages of marine, estuarine and coastal species. As the vast majority of these species begin their life-cycle as free-swimming larvae and/or spend their juvenile life stages in "nursery" areas, the conservation strategy should provide protected areas that support and maintain the replenishment of populations in the wider marine and coastal environment.

The most successful conservation strategies for preserving key marine and coastal habitats are those that have been implemented as partnerships between resource managers and local communities. The Philippines played a leading role in establishing the principle of community involvement in the management and maintenance of with marine reserves. By the late 1970s, marine reserves established and managed by villagers were demonstrating their effectiveness not only in achieving their conservation objectives, but also in improving the long-term productivity and sustainability of the local fishery, even though the area available for fishing was reduced (Russ and Alcala 1988). This partnership principle has now been applied elsewhere and small-protected marine areas managed by the local community are becoming increasingly common in the region (Box 5.6).

In the late 1990s, the World Bank, in association with the IUCN's Commission on National Parks and Protected Areas (CNPPA), prepared a database containing the location of every marine protected area in the world, including those in the Asian and Pacific Region (Table 5.7). The database included background information on site characteristics, biodiversity, key species, etc. as a basis for determining future priorities for conservation investment.

Amongst the countries of the region, about four per cent of the coastal waters of New Zealand are included within marine reserves where no fishing activities are allowed. The remote Kermadec Islands Reserve, with 748 000 ha makes up nearly 75 per cent of the total area, with the rest scattered in six smaller reserves and two marine parks. The Department of Conservation and other groups are considering some 24 additional sites to provide protection for a selection of New Zealand's marine habitats.

Australia's Marine Policy includes plans to expand the existing network of coastal and marine parks and reserves to provide protection for a cross-section of habitats around the entire Australian coastline. Unlike New Zealand's marine reserves, Australian marine parks and reserves allow a range of activities, including commercial fishing, in some or all parts of the reserves. Zoning is a prime management tool for the Great Barrier Reef Marine Park Authority, and parts of the World Heritage Site are closed to everyone, even scientists, while other areas are essentially open to all use, except mining and drilling for oil.

The countries of the South Pacific subregion have actively promoted the establishment of parks, reserves and conservation areas and collectively established the *Convention for the Protection of the Natural Resources and Environment of the South Pacific Region (SPREP Convention)* (Noumea 1986). Since 1992, when the SPREP focus for conservation shifted to community involvement, the SPREP South Pacific Biodiversity Conservation Programme (SPBCP) has assisted 12 Pacific island countries with the development of 17 conservation areas, almost all of which have a coastal or marine component (see Chapter 18).

CHAPTER FIVE

5. Harmonizing Aquaculture and Environment

Some progress is being made in the establishment of appropriate legal and regulatory frameworks for aquaculture in a number of the region's countries, including Malaysia, Papua New Guinea, India, Sri Lanka, the Republic of Korea (Box 5.7) and Thailand.

The Government of India has set up an Aquaculture Authority to regulate the adoption of new technology and the establishment of new farms within and outside the Coastal Regulation Zone. At the State level, the Tamil Nadu Aquaculture (Regulation) Act of 1995 sets out conditions to improve the siting and management of aquaculture facilities and establishes an Ecorestoration Fund, supported by deposits from aquaculturists, to remedy environmental damage caused by aquaculture farms. In mid 1998, Thailand's Ministry of Agriculture, recognizing the accelerated loss of valuable agricultural land to aquaculture, has banned the conversion of rice paddies to shrimp farms.

Table 5.7 Marine Protected Areas in the Asian and Pacific Region

Country/Area	No. of Protected Area	Country/Area	No. of Protected Area
American Samoa	2	New Caledonia	17
Australia	244	New Zealand	14
Brunei Darussalam	4	Palau	1
PR China	41	Papua New Guinea	6
Cook Islands	1	Philippines	19
French Polynesia	1	Samoa	60
Guam	1	Singapore	1
India	11	Sri Lanka	4
Indonesia	30	Thailand	15
Islamic Rep. of Iran	1	Tonga	6
Japan	113	Tuvalu	1
Kiribati	1	Vanuatu	1
Rep. of Korea	6	Viet Nam	2
Malaysia	38*	Total	477

Source: WRI and United Nations Earthwatch website 1999

* Statistics provided by the Government of Malaysia

Box 5.6 The Funafuti Conservation Area Project in Tuvalu

Since Tuvalu separated from Kiribati as an independent nation, the population tripled and unregulated use of the atoll environment led to a marked decline in the quantity of avifauna and marine life on the atoll. Commercial over exploitation of lagoon fisheries undermined the sustainability of important subsistence and local artisanal fisheries. The abandonment of traditional taboos, marine tenure systems and fishing regulations, which were responsible for relatively sustained-yield production over thousands of years, coupled with widespread use of dynamite, pesticides and small mesh gillnets degraded important fisheries resources. By 1992, species of particular nutritional and cultural importance showing evidence of over exploitation included a range of groupers and snappers, emperors and rabbitfish, giant clams, spider conch, lobsters and crabs.

In October 1995, the people of Funafuti, the capital island of Tuvalu, launched the Funafuti Conservation Area Project (FCA). The Government of Tuvalu and the South Pacific Regional Environment Programme (SPREP) supported the people's initiative through the provision of the technical assistance and material resources that the project needed. The project objective was to conserve the biodiversity of Funafuti atoll through the sustainable use of natural resources for the benefit of the community and their descendants.

The project conception was developed in the traditional government *maneapa* after lengthy discussions between the community elders and officials from the national and regional environment organizations. The designated conservation area covers an oblong shaped area of 150 square kilometres or one-third of the total lagoon-fringed island. The conservation vision of the *maneapa* was based upon the elders' recollection of the abundance of marine and bird life on Funafuti in the 1950s and 1960s.

Three years after the project was launched, there was a marked increase in the abundance of marine and bird life in the conservation area. The area was patrolled daily and scientific assessments were carried out by visiting conservation scientists from SPREP. Offenders of the conservation rules have been dealt with in and according to *maneapa* rules and also mentioned in parliamentary exchanges.

In 1997 a Tuvalu Tourism Marketing and Development Action Plan offered the Marine Conservation Area as the cornerstone for the development of a small-scale eco-tourism business. The project funded SCUBA diving courses with certification to provide the human resources and interest for a Dive Operation. A Tourist information centre, the Funafuti Interpretative Centre, was constructed in 1998 and buoys were deployed marking the boundary of the marine reserve. Community workshops discussed community involvement in resource management, the role of protected areas, and the results of the coral reef baseline survey. A pamphlet on management practices for the Funafuti Conservation Area was produced in Tuvaluan and English and several radio talk shows were aired.

The FCA is a success story to the extent that the community has been fully involved in all stages of the project's planning and implementation. The conspicuous return to the abundance of marine and bird life that existed thirty years ago has served to heighten interest and awareness in the benefits and value of conservation. The project is being replicated in other islands in Tuvalu.

Source: 1. SILIGA A. KOFE, ESCAP POC, Port Vila, and Vanuatu
2. UNCED 1992

> **Box 5.7 Aquaculture and the Environment in the Republic of Korea**
>
> In the Republic of Korea, marine aquaculture expanded rapidly from its start in the mid-1960s such that by 1996 the country's aquaculturists produced 538 990 tonnes of seaweed, 306 738 tonnes of molluscs, 11 402 tonnes of finfish and 382 tonnes of crustaceans. The laver culture is carried out by means of a pole and floating net system, while sea mustard and kelp are cultured by long-line systems. Molluscs are farmed using a long-line system for oysters and mussels and a bottom planting system for clams and arkshells. Most of the culture of finfish is carried out in floating net cages, while the culture of prawns is done in embanked ponds.
>
> All aquaculture farms in the country require licensing by municipal authorities. Additionally, all cage culture and other aquaculture involving more than 1 000 m² in surface area must be registered with the Ministry of Environment and operated according to the Aquatic Environment Protection Law. Provisions seeking to minimize the pollution from cage culture include: the use of low-phosphate foods with a sinking rate that does not exceed 10 per cent within a two-hour period and the installation of feed fences with a height of 10 cm above the sea surface to prevent the dispersal of food outside the cages. Aquaculturists are also required to prevent the difference in oxygen levels within and outside the cages exceeding 20 per cent and to remove dead fish immediately and report incidences of diseased fish to the local fisheries administrations. The use of antibiotics and drugs for the control of fish diseases is regulated under the Aquatic Environment Protection Law, whilst licensing provisions require that the seabed under and immediately adjacent to farms is cleaned of debris with dredges more than once every three years.
>
> The Regulations Governing Sanitary Control of Shellfish and their Growing Areas, administered by the Ministry of Maritime Affairs and Fisheries, provide for the administration of water quality standards and the control of water pollution from aquaculture. The National Fisheries Research and Development Institute monitors water quality within the shellfish culture areas as well as the incidence of contaminants in the flesh of the aquaculture products. This entails routine sampling of sanitary indicator bacteria, nutrient salts (to assess eutrophication levels), pesticides and heavy metals. The median coliform most probable number (MPN) of the water should be less than 70/100 cm³, and not more than 10 per cent of the samples taken should have an MPN greater than 230/100 cm³ during the most unfavourable hydrographic and pollutant conditions. The incidence of red tides is also monitored in association with the early warning of aquaculturists when toxic species are identified.
>
> The Environment Impact Assessment Law requires the preparation of an environmental impact assessment (EIA) prior to the construction of city and industrial complexes, port development, land reclamation and water resource development. The establishment of aquaculture ventures is not currently subject to EIA, although this is planned in the near future. The transport of aquatic animals and plants, including the introduction of new species, the quarantining of imported species and the prevention of infected or recessive exotic species into Korean waters, are all subject to regulation by the Ministry of Maritime Affairs and Fisheries. Regulations under the Marine Pollution Control Law provide for government compenzation to aquaculturists in the event of economic loss owing to abnormal environmental changes such as harmful algae blooms. Compenzation may also be sought from private entities and public utilities arising from pollution (including oil spills), reclamation and industrial activities.
>
> Major pollution control and abatement measures under way or planned since 1991 include: the classification of coastal areas according to intended use (fisheries, recreational, agricultural and industrial); the strengthening of water quality standards and the control of industrial and municipal effluent into coastal waters; a national seawater quality monitoring system (for which 280 sampling sites were designated in 1996); investment in treatment facilities for sewage, industrial wastewater and excretion (for the equivalent of US$3.1 billion during the period 1992-1996); the requirement to undertake EIA for all coastal development activities; and the designation of special conservation areas in which most development activities would be prohibited.
>
> These efforts in the Republic of Korea have greatly helped in the promotion of sustainable aquaculture in the country.
>
> *Source:* Hak Gyoon Kim 1995

Elsewhere within the region, certain states and some NGO's, including producer groups, have developed and implemented codes of conduct and practice for particular aspects of aquaculture. Examples include the Code of Practice for Mangrove Protection by the Global Aquaculture Alliance (GAA); the Code of Practice for Australian Prawn Farmers; the codes of practice for cage culture of finfish and pond culture of shrimp in Malaysia; and guidelines for sustainable industrial fish farming (Anon 1997).

B. International Initiatives

1. The United Nations Convention on the Law of the Sea (UNCLOS)

Amongst its various provisions, UNCLOS provides a legal foundation for the sustainable development of coastal and marine resources. Most coastal countries have signed the Convention, primarily in order that they might benefit from the provisions relating to the national ownership of fishing and mining rights extending 200 nautical miles from their coasts. Through the allocation of exclusive rights for a large portion of the world's oceans, the Convention has curtailed the "free for all" approach that had previously encouraged maximum capacity exploitation and had prevented individual nations from implementing conservation and resource management strategies.

Although the implementation and maintenance of the UNCLOS provisions has been slow and problematic, the Convention has fostered the development of national policy and legislation for protection of the coastal and marine environment. In 1998, for example, Australia became the first country in the world to establish a national ocean policy and New Zealand is expected to introduce similar provisions in the near future (Michaelis 1999).

To assist with the implementation of the provisions of the Law of the Sea, the twenty-eighth session of the FAO Conference formulated a Code of Conduct for Responsible Fisheries that provided guidelines for sustainable fishing activities. The Code of Conduct also recommended a regional approach to: (i) the strengthening of scientists' and administrators' capacities; (ii) the development of timely and reliable fisheries information and statistical data as well as the setting up of a regional network; (iii) research and management considerations for shared or transboundary fish stocks; (iv) the development of methodologies for stock assessments; the prevention and control of degradation; and the monitoring of large ecosystems such as the South China Sea or the Gulf of Thailand (FAO 1999).

2. *Regional Seas Programme*

The action-oriented Regional Seas Programme was established more than 20 years ago and now encompasses a large number of discrete regions worldwide, including five regions that include ESCAP member countries of Asia and the Pacific; the East Asian Seas; the North West Pacific; the South Pacific; the Kuwait region (including the Islamic Republic of Iran); and the Mediterranean Seas Region (including Turkey).

The activities of the individual Regional Seas Programme are endorsed by each of the member countries and have typically included joint approaches to environmental assessment and management, legislation and institutional and financial arrangements. National institutions within each of the regions are responsible for implementing agreed actions with the main funding being provided through trust funds provided collectively by the region's governments. The long-term goal is the implementation of relevant global environmental conventions and other agreements, including the Law of the Sea, the London Convention and International Maritime Organization regulations.

3. *Support to Coastal Zone Management*

In the Asian and Pacific Region, ESCAP, the International Centre for Living Aquatic Resources Management (ICLARM) and SPREP have been actively involved in the promotion of sustainable coastal management and have prepared a range of management guidelines and studies including EIA tools for industrial and urban development in coastal areas, port infrastructures, tourism development, hazardous waste management and industrial pollution control. In addition, UNCED's Agenda 21 and the FAO (Clark 1992) have provided guidelines on the integrated management of coastal resources, including coastal fisheries (Scura 1994).

The 1996 International Workshop on Integrated Coastal Management in Tropical Countries reviewed regional progress in the formulation, design, implementation and extension of integrated coastal management (ICM) and produced a set of Good ICM Practices (IWICM 1996). In addition, there has been recent assessment of the current objectives and methods for evaluating internationally funded coastal management projects (Sorensen 1997).

In its efforts to strengthen the capacity of governments, NGO's and the private sector in coastal zone management, the FAO has collaborated with a range of institutions, including ICLARM, NACA, the United Nations Statistics Division, IUCN and other United Nations agencies sponsoring ICM activities. These international efforts have included pilot project to test alternative management approaches and the publication of guidelines on managing the environmental impact of aquaculture. Table 5.8 illustrates the range of regional organizations that provide support to the sustainable management of coastal and marine resources within the South Pacific subregion.

4. *Control of Pollution from Ships*

The control of maritime pollution is primarily founded on three international agreements: (i) the International Convention on Civil Liability for Oil Pollution Damages; (ii) the International Convention relating to Intervention on the High Seas in Cases of Oil Pollution Casualties; and (iii) The International Convention on the Prevention of Marine Pollution by Dumping of Wastes and Other Matter (the London Convention). The United Nations International Maritime Organization (IMO) administers these agreements and was responsible for introducing subsequent safety and environmental regulations for the oil tanker industry to prevent ocean dumping, ship-based discharges and accidental spills. The new rules required double-hulled construction, improved cargo handling procedures and more cautious operations in port and at sea. As a result, the volume of oil spilled into the oceans has dropped by 60 per cent since 1981, even though the amount of oil shipped has almost doubled (Zann 1995).

IMO also works closely with a range of other international and regional organizations in the development and implementation of new maritime pollution control initiatives including the Regional Programme for the Prevention and Management of Marine Pollution in the East Asian Seas, a programme to raise the safety standards of small ships, craft and passenger ferries which are not within "Convention size" and, in conjunction with the ESCAP and with funding from the Netherlands, IMO has been conducting workshops on the adoption of the Convention on the Facilitation of International Maritime Traffic.

IMO's Pacific programme assists with legal, port and safety issues related to maritime transport and has been focussed on upgrading the South Pacific Maritime Code and on the implementation of the 1993 Strategy for the Protection of the Marine Environment prepared jointly with the South Pacific Regional Environment Programme (SPREP) and the Marine Division of the South Pacific Forum.

5. *Regulating By-catch*

The 1995 United Nations Agreement for the Conservation and Management of Straddling Fish Stocks and Highly Migratory Fish Stocks seeks to minimize pollution, waste, discards, catch by lost or abandoned gear and catch of non-target fish and non-fish species. These objectives were reiterated in the Plan of Action produced by the International Conference on the Sustainable Contribution of Fisheries to Food Security, held in Kyoto, Japan, in 1995 (FAO 1999).

The FAO's Code of Conduct for Responsible Fisheries requires: "States, with relevant groups from industry, should encourage the development and implementation of technologies and operational methods that reduce discards. The use of fishing gear and practices that lead to the discarding of catch should be discouraged and the use of fishing gear and practices that increase survival rates of escaping fish should be promoted. Where selective and environmentally safe fishing gear and practices are used, they should be recognized and accorded priority in establishing conservation and management measures for fisheries." (FAO 1999).

The Technical Consultation on the Reduction of Wastage in Fisheries, held in Japan in October 1996, concluded that there had been significant improvements in reducing discarded by-catch in the previous decade. This had come about as a result of less fishing effort, time and area closures of fishing grounds, the use of more selective gear, the utilization of previously discarded by-catch, enforced prohibitions on discarding and consumer-led actions.

Some types of fishing gear, especially long drift nets, had such extensive and damaging effects on target and non-target species that they were banned, either nationally or regionally. For example, the countries of the South Pacific joined together to successfully ban long drift nets from the entire South Pacific subregion, even on the high seas beyond national EEZs (see Chapter 18).

Public concern over by-catch of marine mammals, especially dolphins, resulted in public boycotts of tuna and the initiation of a "dolphin safe" labelling programme. This, in turn, resulted in the development of purse seines and fishing techniques that minimized by-catch of dolphins.

6. *International Conventions on Marine Wetland*

The only international convention to focus specifically on wetlands is the Convention on Wetlands of International Importance (The RAMSAR Convention 1975) (see Chapter 3). Member governments undertake to; (i) designate at least one wetland for inclusion in the List of Wetlands of International Importance; (ii) promote wise use of wetlands; (iii) consult with each other on implementation obligations arising from the Convention, especially, but not exclusively, in the case of a shared wetland or water system; (iv) create wetland reserves. The Convention fosters international cooperation for shared water resources and shared species and provided for the establishment of a Wetland Conservation Fund to provide assistance to developing countries for wetland conservation activities. To date, 67 countries have signed the Convention and 66 have ratified it.

There are 106 RAMSAR wetlands in the Asian and Pacific Region, totalling some 9 698 000 ha. Australia, a founding member of RAMSAR, has designated 40 wetlands, totaling 4 481 346 ha, under the Convention. Other countries that have designated wetlands under the RAMSAR Convention include the Islamic Republic of Iran (18 designated wetlands), Japan (nine), Pakistan (nine), India (six), People's Republic of China (six) and New Zealand (five).

The Convention for the Protection of the World Cultural and Natural Heritage (1975) has designated 36 World Heritage Sites within the region. Of these sites, five are coastal or marine areas: (i) The Great Barrier Reef (Australia); (ii) Fjordland National Park (New Zealand); (iii) Sundarbans National Park (India); (iv) Lord Howe Island (Australia); and (v) Henderson Island (United Kingdom).

Table 5.8 Regional Organizations for the Sustainability of Coastal and Marine Resources in South Pacific

Regional Organization and member states	Primary activities related to marine and coastal issues			
	Coastal Zone Management	Pollution	Marine Resource Use	Conservation
SOPAC The South Pacific Applied Geoscience Commission Australia, Cook Islands, Federated States of Micronesia, Fiji, Kiribati, Marshall Islands, Nauru, New Zealand, Niue, Palau, Papua New Guinea, Samoa, Solomon Islands, Tonga, Tuvalu and Vanuatu	X		X	
SPC – Secretariat of the Pacific Community Australia, American Samoa, Commonwealth of the Northern Mariana Islands, Cook Islands, Federated States of Micronesia, Fiji, French Polynesia, Guam, Kiribati, Marshall Islands, Nauru, New Caledonia, New Zealand, Niue, Palau, Papua New Guinea, Samoa, Solomon Islands, Tonga, Tuvalu, Vanuatu, Wallis and Futuna			X	
SPF – South Pacific Forum Fisheries Agency Australia, Cook Islands, Federated States of Micronesia, Fiji, Kiribati, Marshall Islands, Nauru, New Zealand, Niue, Palau, Papua New Guinea, Samoa, Solomon Islands, Tonga, Tuvalu and Vanuatu			X	
SPREP – South Pacific Regional Environment Programme Australia, American Samoa, Commonwealth of the Northern Mariana Islands, Cook Islands, Federated States of Micronesia, Fiji, French Polynesia, Guam, Kiribati, Marshall Islands, Nauru, New Caledonia, New Zealand, Niue, Palau, Papua New Guinea, Samoa, Solomon Islands, Tonga, Tuvalu, Vanuatu, Wallis and Futuna	X	X		X

Source: IWICM 1996

7. *Ocean Monitoring*

The International Oceanographic Commission (IOC) has established the Global Ocean Observing System (GOOS) to provide information in support of oceanic and atmospheric forecasting, ocean and coastal zone management and research into global environmental change. The system will also serve the needs of the Framework Convention on Climate Change, by underpinning the ability to forecast changes in climate.

The GOOS also includes an integrated, multi-disciplinary, coastal observing system for detecting and predicting change in coastal ecosystems and environments including eutrophication due to nutrient enrichment, toxic contamination, habitat loss, saltwater intrusion, flooding and storm surges, harmful algae blooms and sea level rise.

Databases and sampling methods are gradually becoming standardized for a wide range of coastal and marine ecosystems. More than 230 ichthyologists from 54 countries assembled key information on all species of fish in the world into a single "FishBase" now available on the Internet (www.fishbase.org) and on annually updated CD-ROMs. FishBase eliminates confusion over taxonomic identification of species between scientists from all countries and provides rapid access to the world's knowledge of the ecology, biology and use of fish resources (Froese and Pauly 1997).

The revolution in ocean monitoring, combined with the facilitation of information exchange over the Internet, assists collaboration between developed and undeveloped nations. For example, the Hong Kong University of Science and Technology initiated and coordinated a global "Reef Check" project over the Internet to ascertain the current state of 400 coral reefs in over 40 countries. The Internet not only helped test and standardize ways to identify and measure important indices of ocean ecology, but also enabled the participation of more scientists, and even non-scientists, in the monitoring process.

CONCLUSION

The vulnerable coastal and marine ecosystems of the Asian and Pacific Region provide major resources to the region's peoples supporting a diverse and stabilizing natural system. However, the pace of coastal development and increasing pollution loads threaten the sustainability of the marine and coastal resources and the continued exploitation of the significant reserves of offshore oil and gas provide the potential for both economic prosperity and an increased risk of environmental degradation.

Coastal wetlands, seagrass beds, coral reefs and the sea surface microlayer are key habitats that are especially vulnerable to physical damage and chemical pollution. The reproductive cycles of marine organisms, including finfish, are linked to these key habitats and damage to the habitats reduces the ability of fish and invertebrates to withstand fishing pressure. Many of the region's major fishing areas are showing signs of overfishing; for example, the key species of the Northwest Pacific fishery, the second largest fishery in the world, are currently fished at maximum capacity, rather than at sustainable levels, and, as a consequence, stocks are declining. Although reducing fishing capacity has proved to be a lengthy and politically unfavourable issue, governments have begun the process by reducing fishing subsidies and regulating fishing access rights. The South Pacific Tuna Fishery offers a model of international cooperation for open sea fishing that may prove to be the first sustainable, multi-national ocean fishery in the world.

International agreements and treaties calling for protection of the marine and coastal environments, such as Agenda 21, the United Nations Convention on the Law of the Sea, the FAO Code of Conduct for Responsible Fisheries, have forged a new international awareness of the need for ocean conservation. Although countries in the region are making progress in meeting their obligations under these international agreements, progress is slow and many countries may find the requirements beyond their economic or political reach.

A new consensus seems to be emerging with regard to institutional approaches to the management of marine and coastal resources involving coordination among national government sectors, ministries and departments (fisheries, forestry, agriculture, environment, etc.) as part of a co-operative network of national, state, and local management with active participation of the civil society. Typically, this involves the national government preparing common guidelines and standards and then supporting implementation through the provision of technical and financial assistance to local state, regional or provincial government. Local government then works directly with community members to design long-term coastal plans and to establish community supported monitoring and enforcement activities. This combines both top-down and bottom-up elements with a willingness to form mutually advantageous partnerships between all levels of governance. As success depends on an understanding of the biological necessities of maintaining coastal and marine ecosystems, this approach to integrated management is typically supported by an open system of education and communication.

The control of marine pollution and the establishment of integrated coastal management programmes are legally and institutionally complex and progress has only been possible through international cooperation and, at the local level, the active involvement of key stakeholders and particularly locally affected communities. Future programme success will require the evolution of new mechanisms of open communication and education to enable partnerships between the civil society and government for the protection and sustainable development of the coastal and marine environment.

Chapter Six

The atmosphere in Kitakyushu, Japan: before and after the clean up.

6

Atmosphere and Climate

INTRODUCTION
AIR POLLUTION
REGIONAL ISSUES-TRANS-BOUNDARY AIR POLLUTION
GLOBAL ISSUES
REGIONAL CONTRIBUTION TO CLIMATE FLUCTUATION
REGIONAL POLICIES AND RESPONSES
RESPONSES TO GLOBAL ISSUES
CONCLUSION

CHAPTER SIX

INTRODUCTION

The composition of the atmosphere has been gradually changing over the past millions of years, it is only during the last two to three hundred years, since the beginning of the industrial revolution in Europe and North America, however, that man has begun to affect this change. The process has accelerated over the past 50 years as more countries have also embarked on rapid economic development.

This chapter reviews some of the important trends dealing with local air quality and regional concerns such as haze and acid rain, as well as the regional contributions to, and implications of, global climate change and stratospheric ozone depletion. The policies and response strategies that are being formulated or implemented in the Asian and Pacific Region to address these concerns are also discussed.

AIR POLLUTION

A. Types of Air Pollution-Indoor and Outdoor

While this chapter deals with the issues of pollution to the atmosphere and climate locally, regionally and globally, it is also important to note the significant hazards posed by indoor air pollution. For example, where a large part of the population still depends on traditional biomass fuels for cooking and heating, indoor air pollution may be a larger health hazard than outdoor pollution. The burning of such fuels in a confined space usually produces high levels of smoke and other pollutants. Estimates (for the Asian and Pacific Region) indicate that the concentrations of particulates may exceed WHO guidelines by factors of ten or more (WHO 1997) in many households, particularly in South Asia. The data in Table 6.1 suggest that tens of millions of people in Asia and the Pacific are being exposed to indoor levels of air pollution comparable to the notorious outdoor levels during the "London smog" of 1952, in which about 4 000 deaths occurred due to respiratory diseases.

B. Sources of Air Pollution

The combustion of fossil fuels (coal, oil, and natural gas) is the principal source of air pollution in all urban areas, along with the burning of biomass such as firewood, agricultural wastes and animal wastes in rural areas and some cities. Most of the combustion of fossil fuels takes place in industries, homes, for transportation, and for the generation of electricity. However, in the vast majority of Asian cities, transportation is the largest source of air pollution.

The number of vehicles in Asian cities has been growing exponentially over the last two decades. In Delhi and Manila, for example, they have been doubling every 7 years (ADB 1999). A large portion of vehicles in most Asian cities use diesel fuel, and contribute greatly to the emissions of particulates, especially those that are less than 10 microns in size and are respirable (PM_{10}). Several countries, including India, Pakistan, and the Philippines, still subsidize diesel fuels. In many countries, transportation fuels contain lead and high amounts of sulphur and use older engine designs that emit more pollution than modern ones. Since vehicles in developing countries are typically kept for longer periods than in the industrialized countries, they continue to contribute a substantial share of the air pollution in urban areas as their engines become increasingly less efficient. The situation is compounded by the region's reliance on motor cycles and three-wheel vehicles which frequently use two stroke engines and consequently produce up to 10 times more hydrocarbons than normal 4-stroke engines (ADB 1999).

Table 6.1 Indoor Concentrations of Particulate Matter due to Biomass Combustion

Location	Number of Studies	Duration	Concentration	Size
PR China	8	Various	2 600-2 900	All
Pacific	2	12 hours	1 300-5 200	All
South Asia	15	Cooking	630-820	All
		Cooking period	850-4 400	<10 microns
		Non-cooking	880	<10 microns
		24 hours	2 000-2 800	<10 microns
		Various	2 000-6 800	All
		Urban infants, 24th	400-520	<10 microns

Source: Adapted from WRI 1998

C. Current Levels of Air Pollution

During the process of industrialization in Europe and North America, air quality declined significantly. The same pattern is currently being observed in Asia and the Pacific, where in many urban areas, air pollution greatly exceeds levels considered safe by the World Health Organization (WHO).

Of the 15 cities in the world with the highest levels of particulate matter, 12 are located in Asia (ADB 1999). Furthermore, 6 of these cities also have the highest levels of atmosphere sulphur dioxide. Figure 6.1 depicts levels of total suspended particulates (TSP), sulphur dioxide (SO$_2$), and nitrogen dioxide (NO$_2$) against WHO guidelines, in selected Asian cities. The levels of TSP in several cities are three to four times those recommended by WHO although the situation for SO$_2$ and NO$_2$ is better, with only a few large cities greatly exceeding the recommended safe level.

D. Trends in Outdoor Air Pollution

Air pollution problems resulting from industrialization tend everywhere to be treated in a curative rather than preventative manner. Pollution problems are thus treated when they become high enough to present a health risk, and/or when countries achieve a certain degree of affluence.

Japan was the first Asian country to industrialize, and thus it was also the first to face air quality problems serious enough to encourage the formulation and implementation of policies to address the situation. As demonstrated in Figure 6.1, the ambient levels of TSP and sulphur dioxide in Tokyo are well within the guidelines suggested by WHO, while the NO$_2$ level in 1995 slightly exceeded the Guideline in Tokyo and Osaka, a situation that is commonly attributed to the cities with growing levels of private transport. Newly industrialized countries of the region such as, the Republic of Korea, Singapore, and Malaysia are at a stage in their development where they are also beginning to reduce the ambient levels of the major air pollutants.

E. Economic & Health Implications

Air pollution has economic impacts due to increased mortality and illness, the degradation to crops and property and due to tourists avoiding or shortening visits to cities that are heavily polluted. Estimating a monetary value for air pollution impacts is difficult, as it involves estimating non-market costs and values (e.g., health). However, a number of estimates have been produced, for example, damages caused by particulates and lead emissions in Jakarta have been estimated to be as high as US$ 2.1 billion (ADB 1999). These costs are primarily as a result of premature mortality and the impact of lead emissions on child intelligence, as detailed in Table 6.2.

In regard to health, air pollution is now the principal cause of chronic health problems in many Asian cities. Table 6.3 lists the common air pollutants, and the associated health concerns (ADB 1999). It is estimated that, globally, 200 000 to 570 000 deaths each year are due to outdoor air pollution (WHO 1997 and WRI 1998). According to an estimate by the World Bank (1992), about two to five per cent of all deaths in urban areas in the developing world are due to high exposures to particulates. For example,

if particulate levels in Jakarta were reduced to the WHO standards, an estimated 1 400 deaths, 49 000 emergency room visits and 600 000 asthma attacks could be avoided each year (Ostro 1994).

Figure 6.1 Ambient Levels of Air Pollutants in Selected Large Asian Cities

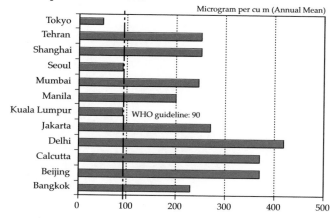

Source: Toufiq Siddiqi 1998

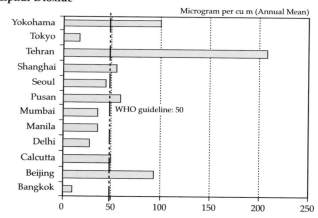

Source: Toufiq Siddiqi 1998

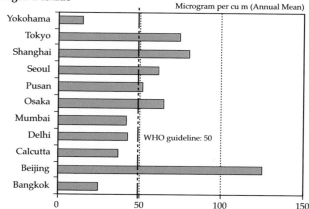

Source: Toufiq Siddiqi 1998

CHAPTER SIX

Table 6.2 Estimated Health Benefits of Reducing Air Pollution in Jakarta

Health Effect	Health Benefits (1989)	Unit Value US$ (1989)	Total in US$ Millions	Indonesia total adj. For PPP[1] (US$M)
Particulate matter				
Premature mortality	1 400	4.0 million	5 600	728
Hospital admissions	2 500	6 306	15.7	2
Emergency room visits	48 800	178	8.6	1.1
Restricted activity days	7 595 000	51	387	50
Lower respiratory illness (children)	125 100	132	16.5	
Asthma attacks	558 000	11	6	0.78
Respiratory symptoms	37 331 000	6	224	29
Chronic bronchitis	12 300	210 000	258	33.5
Subtotal				843
Lead				
Hypertension	135 600	442	60	7.8
Non-fatal heart attacks	190	28 300	5.4	0.7
Premature mortality	158	3.7 million	584.6	76
IQ loss (points)	2 073 205	4 588	9 511	1 236
Subtotal				1 320
Total				2 164

Source: ADB 1999
Note: [1] Purchasing Power Parity

Table 6.3 Health Effects Associated with Common Air Pollutants

Pollutant	Population at risk	Health Impact	Exacerbating Factors
Particulate emissions	Entire population, especially motorists and pedestrians.	Increase in illness, cancer and death from respiratory illness and decrease in lung function.	Especially PM10[1] of if there are high concentrations of diesel emissions.
Lead	Children, motorists, and pedestrians.	Acts as an acute toxin, damaging the kidneys, nervous system, and brain.	Chronic exposure to lead also increases death rates from stroke and heart disease.
Carbon monoxide	Pedestrians, roadside vendors, and vehicle drivers.	Shortness of breath, increased blood pressure, headaches, and difficulty in concentration.	Most significant in pregnant women, young children, and those suffering from heart and respiratory diseases.
Nitrogen dioxide	Urban commuters and dwellers.	Respiratory infection, increased airway resistance, and decreased lung function.	Most significant effects in children and asthmatics.
Ozone	Urban commuters and dwellers.	Irritation of the eyes and respiratory tract and reduced lung function.	Long-term exposure may cause irreversible deterioration in lung structure.
High BOD[2]	Users of untreated public water supplies.	Gastro-intestinal illness.	Greatest impact through dehydration and diarrhoea in young children.
Heavy metals	Ingested through water supply or from exposed foods.	Poisoning, increased child morbidity, and mortality.	Populations on watercourses close to gold mining at risk to mercury poisoning.

Source: ADB 1999
Note: [1] Particulate matters smaller than 10 microns in size
[2] Biological Oxygen Demand

REGIONAL ISSUES – TRANS-BOUNDARY AIR POLLUTION

A. Haze and Smog

Incidents of haze (severe smoke pollution) have occurred from time to time in many parts of the region. However uncontrolled forest fires, mainly in Indonesia, resulted in a particularly lengthy and severe episode affecting several countries in Southeast Asia from late July to early October 1997 (see Chapters 2 and 17). Substantial adverse health effects associated with the high levels of particulates occurred (WHO 1998). During the peak period of the haze in September 1997, air pollution levels considerably exceeded the WHO recommended levels. An estimated 20 million people in Indonesia suffered from respiratory problems, with levels of total suspended particulates (TSP) exceeding the national standard by 3-15 times. Visits to the Kuala Lumpur General Hospital due to respiratory problems increased from 250 to 800 persons a day. The economic costs associated with the haze have been estimated at US$ 6 billion for all the countries affected (WWF 1998). These include direct costs, such as losses to agriculture, as well as indirect costs such as medical expenses and a decline in tourism.

Information on the extent and the impacts of the haze were presented at a Workshop organized by the WHO's Regional Office for the Western Pacific in June 1998. For example, in Brunei Darussalam, measurements taken during the dry weather period February-April 1998 showed that the Pollution Standard Index (PSI) readings exceeded 100, and were sometimes as high as 250. This caused the disruption of daily activities, closure of schools, and changes in government working hours. The PSI in Singapore exceeded 100 for 12 days, reaching a maximum of 138. About 94 per cent of the haze particles were found to be PM_{10} with a diameter less than 2.5 microns. Hospital visits for all haze-related illnesses increased by about 30 per cent. In the Southern provinces of Thailand, the PM_{10} concentrations in the city of Hat Yai also increased significantly. In Papua New Guinea, about 50 per cent of commercial flights were cancelled due to poor visibility. In the city of Port Moresby, visibility during the peak haze period was limited to about one kilometre, and in the southern islands of the Philippines, four to five kilometres.

Besides haze, photochemical smog is also becoming a problem in the region. Air pollution generated in some countries is being carried by winds to neighbouring countries. Under certain weather conditions (primary sunlight) a photochemical smog is formed, when nitrogen oxides from fuel combustion react with volatile organic compounds (VOC) such as unburned petrol. Ground-level ozone (O_3) is the major component of the smog, and it can cause several respiratory diseases. Many cities in Asia are believed to have high levels of ground level ozone, but data are still not generally available.

B. Acid Rain

Acid rain has become a concern in several parts of Asia during the last decade, particularly in Northeast Asia. In People's Republic of China, for example, about half of all the cities monitored had average annual precipitation with pH values less than 5.6, the threshold for acid rain (UNEP 1999). Central and south western China were the areas most affected, with average pH less than 5.0, and acid rain frequency higher than 70 per cent. Of the cities south of the Yangtze River, Changsha, Zunyi, Hangzhou, and Yibib had pH values lower than 4.5.

The total emissions of SO_2 in Northeast Asia were estimated at 14.7 Tg (teragrams, or million metric tonnes). About 81 per cent of these originated in Northeast China, 12 per cent in the Republic of Korea, 5 per cent in Japan, and 2 per cent in the Democratic People's Republic of Korea (Streets et al 1999). The emissions are concentrated in the major urban and industrial centres such as Shanghai, Beijing, Tianjin and Pusan. With the anticipated continued economic growth for this subregion, continued reliance on fossil fuels, and no additional environmental controls, emissions of SO_2 are expected to increase to 41 Tg in 2020. However, Japan and the Republic of Korea currently enforce treatment of flue gas from large coal burning facilities for SO_2 removal, and consequently emissions in both countries have stabilized.

An important step in addressing the problem is the identification of the regions that might be subject to large depositions of acid rain. High deposition areas are being studied by an international group, supported by the World Bank and the ADB, using a Regional Air Pollution Information and Simulation (RAINS-Asia) model developed at the International Institute for Applied Systems Analysis (IIASA). The output from one of the simulations (IIASA 1995), assuming continued large increases in coal use, is shown in Figure 6.2. Japan, the Korean Peninsula, Eastern China, Eastern India, Central Thailand, Northern Philippines, and Eastern Sumatra are amongst the more likely regions to be affected by acid rain.

Additionally, the contribution of nitrogen oxides (NO_x) to acid rain is increasing rapidly in the region and will increase even faster in the future due to the growth of the transportation sector. The increased emissions will contribute to acid rain but will also lead to a rise in ambient levels of ozone. For example, India has large emissions of SO_2 and NO_x, with estimated emissions in 1987 of 3.1 Tg of SO_2 and 2.6 Tg of NO_x (Kato and Akimoto 1992; Wang and Soud 1998). Unless stringent emission controls are introduced, SO_2 emissions alone could increase to more than 18.5 Tg by 2020 (Elvingson 1996).

Emissions of ammonia (NH_3), associated with livestock and the increasing use of fertilizers, are also of concern in Asia. It has been estimated (Zhao and Wang 1994) that present emissions of NH_3 in the

Figure 6.2 Excess Levels of Acid Deposition Projected by the RAIN-Asia Model

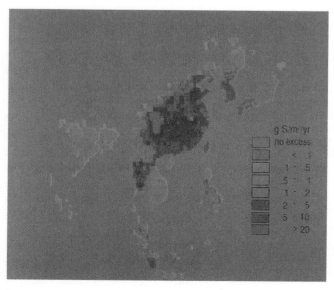

Source: IIASA 1995

CHAPTER SIX

Asian region are about 25 Tg per year. Ammonia is a strong acidifying agent in soils, and exacerbates the effects of acid rain.

Traditional biomass fuels also contribute to the emissions of sulphur and nitrogen oxides. An estimated 4.9 per cent of the total SO_2 emissions and 7.7 per cent of total NO_x emissions in Asia are due to the combustion of biomass (Streets and Waldhoff 1998). In some countries, such as Bhutan, Lao People's Democratic Republic, and Nepal, more than half of the emissions are from the use of biomass.

GLOBAL ISSUES

A. Depletion of the Ozone Layer

The appearance of an "Ozone Hole" over the Antarctic (later observed over the Arctic) during the early 1980s was quickly traced back to the rapid increases in the emissions of gases containing chlorine and bromine during preceding decades. These gases primarily originate from halocarbons arising from human activities. Since stratospheric ozone absorbs much of the ultraviolet radiation reaching the surface of the earth, its depletion exposes people living in the affected areas to higher radiation, resulting in higher incidences of skin cancer and related illnesses.

The global response to this environmental threat is generally considered to be one of the great successes in international cooperation (Box 6.1). The rapid decline in the emissions of ozone depleting substances (ODS) on a global basis is shown in Figure 6.3. Most of the emissions of ODS originated in the industrialized countries, however since 1995, emissions of CFCs from these countries have declined significantly.

B. Climate Fluctuations

1. Long-term Cyclic Variations

There have been changes in the earth's climate since the formation of the planet about 5 billion years ago. It was only after a period of hundreds of millions of years that these changes permitted the evolution of life. Change has continued, with ice ages alternating with warmer eras, on a time scale of tens of thousands of years. Samples of ice cores taken in Antarctica show a good correlation between Antarctic temperature, as deduced from the isotopic composition of the ice, and levels of carbon dioxide (IPCC 1995). An increase in the concentration of the latter from about 190 parts per million by volume (ppmv) coincided with an increase in the mean surface temperature of the earth of about 4 degrees Celsius. The concentrations of methane, another greenhouse gas, also show a similar correlation. The

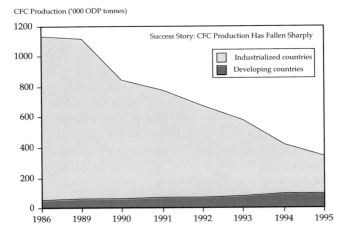

Figure 6.3 Decline in the Global Production of Ozone Depleting CFCs

Source: GTZ 1997

Note: ODP tonnes is a measure by which ozone depleting substances are weighted according to their ability to destroy ozone.

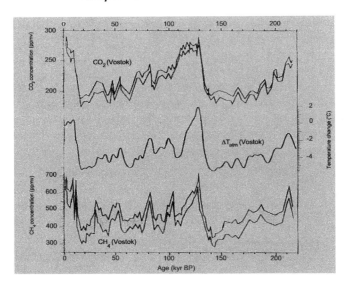

Figure 6.4 Correlation between Levels of Carbon Dioxide, Methane and Surface Temperature

Source: IPCC 1995

changes in concentration and temperature are shown in Figure 6.4.

2. Emissions of Greenhouse Gases due to Human Activities

The majority of scientists believe that emissions of greenhouse gases from human activity are the principal reason for the warming of the earth in recent decades. Nine of the ten hottest years on a global

128

ATMOSPHERE AND CLIMATE

> **Box 6.1 The Montreal Protocol: A Successful Example of International Cooperation**
>
> It is seldom that an environmental threat is recognized and action taken quickly to deal with it. The depletion of the ozone layer over the Antarctic "the ozone hole" was recognized and its link to the use of CFCs established during the 1970s. The Vienna Convention was followed by the Montreal Protocol in 1987, which committed the signatories to phasing out the production and use of CFCs and other ozone depleting substances (ODS). In most international agreements, the schedule for implementation is stretched out beyond the original date, however, in this case, as scientific measurements showed that the size of the ozone hole was increasing at a fast rate, the time table for the phase-out was actually accelerated through the signing and implementation of the London Amendment and the Copenhagen Amendment.
>
> The global production of CFCs in 1998 was only about 7 per cent of the peak reached in 1988. The industrialized countries have almost completed their phase out of these chemicals, and the remaining production is only in the developing countries, which were given an additional grace period of 10 years, up to 1999, to start implementing the provisions of the Montreal Protocol and its amendments. A Multilateral Fund was set up under the Protocol to assist the developing countries to help pay for new technologies, equipment conversion projects, and training of personnel. The Fund has contributed over US$ 600 million to about 1 800 separate projects in more than 100 developing countries. These projects will result in the phasing out of an equivalent of more than 80 000 metric tonnes of CFCs.
>
> In the Asian and Pacific Region, People's Republic of China and India are the largest producers and users of CFCs. China's consumption of ODS increased more than 12 per cent per year between 1986 and 1994. At the end of 1994, the country produced about 60 000 tonnes of ODS and consumed 84 000 tonnes annually. People's Republic of China has made a commitment to phase out consumption of these substances completely by 2010. It has already banned the establishment of new CFC and Halon – related production facilities, and developed general and sector – specific phase out plans, with the assistance of the World Bank and the Multilateral Fund. It was expected to meet the 1999 target for freezing the consumption of CFCs.
>
> India is the second largest producer and fourth largest consumer of CFCs in the world. Its production of 23.7 million tonnes (MT) during 1997 accounted for 16.4 per cent of the world total. During the same year, it consumed 6.7 MT, amounting to about 5.3 per cent of the consumption worldwide. The Multilateral Fund for the Implementation of the Montreal Protocol has approved a World Bank project which will assist India in completely phasing out of CFC production by 2010, with production ceilings set for each of the earlier years.
>
> The successful implementation of Montreal Protocol shows that it is possible to overcome a serious environmental threat with international cooperation particularly through sharing of responsibilities and resources. In fact the establishment of a fund was a major contributor to success of the Montreal Protocol.
>
> Source: UNEP 1998; World Bank 1997; and WRI 1998

basis since measurements began have occurred during the past decade. The gases, which contribute most to climate change, are carbon dioxide, methane, nitrous oxide, and halocarbons.

The Intergovernmental Panel in Climate Change (IPCC) reports (1995) provided details about the natural, as well as anthropogenic, sources of greenhouse gas emissions (Box 6.2). The burning of fossil fuels and biomass is the largest single source of emissions for carbon dioxide, the gas contributing the largest share to the greenhouse effect. Changes in land use, such as clearing of forests for agriculture or residential development is also a major source of carbon dioxide. The keeping of livestock, growing of paddy rice, urban garbage dumps, and the production of fossil fuels are major sources of methane (which although released in relatively small quantities taking residence time into account, its impact as a greenhouse gas is 44 times that of CO_2 on a weight by weight basis). Chlorofluorocarbons, used as aerosols and for refrigeration, were also major contributors to the greenhouse effect, until the emissions were drastically reduced after the signing and implementation of the Montreal Protocol.

C. Impacts of Climate Fluctuations

1.   Changes in Precipitation and Availability of Water

Global climate change is frequently referred to in the media as "Global Warming". This tends to focus attention on the anticipated increase in average surface temperature of the world, estimated to be of the order of 2-4 degrees celsius, if present trends continue. This is important as far as the causes of the change are concerned (for which there exists no consensus), but more importantly in terms of impacts, changes in rainfall patterns and in the location and frequency of extreme weather events, such as cyclones (Box 6.3).

Existing computer models (GC-MS) are not yet able to predict, for example, exactly how the precipitation in each state or province might change. The Commonwealth Scientific and Industrial

CHAPTER SIX

Box 6.2 Anthropogenic Emission and Climate Change

The Rise of Greenhouse Gas Concentrations

- Atmospheric concentrations – the accumulation of emissions – of greenhouse gases have grown significantly since pre-industrial times as a result of human activities.
- Carbon dioxide concentrations – the most important greenhouse gas apart from water vapour – has increased more than 30 per cent from 280 ppmv (parts per million by volume) in the pre-industrial era to 365 ppmv by the late 1990s. The current rate of increase is around 1.5 ppmv per year. Unfortunately, a large proportion of the carbon dioxide put into the atmosphere remains there, warming the planet, for around 200 years.
- Methane – on a weight-per-weight basis some 20 times more powerful as a greenhouse gas than carbon dioxide has more than doubled its concentration, from 700 to 1 720 parts per billion, by volume, (ppbv), primarily because of deforestation and the growth in rice and cattle production. Natural gas leaks are another source. Methane's residence time in the atmosphere is relatively short approximately 12 years.
- Nitrous oxide, associated with modern agriculture and the heavy application of chemical fertilizers, has increased from pre-industrial levels of 275 ppbv to 310, with a current annual growth rate of 0.25 per cent. On a weight-per-weight basis it is more than 200 times more powerful as a greenhouse gas compared with carbon dioxide. Its residence time in the atmosphere is around 120 years.
- The Chlorofluorocarbons, CFC11 and CFC12, both with growth rates of 4 per cent per year during the past decade, have now reached levels of 280 parts per trillion by volume (pptv) and 484 pptv respectively. They have a 'greenhouse gas potential' that is many thousands of times greater than carbon dioxide on a weight-per-weight basis, and they remain in the atmosphere from several thousand years.
- Taking the residence time in the atmosphere of the different gases and their specific effectiveness as greenhouse gases into account, carbon dioxide's contribution is some 55 per cent of the whole, compared with 17 per cent for the two CFCs and 15 per cent for methane. Other CFCs and nitrous oxide account for 8 and 5 per cent respectively of the changes in radiative forcing.

Impacts of anthropogenic emissions on climate change:

- Increases in greenhouse gas concentrations since pre-industrial times (i.e., since about 1750) have led to a positive radiative forcing of climate, tending to warm the surface of the Earth and produce other changes of climate.
- The atmospheric concentrations of the greenhouse gases carbon dioxide, methane, and nitrous oxide (N_2O), among others, have grown significantly: by about 30, 145, and 15 per cent, respectively (values for 1992). These trends can be attributed largely to human activities, mostly fossil fuel use, land-use change, and agriculture.
- Many Greenhouse gases remain in the atmosphere for a long time (for carbon dioxide and nitrous oxide, many decades to centuries). As a result of this, if carbon dioxide emissions were maintained at near current (1994) levels, they would lead to a nearly constant rate of increase in atmospheric concentrations for at least two centuries, reaching about 500 ppmv (approximately twice the pre-industrial concentration of 280 ppmv) by the end of the 21st century.
- Tropospheric aerosols resulting from combustion of fossil fuels, biomass burning, and other sources have led to a negative radiative forcing, which, while focused in particular regions and subcontinent areas, can have continental to hemispheric effects on climate patterns. In contrast to the long-lived greenhouse gases, anthropogenic aerosols are very short-lived in the atmosphere; hence, their radiative forcing adjusts rapidly to increases or decreases in emissions.
- The scientific ability from the observed climate record to quantify the human influence on global climate is currently limited because the expected signal is still emerging from the noise of natural variability, and because there are uncertain ties in key factors. These include the magnitude and patterns of long-term natural variability and the time-evolving pattern of forcing by, and response to, changes in concentrations of greenhouse gases and aerosols, and land-surface changes. Nevertheless, the balance of evidence suggests that there is a discernible human influence on global climate. The IPCC has developed a range of scenarios, IS92a-f, for future greenhouse gas and aerosol recursor emissions.
- The IPCC has developed a range of scenarios, IS92a-f, for future greenhouse gas and aerosol precursor emissions based on assumptions concerning population and economic growth, land use, technological changes, energy availability, and fuel mix during the period 1990 to 2100. Through understanding of the global carbon cycle and of atmospheric chemistry, these emissions can be used to project atmospheric concentrations of greenhouse gases and aerosols and the perturbation of natural radiative forcing. Climate models can then be used to develop projections of future climate.
- Estimates of the rise in global average surface air temperature by 2100 relative to 1990 for the IS92 scenarios range from 1 to 3.5'C. In all cases the average rate of warming would probably be greater than any seen in the last 10 000 years. Regional temperature changes could differ substantially from the global mean and the actual annual to decadal changes would include considerable natural variability. A general warming is expected to lead to an increase in the occurrence of extremely hot days and a decrease in the occurrence of extremely cold days.
- Average sea level is expected to rise as a result of thermal expansion of the oceans and melting of glaciers and ice-sheets. Estimates of the sea level rise by 2100 relative to 1990 for the IS92 scenario range from 15 to 95 cm.
- Warmer temperatures will lead to a more vigorous hydrological cycle; this translates into prospects for more severe droughts and/ or floods in some places and less severe droughts and/or floods in other places. Several models indicate an increase in precipitation intensity, suggesting a possibility for more extreme rainfall events.

Source: IPCC 1995

> **Box 6.3 El Nino and Climate Change: Extreme Natural Events**
>
> While climate change is regarded as a gradual phenomenon, it may largely manifest itself in the changing frequency of extreme meteorological events – unexpected droughts and floods, record heatwaves and snowstorms – that will trigger human disasters. One model for these likely events is provided by the record El Nino caused round the world during 1997 and 1998. The name El Nino – Spanish for the Christ Child – comes from Peruvian fishermen, who named it generations ago for the timing of its usual peak around Christmas. Historical records show the phenomenon has been occurring every two to ten years for at least the last five centuries. Since the turn of this century 23 El Ninos have affected the earth.
>
> El Nino is a fluctuation in the distribution of sea-surface temperatures and of atmospheric pressure across the tropical Pacific Ocean, leading to worldwide impacts on regional weather patterns. No one knows exactly why it takes place, but recent computer climate modelling suggests the frequency and strength of both El Nino and its sister effect La Nina are increased by global warming – 1998 was by far the warmest year since world wide records began 150 years ago. Despite doubts over the precise relationships of climatic cause-and-effect, the mechanisms are well documented. In normal conditions, trade winds blowing west along the equator push warmer surface waters towards Southeast Asia, where they accumulate, evaporate and fall as heavy tropical rains. Meanwhile, off the Pacific coast of Latin America, cooler nutrient-rich waters well up from the ocean depths, causing dryer conditions along the shores of Peru and Chile, and making their fishing grounds among the most fertile in the world. During El Nino, trade winds weaken or reverse, and the warm surface waters of the western equatorial Pacific shift east. This generates unseasonable rain and storms over the Pacific coast of the Americas, while leaving drought to afflict Southeast Asia and the western Pacific.
>
> For the 12 El Nino months from the summer of 1997 to the summer of 1998, Asia and the Pacific experienced some of the most intense and widespread fires ever recorded. Indonesia's rain forests got no rain and the months of dry weather turned the forests into the world's largest pile of firewood. Similarly, the South Pacific sweltered under cloudless skies. As west-blowing trade winds weakened and atmospheric pressure decreased over the central Pacific, warm seas and rain-clouds moved east, radically reducing precipitation levels in the south-west Pacific. Droughts blighted many countries in the region including Australia, Indonesia, New Zealand, Papua New Guinea, Fiji and the Solomon Islands, hitting hard states that rely on arable crops for domestic consumption and export revenue.
>
> El Nino is a periodic natural event, but it has become more intense and frequent in the past 20 years and there is some evidence to suggest that this may be a consequence of global warming, if this is the case, then El Ninos could become semi-permanent features of the world's weather system. Even if not, recent events demonstrate the instability of the world's weather systems and its capacity to switch modes, unleashing extreme weather on unsuspecting communities, and raise the need for further investigation on inter annual climate variability.
>
> Source: Red Cross 1999

Research Organization of Australia (CSIRO 1995), among others is developing computer models to assess regional changes in temperature and rainfall. However, it is clear that a "Permanent" reduction in the rainfall (or snowfall) could have enormous implications for the availability of water, especially in densely populated locations.

2. *Impact on Agriculture*

A change in the average temperature and precipitation is likely to have a significant affect on crop yields either increasing or decreasing them depending on crop types. In some areas, yields might increase, whilst in others they would decline depending on. Countries in the region are beginning to assess the likely implications of such changes on their food production. Initial simulations for studies sponsored by ADB (1994) and the World Bank (Dinar *et al* 1997) suggest that an increase in mean temperature might for example, reduce rice yields in Bangladesh, India, Philippines, and the Republic of Korea. The simulations also showed that the yield of wheat in India might increase due to higher increased carbon dioxide in the atmosphere, but could also reduce due to higher temperatures.

3. *Sea-Level Fluctuation*

Although recently revised downward, the expected rise in sea levels due to climate change are still anticipated to range from 0.3 to 0.5 metres by the year 2100 (IPCC 1995) and could present a big challenge to most countries of the region. Concerns have been expressed by the leaders of many small island nations such as, the Maldives, Tuvalu, Kiribati and Tonga, where most land is only a few meters above sea level. Changes in the sea temperature are also likely to have serious impacts particularly on coral reefs and migratory species of marine life (Box 6.4)

Other countries such as People's Repblic of China, India, Indonesia and Bangladesh have substantial parts of their population living close to river deltas, including many of the Megacities of the region, such as Calcutta and Shanghai. A rise in sea

CHAPTER SIX

level could thus affect at least a hundred million inhabitants and cause large economic losses (see Box 6.5).

4. *Frequency of Storms*

Changes in temperature are likely to be accompanied by changes in the frequency and intensity of storms. Although it may be too early to predict how countries may be affected, there are some indications that a few countries in the region are already being affected by a larger number of destructive cyclones and storms.

5. *Impact on Health*

The health of humans and other species is affected by a number of environmental factors, including the quality of the air and water, temperature ranges, rainfall, and the presence of organisms and vectors that transmit diseases. Since the precise impact of global climate change on these cannot yet be predicted, it is only possible to provide a general indication of the types of health implications that can be expected. For example, Table 6.4 depicts the likelihood of alterations to distribution of vector-borne diseases. One particular concern to the Asian and Pacific Region is the likely change in the distribution of malaria–carrying mosquitoes as a result of warmer surface temperatures, possibly placing several hundred million people at risk every year (WRI 1998).

Box 6.4 Global Warming: Threat to Coral Reefs

According to a recent report, "Climate Change, Coral Bleaching and the Future of the World's Coral Reefs," global warming of 1 to 2 degree celsius over 100 years would cause devastating bleaching events to occur on large tracts of Australia's World Heritage-listed Great Barrier Reef. The reef could die from coral bleaching within 30 years.

Coral reefs are highly sensitive to the first signs of danger. In the late 1980's large areas of coral reefs were damaged or killed by a phenomenon called coral bleaching. The process takes place by the disturbance of plant-like microbes called zooxanthellae, which live in association with coral cells. The zooxanthellae use the waste phosphates, nitrates, and carbon dioxide from coral cells to photosynthesize oxygen and sugars that, in turn, are metabolized by the coral cells. When the water temperature exceeds the normal maximum by more than two degrees Celsius, the photosynthetic process of the zooxanthellae breaks down and the corals begin to eject them. Because the plant cells have pigments, and the corals don't, the coral colonies turn white, as if they were bleached. The corals die if the hot water continues.

In 1998 and 1999, anomalies greater than one degree above the maximum monthly climatological Sea Surface Temperature (SST) were recorded. A network of coral reef researchers and observers examined coral reefs in the hotspots posted on the Internet. The 1998 satellite images showed very hot ocean water in the Indian Ocean and South-West Pacific. This water, sometimes 5 degrees above the normal temperature tolerance of reef building corals, remained over known coral reef habitats for several weeks. The bleaching events that coincided with the distribution of hot water caused the most extensive bleaching of coral reefs ever recorded, affecting formerly lush and healthy coral reefs in Australia, Viet Nam, Thailand, Malaysia, Singapore, the Philippines and Indonesia. Predictions were made for coral bleaching based on the location and duration of hotspots, providing convincing proof that elevated sea temperatures were the primary cause of the bleaching.

In 1998, the Great Barrier Reef, which stretches 2 000 km down the coast of North Queensland, experienced its most serious episode of bleaching on record, with 88 per cent of reefs close to shore affected. According to the principal research scientist at the Australian Institute of Marine Science, while bleaching does not always kill coral, bleaching in 1998 resulted in a high coral mortality rate that left some parts of the reef dead. The same scientist supporting the report said that its prediction of an increased frequency of coral bleaching due to global warming appears very credible and of great concern. This has enormous implications for the health and wealth of tropical and sub-tropical marine dependent societies of Asia and the Pacific. The economic impact of severe coral bleaching would be enormous, especially affecting fisheries and tourism, which form the backbone of the economy in many developing countries of the region.

Source: 1. Kyodo News Service 1999
2. Strong A.E., T.J. Goreau and R.L. Hayes 1999
3. Guch 1999

ATMOSPHERE AND CLIMATE

Box 6.5 Illustrating the Impacts of Sea-Level Rise: Bangladesh

A major concern related to a change in the global climate is the potential of rising sea levels. Bangladesh is one of the most densely populated countries of the world, with a large population subject to frequent flooding and storms. A rise of about 0.45 metre in sea level could inundate about 11 per cent of the total land area of the country, displacing about 5 per cent of the present population, i.e. about 7 million people. If the rise in sea level reaches 1 metre, approximately 21 per cent of the land could be inundated, affecting about 20 million people.

In addition to the physical hardship on the population due to the loss of land, agricultural output would also suffer considerably. The loss in rice production alone is estimated to be in the range of 0.8-2.9 million tonnes per year by 2030, and would exceed 2.6 million tonnes per year by 2070.

Loss Estimates of Rice Output Due to Sea Level Rise (metric tonnes)

Year	SLR 45 cm	SLR 1 m
1995	209 (0.01)	740 (0.01)
2000	950 (0.02)	2 711 (0.04)
2010	11 458 (0.23)	35 192 (0.42)
2020	125 268 (2.23)	412 042 (3.98)
2030	827 212 (13.19)	2 875 351 (23.56)
2040	1 749 582 (25.25)	6 294 330 (45.08)
2050	2 121 854 (27.70)	7 708 359 (49.48)
2060	2 367 394 (27.97)	8 600 366 (49.95)
2070	2 618 802 (28.00)	9 513 691 (50.00)

Source: ADB 1994b
Note: Figures in parentheses show losses as percentages of total potential output in the coastal zone likely to be inundated based on Consultants' estimates.

Forestry, too, is likely to be severely affected. The world famous Sundarbans, one of the largest single-tract mangroves in the world, might be completely inundated if the sea level were to rise by one metre. The rich biodiversity in that area would be lost, as well as a continuing supply of biomass fuel for the area. The economic losses associated with such a sea level rise would also be significant, amounting to several billion dollars annually.

Table 6.4 The World's Major Vector-borne Diseases Ranked by Population Currently at Risk

Disease	Causative agents	Vectors	Population at risk (millions)	Population infected (millions)	Likelihood of altered distribution with climate change
Detigue fever	Viruses	Mosquitoes	2 500	50 per year	++
Malaria	Protozoa	Mosquitoes	2 400	300-500 per year	+++
Lymphatic filariasis	Nematodes	Mosquitoes	1 094	117	+
Schistosomiasis	Flatworms	Water snails	600	200	++
Leishmaniasis	Protozoa	Sandflies	350	12	+
River blindness	Nematodes	Blackflies	123	17.5	++
Trypanosomiasis (sleeping sickness)	Protozoa	Tsetse flies	55	0.25-0.3 per year	+

Source: WHO 1996
Abbreviations: + likely   ++ very likely   +++ highly likely

CHAPTER SIX

REGIONAL CONTRIBUTION TO CLIMATE FLUCTUATION

A. Greenhouse Gases

1. Rising Share of Carbon Dioxide Emissions from Energy Use

While Europe and North America had been the largest emitters of carbon dioxide until the middle of the 1990s, Asia has now assumed this role. The relative contributions of the different world continents to carbon dioxide emissions from fossil fuels in 1996 (Siddiqi 1999) are shown in Figure 6.5.

While in aggregate terms, the emissions of carbon dioxide from People's Republic of China and India are amongst the largest in the world, however in terms of per capita emissions, the USA has the highest contribution, as shown in Figure 6.6. This has implications for designing a universally accepted Protocol for reducing emissions of greenhouse gases. The developing countries feel that they cannot accept limits on their emissions in the near future in view of the close link between economic growth, energy use, and carbon dioxide emissions. Improving the efficiency of energy use, and the use of renewable or other low or no carbon energy sources, are ways of overcoming this dilemma.

A substantial number of studies on greenhouse gas emissions have been undertaken in Asian countries during recent years, many of them with the support of the Global Environment Facility (GEF), the United Nations Development Programme (UNDP), the ADB, and the World Bank. A major regional project, ALGAS, funded by GEF through UNDP, and implemented by ADB, provided a comprehensive picture of the emissions of various greenhouse gases and the options for reducing the rates of growth of emissions in the participating countries. The results for emissions of carbon dioxide from energy use for the years 1980, 1990 and 1995 are shown in Table 6.5.

In many of the Asian countries, traditional fuels such as firewood, and animal and agricultural wastes provide a substantial share of the total energy, as shown in Figure 6.7. The combustion of these fuels also results in sizeable amounts of carbon dioxide, as well as other air pollutants. In conducting emission studies, many countries assume that these emissions are balanced by the absorption of an approximately equal amount by the new growth in forests and

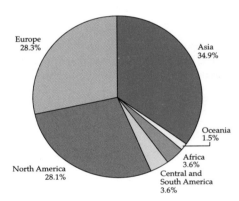

Figure 6.5 Regional Shares of Carbon Dioxide Emissions from the Use of Fossil Fuels

Source: Siddiqi 1999

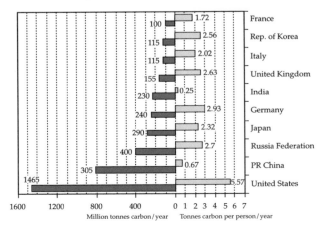

Figure 6.6 Total and Per Capita Emissions of Carbon Dioxide from Fossil Fuels in the Largest Emitting Countries, 1996

Source: Siddiqi, GEE-21

Table 6.5 Emissions of Carbon Dioxide from Energy Use in 11 Asian Countries (in Million Tonnes of Carbon Dioxide)

Country	Emissions (1980) (World Bank 1998)	Emissions (1990) (ALGAS study 1998)	Emissions (1995) (World Bank 1998)
Bangladesh	7.6	21.2	20.9
PR China	1 476.8	2 325.3	3 192.5
India	347.3	565.2	908.7
Indonesia	94.6	156.9	296.1
Mongolia	6.8	13.8	8.5
Myanmar	4.8	6.1	7.0
Pakistan	31.6	69.5	85.4
Philippines	36.5	43.5	61.2
Rep. of Korea	125.2	248.1	373.6
Thailand	40.1	79.7	175.0
Viet Nam	16.8	27.5	31.7

Source: ADB 1998 and World Bank 1998
Note: *The methodologies adopted by the ALGAS study and the CDIAC work cited by the World Bank are slightly different, and caution should be exercized when comparing the data for the different years*

ATMOSPHERE AND CLIMATE

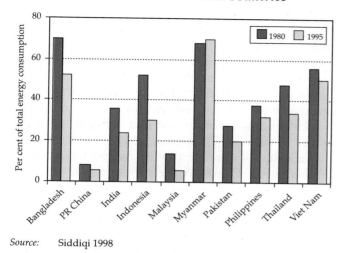

Figure 6.7 Share of Energy Supplied by Traditional Fuels in Selected Asian Countries

Source: Siddiqi 1998

Table 6.6 Carbon Dioxide Emissions from Land Use Changes in the ALGAS Participating Countries

Country	Emissions (1990) (ALGAS study 1999) (in million tonnes of Carbon Dioxide)
Bangladesh	19.8
PR China	-281.2
India	1.5
Indonesia	-334.9
Mongolia	5.5
Myanmar	6.7
Pakistan	9.8
Philippines	82.1
Rep. of Korea	16.2
Thailand	78.1
Viet Nam	31.2

Source: ADB 1999

Note: Negative numbers imply that more carbon dioxide is being absorbed than emitted

Table 6.7 Emissions of Methane and Nitrous Oxide in ALGAS Participating Countries

Country	Methane Emissions (1990) (in million tonnes)	Nitrous Oxide Emissions (1990) (in million tonnes)
Bangladesh	1.7	4.5
PR China	29.1	360.0
India	18.5	255.0
Indonesia	4.9	19.7
Mongolia	0.3	0.1
Myanmar	2.1	8.4
Pakistan	2.7	0.2
Philippines	1.5	30.4
Rep. of Korea	1.4	12.0
Thailand	2.7	11.3
Viet Nam	2.6	14.6

Source: ADB 1998

agricultural crops each year. Land use changes are a major source of carbon dioxide emissions and in some countries, the forests may be a net sink for carbon dioxide, i.e. the removal of carbon from the atmosphere might be larger than the emissions. According to the ALGAS study, two of the 11 countries with completed reports indicated that land use changes represented a net sink for carbon. These data are shown in Table 6.6.

2. Emissions of Methane & Nitrous Oxide

Agriculture and livestock are the largest sources of anthropogenic methane in most countries, although coalmines, the production of oil and gas, and transmission of natural gas are also large contributors. Aggregate emissions are subject to somewhat larger uncertainties than emissions from fossil fuel combustion, since there is a considerable difference in the emissions from paddy rice fields in different areas, and in emissions from different animal species. However, the estimated emissions of methane from the countries participating in the ALGAS study are given in Table 6.7 which also provides the data on nitrous oxide emissions of which agriculture and fuel combustion are the principal sources.

REGIONAL POLICIES AND RESPONSES

A number of policy measures and technologies are available to improve the quality of the atmospheric environment. Many of these approaches are applicable to reducing local air pollution, as well as the emissions of particulates and gases that contribute to regional and global atmospheric pollution. Frequently, a combination of several approaches is used to achieve policy goals. This section highlights some of the key steps taken in combating climatic pollution in the region.

A. Local Air Pollution

1. Ambient Air Quality Standards

A crucial step in improving air quality is the establishment of standards that define what constitutes acceptable levels of particular pollutants in the ambient air. The World Health Organization (WHO) has suggested ranges of acceptable ambient air quality standards (WHO/UNEP 1992), with the middle point of the range normally used to compare the current status of air quality in different locations.

In view of their widespread use, the standards recommended by WHO are reproduced in Table 6.8.

A large number of Asian countries have developed their own ambient air quality standards for the principal pollutants, as well as emission standards for power plants, selected industries, and vehicles. Reflecting their sizes and the diverse situations, People's Republic of China and India have taken the approach of setting different air quality standards (see Table 6.9) for different types of locations, with more stringent standards for conservation and tourist areas, somewhat less strict for residential urban and rural areas, and more permissive standards for industrial areas or areas with heavy traffic.

2. *Emission Standards*

Emission standards place restrictions on the amount of emission industrial facilities such as power plants and industrial boilers are allowed to discharge into the atmosphere, and may also be enforced for vehicles. Although ambient and emission standards have been in place for a number of years, air quality continues to deteriorate as enforcement is lacking in many countries. Nevertheless, some countries are making headway, for example, Malaysia introduced Environmental Quality (Control of Emission from Petrol Engines) Regulations on 1st November 1996. These new regulations focused on a preventive approach toward emissions control by control of vehicular emissions in the manufacturing or assembly stage. Beginning 1st January 1997, new models of motor vehicles were required to comply with certain emission standards before they could be sold (Malaysia Environmental Quality Report 1996).

3. *Fuel Quality and Use Regulations*

A number of countries in the Asian and Pacific Region have set limitations on the amount of specific pollutants allowed in fuels. For example, the sulphur content of coal or oil, or the benzene, lead, or volatility levels of fuel may be limited. Many of the

Table 6.8 Ambient Air Quality Guidelines Recommended by the World Health Organization

Parameter g/m^3	10 minutes	15 minutes	30 minutes	1 hour	8 hours	24 hours	1 year	Year of standard
SO_2	500					125[a]	50[a]	1987
SO_2						100-150	40-60	1979
BS[b]						125[a]	50[a]	1987
BS[b]						100-150	40-60	1979
TSP						120[a]		1987
TSP						150-230	60-90	1979
PM10						70[a]		1987
Lead							0.5-1	1987, 1977[b]
CO		100	60	30	10			1987
NO_2				400		150		1987
NO_2				190-320[c]				1977[b]
O_3				150-200	100-120			1987
O_3				100-200				1978

Source: WHO/UNEP 1992

Notes:
a) Guideline values for combined exposure to sulphur dioxide and suspended particulate matter (they may not apply to situations where only one of the components is present).
b) Application of the black smoke value is recommended only in areas where coal smoke from domestic fires is the dominant component of the particulates. It does not necessarily apply where diesel smoke is an important contributor.
c) Not to be exceeded more than once per month.

Suspended particulate matter measurement methods

BS = Black smoke; a concentration of a standard smoke with an equivalent reflectance reduction to that of the atmospheric particles as collected on a filter paper.

TSP = Total suspended particulate matter; the mass of collected particulate matter by gravimetric analysis divided by total volume sampled.

PM10 = Particulate matter less than 10 in aerodynamic diameter, the mass of particulate matter collected by a sampler having an inlet with 50 per cent penetration at 10 aerodynamic diameter determined gravimetrically divided by the total volume sampled.

TP = Thoracic particles (as PM10).

IP = Inhalable particles (as PM10).

Table 6.9 Ambient Air Quality Standards in People's Republic of China and India (micrograms per cubic metre)

Pollutant	Averaging time	China			India		
		Class 1	Class 2	Class 3	Sensitive Areas	Residential & Rural	Industrial
Total Suspended Particulates (TSP)	Daily	80	200	300	100	200	500
	Annual				70	140	360
TSP < 10 microns	Daily	50	150	250	75	100	150
	Annual	40	100	150	50	60	120
Sulphur Dioxide	Daily	50	150	250	30	80	120
	Annual	20	60	100	15	60	80
Nitrogen Oxides	Daily	100	100	150	30	80	120
	Annual	50	50	100	15	60	80
Carbon Monoxide	Daily*	4 000	4 000	6 000	1 000	2 000	5 000
Lead	Annual		1 000				

Source: WHO and UNEP 1992; CPCB 1994; World Bank 1997; TERI 1998
Note: * the averaging time for India's CO standard is 8 hours.

industrialized countries of the region, including Australia, Japan, Republic of Korea, New Zealand, and Singapore, have phased out lead content in vehicle fuels. Other countries have set standards at the lowest possible level at which older vehicles can still operate.

4. *Licensing and Zoning*

Most industrial facilities require licenses to operate, and their construction frequently requires an environmental impact assessment (EIA) to determine whether and where facilities can be built, and what steps they need to take before being allowed to operate. However, it is important to take into consideration not only the particular facility being licensed, but the cumulative impact of associated or nearly industrial facilities. The large numbers of coal-fired power plants in the Mae Moh area of Thailand, around Shanghai, and in Singrauli (India) for example, have resulted in high concentrations of air pollutants over vast areas of those countries. In response, zoning is used by many countries in Asia and the Pacific to designate locations for polluting industries. These areas are usually designated at some distance from residential areas, but in many developing countries workers and their families frequently live in close proximity.

5. *Enforcement Mechanisms*

Without effective enforcement mechanisms, even the best regulations are unlikely to improve air quality. Several legal instruments are available for enforcement, such as the revocation of licenses to operate facilities and prosecution of polluters. In Asia and the Pacific, Singapore has been particularly strict in the enforcement of environmental regulations. New Delhi is amongst the cities that have begun to take action against highly polluting vehicles, and People's Republic of China has started levying fines to industry for non-compliance.

6. *Economic Instruments*

During the last few years, the role of the private sector in the economic development of the region has increased considerably (see Chapter 13). This provides opportunities for supplementing traditional regulatory instruments with economic instruments such as tax deduction on unleaded gasoline (Thailand) and providing subsidies for the purchase of electricity from wind turbines and other environment friend technologies.

B. **Transboundary Air Pollution**

1. *Haze and Smog*

The severe haze from forest fires during 1997 led the ASEAN Ministers of Environment to endorse (ASEAN 1998) the Regional Haze Action Plan (RHAP). The Plan sets out cooperative measures needed among ASEAN countries for addressing the problem, aiming at three priority areas:

- prevention of forest fires through better management policies and enforcement;
- establishing operational mechanisms for monitoring;
- strengthening regional land and forest – fire fighting capability, as well as other mitigation measures.

CHAPTER SIX

Under the RHAP, lead countries have been appointed to act as focal points for the implementation of the Plan; Indonesia is taking the lead in mitigation, Malaysia in prevention, and Singapore in monitoring. The ADB is providing financial and technical assistance for the implementation of RHAP (see Chapter 13). The countries that were affected by the haze episode during 1997 and early 1998 have already formulated a number of policy responses with a key focus on health, and have also initiated implementation of several actions. For example, Brunei Darussalam has developed a National Action Plan, which includes health guidelines; installation of one fully equipped air monitoring station, and eight PM10 stations; more stringent laws on open burning; and provision of public education through a Haze Information Centre. A National Committee on Haze was also set up, and is co-operating with other neighbouring countries through regional coordinating mechanisms.

In Indonesia, provincial health officials monitor health quality, strengthen the surveillance of haze-related diseases, distribute masks to high risk groups, develop guidelines for health personnel to respond to haze related emergencies, and have established an early warning system for future haze disasters. Long-term effects of haze are also being assessed.

Haze Task Forces/Committees have also been established in countries such as the Philippines, Singapore and Malaysia. The Task Forces are responsible for monitoring the movement of haze, determining the related health hazards and serving as official sources of information. In Singapore a Haze Action Plan has been formulated and is activated when the 24-hour PSI level exceeds 50, and is stepped up when the PSI level reaches 200.

Due to the relatively minor impact of the haze episode on Papua New Guinea, the mitigation measures undertaken by the government have focused on education and information dissemination activities designed to minimize slash and burn practices amongst subsistence farmers.

Responding to the public demand for local air quality data, the initial emphasis in Thailand has been on monitoring air quality rather than on prevention and mitigation. The Ministry of Health have also set up a coordinating centre for public support, and have produced a set of guidelines for public support during haze episodes, covering such aspects as air quality monitoring, health risk communication and public advice on protection measures. Subsequent activities have included the generation of air quality monitoring and meteorology data for a haze warning system.

2. *Acid Rain*

In view of the large use of coal in the Asian and Pacific Region, and rising environmental concerns worldwide, there has been a great deal of interest in the use of "Clean Coal Technologies". Prominent among these are various approaches to Fluidized-Bed Combustion, which not only reduce the emissions of sulphur oxides, but permit poorer grades of coal to be used as fuel. Australia and People's Republic of China are amongst the world's leaders in the development of this technology. Another promising area is that of gasifying coal and then using the gas as a fuel. Cost considerations have however been an obstacle in the use of most of these technologies in the region to date.

The Asia Pacific Economic Cooperation (APEC) has established several Working Groups dealing with Regional Energy Cooperation, including one that is focusing on "Clean Fossil Energy". The reports of these working groups (e.g. APEC 1997) are a useful source of information for the countries of the region.

As mentioned earlier, the emissions of SO_2 in Asia could reach 41 Tg per year by 2020, if no additional measures are taken. The use of basic control technologies (BCT) could reduce these emissions to about 26 Tg, and using advanced control technologies (ACT) to 21 Tg (Streets et al 1999). Moreover, using the best available technology (BAT), which implies the installation of state of the art Flue Gas Desulphurization (FGD) systems, could reduce SO_2 emissions in 2020 to 4.7 Tg, a reduction of 69 per cent from current levels. The annual costs associated with the emission reductions under these scenarios are estimated at US$12 billion for the BCT scenario, US$ 14 billion for the ACT scenario, and US$36 billion for the BAT scenario.

People's Republic of China has adopted a number of measures to reduce acid rain such as the use of clean coal, desulphurization technologies, and the imposition of a levy on sulphur dioxide emissions (UNEP 1999). In addition, People's Republic of China has considerably improved the efficiency of energy use and switched to less polluting energy sources.

Efforts are also increasing to promote subregional cooperation on transboundary pollution in Northeast Asia through the Northeast Asia Subregional Programme of Environmental Cooperation (NEASPEC) (see Chapter 19). In South Asia, the Republic of Maldives (amongst others) has recently ratified the *"Malé Declaration on Control and Prevention of Air Pollution and its Likely Transboundary Effects for South Asia"*. The aim of the Declaration is to achieve inter-governmental cooperation to address the increasing threat of transboundary air pollution and consequential impacts due to concentrations of

ATMOSPHERE AND CLIMATE

pollutant gases and acid deposition on human health, ecosystem function and corrosion of materials. Besides laying down the general principles of inter-governmental cooperation for air pollution abatement, the Declaration sets up an institutional framework linking scientific research and policy formulation. The Declaration also calls for the continuation of this process in stages, with mutual consultation, to draw up and implement national and regional action plans and protocols based on a fuller understanding of transboundary air pollution issues.

RESPONSES TO GLOBAL ISSUES

A. Ozone Depletion

The Multilateral Fund set up under the Montreal Protocol, and the Global Environment Facility, has been assisting the region in meeting the goals of the Protocol. People's Republic of China, the largest producer and consumer of CFCs and halons, has banned the establishment of new facilities producing these compounds, and was expected to meet the 1999 consumption target (UNEP 1998). The countries of Central Asia have also made considerable progress in reducing the consumption of ozone depleting substances (ODS). The consumption figures for recent years, as well as projections up to 2001 (Oberthur 1999), for three of these countries are provided in Table 6.10. The consumption of the ODS is expected to be phased out completely in these countries during the period 2001-2003.

B. Climate Change

Many of the response strategies for addressing global climate change are also effective for reducing the "traditional" air pollutants such as particulates and sulphur dioxide. The adoption of strategies, primarily involving improvements in the efficiency of energy use, and greater use of energy sources other than coal, automatically result in lowering the emissions of carbon dioxide (CO_2), the largest contributor to climate change.

1. *Improving the Efficiency of Energy Use*

There is usually a linear relationship between the amount of energy used and the emission of air pollutants. Utilizing equipment that produces the same output while requiring less energy is frequently one of the most cost-effective approaches to improving air quality. Newer refrigerators and computers, for example, use substantially less energy than their predecessors. Compact fluorescent lighting, increasingly used in offices, is another example of a technology that can reduce energy use by more than 50 per cent, while providing the same amount of lighting. One measure of energy efficiency is the ratio of energy use to Gross Domestic Product (GDP). People's Republic of China is an example of a country that has been able to increase its GDP during the past 20 years at almost twice the rate of increase of energy use.

The Global Environment Facility and UNDP are promoting a number of projects to assist countries in the region to assess their emissions and to formulate strategies to reduce them. For example, the countries participating in the ALGAS project identified a number of mitigation options in the energy sector, as shown in Table 6.11. Most of these represent opportunities for improving energy efficiency.

2. *Reducing Emissions of Carbon Dioxide*

For many of the air pollutants, such as suspended particulates and sulphur dioxide, it is possible to install devices like electrostatic precipitators and scrubbers that can reduce emissions by 90 per cent or more. At present, such an option is not economically feasible for reducing carbon dioxide emissions. The two most effective strategies for reducing emissions of CO_2 are, improving the

Table 6.10 Consumption of Ozone Depleting Substances in Azerbaijan, Turkmenistan and Uzbekistan

ODS	Azerbaijan			Turkmenistan			Uzbekistan		
	1996	1998*	2000*	1996	1998*	2000*	1996	1998*	2000*
CFCs	459.4	173.0	68.0	29.6	n.a.	12.0	260.3	233.0	156.0
Halons	501.1	0.0	0.0	0.0	0.0	0.0	0.0	0.0	0.0
Carbontetrachloride	0.0	0.0	0.0	0.0	0.0	0.0	11.5	11.5	7.5
Methyl chloroform	0.0	0.0	0.0	0.0	0.0	0.0	0.4	0.4	0.2
HCFCs	5.1	n.a.	n.a.	1.5	n.a.	n.a.	n.a.	n.a.	n.a.
Total ODS	**965.6**	**n.a.**	**n.a.**	**31.1**	**n.a.**	**n.a.**	**272.2**	**244.9**	**163.7**

Source: Oberthur 1999
Note: * indicates projections
n.a. indicates data not available

CHAPTER SIX

Table 6.11 Options for Reducing Emissions from the Energy Sector in 11 Asian countries

Category of Mitigation Option	Countries Considering the Option	Specific Options Considered
Energy and Transformation		
Improving energy efficiency of existing facilities	PR China, Mongolia, Myanmar, Pakistan, Philippines, Viet Nam	Power plant renovation, electricity T&D loss reduction, coal beneficiation
Adopting more energy efficient techniques in new capital stock	Bangladesh, PR China, India, Indonesia, Republic of Korea, Myanmar, Pakistan, Philippines, Thailand	Combined-cycle generation, Advanced coal technologies
Utilizing low/zero emission energy sources	PR China, India, Indonesia, Mongolia, Myanmar, Pakistan, Philippines, Thailand, Viet Nam	Hydropower, wind power, biomass-fired power, nuclear power, geothermal power, natural gas
Industry		
Improving energy efficiency of existing facilities and equipment	Bangladesh, PR China, India, Mongolia, Myanmar, Pakistan, Philippines, Thailand	Boiler efficiency improvement, housekeeping/energy management, industry specific process improvements
Adopting more energy efficient techniques in new capital stock	Bangladesh, PR China, India, Republic of Korea, Mongolia, Myanmar, Pakistan, Philippines, Thailand, Viet Nam	Efficient motors, efficient boilers, higher efficiency process in specific industries
Residential/Commercial		
Improving energy efficiency of existing facilities and equipment	Bangladesh, PR China, India, Republic of Korea, Mongolia, Pakistan, Thailand	Lighting system improvements, building insulation
Adopting more energy efficient techniques in new capital stock	Bangladesh, PR China, India, Republic of Korea, Mongolia, Myanmar, Pakistan, Philippines, Thailand, Viet Nam	Improved biomass stoves, CFLs and other efficient lighting systems, efficient refrigerators, efficient air conditioners, efficient boilers
Utilizing low/zero emissions energy sources	Bangladesh, PR China, India, Indonesia, Pakistan	Solar water heaters, PV lighting, biogas
Transportation Sector		
Improving energy efficiency of existing vehicles and systems	Bangladesh, PR China, Republic of Korea, Mongolia, Myanmar, Pakistan	Improved vehicle maintenance, improved bus service, rail system improvements
Adopting more energy efficient techniques in new capital stock	Bangladesh, PR China, India, Republic of Korea, Pakistan, Philippines, Thailand	Efficient trucks, efficient 2/3 wheelers, efficient automobiles
Utilizing low/zero emissions energy sources	Bangladesh, India, Indonesia, Republic of Korea	CNG vehicles, electric vehicles, ethanol vehicles

Source: ADB 1998

efficiency of energy use and increasing the use of low carbon or non-carbon energy sources.

3. *Increasing the Use of Low Carbon or Non-Carbon Energy Sources*

Emissions of CO_2 as well as of traditional air pollutants can be reduced by fuel substitution. Replacing coal and oil with natural gas where feasible can improve air quality. Japan was the first country in Asia to make a policy decision to use liquefied natural gas for electricity generation in areas where air pollution was already high. In addition, the Government of Japan has initiated actions in local government to combat global warming (Box 6.6). In more recent years, the Republic of Korea has also become an importer of liquefied natural gas. Moreover, Indonesia, Malaysia, Brunei Darussalam, and Australia have benefited from this trend, and are four of the five largest exporters of the fuel in the world.

With the exception of biomass, none of the renewable sources of energy such as solar (photovoltaics – PVs), wind farms, and hydropower emit greenhouse gases when operating. Starting from a small base, the production and installed capacity of PVs and wind power have been increasing rapidly in recent years. Japan is now the second largest producer of PVs in the world, and during the 1990s, India has emerged as a major producer of electricity from wind power. The installed capacity for wind power in the different states of India is shown in Table 6.12.

Without the large scale implementation of mitigation options in the energy sector, the emissions

Box 6.6 Measures by Local Government in Japan to Combat Climate Change

Since 1993, local governments in Japan, both at prefecture and municipality level, with the provision of subsidies from the Environment Agency of Japan have established local plans for implementing measures against global warming. The plans are divided into two parts. The first part of the plan provides for facilitating measures against global warming and the second part provides for the implementation of measures against global warming directly by local governments themselves. Elements of the plan include: assessment of the current situation of emissions and removal of Greenhouse Gas within the prefecture/towns; listing of possible measures against global warming by sectors and by actors; setting targets for emission reduction (if possible); and identifying challenges, opportunities and relevant conditions for the implementation of measures set forth.

In addition to the subsidies for establishing plans, the Environment Agency since 1997, has also started to provide additional subsidies to local governments for implementing model measures against global warming. The model measures that qualify for subsidies are: bicycle promotion projects; greening of Government office operations with both hard and soft measures; introduction of CNG operated buses for public transportation; and R&D for coal fired boilers for Activities Implemented Jointly (AIJ) under UNFCCC.

As of July 2000, 29 (of 47) prefectures and 19 (of more than 3 000) municipalities have established plans to promote measures for the control of greenhouse gases. Moreover, the local governments have been formulating their own Action Plans for reduction of greenhouse gas emissions. The formulation of such action plans are mandatory in the provision of the Law Concerning the Promotion of Measures to Cope with Global Warming, which entered into force in April 1999, as part of the initial response to the adoption of the Kyoto Protocol of COP3/UNFCCC.

With their increasing involvement in such activities, local governments in Japan are not only expected to become climate conscious in their overall operations but are also expected to promote the activities of other stakeholders such as citizens, factories and offices within their own jurisdictions.

Source: Government of Japan

Table 6.12 Installed Capacity of Wind Power in Indian States, 1997-1998

Wind power installed capacity (MW)

State	Commercial Projects		
	Additions in 1997/1998	Total	Cumulative Capacity
Tamil Nadu	31.14	687.94	707.30
Gujarat	20.10	149.57	166.91
Andhra Pradesh	1.50	52.74	55.79
Karnataka	11.17	14.44	17.01
Kerala	–	–	2.03
Maharashtra	0.23	0.99	5.60
Madhya Pradesh	2.70	11.70	12.29
Orissa	–	–	1.10
Others	–	–	0.47
Total	66.83	917.38	968.48

Source: MNES 1997

of carbon dioxide from the use of fossil fuels in the eleven ALGAS countries are expected to more than triple from 1990 to 2020. Implementation of the mitigation options could reduce these future emissions by 30-50 per cent (ADB 1998).

4. *Enhancing Sinks of Carbon Dioxide*

Trees and plants absorb carbon from the atmosphere through photosynthesis and release it through respiration and decay. However, deforestation, mainly in the tropics, currently offsets the absorptive capacity by about 2 billion tonnes of carbon each year. This amount can be substantially reduced by increasing the land brought under forestry, and by planting more trees (where feasible) in existing forests. This is already beginning to happen in many countries of Asia and the Pacific, including People's Republic of China, India, and Japan. The ADB, the Global Environment Facility, UNDP, the World Bank, and other development assistance agencies are providing funding for many of the reforestation projects.

5. *Reducing Emissions of Methane and Other Greenhouse Gases*

The degree of difficulty in reducing emissions of any greenhouse gas depends on the source of that gas. It is much easier to reduce emissions of carbon dioxide from a power plant, for example, through improving the efficiency of the boiler, than to reduce emissions of methane from individual livestock animals grazing in the fields. Improved feedstock has been developed to reduce such emissions, but it is still too expensive in most cases to transport it over large distances (which can cause subsidary impacts). Methane from landfills, coal, mines, and oil production, is beginning to be utilized as energy in a number of countries in the region, including People's Republic of China, and India. The options

Table 6.13 Options for Reducing Future Emissions of Methane in the Agricultural Sector in Selected Countries

Category and Type of Mitigation Options	Countries Considering the Option	Features	Mitigation Potential CH_4 Kg/ha or animal/year
Livestock Sector Providing mineral blocks/MNB	Indonesia, PR China	10-30 per cent increase in milk yield (only for dairy cattle) Enhances protein use efficiency Enhances feed conversion efficiency	15.4 (3.8-27)
Molasses-Urea block	Indonesia, Bangladesh, Myanmar, India	Increases feed conversion efficiency 25 per cent increase in milk yield CH_4 reduced by 27 per cent 60 per cent increase in animal productivity	14.0
Urea treatment of straw	PR China, Indonesia, Myanmar, Viet Nam	Rice straw soaked in 2 per cent urea for 15d, improves digestibility up to 25 per cent, 15-20 per cent achievable in field, milk yield increases by 20-30 per cent	6.1 (3.8-8.3)
Chemical/Mechanical feed treatment	Viet Nam, Republic of Korea	Improves digestibility by 5 per cent Enhances weight gain (6 kg/yr) 10-30 per cent reduction in CH_3	10 (5-15)
Genetic improvements	Indonesia	10 per cent reduction in CH_4 (IPCC) 160 per cent increase in milk yield	8.3
Manure Management Biogas plants	Indonesia, Republic of Korea, PR China	70 per cent reduction in CH_4 emissions (where lagoons are utilized)	2-39
Rice Production Intermittent drainage (3-4 times per season) Low CH_4 emitting varieties	Philippines, PR China	Transports less CH_4 from soil to air Tested in few countries only e.g. IR-64	3.7-38
Livestock Sector Using composted organic matter	China, Philippines	Estimated at 50 per cent CH_4 reduction (NR)	48-128 24-62%
Dry-seeded nursery	PR China	Reduces period of flooding	14.4 5-23%
No tillage	Indonesia	Brings about 12 per cent reduction in CH_4 emissions	22.9
Ammonium sulphate usage	Philippines, Indonesia	Competes with methane bacteria and suppresses CH_4 production by about 20 per cent (IPCC)	5.5 (1-10)

Source: ADB 1998

available for reducing future emissions of methane from the agriculture sector are shown in Table 6.13.

6. *Adaptation Strategies*

While efforts of the global community to reduce emissions of greenhouse gases continue, it is necessary to acknowledge that it may be several decades before the emissions stabilize and subsequently decline. Due to the long life in the atmosphere of many of the greenhouse gases, including carbon dioxide, a degree of climate change is very likely to occur. Some countries are therefore suggesting that it may be prudent to begin planning to minimize the possible adverse effects in some areas that may be particularly vulnerable, such as coastlines, areas already suffering from drought, and small island states.

7. *International Agreements on Global Climate Change*

An important outcome of the United Nations Conference on Environment and Development, held at Rio de Janeiro in June 1992 was the signing by world leaders of the United Nations Framework Convention on Climate Change (UNFCCC). More than 165 countries have already signed. By definition, UNFCCC provides a general framework for steps that individual countries might take to address the problem of global climate change. It does not set specific targets and timetables for reducing emissions;

rather *the Conference of the Parties* (COP) has an annual meeting to determine the details for implementing the Framework.

At the third meeting (COP3) in Kyoto at the end of 1997, the countries agreed to a Protocol that commits the industrialized countries to reduce their combined emissions of greenhouse gases from their 1990 levels by the period 2008-2012. Specific targets were agreed to by each of the industrialized countries. Of the countries in the ESCAP region, Japan agreed to a reduction of 6 per cent, New Zealand agreed to keep its emissions at the same level as in 1990, and Australia negotiated an agreement to be able to increase its emissions by 8 per cent above its 1990 level.

Although the Kyoto Protocol represented a major step forward, a great number of issues still remain. For example, at present there is no binding requirement for the developing countries to reduce even the rate of growth of their emissions.

CONCLUSION

Deteriorating quality of urban air, transboundary pollution including haze and acid rain, and the greenhouse effect are the major problems facing the atmospheric environment in the region. Urban air quality has deteriorated in the wake of rapid growth in urbanization, increasing traffic, rapid industrialization and increased energy consumption. Whether as a result of the effect of earth's long-term weather patterns, or a factor of atmospheric pollutants, the threat from global warming has several long-term implications including the potential for sea level rise which would be catastrophic for many coastal areas of the region.

Depending on the rate and extent of climatic fluctuations, the global sea level may rise by as much as 0.95 metres by 2100, up to five times as much as during the last century. The human cost of this could be enormous in Asia and the Pacific because the region has vast coastlines, large amounts of productive land in low-lying areas, and large concentrations of people in coastal cities or near the sea. The densely populated river deltas of Bangladesh, People's Republic of China, Indonesia, Viet Nam as well as Small Island developing states in the South Pacific, are particularly vulnerable. Besides sea level rise, other important consequences of global warming are an increase in climate-related natural disasters (floods, droughts, and storms) and the disruption of agriculture and biodiversity due to changes in temperature, rainfall and winds. The effects may be quite severe on coastal mangrove forests, wetlands and coral reefs. For example, it has been observed that a 0.25 metre rise in sea levels could destroy about half of Asia's remaining wetlands.

The threats posed by haze, acid rain and transboundary pollution have also increased substantially in recent years. Incidents of haze have occurred from time to time, but the most serious episode occurred in 1997 and 1998, where forest fires affected 12.4 million people in Indonesia alone, and extended to neighbouring countries such as the Philippines, Papua New Guinea, Brunei Darussalam, Singapore and Thailand. Acid rain has also become a major concern in several parts of the region, particularly Northeast Asia. At least two thirds of acid depositions in the region are caused by coal-fired power plants with minimal or outdated pollution control equipment. Moreover, sulphur and nitrogen oxides emissions can have serious transboundary impacts, as they can be carried for hundreds of miles. Given the projected growth of energy consumption in the region, the need for urgent and effective emissions controls is therefore paramount.

Part Two

CHAPTER 7-10

PART I

CHAPTER	1	LAND
CHAPTER	2	FORESTS
CHAPTER	3	BIODIVERSITY
CHAPTER	4	INLAND WATER
CHAPTER	5	COASTAL AND MARINE
CHAPTER	6	ATMOSPHERE AND CLIMATE

PART II HUMAN ECOSYSTEMS-EMERGING ISSUES

CHAPTER	7	URBAN ENVIRONMENT
CHAPTER	8	WASTE
CHAPTER	9	POVERTY AND ENVIRONMENT
CHAPTER	10	FOOD SECURITY

PART III

CHAPTER	11	INSTITUTIONS AND LEGISLATION
CHAPTER	12	MECHANISMS AND METHODS
CHAPTER	13	PRIVATE SECTOR
CHAPTER	14	MAJOR GROUPS
CHAPTER	15	EDUCATION, INFORMATION AND AWARENESS

PART IV

CHAPTER	16	SOUTH ASIA
CHAPTER	17	SOUTHEAST ASIA
CHAPTER	18	SOUTH PACIFIC
CHAPTER	19	NORTHEAST ASIA
CHAPTER	20	CENTRAL ASIA

PART V

| CHAPTER | 21 | GLOBAL AND REGIONAL ISSUES |
| CHAPTER | 22 | ASIA AND THE PACIFIC INTO THE 21st CENTURY |

Chapter Seven

The growth of large cities has been accompanied by the expansion of slums.

7

Urban Environment

INTRODUCTION
STATUS AND TRENDS IN URBANIZATION
URBANIZATION PROCESS AND ENVIRONMENTAL QUALITY
POLICIES AND PROGRAMMES FOR URBAN ENVIRONMENTAL MANAGEMENT
CONCLUSION

CHAPTER SEVEN

INTRODUCTION

Cities have played a central role in the cultural, economic and political evolution and development of the Asian and Pacific Region. However, cities have also been significant contributors to the degradation of the region's physical environment both as vociferous consumers of resources and relentless emitters of pollutants. This chapter describes the state and trend of urbanization in Asia and the Pacific and provides an analysis of the policies and programmes on urban environmental management at the national, regional and international levels that have been undertaken to cope with the environmental problems and emerging challenges.

STATUS AND TRENDS IN URBANIZATION

Overall some 37 per cent of the people of the Asian and Pacific Region live in urban areas (United Nations 1998), although there is considerable range in the extent of the urbanization, both across and within the sub-regions. For example, the higher degree of urbanization in Australia and New Zealand contrast markedly with the much less urbanized Pacific Island countries and is sufficiently high to ensure that the South Pacific sub-region is on average the most urbanized in the region, whilst South Asia is the least urbanized followed by Southeast and Northeast Asia (ESCAP 1999). The diversity of countries in the region in terms of levels of economic development has resulted in even greater variation in the level of urbanization. For example, urbanization ranges from a minimum of seven per cent in Bhutan to 100 per cent in Singapore (Table 7.1). It is projected that some of the big countries of the region like People's Republic of China, Indonesia and Pakistan where current urbanization levels are well below 50 per cent, will cross this figure by the next quarter of the century, whilst India's urbanization level will also approach 50 per cent by the year 2030.

A. Degree and Growth of Urbanization

The degree of urbanization in the countries of the Asian and Pacific Region is presented in Table 7.2, Each of the eight highly urbanized (75 per cent and above) countries are industrially advanced nations, (with the sole exception of Brunei Darussalam, which is very rich in oil resources) whilst the eight with low urbanization (25 per cent and below) comprise countries where the levels of economic development and per capita income are low.

In the region as a whole, urbanization continues to grow at an average rate of 2.2 per cent per annum with higher rates being experienced in Southeast Asia (Figure 7.1). As one might expect, those countries with existing high levels of urbanization tend to be experiencing slower rates of urban growth (Table 7.3). In general, urban growth is expected to continue rapidly in most developing countries of the region, although, in the longer term, growth may be expected to slow as high levels of urbanization and socio-economic development are achieved.

The major reasons for increasing urban population are rural to urban migration (and, to a lesser extent, international migration) and the re-classification or expansion of existing city boundaries to include populations that were hitherto classified as being resident outside the city limits; these are estimated to contribute about 60 per cent of the region's urban growth in the near future (Figure 7.2). During the 1990s, Northeast Asia had the highest rate of urban growth due to migration and re-classification primarily as a result of the major changes in human settlements in People's Republic of China and Republic of Korea. Southeast Asia followed with more than half of urban growth accounted for by migration and re-classification especially in Thailand, Indonesia, Malaysia, Lao People's Democratic Republic and Cambodia.

Urban growth due to natural increase is most prominent in South Asia and the South Pacific and in those countries, such as the Iran (Islamic Republic of), New Zealand and Sri Lanka, where the capital cities are the main urban centres. The contribution of natural increase to urban growth is a complicated issue. In the case of the developed countries, i.e. areas such as Japan, Singapore, Australia and New Zealand, the rates of natural increase have generally been falling rapidly over the last twenty years, owing to the better public health infrastructure, and wider acceptance of family planning practices. These countries have already experienced the transition to an urbanized society, and their cities are growing slowly. On the other hand, the rates of natural increase rose in many developing countries, mainly due to improved housing, sanitation and medical delivery systems resulting in higher life expectancy among adults and lower infant mortality rates.

B. Urbanization and the City Size

The cities of Asia and the Pacific Region can be classified into four categories according to their population: 10 million plus (megacities); 5-10 million (large); 1-5 million (medium); and 0.5-1 million (small). In 1990, seven of the world's fourteen megacities were located in the region, by 1996 the

Table 7.1 Urbanization Trends in the Asian and Pacific Region, 1999-2030

Region/ Country	Degree of Urbanization						
	1999	2005	2010	2015	2020	2025	2030
Northeast Asia	40						
People's Republic of China	34	38.4	42.3	45.9	49.1	52.2	55.2
Democratic People's Republic of Korea	63	64.7	66.7	68.9	71.0	72.9	74.7
Hong Kong, China	96	96.1	96.4	96.7	96.8	97.0	97.2
Japan	79	79.8	80.9	82.0	83.2	84.3	85.3
Mongolia	63	66.0	68.4	70.5	72.4	74.3	76.0
Republic of Korea	85	89.3	91.2	92.2	92.7	93.1	93.6
Russian Federation	77	79.3	80.7	82.0	83.1	84.2	85.2
Southeast Asia	38						
Brunei Darussalam	72	74.8	76.9	78.7	80.1	81.4	82.6
Cambodia	23	26.6	29.7	32.9	36.2	39.5	42.8
Indonesia	39	44.7	48.9	52.4	55.4	58.3	61.0
Lao People's Democratic Republic	23	26.4	29.5	32.7	36.0	39.3	42.6
Malaysia	57	60.6	63.6	66.2	68.5	70.6	72.5
Myanmar	27	30.2	33.4	36.7	40.0	43.3	46.6
Philippines	58	62.4	65.5	67.8	69.9	71.9	73.8
Singapore	100	100.0	100.0	100.0	100.0	100.0	100.0
Thailand	34	23.7	26.2	29.3	32.5	35.8	39.1
Viet Nam	20	20.6	22.1	24.3	27.3	30.4	33.7
South Asia	31						
Afghanistan	22	24.2	27.0	30.1	33.3	36.6	39.9
Bangladesh	21	24.3	27.5	30.8	34.0	37.3	40.6
Bhutan	7	8.4	9.9	11.6	13.5	15.6	17.9
India	28	30.5	33.0	35.9	39.2	42.5	45.8
Iran, Islamic Republic of	64	64.1	66.5	68.8	70.9	72.8	74.6
Maldives	28	30.4	33.1	36.3	39.7	43.0	46.2
Nepal	11	13.7	15.8	18.1	20.7	23.4	26.4
Pakistan	33	40.1	43.4	46.7	49.8	52.9	55.9
Sri Lanka	23	25.8	28.9	32.0	35.3	38.6	41.9
Central Asia	68						
Armenia	70	71.5	73.2	75.0	76.6	78.2	79.6
Azerbaijan	57	59.2	61.5	64.0	66.4	68.6	70.7
Kazakhstan	55	63.9	66.1	68.4	70.5	72.4	74.3
Kyrgyzstan	40	42.0	44.6	47.9	51.0	54.1	57.0
Tajikistan	33	34.4	36.8	40.1	43.4	46.7	49.9
Turkmenistan	45	47.0	49.3	52.4	55.4	58.2	60.9
Turkey	74	79.7	82.6	84.5	85.5	86.4	87.3
Uzbekistan	42	44.3	46.9	50.1	53.1	56.1	58.9
South Pacific	70						
American Samoa	52	55.3	57.9	60.6	63.2	65.6	67.9
Australia	85	84.8	85.3	86.0	86.9	87.7	88.5
Cook Islands	63	65.5	67.9	70.0	72.0	73.9	75.6
Fiji	42	44.6	47.3	50.5	53.6	56.5	59.3
French Polynesia	57-	58.2	60.2	62.8	65.3	67.6	69.7
Guam	39	40.9	43.4	46.7	49.9	52.9	55.9
Kiribati	37	39.3	41.7	44.6	47.8	51.0	54.0
Marshall Island	71	74.2	76.2	77.8	79.3	80.6	81.9
Micronesia (Federation State of)	29	32.3	35.6	38.9	42.2	45.5	48.7
New Caledonia	64	66.3	68.5	70.6	72.5	74.4	76.1
New Zealand	87	87.8	88.7	89.4	90.1	90.7	91.3
Niue	29	29.8	31.1	33.3	36.6	40.0	43.3
Northern Mariana Island	55	56.5	58.7	61.4	63.9	66.3	68.6
Palau	73	74.5	76.1	77.7	79.2	80.6	81.8
Papua New Guinea	17	19.1	21.2	23.7	26.7	29.8	33.0
Samoa	21	22.6	24.4	26.7	29.8	33.1	36.3
Solomon Island	19	22.5	25.5	28.6	31.7	35.0	38.3
Tonga	45	51.1	54.9	57.7	60.5	63.0	65.5
Tuvalu	51	57.0	60.9	64.1	66.5	68.7	70.8
Vanuatu	20	21.7	24.0	27.0	30.1	33.4	36.7
Asia and the Pacific	38						

Source: United Nations 1998 and ESCAP 1999

CHAPTER SEVEN

Table 7.2 Degree of Urbanization in the Asian and Pacific Region, 1999

Degree of urbanization	Number of countries	Countries with degree of urbanization
Less than 25%	8	Afghanistan (22), Bangladesh (21), Bhutan (7), Nepal (11), Sri Lanka (23), Cambodia (23), Lao People's Democratic Republic (23), Viet Nam (20),
25-50%	11	People's Republic of China (34), India (28), Kyrgyzstan (40), Maldives (28), Thailand (34), Pakistan (33), Tajikistan (33), Turkmenistan (45), Uzbekistan (42), Indonesia (39), Myanmar (27)
50-75%	9	Democratic People's Republic of Korea (63), Mongolia (63), Islamic Republic of Iran (64), Kazakhstan (55), Malaysia (57), Philippines (58), Armenia (70), Azerbaijan (57), Turkey (74),
75% and above	8	Hong Kong, China (96), Japan (79), Republic of Korea (85), Brunei Darussalam (72), Singapore (100), Australia (85), New Zealand (87), Russian Federation (77)

Source: ESCAP 1999

Note: Pacific Small Island Developing States have not been classified as the quantitative data gives very erroneous picture compared to countries of Asia.

Table 7.3 Annual Urban Growth Rate in the Asian and Pacific Region, 1999

Category	Per cent Annual urban growth	Number of countries	Name of countries with annual urban growth
Very Slow	Less than 1%	4	Kazakhstan (0), Russian Federation (0.1), Japan (0.4), Armenia (0.7)
Slow	1-2.0%	7	Democratic People's Republic of Korea (2.0), Republic of Korea (1.9), Singapore (1.3), Azerbaijan (1.4), Kyrgyzstan (1.3), Australia (1.1), New Zealand (1.3)
Medium	2.1-3.0%	12	Mongolia (2.9), Brunei Darussalam (2.8), Thailand (2.4), Viet Nam (2.1), India (2.8), Islamic Republic of Iran (2.9), Sri Lanka (2.5), Tajikistan (2.4), Turkmenistan (2.3), Turkey (3.0), Uzbekistan (2.6), Hong Kong, China (2.8)
High	3.1-5.0%	10	People's Republic of China (3.3), Cambodia (4.8), Indonesia (3.8), Malaysia (3.2), Myanmar (3.2), Philippines (3.5), Bangladesh (4.5), Maldives (4.6), Nepal (4.6), Pakistan (3.5)
Very High	5.1% and above	3	Lao People's Democratic Republic (5.5), Afghanistan (6.5), Bhutan (6.1)

Source: ESCAP 1999

Note: Pacific Small Island Developing States have not been classified as the quantitative data gives very erroneous picture compare to countries of Asia.

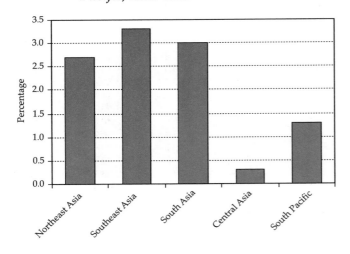

Figure 7.1 Rate of Urbanization in Asia and the Pacific, 1995-2030

Source: United Nations 1996 and ESCAP 1999

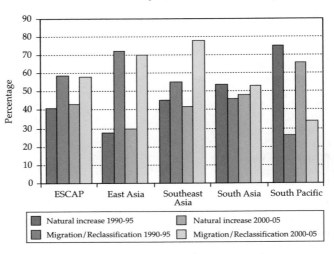

Figure 7.2 Components of Urban Growth in Asia and the Pacific, 1990-2005

Source: United Nations 1996 and ESCAP 1999

number of megacities in the region rose to nine and 1996 projections for early 2000 predicted that there would be twelve megacities in the region (Beijing, Calcutta, Delhi, Dhaka, Karachi, Metro Manila, Mumbai, Osaka, Seoul, Shanghai, Tianjin and Tokyo), although figures recently made available give the number as eleven, with the removal of Seoul and Tianjin and the inclusion of Jakarta. The projected number, size and rankings of cities in the Asian and Pacific Region are presented in Figures 7.3a, b and 7.4. Tokyo continues to be the largest urban agglomeration, both in the region and in the world, a position it has held since 1970 and which it has been projected to retain until 2015 (ADB 1996).

Figure 7.3a   Number of Cities by City Size in Asia and the Pacific in 2000 and 2015

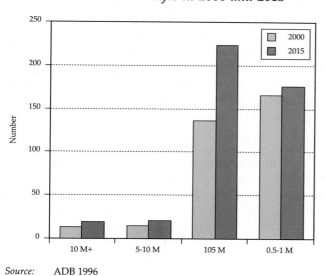

Source: ADB 1996

Figure 7.3b   Percentage of Cities by City Size in Asia and the Pacific in 2000 and 2015

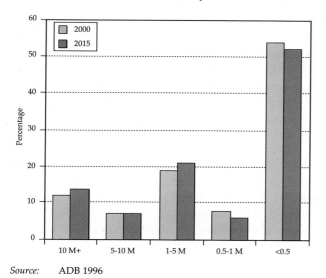

Source: ADB 1996

Figure 7.4   Annual Growth Rates of the Present or Likely to Become Mega-cities During 1975-1995 and 1995-2015 in the Asia and the Pacific

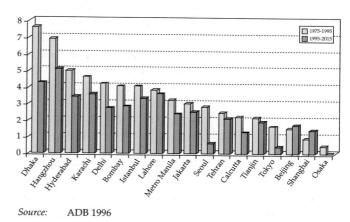

Source: ADB 1996

In addition, there are over 300 cities within the region that may be categorized as small, medium or large. Whilst the 164 small cities (with populations of between 500 000 and 1 million) accommodate only eight per cent of the urban population, the megacities are home to twelve per cent of the region's city dwellers, and the 137 medium sized cities accommodate about a fifth of the region's urban population.

Primacy, or the domination of a country by a single large city, in terms of concentration of population and economic activities, is relatively common in the region. In most cases, the dominant city is also the national capital and has risen to its position of primacy due to a range of factors including, *inter alia*, history, geography, stage of development, political system and government policies (Pernia 1998). There are certain countries, such as Singapore, the Maldives and the island nations of the South Pacific sub-region, where the national capital is the only urban centre.

URBANIZATION PROCESS AND ENVIRONMENTAL QUALITY

A.   Spatial Growth and Urban Environmental Quality

The population of cities in the developing countries of the Asian and Pacific Region has been estimated to have doubled between 1980 and 2000. In certain cities, urban land expansion has been even faster. Jakarta expanded four per cent annually in the first half of this decade, the population of several areas on its fringe grew as much as 18 per cent per year (Planet Earth 1999). According to one estimate some half a million hectares of arable land is being taken by urban developments each year in the developing countries of the region (USAID 1994). In India, it is estimated that on average, about 75 000 ha of agricultural land are lost every year due to urban expansion (Pathi 1992). In Pakistan, the Lahore Metropolitan Development Plan alone envisages transformation of 222 500 ha of agricultural land in Lahore and Sheikhupura districts to an urban locality by the year 2010. According to one estimate, Indonesian cities expanded by 376 000 ha over the period 1985-1995, which is likely to accelerate to 40 000 ha per year during the early part of this millennium. To compensate for this loss of land, crop production on the remaining land may become more intensive and potentially more environmentally damaging (WRI 1995).

Records of long term land-use change show that the urbanized area of countries of Asia and the Pacific has increased from about three (Pakistan) to 11 (Brunei Darussalam) times. The trend was particularly high between the 1980s and 1990s when buildings and infrastructure doubled in some countries and, in the case of Brunei Darussalam, even quadrupled. The increases in the built-up urban areas of some Asian and Pacific cities varied from 180 ha annually for Hong Kong, China, to 2 900 ha annually for Bangkok and Karachi. People's Republic of China has utilized, during the past three decades, about six million hectares of land largely for the construction of factories, public buildings, housing and roads. In the case of Zhangjiagang City, the built-up area increased from 5.6 km^2 to 12 km^2 with per capita increase in housing floor space from 13 m^2 to 18.5 m^2 Similarly, built-up area in Foshan City increased from 22.8 km^2 to nearly 3 000 km^2.

Besides agricultural land, natural areas such as forests, wetlands, and other fragile ecosystems are also lost to residential, industrial and tourism developments. Coastal areas with sensitive ecosystems are under intense pressure from urbanization in the region. Some coastal cities are expanding at a considerably high rate, and reclamation is taking place from the sea to satisfy the soaring demand for new urban land. Land reclamation activities range from draining and filling of marshes and other wetlands and constructing homes or resorts on beaches or dunes, to building seawalls and undertaking large-scale reclamation projects that extend the shoreline into the sea. According to a recent study by the WRI, roughly half of the world's coasts are threatened by development-related activities (Bryont et al 1995). In Singapore, for instance, the demand for land is so great that the island nation has added 6 000 hectares to its land area by filling along the shoreline, increasing its area by some 10 per cent in the last three decades (Sien 1992).

Similarly, over 450 000 hectares of coastal wetlands have been identified for reclamation in the National Plan for Land Development of the Republic of Korea and in Pakistan, mangroves near Karachi are also under considerable pressure as the urban population cuts trees for fodder, fuelwood, construction timber and clear felling to provide land for development.

B. Ambient Environmental Conditions

For most of the region's urban centres, the key environmental challenges are those associated with deteriorating air and water quality, persistent noise pollution and the management of municipal, industrial and hazardous waste. The status, trends and constraints associated with urban generated solid waste forms part of a wider examination of solid waste management, which is presented in Chapter 8.

With a few exceptions, ambient environmental conditions in the cities of the Asian and Pacific Region are far from satisfactory and urban dwellers face worse air and water quality, greater noise pollution and worse living conditions than their rural counterparts.

1. Air Pollution

The deterioration in air quality in urban areas is partially the result of increases in industrial and manufacturing activities, but primarily relates to the growth in the number of motor vehicles in the region. Motor vehicles are particularly concentrated in the urban areas and contribute significantly to the production of various types of air pollutants, including carbon monoxide, hydrocarbons, nitrogen oxides and particulates. For example, it is estimated that in a single day around 56 tonnes of carbon monoxide, 18 tonnes of hydrocarbons, 7 tonnes of nitrogen oxides, and just under one tonne each of sulphur dioxide and particulate matter are discharged through the tailpipes of vehicles in Kathmandu alone (Government of Nepal 1998). In Shanghai, the contribution of carbon monoxide, hydrocarbon, and nitrogen oxide emission by automobiles to the air was over 75, 93 and 44 per cent, respectively (Miankang et al 1997). By 2010, these emissions are estimated to increase further to 94, 98 and 75 per cent, respectively.

In Bangkok, Jakarta and Kuala Lumpur, the annual costs from dust and lead pollution are estimated at US$5 billion, or about 10 per cent of combined city income (World Bank 1996). Air pollution also pushes up the incidence and severity of respiratory-related diseases and cardiovascular disease, particularly amongst children and the elderly (see Box 7.1).

Cities in the more developed nations of the region have recorded improvements in ambient air quality in recent years. In Seoul, for example, sulphur dioxide (SO_2) pollution has fallen from a peak of 0.094 ppm in 1980 to 0.008 ppm in 1998 (Green Korea). These improved levels of pollutants are in line with a number of other cities in high-income countries, including Tokyo, Osaka, Melbourne and Sydney, although the cities of developing countries, such as Shenyang, New Delhi, Tehran and Jakarta, invariably exceed WHO Guidelines on concentrations of particulates and sulphur dioxide.

In addition to particulates and SO_2, nitrogen oxides (NO_x) are increasingly being recognized as a persistent and unhealthy component of urban living, even in the cities of developed countries. In the Republic of Korea, whilst levels of sulphur dioxide and total suspended particulates (TSP) have been declining, slight increases in concentration of other pollutants such as nitrogen oxides, ozone and carbon dioxide have been recorded. Air quality in Singapore has also significantly improved with the adoption of various strategies to prevent air pollution at source (Government of Singapore 1997). In particular, several countries of the region are now promoting the use of unleaded petrol.

2. *Water Pollution*

In many cities in the region's developing countries, the principal water bodies have become heavily polluted with domestic sewage, industrial effluents, dumped chemicals and solid wastes. As many of these water bodies are also relied upon as a source of domestic and industrial water, the available of clean, safe potable water and uncontaminated process water has become a major challenge to many city authorities (see Chapter 4).

The principle rivers flowing through Karachi, Lahore, Kabul and Peshawar are all heavily polluted, as is the Chao Phraya and numerous *klongs* (canals) in Bangkok, the Pasig and Tenajeros-Tullahan Rivers in Metro Manila, and the Ganges in India. Hong Kong, China's Victoria Harbour has also become heavily polluted due to daily dumping of 1.5 million tonnes of untreated sewage (EPD-Hong Kong 1999), whilst the Beira Lake in central Colombo is currently undergoing a massive clean-up operation after decades of receiving untreated industrial effluents and raw sewage. In People's Republic of China, organic pollutants have been the main cause of

---

Box 7.1 Costs of Urban Air Pollution

The rapid economic growth of Bangkok, Thailand, generated levels of pollutants and traffic congestion that carry significant costs in terms of both health and productivity.

A recent World Bank study identified air pollution from particulates and lead, surface water pollution due to micro-biological contamination, and traffic congestion as Bangkok's most serious urban environmental problems, and indicated that even moderate reductions in air pollution and congestion could provide significant benefits. Reducing ambient concentrations of key pollutants by 20 per cent from current levels, for example, would provide health benefits estimated at between $400 million and $1.6 billion for particulates and between $300 million and $1.5 billion for lead. For congestion, the study estimated that a 10 per cent reduction in peak-hour trips would provide benefits of about $400 million annually.

Growth in air pollutants displayed an alarming situation both for particulates and for lead. From 1983 to 1992, concentrations of particulates were up at all six monitoring stations in Bangkok; annual standards were violated at every station in every year since 1988. The World Bank study found that the Bangkok economy operates quite efficiently and that there are therefore few opportunities for "win-win" initiatives that would improve environmental quality without slowing economic growth. There are nevertheless some cost-effective initiatives that deal with the highest-priority problems.

In the area of energy-related air pollution, the study recommended managing demand and imposing emissions standards and taxes. Demand-side management initiatives include the use of energy-efficient lighting and appliances for residential and commercial users, improved building designs, and the use of more energy-efficient motors and production processes in the industrial sector. One way to reduce particulate emissions was identified as the development of incentives to reduce the use of lignite, a fuel that emits more particulates and sulphur dioxide than hard coal or fuel oil. Emissions standards that require new power plants to be fitted with low-sulphur control or combustion technologies and precipitators or a switch to hard coal (instead of ignite) and an increase in taxes on lignite also would be cost effective.

Air pollution should not only be viewed in economic terms as the costs to human health, but also include the issues of quality of life, suffering and death. Air pollution problems impact disproportionally on the poor and marginalized sectors of society, particularly children and the elderly who face the highest risk. Indeed, one of the region's highest causes of child fatality is respiratory disease caused through exposure to both indoor and outdoor pollutants.

Source: World Bank 1996

pollution in rivers passing through urban areas, especially in the northern part of the country (Government of China 1997). In Republic of Korea, although overall water quality in the rivers recovered slightly in 1996, some points along Nakdong and Youngsan rivers are highly polluted (Government of Korea 1997). Most rivers in Nepal's urban areas have also been polluted and their waters are now unfit for human use (Government of Nepal 1998).

In the developing countries of the region, unsafe water is responsible for a large percentage of diseases and a significant proportion of mortality (AMCB Bulletin 1996). Fifteen out of 1 000 children born in the developing countries die before the age of five from diarrhoea caused by drinking polluted water. The microbial diseases endemic to the poorer parts of many cities in the developing countries cost billions of dollars each year in terms of lost lives and poor health. In Jakarta alone, the cost of impaired health from unsafe drinking water is estimated at US$300 million a year (World Bank 1996).

3. *Noise Pollution*

Urban noise is becoming an issue of increasing concern to municipal authorities and residents in many of the region's cities. The extent to which a person is disturbed by a specific noise level varies and is influenced by a range of factors, including the individual's levels of tolerance, the cultural and socio-economic context and the frequency and persistence of a particular noise source. However, standard indices [dB(A)] have been developed for the measurement of sound power levels and these are typically employed as a numerical guide to determining the acceptability of noise.

Few cities in the region routinely measure urban noise levels, primarily due to resourcing constraints and the prioritization of air and water pollution. Where studies have been undertaken, however, the primary sources of urban noise have been identified as motor vehicles, aircraft, railways, construction activities, industrial activities and a range of neighbourhood activities and sources. Of these, traffic noise is often identified as the principle source of noise disturbance, such as in Hong Kong, China, where it is estimated that over one million people are living with unacceptably high levels of noise, primarily from road traffic (EPD-Hong Kong 1999). A similar number of urban residents in Australia live in areas that rank as "acoustically unacceptable" when assessed against OECD standards (Paboon et al 1994). In People's Republic of China, the measurement of traffic noise has indicated an average of 71 dB(A), which is well above international standards of acceptability (Government of China 1997). In summary, however, the lack of robust data for many urban areas in the region precludes an assessment of the extent of noise pollution throughout the region, although the limited data that are available would seem to indicate that the problem is widespread.

Some highly urbanized countries have embarked upon comprehensive programmes to monitor and control noise pollution. For example, in Hong Kong, China, a range of strategies and regulatory mechanisms have been used to reduce noise pollution, including the allocation of significant resources for noise control enforcement, the introduction of requirements for less noisy construction equipment, the strict regulation of out-of-hours construction activities and, in an effort to control traffic noise, the re-paving of roads and the construction of noise barriers alongside highways and surface railways (Box 7.2). Similar control mechanisms have also been adopted in Singapore, whilst in other cities measures have been adopted in specific locations to counter acute noise problems (Best Practices Data Base: Weihai 1999).

C. Shelter and Dwellings

There are two major challenges associated with the provision of adequate dwellings to the residents of the region's cities: the ever-increasing backlog in housing provision; and the inadequacy of utilities and infrastructure including water supply, sanitation and waste disposal. The lack of purpose-built housing or supporting infrastructure impacts directly upon the urban poor, who respond by encroaching on unused land and constructing temporary shelters, which over time have grown into the seemingly permanent shanties and squatter towns that are found in many of the region's cities.

1. *Slums and Squatter Settlements*

The squatter settlements of the Asian and Pacific Region are typically characterized by temporary structures and the absence or severe lack of basic infrastructures and services such as water supply, sewerage, drainage, roads, healthcare and education. The dwellings in these areas are generally made of discarded materials, such as used wooden planks, plastic, corrugated metal, asbestos sheets and cardboard. The population density of the settlements is typically high and inadequate water supply and sanitary facilities result in high incidences of disease.

In contrast to the illegal occupation of land by squatter communities, slum dwellers have legal access to their dwellings through formal ownership or through the payment of rent. However, rents in the region continue to rise faster than the average income and the provision of legally established low cost housing is becoming increasingly constrained by the entry of larger, commercial housing developers and the increasingly complex administrative mechanisms introduced to regulate the market. The commercialization of land and institutionalization of the housing provision sector have left slum dwellers and squatters with fewer opportunities to improve their situation. Unscrupulous developers have also increasingly bypassed the planning and administrative systems such that, in some urban areas, it is estimated that 70-95 per cent of new housing is technically outside the law (O'Meara 1999).

Estimates for the proportion of the population that inhabit squatter settlements in Mumbai, Delhi, Jakarta, Istanbul and Metro Manila range from 15 to over 50 per cent (Fernandes and Varley 1996; Habitat 1996). In Dhaka, about half of the city's population has been estimated as living in slums and squatter settlements with a population density of over 2 500 persons per ha.

Within the low-cost rental sector a range of housing is provided including inner city tenements, houses and apartments with the cost of rental being

Box 7.2 Urban Environmental Protection: The Case of Hong Kong, China

Hong Kong, China's urban areas are besieged with numerous environmental problems such as air, water and noise pollution, and high accumulation of waste. To tackle these problems, the Government has undertaken various environmental protection measures ranging from legislation and regulatory control to implementation of numerous programmes and services. The ultimate responsibility rests with the Environmental Protection Department (EPD) which implements pollution prevention and control measures and help formulate environment-related policies, including planning of Hong Kong, China's sewage and waste management programmes.

The backbone of environmental protection and control are the strategic environmental assessments that have been increasingly used in land-use development and sectoral planning and appropriate studies are integrated in the formulation of district plans. All designated projects must follow the statutory environmental impact assessment (EIA) process and carry out mitigation measures or other actions recommended. Appropriate legislation is also in place governing air, water, waste and noise pollution. Air pollution is controlled through the Air Pollution Control Ordinance, which was extended in 1998 to tighten emission standards for diesel cars and light duty vehicles other than taxis. The Ozone Layer Ordinance bans the import and manufacture of substances that deplete the ozone layer and enables Hong Kong, China to meet its obligations under the 1987 Montreal Protocol on the issue. Water pollution is controlled through Water Pollution Control Ordinance, which manages discharges through a licensing system. The Dumping at Sea Ordinance, which enables the country to fulfil its obligations under the London Convention, prevents dumping of waste into the marine environment. Pollution caused by livestock waste is controlled under the Waste Disposal Ordinance, which also covers the import and export of waste. Shipments of hazardous waste from developed countries to Hong Kong, China were banned in 1998. The 1997-amended Noise Control Ordinance tightened control of percussive piling noise as well as vehicle burglar alarms.

Alongside legislation, a number of programmes and services have been introduced to combat pollution problems, with special emphasis on waste. Three strategic landfills were built, a Chemical Waste Treatment Centre on Tsing Yi Island began its operation, and a Waste Reduction Plan, which aims to reduce waste through incineration, recycling and other measures, was unveiled. The Strategic Sewage Disposal Scheme aims to collect and treat the sewage discharged at Victoria Harbour via an outfall off East Lamma by 2008. Other initiatives on waste include the clearing up of 13 major blackspots in the New Territories and extending services at several refuse transfer stations to the private sector.

A one-year trial of liquefied petroleum gas taxis was successfully completed in 1998 and it is planned that beginning year 2000, all taxis will run on the gas. Tighter emission standards for new light duty diesel vehicles were introduced and a more effective test for smoky vehicles was implemented. In 1998, around 380 000 people exposed to heavy noise from aircraft got relief when the airport was transferred from the highly populated Kowloon to Chek Lap Kok. A programme to phase out noisy diesel, steam and pneumatic hammers has been completed. Technologies for quieter surfaces for low-speed roads and more silent equipment for piling and construction were also tested.

Community awareness and participation is encouraged through the Green Living Campaign, under which a Waste Recycling Competition is undertaken. Secondary and primary schools are provided with environmental education packages and environmental programmes are extended to universities. The territory also celebrates World Environment Day and Environmental Protection Festival, during which most of the talks and workshops are organized for various sectors of the community. Cross-border cooperation continues through the Hong Kong-Guangdong Environmental Liaison Group which at present, concentrates on addressing pollution problems in Deep Bay and conserving Mirs Bay.

The package of actions including strategic and regulatory planning supported by appropriate legislation, promotion of cleaner technology and community awareness and participation have enabled Hong Kong, China not only in cleaning its environment but also safeguarding it from further deterioration.

Source: EPD-Hong Kong 1999

CHAPTER SEVEN

influenced by the standard property market factors of, among other things, location, size, physical conditions and forms of tenure. However, at the low cost end of the market, much of the property is in poor physical condition due a lack of regular maintenance or repair. Typically located in the older sections of the cities, these properties are characterized by a low standard of infrastructure and high person-to-floor space ratios.

Various studies have documented the scale and range of housing sub-markets within cities, particularly those used by low-income groups. For instance, in Dhaka, the major housing sub-markets where the poorest two-thirds of the population live are: squatter settlements; refugee rehabilitation colonies and squatter resettlement camps; 'bastis' (cheap rental accommodation in one or two-storey buildings); inner city tenement housing; and employee housing (including accommodation provided by government agencies for some of their staff and accommodation provided by middle or upper income households for servants). Aside from these, about three per cent of the city's poor live in other accommodation including, for example, boats, vehicles or multiple occupancy rooms that are widely used by single women shift workers.

Some countries, including Indonesia and Pakistan, have managed to integrate most illegal settlements into the wider city by, for example, introducing tenure legalization. Elsewhere, however, the adoption of such pragmatic approaches has been limited and in most of the large cities in the developing countries, little or no action has been taken to resolve the issue of illegal settlements (Durand-Lasserve and Clevc 1996). As squatter settlements are typically established on unused, marginal land, many are to clean drinking water and that 50 per cent lack adequate toilet facilities (Planet Earth 1999). The level of provision with respect to solid waste collection and disposal, transportation and other infrastructure is often much worse.

(a) Water Supply

The declaration by the United Nations of the 1980s as the International Drinking Water Supply and Sanitation Decade aimed to focus attention on the improvement and expansion of water supply, but, as indicated in Figure 7.5, many people in the developing countries of the region still do not have access to a safe and reliable supply of water (AMCB Bulletin 1996; UNCHS 1996; Planet Earth 1999). Whilst the extent of water supply provision varies across the region (see Chapter 4), even within different urban areas of the same country significant differences in the level of supply occur. For example, in Pakistan availability of piped water ranges from 35 per cent in Faislabad to 92 per cent in Karachi, whilst in Thailand, it varies from 78 per cent in Bangkok to less than 10 per cent in Nakhon Si Thammarat. Among the megacities of the region, only Seoul and Shanghai have 100 per cent water service coverage with 24 hours of water supply (ADB 1996). Most megacities have 65-83 per cent water service provision, although the daily duration of water supply varies from 10 to 19 hours daily, except for Dhaka and Karachi where water is supplied for only 6 and 4 hours, respectively.

A major issue in urban water supply is the high rates of unaccounted for water, due to leakages and illegal connections. Rates of unaccounted for water are particularly high in the cities of Dhaka, Hanoi, Mandalay, Metro Manila, Phnom Penh, Calcutta and Apia (Figure 7.6).

Studies have also shown that the reduction of water lost during transmission enables the servicing of a larger number of people at a lower per capita cost, when compared to the costs of providing new water supply capacity. In response, a number of cities have already undertaken actions to reduce losses. For example, the Metro Manila's Metropolitan Waterworks and Sewerage System (MWSS) has already begun repairing and replacing aged and broken water distribution lines in the city, testing and replacing metres and removing illegal connections to reduce the quantities of non-revenue water (ADB 1996). Improvements in the security of water supply systems have been such that cities such as Jakarta, Metro Manila and Bangkok, all of which

Figure 7.5 Urban Population without Access to Safe Water Supply in Selected Countries of Asia and the Pacific

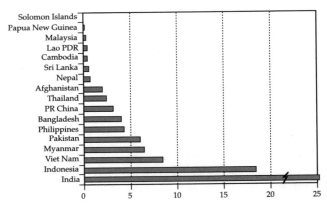

Source:
1. ADB 1997b
2. UNDP 1997
3. UNICEF 1998
4. World Bank 1997

URBAN ENVIRONMENT

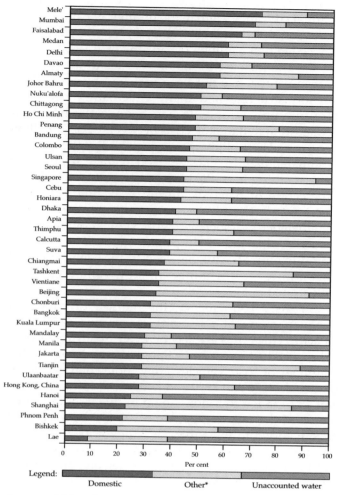

Figure 7.6 Water Use in Selected Urban Areas in Asia and the Pacific

Source: McIntosjh and Yoiguez 1997
* Other use includes industrial, commercial and institutional

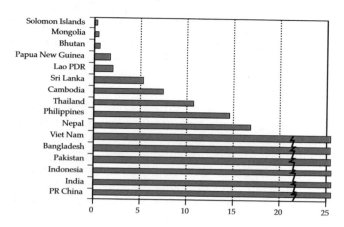

Figure 7.7 Urban Population Without Access to Sanitation in Selected Countries of Asia and the Pacific

Source: World Bank 1997

were heavily reliant on groundwater extraction, are now drawing less than five per cent of their water supply from underground aquifers (UNCHS 1996).

(b)   Sanitation

The provision of wastewater and sewerage infrastructure is also very poor in the cities of the low and middle income countries of the region, with over a third of all city residents still lacking adequate sanitation services (AMCB Bulletin 1996). In total, only some 10 per cent of the region's population is connected to public sewers, with the remainder either without facilities at all, or served by septic tanks or illegal connections to stormwater drainage systems and open water bodies.

In response, in recent years the provision of sanitation infrastructure has been the focus of considerable multi-lateral and bi-lateral investment and many urban centres are in the process of improving their wastewater and sewerage facilities. Non-traditional approaches using natural biological processes and appropriate technology are also being increasingly promoted. For example, an approach has been adopted in Fiji whereby secondary effluent is dispersed through mangrove areas as an effective means of filtering nutrients and avoiding eutrophication in the receiving water bodies. In Central Asia, Uzbekistan is planning a sewerage system that will provide 60 per cent urban coverage by the year 2005 as well providing 100 per cent access to clean drinking water.

Amongst the countries of the region, People's Republic of China has the highest numbers of people without access to adequate sanitation, followed by India (Figure 7.7), although, relative to the total urban population, Afghanistan, Myanmar, Viet Nam and Bhutan have a significantly high proportion (>50 per cent) of urban population without access.

In the absence of basic sanitation infrastructure, much of the untreated wastewater and raw sewage is discharged directly into the lakes, rivers, streams, canals, gullies and ditches of the region's urban areas. It is estimated that over one million tonnes of sewage is produced daily in the cities of the region and that less than two per cent of this quantity is adequately treated before being discharged into watercourses (UNEP 1997).

(c)   Transport

In many cases, the expansion of the cities in the Asian and Pacific Region has occurred with little or no development planning or strategic overview. As a consequence, the provision of transportation infrastructure has lagged far behind the development process and the subsequent provision of road or rail networks has had to be accommodated within an existing urban structure. The greater capital investment required to develop rail infrastructure has delivered a preference for road-based forms of transportation, with consequential impacts associated with congestion, longer and inefficient journey times, less efficient fuel consumption and greater levels of noise and air pollution.

Although in recent years, primary road improvements have been undertaken or planned in most cities of the region, few cities have well-integrated primary, secondary and tertiary road systems. In the cities of the high income countries, road and rail developments are constrained by the effects of land price speculation that often leads to prohibitive costs for the provision of transport infrastructure. For example, by 1987, land prices had become so high in Tokyo that the cost of a seven kilometre road connecting the city centre to a new urban satellite would have been over one trillion yen, 95 per cent of which was for real estate acquisition (Douglas 1989).

The impact of poor transport planning manifests itself in the evident indicators of urban dysfunction particularly severe traffic congestion with uncontrolled mixes of traffic types, long journey times, lack of traffic management, accidents, poor environmental conditions and high costs for the movement of goods.

The transport challenge for the developing nations of the region is to improve the mobility of urban residents while enhancing the efficiency of transportation systems. In cities such as Singapore, Hong Kong, China, Tokyo, Sydney, Kuala Lumpur and Bangkok, light rail, tramway or mass transit systems have been developed to alleviate pressures on the road systems and to provide an opportunity to re-appraise aspects of the city's transportation system. The second Railway Development Study (Hong Kong, China, Government 1999) and third Comprehensive Transportation Study (Hong Kong, China, Government 1999) have each identified a preference for rail-based strategies in the future planning of Hong Kong, China's transportation system, whilst the introduction of measures to curb road traffic in city centres has been considered in a number of cities.

However, the development of alternatives to road transport is not a short or medium term option for many of the cities of the region, due to the capital investment required and, in some cases, due to the political difficulties associated with seeking to counter the current trend toward private vehicle ownership.

Most of the growth in motor vehicle fleets in the developing countries is concentrated in large urban areas. For instance, in Iran (the Islamic Republic of), the Republic of Korea and Thailand, about half of the countries' motor vehicles are in the capital city. In Shanghai, the number of cars doubled between 1985 and 1990 and, in recent years, has exceeded half a million (Miankang et al 1997).

The increase in motor vehicles in the region's urban centres has not been matched by investment in infrastructure and many cities suffer persistent traffic congestion. It has been estimated that congestion in Bangkok costs between US$272 million and over US$1 billion per year (UNEP 1997) and it is estimated that drivers spend, on average, 44 full days a year sitting in traffic jams. Average travel speed in the centre of Bangkok during peak hours is about 12 km/hr (World Resource Institute 1997). The inefficiencies associated with slow moving urban traffic also have implications for economic productivity; the link between traffic congestion and economic performance in key Asian cities is illustrated in Figure 7.8.

As a consequence, the number of passenger cars in many of the more prosperous Asian and Pacific cities, has tripled or even quadrupled over the past 10-15 years. In Bangkok, for example, the number of road vehicles grew more than sevenfold between 1970 and 1990 and, whilst the economic crisis of 1997-98 led to a slow down in this trend, for each year of the last decade some 300 000 new vehicles were added to the streets of Bangkok (UNEP 1997). In People's Republic of China, it is projected that by 2015, there will be 30 million trucks, and 100 million cars (Livernash 1995).

Figure 7.8 Annual Cost of Time Delay (in US$ million) in Some Asian Cities Due to Traffic Jams

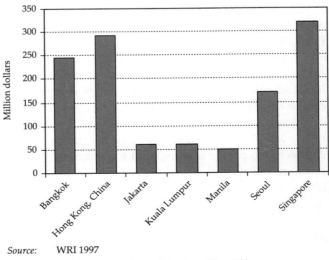

Source: WRI 1997
Note: Bar shows cost of time delay in million US$

In many cities of the low and middle income countries, the growth in the vehicle fleet results primarily from increases in the numbers of motorized two-wheel and three-wheel vehicles, which are more affordable for large segments of the population and often serve as a stepping-stone to car ownership. In Thailand, Malaysia and Indonesia, for instance, two- and three-wheelers make up over half of all motor vehicles. The number of two- and three-wheel vehicles is expected to grow most rapidly in People's Republic of China, India and in other densely populated low-income countries. In People's Republic of China, it is projected that there will be 70 million motorcycles by 2015. Production of motorcycle and cars in India is also increasing by 20 per cent annually, outstripping that for buses which is growing at three per cent per year (Livernash 1995). In Nepal, the number of registered two-wheelers in 1998 was over 100,000, with more than half of these concentrated in Kathmandu.

In many cities of the region, alternatives to road-based forms of transportation are limited. Rail infrastructure, where it exists, is often run down and starved of available investment, which is invariably directed toward expanding the road network. A recent study (EPD 2000) analysed the environmental and social benefits of rail over road-based forms of transportation and found that rail was preferred under each of the comparative assessment criteria, including levels of air and noise pollution, impacts to existing landuses and implications for societal and passenger risk. As stated above, the provision of new or improved railway infrastructure currently lies outside the financial and political reach of many of the region's municipal authorities; in the interim, the rise in motor vehicles, and the associated environmental, economic and social disbenefits, is set to continue.

POLICIES AND PROGRAMMES FOR URBAN ENVIRONMENTAL MANAGEMENT

A. National Actions

In the Asian and Pacific Region, experience has shown that the management of urban areas by central government has simply not been effective (Stubbs 1996), whilst a reliance upon the public sector to provide environmental services (financed by taxation) has frequently resulted in unsatisfactory standards of delivery. In response, several governments in the region have embarked on programmes to reform their urban environmental management policies and promote decentralized and participatory development. This has enabled the mobilization of resources for the provision of improved urban infrastructure at a cost that imposes a lesser burden on scarce governmental finances. In general, urban management policies in countries of Asia and the Pacific have concentrated around five principal areas: i) enhancing urban management through decentralization and institutional and capacity strengthening; (ii) improving financial administration and mechanisms; iii) improving housing and shelter stocks; iv) funding urban infrastructure improvements, such as water supply, sanitation, solid waste management, transport, health, parks and playground, etc.; and v) enacting and improving legislation and regulatory standards for urban environmental management.

1. Enhancing Urban Environmental Management

A number of countries in the region have strengthened local administration through the devolution of functions and responsibilities. In the Philippines, one of the most influential initiatives in this direction was the implementation of the Local Government Code of 1991, under which local governments were given increased autonomy, more responsibilities for provision of services and greater access to financial resources (ADB 1996). Following the code's enactment, Metro Manila's local government has substantially increased its development activities. In the Republic of Korea, environmental management has also been entrusted to local government, especially the management of pollutant-discharging industries; the regulation of waste collection, transportation and recycling; the reduction of noise and vibration; and the control of vehicle emissions (Whang 1999).

As emphasized in the Habitat II Report (see Box 7.3), policy approaches aimed at effective decentralization are needed to provide a framework and a new model for addressing human settlement development issues. Such decentralization should strengthen democracy and provide local authorities with the institutional structures that are accessible to local communities, with the legitimacy to voice the specific concerns and aspiration of their citizens. However, evidence shows that efforts within cities themselves also help in decentralization and the promotion of good governance. The case of the Japanese city of Kitakyushu (see Box 7.4) provides a good example of the benefits of decentralization in providing effective remedies to counter metropolitan pollution (Kojima, et al 1995).

Although urban governments are gradually increasing their powers in the Republic of Korea, People's Republic of China, and the Philippines, in most countries of the region national authorities remain responsible for establishing environmental policy and for developing appropriate legislation and standards for its implementation at the local government level. In recent years, provincial and state governments have also been increasingly entrusted with carrying out urban environmental management, requiring considerable investment in planning, institutional strengthening, capacity-building and community participation. However, few local and city authorities, particularly in the developing countries, are granted the financial resources or provided the revenue raising powers necessary to effectively implement and maintain environmental management activities. The process of decentralization is thus severely constrained by a lack of institutional capacity among local governments, limited resource mobilization at the local level and limited access to long-term financing for investment programmes (World Bank 1998).

The concept of self-government in local affairs has long been in existence in some countries, for example, in India where the Collector remains the dominant figure in local decision-making and often holds sway over village authorities ('*panchayats*') and district boards (Turner and Hulme 1997), whilst in Thailand and Indonesia the adoption of a "mixed authority approach" has effectively enabled central government to retain control while creating an impression of some local autonomy and participation. Although the process of decentralization and devolution is slow it is, nevertheless taking place, for example, in Pakistan, where the devolution of the responsibility for housing and facilities from federal to local governments, and from public to private and community-based organizations is underway (Qadeer 1996). In addition, Papua New Guinea has embarked on a radical decentralization programme, although the success of the process has been severely hampered by the initial lack of administrative capacity, within existing local government.

Another common barrier to effective decentralization is the conflicts of interest both within and between local authorities regarding the benefits of economic development versus environmental protection. With little or no legal accountability, this has led to serious consequences for the environment. For example, in the Republic of Korea local governments seem to have a tendency to favour rapid development over sustainable development (Whang 1999).

2. *Improving Financial Administration and Mechanisms*

It has been estimated (Hardoy, Mitlin & Satterthwaite 1992), that city governments in low and middle income countries often have one hundredth or (in the most extreme cases) one thousandth of the revenue per capita available to most cities or municipal governments of the high income, developed countries and yet their range of responsibilities are broadly comparable.

The implications of the disproportionately high costs imposed by the subsidization of urban services are becoming all too clear within the region. In Dhaka, Hyderabad and Shenyang, for example, it is estimated that the future costs of obtaining water supplies will be three times the current costs, and the prohibitive costs associated with improving solid waste management in Colombo have prevented the implementation of planned actions as the municipal authority examines different financing options.

Box 7.3 Policy Objectives of the Habitat II Agenda

i) mobilize private and collective actors from all members of the society including informal sector in housing and urban development, by adopting special measures to strengthen people's participation in decision making and to improve the general level of education and training,

ii) mobilize local resources for the production of housing infrastructure and services at a sustainable cost, including cost recovery, public-private partnerships, elimination of legal barriers, setting up of financial and credit mechanisms for the whole spectrum of housing needs, promotion of the use of local building materials and application of low-cost technologies,

iii) changes in and revision of inappropriate standards, property regulations, land registration, legalization and recognition of informal settlement,

iv) increase the capacity of local authorities by providing useable and timely information, improve management of services and infrastructure, improve research and local planning, promote inter- and intra-urban networks of actors, improve monitoring and assessment of conditions and trends, and help mitigate the potential effects of natural and human-made disasters, and

v) strengthen the policy making and enabling role of central governments by promoting and integrating national urbanization policies with macro-economic policy; promote political, administrative and fiscal decentralization; promote the role of governments as facilitators; and reform subsidy systems.

Source: UNCHS 1996

Nevertheless, the range of financing options is expanding with the region-wide trend of providing local governments with greater discretion in the levying of taxes, fees and service charges. In recent years, for instance, Metro Manila has awarded the city's water supply and sewerage services to private concession contractors, whilst Singapore has shown that charging road tolls can work and that the advantages are multiplied when licensing entry into the city's central business district is coupled with an efficient MRT system. A range of pollution fees, fuel taxes and elevated vehicle taxes are all being examined as a means of targeting traffic congestion in the urban areas of the region.

> **Box 7.4 Environmental Responsibilities in Local Government: The Case of Kitakyushu**
>
> The Government of Japan is composed of a three tiered structure, operating at the national, prefectural and municipal levels. Central Government policies are also implemented through this structure. The Central Government, prefectures and municipalities, depending on their characteristics hold various powers, authorities and licensing rights. However, unlike the general municipalities, large cities have a special autonomous system. At present, 12 cities are entitled to this status including Kitakyushu.
>
> Kitakyushu was born from the union of five neighbouring cities in 1963, and is a city that has developed as one of Japan's prominently heavy chemical industrial areas. The well known "seven coloured smoke" which formerly symbolized prosperity, rich in dust and sulphur dioxide, was emitted from many large iron and steel, chemical, ceramics, and electric power corporations in Kitakyushu, and has long been a major source of pollution. Water pollution had also started even before the Second World War. The post-war reconstruction period, followed by high level economic growth during 1955-1965, further increased air and water pollution. Many residents in districts surrounded by large factories involved in ceramics, chemicals, iron and steel, etc. suffered from large quantities of dust fall, smoke and offensive odours. Moreover, in Dokai Bay, fish catches dwindled to nothing from 1950 onward. This was caused by the large quantities of industrial wastewater and sewage that flowed into the bay. Alarmed by the worsening situation, the residents raised demands for improvements to the local industries and submitted petitions to the administration.
>
> In subsequent government responses, various policies were devised in Kitakyushu, including monitoring air pollution and institutional development, which lead to the establishment of a pollution administration organization and a Pollution Control Council in the city. In 1967, the first pollution control agreement was concluded between the city government and the industrial corporations. Following the enactment of pollution-related laws by the "Pollution Diet" at the end of 1970, the pollution countermeasures of the city were markedly reinforced, and were implemented in a comprehensive, and systematic manner.
>
> The history of pollution countermeasures in Kitakyushu is characterized by four significant policy measures. Firstly there was a transfer of the authority of the prefectural governor to the city for the purposes of issuing "smog alarms". Throughout Japan, this transfer of authority was made to Kitakyushu alone, and was permitted in view of its distance from the city of Fukuoka, which was the seat of prefectural government. Secondly, a cooperative system of industry and government was initiated, strongly supported by the business community. At the city level, the obligation of the co-operative local self-governing body was to seek industrial development while protecting the health of the residents, and a comprehensive administrative management was required which did not lean toward selection of either "industrial development" or "environmental protection" as one of two alternatives. Thirdly, pollution prevention technologies were introduced by the corporations centred on cleaner production (CP), with pollutant removal equipment playing a supplementary role. For example, in the iron and steel industry, this included the development and introduction of the pre-combustion desulphurization system for coke oven gas, and of the so-called OG system, which conducted dust removal by a non-combustion system for converter gas. Lastly, the adoption of Japanese anti-pollution policy started promoting a "non-economic approach", for instance, the sludge dredging project of Dokai Bay. This project was not conducted on economic principles but on a crisis/risk-management policy designed to cope with future dangers and the actual sense of crisis among local citizenry.
>
> The smog alarm issuance authority which was given to Kitakyushu in 1970, raised the consciousness of both the local administration and a wide range of local citizenry, including the corporations, resulting in the heightened promotion of local pollution prevention initiatives. Meanwhile, the existence and effective operation of a co-operative system of government and industry became extremely effective with regard to industrial pollution prevention. Consultative organs were formed to allow for a full exchange of views and discussions, a process that guaranteed implementation of the concluded pollution control agreements without enacting laws or issuing ordinances accompanied by strict regulations. Pollutant emitting corporations adopted voluntary countermeasures such as improvement of the manufacturing equipment and processes, as well as raising of productivity while striving for resource and energy conservation. The use of cleaner production (CP) technology reduced consumption of raw materials and fuels and lessened generation of by-products that constituted sources of pollution. Finally, the sludge dredging project in Dokai Bay allowed for rehabilitation of the bay and restored the safety of products taken from the bay.
>
> Kitakyushu's experience indicates that appropriate planning and adoption of preventive measures can solve even severe pollution problems in a highly industrialized city with the cooperation of stakeholders and city government.
>
> *Source:* MEIP Report, no date

Property and other land-based taxes remain the mainstay of many local governments, even at relatively low levels of collection. However, such taxes need to be better structured to fully capture the potential revenues which exist, and need to be accompanied by more efficient systems of administration. For example, in Mumbai, which has some of the world's highest property prices, the local government is unable to share in the profits accruing to landlords because of weaknesses in the property tax system. Whilst some progress is being made in a number of cities, for example through World Bank assistance in Dhaka and Jakarta, in most cities there are considerable institutional, administrative and political barriers to reform.

In many of the urban areas of the region, the current mix of locally based revenue sources are inadequate to meet the demands of rapid urban development (Stubbs 1996). The mix includes under-utilization of property taxes and user charges, and the extensive use of cross-subsidies, which often distorts policies to support the poor. User charges are given little emphasis in the region's urban centres, despite their advantages in raising revenues in a fair and equitable manner and their utility in guiding investment decisions. Learning from experience, various cities in the region have utilized commercial capital markets for basic capital investment, which has provided the additional advantage of enabling leverage to be gained in raising funding from conventional public sources.

3. Improving Housing and Shelter Stocks

During the past two decades, major policy changes have occurred with regard to the role of government in the provision of housing. Most governments in Asia and the Pacific have moved away from the role of housing developer towards that of facilitators, shifting the emphasis of housing provision from the public to private sector (UNHCS 1997). Indeed, many governments now limit their direct involvement to the provision of low-cost housing.

Cognisant of the urgent need to re-focus its housing policies, many countries of Asia and the Pacific are adopting "enabling" policies to support individual households in providing for their own shelter (UMP 1996). Measures such as deregulation, changes in credit mechanisms, and lowering of housing and subdivision standards to promote affordability have captured the interest of private housing developers. Moreover, urban renewal policies now focus both on the physical aspects of neighbourhoods as well as social aspects such as employment, education and health (UNHCS 1997).

The importance of mobilising financial resources has also been given emphasis through the privatization of housing-finance institutions; encouraging commercial banks and private developers to invest in the low-cost housing sector with bank quotas, subsidized loans and tax exemption; the promotion of housing mortgages and the secondary mortgage market; and the provision of encouragement and support for community-based finance systems and housing co-operatives (ESCAP 1996).

In India, the National Slum Development Programme is an important post-Habitat II initiative to offer sustainable housing to the urban poor. Other initiatives include rationalising previously complex legislative requirements, such as the Urban Land Act and rental legislation that affects housing and land markets, and the implementation of Constitutional Amendment Acts to install elected local governments, confer land title or tenurial status to squatter settlements, and facilitate the flow of credit to poorer segments of the housing market.

Three main policy solutions have been adopted in the region to address the problems associated with illegal settlements (Fernandez and Varley 1998). Firstly, the review of legislative provisions, which, for instance, in Turkey led to an official tolerance of illegal settlements followed by periodic "amnesty" regularization; although it was widely acknowledged that this approach, in itself, does not solve the problem of access to infrastructure and services. The second form of policy intervention has sought to promote settlement improvement through the relocation of illegal settlers, thereby releasing land for commercial use. Land vacated by illegal settlers is sold at market prices to real estate developers, businessmen, and other parties from outside the settlement and the proceeds are used to subsidize the installation of services within the relocated settlements. This approach has been adopted in a few countries in the region, including the Philippines and Pakistan. Experience in the Philippines, however, shows that relocated illegal settlers often return to the old settlement or seek a new illegal settlement in a location where employment opportunities are more favourable. The third form of policy intervention seeks the regularization of illegal settlements, including the incorporation of such areas within the formal services and infrastructure systems. Such an approach is prevalent in a large number of countries of the region and has been successful in normalising squatter settlements and providing residents with access to minimum standards of service provision, including drinking water supplies, sanitation and street paving.

4. Funding the Improvement of Community Facilities, Services and Infrastructure

Many cities in the region are currently seeking funds to expand and maintain the infrastructure and services required to support growing populations and increasing economic activity (ESCAP 1996). The principle mechanism that is employed to mobilize the considerable investment needs of the developing world's cities and towns is through increased reliance on the private sector. Several cities in the region are now privatising or contracting out the delivery of services such as water, power, solid waste collection and transportation, etc. (World Bank 1996). In addition to the private sector, some success has also been achieved by the countries of the region in the creation of special-purpose agencies able to operate in a market environment, including the water distribution systems in the major cities of Thailand and Indonesia.

In successful public private partnerships, it is recognized that the public and private sectors should have clear and distinctive roles (Stubbs 1996). The public sector takes responsibility for planning, regulation and community protection, whilst the private sector manages the direct implementation and operation of services. In Malaysia, Thailand and the Philippines, for example, the Build-Operate-Transfer (BOT) laws allow private corporations to construct highways and power plants and to operate public sector projects before transferring the assets back to the government, usually after a 20-25 year period (Turner and Hulme 1997). Through these arrangements, governments have secured private sector participation in infrastructure development, whilst reducing the public sector fiscal burden and encouraging the inflow of foreign capital, expertise and technology.

In the Philippines, this approach has been particularly successful in the critical area of under capacity in the power generation sector. However, for other sectors the government has reviewed the BOT law and has sought to offer other options, including Build-Own-Operate, Build-Lease-Transfer, Build-Transfer and Rehabilitate-Own-Operate. The last option was adopted for the Metropolitan Waterworks and Sewerage System (MWSS) (ADB 1996). Whilst the success of such initiatives in the region seems to indicate an increase in private sector involvement in service delivery, it has been recognized that government agencies overseeing the private sector firms need to be strengthened to ensure that these firms are operating in a truly competitive environment (ESCAP 1994). Moreover, governments have to ensure that equity considerations are met, as market-led private firms are seldom motivated to service poorer areas where the potential profits are limited.

In order to ensure that equity and social issues are addressed, over the last few decades, many cities have experimented with the development of neighbourhood or community organizations, which are consulted on the planning of new development, implementation of infrastructure improvement and implementation of tariff or tax increases (ADB 1996). In many countries, programmes are being developed under which community organizations can be responsible for their own infrastructure development, with some notable successes including the delivery of services in sanitation (Karachi), public health (Calcutta) and environmental protection (Metro Manila) (Stubbs 1996). Aside from formal public sector community organizations, informal or private sector community organizations exist, which may participate in urban development. These informal or private sector-led organizations include chambers of commerce and industry, religious associations, and associations of slum dwellers.

5. Enacting and Improving Legislation and Regulatory Standards

A number of countries in the region have developed greater integration in their development and environmental policy making and are able to consider the wider issues of metropolitan scale land management, infrastructure investment, financing mechanisms and governance in an integrated manner. Examples include the Klang Valley Environmental Plan in Malaysia and the Ho Chi Minh City Environmental Planning Project in Viet Nam (Stubbs 1996). In the case of land management, various integrated planning and regulatory systems are now linked to institutional, sectoral investment and fiscal policies within improved urban management systems. Planning and regulatory tools are also being improved, such as the "broad brush", structure planning approaches used in the JABOTABEK Metropolitan Development Plan, the Metro Manila Capital Investment Folio and the current development of plans in Dhaka. In People's Republic of China, the Beijing Environmental Pollution Control Targets and Countermeasures was formulated in an effort to improve the environment of the city (Jiachen, 1999), whilst in Changchun City, environmental protection is integrated as part of the overall planning of National Economy and Social Development (Defu 1999). In lieu of traditional zoning procedures, new techniques such as strategic environmental assessment are being adopted, most notably in Hong Kong, China, as a means of integrating potential environmental considerations at the early stages of strategic policy formulation.

One notable example of improved legislation is Japan's Basic Law on Environment, enacted in 1993, which integrates the two basic laws on Pollution Prevention and Natural Environment and provides mechanisms for responding to the country's environmental problems and emphasizes the local, national and global responsibilities of the State, local government, business and citizens (OECC 1996).

Other cities in the region have also enacted appropriate legislation and standards for environmental improvements (see Chapter 7).

B. International and Regional Programmes

International and regional programmes on urban environmental management have generally focussed on providing support to national programmes. A major development in this direction was the organization of the Second United Nation's on Human Settlements Habitat II held in Istanbul in 1996, which adopted the Habitat Agenda. During the conference, twenty-four countries in the region submitted their national reports with national plans of action outlining priorities for technical cooperation.

The Habitat Agenda provides an operational framework for the implementation of policies and programmes on urban environmental management in the Asian and Pacific Region. Many countries in the region have committed themselves to implementing the Habitat Agenda through local, national and sub-regional plans of action. The Agenda is based mainly on six strategic principles for the implementation of enabling policies for sustainable urban development. These include decentralization, partnership, public participation, capacity building, networking and the use of information and communication technology. The policy objectives of the Habitat II Agenda are presented in Box 7.3.

1. Regional Action Plan on Urbanization

The Ministerial Conference on Urbanization in Asia and the Pacific convened by ESCAP adopted the Regional Action Plan on Urbanization to promote and facilitate economically productive, socially just, environmentally sustainable, politically participatory and culturally vibrant urban areas (ESCAP 1994).

As part of the Regional Action Plan, the Asia-Pacific Urban Forum was established in order to ensure that a regional perspective was maintained and to promote cooperation between national and local urban governments, NGOs, representatives of the formal and informal private sector, the media, academics, research and training institutes and international and regional organizations. Convened once every two years, the Forum has provided a useful mechanism for reviewing on-going regional assistance programmes and their relevance to countries, making these programmes more transparent and demand-driven.

2. Asia-Pacific Initiative 2000

This recently completed project was aimed at capacity building for the sustainable development of urban areas through partnership with private, voluntary and community-based organizations. The programme aimed to support actions that promote improvements in the living conditions of the poor, the empowerment of women, the protection of the environment and the creation of sustainable employment opportunities. Implemented by a partnership of ESCAP, UNCHS and the Metropolitan Environment Improvement Programme (MEIP) of the World Bank, Asia-Pacific 2000 has provided financial and technical support to urban NGOs and NGO coalitions working to strengthen the local resource base of urban poor communities, provided the urban poor with basic affordable environmental services and built the capacity of local community organizations.

3. Land Management Programme (LMP)

The Land Management Programme (LMP) was launched in 1996 as a follow-up to the Habitat II Land Initiative and to capitalize on the various partnership activities which had led to the issuance of the "Global Platform on Access to Land and Security of Tenure as a Condition for Sustainable Shelter and Urban Development: The New Delhi Declaration" (January, 1996). The Habitat II Land Initiative recognized the need to focus on the strategic issues of security of land tenure and enforceable property rights as an essential prerequisite for a nation's long-term success in developing equitable national land policies and practices in support of economic and social development.

The programme aims to produce and disseminate information on a range of issues including best practices focusing on access to land, security of tenure, informal settlements upgrading and regularization, urban land management and land policy reform. The key clients or client groups served include the governments at national and local level, public/private land and property development organizations, private sector at large including the business and professional sector, community-based organizations, and land owning/using sector. Other clients include the landless, homeless, squatters, informal settlers and the urban poor, including head-of-household women and other vulnerable groups.

4. Local Leadership and Management Training Programme

This programme is being implemented by the UNCHS Training and Capacity-Building Section (TCBS), in partnership with the Government of The Netherlands and a range of local government focussed agencies and organizations. The principal aim of the programme is the improvement of living and working conditions through the promotion of effective human resource development and institutional capacity-building for management and development of human settlements. This is achieved through three principal components comprising Settlement Management Training, Local Leadership Training and direct support to National and Local Training Institutions. In general, these activities aim to improve effectiveness and efficiency of local government management, enhance and maintain the quality of services provided to citizens, improve the capabilities of elected local government officials and other local leaders, and strengthen the national and local capacity building institutions. The key clients or client groups served are urban management, municipal development and local leadership training institutions; municipal and local governments and their associations; elected and appointed local government officials and other local leaders; and, local development NGOs and CBOs.

5. *Localizing Agenda 21: Action Planning for Sustainable Urban Development (LA 21)*

The programme aims to localize the Agenda 21 Programme through the enhancement of local sustainable urban planning and management capabilities in a number of selected medium-sized cities worldwide and to support the development and implementation of broad-based action plans for municipal planning and management.

In the Asian and Pacific Region, Vinh City (Viet Nam) has been selected for this programme, although it should be noted that a number of other cities, in People's Republic of China, Japan, the Republic of Korea and elsewhere in the region, have developed their local Agenda 21 independent of this programme (see Chapter 12).

The localization of Agenda 21 has three main elements: the development of a long-term development vision for the city; the formulation of strategies for implementation and the removal of obstacles, and communication and stakeholder participation in decision making and dispute resolution. Specific areas of focus include urban revitalization, buffer zone development between city and fragile nearby ecosystems, waste management, infrastructure improvement, municipal finance and local economic development.

6. *Best Practices and Local Leadership Programme*

The Best Practices and Local Leadership Programme (BLP) is a global network of institutions dedicated to the identification and exchange of successful solutions for sustainable development. Between 1996 and 2000, over 700 good and best practices from more than 100 countries have been documented by UNCHS, establishing a unique data set on how enabling principles are being implemented at the local level.

A regional analysis indicates that innovative practices and policy responses are most prevalent in those sectors where the transition from the direct to the enabling approach is most recent. Where globalization is having the strongest impact on people and their communities, e.g., in Asia, the major thrust appears to be on infrastructure development and the reform of local government. The Asian and Pacific countries where best practices have been documented are Australia, People's Republic of China, India, Pakistan, Philippines and Sri Lanka.

7. *Urban Management Programme for Asia and the Pacific (UMPAP)*

The Urban Management Programme for Asia and the Pacific provides guidance to national governments on ways in which they can improve the management of urban development in their countries (Kendall, 1997). The programme assists in organising city consultations, providing for stakeholder participation in the implementation of urban management policies and techniques.

Among the programme's significant achievement are the establishment of the Regional Network of Urban Experts in Asia and the Pacific (URBNET-Asia) with members from Australia, Bangladesh, People's Republic of China, Hong Kong, China, India, Indonesia, Japan, Korea, Malaysia, Nepal, Pakistan, Philippines, Singapore, Sri Lanka, Thailand and Viet Nam, and the establishment of formal partnerships with regional networks of local authorities such as CITYNET and IULA-ASPAC for information exchange as well as technical services for their capacity building and other operational activities.

8. *Urban Management Programme-Asia*

The Urban Management Programme (UMP) is a global technical cooperation Programme of the United Nations, executed by the United Nations Centre for Human Settlements (UNCHS), with core funding from the United Nations Development Programme (UNDP), and several bilateral agencies. UMP gives advice to local and national governments on ways in which they can improve the management of urban development in their countries. The Programme operates through four regional offices, one of which is in Asia and the Pacific, and a Programme Coordinator at UNCHS (Habitat) headquarters in Nairobi, Kenya. The main focus of the current UMP is to build and strengthen the capacity of governments and other stakeholders to specifically address urban poverty reduction; urban environmental management and participatory urban governance.

9. CITYNET

The Regional Network of Local Authorities for the Management of Human Settlements (CITYNET) was established to promote the exchange of expertise and experiences among local authorities and NGOs in the cities of the Asian and the Pacific Region. Through technical cooperation at local and grassroots levels, CITYNET's major activities include technical advisory services; training activities and study tours; joint applied research; documentation and dissemination of urban development experiences. A regional data bank on the management of human settlements in Asia and the Pacific is currently being developed.

10. LOGOTRI

LOGOTRI is the network of Local Government Training and Research Institutes in Asia and the Pacific. Its members are both governmental, autonomous and private sector institutions and organizations involved primarily in local government training and research. The network was initiated by the UNESCAP with the objectives of establishing technical cooperation among local government training and research institutes in Asia and the Pacific and to strengthen the institutional and technical capacities of local government training and research institutes. To achieve its objectives, LOGOTRI: organizes advisory services, training workshops, study tours, research studies, documentation and information and staff exchange; develops and maintains a regional information resource centre and issue a newsletter; and organizes any other activities, as may be deemed necessary, from time to time, for the purpose of attaining its objectives.

11. Other programmes

The Safer Cities Programme and the Women and Habitat Programme of the UNCHS are among the other urban environmental programmes being implemented in the region. The Safer Cities Programme aims to address urban violence by developing a range of prevention strategies in consultation with the local authorities. It further aims to undertake a local safety appraisal; set up a dynamic coalition between key agencies and actors from civil societies; strengthen the local authority's capacity in dealing with crime prevention and delinquency; and building a methodology for crime prevention.

Meanwhile, the Women and Habitat Programme aims to promote women's equal participation in urban development planning as well as monitoring other UNCHS programmes, projects and activities to ensure that an appropriate gender perspective is maintained. The mechanisms of the programme include policy formulation and development, capacity building, applied research, development of training materials and networking; these are applied with specific emphasis on poverty alleviation, the promotion of equity and diversity and the recognition of women's rights.

CONCLUSION

The shifting trend from agrarian to industrial economies in the Asian and Pacific Region is being accompanied by high rates of urbanization, the rapid growth in the number of megacities and enhanced primacy.

Despite their potential to offer a better quality of life, the cities in the region are beset by growing problems of environmental deterioration relating to the loss of natural resources, the lack of adequate shelter and dwelling provision, deteriorating ambient air and water quality conditions, an increasing backlog in urban service provision and inadequate infrastructure. These conditions impact directly upon the residents of urban areas in particular the poor, and contribute to an acceleration in the deterioration of the urban environment.

The most urgent actions required relate to the improvement of environmental conditions and the strengthening of overall management capabilities. Despite growing trends towards decentralization, most of the city governments in the region have insufficient resources or capacity to effectively address the range and magnitude of the problems facing them. In terms of resources, city governments in the developing countries of the region have only a fraction of the revenue per capita available to the cities of the developed world, and yet they face similar administrative and management burdens. It is imperative, therefore, that the institutional capabilities of city authorities are enhanced, both in terms of human and financial resources.

Chapter Eight

Wastes at Smoky Mountain in the Philippines for hauling by rag pickers.

8

Waste

INTRODUCTION
TYPES OF WASTES
WASTE PROCESSING AND CONTROL
WASTE MANAGEMENT POLICIES AND STRATEGIES
CONCLUSIONS

CHAPTER EIGHT

INTRODUCTION

Waste is an unavoidable by-product of most human activity. Economic development and rising living standards in the Asian and Pacific Region have led to increases in the quantity and complexity of generated waste, whilst industrial diversification and the provision of expanded health-care facilities have added substantial quantities of industrial hazardous waste and biomedical waste into the waste stream with potentially severe environmental and human health consequences. The Chapter discusses the generation, treatment, disposal and management of the growing volume of waste, which poses formidable challenges to both high and low-income countries of the region.

TYPES OF WASTES

A. Generation and Characteristics

A clear appreciation of the quantities and characteristics of the waste being generated is a key component in the development of robust and cost-effective solid waste management strategies. Although amongst some of the more developed countries within the region the quantification and characterization of waste forms the basis for management and intervention, elsewhere little priority is given to the systematic surveying of waste arisings and the quantities, characteristics, seasonal variations and future trends of waste generation are poorly understood. Although there is a lack of comprehensive or consistent information, at the country level, some broad trends and common elements are discernible.

In general, the developed countries generate much higher quantities of waste per capita compared to the developing countries of the region. However, in certain circumstances the management of even small quantities of waste is a significant challenge. For example, in the small islands of the South Pacific subregion, small populations and modest economic activity have ensured that relatively low quantities of waste are generated. However, many of these countries, particularly small atoll countries such as Kiribati, Tuvalu and the Marshall Islands, face considerable waste management challenges due to their small land areas and resultant lack of disposal options.

Throughout the region, the principal sources of solid waste are residential households and the agricultural, commercial, construction, industrial and institutional sectors. A breakdown of solid waste types and sources is provided in Table 8.1. For the purposes of this review these sources are defined as giving rise to four major categories of waste: municipal solid waste, industrial waste, agricultural waste and hazardous waste. Each of these waste types is examined separately below.

1. *Municipal Solid Waste*

Municipal solid waste (MSW) is generated from households, offices, hotels, shops, schools and other institutions. The major components are food waste, paper, plastic, rags, metal and glass, although demolition and construction debris is often included in collected waste, as are small quantities of hazardous waste, such as electric light bulbs, batteries, automotive parts and discarded medicines and chemicals.

Table 8.1 Sources and Types of Solid Wastes

Source	Typical waste generators	Types of solid wastes
Residential	Single and multifamily dwellings	Food wastes, paper, cardboard, plastics, textiles, leather, yard wastes, wood, glass, metals, ashes, special wastes (e.g. bulky items, consumer electronics, white goods, batteries, oil, tires), and household hazardous wastes
Industrial	Light and heavy manufacturing, fabrication, construction sites, power and chemical plants	Housekeeping wastes, packaging, food wastes, construction and demolition materials, hazardous wastes, ashes, special wastes
Commercial	Stores, hotels, restaurants, markets, office buildings, etc.	Paper, cardboard, plastics, wood, food wastes, glass, metals, special wastes, hazardous wastes
Institutional	Schools, hospitals, prisons, government centres	Same as commercial
Construction and demolition	New construction sites, road repair, renovation sites, demolition of buildings	Wood, steel, concrete, dirt, etc.
Municipal services	Street cleaning, landscaping, parks, beaches, other recreational areas, water and wastewater treatment plants	Street sweepings, landscape and tree trimmings, general wastes from parks, beaches, and other recreational area, sludge
Process	Heavy and light manufacturing, refineries, chemical plants, power plants, mineral extraction and processing	Industrial process wastes, scrap materials, off-specification products, slag, tailings
All of the above should be included as "municipal solid waste."		
Agriculture	Crops, orchards, vineyards, dairies, feedlots, farms	Spoiled food wastes, agricultural wastes, hazardous wastes (e.g. pesticides)

Generation rates for MSW vary from city to city and from season to season and have a strong correlation with levels of economic development and activity. High-income countries (such as Australia, Japan, Hong Kong, China, Republic of Korea, and Singapore) produce between 1.1 and 5.0 kg/capita/day; middle-income countries (such as Indonesia, Malaysia and Thailand) generate between 0.52 and 1.0 kg/capita/day, whilst low-income countries (such as Bangladesh, India, Viet Nam and Myanmar) have generation rates of between 0.45 and 0.89 kg/capita/day. Figure 8.1 shows MSW generation by the high, middle and low-income countries of the region.

Taken as a whole, the Asian and Pacific Region currently produces some 1.5 million tonnes of MSW each day and this is expected to more than double by 2025 (World Bank 1999). The current estimate for waste generation may be considered as extremely conservative; the actual levels are probably more than double this amount. Figure 8.2 presents the current contribution of the various subregions to the waste generated by the region (United Nations 1995, World Bank 1995 and 1998, UNEP/SPREP 1997).

The composition of municipal solid waste varies significantly across the region (see Figure 8.3) with some middle and low income countries generating waste containing over 70 per cent organic content, with a corresponding moisture content in excess of 50 per cent. Differences in the characterization and reporting of waste types also differ with some municipal authorities including construction and demolition waste and industrial waste as part of the municipal waste stream.

Some inter-urban differences relate to climate and fuel use. The cities where heating is needed in winter such as Beijing, Shanghai, Seoul and Tokyo and where coal is the main source of energy, have much greater amount of ash in the waste in those cold months. The basic infrastructure brings other variations in cities and towns (such as Calcutta, Dhaka, and Hanoi) with unpaved or poorly paved streets that have large amounts of dust and dirt from street sweeping. There are big differences in amounts of organic waste among cities according to the number of trees and shrubs in public places. Large and bulky waste items such as abandoned motorcars, furniture and packaging are found in the higher-income economies such as Brunei Darussalam, Japan, Republic of Korea and Singapore, but not in low-income countries such as Bangladesh, Cambodia, Myanmar, Nepal, Sri Lanka and Viet Nam. Table 8.2 provides an illustration of the quantities and types of MSW generated in selected countries of the South Pacific subregion.

Figure 8.2 Estimated Generation of Municipal Solid Waste in Different Subregions

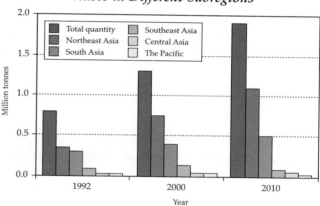

Source: United Nations 1995, World Bank 1995 and 1998, UNEP/SPREP 1997

Figure 8.3 Approximate Composition of Municipal Solid Waste in Selected Cities of ESCAP Member Countries

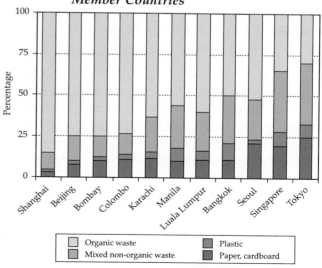

Source: United Nations 1995, World Bank 1995 and 1998, UNEP/SPREP 1997

Figure 8.1 Municipal Solid Waste Generation in Different Groups of Countries in the Region

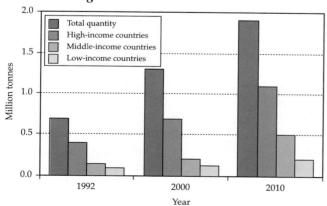

Source: World Bank 1999

Table 8.2 Quantities and Types of MSW Generated in Selected South Pacific Countries

Selected Countries	Types of waste generated	% of total	Total Amount of MSW generated (tonnes per day)
Cook Islands			4.75
	Organic	32%	
	Glass	24%	
	Plastic	12%	
	Metals	10%	
Fiji (Suva, the capital)			35.6
	Metal	10-16%	
	Glass	5-10%	
	Plastic	7-12%	
	Vegetative debris	25-39%	
	Paper	27-34%	
Vanuatu			15
	Vegetative debris	35-40%	
	Wood	25-30%	
	Paper	10-12%	
	Plastic	6-8%	
	Glass/Ceramic	3-5%	
	Metals	2-3%	
	Textile	3-6%	

Source: World Bank 1997

The amount of human faeces in the MSW is significant in squatter areas of many Asian and Pacific cities where "wrap and throw" sanitation is practised or bucket latrines are emptied into waste containers. The latter is common in many cities (such as Calcutta, Dhaka and Hanoi) of the region where sewerage systems are minimal.

2. *Industrial Solid Waste*

Industrial solid waste in the Asian and Pacific Region, as elsewhere, encompasses a wide range of materials of varying environmental toxicity. Typically this range would include paper, packaging materials, waste from food processing, oils, solvents, resins, paints and sludges, glass, ceramics, stones, metals, plastics, rubber, leather, wood, cloth, straw, abrasives, etc. As with municipal solid waste, the absence of a regularly up-dated and systematic database on industrial solid waste ensures that the exact rates of generation are largely unknown.

Industrial solid waste generation varies, not only between countries at different stages of development but also between developing countries (see Figure 8.4). In People's Republic of China, for example, the generation ratio of municipal to industrial solid waste is one to three. In Bangladesh, Sri Lanka and Pakistan, however, this ratio is much less. In high-income, developed countries, such as

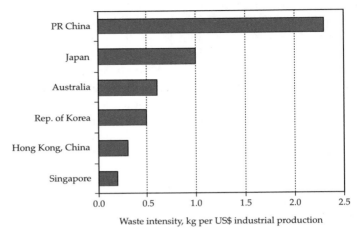

Figure 8.4 Waste Intensity of Industrial Production in Selected Countries in the Region

Source: ESCAP 1997

Australia and Japan, the ratio is one to eight. However, based on an average ratio for the region, the industrial solid waste generation in the region is equivalent to 1 900 million tonnes per annum. This amount is expected to increase substantially and at the current growth rates, it is estimated that it will double in less than 20 years. As the existing industrial solid waste collection, processing and disposal systems of many countries are grossly inadequate, such incremental growth will pose very serious challenges.

3. *Agricultural Waste and Residues*

Expanding agricultural production has naturally resulted in increased quantities of livestock waste, agricultural crop residues and agro-industrial by-products. Table 8.3 provides an estimate of annual production of agricultural waste and residues in some selected countries in the region (ESCAP 1997); the implications of liquid and slurry waste for receiving inland and coastal waters is examined in Chapter 4.

Among the countries in the Asian and Pacific Region, People's Republic of China produces the largest quantities of agriculture waste and crop residues followed by India. In People's Republic of China, some 587 million tonnes of residues are generated annually from the production of rice, corn and wheat alone (see Figure 8.5). Figure 8.6 illustrates the proportions of waste that Malaysia generates from the production of rice, palm oil, rubber, coconut and forest products (ESCAP 1997). In Myanmar, crop waste and residues amount to some 4 million tonnes per year (of which more than half constitutes rice husk), whilst annual animal waste production is about 28 million tonnes with more than 80 per cent of this coming from cattle husbandry.

Table 8.3 *Approximate Estimate of Annual Production of Agricultural Waste and Residues in Selected Countries in the Region*

Country	Annual production, million tonnes		
	Agricultural waste (manure/animal dung)	Crop residues	Total
Bangladesh	15	30	45
PR China	255	587	842
India	240	320	560
Indonesia	32	90	122
Malaysia	12	30	42
Myanmar	28	4	32
Nepal	4	12	16
Pakistan	16	68	84
Philippines	20	12	32
Rep. of Korea	15	10	25
Sri Lanka	6	3	9
Thailand	25	47	72

Source: ESCAP 1997

Figure 8.5 *Proportionate Annual Production of Agricultural Waste in People's Republic of China*

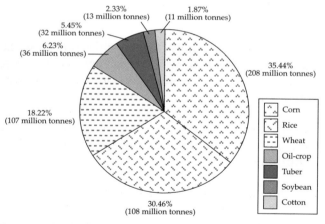

Source: ESCAP 1997

In Pakistan, about 56.22 million tonnes of different crop residues are generated of which 12.46 million tonnes originate from cotton, 2.90 million tonnes from maize, 12.87 million tonnes from sugarcane, 8.16 million tonnes from rice and 19.83 million tonnes from wheat. In addition, Pakistan produces other wastes amounting to some 28 million tonnes of which 58 per cent is animal waste, 40 per cent is sugarcane bagasse and the remaining two per cent comprises a mix of jute sticks, mustard stalks, sesame sticks, castor seed stalks, sunflower stalks and tobacco stalks (ESCAP 1997).

Figure 8.6 *Proportionate Annual Production of Agricultural Waste in Malaysia*

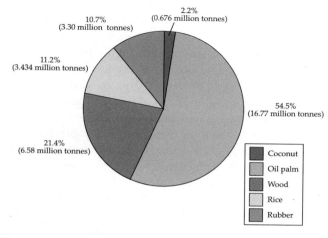

Source: ESCAP 1997

In Sri Lanka, agricultural waste comprises animal waste, paddy husk, straw, coir fibre and coir dust, bagasse, as well as the waste from the timber industry, which comprises sawdust, off-cuts and charcoal. Commercial rice milling generates around 2 million tonnes of paddy husk per annum, whilst coir (the fibres from coconut husks) processing generates an annual 700 000 tonnes of coir dust (ESCAP 1997). Each year, Thailand produces about 4.6 million tonnes of paddy husk, 35 million tonnes of rice straw, 7 million tonnes of bagasse and more than 25 million tonnes of animal waste (ESCAP 1997). Other countries such as Australia, Cambodia, Japan, Lao People's Democratic Republic, Nepal, New Zealand, Republic of Korea, Viet Nam and Small Island States in the South Pacific also generate huge quantities of agricultural waste and residues (ESCAP 1997, UNEP/SREP 1997).

4. *Hazardous Waste*

With rapid development in agriculture, industry, commerce, hospital and health-care facilities, the Asian and Pacific Region is consuming significant quantities of toxic chemicals and producing a large amount of hazardous waste. Currently, there are about 110 000 types of toxic chemicals commercially available. Each year, another 1 000 new chemicals are added to the market for industrial and other uses.

The availability of robust data on the generation of hazardous waste for the Asian and Pacific Region is limited by the reliability of information on the quantities and types of hazardous waste produced at the country level. This is due to a variety of reasons, including the lack of qualified personnel to undertake the necessary assessment, the reluctance of industries to provide process information (including waste arising data) and a poor appreciation of the extent to which generated waste is hazardous. Where data is available, significant difficulties are encountered in seeking to draw international comparisons due to differences in classification and definition of hazardous waste from country to country within in the region.

Most hazardous waste is the by-product of a broad spectrum of industrial, agricultural and manufacturing processes, nuclear establishments, hospitals and health-care facilities. Primarily, high-volume generators of industrial hazardous waste are the chemical, petrochemical, petroleum, metals, wood treatment, pulp and paper, leather, textiles and energy production plants (coal-fired and nuclear power plants and petroleum production plants). Small- and medium-sized industries that generate hazardous waste include auto and equipment repair shops, electroplating and metal finishing shops, textile factories, hospital and health-care centres, dry cleaners and pesticide users.

The principal types of hazardous waste generated in the Asian and Pacific Region, include waste solvents, chlorine bearing waste and pesticide-organophosphate-herbicide-urea-fungicide bearing waste. In particular, solvents are extensively used in the region and, as a consequence, large quantities of waste solvents are produced.

The types, quantities and sources of hazardous waste vary significantly from country to country and are influenced by the extent and diversity of industrial activity. Table 8.4 provides a conservative estimate of the past, current and future hazardous waste generation trends in a number of selected countries (Hernandez 1993, UNEP 1994, United Nations 1995, Nelson 1997). However, it must be stressed that such estimations are founded on data that may be considered incomplete and unverified. In the absence of reliable regional data, a study by the World Bank (WRI 1995) estimated the hazardous waste toxic releases in the Asian and Pacific region and predicted significant increases in hazardous waste production each year in People's Republic of China, India, Indonesia, the Philippines, and Thailand. An even more significant conclusion of the study was that the intensity of hazardous waste generation per unit of output is also set to increase (WRI 1995).

Better and more reliable data are available for the quantities of petroleum waste produced in countries that extract or process crude oil such as in Brunei Darussalam, People's Republic of China, India, Indonesia, Malaysia, Japan and Republic of Korea. In Malaysia, 0.71 metric tonnes of petroleum waste

Table 8.4 Conservative Estimate of Annual Production of Hazardous Waste in Selected Countries and Territories in the Asian and Pacific Region

Country/Territory	Estimated annual production, tonnes x 10^3		
	1993	2000	2010
Australia	109	275	514
Bangladesh	738	1 075	1 560
PR China	50 000	130 000	250 000
Hong Kong, China	35	88	165
India	39 000	82 000	156 000
Indonesia	5 000	12 000	23 000
Japan	82	220	415
Malaysia	377	400	1 750
Mongolia	15	26	45
Nepal	130	260	450
New Zealand	22	62	120
Pakistan	786	1 735	3 100
Philippines	115	285	530
Papua New Guinea	25	45	80
Rep. of Korea	269	670	1 265
Singapore	28	72	135
Sri Lanka	114	250	460
Thailand	882	2 215	4 120
Viet Nam	460	910	1 560

Source: Hernandez 1993, UNEP 1994, United Nations 1995, and Nelson 1997

are generated annually (Malaysia Environmental Quality Report 1998), whilst it is reported that Fiji, Solomon Islands, Papua New Guinea, Federated States of Micronesia, Samoa, Vanuatu, Tonga, Marshall Islands, Nauru, Cooks Islands, Kiribati and Tuvalu collectively generate approximately 10.55 million litres of waste oil per year (UNEP/SPREP 1997).

B. Environmental Impacts of Waste

The economic growth and urbanization experienced in many parts of the Asian and Pacific Region over the past 10-15 years, has significantly escalated the quantities of MSW being generated in many cities, including Bangkok, Beijing, Mumbai, Calcutta, Colombo, Dhaka, Hanoi, Jakarta, Kuala Lumpur, Manila and Shanghai (United Nations 1995, Koe and Aziz 1995). Uncontrolled, open dumping on the peripheries of many of the region's cities has resulting in the degradation of valuable land resources and the creation of long-term environmental and human health problems. The events of July 2000 at the Quezon City garbage dump on the outskirts of Manila, where hundreds of people were killed by the collapse of a "seven storey high" open dump, stands testament to the direct potential consequences of uncontrolled dumping.

Throughout the region, indiscriminate dumping has led to the contamination of surface and groundwater supplies, whilst open burning of waste contributes significantly to urban air pollution. At a global level, the uncontrolled release of methane, which is produced as a by-product of the decomposition of organic wastes, represents a significant proportion of the region's contribution to the greenhouse effect.

The increase in potentially hazardous industrial, biomedical and nuclear wastes has not been accompanied by a commensurate expansion in the provision of waste treatment and management facilities. The uncontrolled dumping of biomedical waste has the potential for transporting pathogens (disease producing organisms), whilst the indiscriminate disposal of oils, used batteries, discarded paints, spent chemicals and carcinogens, such as asbestos, can cause significant adverse impacts on human health and the environment. Various incidents of pollution have also been reported from industrial waste, abattoirs or food processing plants along with biocides and toxic effluents from sawmills and timber processing areas

Table 8.5 Impacts of Various Categories of Wastes on Water, Soil and Air in Selected Countries of Different Subregions

	Agricultural wastes and residues			Municipal wastes			Industrial wastes			Hazardous wastes		
	Water Pollution	Land Pollution	Air Pollution	Water Pollution	Land Pollution	Air Pollution	Water Pollution	Land Pollution	Air Pollution	Water Pollution	Land Pollution	Air Pollution
Australia	⊕	⊕	○	⊕	⊕	○	⊕	⊕	○	⊕	⊕	○
Bangladesh	•	⊕	○	•	⊕	⊕	•	⊕	○	•	⊕	○
Brunei Darussalam	○	○	○	⊕	○	○	○	○	○	○	○	○
Cambodia	•	⊕	○	•	⊕	○	•	⊕	○	•	⊕	○
PR China	•	⊕	⊕	•	⊕	○	•	⊕	○	•	⊕	○
Cooks Islands	⊕	⊕	○	⊕	⊕	○	⊕	⊕	○	⊕	⊕	○
Fiji	⊕	⊕	○	•	⊕	○	⊕	⊕	○	•	⊕	○
Hong Kong, China	⊕	⊕	○	⊕	⊕	○	⊕	⊕	○	⊕	○	○
India	•	⊕	○	•	⊕	○	•	⊕	⊕	•	⊕	○
Indonesia	•	⊕	○	•	•	○	•	•	○	•	⊕	○
Japan	⊕	○	○	⊕	○	○	⊕	○	○	⊕	○	○
Kazakhstan	⊕	○	○	•	○	○	⊕	○	○	⊕	○	○
Kiribati	⊕	○	○	⊕	○	○	⊕	○	○	⊕	○	○
Lao People's Democratic Republic	•	⊕	○	•	•	○	⊕	⊕	○	⊕	⊕	○
Malaysia	•	⊕	○	•	⊕	○	•	•	○	•	•	○
Maldives	⊕	○	○	⊕	○	○	⊕	○	○	⊕	○	○
Marshall Islands	⊕	○	○	⊕	⊕	○	⊕	○	○	⊕	⊕	○
Micronesia Federated States of	⊕	○	○	⊕	⊕	○	⊕	○	○	⊕	⊕	○
Mongolia	⊕	⊕	○	•	⊕	○	⊕	○	○	⊕	○	○
Myanmar	⊕	⊕	○	•	•	○	⊕	⊕	○	⊕	○	○
Nepal	⊕	⊕	○	⊕	⊕	○	⊕	⊕	○	⊕	○	○
New Zealand	⊕	○	○	⊕	⊕	○	⊕	⊕	○	⊕	⊕	○
Pakistan	⊕	⊕	○	•	⊕	○	•	⊕	⊕	•	⊕	⊕
Papua New Guinea	⊕	⊕	○	•	⊕	○	⊕	⊕	○	⊕	⊕	○
Philippines	•	⊕	○	•	•	○	•	⊕	○	•	•	○
Rep. of Korea	⊕	⊕	○	⊕	⊕	○	⊕	⊕	○	⊕	⊕	○
Samoa	⊕	⊕	○	⊕	⊕	○	⊕	⊕	○	⊕	⊕	○
Singapore	○	○	○	○	○	○	○	○	○	○	○	○
Solomon Islands	⊕	⊕	○	⊕	⊕	○	⊕	⊕	○	⊕	⊕	○
Sri Lanka	⊕	⊕	○	•	⊕	○	•	⊕	○	•	⊕	○
Tonga	⊕	⊕	○	⊕	⊕	○	⊕	⊕	○	⊕	⊕	○
Tuvalu	⊕	○	○	⊕	⊕	○	⊕	○	○	⊕	○	○
Thailand	•	⊕	○	•	•	○	•	•	⊕	•	•	⊕
Vanuatu	○	○	○	⊕	⊕	○	⊕	⊕	○	⊕	⊕	○
Viet Nam	•	•	○	•	•	○	•	•	⊕	•	•	○

Source: Anjello and Ranawana 1996, ESCAP 1997, Higham 1998, Hunt 1996, Kiser 1998, Koe and Aziz 1995, Leong and Quah 1995, ST 1995, World Bank 1995 and 1998, ENV 1997

Key: • severe   ⊕ moderate   ○ moderate to negligible

(UNEP/SPREP 1997). The overall impacts of different categories of wastes on water, soil and air in some selected countries of various subregions are given in Table 8.5 (Anjello and Ranawana 1996, ESCAP 1997, Higham 1998, Hunt 1996, Kiser 1998, Koe and Aziz 1995, Leong and Quah 1995, ST 1995, World Bank 1995 and 1998, ENV 1997).

WASTE PROCESSING AND CONTROL

A. Current Waste Management Practices

As indicated in Table 8.6, the current practices employed in the management of solid waste within the Asian and Pacific Region vary considerably between the low, middle and high-income countries. The extent of application and the effectiveness of these practices are reviewed in the subsections that follow.

1. Municipal Solid Waste

(a) Collection and Transfer

In many cities of the region, municipal solid waste (MSW) is gathered in a variety of containers ranging from old kerosene cans and rattan baskets to used grocery bags and plastic drums or bins. In some cities, neighbourhood-dumping areas have been designated (formally or informally) on roadsides from which bagged and loose waste is collected.

Waste collection (and, where appropriate, waste transfer) frequently constitutes the major solid waste management cost for the region's cities. A wide variety of collection systems are used including door-to-door collection and indirect collection, by which containers, skips or communal bins are placed near markets, in residential areas and other appropriate locations. In the high-income industrialized countries of Australia, Japan, New Zealand, Republic of Korea

Table 8.6 Comparison of Typical Solid Waste Management Practices

Activity	Low income	Middle income	High income
Source reduction	No organized programmes, but reuse and low per capita waste generation rates are common.	Some discussion of source reduction, but rarely incorporated in to any organized programme.	Organized education programmes are beginning to emphasize source reduction and reuse of materials.
Collection	Sporadic and inefficient. Service is limited to high visibility areas, the wealthy, and businesses willing to pay.	Improved service and increased collection from residential areas. Larger vehicle fleet and more mechanization.	Collection rate greater than 90 per cent. Compactor trucks and highly machined vehicles are common.
Recycling	Most recycling is through the informal sector and waste picking. Mainly localized markets and imports of materials for recycling.	Informal sector still involved, some high technology sorting and processing facilities. Materials are often imported for recycling.	Recyclable material collection services and high technology sorting and processing facilities. Increasing attention towards long-term markets.
Composting	Rarely undertaken formally even though the waste stream has a high percentage of organic material.	Large composting plants are generally unsuccessful, some small-scale composting projects are more sustainable.	Becoming more popular at both backyard and large-scale facilities. Waste stream has a smaller portion of compostables than low and middle-income countries.
Incineration	Not common or successful because of high capital and operation costs, high moisture content in the waste, and high percentage of inerts.	Some incinerators are used, but experiencing financial and operational difficulties; not as common as high-income countries.	Prevalent in areas with high land costs. Most incinerators have some form of environmental controls and some type of energy recovery system.
Landfilling	Low-technology sites, usually open dumping of wastes.	Some controlled and sanitary landfills with some environmental controls. Open dumping is still common.	Sanitary landfills with a combination of liners, leak detection, leachate collection system, and gas collection and treatment systems.
Costs	Collection costs represent 80 to 90 per cent of the municipal solid waste management budget. Waste fees are regulated by some local governments, but the fee collection system is very inefficient.	Collection costs represent 50 to 80 per cent of the municipal solid waste management budget. Waste fees are regulated by some local and national governments, more innovation in fee collection.	Collection costs can represent less than 10 per cent of the budget. Large budget allocations to intermediate waste treatment facilities. Upfront community participation reduces costs and increases options available to waste planners (e.g. recycling and composting).

Source: World Bank 1999

and Singapore, collection and transfer services are capital-intensive and highly mechanized employing standardized collection vehicles, compactors and containers and providing collection rates in the range of 90 per cent and collection services to most urban and even rural areas. Source separation and subsequent collection of recyclables is governed by regulation and is facilitated by the provision of colour-coded bins or bags or by the establishment of area recycling centres. Whilst a significant number of these cities continue to retain parts of the collection process within their direct municipal control, many others have contracted private sector waste collection firms and have made private sector trade and industrial establishments responsible for the collection and disposal of their own solid waste.

In the middle-and low-income countries of the region, waste collection and transfer tend to be labour-intensive and are undertaken by personnel directly employed by the municipal authorities. Waste collection is undertaken using low-levels of mechanization with handcarts and tractor-trailers being used to collect waste from communal bins and neighbourhood dumping areas. The collection systems are relatively inefficient as the collection vehicles and containers are not fitted with compactors, necessitating the transportation of loose waste and, hence, imposing a constraint on the capacity of the collection system.

In some cities of the lower and middle income countries, such as Dhaka, Calcutta and Hanoi, collection rates are significantly less than 50 per cent, whilst collection rates of well over 50 per cent are achieved in Bangkok, Mumbai, Delhi, Jakarta, Kuala Lumpur, Manila and Shanghai. By comparison, collection rates in Hong Kong, China, Seoul, Singapore, Sydney and Tokyo are in excess of 90 per cent. There are, of course, disparities in collection services between the rich and poor areas and in a number of cities collection services are not extended to the poor, especially those in illegal settlements.

In some cities, decentralized pre-collection has proven effective in achieving increased collection rates. For example, the kampongs (villages) of Indonesian cities have formal responsibility for primary collection, the waste from each kampong being delivered to a transfer station or temporary storage point for collection by the city service. Delhi and Chennai employ similar systems and have achieved reasonably good collection systems as a result. Elsewhere, the lack of efficient transfer facilities represents a weak link in the MSW collection and transportation system. In cities such as Tokyo, Singapore and Sydney, transfer stations are used as a means of gathering waste from a sub-division of the city, compacting the waste to maximize transportation efficiency and then transferring the waste to larger haulage vehicles for delivery to the city's disposal sites. In addition, transfer stations often serve as material recovery centres where recyclables are separated for reuse/recycling. In developing countries, few cities have established well-designed transfer stations with sufficient facilities, equipment and vehicles to manage and process their collected waste.

Increasingly, collection services are being privatized. In the region as a whole, more than 20 per cent of the collection services are now contracted out to private waste collection companies. This practice is gaining momentum, especially in Australia; Hong Kong, China; Malaysia; Republic of Korea; Singapore and Thailand. In Singapore, as elsewhere, the main motivation for privatization is cost saving; the cost of collection and disposal of refuse has tripled during the last decade to more than US$700 million. In 1994, the Ministry of the Environment (ENV) with an authorized capital fund of US$250 million created a private limited company (SEMAC Pte Ltd.). In April 1996, SEMAC took over the collection unit of the ENV, allowing the Ministry to concentrates on its regulatory role of safeguarding public health and environmental standards through legislative and licensing controls (ENV 1997).

Financial constraints and the lack of technical expertise severely limit the effectiveness of solid waste management in the cities and towns of the poorer developing nations. Shortages of storage bins, collection vehicles, non-existent and/or inadequate transfer stations, traffic congestion and a lack of public compliance are factors affecting collection efficiency, resulting in low waste collection rates. In some cities, heaps of refuse are routinely left uncollected and there are illegal deposits on open land, drains and canals. The lack of coordination and overlapping of responsibilities among various government agencies and different levels of local government also contribute to the problem (UNEP/SPREP 1997).

(b)   Material Recovery, Reuse and Recycling

In many countries of the region, including Japan, Republic of Korea and Singapore, the rate of recovery of recyclable materials from MSW has improved significantly in recent years (ENV 1997, Hara 1997). Within the region, overall resource recovery has grown from less than 10 per cent of all MSW in 1988 to 30 per cent in 1998, with much of the increase attributable to greater rates of recovery of paper and paperboard, plastics, glass and metals.

In terms of the total tonnage of materials recovered, paper and paperboard represent the largest category (almost 60 per cent of the total) and this often masks the importance of recovery rates for other materials. For example, recovered aluminium represents only about 3 per cent of the total tonnage of recovered materials, yet in terms of its economic value, recovered aluminium far exceeds the paper product category.

Among the Asian and Pacific countries, Japan recycles huge quantities of materials from MSW stream. Almost half of the waste paper is retrieved or recycled in Japan and that the retrieval rate increased from about 48 per cent in 1990 to about 56 per cent in 1997. Similarly, between 1990 and 1997 the recycling rates of aluminium and steel cans increased from 40 per cent to 60 per cent and 45 per cent to 70 per cent respectively, whilst during the same period glass recycling rates increased from 48 per cent to 57 per cent (Hara 1997). Other countries of the region, such as Australia, Bangladesh, People's Republic of China, Malaysia, New Zealand, Philippines, Republic of Korea, Singapore and Thailand, recycle significant quantities of paper and cardboard, plastics, glass and metals; as an example the categories of materials recycled in Singapore in 1997 are shown in Table 8.7 (ENV 1997). In many cases, particularly in the more developed economies, recycling is undertaken at source (i.e. at the household, business and industry level) and is actively promoted by governments, NGOs and the private sector (United Nations 1995). Elsewhere, such as in Viet Nam (Hebert 1995, World Bank 1995), informal recycling networks have flourished despite the lack of formal promotion or support of the Government.

Waste recycling in developing countries relies largely on the informal recovery of materials by scavengers or waste pickers. In cities of the Asian and Pacific, it has been estimated that up to two per cent of the population survives by recovering materials from waste to sell for reuse or recycling or for their own consumption. In some cities these waste scavengers constitute large communities: approximately 15 000 squatters make their living by sifting through the Smoky Mountain municipal rubbish dump in Philippines (Anonymous 1995); it is estimated that in Bangalore there are between 20 000 to 30 000 scavengers (Hunt 1996); and Jakarta is served by between 15 000 and 20 000 waste pickers (Wahyono and Sahwan 1998). Some of these communities have high levels of organization and the creation of scavenger co-ops has gained momentum in some countries of the region including the Philippines, India and Indonesia (Medina 1998). The role and lifestyle of waste scavengers are highlighted in Box 8.1 (Hunt 1996, Pitot 1996).

(c) Disposal Methods for MSW

Various disposal methods of municipal solid waste in selected countries/territories in the region are given in Table 8.8.

(i) Open Dumping

Open dumping is the most widespread method of solid waste disposal in the region and typically involves the uncontrolled disposal of waste without measures to control leachate, dust, odour, landfill gas or vermin. In some cities, open burning of waste is practised at dumpsites. In many coastal cities, waste is dumped along the shoreline and into the sea, such as Joyapura in Indonesian or dumped in coastal and inland wetlands and ravines as is being practised in Mumbai, Calcutta, Colombo, Dhaka and Manila (UNEP/SPREP 1997).

The scarcity of available land has led to the dumping of waste to very high levels; waste thickness is often over 12 metres and may be over 20 metres, which was the case of the Quezon City dumpsite in the Philippines.

An additional hazard on uncontrolled dumpsites arises from the build-up of landfill gas (predominantly methane), which has led to outbreaks of fire and to adverse health effects on workers and adjacent residents (Perla 1997, Wahyono and Sahwan 1998).

The scarcity of available land has also become a major problem for the disposal of solid waste in Small Island Developing States in the South Pacific subregion. Dumping at sea has frequently been

Table 8.7 Various Categories of Materials Recycled from MSW in Singapore in 1997

Waste type	Estimated quantity in tonnes in 1997			Recycling rate (per cent)
	Total waste disposed	Total waste recycled	Total waste output	
Food waste	1 085 000	24 700	1 109 700	2.2
Paper/cardboard	576 000	324 000	900 000	36.0
Plastics	162 000	35 300	197 300	17.9
Construction debris	126 000	188 000	314 000	59.9
Wood/timber	249 000	34 800	283 800	12.3
Horticultural waste	75 400	67 600	143 000	47.3
Earth spoils	75 400	–	75 400	–
Ferrous metals	75 400	893 000	968 400	92.2
Non-ferrous metal	14 000	76 000	90 000	84.4
Used slag	120 000	135 000	255 000	52.9
Sludge (Industry/PUB)	50 200	–	50 200	–
Glass	30 800	4 600	35 400	13.0
Textile/leather	25 200	–	25 200	–
Scrap tyres	5 600	5 700	11 300	50.4
Others	126 000	1 300	127 300	1.0
Total	2 976 000	1 790 000	4 586 000	39.0

Source: ENV 1997

> **Box 8.1 Recycling: Fortunes and Costs**
>
> There are dozens of recycling enterprises in Hanoi, the Vietnamese capital, despite the fact that it remains one of Asia's poorest cities. The primary reasons for the scale of waste recycling are resource scarcity and poverty. It is estimated that scavengers in the city collect a daily average of 250 tonnes of waste materials, or more than *one third* of the 830 tonnes of refuse produced each day by the capital's 3 million residents. Amongst the materials collected are bottles, paper, metals and plastics.
>
> As Hanoi is a resource-poor, labour-abundant city, nearly everything of value within the City's solid waste is extracted for recycling. The Hong Tien Industrial Cooperative, based in the Hai Ba Trung district, is typical of the enterprises that participate in the recycling of waste materials. The Cooperative specialises in transforming discarded plastic materials into marketable synthetic leather. Each week, during the four months preceding the start of the school year, the workers at Hong Tien take one tonne of old plastic sandals and turn them into 3 000 new red-and-black school bags. During the remaining eight months of the year, Hong Tien produces plastic sheeting that is used for many purposes. Hog Tien and numerous other similar enterprises form an extensive waste recycling network that has developed without government assistance and without the formation of commercial monopolies, as have emerged in many other big Asian cities including Bangkok, Dhaka, Calcutta, Jakarta, Manila and Beijing.
>
> The waste scavengers of Hanoi operate at no cost to the city's municipal authority and provide both financial benefits to the society in the form of avoided costs (such as landfill space, collection and transportation, energy, employment generation, protection of public health) as well as ecological benefits in the form of resource conservation and environmental protection. The recycled materials from wastes work their way from the 'waste economy' back into a productive economy through an elaborate system of buyers. A network of scavengers and junk buyers (estimated to comprise some 6 000 people during the August peak season) collect discarded goods for onward sale to junk dealers, who in turn re-sell the materials in bulk to factories and exporters.
>
> A large number of waste scavengers shuffle from house to house along the streets of Hanoi offering to buy empty beer cars, worn-out plastic sandals, old bottles and used newspapers. On productive days, each scavenger can earn up to Dong 20 000 (US$1.90), although on bad days they may take home almost nothing. It is estimated that some 1 500 families make their living by buying and selling waste materials and a trade network has emerged with clients from Hanoi and the surrounding provinces regularly visiting individual junk dealers to buy, and pre-order, specific types of recycled materials.
>
> However, the business of waste scavenging is not without its human health costs and the rewards for some engaged in extracting materials from waste are inadequate to alleviate their poverty. In many cases, the scavengers picking over the mixed waste of the dumping grounds do not wear protective clothing nor do they have access to washing facilities. The majority of dumpsite scavengers are women and children, who live in overcrowded, poorly ventilated temporary huts, often on the peripheries of the waste dump. The scavengers seldom have access to public or private latrines, are malnourished and suffer from a range of illnesses including worm infections, scabies, respiratory tract infection, abdominal pain, fever and other unspecified diseases.
>
> Source: 1. Hebert 1995
> 2. World Bank 1995

adopted as a solution with old cars and refrigerators being dumped into the lagoons of French Polynesia and municipal waste being bundled into wire gabions for use in sea wall construction in the Marshall Islands. In the latter case, the gabions allowed leachate and loose waste items to pass directly into the ocean water.

(ii) Landfilling

In the Asian and Pacific Region, the disposal of solid waste at a semi-engineered or full sanitary landfill has been adopted by cities from both low and high-income countries as the most attractive of disposal options. Bandung, Singapore, Hong Kong, China, Seoul, Chennai and Tokyo do have well-designed and reasonably operated sanitary landfills, whilst other cities in Australia, People's Republic of China, Japan, Republic of Korea, Malaysia and Thailand have adopted controlled tipping or sanitary land filling for solid waste disposal. Kuala Lumpur employs disused tin mines for MSW landfills around the city.

The generation of landfill gas has been turned to advantageous use at a number of landfills in the region through the development of electricity generation facilities. A landfill/biogas power generation facility is currently commencing construction in Ho Chi Minh City and others are planned for Chennai and, possibly, Colombo.

In the densely populated cities and towns of the region, the land availability for landfill siting is a major constraint. For example, in Hong Kong, China and in Singapore severe land constraints have led to complex engineering infrastructure solutions being developed to ensure high standards of operational and maintenance control and have enabled the development of acceptable landfill solutions in coastal areas, offshore islands and mountainous terrain. In Singapore, the two existing landfill sites are nearing their capacity and an offshore landfill at Pulau

CHAPTER EIGHT

Table 8.8   Disposal Methods for Municipal Solid Waste in Selected Countries of the Region

Country/Territory	Disposal methods				
	Composting (per cent)	Open dumping (per cent)	Land filling (per cent)	Incineration (per cent)	Others* (per cent)
Australia	10	–	80	5	5
Bangladesh	–	95	–	–	5
PR China	10	50	30	2	8
Cook Islands	–	60	30	–	10
Fiji	–	90	–	–	10
Hong Kong, China	–	20	60	5	15
India	10	60	15	5	10
Indonesia	15	60	10	2	13
Japan	10	–	15	75	–
Kazakhstan	–	85	–	–	15
Rep. of Korea	5	20	60	5	10
Maldives	–	90	–	–	10
Malaysia	10	50	30	5	5
Mongolia	5	85	–	–	10
Myanmar	5	80	10	–	5
Nepal	5	70	10	–	15
New Zealand	5	–	85	–	10
Pakistan	5	80	5	–	10
Philippines	10	75	10	–	5
Papua New Guinea	–	80	–	5	15
Samoa	–	80	–	–	20
Singapore	–	–	30	70	–
Sri Lanka	5	85	–	–	10
Thailand	10	65	5	5	15
Viet Nam	10	70	–	–	20

Source: ENV 1997

*Animal feeding, dumping in water, ploughing into soil, and open burning.

Semakau is nearing completion at a cost of S$840 (US$500) million. The landfill consists of a 7 km long bund enclosing 350 hectares of sea that will take care of waste disposal needs of Singapore up to 2030 and beyond. The waste is put into cells and this will eventually rise to 15 metres above sea level (ST 1999).

(iii)   Composting

Whilst small-scale composting of organic waste is widespread in the region, attempts to introduce large-scale composting as a means of reducing the quantities of municipal solid waste requiring disposal, or with the intention of creating a revenue stream from the sale of compost, have been met with limited success. Most of the composting plants in the region are neither functioning at full capacity nor do they produce compost of marketable value. The high operating and maintenance costs results in compost costs that are higher than commercially available fertilisers, whilst the lack of material segregation produces compost contaminated with plastic, glass and toxic residues. Under such circumstances, little of the compost produced is suitable for agriculture application.

The forced-air composting plant in Hanoi is a typical example. The plant is currently operating at 20 per cent of its design capacity, whilst the municipal authorities have been unable to persuade local farmers to take the product free as it is too contaminated with plastics and glass.

Elsewhere, small-scale neighbourhood composting is actively promoted through research and pilot projects. In Indonesia, such schemes have been underway for over a decade and small private enterprises have been established in Cipinang Besar and Watam (East Jakarta) that supply compost to estate gardens and golf courses. In Bandung, a box type windrow composting plant has been established alongside and existing dumpsite (Perla 1997), whilst Ho Chi Minh City has two small composting plants (World Bank 1995). Small-scale vermicomposting (a process that uses worms and micro-organisms to convert organic materials into nutrient-rich compost) of organic waste is carried out in open boxes or containers and is practised in People's Republic of China, India, Indonesia and Philippines (Perla 1997, Thom 1997).

At a slightly larger scale, the composting of organic MSW with agricultural waste and sludge from municipal sewage treatment plants is being piloted in Australia, Bangladesh, People's Republic of China, India, Philippines and Thailand. However, land availability, high operational, maintenance and transportation costs and incomplete waste material segregation remain major constraints to the adoption of co-composting.

(iv)   Incineration

For much of the Asian and Pacific Region, the incineration of MSW remains an expensive and technically inappropriate waste disposal solution. The development of waste incineration facilities has been constrained by the high capital, operating and maintenance costs and by increasingly stringent air pollution control regulations (UNEP 1998). In addition, the combustible fraction of much of the MSW generated in the low and middle-income countries of the region is relatively low, with high organic and moisture contents. For example, the Indonesian city of Surabaya imported an incinerator that is currently operating at two-thirds of its design capacity as the waste needs to be dried on-site for five days before it is suitable for combustion. Even without the cost of air pollution control mechanisms, it is estimated that the cost of waste incineration in this instance is roughly 10 times greater than the cost of open dumping/land filling in other Indonesian cities.

Up-to-date, full-scale incinerators are currently in service only in countries such as Australia, People's

Republic of China, Hong Kong, China, Indonesia, Japan, Singapore and the Republic of Korea, where the combustible fraction of MSW is high and in some instances has been raised by moisture-reducing compaction at transfer stations.

The three incinerators operating in Singapore burn more than 75 per cent of the 6 700 tonnes of MSW that is collected each day (ENV 1997) and a fourth incinerator with a capacity of 3 000 tonnes per day is expected to become operational during 2000. The total electrical generated by the existing plants is about 60 megawatts (250 to 300 kWh/tonne MSW incinerated), a portion of which is used to run incinerator operations, and the balance is sold to the national electricity grid. The planned new plant, costing S$1 billion, will generate 80 megawatts of electricity of which 20 megawatts will be consumed by the plant and the remainder will be sold to Singapore Power (ENV 1997).

The Republic of Korea plans to raise the incinerated portion of their MSW from 8.9 per cent in 1998 to 20 per cent by 2001 (Government of the Republic of Korea, 1999), whilst Japan has 1900 existing waste incinerators, of which 1584 incinerators are operated by local governments with the balance run by private companies (ASIAN WATER 1999). In People's Republic of China, Beihai, Shenyang, Guangzhou, Beijing and Shanghai have all begun constructing incineration plants for MSW with foreign assistance. Hong Kong, China has closed its incinerators because they could not meet air pollution standards, but new plants are under consideration (Wan *et al* 1998).

2. *Industrial Solid Waste*

The methods employed in the disposal of industrial solid waste are broadly the same as those used to dispose of MSW and comprise open dumping, land filling (both semi-engineered and sanitary landfilling) and incineration.

In many countries, including Bangladesh, People's Republic of China, India, Indonesia, Malaysia, Philippines and Thailand, non-hazardous industrial solid waste is accepted at either open dumps or landfills along with municipal solid waste (although where facilities are available potentially hazardous industrial solid waste is disposed of either in secure landfills or is incinerated). In those developing countries with few waste management facilities, industrial waste is often dumped on private land or is buried in dump pits within or adjacent to the site of the industrial facility from which it has emanated.

3. *Agricultural Waste and Residues*

The principal disposal methods for agricultural waste, in a number of selected countries within the region, are presented in Table 8.9.

Table 8.9 *Disposal Methods of Agricultural Waste and Residues in Selected Countries in the Region/Area*

Country	Disposal methods of agricultural waste and residues						
	Land application	Fish farming	Composting	Biogas production	Utilization as		
					Fuel	Animal feed	Building materials
Australia	○	□	●	○	□	○	□
Bangladesh	○	□	□	□	●	○	●
Cambodia	○	○	○	○	●	○	●
PR China	●	●	●	●	●	●	●
India	●	○	●	●	●	●	●
Indonesia	●	●	●	○	●	●	○
Japan	○	○	●	○	○	○	○
Lao People's Democratic Republic	○	○	○	□	●	●	●
Malaysia	○	○	●	○	●	●	○
Myanmar	○	○	○	○	○	○	●
Nepal	□	□	□	□	○	○	□
New Zealand	○	□	○	○	○	○	□
Pakistan	□	□	○	○	○	○	□
Philippines	○	○	○	●	●	●	●
Rep. of Korea	○	○	●	○	○	○	●
Sri Lanka	□	□	○	○	○	○	○
Thailand	●	●	●	●	●	●	●
Viet Nam	●	●	●	●	●	●	●

Source: ESCAP 1997
Legend: ● High   ○ Moderate   □ Low

In most traditional, sedentary agricultural systems, farmers use the land application of raw or composted agricultural wastes as a means of recycling of valuable nutrients and organics back into the soil and this remains the most widespread means of disposal. Similarly, fish farming communities in Bangladesh, People's Republic of China, India, Indonesia, Malaysia, Philippines, Thailand and Viet Nam commonly integrate fish rearing with agricultural activities such as livestock husbandry, vegetable and paddy cultivation and fruit farming (Fauzia and Rosenani 1997, UNEP 1997).

Many countries with agricultural-based economies use agricultural wastes to produce biogas through anaerobic digestion. The biogas (approximately 60 per cent methane) is primarily used directly for cooking, heating and lighting, whilst the slurry from the anaerobic digesters is used as liquid fertiliser, a feed supplement for cattle and pigs and as a medium for soaking seeds prior to germination (Hendersen and Chang 1997).

Purpose-built sanitary landfills have been developed to receive hazardous waste in Australia, Japan, Malaysia, New Zealand and Republic of Korea, whilst hazardous waste incinerators have been developed in Australia, Japan, Hong Kong, China, Malaysia, Republic of Korea, Singapore and Thailand (ASIAN WATER 1998, World Bank 1998). Other countries such as Bangladesh, People's Republic of China, India, Mongolia, the Philippines, Pakistan, Sri Lanka and many Island States in the South Pacific subregion usually co-dispose hazardous waste along with MSW in open dumps or seek to store particularly toxic wastes in sealed containers (United Nations 1995, UNEP/SPREP 1997).

In some countries, including Australia, Japan, Hong Kong, China, Republic of Korea and Singapore, progress has been made on methods for detoxification of hazardous waste and subsequent immobilization by fabrication into bricks and other usable materials.

In Thailand, a major programme of hazardous waste management is underway along the Eastern Seaboard where petrochemical, chemical and non-ferrous industries produce some 250 000 to 300 000 tonnes of commercially viable hazardous industrial waste each year. A hazardous waste treatment plant, managed by the Industrial Estate Authority of Thailand, has been established at the Map Ta Phut Industrial Estate, a focal point of the country's petrochemical and chemical industries.

In Malaysia, the Bukit Nanas Integrated Waste Treatment Facility is the country's first comprehensive treatment plant possessing various facilities including high-temperature incineration, physical and chemical treatment, stabilization and a secure landfill (Malaysia 1998). A centralized hazardous waste treatment facility has also been developed in West Java (Indonesia) to treat hazardous waste from JABOTABEK (Jakarta, Bogor, Tangeran, and Behasi) industrial area. Between 1994 and 1997, the facility increased the quantities of treated hazardous waste from 9.7 tonnes to 29 tonnes, although the economic and political crisis of 1998-1999 saw industrial production slump with a commensurate decline in the quantities of waste treated to 16.6 tonnes in 1998, before increasing to 18.8 tonnes in 1999.

Japan possesses well-developed systems for treating and disposing of the 500 million tonnes of hazardous waste produced by its industries each year. Recycling and material recovery are encouraged to reduce the net amount of wastes requiring treatment and disposal and purpose-build landfills have been developed to receive hazardous waste. However, the most widely practised disposal options is incineration with some 3 840 hazardous industrial waste incinerators across the country (ASIAN WATER 1999), many of which have energy recovery facility to provide heating or for electrical power generation.

In Hong Kong, China, the Chemical Waste Treatment Centre (CWTC) receives most of the hazardous wastes generated by industries (Chua et al 1999) with some solid chemical wastes, including asbestos, tannery off-cuts and treatment residues being co-disposed at landfills.

4. *Biomedical Waste*

The number of hospitals and health care institutions in the Asian and Pacific Region has been increasing to meet the medical and health care requirements of the growing population. Although city planners have long taken into consideration the provision of medical and health care institutions and services, until recent years, they, and even municipal waste management authorities, have paid very little attention to the wastes generated from these facilities, which are potentially hazardous to human health and the environment.

In recent years, however, serious concern has arisen regarding the potential for spreading pathogens, as well as causing environmental contamination due to the improper handling and management of clinical and biomedical waste. Whilst regulatory programmes and guidelines to control waste from such institutions have been introduced in most developed countries, including Australia, Japan, New Zealand, and Singapore, in developing countries, such as Bangladesh, People's Republic of China, India, Indonesia, Pakistan and the Philippines, such programmes have yet to be fully developed (Ogawa 1993, WHO 1996, UNEP/SPREP 1997).

In Australia, the National Health and Medical Research Council has published national guidelines for management of clinical and related wastes and similar biomedical management guidelines have also been produced at the state level. In Japan, the Ministry of Health and Welfare has established a working group who has prepared guidelines for medical waste management. The Standards Association of New Zealand has published the "New Zealand Standards on Health Care-Waste Management" to rationalize and recommend methods for the management of health care wastes within the country. In Singapore, guidelines were drafted for the management and safe disposal of hospital wastes in July 1988 and the Ministry of the Environment subsequently produced the "Hospital Waste Management Manual," which included detailed guidelines for hospital waste handling and disposal and a standard format to assist hospitals in preparing their written policies and procedures. Similarly, in Malaysia the Ministry of Health prepared preliminary guidelines for the management of hospital waste in 1988, whilst the Philippines' Department of Health prepared guidelines in 1990 on effective and efficient methods of collection, storage, and disposal of hazardous waste by hospitals, clinics, and research laboratories.

However, whilst China's National Environmental Protection Agency has recently formulated the Solid Waste Pollution Prevention and Control Law and the Regulations on Management of Hazardous Wastes, hospital waste is generally collected and disposed of together with other domestic wastes and the hospital and waste management authorities have low levels of awareness regarding the dangers associated with infectious biomedical waste (WHO 1996).

In many of the countries of the region, individual hospitals have installed on-site incinerators for the disposal of clinical wastes, although these are often poorly designed and operated and the level of awareness of the dangers by workers is low (UNEP/SPREP 1997).

5. *Radioactive Waste*

Information regarding disposal practices for radioactive waste is not extensive and few systematic country surveys have been conducted. In Japan, low level radioactive waste generated from 46 operating nuclear power plants is packed into 2 000 litre drums and temporarily stored in on-site storehouses. Special enclosed containers are used to package eight drums together and these are then sea and land transported to the Rokkasho-mura Burial Centre in Aomori Prefecture for permanent storage (Tanaka 1993). In Indonesia, low level radioactive waste generated from four nuclear research centres is conditioned into cement matrices in blocks and the embedded waste blocks are transported to the RWMC (Radioactive Waste Management Centre) at Serpong for permanent burial (Suyanto and Yatim 1993). Other countries of the region such as India, Pakistan and Republic of Korea uses permanent land burial methods for the disposal of radioactive waste (Greenpeace 1998).

6. *Transboundary Movement of Hazardous Waste*

The Asian and the Pacific Region is under considerable pressure as a favoured dumping ground for hazardous waste, particularly as domestic pressure has been exerted on industries operating in the industrial nations to dispose of their hazardous waste in a controlled, and hence expensive, manner. Between 1994 and 1997, the industrialized nations sent a total of 3.5 million tonnes of hazardous waste to countries in the Asian and Pacific Region. The first documented case of such imports to People's Republic of China occurred in September 1994 and by the first quarter of 1995, Chinese customs identified 22 separate incidents involving some 3,000 tonnes of foreign hazardous waste. From 1995 to 1996 Chinese customs uncovered almost one case per week of mislabelled hazardous waste, mostly from United States, Republic of Korea, and Japan in particular (Greenpeace 1997). In June 1998, 640 tonnes of Californian waste was found dumped in a Beijing suburb; the waste included toxic sludge, used syringes and decomposing animal bodies (Greenpeace 1997).

Over the same period, India has also seen an increase in the dumping of hazardous waste from industrialized nations (Anjello and Ranawana 1996, Agarwal 1998). Thousands of tonnes of toxic waste are being illegally shipped to India for recycling or dumping, despite a New Delhi court order banning imports of toxic materials. In 1995, Australia exported more than 1 450 tonnes of hazardous waste, including scrap lead batteries, zinc and copper ash, to India, whilst some 569 tonnes of lead battery waste were brought in through the main seaport of Mumbai between October 1996 and January 1997 (Greenpeace 1998).

Despite international agreements, substantial quantities of PVC waste is still exported to Asia as shown in Box 8.2 (Greenpeace 1998).

Various attempts by industry to use the islands of the Pacific as dump sites for hazardous waste (in association with power co-generation) have not been successful largely due to heightened awareness created through the negotiation of the subregional Waigani Convention on transboundary movement of hazardous and radioactive wastes.

However, other countries of the region like Bangladesh, Pakistan, Indonesia, Philippines and Thailand have become dumping grounds of huge

CHAPTER EIGHT

Box 8.2 PVC Waste Export to Asia Despite International Agreement

Between 1990 and 1998, over 100 000 tonnes of PVC waste produced in the Netherlands have been exported to Nigeria, Pakistan and the Philippines. For example, investigations undertaken by the environmental activist organization, Greenpeace, have identified the Dutch export company, Daly Plastics BV, as the holders of a permit for shipping 3 500 tonnes of PVC (approximately 140 truckloads) to one single company in the Philippines in 1998. This quantity of PVC waste is more than the post consumer waste that is recycled in the Netherlands every year.

The Dutch plastic waste exported to Asia is recycled into a range of often poor quality products for which there is no demand in the Netherlands or elsewhere in the industrialized world. For example, the pipes manufactured in the Philippines from recycled plastic are suitable only for temporary projects or low-cost housing. Furthermore, the recycling of PVC in countries with few or unenforced environmental regulations results in workers and nearby residents being exposed to chlorine and other toxic additives that are released during the recycling process. Workers in recycling factories in Pakistan and the Philippines often operate under unhealthy working conditions and respiratory problems, allergies, skin and eye irritations are common.

The primary reasons for PVC waste not being recycled in the country of origin, are the higher wages and production standards of the home market. Unfortunately, even recycled PVC from the industrialized nations often ends up on an Asian dumpsite; such plastics are frequently incinerated in open fires on the dumpsites, thereby releasing dioxin and other toxic substances into the surrounding environment.

In 1994, the Parties of the Basel Convention agreed to ban the export of hazardous waste from OECD to non-OECD countries for dumping and recycling purposes. The ban, which took effect in January 1998, was ratified by the European Union in September 1997. European legislation has been amended accordingly. Since it came into existence, the ban has faced fierce opposition from countries that wanted to keep exporting hazardous waste.

Greenpeace continues to oppose the export of PVC waste and recommends that countries that cannot fully manage the lifecycle of PVC should not be producing it and that instead of assessing poor countries for their capacity to treat PVC waste, the countries producing such waste should rather assess their own production technology, so as to promote non-toxic alternatives.

Source: Greenpeace 1998

---

quantities of hazardous waste for the exporters of industrialized countries both within and outside the region (Greenpeace 1998).

B. Critical Problems and Shortcomings

Many countries in the Asian and Pacific Region face critical problems with regard to waste management. A range of common shortcomings has been identified, including insufficient government priority and political support for action; lack of finance; inadequate long-term planning, indiscriminate disposal of waste; poor handling and disposal of hazardous and biomedical wastes; insufficient recycling and reuse; ineffective legislation and institutions; lack of skilled personnel; and poor monitoring and enforcement.

The prevailing view in many countries, particularly in respect of industrial waste, is that it is not possible to constrain the growth of economy by forcing industry and municipalities into introducing sophisticated and expensive waste treatment and disposal technologies. The short-sighted nature of such a policy, with implications for the long-term problems and costs of waste treatment has yet to be realized or understood in many countries in the region.

A number of problems lie in the political structure of the countries and the government authorities and in the inadequate enforcement of environmental legislation. Roles and responsibilities of many government agencies dealing with waste involve a complex mixture of operational, industrial, commercial and administrative functions. These agencies suffer from a high degree of inefficiency caused by: highly bureaucrat structure; lack of transparency and accountability in decision-making; low salaries; corruption; nepotism and/or selection of inadequate qualified personnel; difficult and complicated methods of procurement; and strong influence of political authorities in technical decisions regarding waste management. Very often, this is connected with the lack of appropriate management systems and a high level of dependency on the budget allocated. The often extremely low levels of salary paid to the government officers forces them towards corruption.

Private sector incentives and initiatives in waste management in many countries are still rare and the responsible authorities are seldom willing to see the provision of waste management services given into private hands. There is a general lack of funding which may be used to establish a waste management system operated by private contractors. In addition, there is also a very low level of public awareness and participation regarding waste management. This is because people are not sufficiently informed about the health and economic benefits of proper waste collection, treatment and disposal.

Much of the existing infrastructure and facilities for the waste collection, treatment and disposal have not kept pace with the economic development in recent years. It is not the lack of knowledge but the lack of finance and administration that is the main reason for the growing inefficiency of waste management practices in the region. Particularly in smaller cities and rural areas (where the patterns of consumption have also changed), the existing standards of waste collection, treatment and disposal remain very low. Adoption of inappropriate technologies creates many problems in the region. There are countless examples of plant failure.

Many problems exist especially for municipal solid waste collection, processing and disposal in the cities of Asia and the Pacific. Waste collection services are often sporadic as they rely upon insufficient numbers of vehicles, which are often old, under-maintained and unreliable. Open vehicles lose part of their load during their trips to the disposal site. Another severe problem is the lack of spare parts of collection vehicles. Collection workers try to earn extra money (sometimes they even have contracts with junk dealers) by sorting out materials or other items from the waste awaiting collection, i.e., they devote much time to this activity and neglect their main duty. Generally, a great deal of time is lost in transporting waste; collection vehicles sometimes need several hours just to travel from the city to the dump site/landfill site because of the heavy traffic and crowded streets. Collection is irregular: again open dumping at the roadside, along and in water channels, rivers and along railways is quite normal. Disposal is often to uncontrolled open dumping sites and, in many countries, industrial hazardous waste and biomedical waste are brought to the same dumpsites.

Waste management practices are most effective where they form part of a robust and integrated approach to the collection and disposal of all generated wastes. At present, however, waste management is given relatively low priority in many countries of the region despite increasing loads that stretch the already limited resources of waste collection and disposal agencies.

WASTE MANAGEMENT POLICIES AND STRATEGIES

Waste management, like many other environmental issues, is multisectoral in nature and encompasses policy making, strategies thinking, the development of legal-institutional-financial-and-administrative frameworks as well as the functional design, implementation, operation and management of waste handling facilities. Although within the region there are excellent examples of integrated waste management systems (including the policies and strategies, developed in Australia, Japan and Singapore, designed to manage waste using a cradle to grave approach), most countries have not developed the necessary waste management policies and strategies, legislation and institutional frameworks. For example, whilst environmental legislation appears on the statute books, few countries have introduced specific waste management regulations relying instead upon outdated unspecific legislation (e.g. public health acts, litter laws) which are seldom or poorly enforced.

Waste management is often hampered by a lack of national policy direction with no clear allocation of responsibilities and little or no national level planning to develop integrated waste management policies and strategies.

Financing remains a critical issue in most regional waste management operations. No sustainable funding plans have been developed or are in place in many countries. Of great concern is that most of the recent documentation on the regional waste management contains little or nothing on this key issue. In addition, there seems to be reluctance or lack of initiatives to move to commercialising waste management activities and few realistic ideas have been tested in raising revenues in the region.

A. National Policies and Strategies

1. Stakeholders, Institutions and Legislation

Many groups of stakeholders, including waste producers, regulators, legislators, consultants, contractors, process and equipment suppliers, educators, NGOs, media and the general public, are involved in national waste management policies and strategies in the region. Although each of these stakeholders plays a potential role, three groups (municipalities and industry (generators), governments (regulators) and legislators provide the key to effective national waste management policies and strategies that integrate the responsibilities of all stakeholders in making waste management a successful venture.

Institutions and legislation at the national level generally provide the basic infrastructure for the implementation of policies, strategies and actions for waste management. In recent years, three general trends in waste management institutions and legislation have been evident in the region. These are the creation of institutions for the strengthening of environmental policies and strategies, the development of more focussed environmental legislation, and the increase of manpower capabilities through education and training.

There has been an upward trend in the status of the above three aspects of waste management, as government ministries and high level agencies have been established specifically to control such activities. However, the lack of funds impedes implementation and enforcement actions and sometimes a lack of community involvement and community participation is a major constraint on improving the standard of waste management services. In some countries, there is an encouraging trend in increased budgetary resources and manpower capabilities for the waste management sector. The current status of national institutions, legislation and manpower capabilities for overall waste management in selected countries of the region are given in Table 8.10 (ESCAP 1994, United Nations 1995, 1996; Blum 1995; CITYNET/UNDP 1996, UNEP/SPREP 1997; World Bank 1998). However, despite these advances, solid waste management in many countries remains diffused due to parallel and over lapping responsibilities.

2. *Waste Minimization and Recycling*

Minimizing the quantities of waste requiring disposal, through source reduction, material recovery and reuse and recycling, is increasingly being realized as the central basis of an integrated approach to waste management. In some countries, such as Japan and Singapore, a reduction in the quantities of waste generated at source has been promoted through the regulation of industry, economic instruments to encourage plant modification or redesign and the education of consumers in the benefits of environment-friendly products. However, the ultimate success of waste minimization depends on cleaner production, which is increasingly being advocated in many developed and developing countries in the region as a more efficient and modern practice than conventional waste management practices (ASEAN/UNDP 1998, World Bank 1998). In some countries, the adoption of cleaner production programmes has reduced the need for end-of-pipe investments in waste treatment in industries and has therefore provided both financial and economic net benefits; these are discussed further in Box 8.3 (United Nations 1995, Aziz and Ng 1998, ASEAN/UNDP 1998, World Bank 1998).

Waste minimization by waste exchange is another option which is practised in some countries. The Industrial Waste Exchange of the Philippines (IWEP) serves as a link between companies that mutually benefit from a waste exchange. At least 600 industrial waste products (including organic and inorganic chemicals, solvents, oils, greases, waxes, acids, alkalis, metals, metallic sludges, plastics, textiles, leather, rubber, wood, paper, and glass) are advertised for exchange with other industries and the IWEP catalogue lists over 130 further waste products that are sought for exchange. Each product is assigned a code to ensure that the producing company's identity and location remain confidential and technical information, such as pH and the presence of any contaminants present is indicated. When two companies come to an agreement, the IWEP withdraws and leaves the producers and users to negotiate directly.

In addition to achieving reductions in the quantities of materials disposed of as waste, the waste exchange scheme has provided substantial benefits to a variety of companies through providing savings in disposal and raw materials costs and in improving the company's public image. For example, Del Monte Philippines, Inc. once spent over P 1.5 million a year to dispose of waste pulp generated by the processing

Table 8.10 Current Status of Overall Waste Management in Selected Countries of the Region

Country /Territory	Legislations	Institutions	Manpower capabilities
Australia	XXX	XXX	XXX
Bangladesh	XX	X	X
Brunei Darussalam	XX	XX	XX
PR China	XX	XX	XX
Cooks Island	XX	XX	XX
Fiji	X	X	X
Hong Kong, China	XXX	XXX	XXX
India	XX	XX	XX
Indonesia	XXX	XX	XX
Japan	XXX	XXX	XXX
Kazakhstan	X	X	X
Rep. of Korea	XXX	XXX	XXX
Maldives	X	X	X
Malaysia	XXX	XX	XX
Mongolia	X	X	X
Myanmar	X	X	X
Nepal	XX	X	X
New Zealand	XXX	XXX	XXX
Pakistan	XX	XX	XX
Philippines	XX	XX	XX
Papua New Guinea	X	X	X
Samoa	XX	XX	XX
Singapore	XXX	XXX	XXX
Sri Lanka	XX	XX	XX
Thailand	XXX	XX	XX
Viet Nam	XX	X	X

Source: United Nations 1995, Blum 1995, CITYNET/UNDP 1996
Legend: XXX Extensive coverage
XX Moderate coverage
X Minimal coverage

> **Box 8.3 Waste to Profits: Some Success Stories of Waste Minimization through Cleaner Production (CP) Programmes**
>
> **People's Republic of China** – At the request of China's National Environmental Protection Agency (NEPA), a US$6 million cleaner production component was included in the World Banks' Environmental Technical Assistance Project, approved in 1993. Under this Cleaner Production Programme, studies were undertaken of waste arisings in 18 industries. A large distillery was one of the plants involved and an initial assessment of the bottling plant identified good housekeeping options that costs less than US$2 000 to implement and resulted in savings of over US$70 000. This initial success was followed by a detailed study of an alcohol plant that identified a number of equipment optimizations, which produced nearly US$700 000 in savings. Three technology replacement options were also identified, costing up to US$500 000 and with paybacks of between one and a half to four and a half years.
>
> **India** – In 1993, a cleaner production demonstration project targeting small and medium sized enterprises (SMEs) was initiated by UNIDO, in cooperation with the Indian National Productivity Council and other industry associations. This DESIRE (Demonstration in Small Industries for Reducing Waste) project focused on three sectors: agro-based pulp and paper, textile dying and printing, and pesticides formulation. Collectively, the 12 companies spent US$300 000 on the implementation of pollution prevention options through cleaner production and saved US$3 milion in raw materials and wastewater treatment costs. The most impressive savings were in the pulp and paper sector, where the Ashoka Pulp and Paper Company invested a total of US$95 000 in the implementation of 24 recommended production changes and achieved a net annual saving of about US$160 000. In this case, the overall payback of the investment was less than seven months.
>
> **Philippines** – Through the Philippines Clean Technology Initiatives, companies such as Del Monte Inc. Philippines, Peter Paul Philippine Corporation, Central Azucarera Sugar Milling and Refining (the Philippines), and Pilipinas Kao Inc., adopted cleaner production systems. Each of these companies obtained significant economic benefits through measures such as water saving, reduction in waste loads, and cost saving in waste treatment and disposal.
>
> **Other Countries** – ICI P Paints, PT Unilever, PT Tifico, PT Semen Cibinong, PT Indah Kiat Pulp and Paper Corporation of Indonesia, Golden Hope Plantations Berhad of Malaysia. Chartered Metal Industries Pte Ltd. of singapore and Cheng Sang industry Co., Ltd. of Thailand adopted cleaner Production techniques which resulted in economic benefits in the form of increased productivity, savings in chemicals, water and fuel, and reductions in waste load and the cost of waste treatment and disposal.
>
> *Source:*   UNIDO 1997, World Bank 1998, and ASEAN/UNDP 1998

and canning of pineapples. The waste pulp is now sold to the Philippines Sinter Corporation (PSC) for P 1.4 million a year for a total of nearly P 3 million in savings and additional revenue. The PSC dries the pulp and exports it to Japan to use as cattle feed. Similarly, Maria Christina Chemical Industries Corporation (MCCI) now sells its carbide sludge wastes to the National Steel Corporation (NSC) for use as a neutralizer in its waste treatment plant at P 330 per tonne – compared to the P 1 500 per tonne that the NSC previously paid for fresh reagent. For its part the NSC produces about 15 000 tonnes per year of mill scale waste which is sold to MCCI for use in the production of ferrosilicon alloys (United Nations 1995, ASEAN/UNDP 1998).

The rate of recycling materials from waste has increased dramatically in recent years in the Asian and Pacific Region. Recycling of waste materials grew from less than 10 per cent in 1990 to 22 per cent in 1998. Most of that increase is attributable to greater rates of recovery of paper and paperboard, plastics, aluminium cans, glass, etc. The paper and paperboard category is dominated in terms of total tonnes of material recovered (almost 60 per cent) followed by plastics, aluminium cans, and glasses. The informal sector plays a significant role in waste recycling in the region. Waste pickers perform the recycling operations in many cities of the region and in resource-poor and labour-abundant economies, such as those in Bangladesh, People's Republic of China, India, Indonesia and Thailand, material recovery and recycling assume particular economic significance. Recycling not only reduces the volume of wastes to be disposed, but also saves these countries valuable foreign exchange which would otherwise be used to import raw materials. Waste reduction through recycling and reuse in People's Republic of China and Singapore has emerged recently as an environmental priority. The governments' goal is to increase recycling of waste from present 10 per cent to 25 per cent in 2002. Republic of Korea recycled 59.5 per cent of its waste in 1998 and, the following year, introduced a system for controlling the use of disposable goods, such as disposable cups and containers, plastic bags, and disposable razors and toothbrushes. Recycling in Australia, People's Republic of China, India and the Philippines has improved dramatically over the last decade and such improvements are likely to continue in the foreseeable future.

3. *Private Sector Participation*

The rationale for the privatization of waste management services is mainly economic. Evidence seems to indicate that public provision is more costly and frequently unsatisfactory due to the inefficiency and rigidity of public bodies. Privatization basically involves the transfer of management responsibility and/or ownership from the public to the private sector and has proven to be a powerful means of improving the efficiency of some waste management services such as collection, haulage, and disposal. There are even examples where such initiatives have been led by direct partnerships between the local community and the private sector in the management of urban waste (see Box 8.4).

A number of countries in the Asian and Pacific Region have introduced at least partial privatization into their waste management systems. Hong Kong, China has entered into long term DBO (design build and operate) contracts for three major sanitary landfills and four large transfer stations as well as for the collection and treatment of chemical waste (Fernandez 1993, United Nations 1995). It is also expanding the participation of the private sector through the intended placing of contracts for the collection and treatment of medical waste, the storage of low-level radioactive waste and the remediation of its closed landfills. Macao, China has let two fourteen-year contracts for the operation and maintenance of its waste-to-energy incinerator and for waste collection, street cleansing and beach cleaning services.

In Malaysia, the privatization of solid waste management commenced in 1997 with a privatization policy oriented towards reducing the Government's financial and administrative burden; promoting competition, increasing the role of the private sector in nation building and providing opportunities to meeting the targeted new economic policy. Privatization has also resulted in the growth of private companies specialising in the waste business and these often complement the services that are mainly provided by the Local Authorities. Several municipalities in Malaysia have let smaller-scale solid waste collection contracts. Selangor and Penang are currently using private sector landfill arrangements and the Federal Government is planning to extend these schemes nationally. Singapore has already engaged private contractors to collect municipal solid waste and private contractors will have roles in its Tuas transfer station and Pulau Semaku landfill. Several cities in People's Republic of China are finalising deals both from local and foreign contractors, mainly in the waste-to-energy field.

In Thailand and Indonesia, there are limited attempts to contract-out the disposal of hazardous wastes, but there are some contracting-out

Box 8.4 Private Sector Initiative towards Urban Waste Management in Pakistan

Rotting garbage creating a health hazard is a common sight in many parts of Karachi. It is also a civic menace for city-dwellers. Municipal authorities have failed to address the issue of solid waste disposal due to lack of capacity. Once it leaves the house, waste is often dumped on any vacant plot of land, or on streets, for want of a proper neighbourhood dumpsite. Where a site exists-usually a low four-wall structure open to the air-waste is more likely to be found lying outside rather than within this makeshift "receptacle". Scavengers rummage there for recyclables, but a large part of garbage remains because there is no regular waste collection service to ensure that the waste is cleared away daily. Waste Busters, a private enterprise has now become active to offer a solution to the poor.

Waste Busters began life three years ago as the Lahore Sanitation Programme. They aimed at providing solid waste disposal services through recycling. They are now called Waste Busters and have branches in Islamabad and Karachi. For Rs100 a month, Waste Busters provide a daily collection service to households who share a concern for the environment. In Lahore, Waste Busters service 10 000 eco-conscious households in Gulberg, Shadman, Model Town, Muslim Town and Cantonment areas. They employ 200 people and an average 50 tonnes of waste is collected and disposed of daily.

In order to manage waste properly, collection isn't enough. Waste Busters now sorts out materials like plastic, glass, paper and organic waste retrieved for recycling purposes. The enterprise divides the city into zones and each zone requires a transfer station where the waste is taken after being collected, for sorting.

In Lahore, organic waste is being efficiently sorted and turned into compost which is sold to farmers and nurseries to be used as fertilizer. The sorting is done at transfer stations set up by Waste Busters at sites allocated by the local municipal administration.

Unfortunately, sorting at source, the mode employed in the West, doesn't work in Lahore. The Waste Busters tried getting households just to separate the organic waste from other household waste but it didn't happen.

Waste Busters are not keen to incur the wrath of the big waste dealers, nor do they want to rob scavengers of their livelihood. "In fact, in Lahore they invite the scavengers to their transfer stations to sort the waste for them and buy it off them."

Eventually the Waste Busters would like to progress from a self-sustaining to a profitable operation. That has already begun to happen in Lahore where the daily production of an average 500 bags of the organic fertilizer, along with the sale of other recyclable material to recycling industries, has brought Waste Busters out of the red.

Source: Sahar Ali 1999

arrangements for collection and disposal of municipal solid wastes and non-hazardous industrial waste generated in Bangkok (Kiser 1998). The Bangkok Metropolitan Administration (BMA) strongly supports privatising medical waste management services. There are some private arrangements for solid waste collection in Japan, Republic of Korea, the Philippines and in Sri Lanka and contracting out of hazardous waste disposal is being pursued in Malaysia and Thailand (Sandra 1994).

Throughout the region, there is a discernible trend towards private sector participation in solid waste management. The examples of Hong Kong, China and Macau China are extremely useful model for cities that have reached the point at which they were ready to improve their waste management arrangements and which carefully considered why and how contracting-out to the private sector was the best means of accomplishing their objectives. The results, in terms of environmental improvement and financial savings, are amply documented in these two cities.

4. *Economic and Financial Strategies*

In some countries of Asia and the Pacific, including Australia, Japan, the Philippines, Republic of Korea and Singapore, a number of different economic tools have been integrated into their strategic waste management plan to ensure that waste in all its forms is minimized, that revenues for waste management are raised, and that, wherever possible, the polluter/user pays.

Different stages of the production and consumption process have produced different forms of waste in the region. The challenge has been to choose the right economic tool given the stage at which the waste has been produced. For example: licences, permits and extraction charges have been used to ensure that excessive use and waste of natural resources inputs does not occur; tax deductions, pollution taxes, and input and product taxes have been used to ensure that clean production practices are encouraged and rewarded; refundable deposits have been used to ensure the recycling of end products when it is economically viable; and performance bonds have been used as an incentive for enterprises to manage their affairs in an environmentally sound manner (Keen et al 1997). However, the use of economic measures to assist in waste management in the region is minimal and sparsely spread throughout a limited number of sectors. The use of economic tools in the overall environmental policy setting in many countries is almost non-existent.

A number of criteria/options have been used in choosing between various policy instruments and strategic alternatives with respect to economic and financial aspects of waste management services in the region (Leong and Quah 1995). The chosen criteria/options have been compatible with the national regulatory objectives and existing legislation, as well as the long term plans of the national environmental protection plan (such as Singapore Green Plan). In addition, this approach has ensured that selected policies are credible substitutes for, or supplements to regulatory legislation, and that they conform to the principle of institutional concordance. Some of the economic and financial criteria/options employed in the region are examined below.

(a) User or Waste-end Fees

These fees are based on the weight or volume of waste generated. They are meant to encourage more recycling as disposal becomes more expensive. There are some cities such as Canberra, Tokyo and Seoul that have successfully implemented kerbside charging schemes (see Box 8.5). Residents are charged per bag which appears to be more effective than the can system in which residents are charged monthly for the use of a specific number of garbage bins per week. These bags are provided for non-separated waste and aluminium cans, glass, cardboard and newspapers are collected separately. Bag users can save money by putting out fewer bags in a given week. Households have to use garbage bins with differentiated charges according to the size and number of bins that are used. Residents are not charged for the removal of various types of separated waste. For example, in Canberra, the cost per week per household (of which there are 94 000) is between A$1.32-1.43 which includes weekly collection of a 140-litre bin and fortnightly collection of a 240-litre-recycling bin (ACT 1996). In some Asian and Pacific households, charges per week range from nothing at all to top figures of less than one US dollar. Higher charges for waste collection are unlikely to be a socially acceptable solution in most Asian and Pacific countries except when the increase in costs is small given the high social benefits.

(b) Waste Disposal Fees

Certain fees (termed *tipping price* or *gate fees*) are payable for disposal of waste at dumping grounds, landfill sites or incinerating plants. Some countries such as Australia, Japan and Singapore impose fees for disposing waste into the designated waste disposal facilities. Table 8.11 shows fees payable for disposal of waste at dumping ground/incinerator in Singapore (ENV 1997).

CHAPTER EIGHT

Box 8.5 Volume-based Collection Fee System for Municipal Waste in the Republic of Korea

The volume-based Collection Fee System for Municipal Wastes was introduced in the Republic of Korea to minimize the generation of waste and encourage households to separate their wastes for recycling. The system was put into effect nationwide on January 1, 1995. Until that time, waste collection fees had been calculated for each residence based on the level of property taxes imposed on houses or apartments, or the size of buildings regardless of the actual volume of wastes that residents generated. The volume-based collection fee system however strongly adhered to the "Polluter-Pays Principles".

Under the system, household waste was to be discarded in the officially designated plastic trash bags, which were manufactured and sold by city, county, and district governments. These regulations, however, did not apply to the discharge of burned coal briquettes and recyclable wastes including paper, waste iron, metallic cans, bottles, and plastics. These were collected at no charge if discarded properly at designated locations as determined by the local governments. Local governments also were given the discretion to set the collection fees for discarded furniture and major home appliances. Waste collected during street cleaning and park cleaning was to be discarded in trash bags for public purposes provided free of charge. The prices of official trash bags were to be determined by ordinance of the local municipal and county governments after consideration of waste treatment costs and the financial state of the local government in question.

The results of a two-year study on the performance of the system in 15 cities and provinces revealed that after the system was introduced, the volume of waste discarded decreased 29.4 per cent to 34 726 tonnes a day from 49 191 tonnes a day previously. The availability of recyclable goods increased 28.5 per cent to 11 468 tonnes a day after the introduction of the system from 8 927 tonnes a day beforehand. By region, the rate of reduction in waste generation was most pronounced in small and medium cities and rural areas than in large cities. Per capita waste generation dropped to 1.01 kg a day.

Performance of the Volume-based Collection Fee System

	Reduction in waste generation (%)	Increase in recyclable wastes (%)	Per capita waste generation (kg/day)
National average	>29.4	28.5	1.01
6 major cities	>24.9	33.2	1.10
Provinces	>35.6	22.6	0.90

Residents of large housing units such as apartment complexes were found to be more conscientious about separating wastes before disposal than residents in areas of individual houses. Wastes such as paper, waste metals, cans, and bottles which are discarded separately for recycling purposes are thoroughly treated by private sector recycling agencies, these agencies account for 30-50 per cent of the volumes of these goods recycled. Only 13 per cent of plastics are collected, however, due to the lack of plastic recycling facilities. Recyclable plastics are therefore stocked in collection sites of local governments and the Resource Recovery and Reutilization Corporation.

The implementation of the Volume-based Collection Fee System served as an opportunity to heighten the general public's awareness of the environment in addition to producing visible benefits such as a meaningful reduction in the volume of wastes generated, an increase in recycling, and an improvement in waste administration service. The entire process of production, distribution, and consumption of goods was shifted to a more environment-friendly paradigm. That is, when consumers buy goods at retail establishments, they have come to prefer goods which entail less waste generation in their production, distribution, and consumption and also good in refillable containers. Enterprises, for their part, have shifted to more efficient production processes to reduce the volumes of waste generated. Consumers cannot be expected to bring packaging materials such as Styrofoam to retail establishments when they buy goods, and in response to changing consumer attitudes, enterprises have been making efforts to develop less voluminous packaging materials. The launching of the system served as an opportunity for local governments to become business minded regarding waste management administration. With the supply of recyclable goods increasing, the recycling industry is also beginning to flourish.

Source: Government of the Republic of Korea

(c) Deposit-Refund System (DRS)

A combination of a tax and subsidy, under a DRS the consumer has to pay a deposit at the time of purchase, usually as part of the product price. The consumer is given a refund if the waste product, such as empty bottles and aluminium cans, is returned to the seller or to an authorized recycling/reuse centre.

The consumer has an incentive to bring back the used item rather than dumping it. For example, in Australia, a glass bottle deposit refund (adopted voluntarily by the soft drink industry) was set at a rate of between 10 and 15 per cent of the value of the item and effected a return rate of over 80 per cent of bottles, whilst a mandatory deposit for PET bottles,

Table 8.11 Fees Payable for Disposal of Solid Waste in Singapore

Disposal site	Cumulative load per vehicle per day	Charges S$
All disposal sites	below half a tonne	Free
Lorong Halus Dumping Ground	first half a tonne	23.50
Senoko Incineration Plant	every additional 0.05 tonne or part thereof	2.35
Tuas Incineration Plant		
Ulu Pandan Incineration Plant	first half a tonne	28.00
a. between 7.30 am & 2.00 pm	every additional 0.05 tonne or part thereof	2.80
b. between 2.00 pm & 5.00 pm	first half at tonne	25.00
	every additional 0.05 tonne or part thereof	2.50
Kim Chuan Transfer Station	first half a tonne	23.50
Refuse disposal fees	every additional 0.05 tonne or part thereof	2.35
Haulage fees	first half a tonne or part thereof	11.00
a. between 8.00 am & 2.00 pm	every additional 0.05 tonne or part thereof	0.55
b. between 2.00 am & 5.00 pm	first half a tonne or part thereof	5.60
	every additional 0.05 tonne or part thereof	0.28

Source: ENV 1997

at a rate of 24 per cent, effected a return rate of 62 per cent. The economic benefits of encouraging recycling can be high (ACT 1996) and a recycling study (Keen et al 1997) found that after allowing for the cost of handling, transport and cleaning, the use of returnable bottles is cheaper in economic terms than non-refundable bottles or cans.

(d) Government Grants/Foreign Aid

In some countries, government grants or foreign aid is arranged for the capital investments necessary to modify existing manufacturing facilities to produce products that generate less waste. Funds are made available for research projects on seeking ways to remove institutional and social obstacles to reduce waste generation and creative ideas for use by non-manufacturing establishments.

(e) Incentives

In some countries, incentives are provided in the form of preferential tax treatment, like tax credits, on the importation of waste treatment facilities that are not locally available. Loans and other assistance programmes are used to encourage compliance with regulatory standards for a limited period of time after new standards are introduced. The common form of incentive is a tax deduction for the installation of anti-pollution equipment by the industrial sector. This has been successfully introduced in the Philippines and in the Republic of Korea and has stimulated the installation of pollution control devices in both countries. In India, a 35 per cent investment allowance (against the general rate of 25 per cent) towards the actual cost of new machinery or plant is provided.

(f) Disincentives

Disincentives are measures to discourage the discharge of wastes into the environment and are based on the Polluter-Pays Principle (PPP). The principle implies that the polluter should bear the financial responsibility (and hence bear the costs) of measures undertaken by public authorities to treat or dispose of wastes in an environmentally acceptable manner. Disincentives take the form of user charges and pollution taxes in some countries of the region, including People's Republic of China, India and Singapore, although their specific application to waste disposal is limited.

(g) Pollution Fines

Pollution fines are prevalent in some countries in the region. Singapore provides a current example of the imposition of pollution fines to waste generators for violating waste management regulations. This measure is particularly effective when the pollution fines are set sufficiently high to make investments in pollution control financially attractive to firms. In this way, the polluters are forced to internalize the environmental costs of their activities.

(h) Economic Sanctions

These deal with the enforcement of compliance with waste management regulations. Although often referred to as economic remedies, these are usually framed as direct penalties that are equal or greater than the costs of compliance. These should not be confused with effluent charges or pollution taxes, but are an economic penalty, which works by imposing on a company a liability that is directly related to the financial savings which results from not complying with waste management regulations. Economic sanctions are prevalent in some countries of the region, including Australia, Japan and Singapore.

(i) Environment Funds

In Thailand, the Government established a five billion baht (US$200 million) Environment Fund to be used for the clean-up of cities and for industrial pollution control (United Nations 1995). In the Philippines, two mining companies were instructed by the Government to create an Environmental Guarantee Fund in the form of security bonds to provide for the rehabilitation and restoration of areas detrimentally affected by mining operations. This scheme has become a part of standard procedure for the issuance of environmental compliance certificates for mining operations. An additional Reforestation Fund has also been instituted as part of the scheme to counter deforestation in areas being mined. People's Republic of China uses Revolving Funds for the purchase of pollution abatement equipment and for introduction of waste minimizing technology (United Nations 1995).

B. Regional and International Initiatives

1. Control of Transboundary Movement of Hazardous Waste

Some 98 per cent of the world's hazardous waste is produced in the industrialized countries and, over the years, international waste traders have increasingly sent hazardous industrial waste from developed countries to the countries of the Asian and Pacific Region. The main impetus of this process is the economic gradient which leads firms to search for the cheapest and easiest dumping grounds, shipping them to those developing countries which have less stringent environmental laws or inadequate enforcement of such laws. The growing concern over the health and environmental implications of the hazardous waste traffic led to the Basel Convention, which was adopted unanimously on 22 March 1989 (Rumel-Bulska 1993). The Final Act of the Basel Convention was signed by 105 States, as well as the European Community (EC), and the Convention entered into force on 5 May 1992. However, by mid-2000, the Basel Convention had been ratified by only some fifty per cent of the ESCAP member countries.

Though the Asian and Pacific Region still remains open to the importation of hazardous waste, progressive governments and activists throughout the region are resisting further imports. In the Southeast Asian subregion, activists met in 1993 to develop and coordinate campaigns to make Southeast Asia a waste trade-free zone and today, Southeast Asian countries are beginning to respond to their citizen's concerns about illegal waste dumping. In November 1992, the Indonesian government prohibited plastic waste imports and also expanded this prohibition to include other types of waste (Indonesia 1997). In August 1992, the government of Malaysia announced a new prohibition on the import of certain hazardous waste (Malaysia 1998). Other countries of the region, including India, Pakistan, Singapore and Thailand, have announced similar prohibitions on the import of hazardous wastes produced outside the region and also on the movement of hazardous wastes generated within the region (Agarwal 1998, Greenpeace 1997). In the South Pacific, waste traders continue to attempt the importation of foreign generated hazardous and radioactive waste into the Pacific Islands such as Marshall Islands, Papua New Guinea, Samoa, Solomon Islands, Tonga, Tuvalu and Vanuatu (UNEP/SPREP 1997). To ban importation and to control hazardous wastes generated in the region, the Waigani Convention was adopted in 1995 (Mowbray 1997). The Waigani Convention is "Convention to Ban the Importation into Forum Island Countries of Hazardous and Radioactive Wastes and to Control the Transboundary Movement and Management of Hazardous Wastes within the South Pacific Region" and it is closely associated with the Basel Convention as a subregional complement.

However, according to some reliable sources (Angellow and Ranawana 1996, Greenpeace 1997, Nelson 1997, UNEP/SPREP 1997, Agarwal 1998, etc.), waste exports to the region and within the region continue despite the adoption of Basel Convention and Waigani Convention.

2. Activities of International and Regional Organizations

Various United Nations Agencies, the World Bank and the ADB have been financing and providing technical assistance for solid waste management services in the countries of the region. The UNCHS and World Bank have undertaken a number of projects on urban waste management with UNDP funding. In addition, some developed countries such as Australia, Japan, USA, Canada, UK, Germany, Netherlands, Denmark, Norway and Singapore are also providing financial assistance to waste treatment and disposal facilities in the region (UNEP/SPREP 1997, ASEAN/UNDP 1998, World Bank 1998).

Similar projects were also undertaken by the United Nations ESCAP, both independently and in close cooperation with UNDP, ADB, CITYNET and other agencies. UN ESCAP was responsible for three projects specifically aimed at waste management and comprising the development of guidelines on the monitoring methodologies for toxic chemical and hazardous wastes; on the managerial and human resources requirements of hazardous waste management in the developing countries of the ESCAP region; and on the legal and institutional frameworks required to prevent the illegal traffic in hazardous products and wastes (ESCAP 1994, United Nations 1995). ESCAP also implemented a project on Capacity Building in Industrial Audit for Waste Minimization which reviewed existing methodologies and guidelines for waste minimization techniques and waste auditing procedures and recommended revised guidelines and procedures (United Nations 1995).

UNEP/EAS/RCU initiated a study on Regional Action Programme on Land-based Activities Affecting Coastal and Marine Areas in the East Asian Seas, which provided technical support to East Asian countries in addressing their coastal and marine pollution problems (Koe and Aziz 1995). Improvement of both hazardous and nonhazardous waste management practices and reduction of waste generation from land-based sources were the prime focus of the study.

CONCLUSIONS

Available data on the quantity and types of solid waste generated, and the methods employed in the treatment and disposal of generated waste, are incomplete, inconsistent and unreliable due to wide variations in data recording, definitions, collection methods and seasonal variations (World Bank 1999). Whilst at a regional level this mitigates against a clear view of the overall status and trends, at the local level the lack of robust data acts as a barrier to the development and implementation of efficient and cost-effective waste management practices.

From the available data, it is clear that, in recent years, there has been a sharp increase in the solid waste generation in the Asian and Pacific Region and estimates indicate that generation rates are set to double over the next 25 years.

Despite the increasing urbanization and industrialization, the economic activity of the region remains predominantly agricultural and, as a consequence, generates substantial quantities of agricultural wastes. However, much of the waste is utilized within rural communities through composting, direct land application, biogas generation or is used as construction materials.

Generation of municipal solid waste from within the region varies enormously from as little as 0.4 kg per person per day to as high as 5 kg per day for people living in the region's high-income, developed countries. Whilst most of the region's municipal waste continues to be indiscriminately dumped on open land, land filling, controlled incineration and composting are increasingly being employed. Industrial solid waste generation rates vary from country to country and even within a country depending upon the nature of the active industries. Open dumping, land filling and incineration are the major disposal processes of industrial solid waste.

Generation of hazardous waste from manufacturing, hospital and health-care facilities and nuclear power and fuel-processing plants is rising and has been estimated to more than double within next 10 to 15 years time. The region's capacity for adequately managing the disposal of such wastes is extremely limited, particularly when additional wastes enter the region through the dumping activities of some industrialized nations.

Over the forthcoming five to ten years, the region faces many waste management challenges that will require clear and effectively implemented policies and strategies. Efforts to develop inter-regional discussions on joint strategies and common approaches are continuing (see, for example, the proposed common strategic considerations presented in Box 8.6), although individual countries will need to address specific issues associated with the current lack of long-term planning, proper policy formulation, insufficient government priority, lack of finance, lack of skilled personnel; lack of public awareness and public participation, inadequate legislation and institutions and poor monitoring and enforcement. These have led to escalating environmental pollution and health problems in the region and, in the long term, may have implications for the continued economic development of the region.

However, evidence from the region indicates that environmental awareness and consciousness of waste is taking roots in industry and business. Australia, Japan, Republic of Korea and Singapore are leading in the waste management area by not only actively pursuing environmental protection through proper institutions and legislation but also developing new and innovative technologies for waste disposal. Countries like People's Republic of China, India, Indonesia, Malaysia, Pakistan, Philippines and Thailand have also made good progress in waste management practices.

Multi-lateral and bi-lateral development agencies (including various United Nations agencies (UNDP, UNEP, UNIDO, WHO, ESCAP), World Bank, ADB, and some donor countries) are offering both technical and final assistance for waste disposal in countries of the region. There is however, a need to intensify efforts towards development of indigenous capabilities in the countries of the region in terms of expertise, equipment manufacture, process technology guidelines, design, construction/installation, operation and maintenance of waste treatment/disposal and pollution abatement facilities.

CHAPTER EIGHT

Box 8.6 Integrated Solid Waste Management – Key Strategic Considerations

There is a striking degree of similarity in the solid waste management needs and constraints within the Asian and Pacific Region and policy makers, municipal managers and practitioners may be assisted in the resolution of these needs and constraints through the adoption of a number of key strategic considerations:

1. Developing waste disposal facilities such as landfills and incinerators often generates tremendous concern – both warranted and reactionary. However, it is possible to reduce opposition to new facilities by involving the community and following a technically sound and transparent site selection process and, wherever possible, using local conditions to ameliorate potential environmental impacts and costs, e.g. siting landfills in geotechnically superior locations. Waste disposal facilities, which often have a useful life in excess of 25 years, need to be well integrated within a sound master plan that reflects regional requirements, standard operating procedures, and financing mechanisms. Sound technical justification and a transparent planning process that respects the general public's valid concerns may not eliminate public opposition, but it is the best way to minimize it.

2. Local governments should minimize residential waste collection frequency to a maximum of twice per week, which is adequate from a public health perspective, but requires social acceptance. Citizens should be encouraged to place their waste in containers that enhance collection efficiency.

3. Local governments should focus primarily on residential waste collection, especially from poor and densely populated areas, and empower the private sector to pick up waste from non-residential sources. Commercial, institutional and industrial waste collection can usually be self-financing. Local governments should license private hauliers to generate revenue and to ensure proper collection and disposal.

4. Waste collection and disposal fees should be based on waste generation rates. Direct user charges and waste fee collection should begin with the business community.

5. An integrated approach toward solid waste management needs to be followed. Municipal waste managers should opt for the least technically complex and most cost-effective solution (e.g. limited mechanization and incineration). Waste diversion should be maximized.

6. All levels of government, including multi-national agencies and transnational corporations must play a role in long-term programme development,. e.g. extended product responsibility, life-cycle analysis, waste exchanges, and natural resources tax regimes.

7. Local governments must honestly and respectively gauge the public's willingness and ability to participate in the design and implementation of waste management programmes. Through good partnerships, progressive programmes can be developed in a complementary manner. These programmes include community-based operations, micro-enterprise development, waste separation for increased recycling and composting and reduced collection frequency.

8. All levels of government should promote the hierarchy of waste management (i.e. reduce, reuse, recycle, recover) and encourage waste separation to maximize flexibility to deal with future changes. Wherever appropriate, governments should view solid waste as a resource, rather than just a "local problem".

9. Although waste collection, treatment and disposal costs often place a large burden on local government finances, improper disposal is far more expensive on the long run. With costs accruing over many years.

10. Local governments are usually in the best position to assume key responsibility for municipal solid waste collection and disposal. However, sustainable financing and sustainable service provision still needs to be defined by a broader set of stakeholders. Local governments need the assistance of all levels of government to provide waste management services efficiently. Regional approaches to waste disposal e.g. shared landfills are especially important.

Source: World Bank 1999

Chapter Nine

Children shoulder the burden of survival; a young girl carries fodder from dwindling forests in Nepal.

9

Poverty and Environment

INTRODUCTION
POVERTY AND ENVIRONMENT: STATUS AND TRENDS
ENVIRONMENTAL CONSEQUENCES OF POVERTY
POVERTY AND NATURAL DISASTERS
POVERTY AND ENVIRONMENTAL HEALTH HAZARDS
POLICIES AND PROGRAMMES ON POVERTY ALLEVIATION AND ENVIRONMENT
CONCLUSION

CHAPTER NINE

INTRODUCTION

Poverty and environment are closely interrelated. Whilst people living in poverty are seldom the principal creators of environmental damage, they often bear the brunt of environmental damage and are often caught in a downward spiral, whereby the poor are forced to deplete resources to survive, and this degradation of the environment further impoverishes people. When this self-reinforcing downward spiral becomes extreme, people are forced to move in increasing numbers to marginal and ecologically fragile lands or to cities.

This chapter examines the interrelationships of poverty and environment and discusses their status and trends. It also highlights the impact of poverty on environmental health and the vulnerability of the poor to natural and man-made disasters.

POVERTY AND ENVIRONMENT: STATUS AND TRENDS

A. Definition-Forms of Poverty

Usually referred to in terms of income or deprivation, poverty has been measured in many ways. Increasingly, equal emphasis is being placed on social definitions (life expectancy, literacy rate etc.), along with the traditional economic ones (Gross Domestic Product (GDP) etc.), to indicate the status and trend of poverty, or to develop poverty related indices such as human development index, human poverty index and gender development index (ADB 1999). Whatever the cause, poverty is often concentrated in environmentally fragile ecological zones, where communities face and/or contribute to different kinds of environmental degradation.

Four forms of poverty based on land use and environment are evident in the region. The first form of poverty occurs in areas characterized by active and productive agricultural land, the efficient and equitable utilization of which by the poor is hampered by low levels of access to land, resources or jobs. For example, the Green Revolution, whilst increasing absolute food production, did little to enhance the income of the small subsistence farmers and sharecroppers. In comparison with large landowners, these farmers lacked resources for the requisite capital investment in the new technologies, and so mostly sold their lands to large landowners and became further impoverished. The overall result has been a bi-polar distribution of land-holdings in the region, with fragmentation of small farms upon inheritance, and a parallel increase in the size of the large holdings.

Against this background, the two variables most strongly correlated with rural poverty are unemployment and limited or no access to land. Studies have shown that in the poorest areas, more than three quarters of unemployed are concentrated in just two classes: landless workers and small farmers (TWE 1993). Many small land holdings have been bankrupted due to the absence of co-operative farming, lack of access to credit and the large capital investments required for new technology. In addition, large landholders have in many cases evicted tenants in an attempt to consolidate their land in response to threatened land reforms or amendments to tenancy laws.

The second form of poverty occurs in areas of marginal lands (deserts, uplands, and already degraded lowlands) with few opportunities for increasing agricultural productivity or for economic diversification. Such areas are generally very low in productivity, and problems are compounded by unsustainable agricultural practices. Approximately 60 per cent of the poor in the region are estimated to be living on these marginal lands.

The third group of impoverished people are those which inhabit coastal areas with inadequate or depleted marine resources. People are attracted to such areas due to economic development, but, in many cases, the pace of this development destroys or depletes the very resources that are fuelling its growth.

The fourth form of poverty is experienced by the poor inhabitants of urban slums and squatter settlements, where there is a constant exposure to poor sanitary and environmental conditions (see Chapter 7).

The geographical concentration of the worst poverty causes serious localized degradation. High population density and growth against a background of an inequitable distribution of productive assets make sustainable development more difficult to achieve. Impoverished communities also tend to rely disproportionately on common property resources such as forest and pasture, which are vulnerable to degradation when exploited by growing numbers of people.

At the country level, developing countries tend to be reliant on natural resources products for the overwhelming majority of their exports, which makes their economies extremely vulnerable to environmental changes and pressures. Macroeconomics, trade and sectoral adjustment policies alter incentives governing natural resources by altering aggregate demand, as well as by distorting relative prices of natural resources and related goods and services. Markets at local, national and international levels also influence local-level environmental management, as do market interventions such as the maintenance of food subsidies that benefit net purchasers rather than producers of food (ODA 1991). In addition, the structural adjustment policies adopted to deal with economic disequilibria (caused by a combination of adverse international economic conditions and internal management), have a disproportionate effect on rates of natural resources exploitation, where exports are from natural resource sectors.

B. The Extent of Poverty

A frequently applied criterion for the assessment of poverty is whether the daily income of an individual (or the average per capita income of a community) is less than one US Dollar (or 'a dollar a day'). On the basis of this criterion, some 1.2 billion of the world's people may be considered poor. Of these, 900 million (or 75 per cent) live in the Asian and Pacific Region, concentrated primarily in South and East Asia (Figure 9.1). South Asia has more than half a billion poor people, with 450 million in India alone. The extent of poverty in other subregions include Northeast Asia, which is dominated by the People's Republic of China (275 million), and around 85 million in Southeast Asia. The remainder of the poor are scattered in the small developing island countries of the South Pacific (ADB 1999).

In the early 1970s, half of the population of the Asian and Pacific Region was poor (based upon the 'dollar a day' criterion), the average life expectancy at birth was 48 years and only 40 per cent of the population was literate. Today, one-third of the people are poor, life expectancy averages 65 years and 70 per cent of the adult population is literate. The economic growth and social progress of the past several decades have ushered in an era of untold prosperity and health in most regions of the world. Globally, per capita GDP has jumped from US$2 257 to US$3 168 in the past 25 years; life expectancy has climbed from 57.9 to 65.6 (WRI/UNEP/UNDP/World Bank 1997). Parts of the Asian and Pacific Region have shared in this prosperity. In Southeast Asia, most countries halved the incidence of poverty or more in just two decades. The Asian crisis notwithstanding, these countries have shown how robust economic growth can reduce poverty. The status of poverty in selected countries is given in Table 9.1.

Table 9.2 indicates how South Asia lags behind Southeast Asia, Northeast Asia and the South Pacific in terms of socio-economic indicators. Southeast Asia has shown the most strides in combating poverty; not only has the incidence of poverty lowered, but its decline has quickened in recent years. Absolute food poverty seems to have

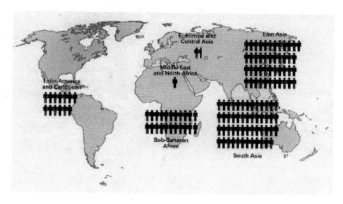

Figure 9.1 Poverty by Developing Regions of the World

Source: World Bank 1998d.
Note: Each figure represents 10 million persons living on US$1 a day or less at 1985 international prices.

disappeared in Hong Kong, China, Singapore and almost disappeared in the Republic of Korea and Malaysia (ADB 1997).

C. Causes and Consequenses of Poverty

1. Causes of Poverty

While the causes of poverty (and their inter-relationships with the environment) are not always discreet, they can be divided broadly into internal (local) and external (global) causes.

(a) Internal Causes

There are many internal, or local, causes of poverty in the region. For example, a lack of essential assets; living in a remote or a resource-poor area; vulnerability on account of age, gender, health, living environment or occupation; and discrimination against an ethnic minority or a community considered socially inferior.

The 'feminization' and 'ageing' of poverty is very visible across the Asian and Pacific Region, and is a result, in part, of population growth and changes in global trade. For example, as the Republic of Korea moved from an agriculture-based economy to one based on heavy engineering, electronics and information technology, young men (and also women) moved into the growing industrial enclaves, leaving the elderly behind, particularly elderly women, to manage the marginal agricultural lands. These trends are also visible in People's Republic of China, India and Bangladesh (FAO 1998).

Often 'population pressure' is given a direct causal role in driving downward spirals of poverty and environmental degradation. However in-depth analyses of population-poverty-environment linkages show that there are numerous factors conditioning

CHAPTER NINE

Table 9.1 *Poverty Measurement Indices in Selected Countries of the Asian and Pacific Region*

	National Poverty Lines			International Poverty Lines		Human Development Index (HDI) 1997		Human Poverty Index (HPI) 1997		Gender Related Development Index (GDI) 1997		Gender Empowerment Measures (GEM)	
	Population Below the Poverty Line (%)			Population Below US$1 a day (%)	Population Below US$2 a day (%)								
	National	Rural	Urban			Rank	Value (%)	Rank	Value (%)	Rank	Value (%)	Rank	Value (%)
Australia	--	--	--	--	--	7	0.922	12	12.5	4	0.921	9	0.707
Bangladesh	35.6	39.8	14.3	--	--	150	0.440	73	44.4	123	0.428	83	0.304
Fiji						98	0.701	30	19.0	79	0.699	40	0.491
PR China	6.5	9.2	<2	22.2	57.8	61	0.763	20	16.3	60	0.749	79	0.327
India	35.0	36.7	30.5	52.5	88.8	132	0.545	59	35.9	112	0.525	95	0.240
Indonesia	15.1	14.3	16.8	11.8	58.7	105	0.681	45	27.7	88	0.675	71	0.362
Lao PDR	--	--	--	--	--	140	0.463	66	38.9	115	0.483	--	--
Malaysia	--	--	--	5.6	26.6	56	0.768	18	14.2	52	0.763	72	0.451
Nepal	--	--	--	50.3	86.7	144	0.463	85	51.9	121	0.441	--	--
Pakistan	--	--	--	11.6	57.0	138	0.508	71	42.1	116	0.472	101	0.176
Philippines	54.0	71.0	39.0	28.6	64.5	77	0.740	20	16.3	65	0.736	45	0.480
Viet Nam	--	--	--	--	--	110	0.664	51	28.7	91	0.662	--	--

Source: 1 World Bank 1998 and 1999
2 UNDP 1999

Table 9.2 *Demographic, Poverty and Economic Indicators for Asia and the Pacific*

	South Asia	East Asia	Asia and the Pacific
Population (million)	1 293.3	1 295.8	486.5
GDP per capita (US$)	521	725	1 063
Human Development Index (HDI)	0.462	0.676	0.683
Human Poverty Index (HPI)	0.430	0.665	0.665
Urbanization (% of total population)	28	33	33
External Debt (% of GNP)	35	17	50
Net foreign Direct Investment (% of GNP)	0.4	3.1	3.3
Export Import Ratio	73	98	96

Source: UNDP 1998

demographic change, including political, social and economic factors (policies, processes and the context within which poverty develops and persists); environmental change (both natural and human-induced changes); and demographic shifts (as a result of both transitions within the resident population and through migration).

Population growth, the shift from rural to urban investment, the growth of urban centres and internal resource exploitation (such that only a few have control of and benefit from economic resources), all lead to repeated cycles of increased poverty and environmental degradation (Figure 9.2). Faster population growth in developing countries, changing age structures, increasing old-age dependency ratios and the pressure on jobs and livelihoods have severely strained vulnerable sections of the population. In the region as a whole, however, population growth rates have declined in recent years and the overall growth rate of 1.2 per cent is now below the world average (1.4 per cent). The highest subregional figures are 1.7 and 1.5 per cent a year for South and Southeast Asia respectively (UNEP 2000).

Historically, migration has tended to be motivated by economic considerations. Rapid population growth, without a corresponding growth in employment or livelihood opportunities, has resulted in people seeking to occupy and exploit marginal lands, or has led to migration to areas that are perceived as offering greater opportunities. Areas hit by natural disasters or suffering long-term ecological decline have engendered "environmental refugees" as declining environmental carrying capacity forces people to seek another place to live, often increasing population pressure on the resources in the receiving areas, therein exacerbating environmental stress and creating a further surge of environmental refugees.

Figure 9.2 Vicious Cycle of Poverty and Environment Degradation in Developing Countries

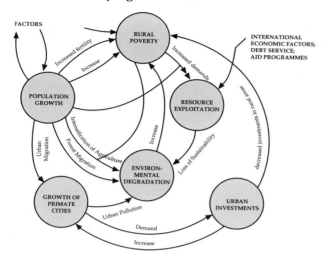

(b) External Causes

The major external (or global) factors facing developing countries in their efforts to alleviate poverty are those associated with the vagaries of the global economic system, a system over which individual countries have little control. The world has, in recent decades, become increasingly interdependent with the emergence of a global economy; for developing countries, access to financial resources is dependent upon a country's participation in this international economy. Moreover, globalization has generally been accompanied by changes of policies and practices whereby the developed countries dictate the terms through which the developing countries participate in the international system (United Nations 1995).

By example, the share of developing countries' exports are 15 per cent of world manufacturing trade. The few value-added products generated in developing countries are hindered by limited market access and the 'new protectionism', which followed the recession of the 1980s. The problems have been compounded by commodity price swings and collapses. Table 9.3 shows the global trade situation of selected countries in the region; despite the reliance upon exports to generate revenue most countries in the region remain net importers.

Debt repayment and servicing form a large part of the annual budget for many of the region's countries, leaving limited financial scope in the funding of social development or environmental remediation programmes (Table 9.4). Although the World Bank's HIPC (Highly Indebted Poor Countries) initiative will provide some assistance, the combined effect of market forces, debt servicing and aid is a net outflow of resources from the developing countries, which provides a key constraint to the alleviation of poverty. While trade liberalization and the removal of tariff barriers have undoubtedly helped some countries, those assisted tend to be the middle-income countries. Exposing the less developed countries to international competition might actually exacerbate their economic problems by restraining their nascent industries and increasing both unemployment and a continued dependence upon external sources of supply.

Many economies of the Asian and Pacific Region have shown robust export-led GDP growth rates in the past decades and have sought to tackle poverty through a range of social, fiscal and economic mechanisms. Yet, much of the economic 'growth' measured in a conventional way looks less impressive when its environmental costs are netted out. Indonesia's rate of growth, for instance, was marked down from seven per cent to four per cent when account was taken of 'depreciation' due to the depletion of its soils, petroleum reserves and forests (Repetto et al 1989).

ENVIRONMENTAL CONSEQUENCES OF POVERTY

Environmental degradation due to pervasive poverty is a matter of great concern in both rural and urban areas in the Asian and Pacific Region. The interaction of poverty and environmental degradation sets off a downward spiral of ecological deterioration that threatens the physical security, economical well being and health of many of the region's poorest people (Figure 9.2).

For example, in the Loess Plateau of People's Republic of China in the 1960s and 1970s the drive to produce more food for the burgeoning population exacerbated land degradation and led to a decline in agricultural productivity and income. By the mid-1980s, more than 5 million people on the plateau's rainfed upland were surviving on incomes of less than $US50 a year. However, consistent government efforts at environmental improvement combining erosion control with improved crop and animal raising practices reduced soil erosion and increased rural incomes. Despite a decrease in the area under cultivation, the improved management of terraces and flatlands has helped to increase total per capita grain production by 30 per cent. Local solutions on environmental adjustments have also been devised in the Rajasthan Desert of India (Box 9.1).

It is often the case that people and countries make an explicit trade off, accepting long-term environmental degradation to meet their immediate needs. In many marginal, rural areas growing populations inevitably lead to daily degradation of

Table 9.3 Balance of Payments and International Reserves for Selected Countries of Asia and the Pacific (millions of dollars) (1996)

Goods and Services	Exports	Imports	Net Income	Net Current Transfers	Current A/C Balance	Gross International Reserves
Australia	78 805	79 568	-15 199	105	-15 857	17 542
Bangladesh	4 508	7 614	-6	--	-1 637	1 609
PR China	171 678	154 127	-12 437	2 129	7 243	146 683
India	42 690	54 505	-4 369	--	-4 601	28 383
Indonesia	51 160	53 244	-5 778	619	-7 023	17 499
Lao PDR	427	787	-4	82	-283	148
Malaysia	83 322	86 595	-4 236	148	-7 362	21 100
Nepal	1 003	1 653	-3	--	-569	627
Pakistan	10 317	15 174	-1 956	--	-4 208	1 790
Philippines	26 795	33 317	3 662	880	-1 980	8 717
Viet Nam	9 695	12 870	-505	1 045	-2 636	1 990

Source: World Bank 1998

Table 9.4 Aid and Financial Flows Relating to Selected Countries in Asia and the Pacific (millions of dollars) (1996)

	Million of Dollars		External Debt		Official Development Assistance	
	Net Private Capital Flows	Foreign Direct Investment	Total millions of dollars	Present value (% of GNP)	Dollars per capita	% of GNP
Australia	--	6 321	--	--	--	--
Bangladesh	92	15	16 083	30	10	3.9
PR China	50 100	40 180	128 817	17	2	0.3
India	6 404	2 587	89 827	22	2	0.6
Indonesia	18 030	7 960	129 033	64	6	0.5
Lao PDR	104	104	2 263	45	72	18.2
Malaysia	12 096	4 500	39 777	52	-22	-0.5
Nepal	9	19	2 413	26	18	8.9
Pakistan	1 936	690	29 901	39	7	1.4
Philippines	4 600	1,408	41 214	51	12	1.0
Viet Nam	2 061	1 500	26 764	123	12	4.0

Source: World Bank 1998

the environment for subsistence, depleting not only the current environment but also future availability. Long-term sustainability of resource use in degraded areas with high populations is an urgent issue that governments of developing countries and international donors have to address through the promotion of appropriate policy instruments.

Poverty and the growing global population are often targeted as responsible for much of the degradation of the world's resources. However, other factors such as the inefficient use of resources, waste generation, pollution from industry and wasteful consumption patterns are key factors in irreversible environmental degradation (UNEP 1997). For example, although developed countries account for only 24 per cent of the population, they consume approximately 70 per cent of the world's energy, 75 per cent of the metals, 85 per cent of the wood, 60 per cent of food, and 85 per cent of the chemicals (United Nations 1995). Whilst practicable solutions remain elusive, the political debate on the "just" entitlement of developing countries and the allocation of responsibility for environmental degradation continues.

The inter-relationship between poverty and environmental damage is complex and is heavily influenced by a range of social, economic, cultural, physical and behavioural factors. These include, the

> **Box 9.1 A Project Addressing Desertification through Local Solutions –
> the "Barefoot Approach" in Rajasthan, India**
>
> The Social Work and Research Centre (SWRC) started work in the Indian province of Rajasthan 27 years ago. It serves an area that has seen two spells of severe drought and famine (each lasting up to five years) over the past 25 years. While working on the provision of safe drinking water, education, health, and awareness raising, the most interesting work of the NGO and the College is on the rehabilitation of the environment, in such a way that the activities also support poverty alleviation. The activities include, the rehabilitation of 600 acres of waste land by planting traditional fuel and fodder species, awareness raising, reinforcing old environment-friendly practices and substituting harmful practices with other alternatives. The greatest achievement of SWRC is the Barefoot College at Tilonia that was started in 1986. Uneducated youth, normally dubbed as unemployable, have been trained in the College to repair and maintain hand pumps in a way that they maintain their recharge rates and do not run dry. 530 "barefoot mechanics" maintain 15 000 hand pumps that serve 3.7 million people. Another 115 "barefoot solar engineers" have installed, and now repair and maintain 75.5 kws of solar panels. The College itself is fully solar powered. The College has also produced "barefoot doctors, chemists, midwives, teachers and civil engineers".
>
> The NGO has a novel approach to managing and recycling waste, and then feeding it into its other programmes, creating employment along the way. For example, waste paper is reused in making glove puppets, which are then used is the 900 plus puppet theatre performances each year. These puppet theatres highlight environmental issues in villages. Kitchen and biological wastes are used in biogas plants that produce fuel for lighting and for supporting laboratories that run medical tests. Agricultural wastes are used for making handicrafts which brings earnings of about US $15/month for each woman involved in the production. Rubber from tires and other waste products are made into educational aides for the 84 night schools run for dropout children, which have annual enrollment of over 1 600 boys and 1 100 girls (children who tend goats and cattle during the day, and cannot attend regular government schools).
>
> Monsoon rains, when they come, are collected in tanks. In areas of extreme scarcity of wood, geodesic domes have been fabricated out of scrap metal to reduce pressure on desert vegetation (this work provides employment for village blacksmiths). The people construct rainwater harvesting structures. They plant trees that are indigenous. They respect the desert, and do not fight it artificially. They know how to use the sun, the air and the wind and the plants and animals. They have converted the area into an ecosystem that is pulsating with vitality, where others see only sand and waste. The success of the project in Rajasthan has already led to expansion of training and capacity building efforts of SWRC all over Indian deserts. The work of SWRC in Rajasthan demonstrate that living in harmony with the environment can improve community's well being. Its barefoot approach also shows that in many cases, problems of poor can be solved by simple technical means adopted to local environment and may not require sophisticated technical training.
>
> *Source:* Roy, B. 1996

ownership of, or entitlement to, natural resources, access to common resources, strength or weakness of communities and local institutions, the individual and community responses to risk and uncertainty, and the way people use scarce time are all important in explaining people's environmental behaviour (Commonwealth Secretariat 1991).

D. Environmental Entitlements and Stress

Environmental entitlements can be defined as, the combined outcome of the environmental resource bundles that people have command over as a result of their ownership, their own production, or their membership of a particular social or economic group; and their ability to make effective use of those resource bundles. These factors mediate the effects of 'poverty' on the use and management of environmental resources and in turn mediate the ways environmental degradation contributes to processes of impoverishment (Melissa & Mearns 1991). Figure 9.3 illustrates this environment and poverty/vulnerability nexus where environmental entitlements are mediated by the "internal" factors of environmental change, environmental management practices and poverty/vulnerability, and the impact of "external" processes.

Although invariably the outcome of various structuring processes, environmental entitlements are dynamic and may improve or decline over time. Entitlement failure either worsens poverty or contributes to environmental degradation because people are no longer able to manage environmental resources effectively. In Bangladesh, for example, agricultural research-led output growth in the highly productive area of Comilla has led to a polarization of income distribution and the out migration of poorer groups to the Chittagong Hill Tracts. Here, through competition with relatively sustainable forms of shifting cultivation, more extractive farming practices have exacerbated the processes of land degradation (ODA 1991).

The management of this relationship between available land entitlement in agro-ecological zones and population densities is a key challenge in

CHAPTER NINE

Figure 9.3 Environmental Entitlements

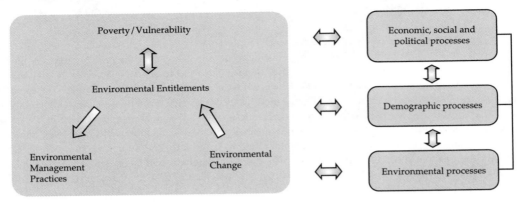

Source: Melissa and Mearns 1991

attempting to combat poverty, whilst preserving environmental resources. The average population density for the Asian and Pacific Region is about 90 persons per square kilometre, and approximately 15 per cent of the total land area is considered arable. For subregional comparison population density varies from 186 persons per square kilometre in South Asia to 3 persons per square kilometre in the South Pacific. Arable land varies from 39 per cent in South Asia, to 9 per cent in the North and Southeast Asia.

1. *Entitlement to Land Ownership*

The distribution of land ownership in most developing countries in the Asian and Pacific Region is characterized by a small proportion of landowners controlling most of the farmland, and a large proportion of farmers controlling small amounts of land. There are also large numbers of landless rural labourers and tenant farmers and farmers with very small land holdings (owned or rented). The South Pacific subregion is particularly affected, for example, with the World Bank estimating that about 90 per cent of men in Tonga are landless (FAO 1998). Gender disparities further distort access to land, with women (who often play a key role in food production) having limited land access. However, in countries such as People's Republic of China, Viet Nam and Lao People's Democratic Republic, land entitlements are centrally allocated by family size and both men and women can cultivate (FAO 1998). This has provided a line of defence against increased poverty.

Lack of secure land tenure acts as a major disincentive to environmentally sound agricultural practices. Farmers without land are likely to be poor and in a 'client-patron relationship' with the landowner. They are in no position to question the technologies they are told to use, and with limited or no access to credit, cannot take the initiative in working in more environment friendly ways. It is, therefore, difficult to break out of the cycle of declining productivity, poverty and increasing environmental degradation.

2. *Entitlement to Common Property*

The poor, especially in rural areas, also tend to be more reliant on 'common property' resources to which everyone has access, for example, forests, rangelands, water points and inshore fishing grounds. In South Asia where the worst poverty is concentrated, families draw heavily from common resources for their fuel, fodder and water. In the dry land areas of India, landless people derive a fifth of their annual income, together with a range of non-marketed goods, from the natural products of common areas. The 'gauchars' (land left uncultivated for grazing in honour of the revered cow and accessible for local people as common property) across the Indian sub-continent provide an ancient method of protecting and rehabilitating environmentally marginal lands. For women across the region, access to common property resources have been crucial in maintaining household food security and micro enterprize ventures. However, the gradual disappearance of common property resources through acquisition and encroachment is exacerbating poverty and environmental degradation. For example, in Bangladesh the clearing of mangroves by commercial shrimping interests has impoverished many coastal households.

3. *Entitlement to Forests*

Villagers or forest dwellers have traditionally managed and controlled deforestation and thus should be entitled to land tenure. Throughout the region, thousands of forest villagers have had an elaborate system of control to ensure sustainable use of their shared forests. In Borneo, for example, generations of Galik people have carefully conserved ironwood trees, and the use of timber from these trees has traditionally been governed by ethics that required broad sharing along kinship lines. The entry of state sanctioned loggers into the Galik area has undermined traditional restraints and, due to fears of losing the ironwood to loggers, the Galik have, themselves, undertaken the unsustainable felling of ironwood trees, endangering both the resource and the village economy (Durning 1994). Similarly, in Pakistan, nationalization of the forests after the change of states of the Swat and Chitral States to districts has undermined traditional management by creating a "Free-for-all" in forest exploitation during the transition period.

4.   Entitlement to Coastal and Marine Resources

With new developments in commercial fishing operations and aquaculture, large traditional communities of fishermen have been marginalized in the region particularly in Southeast Asia, South Asia and the South Pacific. The South Pacific Islands and other coastal areas in the region are especially at risk where communities dependent on marine resources (including reef resources) are severely threatened, as their traditional livelihoods are eroded. Recognition of entitlement and community participation, particularly of the poor, could considerably improve the conservation of marine and coastal resources.

5.   Entitlement to Urban Housing and Shelter

The high price of land and shelter in the urban areas of the developing countries has limited formal access for the poor to these resources. This results in establishing housing and shelters on hazard prone lands or encroaching upon government lands. Due to insecurity of tenure no incentive exists to improve these squatter settlements. However, where the squatter settlements have been regularized, by bestowing property rights, considerable improvements have taken place (see Chapter 7). In such areas, where governments are unable to make adequate provision for housing, services and employment, it is therefore widely argued that the solution lies in self-help. This requires the energy of the urban poor to be channelled in a positive direction so that they may help themselves by building their own shelter and basic services, thus creating employment, organizing their own neighbourhoods and providing for their own needs. A good success story of positive local action example is the Orangi Pilot Project in Pakistan (Box 9.2).

E.   Entitlement and Investment in Environmental Protection

Experience shows that entitlement, or providing poor with access to resources, has greatly helped in ensuring protection and promotion of sustainable development. However, even where entitlement to the land and natural resources can be claimed, it is sometimes difficult to stop their degradation. With few savings and limited access to capital and credit, there is little capacity to invest in environmental conservation measures (such as preventing soil erosion) even when the long-term best interest is recognized. Lack of capital, high implicit discount rates and a foreshortening perspective on the future, explain much of the relationship of the poor to their environment. Yet, many poor communities have evolved models of sustainable behaviour over time, to meet the twin challenges of poverty and environmental degradation.

POVERTY AND NATURAL DISASTERS

The effect of natural hazards on the loss of human lives is directly related to the poverty levels. The overwhelming majority of those worst affected by earthquakes, floods, storms and harvest failures are the poor. All over the region, the poor usually live nearest to natural hazard prone areas, dirty factories, busy roads, and waste dumps that make them vulnerable to environmental disasters.

Disasters have a profound impact on the quality of life through their destruction of food crops and livestock, shelter and other aspects of the built environment, and through the forced dislocation of households and communities, although their most devastating impact is on the direct loss of life and the instant poverty that frequently accompany such events. Forces of nature are no longer the sole cause of natural disasters, such events are often aggravated by the degradation of environment by human activities (Box 9.3).

In the Asian and Pacific Region, losses caused by natural disasters are particularly damaging, depriving countries of resources which could otherwise be used for economic and social development. Since the International Decade for Natural Disaster Reduction began in 1990, the total number of deaths caused by natural disasters in the region has exceeded 200 000. The estimated total damage to property was about US$50 billion prior to the Kobe earthquake in 1995 and very heavy flooding in People's Republic of China in 1995 and 1998. In the 1991 cyclone and storm surge event in Bangladesh, 140 000 people perished, whilst the flood of 1998 affected the lives of 25 million people. This flooding, the worst on record, destroyed part of the rice crop and forced scores of textile factories in Dhaka to close for several weeks, depriving the country of much needed export earnings. The total damage was estimated at US$530 million, resulting

CHAPTER NINE

> **Box 9.2 Community-based Approaches to Improve the Living Environment and Address Poverty in Poor Urban Settlements – The Orangi Pilot Project in Pakistan**
>
> Orangi Township is situated in the Orangi Hills in the western part of Karachi. It has the distinction of being the world's largest katchi abadi or "squatter" settlement. Spread over an area of 8 000 acres, it currently houses about a million people living in about 100 000 housing units, which people have constructed themselves with the help from the informal sector. The Township was originally created by land-grabbers through illegal occupation and subdivision of state land in the 1960s. Bits were then parceled out and sold to poor people who either did not know that they were getting land through illegal channels, or had no other option. The Township was later "regularized" by the provincial government and the city administration. While it was easier to give titles of occupied plots of land to the occupiers, it was most difficult to provide environmental services and take up the task of cleaning up the environment of unplanned huts and housings, with sewerage running across open spaces, lanes, and streets or roads. The extent of morbidity and mortality, was daunting.
>
> The Orangi Pilot Project (OPP) was established in 1981 to meet these huge environmental and poverty challenges. It is one of the first projects in the world that has tackled poverty and environment issues within the same package. In an approach that was very futuristic; OPP recognized right from the beginning that the spiral of poverty-population-environment would have to be tackled as a whole, by the people themselves. Its entry point for intervention through community sanitation and health, therefore, went hand in hand with housing, population programme and entrepreneurship development. At the same time, links were also established with relevant government agencies for regularizing the status of ownership, and for, what was referred to as, "external" development (such as the laying down of trunk sewers).
>
> People were organized to first delineate their "lanes". Everyone living in a lane had to contribute towards its clean up and towards the cost of building sewer lines in those lanes. Households built their own internal disposal system and the connection to the sewer line. These sewer lines were then connected to secondary sewer lines. City agencies put in the main trunk sewers. Local masons were trained to make low-cost bricks used in the construction of these sewers (and also for housing). Community groups had to purchase construction material from these small local producers. The programme, therefore, boosted local economic activity while tackling the living environment.
>
> Credit programmes helped to alleviate poverty, while technical assistance for better housing led to both increased local economic activity, as well as healthier and more sanitary homes. The health and population programme tackled reproductive health and morbidity, as well as preventive measures. The education programme (developed with the private sector through micro-credit) has been so successful, that Orangi has one of the highest literacy rates in the country. A network of over 100 clinics are also involved in preventative, curative and reproductive health care, now managed by an OPP offshoot called OPP-KHASDA. These services have had a direct impact on controlling population growth. The OPP has another very distinctive feature. It does not have "targets" or "time frames" in the conventional sense. The "process"-participation and self-financing for improving the living environment of the poor – was and is paramount.
>
> The outstanding achievements of OPP can be gauged from the fact that under its sanitation programme in Orangi alone 6 000 plus main sewer lines have been established with a total length of 1.5 million rft. 400 plus secondary sewers have been constructed with total length of 165 000 plus rft. In addition 91 000 plus latrines have also been constructed. The community or people of Orangi have invested over Rs79 million (about US$1.6 million) in just this one programme to improve their environment. Having paid for part of this system themselves, they own it and look after it, so that the choked and stinking lanes and unhygienic homes have been "greened" by a functioning system of sanitation. Under its entrepreneurship development programme OPP disbursed loans worth about Rs118 million (US$2.2 million) to the poor in 63 professions between September 1987 and November 1989 alone, helping them to become entrepreneurs and set up their own small businesses. About Rs18 million worth of loans have gone to about 1 200 women entrepreneurs. The entrepreneurs have paid back Rs91 million of the principle and Rs22 million as markup. This programme has had a positive impact on the reduction of poverty.
>
> Over the past two decades, OPP has expanded tremendously and multiplied into many separate institutions across Pakistan, some of which have been handed over to local groups. The flagship operation is now known as OPP-RTI, where RTI stands for Research and Training Institute. Its models have been emulated widely across the country and abroad.
>
> *Sources:*   Hasan A 1997

in a negative effect of five per cent on the growth of gross domestic product (GDP) (ESCAP 1999).

The flood in 1998 in People's Republic of China was the most severe in over 40 years. According to governmental estimates, 223 million people-one fifth of China's population were affected, 3 004 people died and 15 million were made homeless. About 15 million farmers lost their crops. The floods caused severe damage to critical facilities systems as well as industrial facilities and resulted in a total loss of US$12.5 billion, equivalent to 4.5 per cent of the GDP.

In October 1999, the State of Orissa in India was severely affected by a super cyclonic storm in the Bay of Bengal. This resulted in the death of nearly

POVERTY AND ENVIRONMENT

> **Box 9.3 The Interrelationship: Environmental Deterioration, Poverty and Natural Disaster**
>
> The loss of natural resources, for example by deforestation, can generate significant numbers of environmental refugees. For instance, forest-dwellers are being driven from their homes by deforestation and the industrialization of timber in Indonesia, the Philippines, and other tropical nations. Still other communities are being forced out by the disappearance of water supplies, as in Pakistan. The hundreds of people who arrive daily at the bus terminals of cities such as Manila, Jakarta, Dhaka and Karachi seeking a new life are driven, in part at least, by deteriorating environmental conditions in their villages. But they could be the first waves of a much larger tide, who settle in illegal squatter colonies vulnerable to floods, landslides or industrial accidents, in city areas where richer people refuse to live. Being unplanned, they lack the infrastructure, ranging from community health and fire services to dykes and drains, that is needed to cope with disaster and which is available in 'planned' districts. Therefore, once these areas are struck by an extreme natural event – the impact is disasterous in terms of loss of human life and economic damage.
>
> Moreover, the loss of natural vegetation particularly forests, in the Asian and Pacific Region is a major cause of disasters from natural events. The destruction of upland forests, for example, reduces the ability of soils to absorb rainwater. A single tree can absorb 200 litres of water or more per hour. The soil that it holds in place may absorb even more. If the tree is gone, than water runs off the land, pouring down the gullies and into rivers – or, all too frequently, through human settlements. Meanwhile, both the impact of the rain on the ground and the loss of binding tree roots contribute to instability of hill slopes that in turn leads to landslide that end up as silt in rivers, raising river beds and contributing to the floods downstream.
>
> The summer of 1998 saw two major examples of flooding and landslide disasters in the region that were caused by such environmental degradation – in both cases the loss of forested areas in the upland watersheds of major rivers. In the first case along the Himalayan foothills of north-west India, landslides killed more than 300 people within a week in August. Researchers at the Wadia Institute of Himalayan Geology in Dehra Dun reported that landslides had become an increasing feature of the region in the past decade as trees have been cleared for agriculture and road-building.
>
> Meanwhile, in China, government scientists are reported to have found that flooding on the floodplains of the River Yangze – which killed thousands, left millions homeless and wiped out tens of millions of hectares of crops – had been seriously exacerbated by the loss of upland forests and the erosion of their soils. According to the Worldwatch Institute in Washington DC, more than four-fifths of the forests of the Yangze river basin have been chopped down, swelling the river with both flood waters and silt. Following the 1998 floods, the Government of China has banned all logging in the Yangze watershed and decreed that all hillsides logged since 1994 should be replanted.
>
> *Source:* Red Cross 1999

10 000 people, affecting nearly 13 million people in 14 643 villages and causing damage to 1.8 million hectares of crops, 1.65 million houses and loss of 444 000 livestock. In July 1998, the 10-metre tsunami that hit Papua New Guinea took more than 2 000 lives in several coastal villages. In January 1995, the Kobe earthquake killed over 5 000 people in addition to the tremendous damage it caused, and in August 1999, Turkey was affected by a severe earthquake which caused tens of thousands of deaths and injuries, destroyed nearly 100 000 housing units and made some 100 000 families homeless. The total wealth loss was estimated by the World Bank to be in the range of US$3-6.5 billion (equivalent to 1.5 to 3.3 per cent of the gross national product).

Floods and high winds, or cyclones, affect countries in the region on an almost annual basis. They form the predominant type of disaster, accounting for 60 per cent of all natural disasters between 1988 and 1997. Forest and bush fires have also always been a hazard in the region, but recently they reached catastrophic dimensions. During 1997-98, massive fires in Southeast Asia destroyed millions of hectares of forest, and caused more than US$4.5 billion in damage. The fires created serious health problems, accidents on land, at sea and in the air; disrupted transportation systems and resulted in a steep drop in tourism in parts of the region that are heavily dependent upon tourism as a key revenue earner.

The fact that 30 out of the 40 catastrophes that occurred in the world between 1970 and 1997 occurred in Asia and the Pacific (and consequently 87 per cent of the casualties) highlights the significance of natural disasters to the region. In 1997 alone, the region suffered 33 per cent of the worst catastrophes, 67 per cent of the casualties and 28 per cent of the economic losses of the global effects of natural disasters. Only 0.2 per cent of these losses were covered by insurance. National and regional efforts for natural disaster reduction should therefore be closely linked with poverty alleviation and economic and social development activities. It is easy to see how natural disasters have their severest human impact on the poorest communities who inhabit the more vulnerable environments, with limited savings and insurance to help rebuild their lives.

CHAPTER NINE

POVERTY AND ENVIRONMENTAL HEALTH HAZARDS

The WHO has called poverty the world's biggest killer (WHO 1995). Although efforts to reduce poverty and increase disposable income levels continue, the key to identifying new strategies for achieving these objectives lies in an understanding of how poverty affects both the environment and human health.

Environmental health problems emanate from a lack of access to essential environmental resources, primarily sufficient and clean water, enough food, appropriate shelter and fuel and healthy air. At least one in four of the region's population has no access to safe drinking water, and one in two has no access to sanitation (see Chapter 14). Public expenditure on water and sanitation is relatively low at only one per cent of the GDP for most countries. Unsafe water and poor sanitation in developing countries is responsible for a large percentage of diseases and a significant proportion of mortality. In Jakarta, Indonesia alone, the economic cost of impaired health from unsafe drinking water is estimated at US$300 million a year.

Most infectious diseases are "environmental" in origin, as specific environmental conditions increase the biological organisms' ability to thrive or spread. Even diseases such as acute respiratory infections are linked with poor conditions within the household environment, including overcrowding, poor sanitation and indoor air pollution. Since 1998, the Andhra Pradesh government (with donor assistance) has been assessing the overall burden of ill-health associated with lack of water and sanitation infrastructure, and outlining a cost effective strategy for reversing this situation. The burden of disease has been reduced by 17 per cent through the provision of clean water from taps inside the house, private latrines and reducing indoor air pollution through clean cooking fuel.

A. Micro-organisms

Of all the environmental hazards humans encounter, the most formidable adversaries remain micro-organisms-viruses, bacteria, protozoa, and helminthes (parasitic worms). Years of concerted efforts have revealed that while it is very difficult to eradicate microbial threats, it is possible to live in balance with them. However, it is also recognized that human activities can change the environment and disrupt natural ecosystems in favour of the microbes (WRI/UNEP/UNDP/World Bank 1997).

Recent WHO statistics (1999) show that infectious diseases that result from micro-organisms are the major cause of premature death (48 per cent) and of death among children (63 per cent) worldwide. Figures 9.4 to 9.5 demonstrate the widespread incidence of these diseases throughout the region. In particular, six deadly infectious diseases (pneumonia, tuberculosis, diarrhoeal diseases, malaria, measles and more recently, HIV/AIDS) account for half of all premature deaths in the world, killing mostly children and young adults (Figure 9.6). Every day, 3 000 people die from malaria, three out of four of them children. Every year 1.5 million

Figure 9.4 Large Outbreaks

Selected outbreaks of more than 10 000 cases, 1970-1990

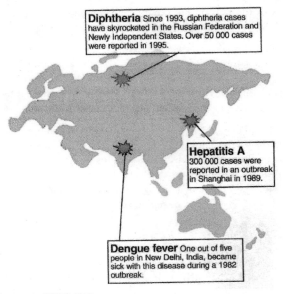

Source: WHO 1999

Figure 9.5 Unexpected Outbreaks

Examples of emerging and re-emerging infectious diseases 1994-1999

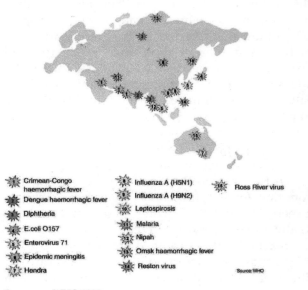

Source: WHO 1999

Figure 9.6 Leading Infectious Killers

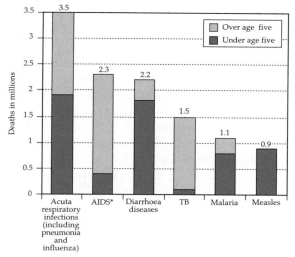

Source: WHO 1999

people die from tuberculosis and another eight million are newly infected. In Afghanistan, one child dies every four days from diarrhoea. Furthermore, families risk being driven into debt through lost earnings and high health-care costs trapping them in a vicious cycle of poverty and ill-health.

B. Poverty and Epidemiological Polarization

Regardless of the overall economic development of a country, health and life expectancy improve with wealth (Stanton 1994). As the income gap increases within a population, the health gap is also likely to grow, leading to what some have dubbed 'epidemiological polarization'. Epidemics are largely concentrated among the poor living in degraded environments. Therefore, one strategy to reduce these risks is to raise incomes and improve the distribution of wealth. Additionally, with wealth comes the increased political leverage required to make the public sector respond to their needs. However, prior to a policy targeted solely at income generation, by targeting policies to reduce environmental threats, incomes may be enhanced improving the ability to work and reducing illness.

Most environmental interventions are very cost-effective as a means of achieving health improvements. In addition, a key development objective of improving people's health requires a holistic approach to mitigating major risks by integrating efforts inside and outside health care systems.

C. Poor Workers and Health Risks

In recent years, a new risk is emerging in the wake of proliferating chemicals and hazardous waste which are being released into the environment through industrial accidents in many countries of the region. Two well known cases are those of Bhopal, India and Khlong Toey in Bangkok, Thailand. When addressing the global problem of occupational accidents and diseases, the International Labour Organization (ILO) estimate that accidents and diseases together cause over 1.2 million fatalities annually. Furthermore, workers face more than 250 million accidents and over 160 million of them become ill due to workplace hazards and exposure to pollution. In addition to workplace accidents, more than 50 per cent of work related fatalities occur in the Asian and Pacific Region. The poorest, least protected, least informed and trained (often women, children and migrants) are generally the worst affected by poor health and safety standards (Communication by ILO, March 2000).

D. Health Costs of Environmental Degradation

The health effects of environmental degradation in selected Asian countries and their related costs are shown in Table 9.5. These costs indicate the scale of the problem and the need for action, particularly in the cities of the region where lead levels are a serious health hazard. According to the World Bank, three types of activities are important in improving the status of environmental health: enhancement in knowledge of environmental health problems; development of appropriate response procedures and integrating critical environmental health issues in the operations of the relevant sectors (World Bank 1999).

POLICIES AND PROGRAMMES ON POVERTY ALLEVIATION AND ENVIRONMENT

Numerous initiatives have been undertaken in the Asian and Pacific Region to tackle poverty and related issues at local, national, and regional levels. However, their cumulative impact is still low mostly because these have been undertaken in an isolated and piecemeal manner. The gravity of the problem demands a more integrated and comprehensive approach to these issues. While faster poverty reduction requires accelerated growth that generates employment and incomes, economic growth alone cannot be relied on to eliminate poverty; complementary well-articulated international, regional and national strategies for poverty reduction are also essential.

A. National Actions

Although the "vicious cycle of poverty" and its intrinsic linkages with the environment, and the

Table 9.5 Health Effects of Environmental Degradation in Selected Asian Countries

Country	Period	Environmental Health Effects	Annual Cost (US$ billion)	Cost as a percentage of GDP
PR China	1990	Productivity losses caused by soil erosion deforestation and land degradation; water shortage and destruction of wetlands	13.9-26.6	3.8-7.3
		Health and productivity losses caused by environmental pollution in cities.	6.3-9.3	1.7-2.5
Indonesia	1989	Health effects of particulate and lead levels above WHO standards in Jakarta	2.2	2.0
Pakistan	Early 1990s	Health impacts of air and water pollution and productivity losses from deforestation and soil erosion.	1.7	3.3
Philippines	Early 1990s	Health and productivity losses from air and water pollution in the vicinity of Manila	0.3-0.4	0.8-1.0
Thailand	1989	Health effects of particulate and lead levels above WHO standards	1.6	2.0

Source: UNDP 1997

urgency to address poverty alleviation in the region is well accepted, little evidence exists to show effective and concerted actions have been taken at the national level. There are few examples of countries linking environmental protection to social investment, such as education, better health-care, and employment generation for the poor, especially women (UNEP 1997).

Many countries have policies and/or legislation relating to poverty, but few have explicit policies on poverty eradication alone, or policies linking environment, poverty, trade and social development. In addition, few environmental policies specifically target equity or poverty issues. Nevertheless, there have been policy initiatives in the social sectors such as a thrust to address poverty directly through employment generation programmes and to improve equity through rural credit. At the same time, many countries have adopted policies to stabilize or moderate population growth rates. The success of efforts directly targeted at poverty alleviation has some notable successes in Southeast Asia, but fewer in South Asia. Success has generally been linked to direct support, which has included the provision of subsidized food or credit and the introduction of micro finance programmes, although such programmes require careful targeting if they are to reach the poorest segments of society.

In order to alleviate poverty, countries of the region have also attempted to tackle the problem of rural landlessness by adopting two broad strategies. In a small number of countries (for example, Indonesia, Malaysia, Sri Lanka) government strategies have involved moving the landless to available arable land as in, for example, the Indonesian transmigration programme. However, in many other countries of the region with inequitable land distribution (for example, India and Pakistan), the demand for cultivable land is to a varying extent being met through land and tenancy reforms.

Agricultural policies (access to credit, pricing policies, inputs and infrastructure) to increase productivity and incomes and thereby alleviate poverty (for example, India, Pakistan, Philippines, Indonesia and Sri Lanka), have been a focus for many official policies, besides the provision of land. Social policies have focused on basic needs of the rural population, especially shelter and safe drinking water. Human resources development has also been emphasized with a high priority on education and training.

Most countries have national policies on health, although they vary considerably in their commitment to the concept of 'Health for All'. The linkage of health, poverty or socio-economic development and environment has not been adequately recognized in many development policies. Health policies and programmes are still mainly formulated and implemented in isolation, with no linkages with related sectors, and focus on curative rather than preventative measures, particularly in terms of environmental issues (e.g. water and sanitation provision).

In recent years, efforts have been made to systematically deal with the predicament of natural disasters. Initiatives have been taken to set up institutions, develop plans, enact legislation and promote programmes and projects to mitigate disasters. Significant progress has been made in environmental rehabilitation, forecasting and early warning, particularly in the case of climatic and volcanic hazards. Post-disaster relief operations have also been significantly improved.

B. Regional and International Actions

Through a series of United Nations conferences, highlighted by the World Summit on Social Development in 1995, the international community agreed to a common set of targets for reducing poverty. In 1996, the Development Assistance Committee (DAC) of the OECD endorsed seven of these as the *Strategy 21* goals. Agenda 21 also emphasized poverty reduction, following on from which, the Fifth Asia and Pacific Ministerial Conference in 1997 resulted in a comprehensive set of targets described in the *Manila Declaration* (Manila Declaration 1997). These targets, for the first time, linked poverty, environment and social development and set a time frame for their achievement.

While the *Manila Declaration* has set out the targets, international agencies have initiated support and have persuaded countries to integrate the targets into their national programmes. In South Asia, the Dhaka Declaration of 1993 enunciated a ten year strategy for the eradication of poverty and all SAARC member countries were directed to prepare national programmes with a view to achieving the objective by the year 2002. In the Delhi Summit of 1995, a SAARC Programme emerged with well-defined subregional objectives to address sustainable human development issues such as poverty, environment, participatory development and empowerment.

The strategy calls for action on three fronts. The centrepiece is a pro-poor development approach, where poverty alleviation programmes lead to increases in savings and capital accumulation. The fundamental premise of the strategy is that the poor should be viewed as engines of growth as opposed to economic liabilities. There is a major shift in poverty alleviation strategy away from the welfare approach to one of facilitating the engagement of the poor in productive activities. Social mobilization alone is recognized as not being a sufficient condition for poverty eradication. The strategies recommend a net transfer of resources to the poor, by removing the "anti poor" bias of many macro policies and correcting policies relating to, *inter alia*, social investment, agricultural development, the informal sector and choice of technology.

Technical cooperation among countries of the region in addressing common health problems has taken place through the assistance of the WHO. The focus of the WHO's support is on improving the quality of national programmes covering protein energy malnutrition, iodine deficiency disorders, vitamin A deficiency and iron deficiency anaemia. The nutrition research agenda has been developed, and technical support provided, for improving the case management of children with severe malnutrition. The WHO's collaborative programme continues to focus on drinking water quality surveillance as well as operation and maintenance of water supply systems. A strategy for sanitation for high-risk communities has also been introduced and the Jakarta Declaration, the outcome of the Fourth International Conference on Health Promotion, provides a crystallized vision for health promotion and development.

In terms of poverty alleviation, various NGOs are also instrumental. The Grameen Bank of Bangladesh is a leading example (Box 9.4). While its credit programme is not aimed at reducing and eradicating the cycle of poverty-environment degradation *per se*, it can be argued that by helping poor landless and marginal farmers, Grameen Bank is also seeking to stabilize the environment and assist poor people to remain on the land.

CONCLUSION

Effective strategies for simultaneously addressing the problems of poverty and resource degradation have focussed on people, with the objective of providing sustainable livelihoods for all as the integrating factor that ensures that policies simultaneously address the issues of resource management, development and poverty eradication.

The ultimate objective is that all poor households are provided with the opportunity to earn a sustainable livelihood, while ecologically-vulnerable areas are handled in an integrated manner encompassing resource management, poverty alleviation and employment generation. A range of activities at national and regional levels, involving governments and citizen's groups and supported internationally, are the keys to achieving better environmental management practices, ensuring the flow of reliable and up-to-date information on demographic and environmental changes, improving training and providing incentives for environmental management. The need to examine the effects of developed countries' macroeconomic and trade policies on environmental management behaviour remains, in particular the consequences for specific groups of poor people.

Individual countries need to take many steps, including investigating the environmental implications of investment in poverty reduction strategies, and examining the effectiveness, practicality and appropriate forms of policy targeting aimed at achieving poverty reduction and sustainable environmental management in different agro-ecological zones and urban environments. The concept of agro-ecological zones (as already used widely in India) would enable different strategies to

CHAPTER NINE

Box 9.4 Fighting Poverty – The Grameen Bank in Bangladesh

The Grameen Bank of Bangladesh is one of the world's leading programmes for poverty alleviation. It was initiated in 1977 and is based on the understanding that the lack of access to credit is the main hurdle in the progress of the rural poor.

The major objective of the bank is to provide institutional credit to landless people for remunerative self-employment. With its specially designed credit programme, it extends loans to women and men living in absolute poverty who cannot otherwise offer collateral for bank loans. Membership to Grameen Bank groups is open to like-minded people having the same social status and whose families own less than 0.5 acre of cultivable land or the value of a family's total assets does not exceed the market price of one acre of average quality land in the locality.

In 1983 Grameen Bank was reconstituted as a specialized financial institution. The operational objectives include: extending banking facilities to the poor especially women; elimination of exploitation of the poor by money lenders; creation of self-employment opportunities for the un-utilized manpower; organization of people to strengthen themselves in socio-political and economic aspects through mutual support: reversing the vicious cycle of poverty – low income, low savings, low investment to 'more income, more credit, more investment'; encouragement of self-reliance among the groups; ensuring better health, nutritional, housing and education facilities for its members.

The total number of borrowers to date far exceeds 2 million, out of which 94 per cent are female borrowers. Grameen Bank's services reach about 40 000 villages. It has extended small credit amounting to US$1 810 million. The small savings of poor villagers to date are over 4 900 million takas (US$130 million). Women are considered more bankable and more trustworthy (this trust has been amply rewarded through a recovery rate of ninety eight per cent on all loans advance to women).

The main reasons for the successes of Grameen Bank are as under:

- Loans are small (average of US$100 each) and carry no interest subsidy
- Loans are given at a much higher interest rate than bank loans in the market, reflecting the extraadministration cost of small loans
- The poor are required to put aside some saving – at least one taka (US2.5 cents) a week, thus encouraging the habit of self-reliance among the poor
- The bank went to the poor, rather than waiting for the poor to come to the bank.

Grameen Bank has also introduced housing loans for the poor. A Grameen Bank member can borrow up to US$640 for constructing a simple tin roof house. It has disbursed housing loans for construction of over 310,000 houses. The GB members are also encouraged to pay attention to their social situations and health conditions. These issues have been documented as the 'sixteen decisions' which are strictly followed by staff and each member. The sixteen decisions certify a change in the attitudes of its clients.

The Grameen Bank is now experiencing with other initiatives, including the creation of a $100 million People's Fund to finance replication of this experiment in other developing countries. Grameen Bank is also working towards reducing environmental degradation. This aspect has also been incorporated in the sixteen decisions. Grameen Bank's approach is total development. The sixteen decisions and the credit are an effective mix to approach alleviation of poverty. This approach has proved that the poor have the capacity to improve their lives. Grameen Bank experiences have been replicated in 40 countries of the world.

Source: World Bank 1997

---

be developed to suit in different zones. For example, in low-potential areas in which the thresholds of ecological sensitivity and resilience have not yet been crossed, social welfare transfers to the poor could be channelled through public works programmes geared at supplementing natural resilience, such as tree-planting by means of rainwater harvesting in drylands.

The world has in recent decades, become increasingly interdependent with the emergence of a global economy. The formation of this economy has been accompanied by policies and practices whereby the developed countries heavily influence the terms under which the developing countries participate in the international system (United Nations 1995), which may result in negative impacts for poverty and the environment. Moreover, as the more developing countries of Asia and the Pacific become dependent on aid, the harder are the terms to which they have to agree for trade.

Efforts to reduce poverty must be comprehensive enough to address all of its many causes. This requires a variety of measures across macro, micro and sectoral levels. Pro-poor and sustainable economic growth is fundamental, but needs to be complemented by social development that permits access by the poor to education, health, social protection and other basic services. These in turn are dependent on sound macroeconomic management and good governance.

Global strategies for sustainable development should integrate economic and environmental considerations of the adverse effects on the environment, which are caused both by marked affluence, and by poverty. This implies that developed countries will also need to play a proactive role in accepting responsibility and should bear a larger burden through the means they have at their disposal, particularly spending their wealth in conserving the environment.

Chapter Ten

The use of agrochemicals helped in enhancing food production but its indiscriminate use is a serious environmental hazard.

10

Food Security

INTRODUCTION

FOOD PRODUCTION AND FOOD SECURITY: STATUS AND TRENDS

ENVIRONMENTAL IMPACTS OF ENHANCING FOOD AND AGRICULTURAL PRODUCTION

POLICIES AND PROGRAMMES FOR PROMOTING FOOD SECURITY AND SUSTAINABLE AGRICULTURE

CONCLUSIONS

CHAPTER TEN

INTRODUCTION

The concept of providing people with food security extends from the individual and local community level to the global level. At the individual level, the concept of food security implies that under all circumstances each man, woman and child has access to sufficient, good quality food to meet the individual dietary requirements consistent with normal active life. At the national and regional levels, food security implies an assured availability of food through production, stock draw down, trade or food aid to meet minimum requirements per capita, and also to meet any unexpected shortfall over a limited period.

The achievement of food security requires the utilization of both renewable and non-renewable agricultural resources and carries the risk of environmental degradation if managed inappropriately. This chapter discusses the food security situation in Asia and the Pacific in terms of food production availability, its inter-relationships with environment, and policy actions undertaken to promote food security.

FOOD PRODUCTION AND FOOD SECURITY: STATUS AND TRENDS

A. Food Resources and Production

Since 1970, overall food production has increased significantly in the Asian and Pacific Region (Figure 10.1), with Asia outstripping the world and developing countries in both total and per capita food production. This trend emerged against the backdrop of a similar performance in total agricultural production and whilst most subregional and individual country performances reflected that of the region, a number of country level performances were more varied in per capita production. People's Republic of China's performance has been particularly noteworthy since its transition to the household responsibility system which gave a boost to food and agricultural production, particularly in the 1980s and early 1990s. In Northeast and Southeast Asia, the total cereal (wheat, milled rice, maize, and coarse grains) production increased, from the 1960s through to the 1980s, at a faster rate than in South Asia (Table 10.1) and, according to the FAO, this differential in the trend of cereal production between the two subregions is likely to continue up to 2010.

1. Crop Production

(a) Cereals

An analysis of production rates for major cereals indicates a slowing down in production growth in the late 1980s and early 1990s from the levels reached in the 1970s (Figure 10.2), with the exception of rice production in South Asia where the average rate of growth in the 1980s was higher than in the 1970s. All projections for cereals and coarse grain production over the next decade show that the average trend of increasing production will be maintained (Figure 10.3), whilst per capita production of starchy roots reveals the opposite trend. South Asia lags behind other regions in terms of per capita production of total cereals and its productivity is projected to decline over the next ten years (Figure 10.4).

Many countries experienced significant variability in food production in recent years, which contributed to food insecurity (Alamgir and Arora 1991; Jazairy *et al* 1992). Cereal production has been less stable than the production of other crops. Several countries have experienced significant production decline (Armenia, Azerbaijan, Bhutan, Fiji, Kazakhstan, Kyrgyzstan, Republic of Korea, Lao People's Democratic Republic, Malaysia, Mongolia, Solomon Islands, Tajikistan, Tonga, Turkmenistan, Uzbekistan and Vanuatu), whilst the food security situation has deteriorated alarmingly in the South Pacific and in the Northeast and Central Asian countries. South Pacific countries are inherently vulnerable to natural calamities and import shortfalls, while countries in Northeast and Central Asia have suffered from breakdown of institutions serving agriculture and food production. While existing centralized institutions dealing with agriculture were dismantled, new ones have yet to efficiently

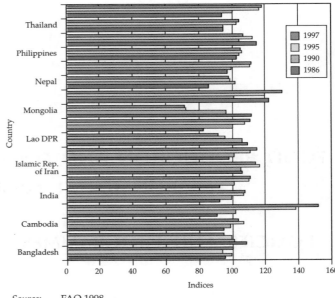

Figure 10.1 Food Security Indices for Selected Countries (1986-1997)

Source: FAO 1998

administer input supply, capital provision and marketing need requirements. Furthermore, Azerbaijan has faced the added problem of a large displaced population (850 000 from conflict over Nagomo Karabakh) and Tajikistan has suffered from the effects of civil war. However, although differences between the countries of the region remain, the food security situation seems to have improved since 1996 (Table 10.2).

The trend of production over latest two years (1996-1998) for which data are available shows a decline in total cereals in most countries except in Northeast and Central Asia, with serious decline being recorded in Indonesia, Democratic People's Republic of Korea and the Philippines. The decline is linked to relatively unfavourable weather including the impact of El Nino (FAO 1998). Wheat production has been better than rice and coarse grains. Rice production is estimated to have declined significantly in Indonesia and the Philippines, coarse grains in Democratic People's Republic of Korea and the Philippines. According to FAO and WFP (1997), grain (cereals and pulses) production improved between 1996 and 1997 by between 3 per cent (Armenia) and 38 per cent (Turkmenistan). Production increased by more than ten per cent in Kyrgyzstan, Tajikistan, Turkmenistan and Uzbekistan. This is attributed to better availability of inputs and improved incentives, although Kazakhstan was affected by financial constraints and fuel and input supply problems, which led farmers to plant seed on unprepared land. As a consequence, it is estimated that in 1997/98 the area sown with grains in Kazakhstan declined by over one million hectares compared with the previous year. In addition to factors identified above, a number of countries in the Northeast and Central Asia, had to deal with the consequences of low prices received by farmers, obsolete machinery, deterioration in the ration of milling to feed quality grain (due to poor quality seed, low level of input use and poor cultivation practices), and a shortage of cash.

(b) Other Food Crops

Among food crops, three other important groups are roots and tubers, pulses and oil. Roots and tuber production in a Northeast, Southeast and South Asia grew at a rate comparable to that for all developing countries crops, and recent performance in the 1990s is better than the historical trend since 1970. Significant increases in production took place in Islamic Republic of Iran, Lao People's Democratic Republic, Myanmar, Nepal, Pakistan and Thailand. Roots and tuber production has declined in the Republic of Korea since the 1970s, while a similar negative trend emerged in several countries in the 1990s, the most pronounced being the Democratic People's Republic of Korea. Thailand and Indonesia gained significant growth in cassava production due to export possibilities to Europe, although this has slowed in recent years.

Table 10.1 Cereals Production (Including Rice in Milled Form) and Growth by Selected Region/Area

Area/period	Production (million tonnes)	Self-sufficiency ratio (per cent)	Growth rates[5]	
			Period	Growth (per cent per annum)
93 developing countries[1]				
1969/71	480	98	1969/71-1979/81	3.0
1979/81	647	92	1979/81-1989/91	2.9
1989/91	863	92	1989/91-1994/96	2.3
1994/96	969	n.a.	1994/96-2010	2.1
2010	1 314	90		
North- and Southeast Asia[2]				
1969/71	216	98.2	1969/71-1979/81	3.8
1979/81	314	94.5	1979/81-1989/91	3.2
1989/91	429	96.2	1989/91-1994/96	2.0
1994/96	474	n.a.	1994/96-2010	2.0
2010	635	96.7		
South Asia[3]				
1969/71	116	97.3	1969/71-1979/81	2.5
1979/81	147	96.0	1979/81-1989/91	3.2
1989/91	203	102.0	1989/91-1994/96	1.9
1994/96	223	n.a.	1994/96-2010	1.8
2010	292	96.3	1969/71-1979/81	
			1979/81-1989/91	
			1989/91-1994/96	
			1994/96-2010	
18 Asian countries[4]				
1969/71	341	n.a.		3.3
1979/81	473	n.a.		3.2
1989/91	647	n.a.		2.1
1994/96	716	n.a.		1.9
2010	952	n.a.		

Source: FAO at 2010 database and Alexandratos 1995

Notes:
1. 93 Developing countries as defined in Alexandratos (1995: 404).
2. North and Southeast Asia includes Cambodia, People's Republic of China, Indonesia, and Democratic People's Republic of Korea, Republic of Korea, Lao People's Democratic Republic, Malaysia, Myanmar, Philippines, Thailand and Viet Nam.
3. South Asia includes Bangladesh, India, Nepal, Pakistan and Sri Lank.
4. Includes East and South Asian countries plus Islamic Republic of Iran and Afghanistan.
5. According to FAO, annual percentage growth rates for historical periods are computed from all the annual data of the period using the Ordinary Least Squares (OLS) method. Annual growth rates for projection periods are compound growth rates calculated from values for the begin- and end-point of the period.

CHAPTER TEN

As for pulses, an important source of protein in many communities in the region, production is likely to grow at a rate faster than in the recent past but it is likely to remain under two per cent per annum. The Asian countries achieved an average of five per cent annual growth in the production of oil crops, higher than the world average, but individual country performance varied widely both over the longer and recent periods. In Southeast Asia, production of oil palm increased rapidly over the past two decades, especially in Indonesia and Malaysia. According to FAO (Alexandratos 1995), over the next decade production of oil crops will grow by 2.8 per cent per annum in Southeast Asia and 2.2 per cent in South Asia.

2. *Present and Potential Crop Yields*

Contribution to crop production increases came mostly from yield increase. The contribution of yield increase to total crop production increase was much higher in South Asia as compared with Northeast and Southeast Asia (Table 10.3). In South Asia

Figure 10.2 Annual Growth Rates of Production of Major Food Crops by Selected Regions/Area

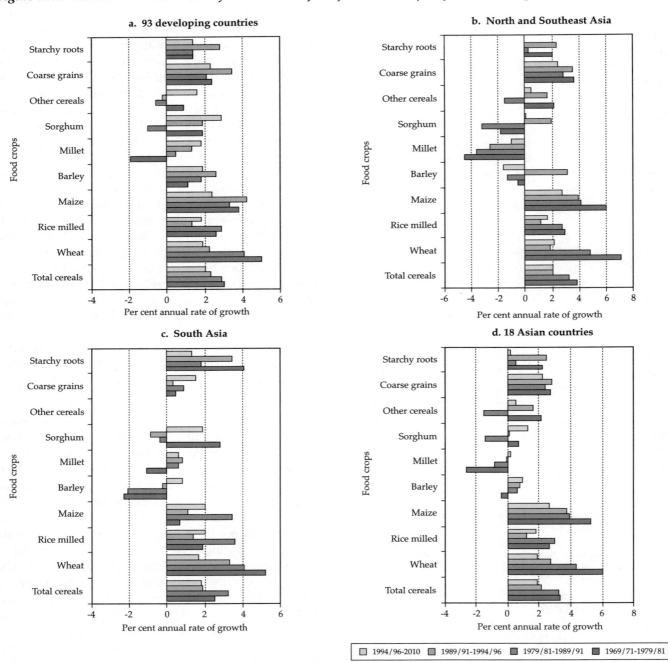

Source: FAO at 2010 database and Alexandratos 1995

increased areas of harvested land accounted for less than a fifth of production increase with the remainder resulting primarily from increases in cropping intensity. The pattern is reversed in Northeast and Southeast Asia and this trend is projected to continue over 1988/90 to 2010 period.

In Central Asia, crop yields declined over the 1982-84 to 1992-94 period except for a modest growth of cereal yields in Kazakhstan and Turkmenistan and roots and tuber yields in Turkmenistan and Uzbekistan. Recent yield trends in the subregion suggest that, along with the larger use of fertilisers and pesticides, improved weather and better outputs were the main contributors to the 19 per cent increase in aggregate grain yields. Nevertheless, average 1997 yield per hectare, although above the 1992-96 average, is still about 4 per cent below that of 1986-90.

The area under irrigation and high yielding varieties of seed has increased across the region and most Asian countries are likely to expand irrigation and the use of modern varieties within the constraints of land suitability and financial resources for investment. According to projections made by FAO, in Northeast and Southeast Asia harvested land under irrigation will increase from 23.2 million hectares in 1989/90 to 27.1 million hectares in 2010, representing respectively 26 and 25 per cent of harvested arable land in use (Alexandratos 1995). The corresponding figures for South Asia are 74.6 and 103.4 million hectares, or 35 and 44 per cent of harvested arable land in use. For Asia as a whole (excluding China), the share of planted area devoted to modern varieties of rice increased from 12 per cent in 1970 to 67 per

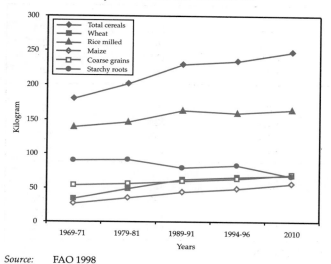

Figure 10.3 Per Capita Production of Major Food Crops in 18 Asian Countries

Source: FAO 1998
Note: Countries as depicted in Table 10.1

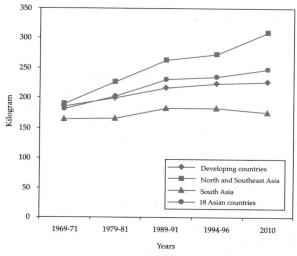

Figure 10.4 Per Capita Production of Total Cereals by Region

Source: FAO 1998
Note: Countries as depicted in Table 10.1

Table 10.2 Recent Trends in Production of Major Cereals in Selected Asian Countries

Area/country	Per cent change between 1996 and 1998			
	Wheat	Coarse grains	Rice (paddy)	Total
Developing countries	1.2%	-0.8%	-1.9%	-0.9%
Asia (12 countries)	4.3%	-4.5%		
Bangladesh	28.6%	0.0%	-5.7%	-4.0%
PR China	-0.5%	-2.9%	-2.0%	-1.9%
India	6.1%	-14.9%	1.1%	0.0%
Indonesia		5.4%	-9.2%	-7.0%
Islamic Republic of Iran	36.4%	2.7%	11.5%	23.8%
Democratic People's Republic of Korea	100.0%	-20.8%	5.0%	-6.7%
Republic of Korea		0.0%	-4.1%	-3.9%
Myanmar	0.0%	25.0%	0.6%	1.1%
Pakistan	10.7%	5.6%	9.2%	-1.2%
Philippines		-9.5%	-8.9%	-7.8%
Thailand		4.3%	-4.0%	-3.0%
Viet Nam		-7.7%	4.0%	3.5%
	Per cent change in production of grains (cereals and pulses) 1997 over 1996			
Armenia	3%			
Azerbaijan	5%			
Kazakhstan	8%			
Kyrgyzstan	19%			
Tajikistan	10%			
Turkmenistan	38%			
Uzbekistan	10%			

Source: FAO 1999 and FAO/WFP 1997

CHAPTER TEN

Table 10.3 Sources of Growth

Region/Area	Contribution to total crop prodcution			
	1970-90		1988/90-2010	
	Harvested land	Yields	harvested land	yields
Developing countries (excl. PR China)[1]	31	69	34	66
North and Southeast Asia[2]	41	59	39	61
South Asia[3]	18	82	18	82
	Contribution to production increase			
	1969/71-1994/96		1994/96-2010	
	area	yields	area	yields
Cereals				
Developing countries (without PR China)	22%	78%	36%	64%
East Asia excl. PR China	31%	69%	43%	57%
South Asia	2%	98%	32%	68%
Total Asia excl. PR China	10%	90%	35%	65%
Wheat				
Developing countries (without PR China)	28%	72%	17%	83%
East Asia excl. PR China	99%	1%	8%	92%
South Asia	37%	63%	12%	88%
Total Asia excl. PR China	33%	67%	6%	94%
Rice				
Developing countries (without PR China)	23%	77%	30%	70%
East Asia excl. PR China	32%	68%	22%	78%
South Asia	10%	90%	15%	85%
Total Asia excl. PR China	20%	80%	16%	84%
Maize				
Developing countries (without PR China)	35%	65%	42%	58%
East Asia excl. PR China	31%	69%	69%	31%
South Asia	25%	75%	15%	85%
Total Asia excl. PR China	27%	73%	54%	46%
Contribution to harvested area increase 1988/90-2010	arable land	cropping intensity		
		38		
	62	18		
	82	78		
	22			

Source: FAO at 2010 database and Alexandratos 1995
Note:
1. 92 Developing countries as in Alexandratos (1995:404) without PR China.
2. In this table the grouping of North and Southeast Asia includes Cambodia, PR China, Indonesia, and Democratic People's Republic of Korea, Republic of Korea, Lao People's Democratic Republic, Malaysia, Myanmar, Philippines, Thailand and Viet Nam.
3. South Asia includes Bangladesh, India, Nepal, Pakistan and Sri Lanka.

cent in 1991. In the case of wheat it increased from 42 per cent to 88 per cent. For maize, the 1990 figure was 45 per cent. High yielding and hybrid rice could expand in India, Pakistan, Bangladesh, and elsewhere, whilst the adoption of semi-dwarf wheat is at its maximum potential in many countries, though further expansion in rainfed areas may be feasible. There might be some movement in adoption of hybrid maize in Asia while the rising trend of hybrid sorghum in South Asia continues. A combination of improved soil management, crop husbandry and cultivars could also raise yield of roots and tubers.

However, raising cropland productivity will be constrained, particularly in Central Asia, where low rainfall and soil erosion will hold back increases in wheat productivity, whilst in South Asia and People's Republic of China the constraints are the slow growth potential for irrigation and soil degradation. Rice yield growth across Asia is likely to be constrained by irrigation, uncertainty of rainfall and natural calamities and lack of high yielding rainfed varieties. However, prospects are good for Bangladesh, Viet Nam and Myanmar and from the new rice variety under development at the International Rice Research Institute in the Philippines (Brown 1998).

FOOD SECURITY

3. Livestock Resources

Livestock production indices show a marked improvement in most of the countries of the region, except in Bhutan and Mongolia. In Bhutan, there is a problem of declining feed/pasture resources and in Mongolia production seems to be linked to a period of downward economic transition, which has witnessed a breakdown of support institutions. Nepal's indifferent performance is again linked to the poor livestock resource base. Among South Pacific countries where data are available, per capita livestock production declined over 1986-96 period in Solomon Islands and Tonga, while it increased slightly in Fiji and Vanuatu. On the other hand, over 1992-97 period, per capita livestock production showed significant decline in all Central Asian countries, except Turkmenistan. Over the 1991-97 period, total meat production declined by 46 per cent and milk production by 33 per cent. The largest reductions in meat production occurred in Armenia (down 40 per cent), Azerbaijan (42 per cent), Kazakhstan (50 per cent), and Tajikistan (47 per cent). The reasons for such drastic reductions include a worsening of the terms of trade for the livestock sector, the lack of competitiveness of the sector due to declining feed conversion rates, the shortage of feed and high transaction costs due to a lack of infrastructure.

Within many of the countries of region, meat and milk production grew at rates well above those of the world (Table 10.4), particularly in People's Republic of China, Indonesia, the Islamic Republic of Iran, the Republic of Korea, Malaysia and Pakistan. According to FAO growth of meat production will slow down over the coming decade up to 2010, although poultry meat is likely to show a rapid rate of growth. Similar projections for milk indicate a significant slow down in growth of consumption and production (Alexandratos 1995).

4. Fishery Resources

Fish is an important source of protein for many countries of the region. Fish production increased by 3.7 per cent annually, from 13 million tonnes in 1970 to 35 million tonnes in 1990 (Table 20.4). Growth in total production has slowed over the years due

Table 10.4 Growth of Production of Meat, Milk and Fish (Per Cent Per Annum) in Selected Countries

Country/Area	Total meat		Milk (cows)		Total fish catch	
	1970-97	1990-97	1970-97	1990-97	1970-80	1980-90
Afghanistan	0.8%	0.2%	-1.9%	0.0%	–	0.0%
Bangladesh	1.6%	5.2%	0.4%	0.5%	0.8%	1.0%
Bhutan	2.1%	2.2%	1.4%	0.0%	–	–
Cambodia	3.0%	5.2%	-0.2%	1.6%	2.8%	6.6%
PR China	7.6%	8.7%	9.1%	6.9%	4.9%	3.6%
India	3.5%	2.2%	5.2%	3.7%	2.9%	1.6%
Indonesia	6.7%	6.6%	8.2%	4.2%	3.4%	1.9%
Islamic Republic of Iran	5.4%	6.6%	4.9%	6.0%	9.1%	7.0%
DPR Korea	2.7%	-2.0%	6.1%	-1.4%	5.0%	0.8%
Republic of Korea	8.7%	7.6%	14.6%	2.4%	5.0%	1.1%
Lao People's Democratic Republic	2.0%	4.8%	2.6%	2.6%	1.3%	0.6%
Malaysia	7.4%	7.1%	2.3%	2.9%	4.1%	1.2%
Myanmar	2.6%	5.5%	5.1%	1.4%	2.0%	0.9%
Nepal	2.1%	2.3%	2.1%	2.8%	7.7%	5.0%
Pakistan	5.6%	7.4%	2.9%	3.7%	3.8%	2.0%
Papua New Guinea	2.4%	1.3%	0.0%	0.0%	2.0%	-1.8%
Philippines	4.3%	6.9%	1.9%	5.2%	2.9%	1.3%
Sri Lanka	2.0%	8.8%	2.4%	1.2%	1.9%	-0.4%
Thailand	4.2%	4.5%	17.3%	13.2%	2.5%	1.6%
Viet Nam	5.2%	5.5%	5.2%	1.8%	1.7%	2.0%
Total 20 countries	6.7%	7.8%	5.1%	4.1%	3.7%	2.2%
Developing countries	5.1%	6.2%	3.6%	3.6%	2.2%	2.0%
Asia and the Pacific	6.7%	7.7%	5.2%	3.9%	3.1%	1.4%

Source: FAO SOFA database

mainly to over fishing, although a few countries show a significant growth in fish catches since the 1970s. Total fish production remained fairly stable in the countries of the South Pacific subregion over the 1986-96 period, with only Samoa experiencing a significant decline in fish catches. The Philippines has witnessed modest growth in fish catches, whilst Sri Lanka experienced a decline in fish catches during the 1990s.

Historically, the development of fisheries has been characterized by several trends, including the depletion of popular stocks, and an increase in the production of freshwater species by aquaculture (e.g. in People's Republic of China). The increased mechanization of fishing fleets has also led to intense conflicts between large and small-scale fisheries, and neighbouring countries have often been entangled in disputes over fisheries jurisdiction.

B. Food Availability Trends

Progress has been good over the past decade in terms of the overall availability of food and nutritional status of the region, although not all countries or population groups have benefited from this trend due to inequities in access, distribution and in the matching of food availability with "food entitlements". The latter is determined by agricultural income, which is dwindling due to

---

Box 10.1 The 1990s Agricultural Crisis in Democratic People's Republic of Korea

The agriculture sector accounts for some 28 per cent of the gross domestic product (GDP) of the Democratic People's Republic of Korea. Land is very scarce, the growing season is short and the climate is harsh, with early frosts, uncertain rainfall in spring and heavy rainfall in July. The total cropland was estimated at slightly over 2 million hectares in 1991-93, an increase of 4.5 per cent over 1981-83, which, on a per capita basis, amounted to only 0.09 hectares, down from 0.1 hectare in 1983. Potential land is estimated at about 5 million hectares and the area currently under irrigation is put at 1.3 million hectares. Although the rate of mechanization is the highest in Asia, 441 tractors per 100 hectares of arable land by 1994-96, many of these tractors were technologically obsolete, incapable of ploughing to proper depths and lack spare parts and fuel. Food (cereals, meat, milk and fish) production increased between 1970 and 1990 but has since declined significantly. Taking the 1970-97 period as a whole, total cereal production (wheat, coarse grains and paddy) declined at annual average rate of 0.6 per cent, although the most significant decline occurred between 1990-97 when production fell by 6.5 per cent. Recent estimate suggests a decline of about seven per cent between 1996 and 1998, from 4.5 million tonnes to 4.2 million tonnes. However, over the 1969/71-1989/91 period, yields of rice and wheat increased annually by over 3 per cent, whilst maize lagged behind with and annual increase of 0.8 per cent.

Policy and institutional weaknesses together with structural problems played a strong role in constraining agricultural development over past decades. The strategy for agricultural development emphasized self-reliance ("juche") in food and feed grain production without due consideration for resource use efficiency. The cooperatives and state farms provided little incentive for farmers to improve efficiency, whilst decision making on production planning, input procurement and distribution and irrigation was centralized through the national Agricultural Commission and its affiliated bodies. With subsidized inputs and a limited role of the market, co-operatives produce some 90 per cent of the country's grain output with surpluses sold to the Government, at a fixed price, with minor transactions taking place through barter arrangements or at local farmers' markets.

Historically, the country's well established trading links with People's Republic of China, the Russian Federation and other socialist block countries provided sources of agricultural inputs, equipment and other assistance. However, the political and economic realignment of many of these trading partners in the 1990s, exposed the Democratic People's Republic of Korea to difficulties in securing the necessary agricultural of inputs and prevented further gains in yield. In response, agricultural production extensified, moving into marginal pasture areas in the hills and, in the absence of mechanical and chemical inputs, employing more labour intensive methods. Despite these measures, food and feed production declined, putting pressure on livestock and poultry production. Efforts to import food to supplement domestic production were frustrated by the Government's declining capacity to import on commercial terms. This situation was aggravated by a series of natural disasters, including hailstorms, high intensity rainfall, typhoon, drought and floods, which adversely affected crop production in 1995, 1996 and 1997. The flooding of large areas of the country left agricultural croplands covered with silt and damaged supporting infrastructure, including irrigation. Flooding and tidal surges led to infiltration of seawater into croplands, whilst specific events, such as Typhoon Winnie in August 1997 and severe rainfall shortages in the critical months of June and July 1999, affected standing crops directly. As consequence, the country suffered from a grave food security crisis in 1997 and 1998. The Public Distribution System (PDS) came under severe stress and delays in the shipment of international emergency food aid threatened an already precarious situation. When food aid did finally arrive, it was the key means by which many lives were saved.

The lessons for the international community regarding the need for adequate and speedy response, irrespective of political considerations, were clear. Similarly, the events of the 1990s were not lost on the Government of the Democratic People's Republic of Korea, whose strategic focus for the avoidance of similar crises is important in enhancing the countries food resources.

Source: 1) IFAD 1997
2) FAO and WFP 1996a, 1997a, 1998b

Table 10.5 Estimated Levels of Undernourishment in Selected Countries

Country/Area	Per cent undernourished			Number undernourished		
	1969/71	1990/92	2010	1969/71	1990/92	2010
Afghanistan	37	73	55	5 079	12 907	18 611
Bangladesh	23	34	21	15 112	39 449	37 096
Cambodia	13	29	36	875	2 469	4 647
PR China	45	16	5	377 264	188 864	75 714
India	36	21	12	199 248	184 473	138 446
Indonesia	34	12	4	41 317	22 133	10 719
Islamic Republic of Iran	32	7	12	9 126	4 219	12 786
DPR Korea	20	9	5	2 905	1 953	1 408
Republic of Korea	2	1	0	750	263	207
Lao People's Democratic Republic	29	24	11	783	1 058	749
Malaysia	14	7	4	1 550	1 266	1 074
Myanmar	34	12	17	9 144	5 215	10 409
Nepal	45	29	27	5 131	5 881	8 311
Pakistan	24	17	7	15 527	20 490	14 232
Philippines	54	21	6	20 138	13 107	5 289
Sri Lanka	21	26	11	2 643	4 605	2 399
Thailand	28	26	3	10 079	14 376	2 096
Viet Nam	24	25	11	10 182	17 232	10 914
East Asia	41	16	6	474 987	267 936	123 226
South Asia	33	22	12	237 661	254 898	200 484
93 Developing countries	35	21	12	916 675	838 709	680 811

Source: FAO at 2010 databases

increases in population, intensification of pockets of poverty, resource depletion/degradation as well as environmental pollution. Additional pressures on the earnings of the poor have occurred through the Asian economic crisis of the late 1990s, which saw several countries confronted with the real spectre of food insecurity. An extreme case in the Democratic People's Republic of Korea where two consecutive years of flooding, drought and typhoon, and the slow response of the international community, combined to create widespread human suffering (Box 10.1). In the countries of Central Asia, food availability contracted significantly in early 1990s. This trend has been reversed in recent years and food aid needs and the number of vulnerable persons needing targeted food assistance have fallen sharply (FAO and WFP 1997).

1. *Per Capita Food Availability*

The growth of per capita calorie supply in the region has slowed down in recent years, except in Fiji, the Maldives, Armenia and Kazakhstan, which each experienced slight increases in daily calorie supply per capita. There are some cases such as Afghanistan, Bangladesh, Cambodia and Democratic People's Republic of Korea where food availability per capita declined between 1970 and 1996, whilst more significant decline was experienced in the 1990s in the Democratic People's Republic of Korea (FAO SOFA database). Although data on average food supply per capita provide a partial view of the food security situation in the region, the average and the distribution of food intake per capita provide indicators of the food security status at the household level. These two parameters were used to estimate the prevalence of under-nutrition in two subregions of the Asian and Pacific Region (Table 10.5) and it is clear that in terms of per capita availability of food, South Asia will remain a difficult region requiring close monitoring and emergency preparedness.

2. *Under Nutrition*

Food insecurity and under-nutrition are linked to agro-ecological zones and their major farming systems (FAO Committee on World Food Security 1993). According to FAO, in dry lands and areas of uncertain rainfall both pastoral systems and upland cereal based system are under stress in Northern China, Indonesia, South Asia, Lao People's Democratic Republic and Mongolia. Within humid

CHAPTER TEN

Table 10.6 Average Annual Net Trade in Food for Selected Countries

Country	Cereals (000 mt)		Oils (mt)		Pulses (mt)	
	1983-85	1993-95	1981-83	1991-93	1981-83	1991-93
Afghanistan	78	190	2 000	3 033	(8 200)	(1 433)
Armenia	n.a.	452	n.a.	n.a.	n.a.	n.a.
Azerbaijan	n.a.	626	n.a.	n.a.	n.a.	n.a.
Bangladesh	1 541	1 412	126 031	298 636	2 470	71 462
Bhutan	n.a.	n.a.	0	249	0	0
Cambodia	94	98	1 800	n.a.	n.a.	n.a.
PR China	10 648	10 887	93 521	1 747 970	(6 378)	(755 453)
Fiji	89	148	(7 839)	2 084	4 178	4 828
India	1 452	(2 690)	1 347 222	241 067	160 845	378 878
Indonesia	2 096	5 406	(344 622)	1 566 942	9 619	53 715
Islamic Republic of Iran	4 457	5 308	280 025	578 378	21 839	22 800
Kazakhstan	n.a.	(4 070)	n.a.	n.a.	n.a.	n.a.
Democratic People's Republic of Korea	131	804	10 717	34 220	n.a.	n.a.
Republic of Korea	6 445	11 907	64 473	315 245	7 729	39 375
Kyrgyzstan	n.a.	426	n.a.	n.a.	n.a.	n.a.
Lao People's Democratic Republic	41	29	2 433	1 523	n.a.	n.a.
Malaysia	2 192	3 612	(2 736 030)	6 010 144	38 163	54 121
Mongolia	19	91	0	1 410	n.a.	n.a.
Myanmar	(839)	(481)	32 667	148 313	(82 833)	(355 000)
Nepal	(3)	61	5 546	27 739	(1 189)	(256)
Pakistan	(718)	1 101	478 900	991 687	78 007	172 212
Papua New Guinea	204	293	(87 012)	(193 198)	65	10
Philippines	1 524	2 560	(944 154)	(919 811)	3 873	27 905
Solomon Islands	13	26	17 043	26 833	18	10
Sri Lanka	810	1 062	(15 561)	28 336	7 438	66 289
Tajikistan	n.a.	536	n.a.	n.a.	n.a.	n.a.
Thailand	(7 767)	(5 420)	42 655	10 256	(211 628)	(102 291)
Turkmenistan	n.a.	588	n.a.	n.a.	n.a.	n.a.
Uzbekistan	n.a.	1 793	n.a.	n.a.	n.a.	n.a.
Viet Nam	288	(1 729)	1 269	13 598	(8 233)	(9 393)

Source: WRI 1997
Note: Positive numbers are net cereal imports, and mt represents metric tonnes.

and peri-humid areas with shifting cultivation, plantations and extensive grazing systems, vulnerable zones are located in Indonesia, Lao People's Democratic Republic, Malaysia, Myanmar, Papua New Guinea, Thailand and Viet Nam. Irrigated and naturally flooded areas, which contain lowland rice based system and irrigated farming system are exposed to threats of waterlogging and salinity compromising food security and degrading land. Many parts of the hill and mountain areas, practising hill farming system and dairy and grazing system are extremely vulnerable and the situation is deteriorating.

FAO (1998) reports a continuation of difficult food supply situation in Afghanistan, Democratic People's Republic of Korea, Indonesia, Lao People's Democratic Republic, Mongolia and Papua New Guinea. In the case of Democratic People's Republic of Korea, 16 per cent of young children currently suffer from wasting, or acute malnutrition, and 60 per cent suffer from long-term malnutrition, placing the country amongst those with the highest malnutrition rates in the world. Interestingly, unlike most other countries, the malnutrition rate is higher among boys than girls (WFP 1998).

Table 10.7 Recent Trends in Cereal Trade in Selected Asian Countries (Million Tonnes)

Country/Area	Wheat				Total cereals			
	Imports		Exports		Imports		Exports	
	1996/97	1998/99 forecast	1996/97	1998/99 forecast	1996/97	1998/99 forecast	1996/97	1998/99 forecast
Developing countries	77.5	73.9	14.6	11.8	149.8	151.1	48.5	50.2
Asia (14 countries)	31.5	24.6	1.5	0.4	60.1	52.5	17.9	20.5
Bangladesh	1.1	2.4			1.1	3.7	0.0	0
PR China	5.2	3.0	0.8	0.3	13.9	11.7	4.0	5.7
India	1.8	0.9	0.6	0.1	2.0	1.1	2.6	2.4
Indonesia	4.2	2.8			6.1	5.5	0	0.3
Islamic Republic of Iran	7.0	3.5			10.1	5.7	0	0
Republic of Korea	3.9	4.4			13.1	11.8	0	0
Malaysia	1.3	1.2			4.3	4.1	0	0
Myanmar					0.0	0	0.1	0.2
Pakistan	3.0	2.4	0.1		3.0	2.4	2.0	2.2
Philippines	2.0	2.0			3.5	3.7	0	0
Singapore	0.3	0.3			0.8	0.7	0	0
Sri Lanka	0.9	1.0			1.2	1.3	0	0
Thailand	0.8	0.7			1.0	0.8	5.4	5.6
Viet Nam					0	0	3.8	4.1
Central Asian countries								
Armenia					0.36	0.34		
Azerbaijan					0.49	0.44		
Kazakhstan					0.02	0		
Kyrgyzstan					0.13	0.12		
Tajikistan					0.26	0.30		
Turkmenistan					0.53	0.56		
Uzbekistan					1.32	0.96		

Source: FAO and WFP 1997c and 1999a

C. Access to Food

1. *Food Distribution and Trade Regimes*

Food distribution channels are mostly market-oriented in Asia and the Pacific, although some countries have government sponsored food distribution schemes including food security reserves, subsidized sale of food and food-for-works programme. India succeeded in raising food grain production ahead of population growth, but in order to improve access of different groups of population to food the Government is involved in grain procurement and distribution, sale of food at less than economic cost and employment promotion schemes for the poor (Rao 1998).

Imports and drawdown from stocks fill national and regional food deficit. Most of the countries of the region are net cereal importers and will remain so in 2010 (Alexandratos 1995). Data on average annual net trade in food by country shows that People's Republic of China, Indonesia, the Islamic Republic of Iran, and Republic of Korea were large importers of cereals from the 1980s through to the early 1990s (Table 10.6). The same is true of Central Asian countries like Armenia, Azerbaijan, Kyrgyzstan, Tajikistan, Turkmenistan and Uzbekistan, whilst Viet Nam became a net exporter for the first time during this period. Myanmar remains a net exporter, whilst Pakistan turned from exporter to importer during the 1980s. People's Republic of China, India and Bangladesh succeeded in achieving near self-sufficiency in cereals, although in the latter two countries this is contingent upon a high level poverty that depresses demand (Alexandratos and Bruinsma 1999).

In recent years, available data have indicated that total cereal imports by 14 Asian countries are projected at 53 million tonnes in 1998/99, down from 60 million tonnes in 1996/97 (Table 10.7). Wheat

CHAPTER TEN

Figure 10.5 Estimated Total Cereal Stock Carryover in Selected Asian Countries

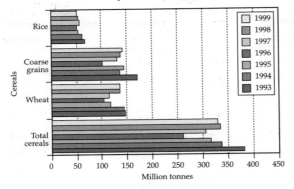

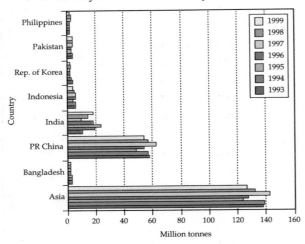

Source: FAO 1999

imports top the list followed by coarse grains and rice respectively. Thailand and Viet Nam will maintain their dominant position in rice exports, while India and Pakistan are expected to improve their position as exporter of fine quality rice. Among Central Asian countries, cereal imports remained the same or were slightly reduced between 1996-97 and 1997-98, which is attributed to better harvests and a reduced demand for bread and feed due to purchasing power limitations. The bulk of the cereal import was in the form of wheat.

2. *Food Stocks*

Stock carry-over (Figure 10.5) is an important factor in stabilising food availability and prices as well as in meeting emergency food requirements. Stock carry-over of cereals in Asia is projected to decline from 137 million tonnes in 1993 to 126 million tonnes in 1999 (Figure 10.5). Due to large population and production base, producers, traders and the government maintain relatively large stocks in People's Republic of China, India, Indonesia, the Republic of Korea, Pakistan and the Philippines. The cost of stock carry-over can be quite substantial, with adequate facilities and financing required to ensure that food stocks are maintained at a minimum required level.

In Central Asia, cereal stocks declined in 1996-97 following lower imports during the previous year, due to high costs in the international market and rising costs of storage as a result of privatization of grain elevators (FAO and WFP 1997). This was combined with higher draw-down of stock in some countries because of lower harvest, as in Uzbekistan and Kazakhstan. Some of the stocks were replenished later through imports. According to FAO, 17-18 per cent of total consumption of stock is an adequate level at the global level (FAO Committee on World Food Security 1999).

Total cereal food aid shipments varied over the years depending on contributions made by producing countries. According to FAO, the 1998/99 forecast of cereal food aid is 9 million tonnes up from 5.8 million tonnes, in 1997/98 (FAO 1999). Much of the food aid is in wheat and coarse grains. Asia received three million tonnes in food aid in 1997/98, it is expected to go up to 3.7 million tonnes when figures are known for 1998/99.

ENVIRONMENTAL IMPACTS OF ENHANCING FOOD AND AGRICULTURAL PRODUCTION

The rising population is placing increasing demands on the supply of food resources, and thereby increasing pressure on the environment (Box 10.2) through the growth of mechanization, expansion of agriculture into marginal or unsuitable lands and intensification of production through chemical inputs (the consequent impacts on land and water resources are discussed in detail in Chapters 1 and 4. The consequent impacts on food security are discussed below).

A. Mechanization

Although increasing mechanization in the region (Table 10.8) has contributed to the rapid increase in food production in many areas, when combined with weak implementation of land and tenurial reforms, it has contributed to increasing marginalization (Box 10.2), landlessness and tenurial insecurity in many countries of the region, particularly in the area affected by green revolution in South Asia. The food security status of small and marginal farmers and landless households in such areas is largely dependent on their ability to rent

Table 10.8 Mechanization of Agriculture in the Selected Asian and Pacific Countries (Number of Tractors)

Country	Per thousand of agricultural workers		Per hundred hectares of arable land	
	1979-81	1994-96	1979-81	1994-96
Armenia	n.a.	64	n.a.	306
Azerbaijan	n.a.	31	n.a.	181
Bangladesh	0	0	5	6
Cambodia	0	0	6	3
PR China	2	1	76	56
India	2	5	24	82
Indonesia	0	1	5	34
Islamic Republic of Iran	17	39	57	133
Kazakhstan	n.a.	51	n.a.	51
Democratic People's Republic of Korea	13	19	275	441
Republic of Korea	1	34	14	563
Kyrgyzstan	n.a.	44	n.a.	238
Lao People's Democratic Republic	0	0	8	11
Malaysia	4	23	77	230
Mongolia	32	22	82	56
Myanmar	1	1	9	10
Nepal	0	0	10	16
Pakistan	5	12	50	144
Papua New Guinea	1	1	699	203
Philippines	112	277	425	923
Sri Lanka	8	9	276	356
Tajikistan	n.a.	37	n.a.	366
Thailand	1	7	11	82
Turkmenistan	n.a.	83	n.a.	347
Uzbekistan	n.a.	59	n.a.	376
Viet Nam	1	4	38	172

Source: IBRD 1999b

land and/or the availability of wage employment giving them the income to buy food. Close to one half of rural population in the region are small-holder populations whose land holding is less than three hectares of cropland per household. In particular, the landless households suffer from food insecurity throughout the year or a part of the year depending on availability of wage employment and food prices.

B. Agricultural Expansion

Impacts of agricultural expansion are determined by the balance of the available cropland not in current use and that which could be brought under new cultivation. The overall scarcity of land is discussed in Chapter 1 of this report. The impact of agricultural expansion on the resource base can be neutral or negative depending on how the land resources are managed.

The impact can be serious if food production expands into areas susceptible to soil erosion such as semi-arid area with cultivation or long-continued grazing, and hill land. Kazakhstan, the largest wheat producer in central Asia, extended cultivation onto marginal land causing serious erosion, which eventually led to contraction of grain areas from 26 million hectares in 1980 to 16 million hectares in 1997 (World Watch Institute 1998). Shifting cultivation with a shortening cycle of fallow has also led to deforestation and loss of soil productivity in certain parts of India, Viet Nam, Malaysia, the Philippines, Thailand, Bangladesh and other countries.

C. Agricultural Intensification

Recent years have seen the continual intensification of agricultural production in the region, through the expansion o irrigation schemes, increased cropping intensities, and increased application of agrochemicals, such as fertilizers and pesticides (Table 10.9). This trend looks set to continue and as discussed in Chapters 1 and 4, will place an increasing burden on the region's land and water resources, with a corresponding risk to future food security.

POLICIES AND PROGRAMMES FOR PROMOTING FOOD SECURITY AND SUSTAINABLE AGRICULTURE

A. National Actions

In most parts of Asia and the Pacific there is some conflict between concerns for household food security and the imperative of resource conservation. Governments have come to recognize that an incentive framework is required to encourage farmers to undertake longer-term investments in land, water and forest conservation. It is not enough to avoid further degradation, land productivity has to be raised to neutralize the current trend which could see population growth outstripping the carrying capacity of the land.

Public and private investments are now encouraged for technology development and adaptation in critical areas of concern such as new crop varieties, which do not use large doses of chemical inputs, new techniques for mountain agriculture, agroforestry and biological control of pests. Policies and programmes have been directed at poverty reduction, improvement of household food security, resource conservation, and reduction of environmental pollution through such measures as

CHAPTER TEN

> ### Box 10.2 Population Growth, Declining Arable Land, and Environmental Degradation in Nepal
>
> From 1970 to 1997, Nepal had one of the highest rates of population growth in the Asian and Pacific Region (2.6 per cent per annum compared with the region's average of 1.9 per cent) and this growing population had a significant influence upon the country's declining levels of arable land per head of agricultural population, (0.14 hectare in 1997 down by 24 per cent from 0.18 hectare in 1970 (FAO SOFA database). Several forces are evident including population pressure in the terai pushing marginal families into areas already heavily deforested and where cultivable land per family is dwindling. The traditional farming system in these areas relies upon a delicate balance between agriculture, livestock and forestry, which includes the cropping of maize, millet and rice on terraced slopes, the use of animal manure for fertilization and the feeding of livestock from communal forest resources. The communal forests also provide the principle source of fuelwood (700 kilogram per capita a year), which requires as much as 3.5 hectares of accessible forest per head – twice as much as that which is currently available. Average cultivated land per household is small (0.85 hectare) and land distribution is such that over 50 per cent of households operate on less than 0.18 hectare. Since these holdings cannot meet family staple food requirements, many families are increasingly relying on livestock raising and waged work to meet family consumption needs. More livestock clearly has adverse implication for the environment since the common forest grazing resources are being depleted at an accelerated pace. If unchecked, it is projected that by 2010 degraded forest, shrub land and wasteland will cover 40 per cent of current forest areas.
>
> Forest over-utilization has been associated with reduced production of biomass, loss of soil cover and increased soil erosion with serious consequences for downstream siltation of rivers and flooding. Much of the flooding problems in Bangladesh originate in the mountains of Nepal. In many areas terraces can no longer be cultivated due to inadequate soil nutrients, whilst increasing fragmentation of holdings, declining land productivity and limited non-farm jobs mean people have to depend increasingly on raising livestock on declining forest fodder stocks. These factors have led to a cycle of environmental degradation, accentuating an already fragile situation of unequal land distribution and insecure tenancy.
>
> The lesson of the past is to seek ways to balance population growth, land ownership patterns and tenancy arrangements with the preservation of the resource base. This is clearly understood by all development institutions working in Nepal. The International Fund for Agricultural Development (IFAD) and other international financial institutions have financed projects drawing upon participatory methodologies for the protection and enhancement of forest resources. These initiatives are building on The Nepalese Government's 1998 Forestry Master Plan, which is promoting community, private and leasehold forestry. The future outcome of this struggle between people and nature will depend heavily on the success of these initiatives.
>
> *Source:* FAO SOFA database, IFAD 1990 and Jazairy, et al 1992

integrated pest management (IPM) and integrated plant nutrition systems (IPNS). Policy makers, governments and NGO's have also joined forces with the private sector to raise awareness at all levels about the importance of maintaining a balance between food security and sustainable agriculture and reducing resource degradation and environmental pollution.

Table 10.9 Selected Indicators of Pesticide Consumption for the Asian and Pacific Region

Item	1983	1988	1993	1998	2003
Consumption (000 tonnes)	885	806	784	819	870
$/kg	6.29	7.48	8.69	10.22	11.94
Consumption (US$ million)	5 571	6 025	6 814	8 370	10 390
Herbicides	1 750	1 970	2 180	2 600	3 150
Insecticides	2 318	2 470	2 790	3 400	4 200
Fungicides	990	1 100	1 260	1 580	2 000
Other pesticides	513	485	584	790	1 040

Source: The Freedonia Group 1999

Under stabilization and structural adjustment programmes adopted by governments of the region in collaboration with IMF, World Bank and the ADB, many countries are moving in a decisive way to eliminate current policy and institutional bias against agriculture and marginal producers. In the past, such policy biases discouraged investment in resource conservation. As a part of the reform process, countries have decentralized decision making and taken steps to promote people's participation which are considered critical to encourage direct involvement of the beneficiaries in protection of the natural resource base and in initiating community based measures to strengthen household food security.

Policies related to incentives and structural adjustment present the greatest challenge to the countries of the region, especially following the Asian economic crisis of the late 1990s. Efforts to encourage private investment in longer term conservation of natural resources have had limited success because of slow start with public investment in rural infrastructure, appropriate pricing policies, adequate availability and access to inputs, equipment, draught animal and services. In policy reform it is understood that allocating use rights of resources to individuals or groups would be an incentive to undertake conservation measures (Box 10.3).

Macroeconomic and sector adjustment policies have also drawn attention to a number of key areas. These include elimination of pricing policies favouring imported food over local crops and capital subsidies that encourage non-sustainable expansion of commercial ranching, logging, fishing, mechanization, and mining. Countries have also initiated reform measures involving gradual reduction of subsidies on agricultural chemicals such as fertiliser, pesticide insecticide, fungicide and others. However, introduction and/or rationalization of water charges have faced resistance. In very few countries in Asia, existing water charges for irrigation cover a fraction of the costs associated with the development, operation and maintenance of water extraction and distribution. In the same way, very little has been achieved in the region by way of reduction of concessions, and imposition of levies in order to contain livestock production and depletion of woodland reserves. Limited progress has been made in the development and implementation of an appropriate combination of taxes/penalty and subsidies to promote sustainable fishery exploitation. Some strategists point to the need for a structural adjustment in the management of natural resources and call for greater priority to be given to rural development and agriculture through a countering of the perceived urban bias in resource allocation.

Following the Third Plenum of the 11th Central Committee of the Communist Party of People's Republic of China, major policy reforms were introduced in Chinese agriculture, a household contract responsibility system was introduced countrywide in the 1980s, investment in agriculture increased and procurement prices increased several times since late 1970s. These factors have led to substantial increase in output and rural income. As for the future, People's Republic of China proposes to hold on to the Household Responsibility System as the cornerstone of its agricultural policy and emphasis will continue to be placed on increased public investment in agriculture, including fertiliser production and research, particularly for improved/high yield breeds.

Box 10.3 Policy Reform and Sustainable Agriculture in Viet Nam

Rice-based agriculture accounts for 26 per cent of gross Viet Nam's domestic product (GDP). However, the agricultural resource base is limited in relation to population as cropland per capita declined from 0.11 hectares in 1983 to 0.09 hectares in 1993. The agricultural population is concentrated in the lowlands, and food security and sustainability of agricultural production is threatened by resource degradation including deforestation, soil erosion, shifting cultivation, and salinity. Over the past two decades, forest cover has decreased from 40 per cent to 25 per cent.

Between 1976 and 1986, the economy of Viet Nam was centrally managed and much of the agricultural sector was collectivized. Since the mid 1980's, the Government's reform programme has resulted in the liberalization of trade, the restructuring of state owned enterprises and the opening up of the manufacturing sector to the private sector. As part of this reform process, farmers have greater autonomy in production decisions and have been invested with long-term use rights to land. Although land is still owned by the state, the use of land was decollectivized in the North of the country in 1989 and was followed by the allocation of use rights to households. Farmers were given tenurial rights for 20 years together with provisions for the selling, leasing and transfer of land through inheritance. These measures encouraged both long-term investment in agriculture and resource conservation as an effective land market evolved across the country.

Rural households met credit needs from the informal sector. The first initiative taken to meet this challenge was the establishment of the Viet Nam Bank for Agriculture (VBA) in 1988. The bank now has 500 branches and 1 500 smaller outlets covering all districts and most subdistricts. The VBA started lending to mass organizations and to the poor and further investment funds were offered by the numerous credit unions established since June 1993. In August 1995, the Viet Nam Bank for the Poor took over poverty lending from the VBA and is supported by the credit made available under the Hunger Eradication and Poverty Reduction Programme.

"Decree 327: Regreening the barren hills" is a major Government initiative to restore deforested hills to sustainable and productive use. People are contracted to protect land and forest and to undertake reforestation. The initiative also encourages the participation of the ethnic minorities of the hill areas, who will be encouraged to reduce shifting cultivation through the provision of improved living conditions. With 9.3 million hectares of forests protected and 2.5 million hectares replanted, the total forest cover is projected to increase to 40 per cent.

Viet Nam stands out as an inspiring example of what can be achieved through a managed transition from central planning to an open and liberal policy regime. The reform measures combined with the rehabilitation of irrigation networks are contributing to the steady growth in agricultural production. Rice production increased by 4.6 per cent per annum over the 1990-1997 period and, from a net importer of rice Viet Nam has transformed itself into a major rice exporter.

Source: 1. IFAD 1993, 1995b and 1996a
          2. FAO 1998b

CHAPTER TEN

Other policy initiatives that would be favourable to the growth of agricultural and food production in People's Republic of China include reform of the grain marketing system, elimination of two-tier pricing, introduction of water pricing (Box 10.4) and deregulation of inter-provincial trade. In 1998, a special fund of US$4.21 billion was used for agriculture. Rural markets will be opened further and foreign investment encouraged to upgrading seeds, technology, agricultural skill and management.

In Thailand, under a recently approved agriculture sector adjustment loan from the ADB, the government has agreed to initiate a wide range of policy and institutional measures. These are designed to increase public and private investment in agriculture, improve production, market efficiency, as well as to streamline agricultural service institutions to address problems of deforestation, land degradation, shortage of water, inefficient use of water and pesticides, and water pollution. Land, labour, capital and natural resources would be appropriately valued allowing production, consumption, saving and investment decisions to be made on the basis of a common appreciation of the environmental and economic trade-offs involved. Improvement of food security and conservation of resources were facilitated by the administrative bodies set at the village and local levels to plan, supervise, monitor and evaluate rural development programmes.

Lao People's Democratic Republic is continuing the process of transition from a centrally planned economy to a more decentralized market oriented economy and the government is taking measures to stabilize the shifting cultivation in hill areas susceptible to soil erosion and degradation. Irrigation expansion together with application of modern inputs is designed to increase cropping intensity, which will reduce pressure on fragile areas.

The Viet Nam Bank for Agriculture (VBA) established in 1996 provides credit to farmers and has made rural development and poverty alleviation a priority policy objective. A larger share of government budget is now allocated to agriculture including irrigation and flood control. Other objectives include increased emphasis on commercial agriculture, diversification, infrastructure (roads and irrigation) and provision of efficient extension, veterinary, and research services (IFAD 1996b).

In South Asia, most countries have taken major initiatives to tackle poverty, malnutrition, food insecurity and environmental degradation associated with food production. For example, Sri Lanka settled farming families on state owned land within the framework of District Integrated Rural Development Projects, under implementation since 1979. In dry regions these projects have given priority to integrated watershed management, including soil and water conservation as in Kirindi Oya, Anuradhapura, Badulla, Kegalle and other regions. Bangladesh, on the other hand, has used a 'food for work' programme to undertake rural investments in infrastructure including flood and salinity control measures which, combined with investments in small scale irrigation, promoted land conservation and improved land productivity. Pressure on land has also been reduced through investments in non-farm income generating activities.

Pakistan has suffered from rapid population growth and increasing poverty-linked degradation of soil in arid and semi arid regions, and waterlogging and salinity. In response, the government has

---

Box 10.4 Agricultural Water Pricing Policies as Market-based Instruments in People's Republic of China

For centuries irrigation has played a crucial role in Chinese agriculture. Since 1949, the area of irrigated land has tripled such that 75 per cent of national food production is provided from irrigated land – a major factor in China's food self-sufficiency.

In July 1985, People's Republic of China took an important first step toward promoting greater efficiency in irrigation water usage. The Chinese government instituted agricultural policy reforms, which invested a greater degree of financial and managerial autonomy in provincial water management agencies, and introduced water charges to cover their operation, maintenance and amortization of capital costs. In general, the pricing structure is differentiated so as to reflect actual costs of water in different uses. For example, charges may vary according to season, and in very dry areas progressive water pricing schemes have been adopted to reflect scarcity. Likewise, irrigation for grain crops is priced according to supply costs without profit, while cash crops may be irrigated for slightly higher cost. The reforms also serve to decentralize authority, making water management authorities more closely tied both to operation and distribution of irrigation water. Management is often further decentralized when a local agency purchases water wholesale and sells in bulk to smaller water user associations responsible for distribution to farmers. These smaller groups strengthen the bond between the water and the supplier who must recover costs.

As a result of these policy reforms, revenues collected by the water management agencies have increased significantly. Farmers have begun to irrigate their crops more efficiently while water use per hectare has declined and crop production has increased.

Source: FAO 1993

introduced initiatives directed at improving the productivity and growth of the agriculture sector with emphasis on rain-fed barani areas, strengthening rural infrastructure and encouraging cottage industries in rural areas.

A major movement in the Central Asian countries was the emergence of smallholder/household production in the 1990s accounting for a growing proportion of food production in the region (FAO and WFP 1997). Livestock productivity has always been higher on private plots than on large state owned and collective farms. In order to facilitate production in a situation where institutional credit is lacking, government and/or the private sector extend commercial credit, for example, as in Turkmenistan and Uzbekistan. Another example of the developments in smaller holder/household production can be found in Turkey where urban agriculture is growing more and more evident in Istanbul (Box 10.5).

B. International and Regional Actions

The right to food is a fundamental human right recognized by, amongst others, the Charter of the United Nations; the Universal Declaration of Human Rights; the International Covenant on Economic; Social and Cultural Rights, 1966; the Universal Declaration on the Eradication of Hunger and Malnutrition, 1974; the Declaration of Principles and Programme of Action of the World Conference on Agrarian Reform and Rural Development, 1979; the Vienna Declaration and Programme of Action of the World on Human Rights, 1993; the Copenhagen Declaration and Programme of Action of the World Summit for Social Development, 1995; and, the Rome Declaration on World Food Security and the World Food Summit Plan of Action, 1996.

1. World Food Summit

Of the international initiatives undertaken to date, the landmark action with regard to world food security was the adoption by the international community of the Rome Declaration on World Food Security and World Food Summit Plan of Action at the World Food Summit in Rome on 13-17 November 1996. The Summit participants (Heads of State and Government or their Representatives) agreed to "ensure an enabling political, social, and economic

---

Box 10.5 Food Security in Cities: Urban Agriculture in Istanbul

In the past four decades, explosive population growth has begun to significantly change Istanbul's cultural fabric. Outlying villages regularly become incorporated into the metropolitan system; at the same time, migrants establish communities within the metropolitan area, bringing with them characteristics of their migratory origins. Throughout Istanbul, informal systems have developed where publicly provided infrastructure is inadequate. For example, informal housing (the gecekondu) and transportation (the dolmus) systems throughout Istanbul are well-known and integral parts of the city's fabric. In a similar spirit, urban agriculture can be thought of as an informal and practical response to inadequate food systems and opportunities.

Urban agriculture is widespread throughout Istanbul, is practised by people from a wide variety of socio-demographic backgrounds, and is practised by poor with the primary motives of basic subsistence, dietary supplement or supplemental income and fungibility (freeing up scarce cash income). Urban agriculture in Istanbul appears to be every bit as cosmopolitan as the city itself, reflecting a wide variety of opportunities, resources and skills. Economically, it ranges from household gardens to commercial greenhouses, from harvesting for household consumption to harvesting for sale. The nexus of two very generalized natural resources (i.e. land and water) and a broadly defined set of social resources (ranging from knowledge, labour and social welfare programmes, to the availability and accessibility of imported or manufactured inputs and capital) determine the viability of urban agriculture.

The availability of a reliable supply of water for crops and livestock is a major constraint faced by urban agriculturists in Istanbul. One question for future research on urban agriculture in Istanbul is to determine the role that water constraints play on such things as crop and site selection and how the people have adopted to these constraints. Our particular adaptation that becomes evident to any visitor who veers even barely off the tourist path is urban agriculture – growing crops, raising livestock, and otherwise harvesting edible produce all over urban space. Throughout Istanbul, many local residents have taken elements of food production into their own hands rather than relying solely on their ability to exchange labour and wages for food.

Istanbul is part of a country which has traditionally been a net food exporter. However, as food exports have increased, so have food imports. As such, locally available foods become relatively more expensive as prices adjust to those offered by international markets and as imported processed foods replace locally grown fresh foods. These events appear to contribute to conditions analogous to what Amartya Sen (1981) might call a boom famine. It is not that food becomes scarce, but that the ability of local populations to command access to food is limited. That is, in the midst of plenty, it is possible for large populations to be hungry, sometimes even to starve. Under such economic conditions, it is reasonable to expect to find creative local strategies of urban agriculture for ensuring household food security, uniquely adapted to the urban resources available in cities such as Istanbul.

Sources: Kadjian, P. 1997 and Tuscon. Sen, A 1981

environment designed to create the best conditions for the eradication of poverty and for durable peace, based on full and equal participation of women and men, which is most conducive to achieving sustainable food security for all" (World Food Summit 1996). The Plan of Action envisages a reduction of the number of undernourished to half the present level by 2015. The Summit emphasized that production increases need to be achieved ensuring sustainable management of natural resources and protection of the environment.

The 1996 World Food Summit also called for improvements to the definition and implementation of the rights related to food as set out in Article 11 of the International Covenant on Economic, Social and Cultural Rights. This task was entrusted with the United Nations High Commissioner for Human Rights, which is actively undertaking consultation with other United Nations bodies, governments, NGO's and experts to develop appropriate recommendations (FAO 1998).

2. *Food & Agriculture Organization (FAO)*

In the spirit of its mandate, FAO has played a vital role in promoting food security and sustainable agriculture in order to raise levels of nutrition and standards of living, to improve agricultural productivity, and to better the condition of rural populations (FAO 1996). FAO has been providing development assistance, agricultural information and support services, advice to governments and a neutral forum for international cooperation. In recent years FAO has reiterated its priority for improvement of food security, management of food emergencies, operation of its Global Information and Early Warning System (GIES) and promotion of sustainable development. A specific priority of the Organization is encouraging sustainable agriculture and rural development, a long-term strategy for the conservation and management of natural resources.

In light of pesticide linked health hazards, Integrated Pest Management (IPM) has been established as the basis of FAO plant protection activities. The approach combines a variety of controls, including the conservation of existing natural enemies, crop rotation, intercropping, and the use of pest resistant varieties. Pesticides may still continue to be used selectively but in much smaller quantities. Five years after IPM was widely introduced in Indonesia, rice yields increased by 13 per cent, while pesticide use dropped by 60 per cent; in the first two years alone the government saved US$120 million that it would spent subsidising the chemicals. (FAO 1995).

For fisheries, the Organization is implementing the Code of Conduct for Responsible Fisheries, supporting aquaculture development, and undertaking monitoring and strategic analysis. FAO has also initiated a special programme of assistance in fisheries to Small Island Developing States.

3. *International Fund for Agricultural Development (IFAD)*

IFAD is a funding agency of the United Nations, which allocates all its resources towards investments in small-holder agriculture and off-farm income generation for the poor (including landless and women) in rural areas. All such investments have a positive impact in terms of improving household food security and protecting the environment. Over 1988-98 period the Fund has committed over US$1 billion to 21 countries of Asia and the Pacific (IFAD 1998). In 1998 the Fund field tested and finalized a comprehensive set of "memory checklists" for issues relating to household food security and gender, also to help with project design. IFAD has introduced an innovative lending instrument, "Flexible Lending Mechanism" (FLM), allowing longer project implementation periods (10-12 years as opposed to traditional five-to-six years) with rolling cycles of design which should be particularly helpful for long-term investment in resource conservation. The Fund has also been selected to house the Global Mechanism of the Convention to Combat Desertification (CCD) which is discussed in greater detail in Chapter 1.

IFAD is continuing with its effort in the region to forge partnerships among governments, NGO's, civil society organizations and other stakeholders with a view to promoting participation and empowerment in decision-making at all stages of project implementation. IFAD is formulating a special programme for Asia with emphasis on community-based rural infrastructure/works programme, micro-credit, environmental regeneration and capacity building. Working in close collaboration with the World Food Programme (WFP), the Fund supports rural public works in order to enhance the productivity of land and water resources.

The Fund has directly intervened through project financing to promote sustainable agriculture in countries in the Asian and Pacific Region, including in Bangladesh, where the Fund has financed an aquaculture development project under which water bodies have been leased to fisherman groups who would now have the incentive to practice sustainable fishing. Similarly, IFAD-financed projects have promoted sustainable farming practices through environmental protection and resource conservation measures and adoption of appropriate technology and cultivation practices in People's Republic of China, Indonesia, Lao People's Democratic Republic,

Pakistan, the Philippines, Thailand and Viet Nam. In the economies in transition in Central Asia, IFAD has taken a number of initiatives to build institutions to promote food security and sustainable agriculture. Two projects in Kyrgyzstan aim to support sheep development and strengthen agriculture support services. In Armenia, food security is being approached through a combination of input supply, financial services, on-farm irrigation, community development and advancement of market-orientation.

4. *Other International Initiatives*

The World Food Programme (WFP) has been involved in 80 countries to fight both hunger emergencies as well as chronic hunger, focussing on the most vulnerable: women, children and the elderly. During the past three decades, WFP has invested about US$7.1 billion and over 12 million tonnes of food in Southeast Asia to combat hunger, promote economic and social development, and provide relief assistance in emergencies. Over 1995-98 period WFP delivered 10.5 million tonnes of food to 223 million people at a total cost of 4.95 billion. According to WFP, food aid is used as a vital catalyst to promote self-reliant development among the poorest of the poor. WFP resources are channelled through three programmes. Under food-for-life, emergency food deliveries are made in a fast and efficient manner to save and sustain life. Food-for-growth projects target needy people (babies, school children, pregnant and breast-feeding women and the elderly) at the most critical times of their lives.

Over 1984-88 period, 29 per cent of the total lending of the ADB (US$29 billion) went to agriculture. For 1989-98, the allocation was US$7.8 billion out of a total of US$52.6 billion, or 14.8 per cent (FAO Committee on World Food Security 1999). ADB-financed projects have focussed on sustainable agriculture through new technology, institutions and investments that protect the environment. Of the 137 projects currently under implementation, 24 projects deal with agriculture and natural resource management and have a strong emphasis on resource use efficiency, environmental protection and resource conservation.

Since 1992 UNCED, UNDP has been assisting countries in developing integrated approaches to managing natural resources to improve livelihood of poor people giving priority to preventive approaches. The primary concern is to ensure that longer-term sustainability is not undermined due to attention to short term requirements. Given that agriculture in the Republic of Korea is one of the most chemical intensive in Asia, UNDP promoted integrated pest management (IPM) techniques through training at all levels including 4 000 farmers, which had significant impact in reducing pesticide use. The Republic of Korea apparently became the first Asian country to adopt a pesticide and fertiliser reduction policy.

Joint action to strengthen regional food security arrangements like those for the Association of Southeast Asian Nations (ASEAN) and South Asian Association for Regional Cooperation (SAARC) has also taken place. The ASEAN Emergency Rice Reserve is perhaps the most advanced scheme, started in 1979.

International research efforts continue to emphasize activities such as the use of traditional crops, agroforestry, balanced use of external and internal inputs in agriculture to preserve the environment, IPM, Integrated Plant Nutrition System (IPNS), agrometeorology, irrigation management and water harvesting, weed control, livestock systems, erosion control, and improved seeds and planting materials. Donors support the research institutes under the Consultative Group on International Agricultural Research (CGIAR) and regional research institutions. In the case of Asia, the International Rice Research Institute (IRRI) and International Centre for Maize and Wheat Improvement (CIMMYT) have made important contributions to improved agricultural technologies. IRRI is working on new technologies for flood-prone rice lands in South and Southeast Asia. The International Centre for Agricultural Research in the Dry Areas (ICARDA) is implementing a programme for the development of integrated feed and livestock production and management technologies in Central Asian countries. The International Crops Research Institute for the Semi-Arid Tropics (ICRISAT) is undertaking research on IPM for pulse-pests. Issues related to increasing and sustaining the productivity of fish and rice in the flood-prone ecosystems in South and Southeast Asia are being addressed by the International Centre for Living Aquatic Resources Management (ICLARM).

International financial institutions have recognized the need to strengthen food security and promote sustainable agriculture. However this understanding is still to be translated into action. IFAD remains the only international funding body to devote 100 per cent of its resources to agriculture and resource conservation. The trends for the World Bank and the ADB funding in this area appear to be declining (Table 10.10).

CHAPTER TEN

Table 10.10 Lending for Agriculture by the World Bank, ADB and IFAD (Current US$ Billion)

Year	Average 1984-88	1989	1990	1991	1992	1993	1994	1995	1996	1997	1998
World Bank/IDA	3.9	3.5	3.7	3.7	3.9	3.3	3.9	2.8	2.6	3.5	2.7
Per cent	23	16	18	16	18	14	19	12	12	19	10
ADB	0.7	0.8	1.2	1.0	0.8	0.4	0.5	0.9	0.8	1.0	0.4
Per cent	29	23	31	21	15	7	13	16	14	11	7
IFAD	0.2	0.3	0.3	0.3	0.3	0.3	0.3	0.4	0.4	0.4	0.4
Per cent	100	100	100	100	100	100	100	100	100	100	100

Source: FAO Committee on World Food Security 1999c

CONCLUSIONS

Although significant progress has been made to improve food security and promote sustainable agriculture within the region, there is evidence of continuing food insecurity and malnutrition, with some countries continuing to suffer cyclical famine or near famine. Available data suggest that the search for food security for an ever-increasing population has depleted the resource base and degraded the environment.

Per capita food production has increased over the past three decades and the prospect is that it will continue to grow over next two decades and perhaps beyond. Increases in yield and cultivation intensity will continue to account for much of the production growth, since per capita availability of land has been on the decline. In absolute terms, many countries are reaching their land potential while in others the cost of bringing new land under cultivation is increasing. Land degradation through desertification, salinity and alkalinity and waterlogging is taking arable land out of cultivation. Shortfall in production to meet minimum dietary requirements has been filled by net trade and food aid. The Green Revolution induced the use of fertilisers and pesticides, which have contributed to environmental pollution in many areas.

Community and national food security arrangements are fragile while regional and subregional food security arrangements are yet to take firm footing. Food aid and emergency food shipments through the World Food Programme (WFP) and bilateral arrangements remain the most potent instrument to avert open food crisis and famine. Pockets of vulnerability exists in most of the countries of the region and the situation is likely to remain so over the foreseeable future as policy initiatives to improve agricultural potential, preserve the resource base and improve access to food for all segments of population continue to show mixed results

FAO makes a number of important observations on the current and prospective world food situation (Alexandratos 1995). Population will grow in the future but at a slower rate than in the past and, although world agricultural growth will slow, progress will continue to be made in improving the availability of food and nutrition in Asia. Global and regional food trade will continue to play important role in stabilising the nutritional status of populations. Many countries will become net importers of food. In many Asian countries agricultural resources have declined to low levels and the trend will continue. A consensus seems to be emerging that food security has to be addressed through poverty eradication and the sustainable use of agricultural resources combined with well functioning markets and reliable food security arrangements at various levels.

Actions need to be taken to identify new sources of food and diversify food consumption habits. Wheat, rice and maize will continue to supply the bulk of energy, protein and vitamin requirements and the production of these crops will need to be strengthened through new varieties, intensive fertilization, disease control and irrigation. Greater support needs to be given to production, processing and marketing of sorghum, millet, barley, rye, and oats, root crops such as cassava, potato, sweet potato, yams, taro, manioc and oilseeds. Pulses and oilseeds are important sources of protein in areas short of meat and fish. Protein present in dry matter of leaves, grasses and waterweeds are potentially of high nutritional value. The sea and water bodies are also important sources of food and protein with the potential of converting trash and fatty fish into fish protein concentrate (FPC).

The key to realising these potentials is sustainable and cost-effective expansion and management of land and irrigation, crop protection measures, mechanization, intensification, and production of livestock, poultry and fisheries. For households without possibility of adequate production, access to food has to be assured by employment and income and market supply. Future prospects for sustainable food security in Asia and the Pacific will depend on how critical policy and institutional questions are addressed. These relate to investments in agriculture and resource conservation, fiscal, commercial, exchange rate and monetary policies, world trade and aid, debt management, land, water and forestry reform, research and extension, markets, client participation, forewarning systems and security reserves.

Food security also recognizes that the demand for food increases over time (in line with population growth) and that an increasing population will put increasing pressure on land, forests and freshwater supply. Therefore, the concept of food security must embody the concept of sustainable agriculture since unsustainable practices will undermine food security at all levels. The challenge for the region, and the world, is how to achieve and to maintain sustainable food security without undermining resources of food production.

PART THREE

CHAPTER 11-15

PART I

CHAPTER 1 LAND
CHAPTER 2 FORESTS
CHAPTER 3 BIODIVERSITY
CHAPTER 4 INLAND WATER
CHAPTER 5 COASTAL AND MARINE
CHAPTER 6 ATMOSPHERE AND CLIMATE

PART II

CHAPTER 7 URBAN ENVIRONMENT
CHAPTER 8 WASTE
CHAPTER 9 POVERTY AND ENVIRONMENT
CHAPTER 10 FOOD SECURITY

PART III NATIONAL AND REGIONAL RESPONSES

CHAPTER 11 INSTITUTIONS AND LEGISLATION
CHAPTER 12 MECHANISMS AND METHODS
CHAPTER 13 PRIVATE SECTOR
CHAPTER 14 MAJOR GROUPS
CHAPTER 15 EDUCATION, INFORMATION AND AWARENESS

PART IV

CHAPTER 16 SOUTH ASIA
CHAPTER 17 SOUTHEAST ASIA
CHAPTER 18 SOUTH PACIFIC
CHAPTER 19 NORTHEAST ASIA
CHAPTER 20 CENTRAL ASIA

PART V

CHAPTER 21 GLOBAL AND REGIONAL ISSUES
CHAPTER 22 ASIA AND THE PACIFIC INTO THE 21st CENTURY

Chapter Eleven

Striving to improve environmental governance: Ministerial Conference on Environment and Development in Asia and the Pacific 2000, Kitakyushu, Japan.

11

Institutions and Legislation

INTRODUCTION
INSTITUTIONS FOR SUSTAINABLE DEVELOPMENT:
STATUS AND TRENDS
NATIONAL INSTITUTIONS
SUB-NATIONAL AND LOCAL INSTITUTIONS
INSTITUTIONAL COORDINATION
ENVIRONMENTAL LEGISLATION: STATUS AND TRENDS
CONCLUSION

CHAPTER ELEVEN

INTRODUCTION

Effective administrative bodies and enabling legal and regulatory frameworks form the corner stones of good governance for sustainable development and enable governments, at all levels, to initiate and enforce critical environmental protection measures. It is now widely recognized that both economic development and the protection of environment are critical for the attainment of sustainable development. In recent years, the focus has increasingly centred on institutional mechanisms designed to promote a balance between environment and development. In addition, environment agencies are also equipped with adequate legal and regulatory powers to conserve natural resources and check environmental degradation. This chapter discusses the institutional framework for environmental protection, and describes the status of environmental legislation that together provides the basic infrastructure for governing sustainable development in the Asian and Pacific Region.

INSTITUTIONS FOR SUSTAINABLE DEVELOPMENT: STATUS AND TRENDS

The Rio Declaration of 1992 and Agenda 21 called for governments to establish effective legal and regulatory frameworks to enhance national capacities to respond to the challenges of sustainable development and underscored the critical importance of fundamentally reshaping the decision making processes to integrate environmental and development concerns. There has been, consequently, widespread recognition of the fact that the social, economic and environmental dimensions of development must be treated in an integrated and balanced way. Individual sectors are no longer to be dealt with in isolation, whilst the efforts of governments and multilateral institutions require multidisciplinary approaches among specialists and inter-sectoral coordination at local, national, regional and global levels.

An encouraging trend, in recent years, has been a shift from highly centralized and compartmentalized bureaucratic structures to decentralized and participatary governance in most developing countries of the region. Promotion of community participation, decentralization of administration, integration of national and local decision making processes, and involvement of NGOs and the private sector in policy making are some of the key changes taking place across the region. Increasingly, Environment Ministries and agencies have been restructured and empowered with greater institutional strength to promote better vertical and horizontal coordination amongst different agencies, to balance conflicting interests and to facilitate integration of environmental and developmental concerns. New legislation, including enactment of comprehensive umbrella laws and the strengthening of existing laws, has enabled judicial institutions to oversee effective enforcement of environmental measures.

NATIONAL INSTITUTIONS

A. Environmental Bodies

National level institutions have a crucial role in formulating and crystallising the strategies, policies and programmes of governments. With respect to sustainable development, their role is ever more critical because it involves translating a fresh and uncharted approach to development planning and policy and the reorientation of existing institutional paradigms and relationships in order to achieve integration of environmental concerns into development processes.

Most countries in the region now have agencies entrusted with the tasks of environmental management. No uniform pattern or model has emerged, as these institutions have tended to be in line with the administrative traditions of the respective countries. Some countries have created a separate ministry for environmental management, while in others environment cells or divisions have been created within an existing ministry. In addition, many countries have also formed a specialized agency under the executive branch of the government. In the design of the institutional structures and the functional domains of the environmental agencies, the institutional policy aspects and the executive policy aspects can be distinguished. Table 11.1 provides a summary of the governmental environmental protection institutions in Asia and the Pacific.

The role of environmental ministries includes coordination with other ministries, implementation of government programmes and projects, monitoring compliance with regulations and providing secretariat services to higher bodies and, thereby, acting as a link between the ministry and the overall policy making body. In federal states, the ministries also coordinate state and provincial activities on the environment. The ministries in their advisory role render technical and political advice and serve as the policy making unit for the government on environmental matters. The experiences of the countries in the region reveal a gradual, yet steady enhancement of capacity in the institutional structures due to increasing importance being accorded to

INSTITUTIONS AND LEGISLATION

Table 11.1 Forms of Governmental Environment Protection Institutions in Asia and the Pacific

Country/Territory/Area	Vision Document	Policy Institution	Executing Agency	Apex National Council
NORTHEAST ASIA				
PR China	China's Agenda 21	State Environmental Protection Administration	State Environmental Protection Administration	
Democratic People's Republic of Korea		Committee of Environmental Protection of Administrative Council	General Bureau of Environment Protection and Land Administration	
Japan	Basic Environment Plan	Ministry of the Environment	Various Departments/bureaus	Central Environment Council
Mongolia	Mongolian Agenda 21	Ministry of Nature and Environment	State Environmental Inspection Department	
Republic of Korea	Green Vision 21	Ministry of Environment (MOE)	Various Commissions/bureaus under MOE	Environmental Conservation Committee
SOUTHEAST ASIA				
Brunei Darussalam		Ministry of Development	Ministry of Environment	
Cambodia		Ministry of Environment		
Indonesia	Indonesia Agenda 21	Ministry of State for Environment	Environmental Impact Management Agency	National Council for Sustainable Development
Lao People's Democratic Republic		Office of the Prime Minister	Office of the Prime Minister	Science, Technology and Environment Organization
Malaysia	Vision 2020	Ministry of Science, Technology and Environment	Department of Environment	Environmental Quality Council
Myanmar	Myanmar Agenda 21		National Commission for Environmental Affairs	
Philippines	Philippine Strategy for Sustainable Development and Philippine Agenda 21	Department of Environment and Natural Resources	Environment Management Bureau	Philippine Council for Sustainable Development
Singapore	Singapore Green Plan	Ministry of Environment (MOE)	Various Divisions under MOE	
Thailand	National Plan	Ministry of Science, Technology and Environment (MOSTE)	Various Departments under MOSTE	National Environmental Board
Viet Nam	• Environmental Vision 2020 • National Strategy for Environmental Protection 2001-2010 • National Environmental Action Plan 2001-2005	Ministry of Science, Technology and Environment (MOSTE)	National Environment Agency	
SOUTH ASIA				
Afghanistan		Ministry of Planning	Various Departments	
Bangladesh	Environmental Conservation Act 1995 and National Conservation Strategy	Ministry of Environment and Forests	Department of Environment	National Environmental Committee
Bhutan	National Environmental Strategy for Bhutan "The middle path" 1998	National Environment Commission Secretariat	Various agencies in the Royal Government of Bhutan	National Environment Commission
India	National Conservation Strategy and National Policy on Pollution Abatement	Ministry of Environment and Forests	Pollution Control Boards and State Departments	
Islamic Rep. of Iran	National Environment Action Plan (NEAP)	Department of Environment	National Committee on Sustainable Development	
Maldives	NEAP 1999-2005	Ministry of Environment		National Environment Council

241

Table 11.1 (continued)

Country/Territory/Area	Vision Document	Policy Institution	Executing Agency	Apex National Council
Nepal	National Environment Policy and Action Plan I & II	Ministry of Population and Environment	Ministry of Population and Environment	Environmental Protection Council
Pakistan	National Conservation Strategy	Ministry of Environment, Local Government and Rural Development	Federal/Provincial Environmental Protection Agencies	Pakistan Environment Protection Council
Sri Lanka	National Environment Action Plan	Ministry of Forestry and Environment	Central Environment Authority	Central Environment Authority
CENTRAL ASIA				
Azerbaijan	NEAP 1998	State committee on Ecology and Utilization of Natural Resources	Ministry of Ecology and Bio-resources	
Kazakhstan	NEAP 1998	Centre for National Environmental Activity Plan for Sustainable Development	Coordinating Working Group	
Kyrgyzstan	NEAP 1996	State Committee on Nature Conservation		
Tajikistan		State Conservation on Nature Conservation		
Turkmenistan	State Programmes of Regional Improvement	Ministry of Nature Protection of Turkmenistan	Ministry of Nature Protection of Turkmenistan	
Uzbekistan	National Environmental Policy	State Committee on Nature Conservation		
SELECTED SOUTH PACIFIC COUNTRIES/TERRITORIES				
Australia	National Strategy for Ecologically Sustainable Development	Department of Environment, Sport and Territories	Federal Environment Protection Agency	
Cook Islands	National Environmental Management Strategy 1993	Interior Affairs	Conservation Service	
Fiji	Sustainable Development Bill	Ministry of Housing and Urban Affairs	Department of Town and Country Planning	
Kiribati	Environment Act, 2000	Ministry of Environment and Natural Resources	Environmental Task Force	
Marshall Islands	National Environment Management Strategy, 1993	Environmental Protection Authority	Environment Task Force	
Micronesia (Federated States of)	National Environmental Management Strategy	Board of Environment and Sustainable Development	Board of Environment and Sustainable Development	
New Zealand	Several Strategies	Ministry of Environment	Various Divisions	
Niue	National Environmental Management Strategy	Conservation Council		
Papua New Guinea		Department of Environment and Conservation	Division of Environment	
Samoa	National Environment and Development Strategy, 1993	Department of Lands, Surveys and Environment	Division of Environment and Conservation	
Solomon Islands	National Environment Management Strategy, 1993	Land, Energy and Natural Resources	Environment Conservation Division	
Tonga	National Environment Management Strategy, 1993	Ministry of Land, Survey and Natural Resources	Environmental Planning Section	
Tuvalu	National Environmental Management Strategy	Office of the Prime Minister	Various Ministries	
Vanuatu	National Conservation Strategy	Ministry of Home Affairs	Department of Physical Planning and Environment	

environmental concerns in governments' overall development policies. However, the environment ministries generally suffer from inadequate financial and human resources, thereby lowering the quality of advisory services and adversely affecting monitoring of compliance. Moreover, in cases where the environmental unit is placed under a multi-functional ministry, conflicts of interest often obstruct the discharge of functional responsibilities.

B. Apex Bodies

Recognizing that the social, economic, cultural and environmental dimensions of development must be treated in an integrated and balanced way, apex bodies, above the level of ministries, have been set up in some countries to guide the sectoral agencies. The mandates for these bodies tend to be extensive and includes: establishing guidelines and mechanisms for sustainable development; formulating policy reforms; recommending new legislation; ensuring linkages with provincial and local governments and NGOs; reviewing and coordinating programmes; and suggesting means for developing capacities, both for human and financial resources.

In Japan, the Council for Sustainable Development is the advisory body to the Government and is mandated to follow up on the progress of measures taken in development planning and facilitate the efforts of different agencies in the field. Similarly in other countries across the region, apex councils have been established and often headed by senior government officials with representation by stakeholders, such as representatives from, industry, private sector, NGOs and the general public. Examples include the Council for Sustainable Development (PCSD) in the Philippines, (Box 11.1), the National Economic and Social Development Board (NESDB) and the National Environment Board (NEB) in Thailand, and the Environment Protection Council in Nepal.

In Iran (Islamic Republic of), the National Committee on Sustainable Development is the key national sustainable development coordination institute. Chaired by the Head of the Department of the Environment, it includes members from the

---

Box 11.1 Philippine Council for Sustainable Development (PCSD): A Mechanism for Integrating Environmental Considerations into Overall Economic Policies in the Philippines

PCSD provides the operational mechanism for sectoral integration and the systematic incorporation of environmental considerations in decision-making in the Philippines. The organizational structure of PCSD, effectively provides the necessary mechanism for integration. The PCSD brings together those agencies which are responsible for the formulation of macroeconomic and sectoral policies, as well as NGOs whose main advocatory interests are environmental and social causes under the chairmanship of National Economic Development Agency (NEDA). This arrangement enhances the effectiveness of PCSD in integrating environmental concerns in economic policy making.

The staff of the NEDA Secretariat serves as the Coordinating Secretariat to PCSD and also provides the necessary technical and administrative support to PCSD. An Interstaff Technical Working Group, supports the unit providing secretariat services to PCSD. The interstaff group relates the policy issues discussed at the council level to existing policies in the sector, and brings them to the attention of PCSD. They also ensure that proposed policy actions are consistent with the overall development objectives in their respective sectors. Since the staff is involved in PCSD work in policy and project evaluation and analysis, it is easy for them to incorporate the environmental considerations discussed at the council level into their review of macro and sectoral policies and programmes.

PCSD recommendations are normally translated into resolutions and Executive Orders which are submitted for the consideration and approval of the President. The requirement of the President for a complete briefing on all policy issues submitted to his office also provides assurance that the concerns of virtually all agencies and sectors are considered. Some of the more notable resolutions and Executive Orders in which the Council has played an important role are: Executive Order 247 on Prescribing Guidelines and Establishing a Regulatory Framework for the Prospecting of Biological and Genetic Resources; Memorandum Order 289 on Directing the Integration of the Philippine Strategy for Biological Diversity Conservation in the Sectoral Plans. Programmes, Projects of the National Government; Memorandum 399 on Directing the operationalization of the Philippine Agenda 21; Executive Order 291 on Strengthening the Environmental Impact Assessment System; and Executive Order 406 on Environmental and Natural Resources Accounting.

To date, PCSD is probably one of the most effective mechanisms for incorporating sustainable development in planning and policy decision-making in the developing countries of the region. Integration of environmental issues into economic decision-making is achieved mainly through dialogue and debate among concerned stakeholders during meetings of PCSD and its committees. Moreover, relevant policy issues and programmes are subjected to the evaluation of member-agencies to ensure that the concerns of all affected sectors are considered.

Source: The Government of Philippines

CHAPTER ELEVEN

Figure 11.1 Role of Planning Commission on Environment in India

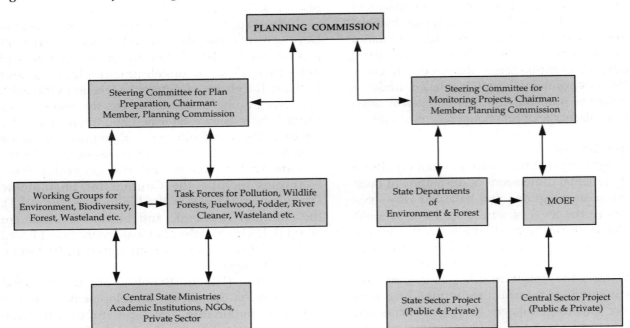

President's Office, the Department of Environment, the Meteorological Organization and the Plan and Budget Organization. In Kazakhstan the Ministry of Ecology and the Centre of National Environmental Activity Plan for Sustainable Development are in charge of integrating environment and development into decision-making.

The creation of high-level apex bodies has the merit of authoritatively reducing inter-ministry conflict of interests and enabling a critical evaluation of the performance of ministries without raising doubts of bias. These bodies guide the sectoral agencies and coordinate their activities, in addition to recommending measures and legislation to effectively implement the national strategies for sustainable development. The representation of major line ministries, the NGOs, corporate sector and community representatives within these bodies recognizes the need for a cross-sectoral approach to sustainable development. The high level of representation, with Prime Ministers heading many such bodies, provide them with a leadership role that cannot be challenged or overlooked by line ministries and agencies. However, the weakness noticeable in these bodies in many countries is the infrequent meetings, which hinders the effectiveness of plans and policies. Often, junior officials, who do not have the requisite authority in policy formulation, represent their departments in the apex bodies or interagency committees, thereby negating the very principles of collective decision-making envisaged in their creation.

C. Integration of Environmental Concerns in Development Process: Planning Institutions

A primary concern of governments in the region has been to integrate environmental issues with the developmental planning processes. A central planning body, which has representation from both the development and environmental streams, and which functions with the holistic vision necessary for sustainable development, is yet to evolve in any of the countries of the region. However, in recent years, there has been a broad awareness amongst policy makers of the interlink between economics and environmental problems, and efforts have been made to link these issues in a decision making framework through the mechanisms of planning.

Bodies such as Planning Commissions (Pakistan, India (Figure 11.1), Bangladesh, Nepal), Boards (NEDA, Thailand) and Authorities (NEDA, Philippines) are crucial macroeconomic institutions that take a long term view of development issues; collate, assess and prioritize the countries public investment programmes; and, coordinate and monitor implementation through line ministries, sectoral agencies, and regional and local governments. Planning bodies can play a crucial role in enhanced promotion of a cross-sectoral and integrated approach to environmental protection and sustainable development.

In Nepal, for example, as part of the regular National Planning Commission's (NPC) mandate, all environmental policies, programmes and projects in the public sector are subject to review and approval by the NPC before they are enforced. There is an Environment Protection Division within the NPC which is responsible for overseeing and coordinating intersectoral activities related to the planning, programme budgeting and monitoring of environment-related actions.

In a number of other countries, environmental units or cells have been created in the planning bodies to integrate environment into economic decision making. For example in India a member of the Planning Commission is in charge of Environment and Forest Sector assisted by a Senior Adviser, a Joint adviser, two Deputy Advisers and three research officers. The Commission performs the coordinating role in the environmental management by having inputs for the environmental component of National Plan.

During the preparation of National Plans, the Planning bodies usually canvass the advice of the ministries of environment and also relevant sectoral ministries through the establishment of ad hoc working groups or committees. These evolving arrangements are advancing the integration of environmental considerations in economic decision-making.

In the island states of the South Pacific, planning processes are still fragmented along sectoral lines though efforts are being made to integrate environmental concerns into the development process (Chapter 18).

Overall, the institutional integration of environment into economic decision-making has seen considerable progress in Asia and the Pacific. However, under-staffing, lack of training of existing staff and conflict of interests with other ministries or agencies obstructs effective integration. In order to bring environmental planning into the mainstream of the planning process developing more effective coordination and integrated institutions and mechanisms is required.

D. Judicial Institutions and Mechanisms

The Judiciary plays a crucial role in promoting the goals of sustainable development. Judicial institutions serve as agencies for interpreting legislation relating to environmental issues, integrating emerging principles of law within the holistic paradigms of sustainable development, providing a coherent and comprehensive strategy for handling diverse sectoral laws into a cross-sectoral approach, ensuring effective implementation of legislation and, in recent years, providing opportunities for people, to canvass for the protection of fundamental rights to a satisfactory environment.

The rule of law becomes particularly important as regulations and procedures which govern human activity serve to limit conflicts arising from competing claims (social, economic and ecological) on scarce resources whilst ensuring sustainable development. Connections and linkages between different forms of activity and their environmental consequences are subject to different interpretations and reflect an inherent complexity of issues. The judiciary, therefore, is called upon to resolve such issues without compromising the fundamental goals of civil society.

The structure of judicial institutions in different countries of the Asian and Pacific Region has not been substantially modified to cater to the requirements of achieving sustainable development. In many countries, the Supreme Courts have taken the lead in interpreting laws and giving directions which have had far reaching impact on environmental management. The Supreme Court in India, for example, in recognizing the role of environmental protection in sustainable development and growth, has been establishing mechanisms for institutionalising judicial dispensation in environmental matters. The Court has adopted and set procedures that become the guiding law for the sub-ordinate courts in the country. The most important innovation has been the Public Interest Litigation that enables individuals and organizations to file a writ petition with the objective of protecting environmental resources and benefiting the affected people. The Supreme Court of India has also established specialized High Court benches known as Green Benches.

Similarly, in Pakistan the superior courts exercise jurisdiction conferred under Articles 184(3) and 199 of the Constitution. The 1997 Environmental Protection Act provides for Environmental Tribunals that will have exclusive jurisdiction to try offences under the Act. Likewise, Nepal's 1997 Environmental Protection Act provides for the designation of a Prescribed Authority before which environmental cases are to be filed. However, appropriate rules for designating such an authority have not yet been formulated and environmental cases continue to be brought before the ordinary courts.

An active judiciary has the potential to establish the rights of people to enjoy certain environmental rights and seek judicial intervention where these are violated. The judiciary may also act as a check on government policies that disrupt fragile ecological balances and generate awareness and consciousness amongst policy makers through court verdicts and orders. However, there is a need for specialist environment courts that can facilitate more consistent and speedier environmental decision-making. These courts would reduce the number of cases brought before the Supreme Courts and High Courts and reduce the administrative costs as a single combined jurisdiction would be cheaper than multiple separate tribunals.

E. **Supporting Institutions**

The role of supporting institutions in environmental management for sustainable development has expanded to bridge the gaps between the vision of sustainable development and the actual implementation of plans and policies. Specialized policy research institutes, R & D institutes, NGOs and private sector institutions have a role in complementing the efforts of governments to prepare plans and programmes, mobilize public opinion, provide research inputs and scientific data, train staff and members of the public and build awareness. Across the region, institutions have emerged that specialize in environmental research and public outreach. In the Republic of Korea, the National Institute of Environmental Research and the Korean Environmental Technology Research Institute have been set up under the Ministry of Environment. In India, the Tata Energy Research Institute (TERI) is working for the standardization of environmental monitoring methods for air, water, soil, wastes, and vehicle emissions and also training personnel from the government and the private sector (see Chapter 15).

At an international level, NGOs such as the World Conservation Union (IUCN) and the Worldwide Fund for Nature (WWF) have been particularly active in the Asian and Pacific Region, and continue to provide technical and advisory support to various country governments (see Chapter 14). Other supporting institutions for environmental management include United Nations agencies such as UNDP and UNEP and bilateral and multilateral organizations, which constitute a major source of financing for sustainable development.

SUB-NATIONAL AND LOCAL INSTITUTIONS

Involvement of sub-national and local governments in environmental management, formulation and implementation of plans and policies and monitoring of environmental laws has increased in the post-Rio years. This has resulted in the devolution of powers and the delegation of responsibilities to provincial governments, local bodies and municipalities to undertake environmental protection measures within the national conservation strategies. At the provincial level, environmental agencies have been set up with the aim of determining critical resources, assessing current resource use, implementing national goals and coordinating the functions of sectoral agencies and local authorities.

Agenda 21 called for local governments to play an important role in its implementation and for the formulation of Local Agenda 21s for effective management. In Japan, the Government provides assistance to the local authorities for their own voluntary and independent environmental activities and for international cooperation at the local authorities level. Local authorities are directly involved in the implementation of laws, regulations and guidelines, and in the observation, measurement and control of pollution, in addition to carrying out various anti-pollution and nature conservation projects, (Box 11.2).

In the Republic of Korea, 159 out of 248 local governments are in the process of establishing or promoting Local Agenda 21s. Through this mechanism local governments, together with citizens and industries, will have their roles and responsibilities for environmental protection strengthened through the establishment and implementation of environmental plans that meet local needs. Provinces, autonomous regions and municipalities in People's Republic of China have also formulated their own local Agenda 21s. For instance, Sichuan (the most populous province in People's Republic of China), Shanxi (a coal-rich province) and Guizhou (the poorest province) have formulated their own local Agenda 21. Provinces in Pakistan such as the North West Frontier Province (NWFP) and, more recently, the province of Boloduta have also formulated their respective Provincial Conservation strategies which focus on a few selected sectors such as industry, agriculture, forestry and wild life, public health, urban environment, education and cultural conservation. Similar efforts to prepare sub-national level environmental plans and strategies are found in Thailand, Indonesia, Malaysia, Viet Nam and the Philippines.

A. **Institutional Arrangements**

Institutional arrangements across the region at regional and local levels have followed national patterns, with an environmental cell or unit responsible for overall environmental management, planning, monitoring and coordination. The environmental cell is vested with regulatory powers to control pollution and check degradation of natural resources. For example, in India, all 25 state governments have either set up a separate Department for Environment or redefined the responsibilities of Forest Department. The state level departments regulate the implementation of laws and regulations within their area and exercise control over the State Pollution Control Boards. In NWFP,

INSTITUTIONS AND LEGISLATION

> **Box 11.2 Synchronization of Environmental Policies, Laws and Administration: A Case Study of Japan**
>
> The Japanese Government responded to an urgent environmental crisis in the late 1960s and early 1970s by enacting stringent emission standards for common pollutants and promoting a policy of working with industry in the widespread application of cleaner technology to combat environmental degradation. Formulation and implementation of environmental policies at national level involved promulgation of a series of environmental laws or acts and their nationwide implementation through the creation of appropriate administrative systems. The Basic Pollution Act (enacted in 1967), for example, laid the foundation for anti-pollution measures and was in effect until it was replaced by the Basic Environmental Act in 1993, which was enacted to take new realities into consideration after UNCED. The Basic Environment Act made it mandatory for the state government to prepare a Basic Environmental Act 1994. The cyclical system of nature and life, the necessity for coexistence, full participation by all sectors and people, and promoting global environment policy, were incorporated as four basic and fundamental principles of the plan. Following the Basic Environment Act, the Environmental Assessment Bill was enacted in 1997 in order to standardize concepts and procedures of impact assessment.
>
> Integration of economic and environmental policy was a major challenge for decision makers in Japan. In this regard, the country's recognition on the economic value of preventive environmental planning and policy, with the issue of waste minimization and environmental compliance in western markets were the positive signals which were also achieved through invoking appropriate legislation. Several acts were promulgated in order to deal with typical environmental pollution problems such as the Air Pollution Prevention Act, the Water Pollution Prevention Act, the Vibration Regulation Act, the Noise Regulation Act, the Odor Prevention Act and the Soil Pollution Prevention Act.
>
> The Japanese Government has also been actively participating in several global environmental conventions. In order to implement these at national level, relevant domestic laws were amended in order to comply with the international treaties (Ozone Layer Prevention Treaty (1985) and its Montreal Protocol (1987), United Nations Framework Convention on Climate Change (1992), the Convention on Biological Diversity (1992), Environmental Protection Protocol to the Antarctic Convention (1997) and the COP3 for the Convention on Climate Change). For instance, manufacturing and use of CFC is now regulated based on a nationally approved Ozone Layer Protection Act. Similarly some amendments were made to the Energy Conservation Act in order to deal with the global warming problem and to achieve the standard of CO_2 emission reduction enforced by the Kyoto Protocol. In 1998, the Global Warming Policy Promotion Act was promulgated to set the basic national policy to address the problem.
>
> Besides the development of synergy between environmental policies, laws and administrations, another noteworthy trend in the implementation of environmental measures at the national level was conflict resolution between institutions on the implementation of environmental pollicies. Occasionally, in the initial stages of implementation of a particular policy arguments arose regarding which agency or ministry was going to assume lead responsibility, especially where jurisdictions were overlapping. Recently for example after the COP3 when amendments to laws and regulations dealing with global warming were considered, there was an apparent conflict between Ministry of International Trade and Industry which favours more of an energy related approach and the Environment Agency which seeks more comprehensive measures. Similar problems developed for initiation of nature conservation, between the Environment Agency and the Forest Agency, when the latter re-orientated its policy towards environmental protection. Such struggles for possession of ruling authority amongst rival ministries and agencies is seen as a source of problems in environmental administration in some countries of the region. However, in Japan, it has had positive effects on environmental governance and is seen as a major step towards integration of policies leading to enhanced environmental management and administration.

Pakistan, the role of the environment unit under the Planning, Environment and Development Department has been streamlined by creating: the Planning Section (project approval, preparation and implementation of parts of the 5 year plan); and the Environment Section (policy formulation, coordination and implementation of the national and provincial conservation strategies; monitoring of regulations and international conventions on environment; and design and implementation of suitable provincial legislation). An Environmental Protection Agency has also been created for checking environmental degradation and pollution control.

Local and municipal administrations have the closest interface with environmental problems such as land degradation, urban waste, air and water pollution. The importance of strengthening institutional arrangements at such levels has been increasingly realized in many countries of the region. Consequently, environmental bodies have been set up in many large and medium sized cities with advanced monitoring systems and assessment capabilities. At the same time, the global trend of decentralized administration has found many countries introducing changes in the administrative structures to empower local bodies by shifting conservation measures and elementary monitoring functions to these bodies.

For example, People's Republic of China has created hundreds of Environmental Protection Bureaux in its cities. These bureaux play a crucial role in formulating and implementing environmental policies and coordinates with the local Planning Commission and other municipal departments in the preparation of joint environmental protection strategies that are inter-institutional and cross-sectoral. In Kuala Lumpur, Malaysia, the task of monitoring urban environment and industrial pollution and enforcing pollution control laws has been delegated to an Environmental Management Unit in the City Hall. However, the all-embracing enforcement powers of the Federal Department of Environment leave the local authorities with weak institutional strength to monitor implementation of environmental measures.

In the South Pacific subregion, institutional mechanisms at the local level for achieving sustainable development have taken different forms. Papua New Guinea, Solomon Islands and Vanuatu have a system of provincial governments in which elected councils have been delegated powers and responsibility over local affairs. The Organic Law enacted by Papua New Guinea identifies two layers of governments: province and district. Amongst other subjects, control over urban facilities and environment, waste disposal, health and water supply have been vested with the provinces. In Vanuatu, a similar system to Papua New Guinea has been adopted, but with a more proactive approach towards public health and urban amenities. For example, the Port Villa Council works closely with the Physical Planning Unit of the Ministry of Home Affairs to regulate use of land and coordinate actively with the Department of Health to improve living conditions in squatter settlements.

In Fiji, provincial administration is relatively undeveloped, with the concentration of powers and responsibilities vested in central government. However, local governments have been delegated urban management responsibilities under the Local Government Act that enables them to promote the health and welfare of the inhabitants. The Department of Health under the council addresses problems of garbage collection and disposal, waste management and drainage control.

However, the devolution of powers and responsibilities to the sub-national, local and municipal bodies has not followed a uniform pattern across the region, or between urban and rural areas. In urban areas there has been greater success in achieving the integration of environmental and development concerns. This is as a consequence of greater and more visible environmental problems in urban and industrial cities and towns, and better awareness of environmental issues amongst the urban populace. In rural areas, where public participation has been greater, the institutional mechanisms have been more focussed on specific environmental management programmes related to resource use.

INSTITUTIONAL COORDINATION

Coordination mechanisms provide harmonizing instruments for close and effective work between different institutions. Lack of effective coordination often results in a dispersed sector-specific policy, orientation and concomitant difficulties in trying to harmonize the various sectoral interests. Ideally, effective coordination mechanisms builds on integrated criteria for development in which conflicting interests get resolved through the mediation of common goals and objectives. Furthermore, each layer of administration is seen less as an adversary determined to confiscate power or privileges and more as an integral component of joint endeavour.

Despite the development of separate environment ministries or advisory councils, the structure of governments in the region continues to be characterized by a high level of fragmentation and differentiation along sectoral lines. The institutions responsible for environmental management have often become another functional line ministry only. In addition, most governments in the region are still highly centralized with devolution of powers and responsibilities to the local and provincial governments being either negligible or progressing in a piecemeal manner. Therefore, the implementation of programmes and plans designed to achieve sustainable development call for establishment of coordination mechanisms that not only bridge the existing gaps in the governmental structures but create avenues for advancing the objectives and goals of sustainable development.

Preparation of strategic environment management plans and the creation of apex bodies/councils above the level of ministries have increased awareness and understanding of sustainable development amongst the multifarious agencies in governments and the public arena. In order to effectively coordinate the activities of ministries, government departments and the agents in the field have all experimented with different approaches. Some countries have created environment units in different ministries; others have liaised with environmental core groups consisting of specialists which act as a roving team moving from ministry to ministry and undertaking specific evaluation and assessment of projects that have environmental implications. The committee system in which members from concerned agencies are drawn into committees to ensure functional integration and coordination of policy has also been introduced in many countries. Vertical coordination among layers of government is particularly important, and depends considerably on the degree of decentralization and the mandate of local governments.

Horizontal and vertical coordination amongst agencies and ministries in People's Republic of China is achieved through four-tier networking amongst the national, provincial, local and sectoral levels. Environmental management departments at each level of government are responsible for supervising implementation of laws and regulations for environmental protection, coordinating the relations among different agencies and guiding their work. The State Environmental Protection Administration is the apex body monitoring and coordinating the activities of the different provincial, local and sectoral agencies.

In the Philippines, Environmental Units (EU) have been created in various sectoral agencies and are coordinated by the Department of Environment and Natural Resources (DENR) which thereby ensures linkages between the environment ministry and the sectoral agencies. The Environmental Units are responsible for preparation of Environmental Impact Statements (EIS), ensuring that agencies meet the EIS norms and facilitating the issue of Environmental Clearance Certificates (ECC).

Singapore set up an Environment Council (SEC) in 1995 as an umbrella organization to facilitate coordination and cooperation between NGOs and other green groups. The SEC has specialized committees to supervise and coordinate the sectoral activities including: the Industry Committee to promote corporate environmental responsibility; the Community Relations Committee to foster greater appreciation of the environment amongst the community; the Education Committee to advise at various levels of education from schools to tertiary institutions; the Research and Publications Committee to initiate, facilitate and publish studies relevant to the environment; and, the Finance Committee to obtain funding from various sources for the committees' projects.

The Environmental Impact Management Agency (BAPEDAL) in Indonesia coordinates with line ministries, especially the Ministry of Home Affairs, and, at the regional level, activities are coordinated by regional government and BAPEDAL-DA (regional offices of BAPEDAL). In the initial stages, community participation was negligible, but this has improved since the participation by groups of universities in the programme.

The NEB in Thailand has a major coordinating role as it has been vested with the authority to prescribe environmental standards and to approve the national plan for Management of Environmental Quality and provincial environmental plans. The presence of most ministers of the government on the board, and the chairmanship of the Prime Minister, makes the NEB a powerful body which line ministries cannot ignore.

In the Republic of Korea, the Environmental Preservation Committee assumes the task of interdepartmental coordination for environmental issues amongst different ministries. It also involves other institutions such as the Korea Chamber of Commerce and Industry, the Seoul National University, the Taegu University, and the Korea News Editors' Association. The NGOs involved include the Korea Saemaul Undong Centre, the Korean Federation for the Environmental Movement, and the Korea National Council of Women. Vertical coordination between the central ministries and agencies and the local governments is maintained by devolving substantial responsibilities, such as performing EIA, monitoring emission standards and protecting nature conservation, to the local governments.

In Sri Lanka, the "Committee on Environmental Policy and Management (CEPOM)" is responsible for integrated environmental concerns in development policies and programmes of the line ministries concerned. It also acts as a mechanism for implementing the National Environmental Action Plan (NEAP). The CEPOMS cover biodiversity, health, industries, coasts, water etc. and are convened by the Ministry of Forestry and Environment and the sectoral Ministry concerned.

Coordination mechanisms in Pakistan are linked to the planning process that leads to the preparation and implementation of five-year plans by the Planning Commission. The Ministry of Environment, Local Government and Rural Development is responsible for coordinating and implementing national environmental policies. At the provincial level, the Annual Development Plan is initiated and coordinated by the Provincial Planning Department. The provincial plan is again vetted at the federal level. The Annual Plan Coordination Committee is chaired by the Finance Minister which deliberates on the proposals before sending it to the National Economic Council, the apex decision making body headed by the Prime Minister. The Ministry of Environment Local Government and Rural Development has established an Advisory Board to involve private sector and NGOs in environmental management.

Among the countries in the South Pacific subregion, most governments have recognized the need for inter-agency coordination in environmental management. In Tonga, the key coordinating agency is the Development Coordination Committee (DCC), headed by the Deputy Prime Minister and comprising senior ministers and officials. With high level and wide-ranging representation, the DCC is expected to ensure that environmental implications of new projects are properly assessed. Unfortunately, the Ministry for Land Surveys and Natural Resources is not represented on this committee and as a consequence, enforcement of laws related to environmental protection suffers. In Fiji, the Department of Environment under the Ministry of Housing and Urban Development only has a peripheral role in the activities of the line ministries and departments. To address this the government has constituted ad hoc committees for specific problems. The representatives on these committees are chosen from different stakeholders, but there are no set procedures for their working. The Environment Management Committee coordinates the activities of different agencies and provides advice on the implications of development proposals. The National Environment Steering Committee oversees the National Environment Project that began in 1991; other committees include: Mangrove Management Committee, Rubbish Dump Committee, National Oil Pollution Committee and Consultative Committee on Ozone Depleting Substances.

The Ministry of Forestry, Environment and Conservation in the Solomon Islands is the lead agency for coordinating environmental issues but coordinates with different sectoral ministries when appropriate. For instance, genetic heritage and bio-diversity projects are undertaken with the Agriculture Ministry. Numerous committees exist for the coordination of policy and activities in different sectors and there is an overlap between the various committees which enables conflicts to be resolved and duplication to be reduced.

Given the segmented and compartmentalized structure of administration in the Asian and Pacific Region, a legacy of fragmentation of concerns and responsibilities preventing effective coordination has been created. Coupled with this is a lack of awareness and appreciation of the interconnections and inter-dependence between various issues, a lack of knowledge about tools and methodologies required for factoring environmental costs and a paucity of trained and skilled personnel. The region's governments will need to continuously review the mechanisms to ensure that responsibilities are well defined, that potential areas of conflict are identified and overlaps in responsibility removed or reconciled, and that potential areas of cooperation are made known to the agencies for mutual benefit and interaction. By making coordination an integral function of the apex environmental bodies in most countries of the region, the governments have signalled their intention to work towards securing integration of environmental and developmental concerns.

ENVIRONMENTAL LEGISLATION: STATUS AND TRENDS

The Rio Declaration and Agenda 21 identified environmental legislation as a critical component in the enhancement of capacity building for sustainable development. Laws and rules to control air, water and marine pollution, to manage municipal wastes and urban settlements, to regulate forests and forest resources and to preserve biodiversity had previously been on the statute books of different countries under separate legislation. In the post-Rio period, many countries in the region have responded to the emerging challenges of achieving sustainable development by reshaping legislation to make it more suited to integrating the environmental and development concerns of the country. The various multilateral environmental agreements have also promoted the development of national legislation.

A. Constitutional Provisions

The Constitutions of many countries in the Asian and Pacific Region, particularly those enacted after the Stockholm Conference in 1972, have enjoined countries to strive towards environmentally responsible development through: the sustainable use of natural resources; the provision of a healthy and clean environment; and the recognition of the rights of individual citizens to a clean and healthy environment. For example, one of the stated goals in the Constitution of Papua New Guinea is: "We declare our fourth goal to be for Papua New Guinea's natural resources and environment to be conserved and used for the collective benefit for us all, and to be replenished for the benefit of future generations." The 1980 Constitution of Vanuatu embodies similar principles, Article 7(d) of the Constitution states: "Every person has the fundamental duty to themselves, their descendants and to others to protect Vanuatu and to safeguard the national wealth, resources and environment in the interests of the present and future generations."

The effectiveness of constitutional provisions on environmental protection depend on the clarity of the legislation, rules and regulations, the administrative and judicial institutions that enforce and implement the laws and the level of awareness in the society regarding the environmental rights of citizens. In countries like India, Bangladesh and the

Philippines, successful judicial interventions through public litigation have established the broad parameters of law that reinforces the provisions of the Constitution. In contrast, the Sri Lankan constitution seemingly states that Articles 27 and 28, which require the state to protect, preserve and improve the environment for the benefit of the community, do not confer legal rights or obligations and are not enforceable in any court. In India, the Directive Principles of State Policy enunciate principles, which, though not enforceable by any Court, are nevertheless fundamental in the governance of the country. Furthermore, the Constitution states that it shall be the fundamental duty of every citizen to protect and improve the natural environment, including forests, lakes, rivers, and to have compassion for living creatures.

B. Environmental Framework Law

Many countries in the region have enacted environmental framework legislation, also known as an 'umbrella law', in order to cover the array of environmental issues. Table 11.2 provides a selection of such framework laws. These framework laws represent a new judicial technique, with a strong environmental policy orientation. Their purpose is to establish an overall coherent policy and provide a basis for the coordinated work of various government agencies with operational responsibility for the environment and natural resources. In some cases, the legislation invests authorities with regulatory powers to address specific issues affecting the environment.

Typical areas covered by these framework laws include: prohibitions on the discharge of oil and waste; prevention of pollution to air, water and soil; prevention of degradation and depletion of natural resources; and the provision of a positive direction towards integrating environment and development.

In the island states of the South Pacific, countries are in the process of formulating comprehensive environmental legislation as the existing laws from the colonial period prove inadequate for dealing with modern environmental problems. Papua New Guinea, for example, has a proposed Environmental Regulation Framework to bring all environmental regulatory work under one authority. Both Tonga and Vanuatu have drafted umbrella laws to facilitate environmental resource management and sustainable development. Fiji has prepared and placed before Parliament, the Sustainable Development Bill. The provisions in the bill include preparation of a natural resource inventory for each sector, a resource plan and a natural resource account, which would be subject to audit. The Bill also seeks to create a self-regulatory mechanism for environmental controls for industries (see Box 11.3).

Many of the framework laws in the region set environmental quality criteria and standards that aim to specify identifiable targets and limits as the basis for government agencies to determine and monitor compliance efforts. For example, under Thailand's 1992 National Environment Quality Act, the Minister of Science, Technology and Environment is empowered to issue standards for emissions, effluents and hazardous wastes. A Pollution Control Committee oversees the environmental protection measures and aids and advises the Minister on the standards. Similar provisions authorising specified government agencies to issue standards and control air, water and waste pollution exist in the legislation of Bangladesh, People's Republic of China, India, Kazakhstan, Kyrgyzstan, Mongolia, Pakistan, Papua

Table 11.2 Some Framework Laws for Environmental Management in Selected Countries in Asia and the Pacific

Sr. No.	Country	Framework Laws
1	Bangladesh	Environmental Protection Act, 1995
2	PR China	Environmental Protection Law of People's Republic of China, 1979
3	Democratic People's Republic of Korea	Law on the Protection of the Environment, 1986
4	India	Environmental Protection Act, 1986
5	Indonesia	Environmental Management Act, 1997
6	Islamic Republic of Iran	Environmental Protection and Enhancement Act
7	Kazakhstan	Law on Protection of the Natural Environment, 1991
8	Malaysia	Environmental Quality Act, 1974
9	Mongolia	Environmental Law of Mongolia, 1994
10	Pakistan	Pakistan Environmental Protection Ordinance, 1997
11	Philippines	Philippine Environment Code, Presidential Decree No. 1152, 1977
12	Republic of Korea	Basic Environmental Policy Act, 1999
13	Sri Lanka	National Environmental Act, 1980 (Incorporating 1988 amendments)
14	Tajikistan	Law on the Protection off Nature, 1993
15	Thailand	Enhancement and Conservation of National Environmental Quality Act, 1992
16	Turkey	The Environmental Law, 1983
17	Uzbekistan	Law on the Protection of Nature, 1992
18	Viet Nam	Law on Environment Protection, 1994

Source: UNEP 1995 (updated by ESCAP)

CHAPTER ELEVEN

> ### Box 11.3 Impact of International Events and Conventions on Institutional and Legislative Development: The Case of Fiji
>
> The United Nations Conference on Environment and Development was a watershed event for environmental policy in many small developing countries, including Fiji. As part of the lead up to this conference, Fiji developed a National Environment Management Strategy that called for the establishment of a Department of Environment (DoE). The strategy also noted that the environmental laws in Fiji required a major overhaul. In 1992, the DoE was inaugurated and soon began to work on new legislation under the Sustainable Development Bill (SDB).
>
> The DoE also coordinates activities to satisfy the requirements of the Government of Fiji as a signatory of the two conventions – United Nations Framework Convention on Climate Change and the Convention on Biological Diversity. The Government of Fiji signed the Convention on Biological Diversity (CBD) in July 1993, and committed itself to developing a set of regulations to provide access for the study of its biodiversity. The process of development of such regulations into the SDB demanded continuous and sustained efforts. In order to initiate the process, an inter-ministry task force was set up in late 1993 to discuss the dimensions of regulatory framework that would be needed for bioprospecting. From the very early stage, government saw its role as regulatory only. It would not be directly involved in bioprospecting ventures but would attempt to ensure that they were carried out under the guidelines of the CBD and other guidelines of best practice such as the Manila Declaration.
>
> The legislation was completed in May 1995 and put out for review as part of the Sustainable Development Bill in 1996. In the legislation, the Conservation and Natural Parks Authority with responsibility for biodiversity conservation, controls the process of granting access to persons wishing to conduct biodiversity research to ensure that: no ecological, social or economic harm is caused by the biological research or exploitation; taking a biological sample does not have an undesirable impact on Fiji's biodiversity; a fair return is provided for commercial exploitation of Fiji's biological resources; and prior informed consent from the resource owners is obtained before any collections can take place.
>
> The proposed draft legislation has been praised for its vision in developing a sound basis for biodiversity research in Fiji in compliance with the Convention on Biological Diversity, and especially for placing this research in the overall framework of conservation in the Sustainable Development Bill.
>
> *Source:* William G. Aalbersberg, University of New Zealand

New Guinea, Saudi Arabia, Sri Lanka, Tajikistan, Uzbekistan and Viet Nam.

Implementation of laws, especially enforcement of standards and environmental quality, is critical for the success of any environmental protection regime. Supplementary legislation on sectoral aspects of environmental management is necessary to bridge the gaps in the framework laws and to provide practical guidelines and criteria to the enforcement agencies. Most countries in the region have laws to cover sector-specific environmental problems and issues. For instance, Fiji has the 1973 Irrigation Act, 1977 Marine Species Act, 1978 Preservation of Objects of Archaeological and Paleontologist Interest Act, 1982 River and Streams Act, and in 1991 Litter Decree. India has a Plant Varieties Act (Box 11.4) whilst Nepal has 69 separate acts and regulations that have a bearing on the environment and natural resources. It is important, however, that these sector-specific acts are rationalized in the context of framework laws and the overall national policies for sustainable development.

C. Pollution Control Legislation

Pollution control legislation has been enacted in almost all countries of the region. When environmental pollution became a public concern during the 1970s, many countries responded by enacting legislation to curb potentially harmful activities, particularly those producing industrial effluent. Pollution control legislation has set standards for emission and effluents and relies on the regulatory agencies for implementation and enforcement of standards.

Anti-pollution legislation covers a range of activities that have the potential to cause pollution and health hazards. Under the Environmental Management Act of Indonesia, there are several regulations to control pollution, including the Water Pollution Control Regulation, the Air Pollution Control Regulation and the Marine Pollution and Degradation Control Regulation. Similarly, under the 1974 Environmental Quality Act (EQA), Malaysia has enacted 22 subsidiary legislative instruments with regulations for *inter alia*: crude palm oil, clean air, raw natural rubber, sewage and industrial effluents, motor vehicle noise, waste treatment and disposal facilities. The regulations set the specified standard for emissions, prescribe penalties for exceeding the limits and specify and empower an authority to enforce the regulations.

The implementing rules and regulations made under the Water Code of the Philippines impose permits for surface and groundwater use, pollution

INSTITUTIONS AND LEGISLATION

Box 11.4 Plant Varieties Act of India

The aim of the Plant Varieties Act (PVA) of India is "to protect the rights of the developers of new varieties to stimulate investment in plant breeding and to generate competitiveness in the field of research and development both in the public and private sectors with the ultimate aim of facilitating access to newly developed varieties and maximizing agricultural production and productivity in the country." Further, the PVA states that "the protection of farmers and researchers rights will strive to balance the need for stimulation and inventive R & D with welfare of the farmers."

The proposed PVA reflects the conflicting pressures between the introduction of plant variety protection driven by developed countries and the strong opposition by farmers against introduction of any form of intellectual property protection in the agricultural sector.

The PVA includes features of the International Union for the 1991 Protection of New Varieties of Plants (UPOV), which sets the minimum standard for plant breeders right (PBR) protection for contracting states. Most contracting countries are implementing the 1991 act into their national laws. The Indian PVA includes elements of both the revised act of 1991 and of the former act of 1978, and has introduced some new features (see table below).

Comparison of India's Plant Variety Act to the UPOV Acts of 1978 and 1991

Issue	UPOV 1978 Act	UPOV 1991 Act	India's Plant Variety Act
Scope of breeders' rights	Production and marketing of propagating material	Production, marketing, exporting, and stocking of propagating material	Production, marketing, exporting and importing of propagating of propagating material
Extent of coverage	Min. 24 species	15 species	All species
Term of protection	Min. 15 years	Min. 20 years	15 years
Exception to rights	Farmers' privilege in practice	Farmers' privilege optional and under conditions	Farmers' rights specifically recognized
Compulsory licensing	In case of public interest (not defined)	In case of public interest (not defined)	In case of public interest, defined as reasonable availability of seeds, and supply of export markets

A major test of the PVA is its consistency with the provisions of the agreement on Trade Related Aspects of Intellectual Property Rights (TRIPs) which stipulates that "members shall provide for the protection of plant varieties either by patents or by an effective sui generis system or by any combination thereof". Whether the 1991 Act, will be considered "effective" is subject to interpretation within the World Trade Organization.

The PVA also covers protection of "essentially derived varieties" which means that when a new variety is genetically too similar to a protected source variety, the marketing of the new variety requires authorization of the breeder of its source, thus ensuring the protection of protected varieties.

Other provisions of the PVA, such as those relating to Farmers and communities rights, and compulsory licensing depart from UPOV. The inclusion of these rights in the PVA bill will help to balance the rights of farming communities with the need to protect intellectual property rights (IPR).

Source: Dhar B. et al 1995

---

control measures and restrictions on disposal of wastewater. The Litter Decree in Fiji imposes stiff fines for disposal of litter in public places. However, it has taken a number of years and considerable public pressure for strict enforcement of the decree. Other notable examples of pollution control laws include China's 1996 Environmental Noise Pollution Prevention and Control Law and Hong Kong, China's comprehensive Noise Control Ordinance and Air Pollution Control Ordinance. Control of illegal transport of hazardous products and wastes has also been receiving the attention of governments in recent years. Thailand has enacted the 1992 Hazardous Substances Act; Indonesia has prepared a complete hazardous waste regulatory programme; Viet Nam has issued a decree on toxic chemicals and radioactive wastes (1995); the Philippines enacted the 1990 Toxic Substances and Hazardous and Nuclear Wastes Control Act; and Malaysia issued the 1994 Guidelines for the Export and Import of Scheduled Wastes.

To complement the traditional command and control regulatory mechanisms for the enforcement of pollution control legislation, some countries have progressively introduced market based instruments to enhance compliance and to create incentives for the agents to implement environmental control

CHAPTER ELEVEN

measures (see Chapter 12). Further measures for abatement rather than control, of pollution are also slowly being introduced.

D. Resource Conservation Legislation

Resource conservation legislation in the region incorporates a wide range of environmental management concerns, including water resources protection and conservation, forest laws, marine resources management, land use management, preservation of natural habitats and conservation of heritage. Most countries in the region have enacted laws specific to these issues and introduced innovations to make the enforcement more effective. However, existing gaps in legislation make dealing with conflicting demands on resources difficult to manage.

In the countries of South Asia, the management of forests and forest resources has been given considerable priority. In India, under the provisions of the 1980 Forest (Conservation) Act, prior permission of the Central Government is essential for the diversion of forest land for non-forest purposes. Linked to this are the provisions in the 1986 Environment (Protection) Act, which restrict the setting up of any new wood based unit, expansion and modernization of such units, renewal of licenses for such units and construction of any infrastructure related to the setting up of new as well as existing wood based units. In Nepal, the government is encouraging user groups and village communities to participate in forest management and it has been made mandatory for industries setting up in forest areas or using forest products to have a detailed environmental impact assessment. Sri Lanka has set up national parks, nature reserves, and sanctuaries to prevent destruction of forest areas.

Viet Nam's 1994 Law on Environmental Protection, explicitly states that the exploitation of agricultural land, forest land and land for aquaculture must comply with land use plans, land improvement plans and ensure ecological balance. It also stipulates requirements governing the use of chemicals, fertilisers and pesticides.

Natural resource-related legislation in the islands of the South Pacific cover a large number of sectors though in their implementation the narrow sectoral concerns still continue to prevail. In Papua New Guinea, for instance, the 1978 Environmental Planning Act is the principal legislation governing mining petroleum sectors, regulating the issue of permits and licenses and defining the total stock of physical, biological and social resources available. The 1978 Water Resources Act, determines the water use pattern in the mines; the 1978 Environmental Contaminants Act, and 1979 Dumping of Wastes at Sea Act regulates environmental contamination and dumping; and the 1978 Conservation Areas Act, provides for safeguarding the natural environment and natural cultural heritage. Natural resources in Tonga, on the other hand, are largely marine or littoral based and include mangrove forests, non-tidal salt marshes, sea grass beds, sea turtle nesting areas, open lagoons and lagoon reef formations. The Water Board Act serves as the main instrument to regulate management and use of scarce fresh water. In Samoa, where fisheries has the greatest potential for providing gainful employment and income, various Fisheries Acts have been introduced to managed the fish stock, control fishing practices and limit exploitation by foreign vessels.

E. Legislation on Environmental Impact Assessment

Environmental impact assessment (EIA) has become an important tool in guiding policy choices by helping to create an awareness amongst project implementation agencies about environmental impact of their actions and the measures required to control negative externalities of the projects. For many countries in the region, environmental assessment methodologies have been made a mandatory exercise through enactment of suitable legislation. Some EIA laws and regulations of developing countries and countries with economic transition are given in Table 11.3. By making EIA compulsory under law, it is envisaged that potential damages to the environment can be minimized or prevented from the project formulation stage itself. This is also seen as a crucial link in integrating environmental concerns into economic decision making process.

Provision for EIA is either in the national framework legislation or in subsidiary legislation. In Malaysia, for example, a handbook of EIA guidelines and 19 separate guidelines have been laid down for EIA in coastal areas, petrochemical projects, industrial estate development and golf course, land reclamation, forestry. In People's Republic of China, under the Environmental Protection Law, the system has been adopted to monitor environmental impact based on design, construction and operation of any particular development project. Nepal has attempted to harmonize sectoral legislation by formulating national EIA guidelines which identify the agencies responsible for reviewing the assessment report. Other countries in the region who have made EIA mandatory include: Thailand, India, Sri Lanka, Bhutan, Indonesia, Viet Nam, the Philippines and the Republic of Korea.

Legislative sanction for EIA has the advantage of introducing greater objectivity in the decision making process. In the context of sustainable

INSTITUTIONS AND LEGISLATION

Table 11.3 Some EIA Laws and Regulations for Selected Countries in Asia and the Pacific

Sr. No.	Country	Key	Title of Separate Legislation
1	Bangladesh	FW	
2	Bhutan	**	Environmental Assessment Act of the Kingdom of Bhutan 2000 (not adopted)
3	PR China	FW	
4	India	X	Environmental Impact Assessment Notification, 1994
5	Indonesia	**	EIA Regulation, 1999
6	Kazakhstan	X**	Provisional Instructions on Procedure for the Conduct of Environmental Impact Assessment on Designated Economic Activity, 1993
7	Kyrgyzstan	FW	
8	Malaysia	X**	Environmental Impact Assessment Order, 1987; Environmental Quality (prescribed Activities)
9	Maldives	FW	
10	Mongolia	X**	Environmental Impact Assessment Reg. 1994
11	Nepal	X	EIA Guidelines, 1993
12	Pakistan	X	Pakistan Environment Act 1997
13	Philippines	X	Pres. Decree, 1987; Rules and Regs. 1979; Proclamation 1989
14	Republic of Korea	X	Act for Environmental, Traffic and Disasters Impact Assessments, 1999 (Environmental Impact Assessment Act, 1993)
15	Russian Federation	X	Environmental Impact Assessment
16	Sri Lanka	FW	
17	Tajikistan	FW	
18	Thailand	X**	Environmental Impact Assessment Notification, 1992
19	Turkey	X**	Environmental Impact Assessment Reg. 1993
20	Uzbekistan	FW	
21	Viet Nam	FW	

Note: (FW) EIA legislation contained in Framework Laws; (X) Separate EIA Legislation; (**) Separate EIA Legislation with some reference in Framework Laws.

development, mandatory EIA also ensures the participation of stakeholders and the public in the EIA process, which brings cross-sectoral ideas and views into perspective and thereby enlightens the decision making process (Chapter 12).

F. Enforcement of Environmental Legislation

Effective enforcement of environmental legislation is contingent upon the availability of adequate staff and financial resources, the administrative and political will of the enforcement agencies and the level of awareness of environmental laws. It is common, however, to find situations where responsibility for enforcement of laws are divided amongst a number of government agencies which pursue conflicting interests, thereby delaying or forestalling their implementation. In response, for enforcement to be effective, developmental planning processes have to be closely coordinated, with powers ideally vested in one apex agency.

Judicial activism and public participation have, in recent years, enhanced enforcement efforts of governments in implementing environmental laws. The courts are not only allowing the public to file public interest litigation for violation of environmental rights, but are also giving directives to the government to take corrective steps for rectifying environmental damage. The imposition of fines and penalties on defaulting industries and closure of polluting units are examples of measures that have been frequently imposed by the courts.

The courts have further expounded on other key concepts of sustainable development. The doctrine of inter-generational equity received judicial sanction in the celebrated Oposa vs. Factoran case (Republic of Philippines, Supreme Court, G.R. no. 10183) whereby the Petitioners, which included a number of minors suing through their respective parents, joined by a national NGO, claimed that the continued issue of permits for timber extraction posed a threat to the natural forests and threatened the survival of the present and future generations. The Court subsequently recognized the right of Filipinos to a balanced and healthful ecology as well as inter-generational responsibility and justice.

The courts have also stressed the "polluter pays" principle and the precautionary principle as critical safeguards for sustainable utilization of natural resources and for environmental balance. Rulings in Sri Lanka, India, Bangladesh and Pakistan on environmental assessments for development projects have provided much needed impetus to the EIA legislation. It is also significant that in most cases the courts have accepted the principle of *locus standi* as a requirement in the promotion of public participation in the judicial process for environmental issues. For instance, a case at the High Court in Kuala Lumpur, held that the natives resident in an area, even though very few in number, had a right to participate in the process of environmental impact assessment for the construction of a proposed dam in their area.

CHAPTER ELEVEN

In the ultimate analysis, the success of the enforcement mechanisms depend on the overall institutional arrangements that have been put in place. The holistic paradigm of sustainable development requires integration of complementary and conflicting concerns that get resolved through the mediation of common goals and objectives. The experience of different countries in establishing an institutional framework for sustainable development does indicate that governments are attempting to rectify the shortcomings of earlier regimes and, with adequate emphasis on developing human potential and awareness the environmental transformation made possible through legislation, countries should be able to advance towards sustainable development.

CONCLUSION

Sustainable development entails the integration and inter-dependence of economic objectives with the protection and conservation of the environment. Institutional mechanisms for achieving this integration are, therefore, of paramount importance. Formulation and implementation of appropriate policies and programmes, enactment of laws, rules and standards for enforcement of policies and creation of delivery mechanisms for implementation of plans and enforcement by institutions provide credibility and stability to the environmental regime. Any institutional framework must be able to provide clear, unambiguous and unidirectional signals and incentives to the agents who must be able to carry out the underlying goals and objectives. The improvements that have emerged in the institutional framework of the region can be summarized as follows.

- Creation of multi-stakeholder agencies at the apex level with the objective of formulating goals, rules and regulations; providing guidance to the executive branches of government; monitoring performance; and, providing policy inputs. These are complemented by the creation of ministries responsible for environment, and empowering them with necessary executive and constitutional authority to undertake environmental protection measures.
- Designing formal and informal coordination mechanisms for integrating cross-sectoral concerns. Though the mode of coordination differs among countries, there is a recognition of the inter-dependence and inter-relationships between various sectors of the government and the economy.
- Recognition that devolution of certain responsibilities and functions to local authorities creates potential for the improved monitoring and management of resources.
- Judicial activism and interventions have increased and made the executive more responsive in meeting environmental standards and safeguards. These have also strengthened public interventions and support for eco-friendly measures.

There is now unanimous acceptance of the need for human action to control environmental degradation, curtail the depletion of natural resources and simultaneously achieve economic objectives which ensure human welfare. These issues need to be resolved through appropriate legislative measures and effective environmental legislation. Governments need to continue to bridge the gap between intent and action, and between policy formulation and implementation. Although appropriate legal and institutional frameworks have been established in most countries of the region, the effective implementation of environmental legislation remains one of the foremost challenges for the achievement of environmentally sustainable governance.

Chapter Twelve

Stringent control on emissions from motor vehicles has helped keep Singapore air clean.

12

Mechanisms and Methods

INTRODUCTION
ECONOMIC AND COMPLEMENTARY POLICIES
METHODS AND TECHNIQUES
MONITORING AND ASSESSMENT
CONCLUSIONS

CHAPTER TWELVE

INTRODUCTION

Whilst institutions and legislation provide the basic infrastructure for systems of environmental governance (see Chapter 11), policy instruments and mechanisms constitute the means by which a system of governance influences the dynamics of social, political, economic and environmental processes. This chapter reviews the policies, approaches and techniques used in the region in environmental governance at the macro- and micro-levels, and discusses the methods that have been applied, including command, control and incentive-based mechanisms. It also reviews the existing and emerging monitoring and assessment techniques that are being adopted in the Asian and Pacific Region to improve environmental governance.

ECONOMIC AND COMPLEMENTARY POLICIES

The merging of environmental and economic decision-making involves a fundamental realignment of developmental policies, with the process of development being viewed as a multipurpose undertaking that includes an implicit and defined concern for the quality of the environment. Significant progress towards that end has been made through the integration of sustainable development principles in the national planning of some of the countries of the Asian and Pacific Region. Increasingly, aspirations for sustainable development are also forming part of the macro- and micro-economic policy making process. National development plans routinely incorporate environmental policies and resource management principles, and sectoral and trade policies are increasingly reflecting environmental concerns. Moreover, a number of countries in the region have initiated action plans and activities to achieve the goals of sustainable development.

A. National Policy Frameworks and Planning Mechanisms

1. *National Strategic Planning*

The formulation and enactment of a national strategic plan or policy document on sustainable development is the first step to the incorporation of environmental issues into the national agenda for development. A common thread underlying national action plans and policy documents is the development of objectives, concerns, goals, policy options and development strategies for sustainable development, which provide a clear focus to the efforts of different agencies, individuals and organizations. The development of national planning mechanisms is increasingly underpinned by long term strategies that outline the institutional mechanisms for inter-sectoral coordination, enforcement and policy making for public participation, and create the climate for altering and shaping behavioural patterns conducive to the goals of sustainable development. Furthermore, the preparation process for such documents generates the cross-fertilization of ideas and views from experts, scientists, politicians and the public at large, which helps to build consensus on the approach to achieving sustainable development.

For example, in the Republic of Korea, the 1995 Green Vision 21 document was prepared as a long-term environmental policy to improve the quality of life by harmonising preservation and development. In order to achieve its goals, the Ministry of Environment has developed annual plans and mid-term action plans. The relevant line ministries, such as the Ministry of Construction and Transportation and the Ministry of Industry and Resources, then developed their own action plans in line with the overall strategy. Viet Nam has also prepared Environmental Vision 2020 as well as the National Strategy for Environmental Protection 2001-2010 and the National Environmental Action Plan 2001-2005 to address both long term strategic, as well as medium and short term, planning goals. In Malaysia, the Second Outline Perspective Plan (OPP2) 1991-2000 calls for the prudent management of resources and the ecosystems, as well as for the preservation of natural beauty and a clean environment to ensure sustainable development for present and future generations. A number of other countries have also prepared and adopted national conservation plans and strategies for sustainable development. A selection of these are listed in Table 12.1.

A major lesson learned from sustainable development strategies, national action plans or sectoral plans throughout the region, is the fundamental importance of understanding and accommodating the interdependence of different concerns-economic, social and ecological.

2. *Centralized Versus Decentralized Decision-Making and Planning*

The decentralization of action plans has also been reflected in the trend towards increased decision-making at the local level. For example, People's Republic of China reported that at the end of 1996, two-thirds of the 30 provinces, autonomous regions and municipalities had organized their respective Leading Groups and established working offices to implement their Local Agenda 21. In

Table 12.1 Governmental Environment Protection Vision in Selected Countries of Asia and the Pacific

Country/Territory/Area	Selected Vision Document(s)
NORTHEAST ASIA	
PR China	China's Agenda 21
Japan	Basic Environment Plan; Action Plan for Greening Environment
Mongolia	Mongolian Agenda 21 (1995)
Republic of Korea	Green Vision 21
SOUTHEAST ASIA	
Indonesia	Indonesia Agenda 21
Malaysia	Vision 2020
Myanmar	Myanmar Agenda 21, Second Outline Perspective Plan, 1991-2000
Philippines	Philippine Strategy for Sustainable Development; Philippine Agenda 21
Singapore	Singapore Green Plan
Thailand	National Plan
	Environmental Vision 2020
Viet Nam	National Strategy for Environmental Protection 2001-2010
	National Environmental Action Plan 2001-2005
SOUTH ASIA	
Bangladesh	Environmental Conservation Act, 1995; National Environmental Management Action Plan
Bhutan	National Environmental Strategy for Bhutan "The middle Path", 1998
India	National Conservation Strategies; National Policy on Pollution Abatement
Islamic Republic of Iran	National Environment Action Plan
Maldives	National Environment Action Plan 1999-2005
Nepal	National Environment Policy and Action Plan I & II
Pakistan	National Conservation Strategy and Action Plan
Sri Lanka	National Environment Action Plan; Forestry Master Plan; Coastal Zone Management Plan
CENTRAL ASIA	
Azerbaijan	National Environment Action Plan 1998
Kazakhstan	National Environment Action Plan
Kyrgyzstan	National Environment Action Plan 1996
Turkmenistan	State Programme of Regional Improvement 2000
Uzbekistan	National Environment Policy
SELECTED SOUTH PACIFIC	
Australia	National Strategy for Ecologically Sustainable Development
Cook Islands	National Environmental Management Strategy, 1993
Fiji	Sustainable Development Bill
Kiribati	Kiribati Environment Act, 2000
Marshall Islands	National Environment Management Strategy, 1993
Micronesia (Federated States of)	National Environmental Management Strategy
Nauru	National Environmental Management Strategy, 1999
New Zealand	Government Energy Efficiencies Leadership Programme; Sustainable Land Management Strategy; Coastal Policy Statement
Niue	National Environmental Management Strategy, 1994
Samoa	National Environment and Development Strategy, 1993
Solomon Islands	National Environment Management Strategy, 1993
Tonga	National Environment Management Strategy, 1993
Tuvalu	National Environment Management Strategy, 1993

Malaysia, each state is empowered to enact laws on forestry and to formulate forest policy independently. The executive authority of the Federal Government only extends to the provision of advice, training, research and technical assistance to the states and the maintenance of experimental and demonstration stations. Indonesia has initiated Local Agenda 21 with the specific objective to prepare practical operational plans for implementing sustainable development at the regional and provincial levels. In New Zealand, an increasing number of local authorities are preparing programmes that can be considered local Agenda 21. Three local authorities have joined the International Council for Local Environmental Initiatives. Environs Australia (formerly the Municipal Conservation Association) has taken the lead in translating Agenda 21 into practical measures to assist its implementation by Local Government. In the Russian Federation, virtually all regions are developing local plans for sustainable development.

In many countries of the region, local authorities and local planning are central to certain environmental activities. For instance, in Thailand, municipalities are encouraged to set up waste management action plans, whilst in Sri Lanka, the issuance of Environment Protection Licensing (EPL) has been decentralized. Similarly, the issuance of Environmental Compliance Certificates in the Philippines has been transferred from the central EIA authority to regional offices for certain projects.

In most countries of the region, land use planning is also locally managed. In Malaysia, local planning authorities prepare development plans, and it is envisaged that by the end of the Seventh Malaysia Plan, each local authority would have at least one local (land use) plan. In Japan, Land Use Master Plans established by prefectural governors, function as a means of comprehensive intra-administration coordination. To reflect the shift towards greater local autonomy, Indonesia's Agenda 21 proposes to grant greater autonomy to Level II governments to facilitate decentralization of land resource decisions.

The trend towards decentralization has led, perhaps inevitably, to occasional problems between central and local governments. For example, the recent introduction of local autonomy in the Republic of Korea has initiated conflict on environmental problems between the central and local governments, and between local governments themselves. To deal with such issues, conciliatory mechanisms such as a "Local Autonomies Association" were activated. Moreover, while devolution often leads to good governance, it can also lead to inter-jurisdictional competitions (Box 12.1).

CHAPTER TWELVE

> **Box 12.1 Devolution of Power and Environmental Regulation in India**
>
> The constitutional amendments in 1992 that became operational in April 1994 provided for the devolution of power to democratically elected local governments. The local governments for the villages (Panchayats), towns and cities have been empowered to undertake environmental activities such as soil conservation, water management, social forestry, water supply, public health and sanitation and solid waste management etc.
>
> Devolution of powers for environmental regulation among three levels of government (Central, State and Local) is intended to enhance participatory planning, and to reduce the cost of regulation, as well as reduce bureaucratic delays and uncertainty in implementation and monitoring. However, in a free and growing market economy, there is a risk that it may contribute to unnecessary competition among the jurisdictions through the relaxation of environmental regulations to attract more business and investment.
>
> The effectiveness of implementation of environmental laws and the degree of compliance by polluters varies significantly among the states, leading to instances where different State agencies have taken different decision over similar environmental issues. For example, the Government of Tamilnadu allowed the establishment of a chemical industry (Nylon-66) while the Government of Goa rejected such a proposal. Similarly, a number of polluting industrial units were closed down in the capital city of Delhi, but were welcomed by the neighbouring States. These examples indicate a lack of coordination among the State agencies. In response, an important safeguard against relaxation of norms has been put in place by the Central Pollution Control Board through the Minimum National Environmental Standards (MINAS). The State Boards can only make these standards more stringent if the local environmental conditions demand, but in no case can they make the local standards less stringent than the MINAS.
>
> The Indian case points out that while devolution of power leads to good governance, it also demands establishment of checks, and balances through a mix of central and state regulations to avoid potential harmful inter-jurisdictional competition in the implementation of environmental regulations.

The lack of local government resource, both technical and financial, is also a constraint to effective decentralization. In Sri Lanka, for example, the option exercised by a Provincial Council to enact its own Environmental Act, and the related institutional structures, created severe operational constraints due to lack of necessary expertise. In such a situation, efficient coordination between the central government (which is often responsible for allocating resources to local government) and provincial administrations becomes very difficult (ESCAP 1999). By contrast, Australia as a developed country has sufficient resources to support local government efforts. The Federal Government has a National Local Government Environment Resources Network and the Local Government Environment Information Exchange Scheme to provide information and support to local government to promote better environmental management.

B. Sectoral Policies

National-level plans and policies on sustainable development are now being increasingly supported by sectoral plans and policies. Indonesia, for example, has initiated the development of Sectoral Action Plans that incorporate sound environmental principles into sectoral planning.

The fulfilment of national obligations under agreed global environmental conventions has also led to a range of sectoral plans, including those related to greenhouse gases, biodiversity conservation and the phasing out of ozone depleting substances. The South Pacific island countries, for example, are developing policies and strategies in response to climate change with the support of the Climate Change Training Programme, whilst countries that are signatories to the Convention on Biodiversity, have completed or are in the process of completing their National Biodiversity Strategy and Action Plans (NBSAPs) (Chapter 3).

Sectoral plans and policies have also been traditionally developed for land, agriculture and water concerns. There are some efforts to integrate environmental considerations and broader sustainable development elements into such sectoral plans and policies (see Chapters 2, 4 and 10).

1. Budgeting

In order to be effective, macro- and micro-strategic planning policies have to be supported by appropriate allocation of budgets. Unfortunately, most countries of the region subscribe to the conventional view that environmental infrastructure is a semi-luxury that contributes to health and quality of life, but little to economic growth (ADB 1997b). Public expenditure on environment in the developing countries of the region, for example, is typically less than 1 per cent, and environmental investments are perceived to be needed only in the later stages of development.

However, a number of countries are advancing economic and fiscal policies, such as privatization appropriate pricing and removing adverse subsidies, that may provide the means of sustaining and/or increasing their environmental expenditure. Improvements in the use of economic instruments may also improve resource-use efficiency and reduce protection and rehabilitation costs by preventing overuse.

Such economic and fiscal policies have already been initiated in the region through the introduction of economic instruments in Japan, Philippines and the Republic of Korea. More recently Fiji, through its Sustainable Development Bill, is working towards providing the use of market-based instruments, such as tradable rights and taxes in almost all sectors where there is an abundance of natural resources, particularly in the management of fisheries and marine resources. This will ensure a move towards market-based incentives which provide the right market signals for the appropriate pricing of resources, and generation of necessary investments.

2. *Trade Policies*

The four key issues of relevance to trade policy in the region are: the international competitive effects of environmental regulations; the trade effects of environment-related product standards, and related ecolabelling and ecopackaging regulations; the use of trade measures to secure international environmental objectives; and the effects of trade on environmental resources.

The developed countries of the region have made efforts to create positive linkages between trade and environment. Australia has called for international action aimed at assessing the environmental effects of trade policies so that reforms can be supported by appropriate environmental policies. Both Japan and the Republic of Korea have taken an active role in international efforts to harmonize and incorporate trade and environmental issues through the WTO Committee on Trade and Environment. However, the position taken by India may well reflect that of many developing countries of the region. India supports an open, equitable, rule-based, cooperative, non-discriminatory and mutually beneficial economic environment. It holds the view that the solution lies not in unilaterally banning trade, but in a two way system of technology transfer, from developed to developing countries, and the pricing of commodities at a level that does not necessitate their overexploitation or jeopardize their development priorities. Trade measures should be applied for environmental purposes only when they address the root causes of environmental degradation, so as not to result in an unjustified restriction on trade.

The dumping of obsolete and polluting technology and products is also a major concern raised by the developing countries of the region. In the absence of appropriate agreements, private enterprises in developed countries with stringent environmental standards are encouraged to shift highly polluting industries to countries with lower environmental standards and regulations. Moreover, there seems to be growing awareness that certain products from the Asian and Pacific countries could be threatened with green bans or may be subject to complex ecolabelling or environmental certification procedures. For example, environmental groups in Australia and New Zealand have proposed a ban on Pacific Island hardwoods because of perceived exploitation of landowners, poor logging practices and the unsustainable nature of the industry. Timber export from Southeast Asia has similarly been the subject of targeting by European environmental groups; in response the countries like Indonesia and Malaysia have developed national systems and capacities for certification. Some countries are also trying to turn the preference for environmentally friendly processes and products to marketing advantage. Fiji's low input small-holder sugar cane production systems, for example, is amenable to conversion to organic production to take advantage of premium prices for certified organically grown and processed products (ESCAP 1999); similar moves are underway in the tea plantation sector in Sri Lanka, with the conversion of existing tea estates to "Biotea" cultivation.

3. *Complementary Social Policies*

(a) Property Rights

Experience has shown that communities with clear property rights with regard to national resources, manage these resources in a much more efficient manner. Australia, for example, supports policies that: define property rights for water; assist in the development of water markets and water trading; and provide irrigators with greater business flexibility. It recognizes that appropriate pricing policies can also help in meeting the long-term infrastructure needs of the irrigation sector.

Property rights can also influence the impact of trade policy on the environment, particularly in cases where exploitation of a natural resource depends critically on the available stock, such as in agriculture or forestry (Lopez 1991). For example, if the trade policy increases the value of timber, there is likely to be more investment in, and maintenance of, the resource with landowners internalizing environmental costs. Many countries of the region provide examples of links between property rights and conservation. In Malaysia, privatization of existing forest plantations has been encouraged, particularly in Peninsular Malaysia; the purpose is similar to that in Sri Lanka where Plantation Sector Reforms (privatization) have improved the productivity, investment profile and efficiency of the sector. In People's Republic of China, the Government formulated an encouraging policy that those who control the area get the benefit, and through this encouraged family contracting, corporate sharing, leasing and auctioning of the usage rights of the land.

India has strongly supported the linking of biodiversity conservation with the need to develop an internationally recognized regime for recognizing the property rights of local communities, particularly in relation to genetic resources. Pending this, all patent applications would be required to disclose the source and origin of the genetic material used; share the knowledge and practices about the use of genetic resources by the local communities and identification of such communities; and give a declaration that laws, practices and guidelines for the use of such material and knowledge systems in the country of origin have been followed.

(b) Public Participation

Public participation is integrated into the governance mechanisms for sustainable development in many countries of the Asian and Pacific region. Communities, NGOs, industry associations, and indigenous peoples are now provided with opportunities for representation in sustainable development and environmental institutions.

The level of such public participation varies between the developed countries, developing countries and economies in transition. Uzbekistan, for example, had no arrangements for the creation and functioning of NGOs, but since independence, their emergence and development became possible and they currently play a substantial role in the formation of public opinion and in governmental decisions on important social issues. In the Russian Federation, NGOs activity has also increased, but due to the lack of supporting legislation and poorly developed systems for information dissemination, the development of dialogue and cooperation between government restitution and NGOs is not yet widespread.

Increased understanding and knowledge at the community level, and recognition of existing local self-help mechanisms, however, have led to increased efforts that directly link governance for sustainable development to the communities. It is recognized that community participation is a long-term process that requires formal procedures supported by the time and effort required to build community consensus, and a willingness by government agencies to recognize this consensus and incorporate it into the planning and implementation stages.

METHODS AND TECHNIQUES

A. Command and Control Instruments

1. *National Level Instruments*

Command and control approaches have been applied to several sectors. For example, the management of forest resources in many countries employs a system of governmental concessions that license the allocation of allowable cuts, with the threat of cancellation or reduction of a concession as a mechanism for ensuring that the resource is protected. Brunei Darussalam's Reduced Cut Policy, for example, provides mechanisms for protecting forest resources and associated biodiversity and has resulted in a 50 per cent reduction in logging area. In commercial fishing, the command and control system is exemplified by New Zealand's Quota Management System (see Chapter 5). A number of countries employ a command and control approach for regulating effluents or emissions, whereby standards are set to a level where the discharge, emission, or deposition of waste, does not affect any beneficial use or result in conditions which are hazardous to public health, safety or welfare or to ecological resources (see Chapters 4 and 6).

In certain cases, building codes or control regulations are utilized instead of standards. In Fiji, where there is no formal legislation on air quality or noise standards, all new factories, dwellings and public entertainment buildings have to be formally approved and the conditions of approval typically require the control of air and noise pollution. In Pakistan, a Building Energy Code has been prepared as a supplement to the National Building Code. It includes specific recommendations for both building design and mechanical equipment, such as fans, lights, and air-conditioning. Improved building designs were found to reduce household energy bills by up to 20 per cent, which could be lowered to 50 per cent with the use of efficient home appliances. In Singapore, the developer submits building plans to the Ministry of Environment for clearance on technical requirements on environmental health, drainage, sewerage, air and water pollution control, and hazardous and toxic wastes.

The achievement of standards is also pursued through the requirement of technology fixes or limits to use. Bhutan has taken a combination of initiatives to curb pollution from traffic and cooking stoves

MECHANISMS AND METHODS

> ### Box 12.2 Control of Air Pollution in Bhutan
>
> The major sources of air pollution in the urban centres of Bhutan are automobiles, and the burning of fuelwood for heating and cooking, as identified by a survey conducted in Thimpu by the National Environment Commission in early 1999.
>
> More than 86 per cent of the vehicle fleet in Bhutan is concentrated in the urban centres of Thimpu and Phuentsholing. The remaining 14 per cent are sparsely distributed in the other districts of Bhutan. As of April 1999, the total number of vehicles registered in Bhutan was around 16 335. In 1997 the total number of vehicles registered was 11 798 and in 1998 the total number was 14 206. Between 1998 and 1999, there was an increase of 14 per cent in the number of vehicles registered in Bhutan. If this trend continues, the number of vehicle in Bhutan will double by the year 2010.
>
> Due to the increasing vehicle fleet in urban centres, air quality is deteriorating and causing concern to the authorities. The likely consequences of deteriorating air quality will be an increase in the frequency of respiratory diseases such as asthma and chronic bronchitis among the urban population.
>
> Some of the main causes identified for high levels of vehicular emission are, inferior fuel quality, aging vehicle population, improper and poor maintenance of vehicles, high altitude resulting in incomplete combustion, import of second-hand vehicles, and the use of two wheelers which often employ 2 stroke engines.
>
> Bhutan has taken major step in its efforts to curb problems relating to air pollution. The following are some of the measures that have been initiated to control air pollution:
>
> - A vehicle emission control programme was initiated in 1995 on a pilot basis with a view to set up baseline data for the purpose of deriving vehicle emission standards. Today, all vehicles must carry an emission test certificate to meet emission test requirements.
> - Bhutan has also started importing unleaded petrol and premium quality diesel (low sulphur content).
> - The import of second hand vehicles has been banned, together with the import of two stroke-engine two wheelers.
> - The Government of Bhutan is also promoting the use of more fuel-efficient stoves for heating and cooking (saw dust stoves).
>
> It is expected that these initiatives will not only help curb air pollution in urban centres but also reduce the pressure on the forests from the increased efficiency in biomass combustion.
>
> *Sources:* Government of Bhutan

(Box 12.2). Vehicular pollution control has also received increased attention in other countries. Indonesia's short-term targets include the reduction of lead content in gasoline by 66 per cent, provision of unleaded gasoline in all urban fuel outlets and regulations on the emission level of road vehicles. In India, the use of unleaded petrol in four-wheel vehicles fitted with catalytic converters has been introduced in four metropolitan cities (Mumbai, Calcutta, Delhi and Chennai), with plans to gradually extend this to other cities. The Committee on Clean Fuels of the Pakistan Environment Protection Council mandated the introduction of catalytic converters in two-stroke engines. In Malaysia, catalytic converters have been required since 1993 for all new cars. Singapore also mandates that all petrol-driven vehicles be fitted with catalytic converters.

Some command and control efforts have focused on particular sectors and priority areas. Indonesia's PROKASIH programme was launched in 1989 with the objective to develop strict policies for industrial wastewater treatment. Its initial focus was on the worst industrial polluters in the 24 most highly polluted rivers with the goal of reducing their pollution load to 50 per cent. In REPELITA VI, the country's Sixth Five-Year Development Plan, the objective has been broadened to reduce wastewater disposed to 50 rivers in 17 provinces by 80 per cent. In Thailand, the Ministry of Science, Technology and Environment (MOSTE) identifies and declares "pollution control areas". The provinces covered by such areas are mandated to formulate a Pollution Control Action Plan that contains remedial and preventive measures. This is then incorporated into the Environmental Quality Management Action Plan for the province.

Bans and phase-outs have been the preferred approach for truly unwanted pollutants, such as ozone depleting substances (ODS). The import of refrigerators using ODS has been banned in Sri Lanka, whilst in Thailand, CFCs have been banned in the production of new domestic refrigerators. Many countries of the region have also developed official phase-out schedules for ODS; for example, Singapore has a Tender and Quota Allocation System that caps the consumption of ODS in accordance with Montreal Protocol guidelines and ensures the equitable distribution of the controlled supply of ODS to registered distributors and end-users.

CHAPTER TWELVE

In the management of toxic chemicals and hazardous substances, countries may identify priority chemicals and substances and subject them to very strict regulation. In Thailand, for instance, about 918 chemicals were named in the Ministerial Announcement (1994) to be controlled by the Ministries of Industry, Agriculture and Public Health. In India, 18 categories of hazardous wastes have been identified and listed for strict regulation. Imports are restricted and exporters are required to communicate details of any proposed transboundary movement of hazardous wastes to the Central Government as well as to the concerned State Pollution Control Boards.

In some countries of the region, penalties are imposed for infringement of regulations. Companies or individuals in Malaysia planning to import chemicals, which are banned or restricted in countries that produce them, have to get prior consent from the Government. Maximum fines of RM 500 000 or up to five years' jail, or both, can be invoked for violations. It is also a serious offence in Australia to move hazardous wastes internationally without the relevant permit. The maximum penalty for offences that are likely to result in injury or damage to human health or the environment is Aus$1 million for a company, or up to five years imprisonment for an individual. In addition, executive officers of corporations may be held liable if they are found to have been negligent.

To combat potential impacts resulting from biotechnology, India has prepared its Recombinant DNA Safety Guidelines and Regulations. The Republic of Korea, New Zealand and Australia have also established systems that deal with these concerns through risk assessment and management. The Russian Federation has also recognized the need to ensure the safety of transboundary technology transfers of genetically modified organisms and their products, and has targeted the creation of a legal and regulatory framework for biodiversity as an urgent national priority. Indonesia has also taken initial steps to develop biosafety procedures for the control of genetic materials resulting from research in the country, as well as for the import of modified living organisms. It is also playing an active role in the negotiations for an international biosafety protocol under the auspices of the United Nations Convention on Biodiversity.

Command and control approaches are also directed towards governmental organizations and employees. In Japan, there is a strong focus on the "greening" of government – controlled industry, as well as overall government operations. Thus, in the area of government procurement, Japan is switching over to recycled paper and is promoting the adoption of low emission vehicles, including electric cars (Box 12.3).

While command and control measures are increasingly being applied in the region, there are several constraints to their effective implementation. For example, with respect to biodiversity conservation, many developing countries have not yet been able to develop comprehensive management plans for their protected areas. In Indonesia, less than five per cent of protected areas (out of a total of over 700 existing and proposed protected areas) have complete management plans and most other sites have not even been accurately surveyed or mapped.

At best, command and control approaches require a willingness to undertake enforcement and the power to enforce regulations. The effectiveness of enforcement is also related to the severity of sanctions, as perceived by the potential polluters. A case study in Sri Lanka revealed that there are problems with regard to enforcement of penalties: the current law is overloaded with command and control mechanisms which cannot be matched by appropriate regulatory resources; lack of relevant regulations and existence of other soft law are making the regulations less effective whilst court delays are resulting in delayed judicial enforcement. The result is that there is a reduced risk of sanction of polluters, and hence the laws are being ignored.

The lack of human resources to enforce legislation is exacerbated by a mismatch between the type of training and experience of available personnel and the critical concerns at hand. In Indonesia and the Philippines, for example, protected areas increasingly include marine and wetland ecosystems. However, the agencies responsible for these areas do not have sufficient experts to manage them, because in the past, the protected areas covered mostly forests. In Armenia, on the other hand, its State Forest Service "Hayantar" is affected by problems of both inappropriate structure and weak capacity, because during the Soviet era various core functions were carried out centrally in Moscow.

Given the above-mentioned constraints, the level of enforcement is often low and an average level of enforcement of 20 per cent is considered to be relatively high for the developing countries of the region (Lohani, et al 1997).

2. *Command and Control at Sub-national and Local level*

The main command and control instruments used at the sub-national level are zoning and land use control. For example, Thailand's Office of Environmental Policy and Planning has initiated a programme of coral protection and rehabilitation by a zoning system based on the classifying of coral

MECHANISMS AND METHODS

> **Box 12.3 Greening of Government Operations in Japan**
>
> In Japan, the national government accounts for 2.2 per cent of gross national expenditure, representing a significant amount of economic activity. Thus if the government conducts its operation in an environment-friendly manner, it can make a major contribution to reducing the burden on the environment.
>
> Realizing this the government of Japan took a lead in adopting environment-friendly actions and encouraged the local governments, corporations and individuals to do the same. The process was initiated with the approval of an Action Plan by the Cabinet in June 1995 for Greening Government Operations (Green Government Action Plan). Details on the implementation of the plan were agreed at a Council of Ministries and Agencies Concerned on Promotion of the Basic Environment Plan.
>
> The Green Government Action Plan provided the actual measures to be carried out in various areas. Some of the most notable initiatives implemented or under way for greening the government operations recently included:
>
> - Environmental considerations are taken into account in construction and maintenance of buildings.
> - Reduce waste generation by 25 per cent or more between 1997 and 2000 particularly reduce the quantity of combustible waste (paper, etc.) by 30 per cent or more in this period.
> - Creation of awareness in government employees through participation and by organizing environmental protection seminars etc.
> - Establishment of a system to monitor progress.
>
> In addition the government makes efforts to save water and energy and promotes recycling and other environmental conservation measures at all its offices. The plan sets out many concrete proposals including 11 numerical targets for achieving objectives on energy savings, use of recycled paper and the introduction of low-emission vehicles.
>
> A survey was conducted to determine the progress in implementing the Green Government Action Plan at all ministries and agencies during 1997. According to the survey, considerable progress was made in many fields however, in some fields performances were still short of targets. Most notably, the amount of waste was still excessive and the diffusion rate of low-emission vehicles (0.13 per cent) was far below the target set for the year 2000 (10 per cent).
>
> The above measures by government of Japan would hopefully further enable greening of the government performance in the country.
>
> *Source:* Environment Agency, Government of Japan 1999

reefs according to conservation and economic development objectives. In Pakistan, the Port Qasim Authority has initiated a number of zoning control activities that involve relocating industries in large industrial estates to clear the coastline. Further examples of coastal zone management mechanisms are provided in Chapter 5.

Land use planning is also used to strengthen the rural-urban continuum and protect prime agricultural lands. India regulates the conversion of agricultural lands to non-agricultural use through land revenue codes, whilst Fiji has a comprehensive "Land Conservation and Improvement Act" where failure to adhere to good land husbandry can result in farmers losing their lease entitlement.

The practice in Singapore is for the planning and development authorities to consult the Pollution Control Department (PCD) on proposed new developments. The PCD not only assesses the impact of such new developments on the environment but also ensures that new industrial and residential developments are properly sited and are compatible with the surrounding land use. Sri Lanka follows a similar strategy. As part of its efforts to ensure proper waste disposal, industries are encouraged to site in industrial estates and the relocation of certain industries that generate hazardous wastes, such as tanneries and the production of pesticides, is ongoing.

The recently passed Clean Air Act in the Philippines provides another example of the application of the land use planning concept to pollution control. It mandates that airsheds be delineated using ecoprofiling techniques and emission quotas are then allocated to specific pollution sources.

The Republic of Korea encourages resource saving and environmentally friendly land development by constructing energy conserving and environmentally sound traffic systems, including networked subway systems, exclusive bicycle roads, a bus lane system and radial and circular transportation networks between metropolitan areas and their suburbs. At the same time, the development of multi-centred cities is promoted to disperse the concentration of human and car traffic throughout the area.

CHAPTER TWELVE

3. *Command and Control at the Project Level*

(a) Environmental Impact Assessment (EIA)

In a large majority of countries, EIA has a statutory basis either in the form of a separate act (e.g. Republic of Korea, Sri Lanka, Philippines, New Zealand) or an amendment to or provision under existing environmental or planning laws (e.g. Indonesia and Malaysia) (see Chapter 11).

Many donor countries and agencies (i.e. multilateral banks) also apply the EIA process to their development assistance projects. In response to their national and international requirement, People's Republic of China prepared its Management Guidelines on Strengthening Loan Projects for EIA, and guidelines have also been issued in countries of the region for EIAs in certain sectors. Thus the Central Pollution Control Board of India has issued guidelines for EIA in key industrial sectors, whilst Bhutan has also developed sectoral guidelines for key development sectors including hydropower, power transmission line, forestry, industries, mining and mineral processing and transportation infrastructure. In addition, guidelines have also been expanded in many cases to include occupational health and safety issues (e.g. in Australia) and the use of environmental health impact assessments of strategic policies and development programmes and projects (e.g. India and the Philippines). The integration of social aspects into EIA has also been initiated, incorporating the potential for cultural disturbances, the impact on cultural sites, the loss of access to subsistence resources, the pollution of water supplies and the development of appropriate mechanisms to facilitate an ongoing relationship between the developer and the landowners for the resolution of future disputes.

In general, at least three major outputs are expected from the EIA process: an identification and analysis of the environmental effects of proposed activities; an environmental management plan which outlines the mitigation measures to be implemented; and an environmental monitoring programme. In some jurisdictions, such as in Indonesia, separate documents are to be prepared for each of these aspects. In others, such as in the Philippines, all three are presented as part of the EIA document.

EIA reports are often reviewed by a designated agency or by a special "Standing Committee" or "Scoping Commission". In most cases a technical evaluation is made by specialists as the basis of the review. In Thailand, EIAs for private projects must be approved by a committee of experts before licenses are granted by the appropriate agencies, although the process has recently been changed to require additional review groups. In Republic of Korea, the Central and Regional Committees for EIA, which consist of academics, engineers and specialists, review all EIA outputs. India follows a procedure whereby the EIA study has to be evaluated and assessed by an impact assessment division (Ministry of Environment and Forests or the State Government depending on the nature and location of the project) who may consult a Committee of Experts, if deemed necessary. In Fiji, all EIA reports are submitted to the Environmental Assessment Administrator of the Department of Environment for an assessment and the recommendations are subsequently submitted to the National Council for Sustainable Development for approval.

In general EIA implementation in the region faces many constraints. Common, though not universal, amongst these are insufficient procedural guidance; inadequate baseline data upon which to base analyses; potential delays in project implementation; the lack of expertise for assessing impacts; inefficient communication of EIA results to decision makers; lack of inter-agency coordination; limited capacity for review of EIA reports; and insufficient commitment to follow up on the implementation of environmental protection and monitoring requirements. Other notable problems include the low status of the agencies in charge of EIA in the overall bureaucracy, the perception that EIA is simply another bureaucratic requirement, and unwillingness of certain governments to open the process up to public debate, the relative weakness of affected interest groups and lack of commitments among government officials. For example, a case study in Malaysia (ESCAP 1999) showed that, while the EIA procedure was supposed to be carried out at the feasibility stage, the lack of adequate studies as well as late submissions of such reports have reduced EIAs to a mere formality.

Pressing issues and emerging challenges include the need to improve the quality of EIA practice (e.g. by establishing monitoring, review, and other control procedures), securing a more cost-effective process, further strengthening public involvement and greater emphasis on training, technical cooperation and professional manpower development.

(b) Environmental Risk Assessment (ERA)

The inclusion of ERA into the mix of environmental management techniques arises from the need to avoid and mitigate the risks and hazards associated with certain forms of development. Past experience such as the Bhopal tragedy in India, (ESCAP 1990), the continued increase in production and movement of toxic chemicals and hazardous wastes and concerns related to new technologies such as biotechnology have made ERA increasingly necessary.

Risk assessment also provides a more accurate presentation of a project's true worth by eliminating investment bias towards projects that promote overuse or degradation allowing comparisons and rankings of different projects or alternatives for a project. ERA has been applied, for example, by the ADB to help make decisions on projects such as the 700-MW coal-fired power plant in Pagbilao Grande Island, Philippines, and the Yunnan-Simao forestry development and pulp mill project in Yunnan Province, People's Republic of China.

Some countries in the region have also developed their own systems for environmental risk assessment. For example, the Republic of Korea has expanded and accelerated its chemical risk assessment capacity and has introduced an Environmental Toxicity Prediction Programme. Australia has established a National Industrial Chemicals Notification and Assessment Scheme and an Existing Chemicals Review Programme to review long established chemicals against contemporary standards.

India applies ERA to industrial units and efforts are underway in the adoption of hazard analysis, off-site emergency plans, the establishment of emergency response centres and poison control centres. India's Ministry of Environment and Forests has also established the National Register of Potentially Toxic Chemicals (NRPTC) for collection, collation, analysis and dissemination of existing national and international information supported by sub-national Registers to create a widespread network throughout the country.

ERA of projects is an emerging issue in many developing countries in the region. It shares similar needs and problems as those involved in the development and promotion of EIA and there is a need to strengthen institutional arrangements and capacities as well as review, monitoring and enforcement powers, with legal provisions for the effective implementation of the process.

B. Self Regulation and Participation of Stakeholders

Self regulation and voluntary or cooperative control mechanisms are being increasingly promoted in the countries of the region, in part due to governments' limited capacities for enforcement through command and control mechanisms. For example, Japanese industries, with the encouragement of the government, have started voluntary emission control measures for some hazardous air pollutants with a pledge and review system. In addition, the Japan Chemical Industries Association has implemented a pollutants release survey and New Zealand's Chemical Industry Council has a Responsible Care programme for the management of hazardous wastes and chemicals. The chemical industry in Australia has also undertaken a voluntary programme to improve the health, environmental and safety performance of its operations. Further examples of private sector self-regulation are presented in Chapter 13.

C. Market Based Instruments

A range of economic instruments has been used in the countries of the region to promote sustainable development (Table 12.2) and there are plans to strengthen these in future. For example "Agenda 21 – Indonesia" frames new programme areas to promote "economic approaches to natural and environmental resource management", "preventive approaches to pollution", and the "development of systems of economic, natural resource and environmental accounting". People's Republic of China also has plans to actively promote shifts in the economic development model and make economic efficiency the core of economic activity. This includes the reduction of energy consumption and increasing production efficiency. Favourable prices, taxation and preferential loan policies are to be adopted in fields that are conducive to society and environment, such as pollution control, clean energy development, utilization of waste material, and natural resources protection. The protection of resources and the environment through economic instruments will be a guiding approach (Box 12.4).

Even in newly emerging economies, such as Mongolia, the policy direction is towards the use of economic incentives for the optimal utilization of natural resources and the protection of the environment. In Uzbekistan, the Cabinet of Ministers approved the concept of the "establishment of scientifically based economic and legal mechanisms for the use of natural resources" on June 26, 1996. Reforms in sustainable agriculture in Armenia include providing the necessary legislative framework for the implementation of a pricing policy, whilst the Russian Federation has created economic machinery to adapt the water industry to market conditions and facilitate adequate financing of water management activities.

The following sections examine the different types of market-based approaches as they are implemented in various countries of the region, including environmental taxes; user fees; targeted subsidies; eco-labels; disincentives; and deposit refund systems.

CHAPTER TWELVE

> **Box 12.4 Environmental Tax Reforms in People's Republic of China**
>
> Environmental taxes are often criticized for being inflexible and difficult to adjust to changing conditions. One difficulty is the growth in number and size of pollution sources and the consequent higher level of emissions. Still another is the increase in the public's willingness to pay for improved environmental quality as incomes rise and environmental sensitivity increases. In a rapidly industrializing and growing economy such as in People's Republic of China, these factors are particularly important. The reformed Pollution Levy System (PLS) in the country therefore includes annual escalation of the rates (proposed at 9 per cent) to deal with these problems. This approach also provides enterprises with information to make optimal investment decisions. The new PLS is designed to be a national system, but People's Republic of China's regions vary widely in industrial intensity, income, population density, and their demand for environmental quality. Accordingly, it provides for regional shift factors that incorporate these considerations into rate determination. As a result, the richer, more heavily industrialized regions will pay a higher rate than less developed areas.
>
> The reformed PLS adopts the "pollution equivalent" (PE) concept to provide an administratively feasible benefits basis for the system. Following this concept, each pollutant is rated according to its harmfulness and assigned a PE index value greater or less than the reference emission, which is assigned the value 1. This factor is then used to adjust the measured quantities of pollutants according to their relative proportions in the overall emissions stream. Because a uniform rate is charged on this constructed tax base, the rate structure itself remains uncomplicated while still accounting for the actual risks of the pollutants in question. The reformed PLS simplifies the task of levying and enforcing the tax applying a flat rate to sources and pollutants where determining actual emissions is impractical. Per vehicle charges for noise pollution, per capita charges for wastewater treatment, and per household charges for garbage are examples. The pragmatic solutions to cover not only large industrial sources but also smaller enterprises and nontraditional emitters, including some non-point sources to use a flat fee only roughly related to emissions.
>
> Projecting revenues from the reformed PLS over time is difficult both because incomplete data on the tax base and because the behavioral responses to the incentive to reduce pollution are uncertain. However, a revenue simulation pilot study covering 77 cities and using 1995 data found that receipts would have been about 10 times the amount actually collected by the old system, assuming no changes in behavior due to the tax itself. On the revenue use side, funds raised by the tax will continue to support environmental administration at all levels, and with the remainder for funds created to fill the investment gap in pollution control caused by capital shortages and by inadequate general government revenue due to the existing tax structure.
>
> Implementation on a pilot basis to test the programme began on 1 July 1998 in three large metropolitan areas in different regions of the country, Hangzhou, Zhengshou, and Jilin. For the pilot programme, applicable tax rates have been set at one-half of those proposed for the reformed PLS. The pilot programme is set to last one year, and it will be followed by an intense evaluation with subsequent required adjustments. Assuming all goes well, nationwide implementation of the new PLS is scheduled for 2000.
>
> One factor in the success of the reform so far is the full participation of field-level personnel, who helped structure a programme responsive to the needs of municipalities, provinces, and the taxed entities. Another factor is the inclusion of persons with political experience. Moreover, participation by key non-environmental agencies avoided provisions that otherwise would have been certain to spawn conflict.
>
> Strengthening and expanding the monitoring, inspection, enforcement, and administrative functions and the performance environmental authorities is a monumental-task and a major challenge to implementation. However, when compared with those of using more intensive command-and-control regulation to achieve the same level of pollution reduction, the tax approach appears better because of its transparency and reduced opportunities for administrative exceptions.
>
> *Source:*   Environment 1999

1. *Environmental Taxes*

Environmental taxes have been used to reduce the use of automobiles and motor fuel and to reduce pollution. Thailand has a tax structure that puts a rate of 100 per cent or higher on automobile imports both as a measure to curb excessive consumption of the item as well as to control a potential source of pollution and congestion. The government also puts a price differential on leaded petrol.

Industries in Singapore may also apply for permission to discharge their effluent containing biodegradable pollutants directly into the public sewers on payments of a tariff, allowing the additional costs of treating the extra pollution load at the sewage treatment works is to be recovered.

The Department of Environment and Conservation of Papua New Guinea and the National Environment Commission of Bhutan have proposed that the costs of environment impact assessment, monitoring and auditing performance should be borne by developers rather than the government. Alternatively, those costs, which are borne by the government, should be recouped through the imposition of fees. This is actually already the practice in the Philippines where developers are required, under the conditions of their Environmental Compliance Certificates, to establish an Environmental Guarantee Fund (EGF) as well as a Multipartite Monitoring Fund.

MECHANISMS AND METHODS

Table 12.2 *Market Based Instruments and Their Sample Applications in Asia and the Pacific*

By sector or theme		Subsidy reduction	Using Markets						
			Environmental taxes on			User fees for natural resources	services	Perform bonds/ deposit-refund	Targeted subsidies
			emissions	inputs	products				
Resource Management	Water resources	Reduction in water subsidy: PR China					• Water Pricing: PR China • Watershed protection charges		
	Fisheries			Fishing input taxes	product taxes	Fisheries Licenses			
	Land management					• Property taxes	Betterment charges: Republic of Korea		
	Forests					Stumpage fees: Indonesia, Philippines, Malaysia	• Park entrance fees • Reforestation taxes: Indonesia		
	Sustainable Agriculture	Reduction in agriculture subsidies: most developing countries							
	Biodiversity/ Protected area	Reduction in land conversion subsidies					• Watershed protection charges: Indonesia • Park entrance fees: Indonesia, Nepal		Habitat protection
	Mineral resources		Fees on mine wastes and tailings: Philippines			Mining royalties: Brunei, Malaysia and other OECD countries			
Pollution Control	Air Pollution	Reduction in energy subsidies: Transition economies, most developing countries	Emission taxes: Republic of Korea, PR China, Kazakhstan, OECD countries	Energy taxes: OECD Differentiated gasoline prices: Philippines	Environment related product taxes: Bangladesh, OECD	Royalties for Fossil fuel extraction			
	Water Pollution		Wastewater discharge fees: PR China, Republic of Korea, OECD, Philippines				Sewage charges PR China, Indonesia, Malaysia, Singapore, Thailand		
	Solid waste						User fees for waste management: Japan, Philippines, PR China	Credit/subsidy policy: Republic of Korea, PR China	Tax relief and subsidized credit for env. investment: PR China, India, Republic of Korea, Philippines
	Hazardous Waste/toxic chemicals	Reduction in agrochemical subsidies in most developing countries	Disposal charges: PR China, OECD, Thailand	Pesticide taxes: OECD				Bond for waste treatment	

CHAPTER TWELVE

Table 12.2 Market Based Instruments and Their Sample Applications in Asia and the Pacific (continued)

By sector or theme		Creating Markets			Environmental Regulation			Engaging the Public	
		Property Rights decentralization	Tradable permit/rights	International offset system	Standards	Bans	Quotas	Information disclosure	Public participation
Resource Management	Water resources		Water Markets: Australia, India, New Zealand	Water trading across borders	Water quality standards			Water efficiency labeling	
	Fisheries	200-mile Exclusive Economic Zone (EEZ)	Tradable quotas/ permits: New Zealand		Fishing standards	Fishing bars	Fishing quota		
	Land management	Land title: Thailand		Tradable conservation credits	Landuse standards/ zoning: PR China, Republic of Korea, OECD, Pakistan	Establishment of environmentally sensitive areas	Land Subdivision		
	Forests	Land Titling	Tradable reforestation credit		• Logging regulations	• Logging/log export bans:	Logging Quota: Malaysia		
	Sustainable Agriculture	• Land ownership Thailand • Participatory irrigation management: India, Philippines, Sri Lanka	Transferable development rights			Ban on use of pesticides: Indonesia		Eco-labeling many OECD Countries	Community self-help groups:
	Biodiversity/ Protected area				Conservation zone: PR China	Establishment of national parks: Philippines, Indonesia			
	Mineral resources				Waste and tailings containment			Energy efficiency labeling	NGO involvements Philippines
Pollution Control	Air Pollution	Environmental liability • Private energy production: Philippines	• Tradable emission permits: Kazakhstan • Auctionable permits for ODS: Singapore		Air Quality and emission standards: PR China, India, Republic of Korea, OECD, Philippines, Singapore	Ban on imports of ODS: PR China	• Emission quotas OECD. Area licensing for vehicles: Singapore	Public disclosure programmes: Indonesia	
	Water Pollution	Environmental liability	Tradable waste water discharge permits		Wastewater discharge standards: PR China, India, Indonesia, Republic of Korea, Malaysia, OECD, Philippines, Singapore		Industrial wastewater Discharge quotas: PR China, OECD	Public disclosure programme: Bangladesh, Indonesia, Philippines	Community pressure: Republic of Korea
	Solid waste	Environmental liability		Tradeable recycled contents	Landfill standards and landfill zoning			Industrial waste exchange programme: Philippines	
	Hazardous waste/toxic chemicals	Environmental liability			Containment/ treatment standards	• Basel convention • Ban on use of some pesticides: Indonesia		Labeling	

272

2. User Fees

Road pricing was introduced in Singapore as early as the 1970s to reduce road congestion and control air pollution. Charging of drivers who want to use roads in the city centre during peak hours, led to a reduction in congestion of 70 per cent (Panayotou, 1994). The Republic of Korea also uses congestion fees on cars entering designated areas. In addition, high parking fees in public parking lots which have previously been open to the public without charge, and a heavy tax for households possessing more than one car has been implemented.

In Singapore, the capital and recurrent costs of building and operating the water and wastewater treatment plants are financed by the government and recouped from users through fees and tariffs. By contrast, dedicated hospital waste incinerators are built and operated by the private sector and users pay full cost fees to dispose of their waste.

The Philippines has recently implemented an environmental user fee for industries discharging wastewater into the Laguna de Bay. In the pilot phase, one pollutant, biological oxygen demand (BOD), was considered. The first batch of industries, including major polluters representing ninety per cent of pollution discharge into the lake, was initiated into the system in 1997. In addition to meeting the existing discharge standards, these industries were required to pay a fee for every unit of pollution they discharge. The polluters were also required to pay a small flat fee in addition to the variable fee to ensure a continuous funding of the system's administrative costs. The environmental user fee system succeeded not only in creating revenue for the implementing agency but also encouraged the larger industries to cut pollution loads by over ninety per cent in some cases. The success of the environmental user fee system in the management of Laguna de Bay supported the passage of further legislation mandating the use of market based instruments. The Clean Air Act of 1999 in the Philippines contains provisions that establish an emission charge system for industrial permitting and in the vehicle registration renewal system. The fees, as well as fines and penalties, go to an Air Quality Management Fund to finance containment, removal and clean-up as well as support research, enforcement, monitoring activities and technical assistance to relevant agencies.

User fees have also been successfully utilized in forest management within the region. For example, by pricing forestry resources, incentives were created for forest protection as well as the raising of revenues: in 1980, Indonesia started a reforestation fee of $4 per m^3 which has now been raised to $22 and earned more than US$1 billion for a fund for forest restoration and conservation. Indonesia also plans to use further innovative and sustainable financing solutions, including charging ecotourists for the use and enjoyment of protected areas to generate revenue to sustainably manage their resources.

3. Targeted Subsidies

Subsidies have historically been targeted at certain natural resource sectors that require particular protection. For example, Malaysia has complemented efforts in sustainable timber production by providing tax exemptions for investment in timber plantations. Japan makes available substantial funds (2.5 billion yen) for low interest loans to private owners of forests for the revitalization of forest improvement activities under the Temporary Fund Law for the Improvement of the Forestry Management Framework.

Other forms of subsidy related to natural resources management include: preferential low interest loans for desertification control (People's Republic of China); reduction in the interest rates of credits given to support forest villagers; and new legal and financial arrangements to encourage the private sector and farmers to be involved in plantation activities (Turkey); incentives for wood substitution, subsidies in use of fuel-saving devices and alternative sources of energy supply like biogas and solar energy (India); provision of kerosene oil depots in places where regular energy supply for people can be ensured without harming the forests (Nepal); and provision of credit-providing financial institutions to coastal communities to finance the development of higher value-added fish-processing businesses (Indonesia).

Taxes have also been exempted or reduced, for example in Thailand, to provide for investment in environmental protection and energy saving equipment. Singapore has tax incentive schemes to encourage owners of trade, commercial and industrial buildings to use energy-efficient equipment and technology and efficient pollution control equipment. The Philippines allows tax deductions for the installation of anti-pollution equipment for pioneer industries. In India, the incentive of a 35 per cent investment allowance, compared to the general rate of 25 per cent, is provided towards the cost of new machinery and plant for pollution control or environmental protection. Sri Lanka also provides some fiscal incentives to industries to use advanced technology for the control or minimization of pollution or waste. In Nepal, a duty concession is provided for environmentally friendly vehicles.

Some countries have set up specialized funding mechanisms to provide support or subsidies. To efficiently prioritize investment and secure new revenue sources, the Republic of Korea introduced the Special Account for Environmental Improvement funded through the various charges imposed on polluters. A Pollution Prevention Fund in Thailand has been established and is expected to serve as catalyst for moving industry, especially small and medium scale industry, towards a more sustainable pattern of development. In Sri Lanka, the Pollution Control and Abatement Fund (PCAF) and the e-friends fund by the National Development Bank, both for investment in effluent treatment machinery, have already shown encouraging results.

4.   Eco-marks and Ecolabels

Eco-marks or Ecolabels are attached on consumer products in some countries of the region to indicate that they are either manufactured by environmentally sound processes, or can be used in environmentally sound ways, or generally over their entire life cycle, impose fewer burdens on the environment than similar products. For a more detailed discussion of ecolabelling schemes operating in the region, see Chapter 13.

5.   Disincentives

Subsidy reduction is another measure that has been effective in encouraging shifts to cleaner production. People's Republic of China, for example, has significantly reduced its subsidy for coal from 61 per cent in 1984 to 11 per cent in 1995. Its total economic subsidy for fossil fuels fell from $25 billion in 1990/91 to $10 billion in 1995/96 (World Bank 1997). Central government budget subsidy to cover operating losses in state-owned coal mines has also been significantly reduced. Coal price controls were removed with growing privatization and about half of the production is now from private mines. This has reduced government spending and through structural adjustment and technological change, has encouraged energy conservation and environmental protection (Gray 1995, Wang 1996).

6.   Deposit Refund System

Japan maintains exemplary deposit-refund systems for solid waste recycling. In Japan's beer bottle recycling system, for example, beer makers levy a fee on the bottles as well as on the containers of beer. The deposit is passed on from manufacturers to wholesale dealers, and then to retail shops and ultimately to consumers. At each distribution stage, the refund is made when the used containers and bottles are collected.

Deposit schemes are also being applied effectively on packaging waste in Turkey and recycling rates of up to 65 per cent are being achieved. Republic of Korea also implements a Deposit-Refund System for products containing toxic materials or discharging mass wastes. Deposit and refund practices have also been quite widespread in the consumer industry of India and Sri Lanka (see Chapter 8)

7.   Constraints

Economic instruments that are self-regulated require fewer inspectors and court actions, and generally result in a reduction in bureaucratic delays. The use of market based instruments, such as pollution levies and environmental user fees, also raise revenue. However, these instruments become less effective when the charge rates are below the marginal costs of pollution control and are not indexed for inflation, and also the imposition of full cost recovery too quickly can creates serious economic problems for industry, as has been experienced in economies in transition including Kazakhstan, Kyrgyzstan and Uzbekistan (HIID 1996a and 1996b). Lack of human resources and expertise is a major constraint to their implementation.

Economic instruments have been found to yield the best results in situations where a strong monitoring system is in place, supplemented by a strict enforcement system. Consultation with industrial associations is essential to minimizing resistance to newly imposed systems, whilst public information and programmes and incremental charge systems are other basic ingredients in achieving success.

D.   International Standards and Mechanisms

Transboundary pollution does not recognize political boundaries and, whether moved by water or air or by trade and other means, eventually affects and imposes external costs on downstream or downwind countries. Moreover, in a world that is increasingly linked by trade, financial flows and technological diffusion, the difficult task of promoting growth while at the same time pursuing sustainable development cannot be pursued by individual governments alone. It is for the same reason that a number of tools and standards such as ISO 14000, ecolabelling, trade related standards, etc. have been developed to check environmental degradation and promote sustainable development (see Chapter 13).

MONITORING AND ASSESSMENT

A.   Environmental and Natural Resources Accounting

Environmental and natural resources accounting provides a methodology for monitoring the sustainability of an economy or the performance of one or more of its sectors. The approach involves measurement of environmental quality and resource stocks at the start and end of on accounting period to determine the changes in natural assets as a result of their use. Physical measurements of change in resources and environmental quality are also translated into monetary units to build a comprehensive picture of the status of environmental and economic resources.

Some countries in the region have already developed systems for environment and natural resources accounting. Japan has developed a System of National Accounts (SNA) that includes the Satellite System for Integrated Environmental and Economic Accounting (SEEA), drawing upon the standards contained in the SNA Handbook on Integrated Environmental and Economic Accounting of the United Nations. Particular importance is being given to quantitative and qualitative changes in Japan's forest and agricultural resources.

In Australia, the valuation of natural assets and expenditure on environmental protection is incorporated into its system of national accounts. Its Bureau of Statistics has developed national account balance sheets to include the market value of natural assets including forests, subsoil assets and land. These estimates are based on resource use values and exclude non-monetary environmental values. A range of environmental accounts, including physical accounts in an input-output framework and financial accounts for environmental protection, are also being developed.

The Republic of Korea has also established a system of Integrated Environmental and Economic Accounting (IEEA) to enable the government (and potentially private firms) to obtain correct information on citizen's welfare. The project to establish the IEEA system is based on the United Nations System of Integrated Environment and Economic Accounts (SEEA).

Environment and natural resources accounting has also proceeded in a more advanced stage in the Philippines, Indonesia, Malaysia and Thailand. In the Russian Federation, there are plans to adopt a more rational approach to account for production in any given region on the basis of strict natural resource accounting. On forestry, for example, production is not only based on the value of the timber but also take into account the economic and environmental consequences of the decisions to cut forests. A land-survey and evaluation of forestland has been started and the values are expressed in roubles per unit of forest area for each type of forest and in each specific region.

B. Environmental Assessment and Indicators

Environmental assessment or an overview of environmental problems of a country or an area is an essential prerequisite not only for environmental accounting but also for overall policy planning. It is for the same reason that State of Environment (SoE) reports are now being prepared at national and local as well as regional and subregional levels. Japan's Environmental Pollution Control Headquarters (1969) produced the first National SoE report in the Asia and Pacific Region and, Osaka City was the first local authority to produce an SoE report. In recent years a large number of reports have been produced in the countries of the region at national level, although local level reports are relatively rare in the developing countries of the region. At the Regional Level, ESCAP produced the first SoE Report for Asia and the Pacific in 1985, whereas the late 1990's saw a number of subregional SoE's by ASEAN, SPREP and the Mekong Secretariat.

Effective assessment needs a comprehensive information system. However, a review of the information base of many countries in the region shows four shortcomings in the assessment and reporting on the State of Environment. Firstly, the lack of quality control in data collection and laboratory analysis of samples. Only in a few cases has the reliability of collected information been verified through ground checks and proofing. Secondly, the basic data available are often not adequate to make a realistic environmental assessment. Several countries in the region often operate on the basis of outdated data. Without updates (e.g. on land use patterns or the utilization of resources), the data is unreliable for decision-making. Third, the statistical compatibility on environmental parameters is far from satisfactory. Collected to serve administrative and economic purposes, they are compiled according to different methods. The collected data are relevant only to the departments doing the assessment. Time series data, where available, are confined to the physical and monetary inputs provided for developing the resource. Data useful for characterizing the sustainability of a resource are hardly collected, because systematic monitoring has begun in most developing countries only recently. Improvements in the measurement techniques call for new benchmark data. Fourth, and finally, environmental assessment becomes subjective in the absence of indicators of sustainable development and there has been no consensus on how to measure (or define) sustainable development. Many countries in the region would like to base the indicators on countries specific goals for achieving sustainability, which give due weight to the particular constraints and challenges of each individual country. For example, several countries would like to adopt a process of pre-testing and review before adopting a set of sustainable development indicators as a standard measurement on their progress.

Other countries, such as Japan and the Republic of Korea, are beginning to apply their long experience with social, economic and environmental indicators towards the development of indicators for sustainable development. These efforts are guided by legislation and national plans, such as the "Basic Environment Law" and the "Basic Environment Plan" of Japan, as well as by institutional needs, ranging from the establishment of a National Commission on Sustainable Development in India to a Council for Sustainable Development in the Philippines.

Environmental indicators are also being developed for specific sectors. For example, Indonesia has applied "criteria and indicators" for sustainable forest management which improved concession performance. The Russian Federation has similarly drafted and confirmed its criteria and indicators of sustainable forest management. Their fundamental purpose is to establish the framework conditions for the functioning of State forest management agencies in such as way as to satisfy the requirements of present and future generations of Russians and to coordinate the efforts for the sustainable development of the forestry sector as a whole.

Most current efforts, however, are directed at developing comprehensive indicator systems. Japan is promoting the development and improvement of an indicator system in which environmental factors are appropriately evaluated in conjunction with the indicators of sustainable development. New Zealand's National Environmental Indicators Programme aims to develop indicators that will reflect the condition of the environment at a particular point in time, show the pressures that human activities place on the environment, and provide measures of the effectiveness of any action in response to these pressures.

Turkey's Ministry of Environment is planning to cooperate with the State Planning Organization and the State Institute of Statistics to integrate environmental and developmental information and develop national indicators on sustainable development. On this, Turkey has already carried out inventories of the existing databases relevant to sustainable development.

In order to coordinate efforts in the region and to link these with regional and international efforts, the Inter-Agency Sub-committee on the Environment and Sustainable Development (ICESD) in Asia and the Pacific agreed to establish a regional project to develop a set of sustainable development indicators, and requested ESCAP to coordinate this project. Accordingly, ESCAP is currently implementing a project to assist in identifying a core set of sustainable development indicators in the Asian and Pacific Region (see Figure 12.1 and Box 12.5).

C. Environmental Quality Monitoring

Environmental quality monitoring provides the basic information for assessing the state of the environment and is central to the measurement of environmental indicators. It focuses on environmental parameters relating to air, water, land and other natural resources through the establishment of monitoring stations as well as utilization of remote sensing and GIS (see Chapter 15). Responsibility for environmental quality monitoring is often under a governmental environmental agency, although other relevant government agencies also participate. In Armenia, for example, the Ministry of the Environment and Natural Resources maintains national monitoring networks all over the country but jointly supervises and monitors background radiation pollution and emissions into the air, water and soil with the Ministry of Health.

Figure 12.1 Step by step approach adopted for the development of Indicators for Sustainable Development (ISD)

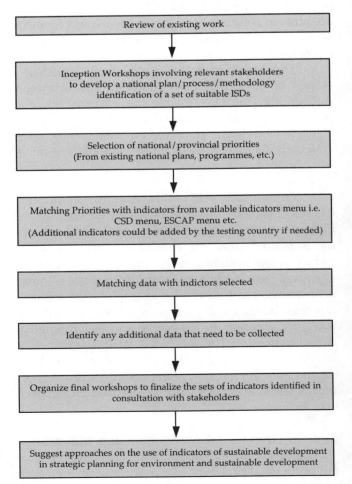

MECHANISMS AND METHODS

> **Box 12.5 Indicators for Sustainable Development in Asia and the Pacific**
>
> Recognizing the importance of the development of indicators of sustainable development (ISD), ESCAP initiated a project on ISDs in Asia and the Pacific with the assistance of the Government of Netherlands. A major objective of this project is to develop a set of indicators through a testing process to help identify trends in the promotion of sustainable development and assist the countries in the region to utilize these towards strategic planning for environment and sustainable development.
>
> The project was initiated by organizing a Regional Meeting on Indicators of Sustainable Development. A very important task of the meeting was to familiarize the participants from countries of the region with the basic concepts and provide the background information on national testing for development of appropriate sets of indicators. The meeting simulated in small groups, the mechanisms for determining priorities and matching them with indicators. It contributed positively to the process of development of ISD's in Asia and the Pacific which was manifested in the commitment of People's Republic of China, Maldives, Pakistan and the Philippines to act as pilot testing countries for the project.
>
> To date, People's Republic of China and the Philippines have developed a large menu of 65 and 80 indicators respectively as well as a more condensed group of 28 and 18 core indicators respectively. Pakistan has identified a menu of 76 indicators and is in the process of identifying core indicators within these. Maldives has identified the priority issues with the selection of indicators ongoing.
>
> A key issue is not only to have a group or set of indicators but also to integrate those into national, provincial and local level strategic environmental and sustainable development planning. This aspect forms the second phase of the project. In order to disseminate the work conducted under the first phase, and to promote the involvement of other countries of the region, a Second Regional Meeting has been planned by ESCAP in October 2000.
>
> The major achievements of the project besides creation of awareness among decision makers on the use of ISD in Asia and the Pacific have been the establishment of methodological approaches and identification of sets of indicators suitable to local conditions on the basis of the pilot study. However, the real success in ISD work will hinge upon the actual use of indicators in strategic planning for environment and sustainable development in the countries of the region.

In some countries, academic institutions are utilized for such monitoring, such as in Turkey where studies on air pollution have been carried out by universities. In addition, a GEF/UNDP Environmental Management and Protection of the Black Sea project is assisted by a pollution monitoring network involving universities and other public institutions. University laboratories can also undertake monitoring, especially when such laboratories include quality assurance procedures and take part in regional or international inter-laboratory calibration programmes. The Institute of Natural Resources (INR) of the University of the South Pacific, for instance, is involved in the monitoring of the Suva harbour area as well as in monitoring programmes for larger tourist resorts.

D. Issues and Problems in Assessment and Monitoring

Environmental monitoring and assessment in the region have improved considerably over recent years, although major constraints remain in some countries. These constraints include inadequate legislative and institutional arrangements. Environmental monitoring and reporting in many countries is distributed among various agencies at the central, state and local levels and the lack or absence of coordinating mechanisms has led to data incompatibility, overlaps and conflicts in enforcement. It should also be noted that in the case of environment and natural resources accounting as well as for environmental indicators, the collection and analysis of data and information could best be approached in an inter-agency manner. Such an approach is used in the Philippines, where environment and natural resources accounting is undertaken through the issuance of a Presidential Executive Order mandating the National Statistical Coordination Board to lead the implementation of the system with the Department of Environment and Natural Resources and the National Economic Development Authority.

It is also important that data should be distributed to decision-makers and the public in a timely manner. This has been done through SoE reports, although these are often constrained by the weaknesses of the existing mechanism for collecting data and their distribution at infrequent intervals. Progress in information technology could help mitigate these problem and initiatives that utilize e-mails, the internet and web pages have shown significant success in the timely dissemination of environmental information.

CHAPTER TWELVE

> **Box 12.6 The Introduction of Economic Instruments is Working Better than Prosecutions in India**
>
> In accordance with the Act for control of water and air pollution, the Pollution Control Board can prosecute defaulting industrial units. In the court, the Board are required to prove the violations of the Act. More often than not, it becomes difficult to establish the nature and impact of violations causes. Hence, a good number of prosecutions launched by the Pollution Control Authorities are dismissed by the Judiciary on technical grounds. Moreover, the legal proceedings are often very long-drawn and defaultees can delay the judicial proceedings and conviction for one reason or the other. Thus, justice is denied by sheer delay.
>
> In order to get over the stalemate, the Board has successfully adopted an innovative method in requiring the defaulting units to furnish time bound action plans for commissioning operational pollution control systems. The industry is also asked to furnish a Bank Guarantee for a sum equivalent to the cost of the pollution treatment system. In cases where the industries do not comply with the time-bound implementation targets, the amount of money assured through the Bank Guarantee is deposited in the account of the Pollution Control Board.
>
> *Source:* Government of India 2000

CONCLUSIONS

Policy structures and incentives for environmental improvements are gradually gaining ground in the countries of the Asian and Pacific Region. The formulation of national strategies for integrating environment and development, improvements in sectoral policies, pricing reforms, participatory management, efficiency of resource use and cost recovery, devolution of powers to local governments and empowerment of local communities are positive steps towards establishing appropriate mechanisms for pursing sustainable development goals.

However, a rigid command and control approach still characterizes environmental policies in most countries of the region. In attempting to control air and water pollution, in particular, countries have favoured emission or effluent standards over more flexible instruments. This inflexible approach has led to high compliance costs and widespread under compliance. Use of market based instruments and privatization is on the increase albeit rather slowly (see Box 12.6).

A new policy model based on a mix of command and control and market based mechanism linked to effective governmental management is showing positive results in countries like Malaysia and the Philippines, and is characterized by: the role of government as a facilitator rather than provider; by a prominent role played by the private sector and civil society; and by pricing reform on environmental goods and services. This model appears to have the greatest potential for developing countries of the Asian and Pacific Region which are deficit in financial resources.

It has been estimated that with such overall improved policies, the region could meet its environmental expenditure at less than 2 per cent of GDP, with the public sector providing less than half this amount. As public expenditure on the environment in developing countries of the region is currently just under one per cent, this scenario will not involve a substantial increase in public expenditures, but a significant redeployment of existing resources toward a more targeted and strategic portfolio, and one that leverages additional capital from domestic and foreign private sources.

Chapter Thirteen

Sustainable development will be possible only with the active cooperation of the private sector.

13

Private Sector

**INTRODUCTION
GREENING OF BUSINESS: STATUS AND TRENDS
BUSINESS COUNCILS, CHARTERS AND ETHICS
PRIVATE SECTOR AND CONCERNS OF THE 21ST CENTURY
CONCLUSION**

CHAPTER THIRTEEN

INTRODUCTION

In the building of a common understanding of the concept of "sustainability" and a practicable consensus on the implications of seeking to achieve "sustainable development", the private sector has a key role to play, both in terms of redefining corporate objectives (and market behaviour) and in using the private sector's considerable influence on economic growth, employment creation and environmental protection to promote and defend the goal of sustainable development.

Like other parts of the world, the role of business and industry in the promotion of sustainable development in the Asian and Pacific Region has enhanced with time and continues to grow with globalization and trade liberalization. This chapter evaluates the status and trends apparent in the interaction of the private sector with environmental issues and of the sector's contribution towards sustainable development in the region.

GREENING OF BUSINESS: STATUS AND TRENDS

In many countries of the Asian and Pacific Region, the proliferation of environmental companies and the increasing number of initiatives being undertaken by both existing and new businesses and industrial enterprises attest to the growing role of the private sector in the provision of solutions to environmental challenges.

In addition to the direct involvement in so-called *eco-business*, (i.e. providing environmental goods and services), the private sector's contribution in cleaner production and environmental research and development (R & D) has expanded considerably. Private sector participation, either directly or through Public-Private Partnerships (PPP), is growing in a range of environmental management sectors, including waste management; water supply and sanitation; integrated resource management and energy planning and development. Elsewhere the private sector is actively promoting energy and/or eco-efficiency; technological substitution of CFC's; management systems and standards for the greening of business and industry; green investment and financing; and new marketing strategies for the promotion of environmental goods and services as well as codes of conduct for business and industry.

A. Eco-business, Environment Markets and Financing

1. Eco-business

In terms of eco-business, companies and corporations in, for example, Australia, India, Indonesia, Japan, Malaysia, Republic of Korea, Singapore, Sri Lanka and Thailand are increasingly providing pollution abatement and remediation technologies, technical assistance and advice on environmental issues and environment-linked engineering and consultancy services (Table 13.1). The governments of these countries have recognized the trade and export opportunities associated with environmental management expertise and have initiated programmes to assist in the development of the export potential of their domestic environmental goods and services firms. In Singapore, for example, tax incentives, research grants, trade missions and exhibitions are promoted and planned to make the city-state a development and distribution centre for environmental equipment and services to other countries in the region. Singapore's Ministry of Environment has also set up a company to market its own expertise and technology to other countries in the region.

The growing trend towards the privatization of environmental services such as water supply, sanitation and solid waste management is also contributing to the development and strengthening

Table 13.1 Companies/Agencies Involved in Environment-Business in Asia and the Pacific

Country	No. of companies	Country	No. of companies	Country	No. of companies	Country	No. of companies	Country	No. of companies	Country	No. of companies	Country	No. of companies	Country	No. of companies
Australia	145	Bangladesh	6	Cambodia	1	PR China	39	Fiji	2	Hong Kong, China	45	India	203	Indonesia	57
Malaysia	55	Maldives	1	Nepal	5	New Zealand	18	Pakistan	20	Philippines	39	The Russian Federation	25	Singapore	78
Islamic Republic of Iran	7	Japan	172	Kazakhstan	2	Republic of Korea	59	Sri Lanka	13	Thailand	54	–	–	Viet Nam	3

Source: Eco Services International 2000

PRIVATE SECTOR

Table 13.2 World Rank of Revenues from Selected Japanese Environment-related Companies, 1995

World Rank	Environment-related Company	1995 Revenue (US$ billion)
7	Mitsubishi Heavy industries (Incineration, Air Pollution Control and Water Equipment)	2 350
9	Ebara Corporation (Water and Incineration Equipment)	2 200
27	Kurita Water Industries (Equipment)	900
44	Hitachi Zosen	602
47	Kubota	558

Source: Eco Services International 2000

of environmental enterprises. Consumer cooperatives have become a powerful force in Japan to popularize green products, while local governments have progressively provided technological and financial support to small- and medium-sized companies. The region's environment industry, however, has yet to play a significant role in the international business markets. For instance, of the top 50 grossing environmental companies worldwide, only 10 per cent are from the region; and all of these are from Japan (Table 13.2). Nonetheless, the growth in numbers of small and medium environment industries in the countries of the region is quite significant, and the current Eco-Services directory shows over 1 000 environmental enterprises in the twenty-four countries within the region that have entries.

2. *The Environmental Market*

With the growth in environmental consciousness and awareness, the region's potential market for environmental goods and services continues to expand. For example, People's Republic of China's direct investment in the prevention and control of environmental pollution was over US$10 billion in 1998, whilst, according to OECD, the regional market in environment-related businesses in 1998 was about US$80 billion and is expected to exceed US$110 billion by 2010. Japan already has the second largest national environmental market in the world with a market of over US$60 billion and the environmental market in the countries of Southeast Asia is expected to grow at the rate of 14 per cent per year.

The potential environmental market may be much larger than the estimates given the underestimation of the potential of some aspects and the exclusion of future "probable markets". For example, the requirements of the Kyoto Conference are likely to result in countries introducing more stringent controls on the management of methane emissions. The two main sources of methane generation (that are amenable to interventive management) are from agricultural wastes and from waste landfills; both of these provide market opportunities for the installation of gas capture facilities and the installation of biogas-fired energy generators. Similarly, opportunities are likely to arise for the enhancement of the quality of treated effluent as organic fertilizer for use in agriculture-irrigation schemes. The size of the potential market may also have been underestimated for analytical services and for instrumentation and information, particularly with the trend towards public sector agencies outsourcing or privatizing their supportive and non-regulatory functions.

The provision of bilateral and multilateral aid, the availability of "soft" environmental loans and the growth of environmentally friendly and ethical investments have also played a significant role in enhancing environmental markets. A large portion of the disbursed loans, grants and donor agency funding has been aimed at improving industrial efficiency, water supply and sanitation facilities, afforestation, waste management and natural resources management. For example, the Asia Sustainable Growth Fund, sponsored by the ADB, aims to raise US$150 million to invest as long-term capital for environmentally sound companies in the developing countries of the Pacific rim and the OECD, ADB, World Bank and international financial markets have provided numerous allocations of development assistance for environmental investments.

3. *Green Investment and Financing*

The "greening" of the financial markets is often seen as a key challenge associated with achieving sustainable development. As financial institutions, such as joint stock companies, banks, insurance companies and financial management firms, begin to take environmental factors into funding considerations, private sector attitude and performance towards sustainability will improve. In the longer term, it is argued, business will see the inherent convergence of investments that achieve both financial and environmental sustainability.

Some initial, tentative steps have been taken by a number of the region's governments to encourage private sector institutions and associations in the adoption of environmentally sustainable business practices.

(a) Private Sector Investment

There is a discernible overall trend in the East Asian and Pacific Region of growth in private sector investment as a proportion of GDP (Figures 13.1 to 13.3). In the wake of increasing investment needs and against a background of declining governmental resources, analysts believe that the demand for private investment will expand significantly in the future. According to the ADB, over the next 30 years the accumulated demand for infrastructure investment within the region as a whole will exceed US$10 trillion and that the countries of Asia will be spending about US$1.5 trillion on infrastructure between 1995 to 2005 alone. The scope for private sector investment in these infrastructure projects is expected to expand substantially in coming years (Box 13.1).

Whilst the proportion of this investment that will be environmentally focussed is unclear, the increasing profile given to environmental assessment during the planning stage of new infrastructure and, more importantly, the requirements for the management of environmental issues during project implementation and operation will ensure that the "green element" in these investments will continue to rise. In addition, the growth in the privatization of environmental services will ensure that the private sector will increasingly be environmentally driven as can be seen from the experience to date of private sector participation in water supply and sewage projects (Figure 13.4).

(b) Green Funds

In recent years, the financial markets have increasingly sought to capture and quantify the environmental risks associated with certain industries and to offer to investors sufficient information to enable the markets to differentiate on the basis of environmental performance. In Japan, for example, five major funds (see Table 13.3.) have been created for this purpose. These funds have so far attracted 190 billion yen (about US$1.75 billion), a rate of growth that far surpasses the rate of emergence of similar eco-funds in the US and Europe. In addition to these funds, several other leading Japanese financial institutions are in the process of starting their own green funds.

Green funds differ from the bulk of the so-called SRI (socially responsible investment) funds that emerged in the western markets in the early to mid 1970s. The Japanese green funds, for example, identify candidate investment ventures on financial criteria, and stringent environmental performance criteria are applied to screen the suitability of stocks for the fund. Unlike the ethical funds of 20-30 years ago, which were marketed and managed by small niche players, the present day green funds are operated by Japan's leading financial institutions. This has had two separate consequences. Firstly, it

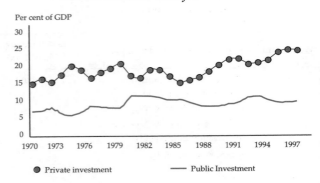

Figure 13.1 **Private and Public Investment in East Asia and the Pacific**

Sources: World Bank 2000

Remarks: The countries in East Asia and the Pacific include American Samoa, Cambodia, People's Republic of China, Fiji, Indonesia, Kiribati, Democratic People's Republic of Korea, Lao People's Democratic Republic, Malaysia, Marshall Islands, Micronesia (Federated States of), Mongolia, Myanmar, Palau, Papua New Guinea, Philippines, Samoa, Solomon Islands, Thailand, Tonga, Vanuatu and Viet Nam.

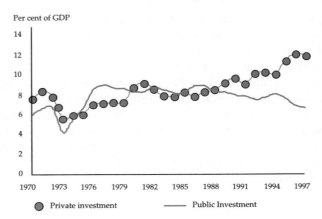

Figure 13.2 **Private and Public Investment in South Asia**

Sources: World Bank 2000

Remarks: The countries in South Asia include Afghanistan, Bangladesh, Bhutan, India, Maldives, Nepal, Pakistan and Sri Lanka

Figure 13.3 **Total Investment in Energy Projects With Private Participation in Developing Countries by Region, 1990-99**

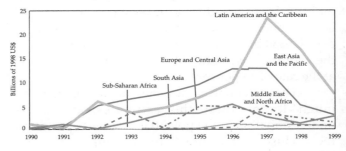

Source: World Bank 2000, PPI Project Database

Box 13.1 Private Investment in Infrastructure

Recent trends have shown encouraging growth in private investment in infrastructure, but not nearly enough to meet the estimated $1.2-1.5 trillion needed over the next decade. Data from 1996 show that the annual private investment flows to East Asian infrastructure projects-from both international and domestic concerns-has reached between $13-16 billion, accounting for 12-18 per cent of all infrastructure investments. This is a noticeable improvement but well short of what is necessary for the region to sustain long term economic growth and competitiveness, improve quality of life and continue the momentum toward a greater role in the global economy. Among countries that received the most investment were Indonesia, the Philippines and People's Republic of China. The lack of bankable, low-risk projects combined with the limits of long-term financing and the weakness of domestic capital markets are the main reasons for the relatively low private sector involvement.

A World Bank study, *Some Choices for Efficient Private Provision of Infrastructure in East Asia*, which came out as a result of a high-level infrastructure conference held in Jakarta in 1996, presented policymakers and the private sector with an overview of the challenges and trade-offs they face in the provision of infrastructure. These challenges include strategic approaches to private involvement, regulatory choices, different methods of contracting private suppliers, management of environmental and resettlement problems when the private sector takes the lead and new ways of financing private infrastructure. Both ministers and private sector executives at the Jakarta meeting agreed that the region can and should aim to increase private sector investment to 30 per cent of the total in the next five years. To address the remaining impediments to private participation, deeper reforms are needed to create a private sector-friendly environment, such as improving methods of contracting with private parties, building regulatory capacity and developing domestic markets. The urgency of undertaking these reforms has assumed greater importance as a result of the Asian economic crisis of the late 1990s.

Private sector involvement in infrastructure financing involves a set of core principles (transparent processes, stable rules, price reforms, maximum competition and incentive-based regulatory structures) that each country, despite the differences in sectors and demands, can adapt to suit to its particular needs. Lessons that have emerged from policymaker discussions show that certain preconditions must exist to ensure the sustainability of private investment in infrastructure projects. These specify that the public must be willing to pay for the proposed projects, that government subsidies or support should be transparent and that private projects must be bankable. Countries need to create the environment to meet these preconditions and may need to implement price reforms, regulatory and legal framework reforms and actions to encourage competition, increased accountability, and capital market development. This will improve the financial viability and profits of infrastructure projects, increase competition and transparency in contracting and reduce the risks of investing in the sector.

Source: www.worldbank.org/html/extdr/extme/ampr_003.htm

Table 13.3   Green Financing in Japan

Top Ten Investments	Nikko Eco Fund	Yasuda Green Open	UBS /JRI Eco-fund	IBJ/Daiichi Life Eco Fund	Partners Investment Eco-Partners
1	NTT DoCoMo (5% of fund)	Sony	NTT DoCoMo	NTT DoCoMo	NTT DoCoMo
2	Fuji Television (4% of fund)	Toyota	Toyota	Sony	Sony
3	Toyota (3% of fund)	Fujitsu	Sony	Toyota	Fujitsu support
4	NTT (3% of fund)	Matsushita	NTT	Seven Eleven Japan	Takura Shuzo
5	Rohm (3% of fund)	Takeda Chemical	Seven Eleven Japan	NTT	Yokogawa
6	Murata Mftg	NEC	Fujitsu	Fujitsu	NTT Data
7	Sony	Rohm	Matsushita	Rohm	Ebara
8	Takura Shuzo	Matsushita	Sumitomo Bank	Murata Mftg	Matsushita
9	Seven Eleven Japan	NTT	NEC	Sumitomo Bank	Seven Eleven Japan
10	Nikon	Kao	Rohm	Matsushita	Kirin Beer
Fund Total (Yen, as of 2/2000)	130 billion	16 billion	9 billion	35 billion	

Source: Eco Services International 2000

Figure 13.4 Private Participation in Water and Sewerage in Developing Countries 1990-97

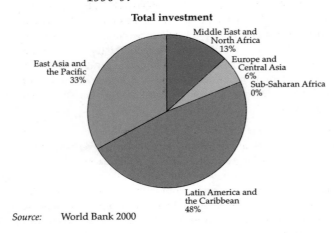

Source: World Bank 2000

explains the swift uptake of the funds in Japan and secondly, the pressure applied to high financial performers to improve their environmental performance rating and thereby attract investors is all the more persuasive as it comes from within the financial markets establishment.

However, none of the Japanese green funds has yet insisted that certain sectors or companies be actively shunned on environmental of health grounds (e.g. tobacco, nuclear power, and forestry). Nevertheless, as the competition in the green fund market heats up and differentiation becomes increasingly difficult, it is likely that negative environmental performance ratings will begin to come to the fore in investment decision-making, thereby adding the punishment of poor performers to the current investment market criteria.

Direct support to the business sector in their shift to sustainable development has been provided through private sector schemes that increase the flow of investments to such industries. In Australia, for example, a superannuation fund is investing exclusively in environmentally friendly companies to encourage corporations to "go green". With assets of Aus$2.2 billion, the Hesta super fund for health and community service workers gives its 400 000 members an investment option that allows them the chance to encourage companies to become "green". Hesta invests in what it considers the best "Eco-performing" companies that are successful and have environmentally friendly practices.

Elsewhere within the region, the financial sector has been slow to bring environmental criteria into the formal investment decision-making system, even at the most basic level. For example, Malaysian banks and insurance companies have no formal policies on the environmental risks (and direct financial liabilities) associated with contaminated land and no assessment is made of land that is offered as collateral on investment loans. Similarly, insurance companies in Malaysia have yet to classify contaminated land as being of a higher risk than other type of land. Whilst Malaysia has been used for illustrative purposes, the same pattern of inadequate or non-existent environmental scrutiny of investments, even on issues that have direct financial implications, is found throughout the region.

(c) Green Loans

There are a number of global green funds that are exclusively directed at environmental programmes. For example, under the Multilateral Fund of the Montreal Protocol, People's Republic of China received US$105.6 million to support 173 projects for its industries to shift to non-ODS (ozone depleting substance) technologies and, thereby, achieve the phasing out of 32 000 tonnes of ODS. Similarly, in the Philippines 51 private sector firms received a total of some US$17.2 million to shift to non-ODS technologies and, thereby, phase out some 1 600 tonnes of ODS.

At the country level, a number of governments have encouraged the utilization of clean production processes by giving incentives which, in Pakistan, has included preferential treatment in loan facilities by banks and development finance institutions and lower import tariffs on pollution abatement equipment. In Sri Lanka, a range of soft loan schemes for environmental improvements, pollution control and resource-efficient technologies are available, such as the *e-friends* soft loans scheme offered to the private sector by the National Development Bank of Sri Lanka. In the Republic of Korea, the government has extended long-term, low-interest loans to companies through the Industrial Development Fund and the Environmental Pollution Prevention Fund for the establishment of facilities to treat, prevent or recycle pollutants (Republic of Korea Report to the CSD).

B. Clean Production: Research and Development (R & D)

According to a recent survey in Japan, 90 per cent of companies with listed stocks and 20 per cent of unlisted companies are developing some form of Eco-business or are funding research and development in the Eco-business sector (Figure 13.5). Similarly, in the Republic of Korea a number of industrial sectors have established research institutes focusing on the development of new environmental technologies. An encouraging trend within the R & D activities reviewed below, is the setting of Eco-efficiency targets by Asian companies, which far exceed the environmental performance standards established by their respective governments.

PRIVATE SECTOR

Figure 13.5 Status of Eco-business in Japan 1996

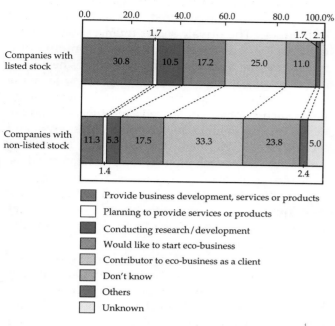

Source:
1) Survey on Environment Friendly Corporate activities
2) Environment Agency of Japan 1996

1. *Waste Reduction and Recycling*

A key focus of the "environmental restructuring" that companies are currently undertaking is the development and implementation of cleaner production technologies, processes and management rationales. One of the main areas that has successfully demonstrated the financial benefits of environmental investments is waste management, where a number of industries have achieved considerable savings in operating, materials and energy costs by adopting a range of measures including reduction at source, maximizing material recovery for reuse or for sale to the growing recycling sector. Waste exchange is also being promoted in a number of countries and territories, including the Philippines and Hong Kong, China, whereby waste from one industry becomes the raw material for another, whilst in Australia and New Zealand Waste-Wise programmes have been initiated (Box 13.2).

Further examples of private sector initiatives in the waste management sector are presented in Chapter 8.

Box 13.2 Construction Companies and the WASTE-WISE Programme in Australia and New Zealand

Australia and New Zealand generate enormous amounts of construction and demolition (C & D) waste; in Melbourne alone, some 4.18 million tonnes of C & D waste is generated each year.

The WASTE-WISE Programme was designed to reduce the amount of the construction waste going to landfill, thereby extending the life of landfills and reducing the costs of wastage to the construction industry. In the first phase of the Programme, five leading Australian construction companies volunteered to work with the Australian and New Zealand Environment and Conservation Council to develop waste reduction, recycling and reuse best practice guidelines.

The Programme identified and addressed the technical and behavioural barriers to efficiently and economically reducing waste. As a result a significant volume of building material waste, including concrete and steel, are now being reused and recycled. The companies participating in phase one of the programme found that waste reduction practices improved their commercial prospects both in Australia and overseas. New jobs were created and increased profits were made through the processing and marketing of recovered construction waste. Participating companies also found that good waste management practices positively influenced the companies work practices and corporate identity. The achievements of the participating companies (Bovis Lend Lease, Multiplex, John Holland, Fletcher Construction and Barclay Mowlem) were as follows:

- Bovis Lend Lease Pty Ltd. recycled 98 per cent of the waste material arising from the State Office Block site in Sydney and 90 per cent of the company's waste concrete and steel was sent to recycling centres.
- Multiplex recycled 60 per cent of site waste at the Homebush Bay Olympic Stadium site between April and August 1997, and established an innovative concrete re-use system where 32 000 cubic metres of concrete was crushed and then reused on site.
- John Holland reused or recycled 760 000 kilograms of building material on John Holland sites in one year.
- Forty-three per cent of waste from the Dandenong Police and Court Buildings site was reused or recycled by Fletcher Construction, thereby saving 55 per cent of its waste removal costs.
- At the Vantage Apartments site in Queensland, Barclay Mowlem, 55 per cent of the timber and wooden formwork was recovered and recycled and, across all operations, waste materials such as concrete, bricks, sand, gravel, soil, steel, metal framing and roofing was collected and sent to recycling facilities.

By reducing the amount of construction waste going to landfill, companies participating in the WASTE WISE Programme made a significant and measurable impact and set the scene for an expansion of the programme to other parts of the Australian and New Zealand construction industry.

Source: WASTE-WISE Construction Programme website www.environment.gov.au/epg/wastewise

CHAPTER THIRTEEN

2. Green Products

Consumer pressure has led to the private sector investigating the production of sustainable products, both for their respective domestic markets and for those products that are exported to the west. However, those industries that operate within markets where little consumer pressure is applied show little sign of restructuring. For example, each day over one million people in Hong Kong, China and billions in Mainland China use expanded polystyrene food containers that are disposed of to landfill and open dumps. In response, a Hong Kong, China-based company, "Join-in Green Products," has developed a range of biodegradable tableware made from a composite of grass and sugarcane pulp. The product decomposes in less than three months and has earned the company a gold prize at Hong Kong, China's first Eco-Products Awards. However, resistance from the restaurant trade and, more significantly, reaction from the existing private sector manufacturers and suppliers of the expanded polystyrene containers have ensured a very slow uptake of the new "green alternative".

In other sectors, greater success has been achieved in introducing new alternative technologies developed by the private sector. For example, solar power is becoming increasingly available in the region with the innovative work done by some enterprising companies, such as Solar Research Design that operates two factories in Malaysia. Its Microsolar heaters channel water through optimally arrayed tubes to produce hot water even on cloudy days and the heated water is also used to compress a coolant, typically an ammonia mixture or lithium bromide, which is then sent to indoor fancoils. The cooling capacity of these systems is similar to a conventional 750 kW air-conditioner. These products have been particularly successful in rural areas that lack connections to electricity transmission systems.

3. The New Generation of Clean Technologies

In the developed countries of the region, the transition is underway from the first-generation environmental technologies (pollution abatement and "end-of-pipe" cleanup technologies) to the new generation of technology systems that includes both the "hardware" of clean technology and the "software" of improved environmental management systems.

Such technology is typically process-controlled and environmentally benign, using less energy and generating less waste. This new technology is linked to, and integrated with, corporate philosophies or covenants that bring together environmental objectives and economic agenda.

Within the region, Japan is leading the way in pursuing policies to encourage cleaner production and developing the required new technologies and techniques. The private sector finances some 60 per cent of all research and development into environmental technology and contributes heavily to a number of government research agencies. Japanese industry is particularly strong in certain clean energy fields such as photovoltaic cells and fuel cells, and in clean motor vehicle technology. Research and technology development for new clean technology cars is aimed at developing electric hybrid cars, whilst similar research is underway in the Republic of Korea where Hyundai is developing a fuel-cell car. These developments are likely to be boosted if policy incentives to control increasing urban vehicular emissions are introduced.

In Singapore, ST Microelectronics is committed to water conservation through an annual reduction in water consumption of 5 per cent. The company has already cut water consumption by 34 per cent over the past six years, despite an increase in production of some 60 per cent. The company has also reduced its annual use of sulphuric acid by 95 per sent through a combination of recycling and reprocessing.

Across all industrial sectors, small and medium-sized enterprises, which contribute substantially to overall pollution levels, are least able to shift to cleaner technology. In India, more than 2 million small enterprises account for about half the country's industrial output and 60 to 65 per cent of its industrial pollution. A similar breakdown of industrial output and pollution contribution emerges from analysis of the small and medium enterprises (SME's) in other developing countries of the region including Bangladesh, People's Republic of China, Indonesia, Pakistan, the Philippines, the Republic of Korea and Sri Lanka. Although operating in all sectors, SME's are particularly numerous in the textile, garment, wood product, food processing, leather product, fabricated metal, component and equipment, rubber and plastic product, pottery, printing and publishing sectors.

4. Technology Transfer

The transfer of clean technology from developed countries to developing countries has been extremely slow. For example, energy efficient technologies, including thermal storage, energy-efficient lighting, distributed generation systems, variable-speed motors and drives, have been available in industrial countries for several decades. However, even within operating factories and subsidiaries of the same company, the transfer of cleaner technology from the developed to the developing countries has been hampered by under-investment and by the transfer of "obsolete" technology (in many cases this may be translated as "no longer complying with environmental standards" of the developed country) from one country to another.

In some developed countries, private sector institutions have been set up to aid the rapid and efficient transfer of environmental technology. For example, Austemex, the export arm of the Environment Management Association of Australia, cooperates with Environment Australia in developing strategies for transferring clean technology to the wider region (Australia Report to CSD). Australia also has a Cleaner Production Case Studies Directory, which has been developed to give industry in other countries easy access to case study information on cleaner production measures used in a wide variety of industries.

5. *R & D Potential*

Although much of the R & D for cleaner production is undertaken by the private sector, collaborative R & D is also taking place between governments and the private sector. In India, for instance, the National Environmental Engineering Research Institute is developing a wide range of environmental technologies to improve pollutant monitoring and the recycling and management of urban and industrial solid wastes. In Indonesia, the government, acting through the Environmental Impact Management Agency, is providing assistance for factories to develop cleaner and less polluting technology, whilst in Thailand the textile, pulp and paper, electroplating, chemical and food industries are all involved with government in promoting cleaner production initiatives.

Despite the growth in R & D, these initiatives have a relatively small effect given the scale of industrial activity of the region. In Indonesia, for example, major industries are using up to 30 per cent more energy per unit of output than comparable industries in Japan and the demand for electricity is increasing at 15 per cent per year. The situation is not much different in other developing countries of the region. Given the perceived lack of resources available, the potential savings from cleaner and more energy efficient technologies are enormous. A study by the Lawrence Berkeley Laboratory estimates that developing countries could avoid spending US$1.7 trillion on new power plants, oil refineries, coal mines and all the attendant infrastructure by spending US$10 billion a year over the next thirty five years to improve energy efficiency and conservation.

C. Use of Environmental Management Systems and Techniques

Environmental management techniques including environmental impact assessment, environmental auditing, adhering to ISO standards, life cycle analyses, environmental monitoring, performance evaluation, and reporting are being used increasingly by the business and corporate sector in the region. Corporate groups and business associations have started to improve their environmental performance either to maximize profitability or in response to public pressure, demands of foreign markets (ecolabelling and trade related standards), or on the instructions of their peers or parent companies.

1. *Environmental Impact Assessment and Environmental Audits*

Environmental impact assessment (EIA) is a required process in many of the counties of the region for the planning and permitting of new projects. Whilst the type and scale of project or development requiring formal EIA varies across the region, many private sector companies choose to commission EIA's, or related studies such as environmental baseline studies, for potentially controversial projects whether these require the formal process or not. The results of these "informal EIA's" are then available to support any public consultation or funding applications that may arise.

Environmental auditing is a tool used to gauge the environmental performance of a facility or company against predefined performance criteria. In some countries of the region, environmental audits are required as part of the pollution control process or as a follow-up to the EIA process to ensure that predictions, recommendations and commitments are fully integrated during the construction, operation and decommissioning phases of the new development. Environmental auditing is also widely used as one of the suite of tools in the development, implementation and maintenance of ISO 14000 environmental management systems (see below).

However, the major growth in environmental auditing has been for the internal use of individual private sector organizations where it has been employed as a tool for determining the environmental risks and potential liabilities associated with specific processes. In particular, environmental auditing has been applied as part of the mergers and acquisitions process to provide potential investors with key data on the current level of compliance of a potential investment venture with a range of local, national, international and corporate criteria. Much of the investment and joint venture decisions accompanying the entry of international companies into People's Republic of China, India, Pakistan, the Philippines, Sri Lanka, Viet Nam and many other countries in the region over the past ten years has therefore been informed, in part, by the results of environmental auditing.

2. *ISO Standards*

Certification to ISO quality and environmental management standards by the private sector in the region is on the increase. The environmental management standard (ISO 14000) covers: (a) environmental management systems; (b) environmental auditing; (c) environmental labelling; (d) environmental performance evaluation; (e) life cycle assessment; (f) terms and definitions; and (g) environmental aspects of product standards. ISO 14000 augments clean technologies and waste reduction techniques by providing a comprehensive and integrated system of environmental management that takes into account managerial commitment, the establishment of environmental policies, objectives and mission statements, data collection, cost-effective technology selection and implementation, employee training and programme monitoring and certification. It is expected that ISO 14000 will create a "pull" factor that will also affect suppliers, contractors, customers, bankers, creditors and other parties. The regional share in worldwide ISO 14000 certification is given in Figure 13.6, whilst progress made towards certification is shown in Figures 13.7 and 13.8.

Since the publication of ISO 14001 standard in December 1996, the number of companies or sites certified with the standard worldwide has grown, of which almost 50 per cent of the top 20 registrations by country (as of November 21, 1999) were from the Asian and Pacific Region.

The process of institutionalizing the requirements of ISO 14001 has begun in many countries of the region. For example, in the Republic of Korea, the ISO 14000 was introduced by the Ministry of Trade, Industry and Energy (MOTIE) under the legal basis of the "Environmentally-Friendly Industrial Structure Promotion Act" and the Korean Chamber of Commerce has been designated as the "Headquarters for Promotion of Environmentally Friendly Industry". In Thailand, the Ministry of Industry has established the National Accreditation Council within which a subcommittee is responsible for ISO 14000 issues. Malaysia and Singapore have also established national organizations to oversee certification to ISO 14000, whilst the Philippines is adopting ISO 14000 standards as part of their national standards. In Fiji, the Ministry of Commerce, Trade and Industry is

Figure 13.6 Share of Asia and the Pacific in Worldwide ISO 14000 Certification

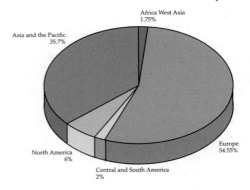

Source: International Organization for Standardization 1998

Figure 13.7 ISO 14000 Certification Growth in the Region from 1995 to end of 1998

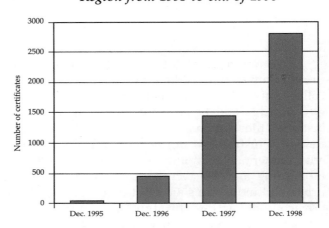

Source: International Organization for Standardization 1998

Figure 13.8 ISO 14000 Certification Growth in Selected Countries of the Region from 1995 to end of 1998

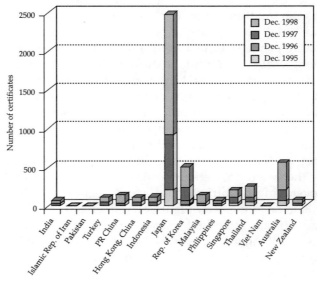

Source: International Organization for Standardization 1998

working on a Memorandum of Understanding with Australia and New Zealand on the effective enforcement of international standards that will address environmental issues, and in China ISO 14000 is receiving increasing interest with an accreditation administration comprising over 30 government bodies (see Box 13.3).

3. *Ecolabelling*

Eco-marks or Ecolabels are attached to, or indicated on the packaging of, products to indicate that they are either manufactured with minimum environmental impact or that they are designed to operate in a manner that reduces impacts to the environment or that over the product life-cycle (i.e. sourcing of raw materials, manufacture, packaging, distribution, use and disposal) they impose less of an environmental burden than similar products.

Following a trend that has been established internationally (particularly in Western Europe), the labelling of products manufactured and traded in the region is increasingly including environment related information. Institutions set up to formalize and standardize ecolabels in the countries of the region, typically include representative of business and industries; some prominent recent examples of formal Ecolabelling are given in Table 13.4. In Republic of Korea, an Eco-Mark Association has been established as a private body composed of representatives from consumer organizations, environmental organizations, businesses and distribution sectors, as well as environmental experts and journalists. Similarly, in Singapore Ecolabelling is through a Green Label Scheme that sets specific guidelines for the manufacturing, distribution, usage and disposal of products. Japan also has an Eco-Mark System,

---

Box 13.3 Sustainable Corporate Management and ISO 14000 in People's Republic of China

Chinese domestic firms follow closely government guidance in environmental management. Most firms with more than 100 staff have at least one full time environmental protection officer, whilst larger firms have departments responsible for environment, health and safety. Until recently Chinese corporate environmental management, however, has been reactive, responding to regulatory pressures only. The recent crackdown on polluting private sector enterprises has forced the environmental issue higher up the corporate agenda.

At both the local and national levels, the Chinese government is increasingly expecting firms with foreign investment to lead the way in responsible environmental management. Expectations of pollution control standards are generally higher from foreign joint ventures and often, the Chinese look favourably on environmental protection investment, training and management by foreign firms as a way of raising the level of domestic industry and transferring skills. Nevertheless, some large domestic Chinese firms are also facing increased pressure to improve their environmental performance. The Jinan Refinery (in Shandong Province) of the state-owned China Petrochemical Corporation is one such large facility in the country and the company is facing strong regulatory pressure to improve its environmental performance over the next five years. It has already invested 13.5 per cent of the capital employed in the period 1990-95 on pollution control and has built a 150 person strong environmental protection unit. Its main issues are air emissions and organic contaminants in wastewater. The company has built emissions control facilities in its asphalt production unit and several wastewater treatment plants.

With growing domestic pressure and the added expectations of the international markets, there is a substantial interest in the Chinese corporate sector towards the development of the ISO 14000 corporate environmental management standards. The country has an opportunity to actively participate in the ongoing ISO 14000 process since many important issues relevant to Chinese trade interests remain to be resolved. In particular, product oriented standards for environmental labelling, life-cycle assessment and environmental characteristics of product policies are still being developed. Further, ISO 14001 provides an excellent opportunity to improve the environmental performance of companies in People's Republic of China. There is growing experience internationally that systematic approaches to improve corporate environmental performance (increasing efficiency, reducing resource use and minimizing wastes and polluting emissions) can improve government relations and public image, reduce costs and expand market opportunities. However, because ISO 14001 does not establish performance standards on its own, environmental performance improvements will depend on the strength of a company's environmental policy and the domestic environmental policy regime. Moreover, People's Republic of China will need to overcome a cost barrier to the implementation of the standard, particularly in small and medium size enterprises.

Since the ISO 14001 standard will involve significant costs which industry cannot incur all at once, a major challenge for governments is to choose industry priorities based on calculations about where best to make domestic environmental improvements, protect export markets and pursue new export opportunities.

In order to assist in this venture People's Republic of China officially adopted the ISO 14001 environmental management standard on 1 April 1997 and has officially designated an organization to administer China's domestic accreditation process. The accreditation body comprises representatives from over 30 government bodies and is headed by a top official from the State Environmental Protection Administration.

Source: ENFO News, December 1997

Table 13.4 Prominent Eco-Labels instigated in Asia and the Pacific

Country	Eco-Label	Year of Establishment
Australia	Environmental Choice Australia	1991
PR China	PR China Environmental Labeling Scheme	1994
India	Eco-Mark Scheme	1991
Indonesia	Ecolabelling Scheme Indonesia	1993
Japan	Eco-Mark Scheme	1989
Republic of Korea	Eco-Mark Scheme	1992
New Zealand	Environmental Choice Scheme	1990
Singapore	Singapore's Green Labeling Scheme	1992
Thailand	Thai Green Labeling Scheme	1994

Source: International Institute for Sustainable Development 1996

which includes a continuous system of surveying and monitoring public opinion with a view to understanding consumer trends and of ensuring that consumer information is in line with public expectations.

Manufactures and suppliers of a range of natural resource products are also increasingly responding to market demands for ecolabelling. There is a growing demand for timber that is certified as originating from sustainably managed forest resources and certification through the Forest Stewardship Council (FSC) and ISO 14000 is increasing throughout the region. Similarly, the increasing international market for organic agricultural products has led to a range of labeling schemes linked to international certification bodies.

4. *Corporate Reporting*

International consumer and shareholder pressure is beginning to call for global accountability and reporting, whereby the private sector is required to measure, reporting and, increasingly benchmark, their social, environmental and economic impact in each country in which they operate. This has led to a number of companies establishing (and most importantly enforcing) corporate environmental standards, which frequently require levels of environmental performance that is higher than the relevant environmental standards of the country of operation.

This process of increasing global accountability has been aided in recent years by the launching of the Global Reporting Initiative (GRI) by the Coalition for Responsible Economies (CERES) in late 1997 (Box 13.4). This initiative draws upon and complements a number of the reporting systems currently being employed within the region, including those of the World Business Council for Sustainable Development (WBCSD), Eco-efficiency Metrics, World Resource Institute (WRI) Greenhouse Gases Metrics and Innovest. A comparative analysis of these and other initiatives on measuring and reporting aspects of corporate sustainability in given Table 13.5.

5. *Environmental Management in SMIs/SMEs*

Efforts to promote environmental management and reporting in small and medium size industries are also being strengthened. In Sri Lanka, the World Bank is funding technical assistance to the Central Environmental Authority to develop environmental codes of practice for small industries. In Thailand, a Pollution Prevention Fund for small and medium scale industry has been established and is expected to serve as a catalyst for moving industry towards a more sustainable pattern of development (Thailand Report to the CSD).

In Japan, the Japan Environmental Corporation (formerly the Pollution Control Service Corporation or PCSC) has a special mandate to provide financial and technical assistance to SMEs which have difficulties due to their limited financial and technological capacity to engage in special projects for environmental clean-ups. The Republic of Korea has also designated 8 research institutions as "cleaner production technology assistance centres" to provide technological assistance to small and medium entrepreneurs (Korea Report to the CSD). The National Cleaner Production Centre project in India assists the National Institute of Small-Scale Industry Extension Training in developing cleaner production courses.

D. **Business and Government Partnerships**

Sustainable development is being actively promoted by public private partnerships in a number of countries. At the policy level, the private sector is often included on government committees and inter-agency councils. In India, industry associations, such as the chambers of commerce, are members of the National Environmental Council, whilst the private sector is represented on Thailand's National Environmental Board and representatives of the business sector in the Philippines are members of the Philippine Council for Sustainable Development. In Sri Lanka, a forum for improving dialogue between government agencies and the private sector for the promotion of sustainable development, known as the Lanka International Forum on Environmental and Sustainable Development (LIFE) has been established.

PRIVATE SECTOR

Box 13.4 The Global Reporting Initiative: An Emerging Tool for Corporate Accountability

Business economic power is expanding in the global economy. Many NGOs argue that with power should come responsibility, and with responsibility there must also be accountability, not only to shareholders, but also to others with a stake in business activities. A multiyear project on corporate reporting has been launched by the Coalition for Environmentally Responsible Economists (CERES). The Global Reporting Initiative's (GRI) core mission is to establish, through a global, voluntary and multi-stakeholder process, the foundation for uniform corporate sustainability reporting worldwide. Over time, GRI seeks to elevate corporate sustainability reporting practices to a level equivalent to and as routine as financial reporting and to design, disseminate and promote a globally applicable standardized reporting format. Drafted in March 1999, the GRI issued a draft set of Sustainability Reporting Guidelines for comment and pilot testing.

The draft Reporting Guidelines document has nine points comprising CEO Statement (describes a company's understanding of and commitment to sustainability); Key Indicator (provides a summary of key social, environment, and economic performance indicators); Profile of Reporting Entity (provides an overview of the reporting entity and scope of the report); Policies, Organization, and Management Systems (a statement of the organizational and management processes intended to met sustainability); Stakeholder Relations (information on the processes and methods for engaging stakeholders); Management Performance (indicators of operational performance in key aspects of sustainability); Operational Performance (indicators of operational performance in key aspects of sustainability); Product Performance (indicators of product performance in key aspects of sustainability); Sustainability Overview (description of how sustainability is integrated in decision making and performance).

The draft GRI reporting guidance proposes three complementary reporting styles to help understand the interrelations and balances among the three dimensions of sustainability:

- Systematic Accounting and Reporting – A total approach comprising economic, social, and environmental accounts including flows and accumulated balances of values. Systematic accounting and reporting requires a sophisticated knowledge of causes and effects within and among the environmental, economic and social dimensions. Though not yet well developed, it presents a model for future sustainability reporting.

- Thematic statement – The enterprise's interpretation of a conceptual theme – a core issue or challenge – that relates in some way to all dimensions of sustainability. Sample themes include diversity, added value, productivity, integrity, health, and development.

- Case Studies – An exploration of an enterprise's decision making in a particular situation, treating the interrelations of environmental, economic and social aspects in terms of how it manages specific situations in the past, present, or future.

On face value, the GRI reporting system appears quite pragmatic however, the real test of its success in the Asian and Pacific Region will come with its applications particularly by the firms in the developing countries of the region.

Source: GRI's website: http://www.globalreporting.org/

Table 13.5 *A Review of Various Initiatives on Measuring and Reporting Aspects of Corporate Sustainability*

Initiatives	M	R	3rd Party Verification	Social	Env.	Econ.	Inter. Sustainability	Constituency
WBCSD Eco-Efficiency Metrics	✔				✔			Business
CERES	✔	✔			✔			Multi-stakeholder
Social Accountability SA 800	✔		✔(C)	✔				Multi-stakeholder
UK Ethical Trading Initiative	✔	✔	Not Yet	✔				Multi-stakeholder
European Eco-Management & Audit Scheme (EMAS)	✔	✔	✔					Inter-Government
ISO 14031 EPE	✔				✔	(fin)		Multi-stakeholder
NEF Quality Scoring Framework	✔	✔		✔				NGO
WRI/WBCSD GHG Protocol	✔	✔			✔			Multi-stakeholder
UNCTAD	✔	✔			✔	(fin)		Inter-Government
CEFIC	✔	✔			✔			Business
VFU	✔	✔			✔			Business
sustainability/UNEP		✔		✔	✔			Business/Inter-Government
NRTEE	✔				✔			Multi-stakeholder
GRI	✔	✔	Not Yet	✔	✔	✔	✔	**Multi-stakeholder**

M = Measurement; R = Reporting; C = Certification; fin = Financial

CHAPTER THIRTEEN

The private sector has also been included in the Inter-Agency Committees which formulate, review, and revise EIA activities, regulations and other policy elements in a number of countries in the region. For example the Business Council of Australia, Minerals Council of Australia, and the National Association of Forest Industries regularly interacts with the government through the Australian Intergovernmental Committee on Ecologically Sustainable Development.

Across the region, public private partnerships have aided the initiation and successful implementation of new environmental management schemes. For example, such a partnership launched the Industrial Ecowatch Project in the Philippines, under which factories are visited and rated on environmental performance and the results of this process are made public. The initiative is founded on an agreement between the government and 23 industry associations representing 2 000 companies across the country. In the Republic of Korea public private partnerships have allowed a sharing of expertise that has led to greater cost-effectiveness in improving environmental performance. In New Zealand, the oil industry, in conjunction with relevant government agencies, has produced a guide to the installation of underground storage tanks. This work will be extended to address the removal of old tanks, site remediation and sampling standards (New Zealand Report to CSD).

Public private partnerships have proven particularly effective in situations where existing service delivery is either too costly or inadequate and where private sector participation provides an opportunity for enhancing efficiency (and thus lowering costs) and mobilizing private investment (and thus expanding the resources available for urban infrastructure and equipment).

BUSINESS COUNCILS, CHARTERS AND ETHICS

A. World Business Council for Sustainable Development (WBCSD)

The World Business Council for Sustainable Development (WBCSD) was formed in January 1995 through a merger between the Business Councils for Sustainable Development (BCSD) and the World Industry Council for the Environment (WICE). The WBCSD has 120 international companies from 30 countries comprising of 20 major industrial sectors within its coalition and also benefits from a global network of national and regional Business Councils for Sustainable Development (BCSDs) such as those established in Indonesia, Malaysia and Thailand (see Box 13.5). Elsewhere, in lieu of BCSDs, are the Confederation of Indian Industry in India, the Philippine Business for the Environment in the Philippines and the Vernadsky Foundation for the Russian Federation.

The aims of the WBCSD are to promote: *Business leadership* – to be the leading business advocate on issues connected with environment and sustainable development; *Policy development* – to participate in policy development in order to create

---

Box 13.5 Business and Environment Programme in Thailand

The Business and Environment Programme is designed to encourage the business community in Thailand to actively participate in meeting the country's environmental challenges through the adoption of sustainable practices i.e. minimizing the use of natural resources, preventing pollution, and developing green products with a minimal impact on the environment.

The programme is being implemented by the Thailand Environment Institute (TEI) and the Thailand Business Council for Sustainable Development (TBCSD). The Council consists of fifty-five members from varied sectors of the business community. The objectives of the TBCSD are to promote the concept of "Sustainable Development" amongst business leaders and to disseminate information on sound environmental practices to the business community and the general public. In order to ensure business leaders take a leading role in preventing and solving environmental problems, the TBCSD encourages the integration of the "Sustainable Development Approaches" into its members corporate operation and management. It is believed that these approaches will effectively improve the environmental performance of commercial organizations and also encourage other members of the business community to recognize the advantages of sustainable development.

During the past year, the Programme continued to act as the Secretariat for the Thailand Business Council for Sustainable Development. In this capacity, it managed the implementation of the "TBCSD Sustainable Development Approaches" and other projects sponsored by members of the Council. One such TBCSD sponsored project, "Coal and its Impact on the Environment", undertaken by the Policy Research Programme, in consultation with the TBCSD Working Group, has been completed and a number of other projects are in the pipeline. The on-going efforts and actions on waste reduction, energy efficiency and pollution prevention are likely to provide a competitive advantage, help strengthen business community's/corporate citizens reputation regarding environmental issues and enable the risk of potential liabilities to be managed proactively.

Source: www.tei.or.th/bep/Bep.cfm

a framework that allows business to contribute effectively in sustainable development; *Best practice* – to demonstrate progress in environment and resource management in business and to share leading edge practices among members; and *Global outreach* – to contribute through its global network to a sustainable future for developing nations and nations in transition.

The WBCSD has also developed some consensus in the definition of Corporate Social Responsibility (CSR) as "the commitment of business to contribute to sustainable economic development, working with employees, their families, the local community and society at large to improve their quality of life." From WBCSD members in the region, a variety of emphases have emerged. In the Philippines, the focus is on determining the real needs of stakeholders, defining ethical behaviour, building partnerships and having a visionary and leading role. In Thailand, the emphasis is on the concept that the bigger the company, the bigger the obligation; the importance of environmental mitigation and prevention; the need for transparency; the importance of consumer protection; awareness of and change in people's attitudes towards the environment; and the relevance of youth and gender issues.

It is important to note, however, that WBCSD itself recognizes that CSR is not yet high on the business agenda of many companies. For example in Thailand, owing to the vulnerability of the private sector, CSR is seen more as a luxury item particularly by businesses that have their survival at stake. Nevertheless the importance that Thai society places on the environment has led to recognition of CSR and a development of the desire to increase CSR awareness and build partnerships between NGOs, government and business. Companies in the Philippines have also started to see CSR as helping meet their business agenda since the private sector has traditionally played an important role especially in filling the gaps where government is weak or unable to provide effective leadership.

Those who subscribe to the concept of CSR have already realized its link to long term profits. In the Philippines many companies have noted that "sustainable business cannot be achieved by following the traditional thinking that the only thing business has to do is to make a profit ... it now means not only making a profit but also managing issues that concern many stakeholders. Managing these concerns will enable the company to continue to make a profit." and that "CSR is the backbone of sustainability in the long run." These developments illustrate the growing realization within the private sector of the need to internalize the environmental costs associated with their business activities.

B. Business Charters and Ethics

A recognition of the private sector's responsibility to the environment has also led to the development of business charters and ethics statements including the Business Charter for Sustainable Development of the International Chamber of Commerce and the Principles of the Coalition for Environmentally Responsible Economies (CERES). The signatories to these documents in the Asian and Pacific Region have pledged to put into practice the principles of environmental management for sustainable development which include recognizing the need for long-term economic growth and environmental sustainability; developing and providing products and services that have no undue environmental impact, are safe, are efficient in consumption of energy and natural resources, and can be re-used, recycled, or disposed of safely; protecting the biosphere, preserving biodiversity and safeguarding habitats; informing and working with the public and creating dialogue with stakeholders; working towards self-regulation and a demonstrated environmental commitment by corporate management; and, undertaking environmental auditing to measure progress.

Developed countries of the region have also devised country-specific business charters, ethics statements and programmes for sustainable development. For example, the Business Council of Australia has an environmental committee and has developed position papers on a business perspective on sustainable development. Similarly, in Japan, the Federation of Economic Organizations (Keidanren) compiled the Keidanren Global Environment Charter that prescribes action standards for Japanese enterprises. In the Republic of Korea, the environment committee of the Federation of Korea Industries (FKI) comprises seven member companies and environmental experts and has developed and promoted a code of environmental conduct.

A good example of a sectoral code of practice is tourism, where a variety of codes have been developed. For example the Ecotourism Association of Australia, in consultation with the Australian Tourism Operators Network and Tourism Council Australia, finalized the National Ecotourism Accreditation Programme (NEAP) in 1996. The Federal Government provided funding for the development of the NEAP, including a feasibility study, development of criteria and a pilot project, although since the initial establishment funding, the programme has been entirely industry managed and funded.

In developing countries such charters and codes of conduct do exist but are not so common. An example is the Bangladesh Textile Mills

CHAPTER THIRTEEN

Associations and the Bangladesh Garment Manufacturers and Exporters Association which have agreed to develop a social and environmental code of conduct for their members.

C. Voluntary Action and Agreements

Community pressure and increasing legislative scrutiny have led to companies or industry associations taking "voluntary" action on behalf of environmentally sustainable practices. For example, the Bac Giang Chemical and Fertilizer Company in Viet Nam took voluntary action to prevent emissions of ammonia and other gases following concern that leaks from the factory's outdated plant were threatening the company's long term viability and the local environment. The result was a reduction of costs by 40 per cent and an environmental award in 1997 by the Viet Nam Labour Union for being the 'cleanest and greenest' company in the sector.

The chemical industry in Australia has undertaken a voluntary programme to improve the health, environmental and safety performance of its operations. The chemical industry in India has also supported the development of a voluntary code of ethics on the international trade on chemicals.

D. Recognition and Awards

Industry and business are also supporting and promoting improved environmental performance through the funding of awards, competitions and conducting performance surveys. For example, the Singapore Environment Council awards an annual Environmental Business Award. Another organization that helps to support environmental awards is the Japan Shipbuilding Industry Foundation, which has made an endowment to finance the Sasakawa International Environment Prize (which is awarded to individuals who have made outstanding contributions to the management and protection of the environment). Similarly, the Asahi Foundation Blue Planet Award goes to best researchers in the field of environment and the foundation also conducts surveys on environment to analyse new trends and needs in environmental management.

PRIVATE SECTOR AND CONCERNS OF THE 21ST CENTURY

A number of established and emerging environmental foci can currently be discerned through the actions of the private sector in the Asian and Pacific Region including forest management, genetic engineering, growing tourism and meeting the backlog of environmental services. For example, the private sector is increasingly being brought into plantations and forestry management. Some sectors of the business community have also shown a level of maturity in their relationship with consumers through their disclosure and neutrality on the issue of genetically modified food. For example, Jusco, Japan's leading supermarket chain introduced a line of GM-free foods and the resulting increase in the sales of these foods illustrated the importance of disclosure to consumers. Suppliers of corn snacks such as Tohato and beer brewer's Kirin, Asahi, and Sapporo also followed suit. A major supermarket chain in Hong Kong, China, Park'n' Shop plans to offer its customers a choice by testing store-brand items and labelling them. On this issue, the Rockefeller Foundation which has contributed US$100 million to bio-engineering research now advocates a new corporate culture that includes supporting more careful monitoring, open reporting and public participation, to win back public trust.

The tourism sector is an important source of revenue to many of the economies of the Asian and Pacific Region. Whilst an international industry code of conduct for tourism has been established, the World Travel and Tourism Centre's Environment Charter and Environmental Guidelines, the challenge is persuading the sectors many travel agencies and tourism related companies to accept and implement the provisions of these codes (Box 13.6). Within the region, the Pacific Asia Travel Association (PATA) supports the promotion of environmentally friendly tourism and has developed a regional Code for Environmentally Responsible Tourism that demands the adoption of necessary practices to conserve the environment as well as consider cultural values, customs and beliefs and promote wide community participation in tourism planning. The PATA Green Leaf ecolabelling scheme was launched as early as 1994 and the associated Green Leaf programme provides checklists that assist companies in assessing their environmental practices and in the implementation of the PATA Code. PATA also sponsors workshops and conference for raising awareness on the importance of sustainable tourism and has a foundation that awards scholarships and grants, including custom-made programmes for managers from the Asian and Pacific Region. The hotel industry, through the International Hotels Environment Initiative (IHEI), has published *Environmental Management for Hotels – The Industry Guide to Best Practice* and established its first regional chapter, the Asia Pacific Hotels Environment, intended to promote environmental awareness and action in hotels in the region. IHEI has also prepared an environmental toolkit for small and medium-sized

> **Box 13.6 International Hotels Environmental Action Plan for the Asian and Pacific Region**
>
> Hotels are at the very heart of the travel and tourism business, the largest and fastest growing industry in the Asian and Pacific Region. In 1995, the global industry generated an estimated US$3.4 trillion in gross output, employed 211.7 million people and produced 10.9 per cent of world GDP. Industry success relies upon the preservation of a high quality environment, hence the identification of the environment agenda as being of utmost and immediate importance by many industry leaders. By convening this meeting in association with UNEP, the hotel industry is continuing to position itself as a leader on environmental management and sustainable development within the international tourism industry.
>
> The International Hotels Environment Initiative (IHEI) is a network of 11 of the world's leading hotel chains which aims to improve environmental performance throughout the industry by promoting the business benefits of action. The Asia-Pacific Council is the campaign's regional body and is governed by 16 leading regional hotel chains. The IHEI is coordinated from London by The Prince of Wales Business Leaders Forum (PWBLF), a group of the world's leading multinational companies established to take up the challenge of sustainable development and corporate responsibility in the community.
>
> The meeting on the Environmental Action Plan for Asia and the Pacific was organized by the IHEI and its regional Asia Pacific Council, the International Hotel Association (IHA) organized a meeting in October, 1996 in collaboration with the United Nations Environment Programme (UNEP) to explore potential collaboration between the hotel industry, national hotel associations, government and non-governmental organizations. The meeting was the first in a series of regional Hotel Industry Environmental Fora, organized by the IHEI, the IHA and UNEP. The key elements of the final resolution of the meeting included: hotel business leaders should drive environmental action internally through their own hotel companies and should also support the efforts of associations at the national level; at the country level, hotel associations should drive environmental education programmes for their members; exchange of experience between national hotel associations will be facilitated through the IHA; government and national associations should work in partnership to encourage action by hotels. The exchange of hotel environmental best practice and promotion of the business benefits of environmental action was acknowledged as vital to any such strategy's success.
>
> Internationally, progress on these fronts are profiled through "GREEN HOTELIER", the recently launched IHEI magazine that networks environmental news, information and contacts in the international hospitality industry. The IHA assist national associations to develop programmes for their members and UNEP will support these efforts through providing technical and educational information through various UN units.
>
> It should be mentioned here however that developing guidelines and code of conduct are not an end by themselves. The real success would lie in their wider dissemination and implementation.
>
> *Source:* ENFO 1996

hotels. It should be noted that the region already has hotels that enjoy international reputations for their excellent environmental practices (Box 13.7).

The role of the private sector has also increased in the provision of environmental services and infrastructure. In India, for example, private sector investment contributes as much as 90 per cent of total investment in housing. However, privatization or the public private partnerships is not without some inherent structural and timeframe constraints. Solid waste management contracts that are attractive to the private sector need to be of sufficient duration to allow depreciation of investment, large enough to allow economies of scale and competitive enough to encourage efficiency. For their part, the public sector needs sufficient autonomy to enter into multi-year agreements that capture economies-of-scale, as well as efficiencies. Unfortunately, many countries of the region have procurement laws that place low monetary ceilings on the sizes of contracts before provincial and central reviews and approval and do not allow contracts to extend beyond one fiscal year.

CONCLUSION

Despite the increasing role being performed by some elements within the private sector in the provision of environmental goods and services, in self-regulation and in the promotion of environmental performance standards, many companies, industries and sectors in the region remain reactive rather than proactive in their approach to environmental issues. Despite the numerous examples of proactive pollution control from within the region, a total of less than five per cent of the region's investment in pollution control is provided by the private sector. The two major constraints to "greening" the private sector, particularly in the developing countries of the region, are weak environmental monitoring and enforcement of existing standards and the lack of green consumerism.

The governments of some countries in the region have begun to recognize the financial opportunities and political advantages that accrue from stricter environmental management and the

CHAPTER THIRTEEN

> ### Box 13.7 The Kandalama Hotel: An Eco Experience
>
> The Kandalama Hotel is the first hotel in Asia to be certified under the provisions of Agenda 21 of the Rio Earth Summit as a Green Globe Hotel. Built next to an ancient irrigation tank and surrounded by tropical dry evergreen forest, the hotel is within Sri Lanka's Cultural Triangle and is flanked by two World Heritage sites. The initial public protest due to the sensitivity of the hotel's location was overcome through the completion of a formal EIA report and the successful implementation of the report's recommendations.
>
> The environmentally sensitive design of the hotel avoids disturbance to the existing landform profile and trees, which needed to be removed during construction, were rootballed and replanted. To replant denuded patches of land, a nursery of 3 000 indigenous plants was established.
>
> The village community was given the highest priority in the provision of employment and many of the services and utilities brought into the remote rural area were extended to local villages including electricity, roadways, deep wells and biogas generators.
>
> The environmental management of hotel operations are categorized into six key elements: waste minimization (reuse, recycling and reducing); energy efficiency conservation and management; fresh water resources management; wastewater management; environmentally sensitive purchasing; and social & cultural development. These areas are in accordance with those laid out in Agenda 21 of the Earth Summit.
>
> In order to obtain total participation of hotel staff, environmental committees have been established to cover key factors and every employee as a member of at least one committee. Environmental conservation is an essential responsibility in everyone's job description.
>
> The hotel's environmental policy is available to the public and the environment factors are deeply integrated into the hotel management. Leaflets and brochures with emphasis on environmental conservation are available for all stakeholders. Guests have access to information on nature, special bird watching trails, the surrounding forest, its fauna, flora and its unique bio-diversity. Sustainable practices such as use of rainwater, solar heating panels are currently being employed, whilst windmill pumps and biogas generation are planned for the future.
>
> In recognition of the standards of environmental management adopted by Aitken Spence, the operators of the Kandalama Hotel, the hotel as won the following environmental awards:
>
> - International Green Globe Award for 1996, 1997, 1998 and 1999;
> - Sri Lanka Association for the Advancement of Science Award-1997;
> - TRAVTALK' Award-1997 by the World Tourism and Travel Council; and
> - PATA Green Leaf (Gold Award) for Environmental Education-2000.
>
> *Source:*   South Asia Comparative Environment Programme

enforcement of regulations. The consequential development of domestic expertise in pollution control also provides opportunity for environmental goods and services firms to develop and export. This has led to the promotion of tax incentives, research grants, trade missions and promotion of exhibitions to this new sector.

If the slow current growth in environmental awareness was to accelerate, the region has the potential to become one of the largest international markets for green goods and services. Already, the group of countries comprising Japan, Australia, New Zealand, People's Republic of China, India and many countries of the ASEAN account for an environmental goods and services market of US$80 billion. Within these countries there is a proliferation of environmental companies and an increasing number of initiatives being undertaken by both existing and new businesses and industrial enterprises.

Corporate groups and business associations have also started to take steps to improve their environmental performance and are employing environmental management tools, such as EIA and environmental auditing, to assist and standardize the process. The adoption of the ISO 14000 environmental management standard is expanding, ecolabelling is becoming established in certain markets and a consensus is emerging on corporate environmental reporting. The new wave of environmental awareness is also resulting in the establishment of business councils and associations for sustainable development, as well as business charters and codes of conduct. As small and medium industries and enterprises dominate output in a number of industrial sectors, there is also a need to apply greater focus in the promotion of environmental management techniques to members of this sub-sector.

The "greening" of the financial markets is often seen as a key challenge on the road to sustainable development. As financial institutions, such as joint stock companies, banks, insurance companies and financial management firms, begin to take environmental factors into funding considerations, private sector attitude and performance towards sustainability will improve. In the longer term, business may see the inherent convergence of investments that achieve both financial and environmental sustainability.

There is a discernible overall trend in the growth in private sector investment as a proportion of GDP in some parts of the region. In the wake of increasing investment needs and against a background of declining governmental resources, analysts believe that the demand for private investment will expand significantly in the future.

Chapter Fourteen

A Pakistani Village Development Committee discusses environmental protection.

14

Major Groups

INTRODUCTION
ROLE OF NGOs & MAJOR GROUPS
NGO ROLES AND ACTIVITIES
THE ROLE OF OTHER MAJOR GROUPS
CONCLUSION

CHAPTER FOURTEEN

INTRODUCTION

There is a growing environmental awareness amongst stakeholders, individuals and communities within the Asian and Pacific Region. This increase in knowledge and awareness has been, by and large, the result of campaigns and education programmes run by major public interest groups concerned with the environment. These include non-governmental organizations (NGOs) at the international, regional and national levels, as well as groups concerned with the empowerment of marginalized sections of society, such as women, indigenous peoples, and youth groups, and other community based organizations. Such organizations have worked to foster grassroots-based approaches to the protection and preservation of the region's environment.

Major public interest groups are increasingly contributing efforts towards sustainable development through participation, advocacy, demonstration projects, monitoring and research, as well as cooperation and networking with other NGOs and government departments. This chapter reviews the contributions made by such groups to the promotion of environmental knowledge, awareness and action in the region.

ROLE OF NGOs & MAJOR GROUPS

A. Status and Trends

Traditionally, community based organizations played an important role in the management of common property resources such as forests and fisheries in the Asian and Pacific Region. Although over successive years their role was reduced by governments in some countries, recent years have seen a re-emergence of community involvement and the development and growth of NGOs, youth, women and indigenous people's groups and associations of farmers and businessmen.

NGOs have, in particular, played an important role in raising environmental concerns, developing awareness of environmental issues and promoting sustainable development. The encouragement of public participation in environmental management through legislation in recent years has also enhanced the role of NGOs and Major Groups. For example, in Thailand, Article 56 of the 1997 Constitution recognizes the rights of people to participate in the protection of natural resources and environment. Similar provisions have been made, for example, in the Philippines, New Zealand (Resource Management Act), Azerbaijan (EPA 1999) and the Australian Landcare and Coast Care programmes.

The roles and activities of major public interest groups in the Asian and Pacific Region are constantly evolving, as the issues they deal with change, and the political and social landscape they work within alters. However, between individual countries there are great differences in both the number and types of major public interest groups that exist and the way they operate, reflecting the diversity of cultures and political establishments, and levels of economic and social development. For example, in India, there are numerous NGOs and community-based organizations using a wide variety of means to raise awareness, and in Singapore, about 40 green groups operate under the umbrella of the Singapore Environment Council, an organization set up by the government to champion environment-related activities (SEC 1998).

In recent years, the range of activities undertaken by environmental NGOs and other major groups has broadened. They now undertake a much wider range of activities than simply raising environmental awareness and/or acting as pressure groups. Their activities now include environmental monitoring; promoting environmental education, training and capacity-building; implementing demonstration projects; conducting advocacy work in partnership with the government; and the promotion of regional and international cooperation on environment. Many also get involved in the practical management of conservation areas, and promote community or individual action and campaign for greater accountability on the part of the government and corporate sector. The majority of NGOs in the region now work concurrently on environment and development, thus acknowledging that environmental problems are embedded in economic and social systems (Oh 1998).

Unfortunately, the Asian financial crisis of the late 1990s has adversely affected the financial position of many major public interest groups, particularly in Southeast Asian countries such as Indonesia, Thailand, and Malaysia and has curtailed much of their work. For instance, the Indonesian NGO *Wahana Lingkungan Hidup* (WALHI) has over 600 groups actively involved in a variety of work, including campaigning against the burning of the country's tropical forest, however, financial restrictions have caused the group to cease the regular publication of its investigative reports.

Despite this financial set back, the work of NGOs has won considerable credibility and appreciation in the region, which can be evidenced by the growing partnerships with government. For instance, NGOs and major groups have substantially increased their involvement in policy related work, playing a key role in assisting government agencies to meet the requirements of environmental management. The conviction that NGOs and government agencies can and should work together in a complementary relationship has become stronger and the credibility that NGOs have acquired from successful campaigns has created strong and growing public support for their new advocacies. For example, in the Central Asian subregion, independence has led to the emergence of NGOs with very strong technical backgrounds, who play a substantial role in the formation of public opinion.

Established nationwide networking has assisted NGO advocacy at the national level to proceed, even during periods of political instability. In the Philippines, the Philippine Federation for Environmental Concern (PFEC) is the earliest network in the country for environmental advocacy, formed in 1979 at the height of the martial law period. Other networks such as the Green Forum have added strength to the overall advocacy work in recent years.

At the regional and international level there is evidence of an increasing networking of NGOs and other major groups. At these levels, they are concerned with the sharing of experiences, influencing discussions and policy-making of regional and international inter-governmental bodies and raising concern over the globalization process. Active regional and international NGOs working on sustainable development concerns include: the International Organization of Consumer Unions (IOCU); the Asian NGO Coalition for Agrarian Reform and Rural Development (ANGOC); Asian Cultural Forum on Development (ACFOD); and the Asian Alliance of Appropriate Technology Practitioners (APPROTECH ASIA). The Third World Network has also consolidated a regional stance especially on issues related to trade liberalization and globalization.

The new electronic forms of communication, principally e-mail and websites (Box 14.1) have greatly assisted in these endeavours. For example, in India, networks have been built between people's organizations, NGOs, and women and children's organizations in efforts to change attitudes towards environmental protection and conservation (ESCAP 1997). Networking efforts among NGOs within countries such as Pakistan, Nepal, the Philippines and Bangladesh have also resulted in expanding and strengthening of their activities.

Box 14.1 Harnessing the Power of the World Wide Web

One key feature of the last five years has been the increasing use by NGOs of information technology to promote conductivity. The rapid convergence of media, the growth of satellite communication and new technologies such as the E-Mail and the World Wide Web have all increased the scope and availability of environmental information resources. This has helped speed the development of networking and liaison groups – both between NGOs and with government agencies.

One example of an NGO network that has a strong presence on the web is the Third World Network, which shares a page with one of Malaysia's leading campaign groups, the Consumer Association of Penang and Pan Asian Networking, an organization which helps researchers and communities in the developing world find solutions to their social, economic and environmental problems. This page is used to bring people up to date with various campaign issues and publications and to link activists.

A website that has been specifically developed to facilitate regional networking is ECANET (Environmental Communication Asia Network, Website 21), developed and operated by AMIC. Support for this website has been provided by the ADB and UNESCO. The website disseminates information on environmental groups in the region, bibliography on environmental information (including websites) and environmental success stories written by Asian journalists. At international level, websites have been developed by UNEP in collaboration with other agencies. Chief among them is Infoterra, one of the most comprehensive environmental resource systems currently available which facilitates the exchange of scientific and technical information. This website has links with over 6 800 national and international institutions, NGOs, industrial and commercial enterprises, academics and experts from around the world.

Small NGOs and CBOs particularly in rural areas of Asia and the Pacific are slow in harnessing the benefits from internet and world wide web. It is important to strengthen their capacities in this respect so that they could take full advantage of the growing information technology which provides tantamount opportunities not only for networking both nationwide and worldwide, but also for strengthening the capacities of major groups, especially NGOs.

Source: http://www.capside.org.sg/ and http://iisd1.iisd.ca/50comm/commdb/list/co8.htm

CHAPTER FOURTEEN

NGO ROLES AND ACTIVITIES

Local, national and regional NGOs have emerged as major players and partners in both development and conservation activities in the region. At the community level, they are in the front line in providing assistance in the acquisition of basic needs and amenities; in identifying issues, raising awareness, and providing information to grassroots communities; in articulating the communities' problems and needs and bringing these to the attention of those who can affect change; in defending both the environmental and developmental rights of communities and building the capacity of communities to manage their natural resources; and in dealing with sustainable development concerns. NGOs that work at the national level focus mainly on policy work, playing a vital role in the identification of the weaknesses and gaps in current policy or legal frameworks; in information gathering and educating the public, private sector and government; and, in certain cases, in activist lobbying and protest movements.

A. Awareness-Raising, Campaigning and Advocacy

Across the region a large array of groups work to raise awareness of environmental issues and push for changes in policy and development programmes. These groups carry out environmental awareness raising and campaigning locally, nationally, and internationally, with some campaigns operating simultaneously at all levels. In India, for example, the *Kerala Sastra Sahitya Parishad* (KSSP) has earned international recognition for its work in mobilising public opinion among people's organizations in the State of Kerala (United Nations 1995). The KSSP is regarded as one of the best-informed and best-organized grassroots movement in India, with over 20 000 members.

In Pakistan, the Society for the Conservation and Protection of the Environment (SCOPE), established in 1988, is particularly successful at national environmental campaigns, whilst giving priority to developing linkages with local NGOs, research institutes, universities and government departments. In addition SCOPE motivates grassroots groups and undertakes public interest litigation and advocacy work (Non Governmental Liaison Service 1997).

Scientific and technical NGOs are assisting in bridging the gap between science, policy makers and the citizenry. Their research and education work is proving a vital addition to the decision and policy-making process. In India, for instance, the Centre for Science and Environment publishes 'Citizen's Reports on the Environment' which focus on specific environmental issues, such as urban pollution, and flood management. Written in non-technical languages, these reports enable the general public to better understand the issues.

Many of the more established NGOs in the region work on major national campaigns using a range of promotion activities, from grassroots awareness-raising, through to lobbying and media campaigns (Box 14.2). Such campaigns are multi-faceted, involving research, awareness-raising, education and lobbying. The Worldwide Fund for Nature (WWF) in Malaysia, for example, has launched the Species 2 000 Campaign to mobilize effective national action to conserve Malaysia's wildlife. (WWF Malaysia, *Website 6*). In doing so, WWF Malaysia has forged partnerships with many groups involved in conservation, from Federal and State government agencies to universities, other NGOs and local community groups. Similar alliances have been made by environmental groups in India, Malaysia and Philippines to raise the awareness of governments and the general public with regard to the loss of fauna and flora species and consequences for biodiversity.

One of the great challenges for NGOs campaigning on environmental issues is to involve as many people as possible and, particularly where religion plays a major role in everyday life, getting the environmental message across to key religious groups. The Alliance of Religions and Conservation (ARC) has been working internationally with many faiths to forge new, practical models of religious involvement with environmental issues. The group espouses the Ohito Declaration of 1995, a declaration on religions, land and conservation that states "for people of faith maintaining and sustaining environmental life systems is a religious responsibility" (Xiamin and Halbertsma 1997). The Ohito Declaration and the work of organizations such as ARC has led to the re-discovery of 'holy ground' and the concept of the need for Man to preserve and protect the environment by all the major religions of the world.

The scope of ARC's network activities is shown in the involvement of the Taoists, who formally joined ARC in 1995; the ninth faith to do so. Following meetings with WWF/ARC staff, the Taoists asked ARC to join them in launching a campaign to protect their sacred holy mountains in China, which were threatened by changes in forestry, agriculture, urban development and, of late, tourism.

Beyond national frontiers, many environmental NGOs have joined forces to campaign internationally. WALHI, Indonesia, for instance, worked alongside international NGOs such as WWF to bring the plight

MAJOR GROUPS

> **Box 14.2 NGOs Working to Improve Media Coverage of Environmental Issues**
>
> The media is a vital conduit for environmental campaigns and programmes, however there is much need for improvement in the way the regional media handle and report the key issues. Many NGOs across Asia and the Pacific are working to support the media and improve coverage of environmental issues.
>
> For example, in Bangladesh, the Nature Conservation Movement (NACOM) and the Society for Environment and Human Development (SEHD) have worked closely together to establish a multimedia centre focusing on sustainable development. The organizations have extensively researched and reported on forests, indigenous people, human rights and other environmental issues. They are building information bases on these issues for the use of public, social entrepreneurs, journalists, and local communities and are also making an effort to improve the standard of investigative journalism – through training on in-depth reporting, providing information bases to journalists, members of other media and undertaking research activities. SHED founded in 1993, is a non-profit organization dealing with environment, development, and multilateral development bank and human rights issues. NACOM-Nature Conservation Movement – is a national nature conservation research and management organization which is particularly experienced in biodiversity research and non-formal environmental education and awareness building projects.
>
> Another NGO that is deeply involved in communication about environmental issues is the Asian Media Information and Communication Centre (AMIC), headquartered in Singapore. AMIC conducts seminars and workshops for media practitioners, publishes books, produces broadcast and audio-visual materials which are used to alert audiences to environmental concerns. AMIC's environment-related activities have mobilized communication educators and media practitioners to promote environmental conservation and protection.
>
> Journalists themselves have mobilized to ensure that the environmental message is given maximum coverage. Leading this initiative is the *Asia-Pacific Forum of Environmental Journalists* (AFEJ), currently headquartered in Colombo, Sri Lanka. This group coordinates the efforts of 15 national forums of environmental journalists. Its work programme includes specialized regional training workshops on environmental reporting, publication of books, technical assistance to members, journalist exchange programmes and awards for excellence in environmental reporting. The AFEJ's constituent members are also very active at both regional and national levels. For example, the Sri Lanka *Environmental Journalists Forum* (SLEJF) has compiled and published 20 success stories written on the environment in Sri Lanka. The SLEJF was also the principal organizer of the *Sixth World Congress of Environmental Journalists*, held in Colombo, Sri Lanka from October 19 to 23, 1998. This meeting resulted in the formulation and adoption of AFEJ's *Five-Year Strategic Management Plans*.
>
> The power of television is not ignored. The Asia-Pacific Institute for the Development of Broadcasting (AIBD), headquartered in Kuala Lumpur, Malaysia has worked with regional and national NGOs in conducting courses for broadcasters on popularizing environmental issues, and the Television Trust for the Environment (TTE), located in Colombo, Sri Lanka, works closely with organizations such as UNEP in producing television and video documentaries about environmental issues. These efforts, contributing to media coverage of environmental issues, are likely to improve the situation even further with their multiplier effects.
>
> Source: Khan 1997 and Wickramaratne 1998

of communities affected by forest fires to international attention. WALHI also brought together the protests of thousands of affected villagers to highlight the involvement of big business in forest destruction, and led a court action against plantation and logging companies implicated in the fires.

NGO groups have also had international success in the campaign against the trade in endangered animal products. Although most Asian and Pacific Region countries have ratified the CITES Convention (see Chapter 3), an active trade in endangered species still exists. NGOs are active not only in lobbying governments for effective enforcement of existing legislation to protect endangered species, but also in raising the awareness of the general public with regard to the importance of conservation. The power of the media has also been used effectively in this arena by the Asian Conservation Awareness Programme (ACAP), using graphic images of animals being slaughtered for their tusks, fur and gall bladders in a series of hard-hitting TV, cinema and poster commercials. Through such international and national campaigns, NGOs in the region have been trying to persuade the general public to change their habits with regard to exotic food and clothing and in getting the government to implement the provisions of CITES.

Increasingly, regional NGO groups are taking part in global campaigns, such as Greenpeace China and its much-publicized campaign against potentially toxic phthalate-containing PVC toys and children's products. In November 1998, following the campaign, Toys R Us announced a worldwide product withdrawal of phthalate-containing teethers, rattles, and pacifiers. Throughout the region, NGOs also work with regional and international agencies to also observe special days – such as *World Water Day* (March 22), *World Environment Day* (June 5) and *World Habitat Day* (October 1).

CHAPTER FOURTEEN

Green political parties are entering the formal political arena in some countries, such as in the Philippines, where the Philippine Federation for Environmental Concerns (PFEC) is organizing a Green Party modelled on those in Europe. During the last election two groups, the Green Philippines and the Philippine Greens, linked efforts to list candidates. Elsewhere in the region, Green Politics are an established strand of the party political system. For example, the world's first recognized "Green Party" was the Values Party of New Zealand which, although subsequently absorbed by the political mainstream, provided a political platform that led directly to the adoption of progressive environmental legislation by the New Zealand governments of the 1970's and 1980's.

B. Environmental Monitoring and Reporting

Many NGO campaigns are based on research that highlights or monitors specific environmental issues. At the local or community level, NGOs are in a good position to keep track of critical issues on a continuing basis and, in many cases; NGOs complement the work of government institutions and cooperate with law enforcement authorities.

1. Environmental Reporting

In Bangladesh, for example, about a dozen NGOs cooperate to produce periodic "State of the Environment" reports, and in India, the Centre for Science and Environment publishes Citizen's State of Environment Report. In many instances, these broad assessments are supplemented by specialized reports that facilitate debate on key environmental issues, promote public awareness and encourage active community participation in environmental protection.

In addition to State of the Environment Reports, Investigative Reports on the specific environmental issues are also increasing in the region. In order to assist in such endeavours, the Asia-Pacific Forum of Environmental Journalists (AFEJ), with ESCAP support, has published a book entitled "Reporting on the Environment: A Handbook for Journalists". AFEJ affiliates are active in a number of countries in the region (including Nepal, India, Sri Lanka, Malaysia, the Philippines and Thailand) and has published Citizens' Reports on the Environment, as well as newspaper reports. AFEJ associate members such as the Asian Institute of Development Communication and the Asian Media Information and Communication Centre have helped conduct workshops on environmental reporting. PINA, the Pacific Islands News Association and its Pacific Forum of Environmental Journalists have established the PINA Pacific Journalism Centre in Suva, Fiji, which runs a regular environmental news service distributed to the 22 countries and territories in the South Pacific.

2. Environmental Journalism

Environmental journalism can play a vital role in creating market reactions in the international stock markets. For example, studies in the U.S. and Canadian stock markets have shown markets react significantly to environmental news. Gains from good news and losses from bad are in the range of 1 to 2 per cent. World Bank researchers examined whether the same trends hold true for developing countries (Argentina, Chile, Mexico, the Philippines) and found that 'good news gains' average 20 per cent whilst 'bad news losses' range from 4 to 15 per cent. (World Bank 2000)

3. Environmental Monitoring and GIS

Environmental monitoring and reporting by NGOs have contributed substantially to environmental protection across the region. For example, a reef monitoring and evaluation project undertaken in Pangasinan in the Philippines (United Nations 1992, *Website 2)* resulted in improved protection of the coral ecosystem and, ultimately resulted in increased catches for local fishermen. It also stimulated community-based action where villagers of Cabacongan (province of Bohol) formed a *Bantay Dagat* ("Sea Watch") group to apprehend illegal commercial fishing boats.

A number of NGOs have embraced Geographic Information System (GIS) technology to help them with environmental monitoring. For example the International Centre for Integrated Mountain Development (ICIMOD) (United Nations 1999, *Website 3)* has established a decentralized network that collects, stores and disseminates key bio-physical and socio-economic data. Other groups doing similar work include the enhancement of GIS training at the University of the South Pacific in Fiji (United Nations 1999, *Website 4)*.

At the national or regional level, NGO monitoring can track the effectiveness of legislation, and investigate issues such as the movement of hazardous waste, the migration of species, the trade in endangered or restricted animals or plants or research on the state of rivers, forests or other ecosystems. NGOs use this type of work to influence practical actions, and to develop campaign strategies or propose policies. For example, Greenpeace China's monitoring of the country's trade in toxic waste led the banning of hazardous waste imports into or through the Hong Kong Special Administrative Region (Greenpeace, *Website 5)*.

The availability of reliable information on the environmental impact of development and economic policies is a critical requirement for robust environment decision making. A number of NGOs and other groups are working to present information to influence government decision-making. Singapore based, the Economy and Environment Programme for Southeast Asia, or EEPSEA, operates across the region, coordinating and supporting a network of researchers who investigate the environmental impacts of policies. It works in close partnership with a range of organizations, including the Vietnamese University of Agriculture and Forestry, the University of the Philippines, Los Banos and the China Centre for Economic Research, Beijing. The research undertaken by EEPSEA and its associated network has influenced the outcome of a range of issues including pollution in People's Republic of China, water supply in Manila, the impact of international law on farmers in Sri Lanka and developing policy for Thailand's National Parks (EEPSEA 1998).

In terms of monitoring the urban environment, a number of NGO initiatives have been launched through the United Nations Development Programme regional offices. These include *Asia-Pacific 2000*, an initiative to help NGOs address the urban environment challenge, and the Urban Governance Initiative. Both these programmes have worked closely with regional and national NGOs in organizing activities to promote knowledge and awareness of urban environmental issues. Various publication have been produced, such as *Our Cities, Our Homes*, which is an A to Z guide on human settlement issues, and a community action guide called *Water Watch*, published in collaboration with a regional NGO Asia-Pacific People's Environmental Network (APPEN).

4. *New Research*

In the last few years a number of areas in water resource; solid and hazardous waste management; and biodiversity (genetic engineering) related fields have gained importance. New NGOs are emerging in the region to promote knowledge and awareness of these issues, while more established NGOs are responding to the new challenges by re-focusing their research efforts. For example, in India, the Centre for Science and Environment (CSE) was instrumental in convincing 28 selected pulp and paper mills to provide data similar to the USEPA's "toxic releases" database. The results of the survey showed that only two companies were in compliance; however, it also showed that a mill with sound ecological sense is more likely to maintain steady profits than other plants. The project prompted nine of the companies to improve pollution controls. The CSE plans to introduce green ratings to two sectors annually with India's automotive industry as the next target. (Asia Week 2000).

In the Philippines, the Pasig River Movement, a coalition of NGOs advocating the clean-up of Metro Manila's major river, regularly gathers data on companies discharging waste in the river. The ten worst polluters of the year are given the *Lason* (Poison) Award, which receives wide media coverage, and has led companies to set-up effective wastewater management facilities. Several "poison" awardees have since redeemed themselves and became "environment" awardees the following year.

C. Education, Training and Capacity Building

An increasing number of NGOs have also been using education to encourage participation in conservation activities (see Box 14.3). One of the most active international NGOs working in the field of environmental education in the region is the International Union for the Conservation of Nature and Natural resources (IUCN). The IUCN exercises this function through its Commission on Education and Communication (CEC), which has sought to develop environmental expertise among teachers in the region by focusing on providing information, training, capacity building and networking (CEE 1997).

Other NGOs have worked extensively with governments to help develop and implement national environmental education strategies (see Chapter 15). For example, in Nepal, NGOs have collaborated on the implementation of the environmental education element of the country's National Conservation Strategy and have provided technical assistance to government agencies in running the programme.

D. Government and NGO Partnerships

Government and NGO partnerships are being encouraged in a number of ways. In some countries laws exist to ensure community and major groups participation in developing regional and national policies and plans. For example, in Thailand, the 1992 Environment Act delegates the work on environmental management to provincial and local authorities, and encourages people's participation through environmental NGOs (Government of Thailand 1992). As countries put into practice formal policies requiring public participation in decision-making, the influence and potential political power of NGOs increases. For example, in the Philippines, the EIA regulations and implementing rules relating to the Mining Act and the Indigenous Peoples Rights Act require the voluntary informed consent of indigenous peoples before projects can proceed. In many cases, people have little experience or access to information on the issues involved, and NGOs undertake to advise on hearings and consultations, and on negotiation and settlement strategies. However, elsewhere within the region, the influence of unaccountable public interest groups within their own political agenda has given cause for concern, and in some cases has led to their loss of credibility and/or sanction. Elsewhere, governments have made moves to formalize the role of independent NGOs within the decision making institutions by giving them representation.

1. *Institutionalization of NGO-Government Partnerships*

The participation of major public interest groups is also being institutionalized through formal representation in decision making and management bodies. For example, in the Republic of Korea, the Environmental Preservation Committee, chaired by the Prime Minister, was established by the 1990 Basic Environmental Policy. The committee reviews all major environmental policy decisions and includes not just government ministries, but also representatives of environmental associations. The Philippines has a similar body in its Philippine Council for Sustainable Development (PCSD) that is chaired by the Director-General of the National Economic and Development Authority and submits high level reports through its meetings with the President. There is equity in the decision-making between government and major group (civil society) representatives in the PCSD; major issues are only deemed resolved when there is a consensus between major group (civil society) representatives and government counterparts.

In India, the National Environmental Council has five representatives from NGOs as well as the National Consumers Federation. Sri Lanka also has inter-agency committees with NGO participation for the formulation of environment policy and in implementing action plans to meet its obligations under international environmental conventions. NGOs in the Republic of Korea also participate in environment-related decision-making processes through the National Council of Environmental Organizations.

The mandate for civil society participation has also been provided through the development of programmes that facilitate direct community-based management of the environment and natural resources. India, for example, promotes people's participatory institutions like *Panchayati Raj* institutions, co-operatives and self-help groups. Similarly, Indonesia recognizes indigenous mutual-help and community participation mechanisms such as the water user's associations *Subak* in Bali and *Mitra Cai* in West Java. Thailand's Water Resource Utilization Promotion Project has organized water resource users to establish Water Resource Utilization Groups for water resource management.

Programmes related to community forestry have been another way to recognize and support community user groups. India's Eco-Development Programme involves local communities in the maintenance of designated buffer regions surrounding protected areas (see Chapter 3). The Hill Community Forestry Project in Nepal hands over accessible hill forests to user groups that are willing and able to manage such forests. Thailand, the Philippines, and Sri Lanka also have similar policies promoting the active implementation of community or village forestry.

Some countries have recognized the role of indigenous peoples in sustainable development. The Philippines has released "Certificates of Ancestral Domain Claims" over more than 2 million hectares of land and waters, many of which are in protected areas, where no project can proceed without the voluntary and prior informed consent of the ancestral landholder. Similarly, Fiji recognizes traditional fishing rights and has a policy that no commercial fishing activities can proceed unless the consent of the chiefs and the people with the right to fish in these areas is obtained.

The Mongolian Government co-operates closely with NGOs, such as the Mongolian Association for the Conservation of Nature and Environment, to coordinate the voluntary activities of local communities and individuals to protect nature and wildlife. The government also works with the Green Movement that promotes public environmental education in support of traditional protection methods (MNE, UNDP and WWF 1996).

In many countries, NGOs also co-operate closely with government organizations and other civil groups in formulating programmes and plans. In Bangladesh, for example, the national environmental management action plan was prepared by the Ministry of Environment and Forests in collaboration with the Forum of Environmental Journalists of Bangladesh (FEJB), environmental NGOs and other civil society groups (ESCAP 1997). IUCN Pakistan helped the government in preparing Pakistan National Conservation Strategy and finalising the draft Pakistan Environmental Act. In Sri Lanka, NGOs are involved in the development of forestry development plans and have provided assistance in enforcing the prevention of illegal logging.

2. *Provision and Validation of Environmental Information*

The provision and validation of information forms the basis for successful collaboration between governments, NGOs, community representations and other stakeholders, including project proponents, the private sector, the scientific community and the press and broadcast media. In circumstances where NGO/major groups are formally involved in inter-agency bodies (i.e. Councils for Sustainable Development) and in public consultation processes, as in the Republic of Korea, India, the Philippines, Pakistan and many other countries, information exchange and dialogue is a continuous process that allows ongoing verification and clarification between parties, thereby facilitating progress and reducing the need for conflict resolution.

3. *Provision of Funding Support*

Many governments in the region provide or facilitate, funding to strengthen NGO activities in environmental protection and natural resource conservation. Governments, donors and NGOs have also joined together to form endowment funds to support environmental projects. Examples are in Bhutan (Bhutan Trust Fund for Environmental Conservation), Indonesia (Indonesian Biodiversity Foundation or *Yayasan KEHATI*), and the Philippines (Foundation for the Philippine Environment). Additionally, Japan's Fund for Global Environment, based on contributions from the government and the private sector, was established to provide assistance for the global environmental conservation activities of NGOs.

The Global Environmental Facility (GEF) recognizes that civil society has become an important force in implementing Agenda 21 (GEF 1998). About 20 per cent of the funds expended by the GEF involve NGOs in the design, planning and/or implementation of these projects. The GEF also supports a Small Grants Programme (SGP), implemented through the UNDP, and has started a Medium-Sized Grants Programme where NGOs can access up to US$1 million for biodiversity conservation projects.

E. **Regional and International Cooperation and Networking**

NGOs in the Asian and Pacific Region are increasingly working together to deal more effectively with transboundary issues and to improve the impact of associated campaigns. One high-profile grouping is the Third World Network (TWN) based in Malaysia, which is an independent non-profit international network of organizations and individuals involved in issues relating to

---

Box 14.3 Using Education to Encourage Participation: Success Stories

Reef Education

In Australia, the Great Barrier Reef Marine Park is managed primarily through the community's understanding and acceptance of zoning and other management practices. An education and information programme run by the Marine Park Authority (GBMPA) provides educational, advisory and information services. The strategy has marshalled strong public support for the marine park concept and high levels of public participation and cooperation in developing zoning plans.

Global Concerns

Global Concerns is an organization within the United World College of Southeast Asia (UWCSEA) which integrates awareness of issues associated with development and the environment into the school's education programme. A key part of its work is providing material support for a wide variety of projects in Asia. Students and staff are encouraged to visit the projects and work with the local community. This not only provides feedback for the group's fund-raisers, but also allows the projects to be included as part of the UWC's normal teaching programme. In this way, the group instills in its students a knowledge of the role they can play in serving society. Projects which the organization has supported include a community-development based conservation programme in Kenong Rimba Park, Malaysia, a self-sufficiency project in a village in Sarawak, Malaysia and an agro-forestry project in Bali, Indonesia.

Recycling Awareness

The city of Bangalore in India faces a massive resource management and waste challenge with hundreds of tonnes of garbage remaining uncollected every day. To help to address this problem, the *Centre for Environmental Education, Southern Regional Cell* (CEE South), has run a major hands-on education programme in the city to create awareness among urban people about the need to reduce wastage of water, fuel and other natural resources. Volunteer Communicators have been trained to run courses on recycling, composting, health issues and anti-littering. The programme has resulted in much increased awareness and participation in waste reduction and litter clean-up.

Source: United World College of Southeast Asia 1997

development (TWN 1999, *Website 10*). TWN conducts research; publishes books and magazines; organizes and participates in seminars; and provides a platform representing the interests and perspectives of developing countries at international fora, such as United Nations conferences.

SUSTRANS (Sustainable Transport Action Network for Asia and the Pacific) is another informal network of interested government officials, transport and urban planners, NGO representatives and academics. It is dedicated to promoting transport policies that foster socially just and ecologically sustainable mobility. The SUSTRANS network is based in Penang, and allows for the sharing of information about current transport plans and problems, for joint campaigning and for cooperation with other NGOs working in related areas such as air-quality and housing rights. Other networks provide links between international counterparts, both within the Asian and Pacific Region and globally (see Chapter 15).

F. Management of Resources and Environment: Community Based Projects

With the revival of community involvement in management and policy issues, many public interest groups have become involved in community-based field projects. For example, the Philippine Federation for Environmental Concern (PFEC) has undertaken community forestry projects in partnership with local communities, whilst the Agha Khan Rural Support Programme (AKRSP) has assisted in promoting forestry and sustainable agriculture in the northern areas of Pakistan. Box 14.4 highlights further examples of community participation in the management of resources and the environment and how the work of religious groups is increasingly involved in finding practical solutions to social problems.

Box 14.4 Community Based Projects

To address the specific needs of local communities, many NGOs are working to directly solve problems at the grassroots level. One example is a programme run in Phang Nga Bay, on the Andaman sea coast of Thailand. Organized by the Andaman Sea Fisheries Development Centre Organization (AFDEC), Phuket, Thailand and the FAO/Bay of Bengal Programme, Chennai, India, the programme addressed the over-exploitation of fish and degradation of the natural environment which was causing a reduction in fish catches and incomes. The project brought representatives of many of the 114 villages of Phang-Nga Bay together for regular monthly meetings to discuss, initiate and monitor management activities. These included the promotion of cage culture of finfish; the banning of the use of trawls and motorized push nets within 3 km of the shoreline, and; the installation of over 40 artificial reefs at the entrance to the Bay. The project has increased resource health and productivity, raised the production of shrimp and crabs and achieved a strong support for the implementation of community-based decisions.

Another similar project in Thailand is run by the Yadfon Association, based in Trang. The Association works with local fishing villages to promote community-based coastal resource management and sustainable fisheries management. Fish in the region thrive on seagrass and mangrove swamps which are increasingly threatened. In response, the Yadfon Association has worked with local fishermen to responsibly manage mangrove forests and develop seagrass conservation projects – while organizing activities to supplement income and reduce the use of destructive fishing equipment. The project has resulted in considerable environmental improvement, which, in turn has generated a positive impact on the villagers' livelihoods. Even after the initial project funding ceased, activities continued through a network of fishing villages involved in participatory management of coastal resources.

NGOs are no less active in cities, where they are often called on to fill in where municipal support services fail. One group which shows the type of work undertaken is The Clean Ahmedabad Abhiyan project. This programme was formed by concerned citizens, women's and voluntary organizations and the municipal corporation to research and find permanent and sustainable solutions to the health hazard and sanitation problems caused by the decomposing garbage on urban roads. Through a concerted public awareness campaign, households were involved with segregating wet and dry garbage. A special bag with three compartments was developed to segregate and store recyclables – paper, plastic and miscellaneous. The programme has resulted in improvement in the health standards of the community as well as providing income and increasing the amount of waste recycled.

One example of a multi-faceted campaign, which embraces awareness raising and action, is a project undertaken in the Punjabi town of Anandpur, which has been set up to reduce the environmental impact of the festival of Hola Mohalla. This festival draws over two million Sikhs to the town and creates many environmental problems. At the heart of the project is a Sikh Heritage Centre stressing the centrality of faith to the whole undertaking. Alternative energy sources, such as solar power, will be used, public transport will be made a priority and low energy housing and small industrial units to manufacture environmentally-friendly goods such as fuel efficient stoves will play a key role.

Source: United Nations 1999, *Websites 12 and 13*; Xiamin and Halbertsma 1997

THE ROLE OF OTHER MAJOR GROUPS

A. Gender

Women play crucial role in helping the communities and societies of the region to improve the condition of the environment and achieve sustainable development. Indeed, as farmers, housewives, mothers or social mobilizers, women are at the heart of issues and often bear the brunt of problems resulting from environmental degradation.

The role of women in sustainable development and the environment has been highlighted since the 1980s. However, over the last decade, three key initiatives have provided an international and regional framework within which women's issue may be viewed: the *Rio Declaration on Environment and Development*, 1992; the *Beijing Declaration and Platform for Action*, 1995, and; *ASEAN's Vision 2020 Statement*, 1998. These documents have set the tone for, and determined the shape of, environment-related programmes at national, regional and international levels. The Beijing Platform, for example, proposes specific actions to promote women's access to education, inheritance, economic resources and decision-making.

Women's NGOs utilized the opportunities arising from these international and regional initiatives to revive, reactivate and strengthen movements at the national and local levels. Follow-up meetings have been organized by women's associations in countries such as Malaysia and Thailand to review and develop initiatives for women to promote sustainable development. The Singapore-based women's organization ENGENDER has actively participated in meetings at national and regional levels to monitor and encourage women's participation in environmental issues. A number of NGOs are also actively involved in projects aimed at empowering women in communities for the management and protection of the environment. Examples of such projects are given in Box 14.5.

Women's organizations have proved themselves as some of the most potent environmental campaigners in the region. One of the most well known success stories of women's participation in environmental protection is the Chipko movement. This grew out of grassroots opposition to the destruction of India's forests, which saw villagers seeking to protect this vital resource through the Gandhian method of non-violent resistance. In the 1970s and 1980s, the movement achieved a 15-year ban on green felling in the Himalayan forests of Uttar Pradesh, stopped clear felling in the Western Ghats and the Vindhyas and generated pressure for a natural resource policy which is more sensitive to people's and ecological requirements.

Women have also played a key role in uncovering the negative health impacts associated with environmental contamination. For example, rural women in Bangladesh have been among the most vocal groups calling for action against arsenic contaminated water, a major and widespread problem (see Chapter 4). During the early 1990s women living in these districts realized that their children's health was being seriously affected. At the April 1997 Session of the Commission on Sustainable Development, and the Special Session of the United Nations General Assembly, the Women's Caucus raised the issue of the arsenic case to increase international attention on the problem. Since then, women's groups have played an important role in working with communities to survey and monitor the effects, generating media attention and mobilizing national and international initiatives.

B. Role of Children and Youth Groups

Taken as a whole, the percentage of children in the Asian and Pacific Region has steadily increased over the last decade. Demographic statistics reveal that children under 15 comprise nearly one-third of the world's total population, with 60 per cent of them living in Asia. Because of this, children and youth groups are becoming major factor for involvement in environmental work. In Bangladesh, for example, one third of the total population is between the age of 15 and 30 with approximately 90 per cent unemployed, illiterate and poor. The National Federation of Youth Organizations in Bangladesh (NFYOB) works to empower youths by providing training in livestock, poultry farming and the establishment of small-scale income generating projects. Many youth groups also work directly on environmental campaign issues aimed at increasing environmental awareness and participation amongst the young. Activities include: workshops in schools and colleges; integrating environmental issues in syllabi of subjects taught in schools; formation of youth environmental clubs; and action projects at grassroots level involving youths in the conceptualization, planning and execution of these projects. For example, in Viet Nam, the Viet Nam Youth Union (VYU) has collaborated with government agencies in implementing an UNDP-funded project on the promotion of environmental awareness and, as 80 per cent of the population live and work in rural areas, radio was used as the main medium for promoting public awareness of environmental issues (ESCAP 1997). Another child-centred initiative that originated in Australia and New Zealand and has been adapted in some Asian cities is "making cities children friendly." This has been initiated by the Asian Network on

CHAPTER FOURTEEN

Box 14.5 Empowering Women

Women's NGO groups are working to empower women and improve their standing in the decision making process. One example is the (Indian) Community Development Society (CDS), Alappuzha (Alleppey). This is a successful model of women in development that has now been replicated in 57 towns and one entire district in Kerala State. The objective of the CDS is to improve the situation of children under 5 and of women age 15 to 45 years. CDS work includes literacy programmes, income generating schemes for women, provisions of safe drinking water, low cost household sanitary latrines, kitchen gardens, food-grain bank, immunization, and child-care. The CDS has resulted in the empowerment of women and the building of community leadership. It is a unique example of community based poverty eradication efforts by women. Since its small start in 1993, the CDS has grown to a large-scale women's movement with membership of 357 000 poor women (20 per cent of poor people in the State) from both rural and urban areas.

Similar work in empowering women to play an active role in environmental improvement and development is done by the Aurat Foundation in Pakistan and Seikastu Club in Japan. The Aurat Foundation works to help women acquire greater control over knowledge and resources; to facilitate women's greater participation in political processes and governance; and to transform social attitudes and behaviour to address women's concerns and development. The Foundation works directly at a grass roots level on environmental issues. It has facilitated meetings between peasant women and policy makers, planners and political representatives, as a result of which the women were able to express their concern about the impact of environmental degradation on their livelihood and their lives. The Foundation has also lobbied with Government about the concerns of peasant women and has championed the demands of rural women to the technology transfer and agriculture extension departments in Punjab. This has led to the development of demonstration and training projects designed to improve the productivity of peasant women.

In Nepal, a local NGO, Women in Environment (WE), attempts to counter both environmental degradation and poverty by getting women actively involved in environmental projects. Working with women social workers, environmentalists, women's rights advocates and other volunteers the organization has successfully mobilized women to work on such projects as National Park buffer zone management, river bank stabilization, kitchen garden development and the creation of revolving loan fund for environmental work. The Sindh Rural Women's Uplift Group in Pakistan owns 108 acres (43 hectares) of fruit orchard in which they use "organic and sustainable cultural practices" to fight against the use of synthetic pesticide and insecticide. The Group believes in maintaining soil and plant health to reduce disease attacks – and to reduce environmental contamination.

Another example of an NGO group which works with women to develop sustainable solutions to environmental problems is the Viet Nam Women's Union (VWU). This is a large organization with over 11 million members, which promotes the role that women play in Vietnamese society. In order to promote energy self-sufficiency for rural families with no access to the electrical grid the VWU has joined in the Rural Solar Electrification Project, in conjunction with the Solar Electric Light Fund (SELF) – an American non-profit NGO which promotes rural electrification. The project has provided electricity – from solar photovoltaic cells – for 240 households and to 5 community centres. This is an especially timely initiative, since Viet Nam is in the process of designing a national rural electrification master plan with the World Bank in order to integrate renewable sources of energy into an overall rural power delivery system.

Source: United Nations 1999 Websites 16 and 17

Promoting Awareness of Environmental Issues among children (See Box 14.6).

In Pimpri Chinchwad, a major industrial city in India with over 2 000 engineering, chemical, rubber, pharmaceutical and automobile companies, environmental education and awareness are deployed through children's programmes. Several NGOs working in the area, (including the "Centre for Environment Education" (CEE), "Regional Cell for Central India", and the "World Wide Fund for Nature" – India (WWF-I)) have long recognized the need for environmental consciousness, especially among those living without adequate housing or amenities. However, with adult members of these households employed in different sectors, it has traditionally proven difficult to develop effective and targeted strategies for environmental education and consciousness. Thus children became the focus of a series of initiatives aimed at 90 municipal schools, which led to the development of Nature Clubs and the beginning of an environmental education resource centre and project newsletter, *"Shrishti"* (Creation) (United Nations 1999 *Website 21*).

In order to enhance interest of youth in environmental protection and management, eco-clubs have been initiated in schools in a number of countries of the region. For example, "Young Zoologist" Clubs in schools are encouraged in Sri Lanka. Similarly, over 3 000 Educational Environmental Clubs have been set up in various parts of Pakistan. Thailand has an Environmental Development Campus to help children understand and implement environmental conservation activities and in Japan, the Junior Eco Club project has been active since 1995. The main regional agency that has worked to strengthen and foster groups involved in

> **Box 14.6 NGO Contribution Toward Making Cities Child-friendly**
>
> Sporadic urban growth in many cities in Asia and the Pacific poses significant risks to the well-being of children. Research commissioned by UNICEF has noted that the health and often the lives of more than half of the world's children are constantly threatened by environmental hazards, in their home and surroundings and in the places where they play and socialize. The research also indicates that 40 000 child deaths occur each year from malnutrition and disease, and that 150 million children a year survive with ill health, with retarded physical and mental development. More and more young people are being admitted to hospital with asthma due to car fumes, while other pollutants are linked with a whole range of other health problems in the young. Shanty town dwellings with inadequate basic facilities exposes children to diseases and dangers, while traffic claims many young lives on a daily basis. Because of such problems, one of the greatest challenge for urban administrations in the new millennium is in the area of child development and protection.
>
> In Malaysia a number of concerned NGOs have got together to try and address this challenge. In September 1996, The Malaysian Council for Child Welfare (MCCW) and the National Council for Women's Organizations (NCWO) organized a National Conference on the Right of the Child in Kuala Lumpur. The Conference was supported by the United Nations Children's Fund (UNICEF), Malaysia and received technical cooperation from Asia-Pacific 2000, which is a Project of the United Nations Development Programme (UNDP).
>
> At this conference, serious concerns were raised about the quality of life of the urban child, who is often caught between his or her own needs and aspirations and that of his parents. Subsequent to this meeting, on 5th July 1997, the MCCW, NCWO and the Management Institute for Social Change (MINSOC), with technical support from Asia-Pacific 2000 and UNICEF, organized a follow-up national workshop on 'The Urban Vision 2020 Initiative: Making Urban Areas Child-Friendly'. Involving over 150 participants from government departments, tertiary institutions, non-governmental organizations as well as interested individuals, the workshop concluded with concrete proposals on improving the socio-economic environment of children, addressing issues that arise within the home, school or community pace and the safety and health of urban children.
>
> Out of these deliberations, there emerged the Malaysian Charter on Making Urban Areas Child-Friendly and its associated Ten Strategic Actions aimed specifically at urban local authorities. The Initiative then commissioned the development of a child-friendly survey instrument – 'The Child's Report Card' as a tool for children to assess the friendliness of their own neighbourhood environments.
>
> The Malaysian Child-Friendly Cities Initiative is a complement of the International Child-Friendly Cities Initiative (CFCI) which was launched during the International Workshop on Children's Rights. The objective of the CFCI is to help translate the Convention on the Rights of the Child (CRC), into concrete actions that can be implemented at the local level, by just about everyone.
>
> Source: Saira Shameem (Compiler) 1998; Satterthwaite 1996

children and youth activities is UNICEF. One of the key follow-up areas in UNICEF's response to Agenda 21 is enhancing partnership with NGOs, research institutions and community groups to initiate innovative activities that take account of the concerns of children and young people. UNICEF has assisted each of the countries in the region in organizing activities whereby children and young people become focal points for environmental awareness raising.

C. Role of Indigenous People

There are still a large number of communities in the region, inhabiting remote areas, in close proximity to nature, practising traditional farming, fishing, agricultural and forestry techniques. Conserving the environment is a part of their way of life. The age-old traditions and experiences of these communities (usually termed "indigenous people") can help improve the efficiency of resource use and it is for this reason that a number of NGOs build on traditional or indigenous knowledge systems. These knowledge systems are researched and disseminated so that the wider public can learn from them.

Major groups in indigenous communities themselves are also active in environmental protection. An example which illustrates how local indigenous groups are actively involved in conservation work, is Soltrust, one of the major local indigenous organizations in the Solomon Islands dedicated to promoting sustainable forest management, where logging operations are a major concern for both the government and the indigenous peoples. Despite many awareness campaigns on sustainable development, both the number of logging companies, and the unsustainable rate of harvesting of timber resources have been increasing. Established in 1986, the group's more recent work has involved the Rarade Community of the Isabel Province, and island province that has been out of reach by loggers until recently. A partnership between Soltrust and the community was created as a model for future eco-forestry activities, not only in Isabel and in the Solomon Islands at large, but also for neighbouring countries facing similar situations (United Nations 1998).

CHAPTER FOURTEEN

In many parts of the region, rapid industrialization, the development of suburbs and the conversion of land for agricultural purposes has encroached upon the traditional homeland of indigenous people. At the same time greater numbers of indigenous people have either become displaced because of development or have moved into urban areas in pursuit of education and/or employment. This has resulted in the reservations and sanctuaries shrinking in size and often being hemmed in by developmental projects, with negative consequences for their once pristine environment. However, indigenous groups are now beginning to organize resistance movements. In Australia, for example, aboriginal communities in states such as Queensland have joined forces with environmental groups to prevent the further depletion of their land and forest reserves by logging and mining concerns. In New Zealand, people of Maori descent have banded together to assert claims to their land and also to protect them from further environmental damage. A number of tribes have petitioned the courts in order to reclaim their tribal lands. In the northern part of Thailand, the increasing mobility of traditional people poses a serious threat to the "sustainability" of the hilltribes distinct cultures. The threat comes from the influx of consumerism, lack of land security and large migrations to the cities. In order to counter these threats the "Inter Mountain Peoples Education and Culture in Thailand Association" (IMPECT) was founded with the intention of supporting, promoting and revitalising the traditional belief systems, agricultural traditions and cultures of the hilltribes. To make the children and youth proud of their culture, the relationship between the traditional lifestyle and the conservation of their natural surrounding has been promoted through a locally developed curriculum. In response there has been an increased feeling of the value of traditional knowledge among the children and youth in the target villages.

The close links between some NGOs and indigenous communities, especially vulnerable groups, also provides for the representation of such groups at the national and international levels. This is important for resolving issues, especially those related to globalization and its homogenizing influences that endangers indigenous cultures and cultural diversity.

D. Role of Farmers and Agricultural Groups

The primary economic activities in most countries in the Asian and Pacific Region are subsistence based. Farmers and agricultural groups, alongside traditional 'hunter-gatherer' and fishing communities, include some of the least advantaged sectors of the population in these countries. Many NGOs and community organizations have been established to champion the cause of these disadvantaged groups, particularly in the South Asian countries of India, Pakistan and Bangladesh. Such NGOs include Development Alternatives, which was set up in India to undertake and support programmes where rural people, primarily farmers and agricultural groups, are consulted on developmental and environmental issues. Among these are projects that adapt new technologies in shelter, textiles, energies and biomass that enable agricultural communities to increase their productivity and simultaneously minimize the negative effects upon the environment. The Bangladesh Rural Advancement Committee (BRAC), provide another example, which has done commendable work in improving the lives of farmers and agricultural groups.

At regional level, the Asian NGO Coalition for Agrarian Reform and Rural Development (ANGOC) has collaborated with national and regional agencies to ensure that the perspectives of farmers and agricultural groups are adequately reflected in programmes that promote sustainable development.

E. Role of Workers, Trade Unions and Business NGOs

Corporate groups and business associations have also started to work on environmental issues (see Chapter 13). At the international level, the International Chamber of Commerce (ICC) has put together a document titled the ICC Charter for Sustainable Development, which has been translated into more than 20 languages and publicized throughout the world. In the industrial sector, the ICC is affiliated to the World Industry Council for the Environment (WICE) and has collaborated in transferring environmental technology from the industrialized countries to institutions in this region.

Private sector-NGO partnerships are also developing in the region. For example, in 1993, United Nations Development Programme approved a project to prevent and manage marine pollution in the seas of North and Southeast Asia. Among the project's major activities is a demonstration project at Batangas Bay in the Philippines. Responsibility for the project activities is given to the Batangas Bay Environmental Protection council, which includes representatives of government, community organizations, local NGOs and an industry-based NGO, the Batangas Bay Coastal Resources Management Foundation (BCRMF) which represents the private sector. The involvement of BCRMF has been cited as a major factor in the success of the project's efforts.

In recent years, workers' organizations have also started to focus on the environment, and as a result, issues such as the awareness of the link between employee health and safety and the environment have come to the fore. Organizations such as the International Confederation of Free Trade Unions (ICFTU) have urged constituent member unions to promote knowledge and awareness of the environment among members. At national level, however, the strength and influence of trade unions varies considerably across the region.

CONCLUSION

It is evident that public concern for the state of the environment in the Asian and Pacific Region is increasing. This has been reflected not only in the increase in the number of public interest and community groups involved in environmental activities, but also in the scope and diversity of such activities. It was reported that there were about 1 500 Asian and Pacific NGOs represented at the Earth Summit in 1992; in recent years the number of active groups has increased to nearly 10 000. In addition, the increase in the range of their activities reflects not only the increasing professionalism with which major groups are fulfilling their obligations and responsibilities, but the greater recognition and credibility accorded to them by national governments, regional and international organizations.

A number of public interest groups have also strengthened their participation at grassroots and community levels, and have played a vital role, not only in awareness-raising and campaigning, but also in education, training and capacity-building. They have made considerable headway in their attempts to promote the concept of sustainable development, particularly among women, children and other NGOs throughout the region. Their activities show that they effectively use all media of communication, traditional as well as the new communication technologies, to disseminate information to the grassroots and to strengthen networking.

The accountability and professionalism of NGOs and civil society groups is crucial if they are to become established as appropriate representatives of the needs and concerns of those members of society who are disadvantaged, disenfranchised and poorly informed. The important role that such groups provide needs to be founded on robust information as well as the direct needs of the 'client' community. Mechanisms that can best be utilized to ensure balanced and equitable networking among NGOs need to be identified, and a focus maintained on the interests of indigenous people, women, children, youth and other disadvantaged sectors.

It is of paramount importance that NGOs in the region strive to build capacity within, and amongst, themselves and to strengthen their capacity to organize dialogue and act as public advocates with governments and regional and international bodies.

Chapter Fifteen

Environmentally friendly habits need inculcation from childhood: A Chinese mother teaching her child to throw trash in a waste bin.

15

Education, Information and Awareness

INTRODUCTION
ENVIRONMENTAL EDUCATION
ENVIRONMENTAL TRAINING
REGIONAL COOPERATION FOR
ENVIRONMENTAL EDUCATION AND TRAINING
INFORMATION AND AWARENESS
ENVIRONMENTAL COMMUNICATION
ENVIRONMENTAL COMMUNICATION BEYOND MEDIA
CONCLUSION

CHAPTER FIFTEEN

INTRODUCTION

The pursuit of sustainable development and environmental conservation policies, objectives and targets requires the public to be sufficiently sensitized about the multiple dimensions of environment and development. Awareness and understanding of environmental issues provide the basis and rationale for commitment and meaningful action towards environmentally sound and sustainable development.

This chapter provides an overview of developments in environmental education, information, awareness and training in the Asian and Pacific region during the 1990s. The trends and patterns in environmental education in formal and non-formal sectors, and initiatives and programmes that have been undertaken by governments, non-governmental organizations, communities and by regional and international organizations are also identified. The underlying needs for environmental education, information and communication are also discussed with a particular emphasis on the constraints and key issues that need attention.

ENVIRONMENTAL EDUCATION

A. Main Trends and Conditions

Education has been identified as a critical driving force for change in the Asian and Pacific Region, and countries and regional organizations have adopted a range of strategies for implementing programmes in environmental education (Fien 1999a). The overall trends in environmental education information and communication in the region reflect the concerns of people and societies in transition.

Environmental education is now being seen as an instrument and a process that enables participation and learning by people of all ages, based on two-way communication rather than the old paradigm of a one-way flow of information, from teachers to pupils. The content and substance of environmental education is also undergoing review and change. Reorienting education as a whole towards sustainability involves the various levels of formal, non-formal and informal education at all levels of society. Environmental education has developed within the conceptual framework that emerged from the first international conference in Tbilisi (1977) and is now seen as education for sustainability. This allowed environmental education to address the broad range of issues and concerns included in *Agenda 21* and others which evolved through the meetings of the Commission on Sustainable Fdevelopment (UNESCO 1997).

The key international conventions on environment place a high value on public awareness, education and training and obtaining information through monitoring as essential elements for the success of the conventions. For example, the Convention on Biological Diversity, emphasizes the importance of public education and awareness through promoting and encouraging measures required for the conservation of biological diversity. Since the convention came into force in December 1995, the contracting parties (countries) have been motivated to address issues related to education and awareness on biodiversity.

In addition countries in the region recognize the immensity of the challenges they face, and of the vital role that environmental education can play in meeting these challenges. There is a growing perception by governments of the need to integrate environmental education information and communication into the country's on-going programmes. As a result of the *Agenda 21*, the level of cooperation and collaboration between environmental and the educational institutions has increased. In some countries, governmental environmental agencies have statutory requirements to engage in activities related to environmental education and awareness. For example, the Malaysian Department of Environment has established an educational division under the Environmental Quality Act, which is actively engaged in promoting and implementing a variety of activities.

In many countries of the Asian and the Pacific Region environmental topics have been included in education courses, through integrating environmental concerns in other subjects and through specific courses for the environment. Government, NGOs, educational institutions and media have undertaken some serious efforts to meet the growing environmental challenges by promoting environmental education, information and communication in their respective countries. Activities such as green bank, green press, eco-lubs, eco-polies, eco-farming and eco-harvesting; are emerging in the region. Special economic incentives (such as subsidy, tax-exemption and other incentives) are provided to schools in some countries where environmental education courses are offered.

There is greater recognition of the role of NGOs and civil society organizations, and the need for meaningful community participation in debates and action programmes aimed at education and training for sustainable development. NGOs and governments are increasingly working together, reinforcing each other's strengths and outreach. Linkages between governmental institutions and

NGOs are improving in most countries of the region, and in some cases, governments are actually depending on mature and experienced NGOs to promote environmental awareness, communication, and training activities (See Chapter 14).

B. Formal Environmental Education

Environmental education is increasingly a prominent part of primary, secondary and tertiary education in Asia and the Pacific. The formal education sector plays a vital role in environmental education and awareness by exposing the younger generation to the information, issues, analyses and interpretations on environment and development.

A number of factors have influenced the development of environmental education in the region. The two over-arching factors are national education policy and national environment and population policy. These policies are a reflection of national cultural values, priorities and socio-economic goals in most countries. The national environmental education policy is usually the result of decisions made in these broader fields (UNESCO-PROAP 1996). For example, environmental education in Australia has seen two major shifts since 1970s. First, there has been a distinct move away from nature and science-based environmental education to a concern with the social, economic and political aspects of sustainable development. There has also been a major shift from schools to adult and community environmental education. (Fien 1999b).

Rather than establishing a new subject, most countries have opted to infuse environmental education objectives and strategies into the existing curricula, while some other countries practice both options. In addition, the focus on practical learning in the real world in environmental education helps schools to address important general educational objectives related to values and to skills development (UNESCO-PROAP 1996).

Nationally determined syllabi often provide for a coordinated programme of environmental topics in both primary and secondary schools in countries of the region. However, there is some lack of coordination at different levels of education in the national framework, which prevents the development of a comprehensive environmental education programme.

1. Primary and Secondary Levels

The diversity of approaches in primary and secondary education seen across the Asian and Pacific Region are based on each country's major and threatened resources, and issues of concern. For example, in the Maldives, environmental education and awareness programmes highlight issues of the marine environment emerging from the National Environment Action Plan of 1989 (IUCN 1998). Whereas in Nepal, the national goals of education are to teach thoughtful protection and wise use of the natural environment and national heritage. In Nepal, the general need of environmental protection from specific problems resulting from population pressure on natural resources, and the links between environment, population and natural resources, were addressed in the environmental education plan (IUCN 1998).

In Republic of Korea, during the 1990s, "Environment" and "Environmental Science" were included in the middle school curriculum as separate courses, attributing to the common understanding that environmental education for young children was the most important means to solve persistent environmental pollution (Kang 1999). In order to raise young people's awareness of environmental preservation, and to cultivate environmental perspectives, the Republic of Korea designated 63 schools as demonstration environmental schools from 1985, and has disseminated their best practices to an additional 26 schools in 1999. These schools are now eligible for subsidies and environment-related educational materials (Green Korea 1999).

In People's Republic of China, environmental education is coupled with the working schedule of public agencies and educational institutions. This was formalized by a regulation issued by the National Bureau in the Conference of National Environmental Education in 1992. At pre-school and primary levels, environmental education is carried out through games, audiovisual means and the study of natural systems. In secondary schools, environmental concepts are infused into the courses of physics, chemistry, biology and geography. At the same time, the teachers use local examples to help develop environmental understanding among the students (UNESCO-PROAP 1997).

In small island countries of the South Pacific subregion, considerable effort is directed towards infusion of environmental education into various subjects within the primary and secondary school systems. The syllabus for the secondary level has a strong emphasis on environmental studies, developed either independently, as in the Solomon Islands and Papua New Guinea, or in collaboration with the South Pacific Regional Environmental Programme, SPREP (Ravuvu 1998).

2. Tertiary Level

Tertiary level education has responded to the increasing demand for environmental managers and experts in the 1990s. Key trends have been observed across the region in relation to environmental education at tertiary level these include:

- basic environmental concepts and elements added to existing courses at undergraduate and postgraduate levels, for all students irrespective of their courses;
- new environmental units or modules introduced into a large number of courses at undergraduate and postgraduate levels, thus increasing the depth and detail of environmental study;
- new non-degree programmes and courses (at foundation, certificate or diploma levels) introduced by tertiary education institutions to cater to the demand for in-service training and upgrading of knowledge and understanding on environmental issues and practices;
- an increase in the publication of relevant textbooks and audio visual material;
- greater emphasis on training the trainers, and in strengthening the tertiary education system and research capabilities;
- more research on environmental education policies and practices;
- a greater dialogue and information exchange between the users of environmental skills and talent, in government, private and NGO sectors etc., and the institutions of tertiary education, ensuring, education and training address prevailing practical needs; and
- increasing emphasis placed on adult and community education, using both formal and non-formal methods to raise the overall environmental literacy levels.

Increase in environmental studies has been a response to the market realities, and also to the growing recognition of the environmental crisis and the management options available. Countries have adopted different mechanisms to cater to their specific needs in tertiary education. For example, in Viet Nam in 1995, the Ministry of Teaching and Education made it compulsory for a course in Environment and Man to be taught in all aspects of natural sciences, social sciences, humanities, agriculture etc., with more specialized environmental science and technical courses in other university courses. Since the early 1990s, short, medium and long-term postgraduate courses for environmental managers and researchers have been conducted in various specialized centres (UNESCO-PROAP 1997).

Specialized degree courses have been introduced in Thailand, at graduate level, with compulsory course introduced and targeted to first year undergraduates. The courses raise awareness and understanding of topical environmental issues and are integrated with other courses, especially science and social science (UNESCO-PROAP, 1997).

In People's Republic of China and Pakistan, many colleges and universities have introduced environmental courses for undergraduates and postgraduates, and have also directed then towards training professionals and officials. In Mongolia, the government has prepared a masterplan for environmental education and awareness, which will cover both formal and non-formal education and communication (IUCN 1998).

However, constraints that prevent countries in Asia and the Pacific from achieving their environmental education and training needs are evident. At a UNESCO sponsored regional seminar on environmental education in 1996, it was noted: "Each country has already initiated a number of education and training programmes at the tertiary or higher level of education that are related to aspects of the environment. In various ways and to varying degrees, each of the countries has responded to the problems of the environment with environmental laws and regulations, and with associated political and institutional initiatives. However, in all countries, the development of persons with the conceptual understandings and skills that these regulatory intentions require for implementation and management lags well behind what is needed" (UNESCO-PROAP 1997).

3. *Materials and Study Aids*

In line with the growing interest and activity in environmental education, awareness and training, the demand for educational materials and study aids has also increased. However, the limited availability of materials may not meet individual country requirements in terms of local language, and in coverage of the most relevant issues to the country.

Many government and non-government institutions have risen to this challenge. For example, in Malaysia, the Academy of Writers was enlisted to produce storybooks that will instill environmental values and attitudes amongst primary school children. A similar initiative has been made in the South Pacific subregion where there have been several efforts to produce locally relevant environmental education material at the primary and secondary school levels.

A widely felt constraint has been the lack of standardization in textbooks and other material on environmental issues and a failure to provide the full information base. While the subject of environment can be interpreted and presented in many ways, and it can be looked at through a scientific or cultural angle, there is a basic need to present facts accurately and discuss issues in a balanced manner. If this is lacking, the wider goals of environmental education cannot be met. An extensive review of environmental textbooks, supplementary readers and other material is required in many countries of the region as part of a process to improve their quality.

Establishment of environmental study centres has helped a great deal in developing materials and study aids for both formal and non-formal environmental education. Such centres also provide students with information, insights and practical activities, and adult visitors gain information and advice. In Indonesia, the Ministries of Environment and Education have been collaborating since 1979 in establishing Environmental Study Centres (PSLs) in all public universities. By 1997, there were more than 72 PSLs covering all 27 provinces of Indonesia. These centres perform three functions: education and training pertaining to environmental management; extension services, fostering public education and awareness; and research and surveys in support of environmental management.

The Department of Environmental Quality Promotion (DEQP) in Thailand has supported the establishment of Provincial Environmental Education Centres (PEEC) in selected schools since 1995 and are expected to expand these to every province by 2001. The main objectives are to develop environmental education materials and tools in relation to local environmental problems, and to strengthen the capability of environmental educators in all parts of the country.

4. *Issues and Constraints*

There are many constraints and barriers to the widespread adoption and practice of environmental education in the region. In the formal education sector, class sizes are often large and teachers lack resources and experience in interactive pupil-centred teaching strategies. School curricula are also dominated by competitive academic curricula which prioritizes end–of–course examinations and discourages the development of locally and personally relevant intellectual skills. Outside the formal education sector, environmental education is often poorly organized and resourced.

Another common constraint is the lack of clear integration of environmental education objectives and programmes with national education and environmental policies. In some countries, the absence of national policies or guidelines for environmental education has resulted in a lack of coherent strategies and long term planning. Even in those countries which do have such policies, the educational systems are often insufficiently dynamic to accommodate the evolving social, economic, political and conservation aspects of sustainable development.

The important pre-requisites for the successful introduction of environmental education in schools include: the existence of clear and well communicated policies; the political will and availability of sufficient resources for implementation; curricula revision; proper preparation of teachers through in-service training; the availability of relevant materials in local languages; networks for exchange of expertise between teachers; and adequate assessment and incentives for teachers development. It is encouraging to note that in spite of many constraints, practitioners and promoters of environmental education have found innovative ways of teaching throughout the region.

C. **Non-formal Education**

Non-formal environmental educational activities exist alongside the formal educational systems, at curricular and extra curricular levels, in occupational training, and through wide public awareness activities through non-formal channels such as mass media, and voluntary organizations. Different communities, institutions and individuals choose methods and practices that best suit their local needs and capacities.

1. *Learning by Doing*

In several countries, there are efforts to get students to relate to local problems, while understanding their global implications. In Bangladesh, an environmental education programme called *Muktangan Siksha*, or open-air education, encourages field programmes related to the surroundings and communities. In Myanmar, an imaginative pre-school and lower primary environmental programme bases its teaching on a study of the surroundings, or *patwinkyin*, without formal textbooks (Kartikeya V. 1995). In Sri Lanka, a WWF-supported innovative environmental education programme involving over 750 schools has been implementing an approach called 'greening of learning'. In this, students are encouraged to beautify the school garden, start a plant nursery or engage in other 'green' activities within the school premises. Due to its success the WWF has started introducing the same concepts and approaches in other countries, such as Viet Nam.

Research and advocacy organizations are increasingly involved in developing non-formal environmental activities. For example, Development Alternatives (DA) located in India, has launched the Community Led Environmental Action Network

CHAPTER FIFTEEN

(CLEAN) which promotes among school children and communities activities based on the "four r" concept: refuse, reduce, recycle and reuse (DA 1998). In several countries, government agencies or NGOs support nature clubs or environmental societies in schools as a means of encouraging and inspiring students to undertake non-formal environmental activities.

2. *Outdoor Activities*

Government agencies as well as NGOs have developed a wide array of outdoor activities that expose youth and adults to different aspects of environmental awareness, action and understanding. For example in Nepal, Environmental Camps for Conservation Awareness (ECCA), a local NGO, has been active for over a decade in organising outdoor environmental activities aimed at children, both able and disabled. These camps are held at places of environmental significance and aim to raise awareness about conservation issues, and to potential careers in conservation.

Similar environmental camps have become a regular feature in many countries. For example, in Malaysia, environmental awareness camps are regularly organized by the Department of Environment. The camps are for children between 14 and 16 years of age, at the Nature Education Centre (NEC) established in 1992 by the Malaysian Nature Society. In Singapore, the Ministry of Environment allocates small portions of the beach to volunteering schools under an 'Adopt a Beach' programme. Students are then responsible for keeping that stretch of the beach clean, and in that process learn aspects of the coastal and marine environment.

The China Association for Science and Technology (CAST) is engaged in the promotion of non-formal science education for children and youth. CAST has 165 natural science societies and is established at county level all over the country. CAST relies on its member scientists and technologists to provide the knowledge base and necessary human resources for various activities organized to promote better understanding of science, technology and environmental issues. Also in People's Republic of China, the State Environmental Protection Administration (SEPA) is playing a lead role in non-formal activities, and has introduced the Global Learning and Observations to Benefit the Environment (GLOBE) Programme, a worldwide network focusing on science and education, which brings together students, teachers and scientists in order to share information on monitoring the global environment.

In Australia, Waterwatch was established in the early 1990s as a national level, community-based water quality monitoring programme under the National Landcare Programme. Monitoring water quality on a regular basis gives communities a greater understanding of the natural environment, and may lead to action which will have local, regional or even national benefits (Palmer J. 1995). Both school children and adults participate in water monitoring activities, and government funding is provided through facilitators who have been appointed in each state attached to a lead state government agency.

3. *Innovative Approaches*

New strategies and innovations for environmental education have been developed and applied throughout the region. For example, in Singapore, the Ministry of Environment in 1996 published the 'Fun and Discovery Through Environmental Clubs', outlining environmental activities and clubs. Similar publications have been developed in Japan, India and Bangladesh.

The observation of National Environment Days and Weeks across the region, provide a focal point for environmental activities, including seminars and exhibitions. In New Zealand, school education kits on sustainable agriculture are also provided to educate young people about agriculture and to survey changes in land management and planning.

Japan, has various examples of innovative education, public awareness and training activities including, Environmental Counselor Registration System, the Environmental Activities Evaluation Programme and various campaigns for conservation of natural resources and energy. In addition to these programmes, the "Junior Eco Club programme" supported by the Environment Agency has been a very effective programme at the elementary and junior high school age children level. Club activities are supported by local governments, and a nationwide festival for the Club is organized at the end of each school year. There are about 4 000 clubs with 70 000 children in Japan and its popularity is increasing (ITO 1999).

Rising to the challenge of going beyond mere awareness raising, the Centre for Environment Education (CEE) in India has produced an exhibition package called 'Act Now'. It focuses on some actions that people can take in their everyday lives to help improve the environment. Another environmental education package produced by CEE consists of a one-hour video story in Hindi called *Dhraki*, which discusses some concepts of drought, conservation of water, land and vegetation, and the management of exotic plant species. The accompanying booklet suggests ways of developing activities that will involve children or other viewers on these issues and themes and has been used widely in classrooms, environmental camps and workshops. In response to the lack of material the CEE has also developed the Environmental Education Bank, providing situation specific material for use in both formal and non-formal instructional situations. Such materials include physical resources like posters, kites, booklets, films, publications and a computerized database of over 800 environmental concepts, 2 500 activities and 600 case studies. The Bank is usually accessed through a five-day workshop, organized by the Centre at the request of small groups.

In Thailand, an example of innovative education and public awareness activities related to sustainable development is a Management of Science and Environment course in Hard Amra Aksornluckvittaya School, in Samutprakan Province. This project used local problems, such as a degraded mangrove, to let the student groups analyse problems systematically, searching for options and solutions and preparing work plans for action. As a result, teachers and students replanted the mangroves in the Asokaram Temple and used the area as a study site.

ENVIRONMENTAL TRAINING

Agenda 21 states that training is one of the most important means to develop human resources and facilitate the transition to a more sustainable world. "It should have a job-specific focus, aimed at filling gaps in knowledge and skill, and would help individuals find employment and be involved in environmental and developmental work. At the same time, training programmes should promote a greater awareness of the environment and development issues as a two-way learning process" (United Nations 1993).

A. **Types of Training**

1. *Teachers in Formal Education*

In a large number of countries in the region, teacher-training programmes at both pre-service and in-service levels have incorporated elements of environmental education. In general, governments of most the Asian and Pacific Region are responsible for training teachers in formal environmental education. In Malaysia, the establishment of National Institute for Environmental Skill and Training (IKLAS) is an important step forward for the Department of Environment to equip its own personnel, as well as staff of other related government agencies and private sector, with the knowledge and skills for pollution control and sound environmental management. The IKLAS is expected to be operational by the year 2001 (Malaysia Environmental Quality Report 1996). Similarly, in Thailand, the Ministry of Education has trained teachers, administrators, educational planners, and non-formal educators. In Sri Lanka, March for Conservation, a university-based NGO, has designed modules for introducing environmental concepts to primary and secondary teachers, and conducts short-term training programmes for teachers.

Regionally, initiatives for training teachers focus mainly on material development and training of trainers. *Learning for a Sustainable Environment: Innovation in Teacher Education through Environmental Education Project* is a joint undertaking of UNESCO's Asia Pacific Centre for Educational Innovation for Development (ACEID) and Australia's Griffith University. This long-term project seeks to expand the range of innovative practices used in teaching education programmes by introducing teachers and teachers-in-training to the curriculum planning skills and teaching methodologies of environmental education.

In the South Pacific subregion, teacher training in formal education is seen to be the key to the success of environmental education in schools. A teacher's guide to environmental education has been produced for adaptation by all countries. Several teacher-training workshops have been held at national and subregional level, where the teachers are encouraged to use outdoor education and investigative learning approaches.

2. *Practitioners of Non-Formal Education*

Training for the practitioners of non-formal education is less systematic in most countries, and on the whole fewer opportunities exist. This is partly due to the enormous diversity of professional backgrounds of the individuals engaged in non-formal environmental education activities. Most non-formal education activities are designed and carried out by NGOs or community organizations, whose large numbers and wide geographical spread makes it difficult to expose them to centralized and long-term training programmes. More effective in their case are short-term refresher courses and skills development seminars and workshops.

The Indian Centre for Environmental Education, offers a training Programme in Environmental Education for Indian and overseas participants and introduces various approaches and methods in communicating environmental messages to different target groups. The CEE in cooperation with IUCN and WWF also offers a Certificate Course in Environment Education (CCEE) as a means of in-service training for professionals already engaged in environmental education work. The Centre for Environmental Concerns (CEC) in the Philippines, offer a course that includes elements of community-based rehabilitation technology, community-based environmental monitoring, and participatory approaches to environmental education.

In Bangladesh, the Environment and Social Development Organization (ESDO) has conducted, with the support of the Advocacy Institute in the US, several national workshops to train local activists, with examples of strategies and practices of the US counterparts. In the South Pacific subregion, organizations like SPREP and the SPC (Secretariat of the Pacific Community) have concentrated on providing communication skills for extension workers in specific fields such as fisheries, coral reef monitoring and conservation area management.

3. *Training Media Professionals*

Training and sensitising journalists on different aspects of the environment remain urgent needs in most countries of the region where the standards of environmental reporting are low. Even journalists and broadcast producers who are well trained in their craft face new challenges in reporting on issues related to environment and sustainable development. The technical nature of most environmental issues requires the ability to grasp these technicalities and then to interpret these in layman's terms. Many environmental stories involve the assessment of risks and the weighing of costs and benefits, all of which require experience, skill and a strong sense of balance in journalists.

In spite of a decade of heightened interest in environmental journalism in the region, the formal training courses and curricular for journalists in many countries do not as yet pay sufficient attention to the specialized needs of environmental journalism. However, some national forums of environmental journalists regularly organize short training activities, including workshops, seminars conferences etc., for the benefit of their members and other environmental journalists. The national forums in Bangladesh, People's Republic of China, Japan, Malaysia, Nepal, Pakistan, Philippines and Sri Lanka have been particularly active in this respect. For example, from 1995 the Japanese Forum of Environmental Journalists (JFEJ) has organized an annual environmental study tour of Japan for selected journalists from both print and electronic media in the region. Journalists and producers are exposed to Japanese environmental policy and practice through a series of lectures, discussions.

At a regional level, several initiatives are noteworthy. UNEP has conducted several workshops, seminars and training programmes under its Environmental, Communication and Information Strategy for Asia and the Pacific (1995-2000). A high-level meeting was held in Beijing, People's Republic of China, in 1996 that brought together editors and managers from leading publications in the region to discuss how to boost environment related coverage in the media. Parallel to this, training workshops were organized for environmental journalists drawn from several countries in the region.

In the Pacific Island countries, SPREP and UNESCO have cooperated on a series of national 'Environment and Media' workshops to train television, radio and print journalists in incorporating environmental issues in their reporting. Five training workshops were held in 1999 (Samoa, Tonga, Marshall Islands, Cook Islands and Fiji) and an additional 4 are proposed for 2000.

One weakness in short-term training activities available for journalists is that most programmes are targeted only at the print media, with relatively few opportunities for journalists and producers working in radio and television. With this in mind, regional organizations have started supporting training for broadcast journalists. Panos South Asia office, based in Kathmandu, is placing particular emphasis on training radio journalists to improved coverage of environment and development. Panos has also organized several field study tours for environmental journalists from South Asia to better understand the complex issues of trans-boundary water resource sharing and management (Panos 1998).

The training of audio-visual communications is another widely felt training need and several organizations are now engaged in offering such training. The Worldview International Foundation regularly conducts training courses for NGO activists and media professionals in the use of video for documenting environmental abuses, issues and problems. Meanwhile, the Television Trust for the Environment (TVE) annually organizes a technical skills development workshop for NGO professionals on using television and video for awareness and advocacy work. It has also conducted national level training programmes on environmental video programme making for producers in People's Republic of China, Bangladesh and Sri Lanka, each tied up to an actual production that was later broadcast. In mid 1999, Panos and TVE collaborated in organising a South Asian level workshop that exposed mainstream print and broadcast journalists to the effective ways of using the Internet for researching environmental stories, and also disseminating them through the World Wide Web.

4. *Specialized and Technical Training*

The number of specialized training programmes at country and regional levels has increased since the 1990s, a reflection of the greater market demand for specialized environmental skills and environmental managers. Traditional areas of environment related specialized training includes protected area management, environmental quality monitoring and environmental information systems. Relatively new areas of training include, environmental economics, environmental impact assessment (EIA) and in the adoption and implementation of environmental standards, especially the ISO 14000 series. In Malaysia, for instance, universities have worked with experts from the SIRIM Industry Standards Committee on the Environment to produce both training materials and training modules for ISO 14000 implementation.

Training activities are also conducted by some of the larger and well-established NGOs working nationally or regionally. For example, the Society for Participatory Research in Asia (PRIA), based in New Delhi, India, regularly organizes national and regional level training courses and programmes on occupational and environmental health. The UNEP International Environmental Technology Centre (IETC), based in Osaka, Japan, organizes specialized training programmes at regional and subregional level under its capacity building initiative. Professional target groups for its post-graduate training interventions are decision-makers in central and local governments, civil society and industry, academia, NGOs and senior trainers attached to regional or national environmental training centres.

Other specialized training includes the Regional Community Forestry Training Centre (RCFTC), Thailand, which runs several courses on areas such as community-based ecotourism for forest conservation and rural development, offered in collaboration with the Institute of Forestry in Nepal. The United States-Asia Environmental Partnership (US-AEP) is a regional initiative that provides technical assistance and training to Asian governments, business and industry, and NGOs in relation to environmental management and technologies. Training focuses on five critical environmental areas: addressing global climate change; providing safe drinking water and wastewater management; reducing urban air pollution; promoting solid waste management (including medical waste and landfill methane recovery); and participating in regional policy projects on performance matrix, environmental management systems, and public disclosure (US-AEP 1998). The Pacific Islands Climate Change Assistance Programme (PICCAP) at SPREP, undertakes Greenhouse Gas Inventory and Vulnerability and Adaptation Assessment (V&A) training for countries who have insufficient capacity to undertake these tasks, in order to fulfil their obligations under the United Nations Framework Convention on Climate Change (Box 15.1).

Box 15.1 PICCAP: A Training Success in the South Pacific

The Pacific Island Climate Change Assistance Programme (PICCAP) is a GEF funded regional climate change project. It involves 10 Pacific Island countries (Cook Islands, Federated States of Micronesia, Fiji, Kiribati, Marshall Islands, Nauru, Samoa, Solomon Islands, Tuvalu, Vanuatu) and is coordinated and executed by the South Pacific Regional Environment Programme (SPREP). PICCAP has been designed in such a way as to strengthen the capacities of participating countries, in terms of training, institutional strengthening and planning activities, to enable them to meet their reporting obligations under the United Nations Framework Convention on Climate Change (UNFCCC).

In less than three years, PICCAP has achieved excellent results. Through its subregional approach, PICCAP has fostered greater sharing of information, built up a qualified pool of climate change experts from within the South Pacific subregion, instituted cross-sectoral climate change country teams with technical and policy-related functions, established a database of climate change information, and assisted with the development of national climate change action plans which have formed the basis for initial national communications and the implementation of the UNFCCC at national level. Much effort has gone into the training and related activities needed to the build in-country capacity that has allowed these countries to prepare and report on their national greenhouse gas (GHG) inventories, GHG mitigation strategies, assessments of vulnerability to climate change impacts and of adaptation options, national climate change action plans, and enabled their initial national communications to the UNFCCC.

One notable achievement under PICCAP has been the design, development, and delivery of a comprehensive university-based vulnerability and adaptation certificate programme for the region. This was done in conjunction with the Climate Change Training Programme implemented by UNITAR and with cofunding from the Government of New Zealand, and included the transfer of entire training course from the originating institution (the International Global Change Institute IGCI, University of Waikato, New Zealand) to a regional institution, the University of the South Pacific in Fiji.

PICCAP is an example of a subregional training and technical and policy-related cooperation activity that has enhanced national capacities to address an environmental issue of critical importance to the small island developing states of the Pacific.

Source: IGCI, New Zealand

CHAPTER FIFTEEN

B. Main trends and Conditions

As countries in the Asian and Pacific Region address environmental management problems and create institutional mechanisms in response, the training needs in environmental related sectors continue to expand rapidly. A variety of training and skills development activities are being pursued at local, national, subregional and regional levels by governments, NGOs as well as by international and inter-governmental organizations in the region. In terms of content, duration, methodology, instruction mechanisms, types of participants and other factors, there is great diversity, but the end result is an overall strengthening of skills and capacity to better manage the environment and natural resources. The need for greater numbers of trained personnel in a wider range of environment related disciplines and skills are required for:

- the implementation of National Environmental Management Plans (NEMPs) or action plans (NEAPs);
- servicing of international environmental conventions and treaties that many countries have become parties or signatories to, each of which places specialized demands and requirements on the participating countries;
- the adoption of international environmental standards, such as ISO 14000, as well as environmental policy and management tools including environmental impact assessment, environmental audits and eco-labelling, which require well-trained and highly skilled personnel in sufficient numbers within government, industry and other sectors; and
- the provision of environmental procedures, systems and checks, demanded by the increased environmental awareness of the public. Market forces are also demanding greater commitment to, and professionalism in, environmental management and compliance.

A question often arises as to who should be trained given the limited available resources. While environment related training is needed across the board in a large number of sectors, professions and pursuits, some prioritization is required to identify areas for immediate support and intervention. In this regard, UNEP's Environmental Education and Training Unit identified the following categories for priority environmental training: teachers, teacher trainers and teachers' curricula developers; students at all levels; policy makers; decision makers; media professionals; and other opinion leaders and multipliers.

In addition, foreign service personnel who negotiate environmental conventions on behalf of their countries; government officials who negotiate donor funding support for development projects; managers and administrators in educational and media institutions, who often act as gatekeepers, deciding what training and exposure trainees will receive, have all been identified as target groups for environmental training.

C. National Networks For Environmental Training

National level networks of tertiary level institutions have been formed in several countries of the region to share resources and efforts in providing environmental training. One such network is the Philippine Association of Tertiary Level Educational Institutions in Environmental Protection and Management, PATLEPAM, which by mid 1999 had linked over 300 higher educational institution. It has conducted training programmes in environmental impact assessment, and a large number of seminars and workshops on environment and education (Supetran 1999). UNEP/NETTLAP has catalysed the formation of two country level networks in Thailand and Malaysia for environmental training and research. These networks, known by their acronyms THAITREM and MATREM respectively, have brought together dozens of tertiary level institutions in each country for undertaking collaborative research and training activities.

The Australian Environmental Education Network (AEEN) is a national network of environmental education and information programmes, materials and publications. The Network includes access to materials and programmes produced within the Federal Environment Portfolio, States/Territories Environmental Education Resources, a number of current school and community environmental programmes, and links with tertiary resources and a bulletin board for the exchange of ideas.

Education Network Australia (EdNA) aims to facilitate the provision of cost effective education to all parts of the education community in Australia. EdNA is founded on cooperation and consultation between representatives of all sectors of the education community including Commonwealth, State and Territory governments, non government schools, the vocational education and training sector, the higher education sector and the adult and community education sectors. The aim is to maximize the benefits of information technology for all sectors in education and to avoid overlap and duplication between the various sectors and systems.

EDUCATION, INFORMATION AND AWARENESS

REGIONAL COOPERATION FOR ENVIRONMENTAL EDUCATION AND TRAINING

Recognition is growing that many environmental issues and challenges are common to more than one country in the region, and that countries and communities can learn from each other in their responses to similar situations and problems. In some cases, the transboundary nature of environmental problems and their impact make it an imperative for countries in the same geographical vicinity to work together to address and cope with environmental trends and conditions.

Regional networking is one of the most cost-effective ways of promoting active cooperation among those engaged in environmental education and training. The Asia Pacific Network for Tertiary Level Environmental Training (NETTLAP) contributes to human resource development and the strengthening of tertiary institutions in the Asian and Pacific Region. Staffs of tertiary institutions are key targets for environmental training for the multiplier effect they can generate; they often act as advisors to government and industry; and because of the high standing university staff members enjoy in many communities as opinion leaders. The roles and objectives of NETTLAP are outlined in Box 15.2.

Another example of regional network is the South and Southeast Asia Network for Environmental Education (SASEANEE). It was initiated in 1993, coordinated by the CEE in India and the Commission on Education and Communication of IUCN. SASEANEE membership includes governmental and non-governmental organizations, academic institutions, as well as agencies and individuals involved in or interested in networking, initiating or supporting environmental education activities in the South and Southeast Asian countries. Activities include: a directory of persons and institutions involved in environmental education; a newsletter called *SASEANEE Circular*; and short training courses in environmental education.

In 1995, the South Asia Cooperative Environment Programme (SACEP) initiated a project to assess environment related training needs in its member countries, with a view to developing a Regional Plan of Action. Country studies were carried out in Bangladesh, Bhutan, India, Maldives, Nepal, Pakistan and Sri Lanka, which were coordinated by CEE, India. Based on the country level assessments, SACEP and CEE prepared a Regional Plan of action that sought to help synergize environmental capacity building within the subregion. It found that courses were well developed and that some institutions had also developed considerable expertise and experience in their areas of specialization and provide appropriate training. However, the assessments underlined the fact that the available training opportunities were inadequate to cater to the growing needs, and in some cases, the quality of training required improvement.

INFORMATION AND AWARENESS

A. **Monitoring Assessment and Reporting**

Environmental monitoring by government agencies and institutions in many countries has focused on certain environmental concerns, such as the quality of air, water and other natural resources. Monitored data are transformed into information that show environmental trends and effectiveness of past mitigation measure which are vital for environmental management. This information is also utilized in the state of environment reports; for example, of Australia; New Zealand; Malaysia; India; Fiji; Japan; Hong Kong, China; Turkmenistan; Azerbaijan; Uzbekistan; and Palau. NGOs and community-based organizations also play a significant role in collecting and disseminating information to the community and to all spheres of government.

1. *National level*

Air pollution is a concern for which extensive monitoring systems have been put in place in many countries of the region. In the Republic of Korea, for example an automatic air pollution monitoring network measures seven atmospheric pollutants which includes TSP, SO_2, NO_2, CO, O_3, etc. Other countries in the region may not have such a sophisticated system but nonetheless have established monitoring systems to cover sites across the country. Indonesia now has 31 air quality monitoring stations nationwide to help check increasing air pollution arising from motor vehicles in large cities. People's Republic of China illustrates a wide array of monitoring stations and also implements a quantitative examination system for urban environmental control. The central and provincial governments have already performed quantitative checks in over 37 major cities and 330 smaller cities.

In the Russian Federation, special attention is given to the development and support of a system for monitoring the condition of the ozone layer as well as a system for monitoring ultraviolet radiation over the whole country and adjoining territories. Information is processed and presented in the form of daily maps showing deviations in the volumes of total ozone content from norms established over many years.

CHAPTER FIFTEEN

In Australia, a number of strategies and plans provide a focus for particular resource issues, including Agriculture Land Cover Change project and the revised National Overview for the Decade of Landcare Plan (the main strategic plan for the National Landcare Programme The government, through its Commonwealth and State agencies, funded a programme to monitor changes in Land cover from 1990-1995, through the use of Satellite data. Information from the project provides as baseline for future monitoring which is vital to land clearing and agriculture development that have major impacts on wide range of country's natural processes.

Box 15.2 NETTLAP: Building Regional Capacity through Environmental Research, Training and Education

In 1993, UNEP's Regional Office for Asia and the Pacific (ROAP) established an ongoing programme to enhance the region's capacity to manage the environment in a sound and sustainable manner. This initiative, called the Network for Environmental Training at Tertiary Level in Asia and the Pacific (NETTLAP), has evolved into a major contribution in helping to achieve both national and regional goals of sustainable development. It has explicitly recognized that tertiary institutions such as universities, technical and training institutes and teacher training colleges, play a major role in building capacity for sustainable development.

Staff of these institutions were identified as "agents of change" for two reasons. Firstly, a large multiplier effect is associated with actions that strengthen tertiary institutions and enhance the abilities of staff to transfer, to their colleagues and students, state of the art understanding and international best practices. The improvements involve many people in a short period of time-graduates are soon improving the environmental management policies and practices in industry, government and the community. Secondly, governments and industry keenly seek the advice and guidance of staff from universities and technical institutes. Industry in particular recognizes the ability of such people to bring innovative solutions to current environmental problems and creative approaches to preventing the occurrence of new problems.

Initially NETTLAP focussed on strengthening key tertiary institutions in 35 developing countries in the region. The early efforts of NETTLAP did more to recognize the enormity of the need and, in relative terms, little to address it. But incrementally NETTLAP made a difference throughout the region. The benefits of the multiplier effect have begun to be seen. This was particularly so in the case of the design, preparation and dissemination of environmental curricula and the supporting instructional methods, materials and tools for use in tertiary and other relevant institutions in the region. These efforts resulted in sets of curriculum guidelines and associated training methods, resource materials and tools cover such topics as environmental economics, hazardous waste management, toxic chemicals management and coastal zone management. These outputs are still widely sought, and extensively used in the region.

By the late 1990s, NETTLAP had matured in several ways. Significantly, it has shifted its target from institutional strengthening and human resources development in the tertiary sector itself to assist developing countries to plan and implement their own activities to build the capacity to achieve effective environmental management, and sustainable development. Countries can take a comprehensive approach that targets other important "agents of change" – politicians, government officials, and leaders from the private sector and NGOs.

Therefore, in recent years NETTLAP has focussed on building national networks that can facilitate the linking of policy makers, development planners, environmental managers from industry and key staff from tertiary institutions. In addition to sharing expertise, experiences and best practices, the networks are also designed to help identify current and emerging needs that can best be addressed through a symbiotic relationship between these key players. Through these national partnerships, NETTLAP is linking research, training and education in order to improve the capacity to prevent or minimize adverse impacts on the environment. This involves identifying and implementing responses that are sustainable, assured of achieving the desired results and identified needs, supportive of related policies (e.g. appropriate economic and social development), innovative but consistent with traditional indigenous practices, add value to other initiatives and encourage complementary activities.

NETTLAP has played a key role by facilitating nationally "owned and driven" environmental capacity building networks in the Philippines, Malaysia and Thailand. The major achievements of the latter two networks for training and research in environmental management have resulted, in part, from significant funding from DANCED, an initiative of the Government of Denmark. Given the success of its current approach of "Regional Cooperation with National Implementation", NETTLAP is in advanced discussions to help develop similar networks in People's Republic of China, Viet Nam, India and the Mekong countries.

Key players in NETTLAP's strategy – "Regional Cooperation with National Implementation" – are such subregional organizations as SACEP, ASEAN and SPREP. NETTLAP is working with these, and similar organizations, to ensure that its actions are supportive of their strategies and action plans related to environmental research, education and training. NETTLAP has evolved as new needs are identified and past needs are addressed. But despite its efforts, and its significant successes, much more needs to be done. NETTLAP has shown that the most effective approach is one that builds synergies between key international, regional, subregional and national players.

Source: NETTLAP office at UNEP/ROAP

Wastewater is also intensively monitored in some countries across the region. Databases in Australia (Australian Waste Database) and New Zealand (Waste Analysis Protocol; WAP) are used for collating national baseline and update information, so as to monitor waste management services and ensure national objectives are met. These data will also eventually form the basis for developing national waste reduction targets.

Environmental quality monitoring has also been focused on natural resources, such as coastal and marine resources. In Thailand, the Fisheries Department of the Ministry of Agriculture monitors coastal zones and estuaries for toxic chemicals, heavy metals and oil pollution. India has a programme on Coastal Ocean Monitoring and Prediction System that is engaged in a systematic monitoring of marine pollution in the country and conducts studies relating to waste assimilation capacity of coastal waters. The Russian Federation monitors and evaluates the environmental impact of activities affecting coastal and marine regions such as in the Caspian Sea where a system is being developed to forecast its fluctuating level over various time spans to predict possible future changes.

Forest resources and biodiversity are extensively monitored throughout the region. For example, National Biodiversity Surveys are carried out approximately every five years in Japan. Indonesia conducted a project with GEF support to establish a Biological Diversity Inventory and develop a User Advisory Group Information System. In Australia, a National Forest Inventory and Wilderness Inventory are carried out to produce a national State of the Forests report every five years. The Republic of Korea has a 10-year periodic forest inventory with site surveys to produce a geographic map that is computerized into digital databases. Monitoring for forest pests and diseases has also been developed in the Russian Federation and the Republic of Korea for early warning systems related to pest and disease outbreak.

Other monitoring techniques include the use of Geographical Information Systems (GIS), which are used to establish nationwide conservation and protected areas (Australia); identification of environmentally sensitive areas (Malaysia, Fiji); forest resource mapping (Indonesia, the Philippines, Myanmar, Solomon Islands); flood action planning (Bangladesh); studying ecosystem changes, river pollution and marine environmental surveys (Republic of Korea); analysis of river change, water logging and salinity, desertification, and agro-ecological zonal maps for water research (Pakistan); and industrial pollution control (Thailand). The significance of disaster prevention and mitigation has also prompted many countries to take initiatives in the areas of disaster forecasting, early warning, risk assessment and mapping of climate and water related hazards.

2. *Regional and Global Monitoring*

Environmental quality monitoring goes beyond country activities and programmes when environmental concerns take on transboundary characteristics. An example is the Mekong GIS project (of the Interim Committee for Coordination of Investigations of the Lower Mekong Basin), which was initiated to evolve a network of data centres in Lao People's Democratic Republic, Thailand and Viet Nam. Regional cooperation in Central Asia has also been initiated on questions relating to radiation safety.

Remote Sensing and GIS are also being applied as a tool for monitoring at regional level. The value of remote sensing in monitoring transboundary pollution was demonstrated recently when ASOEN (ASEAN Officials for the Environment) used the data developed from these for planning inter-country cooperation regarding haze caused by forest fires in Indonesia.

3. *Issues and Problems in Monitoring and Assessment*

The most often cited constraints in relation to environmental quality monitoring in developing countries relate to inadequacy of funds and the lack of manpower and/or training, which leads to low coverage and low frequency of monitoring, particularly in the rural areas. This is especially so when the monitoring cover a wide area, for example, in Mongolia there is a need to monitor about 20 000 bored wells and a similar number of dug wells as it is believed that 70-80 per cent of these are contaminated. However, present capacity only allows chemical analyses for 14 per cent of the bored wells and none of the dug wells. Lack of training and technical capacity has also reduced the capacity of developing countries to actively participate in global monitoring efforts.

Environmental monitoring and assessment in the developing countries of the region still requires strengthening in terms of standardization of, monitoring network system design; sampling and analytical methodologies; quality control in data collection and laboratory analysis; national procedures for harmonizing data collection; and improving the access to data of researchers and interested citizens. There is also a need to establish systems for collecting new benchmark data to assess the sustainability of resources, as well to environment with health, population, and economics.

B. Information Dissemination

Dissemination of environmental information is extremely important in integrated environmental management, since it plays a vital role in sensitising individuals to environmental issues. In the past, dissemination has been done through campaigning using mass media and scientific publications, however, advanced technology, especially the electric media, is now playing an increasing role. However, access to environmental information is not easy in most countries in the Asian and Pacific region. Some region-wide efforts such as the UNDP funded Asia Pacific 2000 initiation have concentrated on promoting such efforts (Box 15.3).

1. National Level

In some countries, there is no legal policy to disseminate environmental information to the public or private sector. Lack of effective coordination makes information exchange more difficult. For example, in the Russian Federation, though environmental information distribution policy does exist by law, there is no legal administrative mechanism available to put it into practice. As a result, the information cannot be obtained easily, even among coordinating agencies.

In Australia, environmental information can easily be accessed from a diverse range of institutions. There are numerous projects being undertaken by government agencies at the Federal and State level and in research and teaching institutions, which are aimed at developing methods of integrating economic, social and environmental information. These include state of environment reporting, the development of indicators of sustainable development, methods for resource valuation and systems of environmental and natural resources accounting.

The Environmental Resources Information Network (ERIN), located in Environment Australia provides environmental information for policy development and decision-makers. ERIN databases

---

Box 15.3 Awareness Raising Campaigns

The UNDP-funded Asia Pacific 2000 initiative (AP2000) seeks to support and strengthen the role of civil society in meeting the challenges of urban environment and poverty. Established in the early 1990s, it works closely with governments, NGOs, researchers and all other stakeholders in the region's unfolding urban drama. Studying how AP2000's partners are mobilizing themselves sheds light on the use of different campaign methods for environmental awareness, advocacy and activism.

Environmental Campaigns in urban areas of Asia and the Pacific have taken various forms. Some have been as simple as making and putting up posters, or conducting slide presentations and exhibits, for example as the Penang Organic Farm (PAF) does to promote the concept of Eco-cities. Eco-cities are designed to revitalize nature in over-exploited areas, while bringing new prosperity to its inhabitants. The Asia Pacific 2000 initiative has involved PAF to become the core of an Asia-wide Eco-cities network that will create awareness of alternative ways of building cities and to create sensitivity toward the environment. There are many other ways of carrying out campaigns. For example, in March 1995, AP2000's partner Waste Wise in Bangalore organized a three-day 'Festival of Recycling'. This event brought together the community, local authorities and private enterprise to promote the concept of reducing, reusing and recycling as a way of life. Now, the organization is spreading its advocacy work beyond Bangalore to the region with Waste Wise Asia Pacific.

Some campaigns target specific audiences. Magic Eye, a Thai community development organization, for example, aimed its 'Do Not Litter' campaign at school children. The programme, which made children aware of the need for proper waste disposal – for instance through separation and recycling – was a success. But Magic Eye's ultimate target was the parents, who are in a better position to put into practice the waste management schemes advocated through the programme.

Sometimes campaigns involve protest and agitation against policies or practices that are felt to be harmful to the environment (see Chapter 14). Across the region, NGOs and other citizens' groups regularly organize such campaigns which employ a variety of means, ranging from letter writing, marches or rallies, to the more active forms such as *satyagraha* or sit-in protests that are often adopted in parts of South Asia. In India, sustained campaign efforts have also used the method of *yatras* or long marches by groups of people – sometimes stretching across hundreds of kilometres – which attract news media interest, and expose thousands of people in communities along the way to the messages of the campaign.

Government agencies also use campaigns to rally public support for specific environmental issues. Singapore has a well established practice of environmental campaigns like the Clean and Green Week, and it's calls to Save Energy and Save Water. In Japan, October of each year is observed as the Recycling Promotion Month, during which all government ministries and agencies concerned with the environment conduct an extensive campaign to persuade the people to recycle resources.

Hopefully these and other campaigns in the cities of Asia and the Pacific will go a long way in improving the urban environment.

Sources: 1. Inter Press Service, Manila 1997
2. Ministry of the Environment, Singapore 1996

store a vast array of information about the environment, ranging from endangered species to drought and water pollution. Information is drawn from many sources including maps, species distributions, documents and satellite imagery and the information is easily accessible through the internet.

Similarly in Japan, public access to data on air and water pollution and natural environment is provided to government agencies, laboratories and outside users under pollution control laws.

The Government of India and the State Governments, through several of their organizations have taken various steps to develop information network capabilities of both the public and private sectors. The Environment Information System (ENVIS) has the joint objectives of building a repository and dissemination centre for environmental science and engineering, and providing national environmental information services to originators, processors, and disseminators of environmental information.

2. *International Level*

Internationally, INFOTERRA of UNEP is an important source of information. Additionally, United Nations system-wide Earthwatch programme initiated in 1994 by UNEP has emphasized environmental education. It aims at coordinating, harmonising and integrating observation, assessment and reporting activities across the United Nations system in order to provide environmental and socio-economic information to interested parties. Some countries also promote dissemination of environmental information for users around the world. The National Resource Information Centre (NRIC) of Australia develops advanced computing systems designed to service policy and community needs for information, for example, on sustainable development. Its information resources consist of more than 50 national and 100 regional spatially maintained datasets, and FINDAR, a software package for interrogating metadata on more than 6 500 databases that it maintains as a directory linked to all other major international directories.

In March 1996, the NIES, Japan, began to provide environmental information from its research activities to the world via the Internet. At the same time, the Centre established a computer communication system for the general public called the "Environmental Information & Communication Network (EICnet)" in order to promote national activities for conservation of the environment, including "EI-Guide" a survey of environmental information including explanations of laws, treaties, and environment-related terms.

3. *Issues and Constraints in Information Dissemination*

Effective communication plays an important bridging role between information and target groups. The first stage of communication, however, which is listening to what people know, think, believe and do, has often been weak. In the formal education sector, timely access to credible information is one of the major constraints for environmental education in the region. Furthermore, in some countries, although data and information are often collected by different agencies, a lack of effective coordination and communication among related agencies make information sharing ineffective. NGOs have been effectively involved in dissemination of environmental information to promote environmental awareness, although there is still a need for improved promotion at the regional level.

ENVIRONMENTAL COMMUNICATION

Environmental communication has now emerged as strong complementary practice to environmental education. A broad definition of environmental communication would be, "the sharing of information, insights and opinions on environmental issues, trends, conditions and solutions using any means of communications, ranging from inter-personal methods to means of mass communication using the modern as well as traditional media". In this sense, environmental educators constantly engage in environmental communication. However, an important distinction is that, while all environmental educators are communicators, not all environmental communicators are necessarily educators. Across the Asian and Pacific Region, individuals and institutions engage in a very wide range of environmental communication activities with varying degrees of sophistication, outreach and impact.

A. **Communication Trends**

In the mid to late 1990s, the main focus in environmental communication was to inspire positive behaviour, on the part of individuals, communities, corporate and industrial bodies and others, to help conserve the environment and achieve sustainable development.

A key issue for many countries in the region is to integrate environment into development policy, and to use communication and education in an integrated way as an instrument of policy. There is, however, a tendency to focus on formal school education target groups, both by governments and NGOs, rather than addressing other groups who can make a difference in a policy issue, or in an immediate practical sense (IUCN 1998).

One of the major channels through which environmental communication is practised is the mass media, which includes both print and electronic sections. The communication media in Asia and the Pacific underwent rapid change during the 1990s, with liberalization of media policies allowing private sector involvement, spread of global media networks owned by trans-national companies and the proliferation of new communication technologies which have removed the barrier of large investments required to enter the media field.

Trans-national television is the most prominent example of the changes taking place in the region's communications scenes. More than 386 million households in the region (more than 55 per cent) are now equipped with television sets. Estimates indicate that by the year 2005, more than 447 million households will have television sets (UNESCO 1997b). The last few years has also seen the rapid expansion in the use of the Internet, with most countries in the region already connected to the global information superhighway. Cellular phones, faxes, email, electronic networks and cable are also expanding countries' outreach, often at unprecedented growth rates.

The Asian and Pacific Region is experiencing a trend where the public service component on television is declining while the number of broadcast hours and channels continues to increase. The medium's potential for non-formal education and for raising public awareness remains largely untapped. While advertisers and sponsors compete to support entertainment, news or sports programmes, the more educational programmes, documentaries, investigative current affairs programmes or in-depth interviews, are having to contend with budget cuts, intense competition for prime-time slots, and an overall decline in the public service spirit in broadcasting (TVE 1999). Media's role in environmental communication remains effective only to the extent that the environmental experts, researchers and activists engage and use the media to influence and shape the accuracy, balance and scope of environmental coverage.

In response, throughout the region, governments and NGOs are adopting strategies to mobilize communications and cooperate with the media, and in some cases, to strengthen the media's capacity to cover environmental issues more effectively (Box 15.4). In 1996, Malaysia's Department of Environment (DOE) launched two major programmes with the cooperation of the electronic media. One was a TV serial called 'Bicara Alam' that discussed environmental issues of current interest, the other was a radio environmental awareness quiz.

B.   Print Media

Environmental reporting is now well-established within the region's print media, newspapers and magazines. Investigative journalism on environmental issues has resulted in exposing environmentally damaging plans, polices or practices. Newspaper exposures and subsequent public pressure have forced governments, local authorities, industries and others to change their plans, to tighten laws and regulations, and to abandon certain development projects whose environmental and social costs outweighed any benefits.

This is illustrated by Thailand's two leading English language newspapers, the *Bangkok Post* and *The Nation*, both of which have been honoured by UNEP with its Global 500 awards for excellence in environmental coverage and for their commitment to the environmental cause. The Nation newspaper has a recognized tradition of covering groundbreaking environmental stories, and the production of a weekly environmental page, Earth Focus. In Malaysia, a leading newspaper group, the *New Straits Times*, has supported and managed a national environmental education programme for students since 1992, which aims to enhance the awareness of the environment, culture and social values, whilst encouraging problem solving, motivation and participation in environmental programmes.

Producing special publications for children is another important strategy. Many conservation organizations produce material meant for children and young adults. The Centre for Science and Environment (CSE) in India, for instance, has published the science and environment fortnightly *Down to Earth* since 1992 and launched a children's supplement in 1998, called *Gobar Times*. It aims to stimulate young minds to question prevailing development patterns, lifestyles and governance systems. The supplement carries news and views on the environment, science and technology, stories from various traditions including environmental movements and inventions, and highlights of social implications of the issues. It uses comic strips, cartoons, quizzes, essay competitions and interactive pages to engage children's minds.

EDUCATION, INFORMATION AND AWARENESS

Box 15.4 New Environmentalism in People's Republic of China

In the 1990s, the environmental movement in People's Republic of China gained considerable momentum at all levels: in the media, among academics and professionals, and at different levels of government. At the official level, the National Environmental Protection Administration (NEPA) has a Centre for Environmental Education and Communication (CEEC) based at the Sino-Japanese Friendship Centre for Environmental Protection in Beijing, a modern facility opened in 1996. The CEEC's main functions include organising and undertaking national communications activities on environmental protection; managing a public inquiry system for information of environmental communications and education; compiling teaching materials on environmental protection; producing TV programmes on environmental issues; conducting training; and providing professional advice to local level CEECs.

Under CEEC's leadership, a vast network of environmental communicators has been mobilized across People's Republic of China, to provide information, clarifications and interpretations on a wide range of environmental issues and concerns. Meanwhile, a growing number of citizens' groups and media professionals are also actively participating in strengthening environmental awareness and advocacy work. A pioneering role is played by Friends of Nature, a citizens' group formed in 1993 whose membership comprises several hundred academics and journalists concerned about the environment. In 1996, Friends of Nature initiated a campaign that prompted the government to ban logging in Deqing country in the south-western province of Yunnan, home to the endangered golden monkey (due to excessive deforestation). Similar campaigns have been launched on saving the natural habitats of other endangered species, and to stop the caging of wild birds. Campaigns are carried out using persuasion as well as pressurizing tactics (media exposure) rather than using directly confrontational approaches.

A concerned and sympathetic media has become one of the biggest forces for change in People's Republic of China. For many years, Peple's Republic of China has had substantial coverage of environmental issues in its media: by the early 1990s, the country's environmental protection institutions at various levels had founded 16 specialized newspapers and 87 special magazines on environment, and there were eight newspapers and magazines with a national readership devoted solely to this subject. A major trend observed in the mid 1990s was the emergence of television as a powerful medium for raising awareness and advocating policy reforms on environment.

The Global Village Environmental Culture Institute of Beijing (GECIB) has been at the forefront in using television for environmental communication. In 1996, it created two successful television series. The first series, called *Green Civilization and People's Republic of China*, is aired on China Educational Television (CETV). It aims to cultivate the people's environmental consciousness through spreading basic knowledge on sustainable development. Encouraged by the response to this series, GECIB created a second series of weekly programmes called *Time for Environment*, which runs on China Central Television (CCTV), the country's national television network. This programme deals mainly with the latest international and domestic environmental events.

GECIB's work symbolizes the commitment of the new environmental movement in People's Republic of China. It is run by professionally skilled volunteers; the actual costs are covered by donations and grants. Encouraged by GECIB's successful television ventures, several television stations have launched their own programmes looking at different aspects of environment and development. While some deal with the traditional aspects of natural history, other programmes adopt a more investigative approach, probing and exposing environmental degradation, violation of existing laws and regulations, and instances of corruption and negligence leading to environmental disasters.

Expansion of operations to engage in book publishing and multi-media ventures that disseminate environmental information in a variety of ways is also on the agenda. Other bodies are already exploiting the Internet for this purpose: for example, the China Green Students Forum, a network of 20 university-based environmental groups, are linked through the Internet. As People's Republic of China expands its presence in the Internet, more groups and organizations are poised to take advantage of the new medium.

Sources: IUCN-CEC 1998; Time magazine 1 March 1999; and Sino-Japanese Friendship Centre 1997

C. Broadcast Media

The broadcast media, radio and television, in all their variations, have established themselves as the most pervasive and powerful forms of mass media in Asia and the Pacific today. They represent a major channel through which information on environment can be conveyed to the people (Box 15.5). Recognising the need for programmes in the regional language and context, major international media groups have associated with regional and national companies.

A wide range of activist groups and media organizations also produce television and radio programmes on environmental issues, using the media as a means to raise awareness. For instance, since 1996, the Nepal Forum of Environmental Journalists (NEFEJ) has been producing a weekly radio programme for Radio Nepal on community forestry. Called "Samudayik Ban", it is produced on behalf of the Federation of Community Forest Users in Nepal, and is designed to support community forest user groups through sharing experiences on

CHAPTER FIFTEEN

Box 15.5 Moving Pictures to Save the Environment

As the audio visual media become the dominant source of information and entertainment in the Asian and Pacific Region, more environmental activists and organizations are moving to take advantage of these powerful media for environmental education and awareness. Using low cost video production equipment spawned by the digital revolution, and responding to the growing demand for programming in most countries, NGOs and small media organizations are carving out their own niche in the region's media landscape. While many use video as a medium to raise public awareness and understanding of environmental issues, some are using it to document instances of environmental crimes and degradation; to lobby for specific policy reforms; or to raise funds for specific campaigns.

Recognizing that their effectiveness depends on having access to skills, information and programming, a number of the region's organizations have come together to form the Asia and Pacific Video Resource Centre (VRC) Network. A 'VRC' is an existing environmental or media organization that commits itself to using television and video for environmental communication. By mid 1999, 18 such VRCs were active in 15 countries in the region, engaged in a range of activities such as: producing new video programmes; versioning or adapting foreign programmes into local languages; and distributing the programmes using means such as public screenings, seminars, and film festivals. The network is affiliated with the Television Trust for the Environment (TVE) which provides them with high quality programmes and technical assistance. VRCs also receive support from UNEP and other agencies.

Members of the VRC network have found innovative means of using the audio visual media to promote environmental awareness. For example:

- WWF Pakistan has made arrangements to screen environmental videos at departure lounges of key international airports in the country;
- Nepal Forum of Environmental Journalists is producing its own weekly magazine programme for local television where environmental issues are regularly covered;
- Environmental Broadcast Circle in the Philippines has integrated video film viewing into the course work at some universities;
- Women's Media Centre in Cambodia produces short duration television 'spots' that convey concise messages on topics such as women and the environment;
- Centre for Environment Education in India uses videos in awareness programmes organized for children.

Several members of the network regularly organize environmental film festivals. Earth Vision has emerged as a leading environmental film festival in the Asian and Pacific Region. Conducted every year since 1992 by the Earth Vision Organization in Japan, it is open to film makers in the region, and offers prestigious awards and cash prizes. Film South Asia is another documentary festival that was started in 1997 by the Kathmandu-based Himal magazine. According to the organizers, the festival is an attempt to counter the domination of satellite television.

Sometimes, individual television programmes can capture the attention of viewers across boundaries and catalyze change. In 1995, TVE Japan produced a 30-minute documentary titled *Japan's Lessons on the Economy and the Environment: Our Pollution Experience*. It traced Japan's industrial pollution problems and effects on public health and the environment, and documented how the Japanese industry was forced to adopt pollution control measures. Because of its clear lessons for the currently industrializing countries in the region, this programme has been versioned into several main Asian languages – such as Mandarin, Hindi, Urdu, Tagalog and Bengali – by members of the VRC network to promote the key message in the film.

Similarly, a UNEP/TVE video called *Saving the Ozone Layer: Every Action Counts*, produced in 1995, has been versioned into several key Asian languages and is used widely for education and awareness purposes. This forms part of the audio-visual material available to organizations marking the World Ozone Day on September 16 each year.

Sources: TVE International and TVE Japan; UNEP-ROAP

---

policies and practices. Similar community radio broadcasts are also taking place from a number of radio stations in the region (Box 15.6).

Other private sector organizations are also engaged in producing environmental material for broadcast. For example, a private television production company named Miditech, based in New Delhi, India, has been producing a weekly television series called *Living on the Edge* which combines news and views on environment with clever visuals and interactive segments, making the programme both educational and entertaining.

D.   Major Constraints and Responses

Even though the quantity of environmental coverage in the region's media has continued to increase since the mid-1990s, several constraints and drawbacks remain.

A major difficulty faced by journalists and producers covering the environment is access to

EDUCATION, INFORMATION AND AWARENESS

Box 15.6 Community Radio: A Growing Cacophony of Local Voices

Community radio stations are designed to encourage participation by a large representative sample of the various socio-economic levels, organizations and minority groups within a community. The stations facilitate the free flow of information and opinions, encouraging freedom of speech and enhancing dialogue.

In recent years, there has been an increase in the number of local or community radio stations, usually characterized by their low transmission power and restricted audience. This could be a university campus, a specific district in a large city, or small stations in rural areas servicing distinct communities. Since the late 1970s, when the technology and expertise for starting community radio stations became widely available, a whole range of community radio experiments has taken place, and today community radio initiatives are found in all regions of the world. Some have been used effectively to promote environmental awareness and to generate community level discussion and debate on sustainable development issues.

Community radio is found in many countries of the Asian and Pacific Region. In the Philippines, parallel to the conventional radio stations (of which there are more than 500), there are some 35 local radio stations headed by community groups, denominational associations and educational organizations. Among the best known is the Network of Tambule Community Radio, comprising eight small community stations located in isolated regions. Set up and managed by local volunteers, the network receives support from UNESCO and the Danish International Development Agency (DANIDA). With basic technical training in radio broadcasting, villagers prepare news bulletins as well as programmes focusing on local issues. Open debates on these stations maintain pressure on the local authorities to find viable solutions to environmental, social and other community problems.

In Thailand, the community radio station in Chiang Mai has been broadcasting programmes on substitution crops as part of a government programme to reduce the poppy crop. This Hill Tribes Radio also seeks to improve the socio-economic conditions of the mountain tribal people living in the northern region of Thailand. In Australia, the 'Public Radio' community network was given legal status in 1974, and now has more than 130 community radio stations.

In Sri Lanka, Mahaweli Community Radio (MCR) was started in 1980 to support the new settlers in the Mahaweli River diversion programme, one of the largest development projects in Asia at the time. Radio has been used to motivate people in the Mahaweli settlements to take on the responsibility of bringing about change in themselves and in their communities. MCR has adopted many concepts of community-based radio programming and production, and has demonstrated that the medium of community radio can influence the thinking and attitudes of rural populations. MCR programmers have incorporated concepts of sustainability and conservation into their programmes over the years.

In 1997, Radio Sagarmatha in Nepal became the first privately owned and managed community radio service in South Asia. The station is run on a non-profit basis by the Nepal Forum of Environmental Journalists (NEFEJ) in collaboration with Himal South Asia magazine, Nepal Press Institute and the Worldview International Foundation. Initially starting with two hours of transmissions every day to the Kathmandu valley, the new station quickly increased its number of broadcast hours as well as its audience share. The core programming covers social, cultural, environment and developmental issues, while the rest of the air time is used to broadcast Nepalese classical music and folk songs, etc.

The growing developments in community radio throughout Asia and the Pacific are facilitating the dissemination of cultural and community specific information and awareness raising with respect to local environmental concerns.

Sources: 1. UNESCO 1997
2. Nepal Forum of Environmental Journalists (NEFEJ) 1998

---

reliable sources of information and having the technical information interpreted by experts. Although identified some time ago, the response to this constraint has been slow. The restriction of access to information by the public and media, still prevails in some countries and inhibits open coverage, discussion and debate of issues that are of public interest. In other cases where information is available, journalists lack credible means of interpreting technical issues and explaining them in layman's terms. In some countries, such as the Philippines, the scientific community has set up media referral services which offer free advice and information to journalists working on a science or environment related story, directing them to relevant experts and institutions.

In the few instances where such collaboration has been established between the environmental community and the media, impressive results have been recorded. In Pakistan, for example, the IUCN has involved the media as a key target group and stakeholder in the implementation of the National Conservation Strategy, resulting in a strengthened capacity of the media to cover the complex issues of sustainable development.

However, in many countries, environmental activists and government agencies handling the subject of environment have not developed fruitful media relations and use the media only to generate publicity for events and individual actions. While such publicity campaigns are useful, the greater interests of communities and sustainable development are better served by permitting investigative, balanced and accurate reporting.

Environmental journalism encounters other problems that are not easily overcome. For example, bottlenecks sometimes exist, where editors and programme managers may not appreciate, and thus pay limited attention to, environmental news. These bottlenecks remain partly because the sensitization that has occurred among reporters and producers has not been extended sufficiently well to cover the media gatekeepers. Another limiting factor is legitimacy of environmental issues and where such legitimacy is derived from. For instance, sometimes major environmental stories are under-reported in the country of origin until the foreign media picks it up and gives international coverage. In other cases, owing to prevailing restrictions on media freedom or due to problems of accessing reliable information, local journalists are unable to cover certain environmental stories. Foreign correspondents, operating at a different level are usually not as restricted by domestic information policies, and are better positioned to cover such stories. However, coverage by global news networks or international features services are tailor-made for a global audience, and as a result only highlights of a complex situation may be covered.

The anomalies of global media are such that sometimes the international media organizations enjoy a disproportionately high degree of influence. A case in point is the mass-scale arsenic poisoning in Bangladesh and West Bengal in India due to naturally occurring arsenic levels in the groundwater. Some fifteen million people in Bangladesh are forced to use the groundwater contaminated by arsenic, and it has become one of the largest cases of poisoning in history. This major environmental story has been reported and covered in the Bangladeshi media for several years since the mid 1990s. But worldwide attention increased only after the influential global media started reporting on it. The BBC and the *New York Times*, between them, covered the story and raised the issue to a new level of recognition among donors, the development community and the public than had been achieved by the previous media coverage in Bangladesh and in the region.

E. The Internet and World Wide Web

Recent years have seen the extremely rapid growth and development of the Internet as an information provider. According to the International Data Corporation, the number of Internet Web users in the Asian and Pacific Region will rise from 6.5 million in March 1998 to 29.3 million by the year 2001 (Panos 1998). However, large disparities in access and use of the Internet exist across the region and in individual nations. Major constraints remain that hinder the rapid expansion in Internet access in developing countries such as: poor telecommunications (lack of working phone lines); an inability to afford computers; and the higher cost of providing Internet services. However it is predicted that the Internet will move from a minority to a mass medium, with wide accessibility in a short period of time.

The earliest users and disseminators of Internet use and technology were academic and research organizations, belonging to the Association of Progressive Communications (APC). These have actively supported or established networks in regions around the world for several years, and often provided countries with their only link to the Internet until commercial Internet Service Providers (ISPs) became operational. Partly because of these initiatives, analysts say the Internet may have a greater social impact in developing countries than anywhere else, as the academic organizations are some of the best-informed and are often campaigning for greater democracy, social equality and protection of the environment (Panos 1998).

During the past few years, an increasing number of environmental organizations in the region have recognized the potential of the Internet and started using it for exchanging information; for advocacy and activism; for public awareness and education; as well as for publicity and promotional purposes. Although the number of websites with a strong environmental content originating from the region is still numerically low, the quality and scope of some of the currently available sites are impressive.

F. Traditional Media

For many years, environmental communicators have recognized the value and power of using folklore and traditional media, such as, dance, songs, drama, puppetry and miming, to take environment and development messages to the public. Historically, traditional or folk media have often played a role in the communication and promotion of new ideas, apart from its traditional role of preserving and teaching established values. Today, in spite of advances in the modern forms of mass media, many people still relate more readily and easily to traditional media, which are closer to their local cultures, and are often more interactive and participatory than the regular forms of mass media.

Various forms of traditional media are being used or adapted to convey environmental messages to children, communities and specific target groups in different countries of the Asian and Pacific Region. For example, the *wayang kulit* or puppet's shadow play is one popular folk art form among Javanese communities in Indonesia. These plays tell stories containing important values relevant and useful to people's lives, both physically and spiritually. The original wayan stories are rich in ethical messages regarding the close relationship between people and nature, appreciation of other living beings, warnings on probable natural hazards due to human actions, etc. In recent years, wayan plays have been used to communicate messages supporting Indonesia's national family planning programme, reaching vast audiences (CEE 1994).

ENVIRONMENTAL COMMUNICATION BEYOND MEDIA

Environmental communication and awareness rising is also taking place through citizen volunteers, through the efforts of the corporate sector, and through law and faith-based approaches. Each of these approaches has wide ranging applications across the region.

A. Citizen Volunteers

Mobilising volunteer action is a time-tested tradition in environmental management and activism. Volunteers may be drawn from different sections of society; some will bring in specialized skills and knowledge; others will provide a donation of labour and time.

For example, in 1987 the Regional Environmental Management Offices in the Republic of Korea launched the Environmental Watchdog System to raise public awareness of the need for environmental conservation and to encourage citizens to monitor environmental pollution. The system has contributed greatly to increasing public awareness of environmental issues through public relations, monitoring, reporting, and by accommodating public views and opinions. While the vast number of people involved was a positive development, it also posed management difficulties. To address this concern, the Ministry of Environment set up the Honorary Guard System for the Natural Environment as a watchdog in accordance with the Natural Environmental Preservation Act. In 1995, there were more than 11 000 honorary environmental watchdogs, representing every walk of life in society (MOE, Republic of Korea 1995).

B. Working with the Corporate Sector

Today, environmental communication initiatives often work closely with the corporate sector. Recognising that business, commerce and industry are key players in achieving sustainable development and equitable growth, environmental educators have started to form partnerships with socially and environmentally responsible corporations (see Chapter 13).

Increasingly, the corporate sector is initiating environmental awareness, communications and education activities on its own, as part of their community outreach or service programmes. In some cases, large corporations underwrite the cost of environmental communications activities initiated by inter-governmental organizations. A prominent example is how the Canon Corporation of Japan has consistently supported the United Nations Environment Programme (UNEP) to organize a worldwide photographic competition entitled "Focus on Your World" in 1992, 1995-96 and again in 1999.

The Environmental Conservation Committee of Sony has been actively involved in environmental work for many years, monitoring regional committees set up in North America, Europe and Asia. These committees identify environmental problems in their regions, and devise responses that can be implemented easily and efficiently. In October 1998, Sony Group in Singapore, in association with the Tertiary Institutions Council for the Environment (TICE) and the Ministry of Environment, established the Sony Environmental Conservation Prize to stress the importance of environmental issues. Open to university and polytechnic students, the competition solicits innovative ideas for environmental conservation. Other Sony companies across the region are working with environmental initiatives and activities, such as protecting mountains and rivers in the Republic of Korea, and assisting with the provision of forestry education at Muhak Mountain in the Masan region (Sony 1999).

In a growing number of countries, business and industry are coming together to form alliances or common fronts for taking corporate environmental responsibility to greater professional levels (see Chapter 13). Such business councils for sustainable development have already been formed in several countries, such as Japan, the Philippines and Sri Lanka, and the chambers of commerce in these countries are actively involved in enabling companies to become more environmentally responsible corporate citizens.

C. Faith and Law based Approaches

Public interest environmental litigation is increasingly popular in the region. In this, a group of lawyers or an NGO would initiate legal action for and on behalf of a local community directly affected by a specific instance of environmental degradation. The strengthening of environmental legislation in most countries during the 1990s (see Chapter 11), and the increased awareness on environmental rights, has fuelled this process. Beyond mere litigation, most environmental law groups also engage in raising awareness on the legal provisions and rights related to the environment.

The Bangladesh Environmental Law Association (BELA) has been at the forefront in using the law as an instrument of environmental education and awareness raising. It works with both the victims of environmental degradation, helping them to assert their rights, and also with law enforcement and judicial officers, assisting them in their official functions. In Sri Lanka, the Environmental Foundation and the Mihikatha Institute are both engaged in similar tasks of advocacy and education. Both organizations operate environmental legal aid clinics which not only offer free legal advice to those affected by environmental problems, but also engage in awareness raising related to legal provisions and rights. The Mihikatha Institute carried out a country-wide training programme that exposed over 2,000 practising lawyers on the aspects of environmental law, mediation and rights (Mihikatha 1997).

Environmental education and the promotion of environmentally friendly life-styles can also be carried out through faith groups (see Chapter 14). In the mid 1980s, the WWF brought senior representatives of the world's major religious faiths together in Italy to discuss each faith's teachings and principles on man and the environment. Having found a considerable amount of common ground among all religions on this subject, it was adopted as a strategy to enlist the support of religious leaders and groups to promote environmental awareness and, in particular, environmentally friendly lifestyles. In the Asian and Pacific Region, a number of initiatives have also been started with support from the UK-based Alliance of Religions and Conservation.

Some countries are actively supporting the development of sustainable lifestyles through religious teachings. For example, Bhutan is actively promoting the Buddhist way of life to ensure that its citizens are sensitive to the conservation of environment and natural resources. The Royal Government of Bhutan made a national commitment in 1995 to uphold its obligations to the future generations by charting a path of development called the Middle Path. This development upholds both environmental and cultural preservation as an integral part of the development process. To ensure that the modern development pressures will not lead to negative impacts on the environment, the Bhutan government is invoking Buddhist teachings to impress upon the people the value of environmental conservation and sustainable lifestyles (Royal Government of Bhutan 1999).

CONCLUSION

As the Asian and Pacific Region enters the new millennium, it is clear that a new surge of interest, enthusiasm and activity is underway on many fronts to place environmental education, training and communication higher on the public agenda in countries of the region.

Although a considerable amount of work has been done and achievements made in the 1990s, many challenges remain. As many countries of the region struggle to overcome the social, economic and cultural barriers placed on them by poverty and under-development, and at the same time face up to the new challenges of economic globalization, the priority assigned to environmental issues and conservation is at risk of being overlooked or traded off for more immediate benefits, and for survival needs. The environmental educators and communicators of the region need, therefore, to be vigilant and active to ensure that governments, industry and other key players in the sustainable development arena remain mindful of their international and national commitments to environmental conservation, in addition to ensuring that sufficient investments of resources, time and attention are made to consolidate the achievements of the 1990s.

Part Four

CHAPTER 16-20

PART I

CHAPTER	1	LAND
CHAPTER	2	FORESTS
CHAPTER	3	BIODIVERSITY
CHAPTER	4	INLAND WATER
CHAPTER	5	COASTAL AND MARINE
CHAPTER	6	ATMOSPHERE AND CLIMATE

PART II

CHAPTER	7	URBAN ENVIRONMENT
CHAPTER	8	WASTE
CHAPTER	9	POVERTY AND ENVIRONMENT
CHAPTER	10	FOOD SECURITY

PART III

CHAPTER	11	INSTITUTIONS AND LEGISLATION
CHAPTER	12	MECHANISMS AND METHODS
CHAPTER	13	PRIVATE SECTOR
CHAPTER	14	MAJOR GROUPS
CHAPTER	15	EDUCATION, INFORMATION AND AWARENESS

PART IV SUBREGIONAL OUTLOOK

CHAPTER	16	SOUTH ASIA
CHAPTER	17	SOUTHEAST ASIA
CHAPTER	18	SOUTH PACIFIC
CHAPTER	19	NORTHEAST ASIA
CHAPTER	20	CENTRAL ASIA

PART V

| CHAPTER | 21 | GLOBAL AND REGIONAL ISSUES |
| CHAPTER | 22 | ASIA AND THE PACIFIC INTO THE 21st CENTURY |

Chapter Sixteen

Irrigation has brought vast tracts of land under cultivation but its mismanagement has led to waterlogging and salinity in South Asia.

16

South Asia

INTRODUCTION
SHARED ENVIRONMENTAL PROBLEMS
CAUSES
POLICY RESPONSE
SUBREGIONAL OUTLOOK

CHAPTER SIXTEEN

INTRODUCTION

The South Asian subregion comprises Afghanistan, Bangladesh, Bhutan, Islamic Republic of Iran, India, Maldives, Nepal, Pakistan, and Sri Lanka. Given the economic, social and cultural context of the countries of South Asia, similar challenges confront them in protecting their environment and natural resources. For instance, high rates of population growth, urbanization, and widespread incidence of poverty are common, although improvements have been witnessed in all major indicators of human development over recent years (UNDP 1998).

South Asia is also home to a significant but shrinking array of terrestrial and marine biodiversity. For example, the Hindu Kush Himalayan belt is home to some 25 000 major plant species, comprising 10 per cent of the world's flora (Shengji 1998). In addition, Sri Lanka is one of the most biological diverse countries in the world (see Box 16.1), and India contains extensive savannah and forest habitats, including many endemic species of international importance. South Asia is also home to around 14 per cent of the world's remaining mangrove habitat, in addition to the highest percentage of threatened wetlands, 82 of which are in Bangladesh. Table 16.1 provides a summary of the key environmental issues and their causes in South Asia.

SHARED ENVIRONMENTAL PROBLEMS

A growing population reduces the per capita availability of land and water, which consequently impairs the people's ability to produce food. Land degradation and water scarcity are thus closely tied to food security concerns. Urbanization and poverty, particularly the concentration of population in a few large South Asian cities, is another important challenge. It poses a strain on the limited resources of these cities, where infrastructure is already overstretched.

In prioritizing shared environmental concerns in the subregion, the most important are:

- **Land degradation** South Asia has the largest area of irrigated land in Asia and the Pacific (nearly 90 million ha). However unplanned and badly engineered irrigation developments have resulted in land degradation through factors such as water erosion and salinity. Irrigated croplands of many countries, such as India and Pakistan, are severely affected by salinity. Wind erosion also presents a serious regional issue. Afghanistan, the Islamic Republic of Iran, Pakistan and India suffer most from desertification (UNEP 1997). The worst sufferers as a result of land degradation are the poor.

- **Water scarcity and degradation** A number of areas within Afghanistan, India, the Islamic Republic of Iran, and Pakistan are suffering from water scarcity. Groundwater depletion has also emerged as a major concern in parts of India, Bangladesh and Sri Lanka in recent years, and depletion of freshwater aquifers is threatening water supplies in the Maldives. Adding to these problems is the cross border degradation of existing sources specifically through uncontrolled release of sewage (the problem of pathogenic water pollution has grown to alarming proportions in the subregion, UNEP 1997), industrial wastes, agricultural run-off, commercial pesticide, and arsenic contamination (see Table 16.2).

- **Deforestation and biodiversity loss** South Asia shares approximately 19 per cent of the Asian and Pacific region total forest cover, providing cross-border habitats for a wealth of species. Within the subregion the Islamic Republic of Iran and Pakistan are the countries with the highest rates of deforestation. The effect of habitat loss on biodiversity has still to be quantified, though overall habitat losses have been the most acute in the Indian sub-continent (UNEP 1997). Diverse habitats such as mangrove have also faced severe pressures through aquacultural developments in Bangladesh, India and Sri Lanka (UNEP 1997) (see Box 16.1).

- **Impacts to the marine environment** Pollution and over extraction of resources are both key concerns in relation to South Asia's marine environment. Unplanned developments are also affecting the coastal zone, and leading to erosion, while changes in the region's seawater temperatures are impacting on its coral systems (see Box 16.2).

- **Atmospheric pollution** Rapid growth in energy demand and the reliance on coal have translated into significant increases in the emission of air pollutants. Urban air pollution has emerged as a major problem in many cities, in addition to fly-ash generated through coal mining (see Box 16.3). Effects of acid deposition in excess of critical loads have been recorded in areas of Northeastern India and Bangladesh (UNEP 1997), and pristine forests of Bhutan (Communication by the Government of Bhutan, May 2000). An increase in slash-and-

Table 16.1 Key Environmental Issues and Causes in South Asia

Country	Key Issues	Key Causes
Afghanistan	Soil degradation; overgrazing; deforestation; desertification; loss of biodiversity; food security risks; natural disasters such as earthquakes and droughts.	Population growth; increased demand for bio-fuels, building materials, and agricultural lands.
Bangladesh	Marginalized populations forced to live on and cultivate flood-prone land; loss of biodiversity; limited access to potable water; water-borne diseases prevalent; water pollution, especially of fishing areas; arsenic pollution of drinking water; urban air pollution; soil degradation; deforestation; severe overpopulation: natural disasters (especially floods and cyclones which kill thousands of people and causes heavy economic losses every year); food security risks; industrial pollution; import of hazardous waste.	High population density and urban primacy; reliance on private transport; urbanization and deficits in urban infrastructure (including one of the world's 30 largest cities – Dhaka); increases in unmanaged marine-based tourism; green revolution/agrochemicals and run-off; high demand for bio-fuels; lack of controls on industrial effluent; over exploitation and/or pollution of groundwater.
Bhutan	Soil erosion; limited access to potable water.	High rates of urbanization.
India	Deforestation; soil erosion; overgrazing; desertification; loss of biodiversity; air pollution; water pollution; huge population base and large growth rate is overstraining natural resources; natural disasters such as floods, cyclones and landslides are common; high death rates and ailments associated with indoor air pollution.	High rates of urbanization and deficits in urban infrastructure (including in four of world's 30 largest cities); reliance on private transport; industrial effluents and vehicle emissions; increases in unmanaged marine-based tourism; green revolution/agrochemicals and run-off; reliance on bio-fuels.
Islamic Republic of Iran	Air pollution, especially in urban areas; deforestation; overgrazing; desertification; oil pollution in the Persian Gulf; inadequate supplies of potable water; food security risks; natural disasters such as floods, earthquakes, and landslides are common.	Excessive pressure on forests and rangelands; high rates of urbanization and deficits in urban infrastructure (including one of world's 30 largest cities – Tehran); inefficient public and private transport; vehicle emissions, refinery operations, and industrial effluents.
Maldives	Climate change; beach erosion; depletion of freshwater aquifers; degradation of marine habitats.	High population densities; increases in marine-based tourism; sea level rise.
Nepal	Deforestation; soil erosion and degradation; loss of biodiversity; water pollution; natural disasters such as floods and landslides in rural areas; food security risks.	High rates of urbanization; reliance on private transport; increased demands for timber; increased population density and cultivation of marginal lands.
Pakistan	Water pollution; seasonal limitations on the availability of natural freshwater resources; majority of the population lacks access to potable water; deforestation; soil erosion; coastal habitat loss and degradation of marine environment; desertification; loss of biodiversity: natural disasters, mainly due to floods.	High rates of urbanization and deficits in urban infrastructure; industrial wastes; population increases in coastal areas and rise in tourism; depletion of mangroves for aquaculture; overfishing; increased demands for timber/bio-fuels; hunting/poaching; green revolution/agrochemicals and run-off.
Sri Lanka	Deforestation; soil erosion; pollution by municipal and domestic waste; loss of biodiversity; coastal degradation; limited access to potable water; water-borne diseases prevalent.	Excessive pressure on forests; increases in marine-based tourism; poaching; sea level rise; deficits in urban infrastructure; water pollution by municipal and industrial waste, and agricultural run-off; extensive mining activities.

Source: Compiled from ADB 1990; WRI 1998 and 2000; Ghimire 1995; Sharma 1993 and 1996; UNFPA 1997; Bhattacharya 1996; Malla and Shreshtha 1996; Singhal and Sachdeva 1997; Satyaramacahandra 1996; Government of Pakistan 1994; Government of Maldives 1994; UNDP 1998; Saxena and Dayal 1999; and World Bank and UNDP 1995.

CHAPTER SIXTEEN

> ### Box 16.1 Sri Lanka and its Biodiversity
>
> Due to its location and topography, Sri Lanka, is one of the smallest but biologically most diverse countries in Asia. Consequently, it is recognized as a biodiversity hotspot of global importance. Among the terrestrial ecosystems are forests varying from wet evergreen forests to dry thorn forests, grasslands, wetlands and freshwater bodies and a complex network of rivers. These together with the coastal and marine ecosystems such as sea grass beds, coral reefs, estuaries and lagoons, and associated mangrove swamps constitute the diverse and complex network of ecosystems in the country. In addition, there are numerous man-made ecosystems related to agriculture and irrigation, which have a direct bearing on the conservation, sustenance and survival of biological resources.
>
> Sri Lanka's high population density, high level of poverty, and wide spread dependence on subsistence agriculture are exerting considerable pressure on the biodiversity of the country. Extensive land degradation and deforestation and the unregulated exploitation of natural resources (e.g. Mining for coral lime, sand and gemstones) are some of Sri Lanka's most pressing problems. In response, the National Conservation Strategy, the National Environmental Action Plan, the Forestry Sector Master Plan, the National Coastal Zone Management Plan and Coastal 2000 are some of the policy instruments that are addressing biodiversity conservation. There are also many Government Institutions whose responsibility is to translate these policy initiatives into action. However, despite the legal, policy and institutional support for its conservation, the country's biodiversity is continuing to diminish. The growth and movement of population, the opening of economic markets, and new trends in industrial development are expected to have a growing adverse impact on biodiversity unless some systematic and stringent corrective measures are taken.
>
> Sri Lanka ratified the Convention on Biodiversity in 1994 and as a response to article 6 of the Convention, the preparation of "Biodiversity Conservation in Sri Lanka- A Framework for Action" began in early 1996. What this plan proposes is a course of action to ensure that the biological diversity within the country is conserved and used sustainably.
>
> *Source:*   Forestry Sector Master Plan of Sri Lanka 1995

Table 16.2 Organic Water Pollution Resulting from Industrial Activities in South Asia

Countries	Emissions of organic water pollutants (kg/day)	Industrial Shares of Emissions of Organic Water pollutants (%)							
		Primary Metals	Paper and Pulp	Chemical	Food and Beverage	Stone, ceramics and Glass	Textiles	Wood	Others
Bangladesh	186 852	2.8	6.8	3.5	34.2	0.1	50.9	0.6	1.1
India	1 664 150	15.5	7.5	8.2	51.5	0.2	11.6	0.3	5.2
Iran (Islamic Rep. of)	101 900	20.6	8.0	8.0	39.7	0.5	17.3	0.7	5.4
Nepal	26 550	1.5	8.1	3.9	43.3	1.2	39.3	1.7	1.0
Pakistan	114 726	14.1	5.8	7.3	39.5	0.2	30.1	0.3	2.7
Sri Lanka	55 665	1.2	8.9	7.2	42.2	0.2	38.3	0.7	1.3

Source: World Bank 1998

Figure 16.1 Access to Safe Water and Sanitation (% of Urban Population) in South Asia

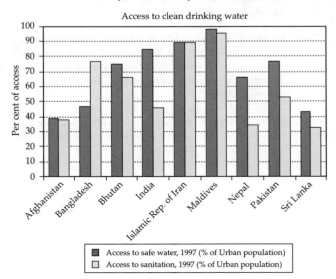

Source: World Bangk 1997

burn agriculture is also contributing to incidents of haze in certain areas.

- **Deficient urban infrastructure** The capacity of urban infrastructure (water supply and sanitation, waste management etc.) has not been increasing at a rate commensurate to urbanization. Surmounting deficits are growing particularly acute in sub-urban sprawls and inner-city squatter settlements, where it is the poor who are the most vulnerable (see Figure 16.1).
- **Natural disasters** The north-eastern Indian subregion represents the greatest area of seismic hazard in the Indian subcontinent. Floods, cyclones and landslides are other common hazards. Floods regularly cause widespread havoc in India and Bangladesh, and also occasionally in Pakistan and Nepal.

> ### Box 16.2 Coral Reef Degradation in the Indian Ocean (CORDIO)
>
> The coral reefs of the Indian Ocean were severely affected by the elevated sea surface temperatures during 1998, often between 3-5 degrees above normal, which bleached and subsequently killed a large percentage of the corals. The coral mortality ranged between 50 and 90 per cent around the Indian Ocean, with some areas facing a mortality close to 100 per cent, for example in large areas along the East African coast and in the central Indian Ocean. This coincided with the strongest El Nino (and the reversing La Nina) ever recorded.
>
> As a response to this, the Coral Reef Degradation in the Indian Ocean (CORDIO) programme was launched in early 1999. The programme's focus was on the ecological and socio-economic effects of the mass coral mortality and the degradation of the coral reefs. Projects also investigated the prospects for restoration of the reefs as well as providing alternative livelihoods for dependants of coral reefs. CORDIO is being implemented in 12 countries and is funded by The World Bank, SIDA (Swedish International Development Agency), WWF Sweden and two other Swedish agencies. The results of the first year of research in the South Asian Region showed that in India the reefs of Gulf of Mannar were severely affected by mortality of coral during 1998. Post bleaching surveys on the coral reefs of the 21 islands in the Gulf show the mean cover of coral is approximately 26 per cent. However, there is considerable variation between reefs with the cover of live coral ranging between 0 per cent and 74 per cent. In addition to reef-building corals, sea anemones and octocorals (soft corals) also bleached as a result of the increased sea temperatures that prevailed during 1998. Subsequently, a decrease in biodiversity of these reefs has been reported. Furthermore, extensive beach erosion on some islands was reported. Initial assessments of recovery processes in Lakshadweep Islands during 1999 indicate that the cover of live coral has increased to 15 per cent to 20 per cent compared with 5 per cent to 10 per cent reported immediately after the bleaching event.
>
> Monitoring of the reef tops conducted during 1999 in the Maldives showed that the cover of live coral has not increased since the post bleaching surveys conducted in 1998 and remains at approximately 2 per cent. At present, the cover of live coral is 20 times lower than that recorded before the bleaching event. However, re-colonization of fast growing branching corals has been recorded, indicating that reef recovery processes are underway. Furthermore, some reefs are abundant with coralline algae providing potential areas for coral recruitment. Nevertheless, despite these reasons for hope it is clear that the reefs of the Maldives were seriously affected by bleaching and subsequent mortality of coral will require many years to recover.
>
> Most shallow coral reef habitats in Sri Lanka were also severely damaged as a result of coral bleaching in 1998. Surveys conducted between June 1998 and January 2000 (7) revealed that many of the dominant forms of reef building corals in the shallow coral habitats have been destroyed. Invasive organisms such as tunicates, corallimorpharians and algae now dominate the dead coral reefs. Furthermore, the dead coral patches were rapidly inundated by sediment thus preventing re-colonization of coral larvae. Also, in every area surveyed thus far, except Trincomalee in the Northeast, the hydrocoral, *Millepora spp.*, which was once common, appears to be completely absent. The impact on fish by the loss of live hard corals is clearly visible in the decreased abundance of several species of fish that depend on live corals for food (e.g. Chaetodonts). Despite the destruction of corals in shallow water (< 8 m), corals growing in deeper waters (> 10 m) have recovered from bleaching almost completely providing a source for new recruits and reef recovery.
>
> The above observations indicate that corals of Indian Ocean need to be closely monitored to observe the impacts of temperature changes on them. The initiation of the CORDIO programme is an important project in that respect. The scientific data gathered during the project will help not only in assessing the damage and recovery of coral reefs but will assist in building national capacity for coral reef monitoring, and promote subregional cooperation through exchange of information and expertise.
>
> *Source:* South Asia Cooperative Environment Programme, Colombo 1997

CAUSES

Average population has been growing at the rate of 1.8 per cent per annum (UNFPA 1997), and widespread incidence of poverty characterize much of the subregion, although improvements have been witnessed in all major indicators of human development over recent years (UNDP 1998). Per capita GNP is lowest in Nepal and highest in Islamic Republic of Iran, and varies from US$210 to US$1 650 respectively. Overall the region is home to more than 500 million poor people living below a 'dollar a day' poverty line (World Bank 1999), representing an average of nearly 34 per cent of South Asia's population. 29 per cent of the population is urban, and urbanization is growing at 3.4 per cent (UNFPA 1997). The economies of countries of the subregion are primarily agricultural although in recent years the pace of industrialization has increased. Nevertheless, industrialization has not been able to completely absorb the growing labour force presented by the region's demographic trends. Major socio-economic indicators of the region are provided in Table 16.3.

CHAPTER SIXTEEN

Box 16.3 Air Pollution In India

Air pollution will intensify substantially by the year 2015 in India under 'a business as usual' scenario. Under this scenario, India will be producing SO_2, NO_x, particulate emissions, and ash at three times the current levels and CO_2 emissions will be 775 million metric tonnes per year, as compared with 1 000 million metric tonnes per year now produced by power generation across the entire European Union.

Emissions on this scale are bound to affect air quality and have major human health impacts. The damages caused by particulate matter to the respiratory system are a particular cause for concern. Consequences include significant increases in mortality, hospital admissions for respiratory infections, emergency room visits for bronchitis and other chronic pulmonary diseases and the number of days asthmatics experience shortness of breath.

The state of air pollution in different cities of India is fairly acute. Although the power sector contributes to the problem, most of the pollution, especially in the urban areas, stems from other sources. These include, residential and commercial stoves; industrial boilers; inefficiencies in the transport sector and the liquid fuel chain; the extensive use of traditional fuels in the city slums; and emissions from non-energy sources. Urban air pollution is, therefore, a cross-sectoral issue that requires a city-wide approach to achieve comprehensive air quality management. Thus, a strategy to address urban air pollution must integrate a range of activities at the municipal level, especially at small sources of energy and power stations located in densely populated areas and in the transport sector. This strategy should also include a major focus on the petroleum subsector.

Currently the power sector in India is on the verge of fundamental and significant reforms that will have profound implications for environmental management. India is moving from publicly owned, vertically integrated, monopolistic system with highly distorted prices for fuels and electricity to a more liberal system with market prices, competition, a greater role for the private sector, and commercial incentives. These changes will effect every aspect of the energy production system: the demand for electricity, the financial viability for all the entities involved, the choice of fuel and technologies, pricing decisions, and the respective roles and relationships among the state, the power sector, regulators, and fuel suppliers. During this time of transition, it is critical to determine how best to take advantage of the opportunities it presents to protect the environment and avert threats to public health. In particular, there is a need to find a more appropriate balance between economic development and environmental concerns.

Source: World Bank 2000

Table 16.3 Major Socio-Economic Indicators for South Asian Countries

Countries	Population (2000) (thousands)	Annual growth rate in population (%)	Population density (person per km^2)	Urban population as % of total	Annual growth rate of urban population (%)	Total GNP (US$ millions) 1998	Per capita GNP (US$) 1998
Afghanistan	22.7	2.9	38	21.9	5.8	–	–
Bangladesh	129.2	1.7	965	21.2	4.0	44 224	350
Bhutan	2.1	2.7	16	7.1	6.1	354	470
India	1 013.7	1.5	330	24.8	2.8	427 407	440
Iran (Islamic Rep, of)	65.6	1.5	38	61.6	2.2	102 242	1 650
Maldives	0.29	2.7	875	28.3	3.3	296	1 130
Pakistan	142.3	2.6	171	37.0	4.2	61 451	470
Sri Lanka	18.8	1.0	291	23.6	2.6	15 176	810
Nepal	22.9	2.4	160	11.9	4.8	4 889	210

Sources: 1. World Bank 1999 and 2000
2. ESCAP 2000
3. United Nations 1998

POLICY RESPONSE

A. National Initiatives

Notable initiatives have been undertaken by the governments of South Asia in recent years to strengthen institutions; improve regulatory systems; implement financial and policy reforms towards sustainable development; and enhance private sector involvement. Many new public sector institutions have been established, including environmental ministries, while independent environment agencies, departments and pollution control boards have also been created to support them.

A number of common limitations are, however, observed in relation to achieving sustainable development in the subregion. These relate to bottlenecks facing the legal, industrial, policy, and NGO communities, and are as follows.

- **Legal** Legislation in several countries of the subregion has failed to respond to changing paradigms of development, and many resource laws and statutes have been rendered obsolete by recent developments. Moreover, there is a weak regulatory framework, and problems of implementation and enforcement are acute. This is manifested in slow adoption of EIA practices, and inadequate public participation in formulation and implementation of laws. The paucity of financial resources to implement laws, inadequate penalties for violation and lack of political commitment are also observed.
- **Industrial** Environmental awareness among the industrial sector is increasing and is substantially better today than a decade ago. However the region is dominated by small and medium scale industry, a significant number of which produce hazardous wastes. The scale of industry in the subregion is analogous with meagre budgets and scant, if any, resources for allocation to research and development. A need for closer ties between industry and the universities is recognized, together with the development of appropriate environmental management capacity.
- **Policy** While it may be difficult to gauge the overall impact of environmental policies on the environment, it appears from the existing literature that local level decentralized approaches are having a far greater impact on resource management than other policy instruments. This is borne out, for instance, by the success of water user organizations in much of India and Nepal and also through similar approaches in forestry. A critique, however, has been that such approaches have focused exclusively on the accomplishment of physical targets, often to the detriment of participatory goals (Khan 1996).
- **NGOs** Several factors hamper the working of NGOs in the subregion, including lack of financial resources, training and strategic goals. In addition, numerous cases have been identified where, even when funds have been available, NGOs remain unable to access them due to insufficient organizational capabilities.

B. Subregional Cooperation

Opportunities for subregional policy cooperation are promoted through two principal programmes:

- The South Asia Cooperative Environment Programme (SACEP); and
- The South Asian Association for Regional Cooperation (SAARC).

1. *The South Asia Cooperative Environment Programme (SACEP)*

Since its inception in 1982, SACEP has initiated a number of projects, which are building national capacity to manage environmental issues. The overall focus of SACEP's activities includes capacity building and institutional strengthening; conservation and sustainable use of biodiversity; ecosystems conservation and management; environmental information and assessment; and education and awareness-raising. SACEP's members include Afghanistan, Bangladesh, Bhutan, India, Iran (Islamic Republic of), Maldives, Nepal, Pakistan, and Sri Lanka.

(a) The SACEP South Asian Regional Seas Programme

The formulation of the Regional Seas Programme was a major achievement under the aegis of SACEP, and it is one of the few major transboundary environmental programmes of South Asia. Under this programme, a South Asian Seas Action Plan was also prepared along with national and regional overviews and action plans. The implementation activities relate to integrated coastal zone management; development of national and regional oil and chemical contingency plans; and protection of the marine environment from the impacts of land-based activities.

(b) Improvement of the Legal and Institutional Framework

Another major programme undertaken by SACEP has been the improvement of the legal and institutional framework in the countries of the subregion with technical assistance from UNEP Regional Office for Asia and the Pacific. Under this programme, national workshops were organized in Bangladesh and Nepal covering environmental law from both national and international convention implementation perspectives. In the Maldives, support was given for a National Planning Meeting to develop Draft National Environmental Legislation. In Sri Lanka, activities were carried out in development of regulations; preparation of a model statute; establishment of environmental standards; preparation of the state of environment report; training programmes; and, an environmental awareness raising programme for children.

(c) Private and Public Cooperation Initiative

SACEP launched this initiative to promote cooperation between governments and the private sector, with support from UNEP and NORAD. Under the initiative, a Regional Seminar on Cooperation for the Promotion of Environmentally Friendly Business

Practices is being convened. The objective of the seminar, which will be attended by representatives of the Governments and business sectors of the seven South Asian Countries, is to share experiences in improving industrial, agricultural and business management and to identify present and future challenges for advancing towards sustainable development goals. The seminar expects to initiate a regional dialogue by promoting networking among national private sector institutions including Chambers of Commerce and Industry; Employers Federations; and, Industrial/Agricultural Research Institutions.

2. *South Asian Association for Regional Cooperation (SAARC)*

SAARC was established in 1983, with its headquarters in Kathmandu, and includes the countries of Bangladesh, Bhutan, India, Maldives, Nepal, Pakistan, and Sri Lanka. SAARC has a particular focus on economic cooperation, although it also covers many aspects of regional cooperation (including environment). SAARC has grown steadily and, as a result of recent coordination initiatives between the two programmes, its environmental activities are complementary to those of SACEP.

SAARC has set up several technical committees in many fields. The Committee on Environment was given the status of a Technical Committee in 1992, in which year a special session of this Committee was held in Pakistan to prepare modalities and programmes of action. The implementation of the recommendations of the Regional Study on Greenhouse effects has also been mandated to this Committee.

3. *International Programmes and Projects*

Countries of the subregion are also participating in four transboundary efforts being promoted by the World Bank in Asia and the Pacific. URBAIR and the Two-Stroke Vehicle Engine Initiative address the rapidly worsening air pollution problem in South Asia's largest cities. The Bay of Bengal Environment Programme funded by GEF and jointly implemented with FAO, addresses fisheries research, environmental emergencies, large marine ecosystems, and coastal zone management in and around the bay. Both South and East Asian countries are involved in this programme. The South Asia Development Initiatives seeks to improve regional cooperation in the poorest part of South Asia (Bangladesh, Bhutan, Nepal and eastern India) in water resource management, energy development and trade, and transport and commerce. Lastly, a programme for the preservation of Cultural Heritage in South Asia is being implemented in Bangladesh, India and Nepal to promote active involvement and financial support of the public, NGO, and private sectors to rehabilitate and protect national heritage sites.

SUBREGIONAL OUTLOOK

A distinctive feature of the South Asian subregion, is that, while the last five years has seen a growth in cooperation and coordination of intra-regional sustainable development initiatives, historically this has been lacking. However, many of the South Asian countries, like those of other subregions, are also party to several international agreements, covering many aspects of economic development and environmental conservation, and as such their intra-regional cooperation on these could be of considerable future mutual benefit.

What is imperative in the region is the continued move toward the decentralization of environmental management, the increased involvement of the public sector in environmental decision making, and importantly, concentration on the key concern of poverty alleviation. In this respect SACEP and SAARC are considering several further proposals. Indeed, it is expected that within the next few years, there will be a substantial portfolio of subregional environmental programmes underway, which will benefit the individual countries as well as the subregion as a whole, and improve the quality of life of the population. It is hoped these programmes will also yield positive results that will spread far beyond the boundaries of South Asia.

Chapter Seventeen

Fires in recent years have caused serious damage to both forests and the atmosphere in Southeast Asia.

17

Southeast Asia

INTRODUCTION
SHARED ENVIRONMENTAL PROBLEMS
CAUSES
POLICY RESPONSE
SUBREGIONAL OUTLOOK

CHAPTER SEVENTEEN

INTRODUCTION

The Southeast Asian subregion comprises the countries of Brunei Darussalam, Cambodia, Indonesia, Lao People's Democratic Republic, Malaysia, Myanmar, Philippines, Singapore, Thailand, and Viet Nam. The subregion remains very diverse in terms of economic development, political systems, ethnicity, culture, and natural resources. Singapore, for example, is an OECD country and Brunei Darusslam, and oil-rich microstate. Myanmar, Lao People's Democratic Republic, and Cambodia are essentially agrarian economies, while Malaysia, Thailand, the Philippines, Indonesia, and Viet Nam are rapidly industrializing. The diversity of the region is also reflected in the Human Development Index of its member countries, which range through high to medium to low. The region has witnessed sharp economic growth and subsequent sharp decline following the financial crisis in 1997. Consequently, there have been great variations in GDP per capita in the subregion. Over the past decade this has ranged from about US$250 in Viet Nam to US31 900 in Singapore, with average growth rates up to 1996 in the range of 6-7 per cent. Both the Asian Development Bank (ADB) and the International Monetary Fund (IMF) agree, however, that the prospects for economic recovery look positive from the beginning of 2000.

Southeast Asia is home to about half of the world's terrestrial and marine biodiversity, which in the tropical forest of the subregion remains largely undocumented (World Bank 1992). Around 30 per cent of the world's coral reefs are situated within the subregion, with the seas around the Philippines, Indonesia, and Malaysia constituting the centre of marine biodiversity. Some of the last remaining intact expanses of mangroves also occur in Southeast Asia. Table 17.1 provides a summary of the key environmental issues and their causes in Southeast Asia.

SHARED ENVIRONMENTAL PROBLEMS

The rapid economic development of recent years has led to a number of shared environmental problems in the region. These include diminishing forests; altered habitats and decreasing biodiversity; land degradation; polluted waters and declining availability of potable water; and the degradation of marine and coastal resources.

In prioritizing shared environmental concerns in the subregion, the most important are:

- **Deficient urban infrastructure** The capacity of urban infrastructure (water supply and sanitation, waste management etc.) has not been increasing at a rate commensurate to urbanization. Surmounting deficits are growing particularly acute in sub-urban sprawls and inner-city squatter settlements, where it is the poor who are the most vulnerable. Water pollution in the urban environment has had the severest impact on human health in Southeast Asia (World Bank 1998), and eutrophication presents one of the most serious problems of the subregion. For example, there is evidence that the rivers of Metro Manila can be considered biologically dead as a result of discharges of industrial (30 per cent) and domestic (70 per cent) effluent. During recent years, the incidence of water-borne diseases caused by contaminated water has been widespread, as well as the diseases caused by mosquitoes (that breed in stagnant water) such as dengue and malaria. Data for Indonesia and the Philippines show that 77 per cent and 56 per cent respectively, of the urban populations of those countries are connected to piped water (WRI 1999). As for flush toilets, recent data is only available for urban Philippines which shows that 37 per cent have such sanitation facilities (WRI 1999). Figure 17.1 provides data on the connection of some major Southeast Asian cities to basic services such as electricity, water and sewage. It demonstrates that the cities have a high percentage of their populations connected to water supplies, but that connections to a sewerage system lag far behind.
- **Deforestation and biodiversity loss** Continuing exploitation of forests and other habitats is

Figure 17.1 Share of Selected Southeast Asia Urban Household's Connection to Basic Services

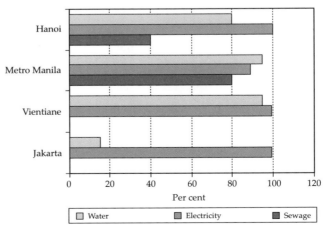

Source: WRI 1999

Table 17.1 Key Environmental Issues and Causes in Southeast Asia

Country	Key Issue	Key Causes
Brunei Darussalam	Seasonal smoke and haze	Transboundary pollution from forest fires in neighbouring countries.
Cambodia	Soil erosion; sedimentation; water pollution, deforestation; loss of biodiversity – threats to natural fisheries.	Unmanaged waste and effluent discharge into Tonlesap lake; destruction of mangrove swamplands through extensive industrial and aquacultural development.
Indonesia	Deforestation; loss of biodiversity; water pollution; air pollution in urban areas; national and transboundary seasonal smoke and haze; land degradation; pollution of Malacca straits.	Deficiencies in urban infrastructure – unmanaged industrial wastes and municipal effluents and waste; vehicular congestion and emissions; extensive land clearance and forest fires for pulp wood and oil palm production; extensive and unmanaged mining activities; national and transboundary industrial pollution (from Singapore and Malaysia); tourist developments in coastal regions beyond existing carrying capacity.
Lao PDR	Deforestation; loss of biodiversity; soil erosion; limited access to potable water; water-borne diseases prevalent.	Land clearance; shifting cultivation; inadequate water supply and sanitation infrastructure.
Malaysia	Urban air pollution; water pollution; deforestation; loss of biodiversity; loss of mangrove habitats; national and transboundary smoke/haze.	Vehicular congestion and emissions; deficiencies in urban infrastructure – industrial and municipal effluents; extensive land clearance and forest fires for pulp wood and oil palm production; unmanaged coastal developments; tourist developments in coastal regions beyond existing carrying capacity
Myanmar	Deforestation; loss of biodiversity urban air pollution; soil erosion; water contamination and water-borne diseases prevalent.	Land clearance; excessive mineral extraction; vehicular congestion and emissions; deficiencies in urban infrastructure – unmanaged industrial and municipal effluents.
Philippines	Deforestation in watershed areas; loss of biodiversity; soil erosion; air and water pollution in Manila leading to waterborne disease; pollution of coastal mangrove habitats; natural disasters such as earthquakes and floods.	Illegal forest cutting; land clearance; rapid urbanization and deficiencies in urban infrastructure – unmanaged industrial and municipal effluents, inadequate water supply and sanitation; tourist developments in coastal regions beyond existing carrying capacity;
Singapore	Industrial pollution; limited natural fresh water resources; waste disposal problems.	Seasonal smoke/haze; limited land availability for waste disposal.
Thailand	Deforestation; loss of biodiversity; land degradation and soil erosion; shortage of water resources in dry season and flooding in rainy season; conflict of water users; coastal degradation and loss of mangrove habitat; urban air pollution; pollution from solid waste, hazardous materials and hazardous waste.	Non-strategic and sporadic developments and destruction of critical watersheds; unmanaged aquaculture developments; growth in tourist industries exceeding growth in tourism carrying capacity; deficiencies in urban and rural infrastructure, particularly central draining – freshwater resources being polluted by domestic and industrial wastes and sewage runoff.
Viet Nam	Deforestation and soil degradation; loss of biodiversity; loss of mangrove habitat; water pollution and threats to marine life; groundwater contamination; limited potable water supply; natural disasters such as floods.	Land clearance for industrial zones; extensive aquaculture and overfishing; growing urbanization and infrastructure deficiencies – inadequate water supply and sanitation, particularly in major cities of Hanoi and Ho Chi Minh.

Source: WRI/GEF/UNDP/IMO 1997; APEC 1997; Great Barrier Reef Marine Park Authority et al 1995

impacting severely on biodiversity in Southeast Asia. The three largest forest areas per capita are in Lao People's Democratic Republic, Brunei Darussalam, and Malaysia, while the highest proportion of rainforests are situated in Malaysia, Indonesia and the Philippines. Latest available figures show an average deforestation rate in the region of 1.8 per cent (WRI 1999), rendering a total loss of original forest cover of around 2.4 million hectares between 1990 and 1997 (Schweithelm 1999). Highest loss rates were in the Philippines, Thailand and Malaysia, and the principal cause was logging, shifting cultivation, and conversion to plantations.

- **Degradation of the marine environment** Coastal and marine resources in the subregion suffer from a high degree of stress due to overfishing, destructive fishing methods (blast fishing and use of cyanide which contributes to the decimation of coral reefs), siltation from soil erosion, marine-based and inland water pollution, and the destruction of mangroves for shrimp ponds and aquaculture. Population

Table 17.2 Projected Coastal Populations

Countries	Population in Coastal Urban Agglomerations (Thousands)	
	1980	2000
Brunei Darussalam	N/A	N/A
Cambodia	50	287
Indonesia	29 166	58 303
Malaysia	3 997	9 158
Myanmar	3 923	7 695
Philippines	17 736	37 181
Singapore	2 414	2 960
Thailand	5 698	13 541
Viet Nam	5 585	14 317
Total	70 549	143 442

Source: WRI 1994 and 1995

growth and unplanned and sporadic coastal developments are also posing serious implications to coastal and marine resources. Table 17.2 shows the rate of increase of coastal population between 1980 and 2000. The subregion also shares the Coral Bleaching problems experienced in South Asia (see Chapter 16, Box 16.2).

- **Forest fires** The indiscriminate clearing of land for pulpwood and oil palm plantations is fuelled by the high demand for paper and palm oil products throughout the world. The traditional way of clearing land in most of the Southeast Asia subregion is by fire. The activity has led to devastating cross border impacts to habitat corridors, and has caused significant transboundary air pollution problems with particulates, smoke, and haze. The haze from forest fires that engulfed Indonesia, Malaysia, Brunei Darussalam, and to a lesser extent the Philippines, in mid 1997, and intermittently hereafter, have been observed as some of the worst episodes of air pollution in recent world history.

- **Atmospheric pollution** Data for air pollution levels in each Southeast Asian country are not available. However, the high concentration of industries in the urban centres of the region especially in the two largest cities of Metro Manila and Jakarta, indicate the high air-pollution potential. Vehicular emissions, particularly in Jakarta, Bangkok and Metro Manila, also contribute largely to the poor ambient air quality of these cities. However, since the 1997 financial decline traffic congestion in the major cities has also been reduced. For example, between 1996 and 1999 automotive production in Thailand fell by 55 per cent and sales fell 63 per cent (Brown 1999). Nevertheless, growth trends in air pollution are being observed in many cities, for example, Malaysia's urban centres have not yet reached critical levels of air pollution although traffic congestion and consequent vehicular emissions is fast becoming a problem in Kuala Lumpur and elsewhere.

- **Land degradation** The subregion suffers from soil erosion and contamination. For Southeast Asian countries, erosion mostly takes the form of surface water erosion, which contributes to the loss of topsoil (Lynden and Oldeman 1997). Problems are most acute in the Philippines, Thailand, Viet Nam, Malaysia, Indonesia, Cambodia, and Lao People's Democratic Republic, where water erosion impacts an average of 20 per cent of the total land areas. Thailand, Cambodia and Myanmar also suffer from the impacts of land contamination, where soil fertility declines have affected a total of 36.5 million hectares (Lynden and Oldeman 1997).

CAUSES

Before 1997, Southeast Asia was already experiencing the environmental costs resulting from the 'grow now, clean later' development policies of the period. Respective governments, influenced by civil society, had started action to address environmental issues and cooperate toward dealing with the region's key environmental problems, which were principally associated with the drive of globalization and trade pressures, coupled with fast track developments and inefficient resource consumption. However, the economic crisis of the 1990s significantly reduced the force of environmental efforts already underway (World Bank 1998).

External development drivers are also compounded by the internal pressures of population growth and poverty. The total land area of the Southeast Asia subregion is more than 435 million hectares, representing about 3 per cent of the total land surface of the earth, although the region is home to about 520 million people, or about 11 per cent of the world population. The average population has been growing at a rate of 1.5 per cent, representing the second highest growth rate in the Asian and Pacific Region. An average of 39 per cent of the population is urban, and urbanization is growing at 3.5 per cent. Projections indicate that by 2150, there will be three megacities (with population of more than 10 million) in Southeast Asia: Jakarta, Manila, and Bangkok. In summary, the region's major socio-economic indicators are provided in Table 17.3.

POLICY RESPONSE

A. National Initiatives

Many national initiatives have been undertaken by the governments of Southeast Asia with a view to achieving environmentally sustainable development. Amongst these are the creation of institutions, development of legislation, use of economic instruments, and development of links with the private sector. A number of common limitations are, however, observed in relation to achieving sustainable development in the subregion.

- **Policy** Although there are often confusing and fast changing jurisdictions of government agencies, the environmental policy initiatives of the region are moderately advanced and adequate numbers of institutions and mechanisms exist for their implementation. National Environmental Action Plans have been developed for many of the Southeast Asian countries, and cooperation and coordination of policy responses is also active. What is often lacking, however, is the capacity to implement and deliver on policy targets and ideals, which can be restrained through budget limitations and weak bureaucratic processes. Moreover, market based instruments remain in operation in a number of countries that do not account for environmental impacts, and indeed in many cases encourage non-sustainable activities. This is a salient feature in respect to the extent of aquaculture developments, and the leverage potential of 'fast grow' cultivation subsidy initiatives involving forest clearance.
- **Legal** A common observation in the region is that there is sufficient legislation in place to protect the environment but enforcement has been weak or non-existent in many cases. Command and control mechanisms are fairly advanced, though many countries lack either monitoring capacity or enforcement actions. For example, the regular sound of explosions is evident due to blast fishing in certain areas, although the practice is officially illegal. Furthermore, while the clearing of forest areas by fire was outlawed by the Indonesian Government immediately following the 1997/98 devastation, the recurrence of the fires in August 1999 showed the practise still to exist.
- **Industrial** The regions industry has traditionally focused on forestry and marine products, driven by the increasing global demand. Chief exports include tropical hardwoods, live fish, crustacea, and 'ornamental' and 'medicinal' animal products. Live fish may be sold for between twice and twenty five times the price of dead fish, and thus represent an extremely lucrative trade. However, the fishing method preferred for live catches by local fishermen is that of cyanide poisoning, posing significant chronic pollution problems (particularly to coral habitats) in coastal and inland waters (APEC 1997). Meanwhile, although modern industrial developments are adopting environmental management practices, and in many cases working toward international environmental accreditation, such approaches are restricted to large or even international conglomerates. It would appear that the SME sector remains reluctant through either perceived constraints of inaffordability or capacity deficits.
- **NGO** The concept of public consultation in decision making is at an early stage of development in most countries of Southeast Asia. The institutional framework therefore remains somewhat *ad hoc*, and the NGO community marginal. Furthermore, the process of advertising development proposals to project stakeholders, for example, can be constrained by minimal distribution and readership of rural and/or regional media. However, autonomous municipalities, including people's representative committees at the local and district levels form the administrative structures of many of the Southeast Asian countries. District committee leaders are themselves usually members of the local population, and are consequently viewed as sharing a firm understanding of local issues and concerns. Imaginative usage of this existing framework may present the subregion with some potentially strong foundations for involving the public in the region's environmental decision making.

B. Subregional Cooperation

Southeast Asian countries have a long history of cooperation on environmental issues and concerns, dating back to the first subregional environment programme in 1977 (ASEP I). The programme was further endorsed by Southeast Asian Governments at the first subregional Ministerial Meeting on the Environment, which was held in Manila in 1981. This assisted in the development and implementation of two further ASEPs, namely ASEP II (1982-1987), and ASEP III (1988-1992). The significant levels of

CHAPTER SEVENTEEN

Table 17.3 Major Socio-Economic Indicators for Southeast Asian Countries

Countries	Population (2000) (thousands)	Annual growth rate in population (%)	Population density (person per km^2)	Urban population as % of total	Annual growth rate (%) (2000)	Total GNP (US$ Millions) 1998	Per capita GNP (US$) 1998
Brunei Darussalam	339	2.4	60	72	2.7	–	–
Cambodia	12 227	2.4	65	16	4.4	2 945	260
Indonesia	212 107	1.3	112	41	3.9	130 600	640
Lao PDR	5 433	2.6	22	24	5.0	1 583	320
Malaysia	23 171	2.0	68	57	3.1	81 311	3 670
Myanmar	48 785	1.8	68	28	2.7	–	–
Philippines	75 967	2.0	252	59	3.4	78 938	1 050
Singapore	4 146	3.5	5 186	100	3.5	95 453	30 170
Thailand	62 320	1.0	120	31	2.4	131 916	2 160
Viet Nam	79 832	1.4	235	24	3.6	26 535	350

Source:
1. World Bank 1999 and 2000
2. ESCAP 2000
3. United Nations 1998

subregional cooperation have stemmed from the realization that environmental problems, especially those that require long-term and difficult solutions, could be better addressed by collaborative efforts through the sharing of knowledge and the pooling of resources. Such unified positions on international environmental issues of common interest have also allowed the subregion to be vocal and effective in international fora. Opportunities for cooperation are presently promoted through three principal means:

- The Association of Southeast Asian Nations (ASEAN) Senior Officials on the Environment (ASOEN);
- Other Subregional Cooperation Programmes; and
- International Programmes and Projects.

1. *ASEAN Senior Officials on the Environment (ASOEN)*

ASOEN was established in 1989, and is composed of Permanent Secretaries who serve as national chairs or focal points for the coordination of all environment programmes in the subregion. Members meet annually and are assisted by three working groups targeting the issues of, conservation; the marine environment; multilateral environment agreements; and a task force on haze. Cooperation was again enhanced in 1994 by the Bandar Seri Begawan Resolution on Environment and Development, which not only allowed for the harmonization of the region's environmental air and water quality standards, but also led to the development of the ASOEN Strategic Plan for Action on the Environment.

(a) ASOEN Strategic Plan for Action on the Environment (1994-1998)

The plan consisted of five objectives, ten strategies, and twenty-seven actions. A number of achievements have been made during the plan's implementation period. These include for example, the adoption of the Cooperation Plan on Transboundary Pollution in 1995, addressing atmospheric pollution, movements of hazardous wastes, and, transboundary shipborne pollution (in-line with GATT principles). The plan also facilitated the development of a Regional Haze Action Plan, following the 1997 forest fires. Under this plan, a technical Task Force was developed chaired by Indonesia, and comprising officials from Brunei Darussalam, Indonesia, Malaysia, and Singapore. Other important achievements in regional cooperation activities under the plan were the establishment of a Regional Centre for Biodiversity Conservation in the Philippines (see Box 17.1), and the promotion of an integrated framework for the management of Southeast Asia's coastal zones.

(b) ASOEN Strategic Plan for Action on the Environment (1999-2004)

The current plan consists of the key activities to be implemented by ASOEN and its supporting bodies over the five-year period. This period includes the likely ratification of the ASEAN Agreement on Transboundary Haze Pollution, following its discussion at an expert working group in May 2000 (see Box 17.2). Programmes relating to the protection of biodiversity are also being progressed, including the protection of heritage parks and reserves, and

> **Box 17.1 ASEAN Regional Centre for Biodiversity Conservation (ARCBC)**
>
> The subregion contains some of the richest and yet most threatened biodiversity in the world. The region contains four major biodiversity "hot spots" and 36 out of a global total of 221 Endemic Bird Areas (see Chapter 3). Indonesia is ranked as one of the top five countries in the world for biological richness and contains more than 15 per cent of all vertebrate species. The marine areas of the region are also the richest in the world.
>
> ARCBC is being established under a European Union (EU) and ASEAN cooperation project to promote biodiversity conservation in the subregion. Its activities will include capacity building; research; training; networking; raising of awareness; collation and analysis of information; data sharing; technological exchange; improved data management procedures; adoption of ASEAN standards; and, formulation of the ASEAN framework on access to genetic resources. ARCBC will serve as a central focal point for the elaboration and coordination of a regional permanent network through National Biodiversity Reference Units (NBRUs) located in the ASEAN countries, maintaining and intensifying the links established by the project beyond the duration of European Union (EU) support.
>
> A regional approach will be promoted by the ARCBC in addressing the major problems of loss of forest cover and other natural habitats; weak protection of biodiversity resources; overuse of biodiversity resources; pollution of natural habitats; loss of endemic species; and loss of agricultural biodiversity through abandoning of old crop varieties. A regional approach is needed simply because species and ecosystem distributions do not follow national boundaries.
>
> *Source:*   ASEAN Secretariat Jakarta

> **Box 17.2 ASEAN Combat Against Forest Fires in Indonesia**
>
> Health threatening smog caused by forest fires in Kalimantan, Indonesia, spread through Southeast Asia during 1997. Estimates have suggested that the smog caused US$4.4 billion in damage and wiped out 5 million hectares of forest, agricultural land, and bush land, equivalent in size to Costa Rica. About 80 per cent of the fires that burnt agricultural land, grassland and rainforest in Indonesia since 1997 are blamed on large owners who illegally used fire to clear unwanted vegetation, however these allegations have not been proven.
>
> A meeting of the subregional Ministers **Transboundary Haze Pollution**, was convened as the main intergovernmental body to plan, organize meetings, establish funds and coordinate regional activities. Since then, the Ministers have set up a special fund to finance the fight against the forest fires. Support has been solicited from donors like the United States, United Nations Environment Programme (UNEP), and other nations and organizations. At the 4th meeting of the Ministers held in Singapore on June 19, 1998, Indonesia proposed an aerial surveillance plan, which would enable fire-fighters to spot fires early and to take prompt attention to check their spread. Along with the Indonesian proposal came a Malaysian offer of providing expertise and training and a pledge from Singapore to provide the necessary communication equipment for the immediate transmission of information to agencies on the ground. Communications between the relevant agencies at the provincial and district levels in Sumatra are also given high considerations. In July 1999, Southeast Asian environment ministers came up with a more comprehensive plan to stop forest fires through a coordinated fire – prevention campaign. The plan piloted in Indonesia included education, fire prevention, fire-fighting, an surveillance techniques. Southeast Asian countries have agreed to strictly enforce and develop laws against open burning on their land. Malaysia and Brunei Darussalam, as a result, adopted a legal sanction called the "presumptive clauses" which presumes or apprehends a landowner to be responsible when a fire breaks out in his or her property.
>
> Combating forest fires in Indonesia is a continuous action locally and regionally, which necessitates further cooperation, pooling of more financial resources, provision of adequate communication and satellite technologies, and training. The efforts of the subregion's governments to avert a recurrence of disaster-level blazes also calls for change in forest policy and greater public cooperation in fire prevention and fire-fighting campaigns.
>
> *Source:*   Jakarta Post, June 20, 1998; Business World, July 8, 1998; and http://www.idn.org/news/0698/df062398-5.htm

the adoption and implementation of the ASEAN Protocol on Access to Genetic Resources and Sustainable Management of Water Resources.

The latter is mostly focusing on developing cooperative approaches to key international environmental agreements and Agenda 21. Publication of Southeast Asia's State of Environment Report is also taking place under the education and awareness raising initiative.

2. *Other Subregional Cooperation Programmes and Projects*

(a)   The Mekong River Commission

The Mekong River Commission is an intergovernmental cooperation initiative responsible for coordination in the use and development of water resources in the Lower Mekong Basin. The Commission's present membership includes

CHAPTER SEVENTEEN

Cambodia, Lao People's Democratic Republic, Thailand, and Viet Nam. Programme activities encompass both policy and planning and technical support, including data collection and resource development.

(b) Greater Mekong Programme

The Greater Mekong Programme was established for the promotion of economic cooperation among the countries of the Greater Mekong subregion, including Cambodia, People's Republic of China, Lao People's Democratic Republic, Myanmar, Thailand, and Viet Nam. It is an ADB funded initiative focused on: harmonization of environmental standards and legislation; capacity building for environmental management planning and assessment; and, technology transfer.

(c) Subregional Projects

A number of independent subregional environmental projects are also underway. These include, for example, conservation of turtle habitats by the Philippines and Malaysia under the Turtle Islands Heritage Protected Heritage programme; cross-border cooperation in the management of a national park between Kalimantan in Indonesia, and Sarawak in Malaysia; and, the potential future development of a 'forest ecoregion' bordering Viet Nam, Cambodia and Lao People's Democratic Republic.

3. *International Programmes and Projects*

Bilateral and multilateral projects and programmes are also contributing to the subregion's sustainable development objectives. A significant number of projects are either planned or underway targeting the region's land and marine resources. Projects include:

- Coastal Resources Management Project (USAID);
- Red Tides and Living Coastal Resources Management projects (AIDAB);
- Assessment of Marine Pollution by Heavy Metals (CIDA);
- Metropolitan Environment Programme (World Bank);
- Regional Study on Global Environmental Issues (ADB);
- Promotion of Market-Based Instruments for Environmental Management (ADB);
- Forest Fires Monitoring and Warning Systems-various (UNEP, GEF, DANIDA, USAID);
- Coastal and Marine Environment Management Information Systems (UNEP, ADB);
- Protection of the Greater Mekong Subregion-various initiatives (ADB, USAID); and,
- Regional Centre for Biodiversity Conservation (EU) (see Box 17.1).

SUBREGIONAL OUTLOOK

Southeast Asian countries have a very strong history of environmental cooperation. A major challenge for the subregion is therefore the continuation and development of cooperation and coordination initiatives, whilst balancing economic development perspective with long-term environmental goals and responsibilities. Economic growth has had its rewards in rising incomes, literacy and life expectancy, but at the same time it has greatly diminished the region's natural capital and destroyed valuable habitats and biological diversity. Future development will need to be guided through the reform of economic policies in such a way that they conserve natural capital, and maintain and protect conservation areas; joint subregional management of shared stocks will also be a priority. For example, concrete activities will be required to reverse the unfavourable trends in the erosion of forest resources, and cooperation in the management of coastal areas and marine fisheries will also be essential to prevent threats to food security and livelihoods within some Southeast Asian countries. The coordination and sustainable management of tourism is also a high priority. The World Bank estimates the cost of Southeast Asia's environmental degradation to be about five per cent of its GDP. It also estimates the cost of pollution abatement per life saved at US$1 000 or less (World Bank 1998). In light of this, a number of countries in the region are at stages in the development of their economies where the introduction of technology and capacity for pollution prevention and abatement in parallel to command and control will be key. Furthermore, major steps to significantly increase the efficiency of resource consumption per unit of output, the region's economic development will be short-lived, and at the increasing cost of the environment.

Chapter Eighteen

The Oceanic Realm of South Pacific with its rich biodiversity needs serious conservation efforts.

18

South Pacific

INTRODUCTION
SHARED ENVIRONMENTAL PROBLEMS
CAUSES
POLICY RESPONSE
SUBREGIONAL OUTLOOK

CHAPTER EIGHTEEN

INTRODUCTION

The South Pacific subregion comprises a total of twenty one countries and territories, which in this chapter, are presented under the following four groupings: Australia and New Zealand; Melanesian Countries (Papau New Guinea, Solomon Islands, New Caledonia, Vanuatu, and Fiji); Mid-sized open islands of Polynesia and Micronesia (Tonga, Samoa, American Samoa, French Polynesia, Palau, Guam, and the Northern Mariana Islands); and the Small island micro-states (Cook Islands, Kiribati, Tuvalu, Federated States of Micronesia, Marshall Islands, Niue, and Nauru). The subregion is quite diverse, politically, economically, geographically, and ethnically. Australia and New Zealand tend to face issues of marine pollution, deforestation, and desertification, while the small island developing states face common environmental challenges in the threat of sea level fluctuation, isolation, exposure to disaster, and shortage of resources.

The South Pacific has the lowest population of all the Asian and Pacific subregions (just over 30 million) and while it has a high rate of population growth, in a number of cases this has been absorbed through migration to the regions larger peripheral islands. The subregion has some of the lowest per capita arable land resources, together with the highest per capita marine resources. The South Pacific also has some of the highest marine diversity in the world – up to 3 000 species may be found on a single reef (SPREP 1993). The many thousand islands are surrounded by a rich complex of coastal ecosystems, including mangroves (around 10 per cent of the world's total habitat), seagrass beds, and estuarine lagoons. Terrestrial diversity is shaped by the endemic island ecology, coupled with the importation and invasion of foreign species. For example, over 75 per cent of the biodiversity of New Caledonia is endemic, including several plant species limited to one small area of one mountain. This chapter highlights the major environmental issues in the subregion with particular emphasis on shared concerns and challenges. Table 18.1 provides a summary of the region's key environmental issues and their causes.

SHARED ENVIRONMENTAL PROBLEMS

The island nations of the South Pacific subregion control Exclusive Economic Zones (EEZ) of 200 nautical miles from their coasts. This represents a significant portion of the high seas fisheries and seabed mineral wealth of the global ocean hemisphere. Shared environmental problems are dominated by fluctuations in sea level, increasing vulnerability to natural disasters, decline in marine resources and erosion of the coastal zone. Other common problems of the subregion include those associated with the degradation and depletion of land and water resources, loss of biodiversity and deforestation. In prioritizing the region's shared environmental concerns, the most important are:

- **Sea level and temperature fluctuations**
Temperature patterns of the marine

Table 18.1 Key Environmental Issues and Causes in the South Pacific Region

Country	Key Issues	Key Causes
Australia and New Zealand	Soil erosion; soil salinity; degradation of in-land and marine waters; depletion of wetlands; desertification; depletion of fisheries; loss of biodiversity.	Overgrazing; poor farming practices; land clearance and deforestation; invasion of exotic species; overfishing; over development of the coastal zone; shipping pollution.
Melanesian Countries (Papau New Guinea, Solomon Islands, New Caledonia, Vanuatu, and Fiji)	Deforestation; land degradation/soil erosion; loss of biodiversity; water degradation and limited access to potable water; local depletion of coastal fisheries.	Commercial logging; land clearance; mining; climate change; population growth and deficiencies in urban and rural infrastructure; over fishing.
Mid-sized open islands of Polynesia and Micronesia (Tonga, Samoa, American Samoa, French Polynesia, Palau, Guam, and the Northern Mariana Islands)	Deforestation; soil erosion; loss of biodiversity; local depletion of coastal fisheries; degradation of in-land and marine waters.	Expansion of commercial agriculture and agro pollution of run-off; population growth and expansion into marginal lands; indiscriminate collection of coral and shells; invasion of exotic species; overfishing; hunting, particularly of native sea turtles.
Small island micro-states (Cook Islands, Kiribati, Tuvalu, Federated States of Micronesia, Marshall Islands, Niue, and Nauru	Vulnerability to natural disasters; water degradation and limited access to potable water; coastal erosion.	Climate change; groundwater salinization; deficiencies in urban and rural infrastructure.

Source: Complied from WRI 1999; UNEP 1999; Counterpart International 1997; ADB 1997 and United Nations 1999

environment regulate the distribution of plants and animals, and slight, short or long-term variations can have dramatic impacts. Problems have been witnessed with coral bleaching (see Chapter 16, Box 16.2), and species migration from traditional fishing areas. Cities, villages, agriculture, and infrastructure are all concentrated in the regions coastal zones, which are especially vulnerable to sea level rise. Determining the severity of this problem is especially complicated by natural and intricate sea level shifts associated with recurring ice ages, however there are many potential impacts which will require adaptive policy responses (see Table 18.2).

- **Water scarcity and degradation** Limited and decreasing supplies of potable waters resources are threatening many of the regions island populations (see Figure 18.1). At the same time, demand is increasing through tourism and agricultural related developments, and drought is a common problem. This is often compounded by pollution of groundwater, extensive leakage and clandestine connections to the existing system. Changing weather patterns are also a factor, with one of the subregions worst droughts on record recorded as a result of the reduced rainfall during the 1997/1998 El Niño event.
- **Land erosion and degradation** Widespread overgrazing in the larger countries, and water and wind erosion in the smaller islands, are common causes of land degradation. The increasing pressures from globalization and the use of chemicals in commercial agriculture are also degrading land resources, in addition to entering the terrestrial water and marine environments. Pressures on land also stem from the rise in urbanization and increasing developments in the coastal regions.
- **Deforestation and biodiversity loss** The biological diversity of the South Pacific subregion is some of the most critically threatened in the world (Given 1992). As the economies of many of the countries remain subsistence based, this is more than an environmental threat. Biodiversity is threatened by large scale deforestation and the pressures on marginal lands imposed through increasing population and shifting cultivation. Land-based sources of marine pollution are also thought to be one of the four biggest threats to marine diversity, together with the introduction of invasive species, and habitat destruction, including dynamiting.

Table 18.2 *Indicative List of Potential Impacts of Climate Change and Sea-level Rise Requiring Adaptive Responses in the South Pacific Subregion*

Coastal zone	• Inundation and flooding of low-lying areas • Coastal erosion • Possible increase in cyclone-related effects • Changes in sediment production due to changes in coral reef systems
Water resources	• Changes in freshwater lenses and other groundwater resources • Salt intrusion of groundwater resources • Changes in surface-water resources • Changes in surface run-off, flooding and erosion
Agriculture	• Changes in commercial crop yields • Changes in subsistence crop yields • Changes in plant pest populations • Possible changes associated with changes in ENSO, drought and cyclone patterns • Changes in soil quality
Fisheries	• Changes in distribution and abundance of offshore fish species • Changes in productivity of inshore fisheries • Changes in fish breeding sites
Ecosystems	• Coral bleaching and coral degradation (also possible increased upward coral growth) • Changes in mangrove health and distribution • Degradation of sea grass meadows • Changes in forest ecosystems • Changes in wetland systems
Human Health	• Increased incidence of vector-borne disease such as malaria and dengue fever • Increased heat stress and heat-related illnesses • Indirect effects on nutrition and well-being secondary to effects in other sectors, such as agriculture and water resources • Deaths, injuries and disease outbreaks related to possible increases in extreme events such as cyclones, floods and droughts

Source: SPREP 1999c in UNEP 1999

- **Degradation of the marine environment** Increasing economic development activities over the last ten years are placing imminent threats on the region's marine environment. Negative impacts have been observed to fisheries, mangrove forests, sea grasses, coral reefs, and surface conditions, including red (toxic) phytoplankton blooms; together with oil pollution, and floating and suspended solid wastes. Increasing levels of UVB penetration are also impacting on fish eggs and plankton

CHAPTER EIGHTEEN

Figure 18.1 Percentage of Population in Selected Pacific Island Countries with Access to Safe Water

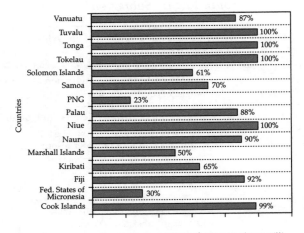

Source: UNDP 1996 in UNEP 1999

species which occupy the surface waters of the South Pacific.

- **Increasing vulnerability to natural disasters**
Physical natural disasters including volcanic eruptions, earthquakes, tsunamis, cyclones and droughts, threaten much of the South Pacific region, and the resilience of the fragile island ecosystems to disasters is increasingly impaired. The region's tropical and sub-tropical climate is punctuated by climatic extremes which have far reaching impacts on land use and serious environmental consequences. Melanesia, Guam, and the Mariana Islands are part of the 'Pacific Rim of Fire' (a region of severe seismic activity) for example. The impact of these disasters can be highly localized, but severe. In many other countries of the subregion, severe tropical storms and cyclones are accompanied by massive rainfall and the low pressure may cause the sea to rise as much as 2 metres, itself stimulating wide spread flooding, coastal inundation, land erosion, destruction of housing and gardens, loss of vegetation, and salinization of water supplies (see Figure 18.2).

CAUSES

Lifestyle changes dominated by the movement away from subsistence and towards consumer lifestyles have characterized much of the increasing pressures on the South Pacific environment. Climate change, population growth, and coastal urbanization are presenting an increasing strain on the region's fragile resources and deficient infrastructure. The average population growth across the subregion is estimated at 1.2 per cent per annum (ESCAP 2000), and indications are that GDP per capita has tended to decline, together with the standard of living (ESCAP 1999). However, while conventional economic and social indicators demonstrate that a significant percentage of the region's population experience a high poverty level, many still enjoy a degree of subsistence affluence which does not form part of the conventional calculation. Nevertheless, poverty is a serious emerging issue, particularly in relation to the growing levels of urban inward migration. Major socio-economic indicators of the subregion are provided in Table 18.3.

Figure 18.2 Estimated Levels of Vulnerability to Specific Natural Hazards in Selected South Pacific Islands

Country	Cyclone	Coastal flood	River flood	Drought	Earthquake	Landslide	Tsunami	Volcano
Fiji	xxx	xxx	xxx	xx	x	xxx	x	
Fed. States of Micronesia	xx	xxx	x	xxx	x	x	xxx	
Kiribati	x	xxx		xxx	x	x	xxx	
Marshall Islands	xx	xxx		xxx	x	x	xx	
Palau	xx	xx		xx	x	x	xx	
Solomon Islands	xxx	xxx	xxx	x	xxx	xxx	xxx	xxx
Tonga	xxx	xxx	xx	xxx	xxx	x	xxx	xxx
Tuvalu	x	xxx		xx	x	x	xxx	
Vanuatu	xxx	xxx	xxx	x	xxx	xxx	xxx	xxx

Source: UNDHA 1996 in UNEP 1999
  x Low   xx Medium   xxx High

Table 18.3 Major Socio-Economic Indicators for South Pacific Countries

Countries	Population (2000) (thousands)	Annual growth rate of population (%)	Population density (person per km²)	Urban population as % of total	Annual growth rate of urban population (%)	Total GNP (US$ millions) 1998	Per capita GNP (US$) 1998
American Samoa	64	2.9	315	48	4.6	–	–
Australia	19	1.0	3	85	1.0	387 006	20 640
Cook Islands	19	-0.5	–	59	0.6	–	–
Fiji	825	1.6	43	49	2.9	1 748	2 210
French Polynesia	23	2.33	62	53	1.3	–	–
Guam	148	1.0	270	39	2.5	–	–
Kiribati	91	2.5	118	37	2.2	101	1 170
Marshall Islands	62	2.7	342	66	1.7	96	1 540
Micronesia (Federated States of)	118	1.9	162	27	0.4	204	1 800
Nauru	12	1.8	–	100	1.8	–	–
New Caledonia	218	2.5	11	77	3.4	–	–
New Zealand	3 869	0.5	14	86	1.1	55 356	14 600
Niue	2	-3.1	–	35	1.2	–	–
Northern Mariana Islands	77	5.5	143	90	5.6	–	–
Palau	19	2.2	40	71	2.9	–	–
Papua New Guinea	4 807	2.2	10	17	3.9	4 104	890
Samoa	169	0.6	60	22	2.4	181	1 070
Solomon Islands	448	3.4	15	20	5.8	315	760
Tonga	100	0.6	137	32	0.8	173	1 750
Tuvalu	10	0.9	–	42	4.8	–	–
Vanuatu	200	3.0	15	21	4.3	231	1 260

Source: 1. World Bank 1999 and 2000
2. ESCAP 2000
3. United Nations 1998

POLICY RESPONSE

A. National Initiatives

Policies and initiatives in the region stem from a mixture of historic social and cultural values coupled with, in many cases, a recent colonial administration. The South Pacific subregion therefore has a wide range of systems of government. Nonetheless, strong national identities exist, together with a history of cooperation on issues of regional concern, and a well-developed legal and institutional framework through which to address them. However, it is only relatively recently that policies have taken account of strategic environmental dimensions, and a major challenge will be for the subregion to meet its emerging economic development needs, while sustaining its subsistence bases and its values of cultural and social cohesion. A number of common themes are observed in relation to present sustainable development limitations.

- **Policy** While capacity for the implementation of environmental policy is high in the developed countries of New Zealand and Australia, one of the most significant causes of unsustainable behavior in the smaller countries of the subregion is the lack of capacity for adequate environmental planning. In the independent South Pacific islands, government environmental units and planning units lack both human and financial resources. In many countries, environmental units have no direct input into the main decision making processes, and in most other countries, environment is marginalized as a department within a multi-function ministry.
- **Legal** While there is a strong legal framework (much of which stems from traditional community structures), there is a lack of enforcement or implementation of many policies or legislation, together with a growing weakness in the protection of the subregion's indigenous property rights. A number of regulatory mechanisms are also becoming out-dated and have failed to respond to changing paradigms of development.

- **Industrial** Although the subregion is perceived as having a small industrial base, growing industrialization is particularly acute in terms of the micro- and small-scale industries. Greater effort is required promote research and development both within and across these industries and academia, of environmentally sound techniques and technologies. Focus is also required on the exchange and use of data and information relating to ocean and fishery sciences. The major agricultural, fishery, forestry, and tourist industries also need to be sustainably managed and more closely integrated into the planning framework.
- **NGO** Grassroots environmental NGOs are a recent development in many of the South Pacific islands, although they have been active in New Zealand and Australia since the 1920's. Recent NGO activity has however witnessed some surprisingly effective success stories. A hard-hitting NGO media campaign was afforded the main responsibility in the rapid and powerful action of governments in the region to curtail the use of drift netting, for example. The full potential of many community-based and indigenous NGOs is still not fully recognized, often due to a lacking capacity in project management and implementation, in addition to weak accountability and monitoring.

B. Subregional Cooperation

Coordination of policy responses facilitates a necessary strategic approach to sustainable development and environmental problem solving in the region. The South Pacific is one of the two subregions (together with Southeast Asia) in Asia and the Pacific to have ratified subregional conventions on environmental protection (this is further discussed in Part V). Opportunities for cooperation in the subregion are promoted through a range of policies and programmes, in addition to national and international projects, which are outlined under the following two headings:

- the Council for Regional Organizations of the Pacific (CROP); and
- Other Cooperation Programmes in the Subregion.

1. *The Council for Regional Organizations of the Pacific (CROP)*

CROP is the formal coordination mechanism for subregional organizations in the South Pacific. Formerly the South Pacific Organizations Coordinating Committee, it aims to ensure complementary mandates, common goals, and synergy of regional initiatives. Member organizations include (UNEP 1999):

- **South Pacific Regional Environment Programme (SPREP)** SPREP was established in 1982 by the government and administrations of the South Pacific countries and four other countries with a direct interest in the region. It is the major inter-governmental organization charged with promoting regional cooperation, supporting protection and improvement of the South Pacific environment and ensuring its sustainable development. With the help of ESCAP and others under the Barbados Programme of Action, SPREP has assisted small island developing states in capacity building through development of National Environmental Management Strategies and legislation on environment. In the last ten years SPREP initiatives have focused on biodiversity conservation (see Box 18.1), waste management, climate change, impact assessment, and environmental assessment and awareness raising.
- **Forum Secretariat** This was established in 1971 from the independent and self-governing countries of the South Pacific. Its fifteen member countries are Australia, Cook Islands, Fiji, Kiribati, Marshal Islands, Federated Stated of Micronesia (FSM), Nauru, New Zealand, Niue, Papua New Guinea, Samoa, Solomon Islands, Tonga, Tuvalu, and Vanuatu. The Forum is responsible for facilitating maintaining and developing cooperation and consultation across it's membership on issues such as trade, economic development, transport, and energy.
- **Forum Fisheries Agency** FFA was established in 1979 with a broad mission to enable the region to obtain maximum sustained benefit from the conservation and sustainable use of its fisheries resources, with a particular emphasis on tuna.
- **South Pacific Applied Geoscience Commission** It was originated in 1972 and its member countries are Australia, Cook Islands, Fiji, Guam, Kiribati, Marshall Islands, FSM, New Zealand, Papua New Guinea, Samoa, Solomon Islands, Tonga, Tuvalu and Vanuatu. SOPAC is involved in the provision of advice on the environmental effects of coastal zone developments, water and sanitation, pollution and health issues.
- **Secretariat of the Pacific Community** It was first established in 1947, and is mandated to

SOUTH PACIFIC

> **Box 18.1 The Vatthe Conservation Area – Big Bay, Espirito Santo, Vanuatu**
>
> The Vatthe Conservation Area is located at the southern end of Big Bay on the island of Espirito Santo in Vanuatu. It is 2 276 hectares of lowland alluvial rainforest owned by the villages of Sara and Matantas. The Vatthe Conservation Area Project was initiated in 1994. The basic idea for the project began in 1993 during a biodiversity survey of the area by the Vanuatu Environment Unit in collaboration with the Royal Forest and Bird Conservation Society of New Zealand. During the study there was some discussion with the community on the idea of setting up a national park.
>
> The project was developed by the Environment Unit and ran into some constraints in its early stages. It has been argued that involvement and therefore ownership was lacking with the local community, and a major, long-term land dispute between the two villages was initially seen as a serious problem for the project. In 1995, as the community became more involved in the work plan and budget, and SPREP arranged a trip to Fiji for the landowners where two conflicting groups were able to share new experiences and view the progress of other conservation areas together. The project planning sessions and discussions acted as a stimulus for peace between the two villagers and in 1995 a traditional ceremony united them to a common cause. By 1996, the communities had formed a Community Management Committee that began to assume control over work plans, activities and budget. Two conservation Support Officers were appointed by the Committee to work with the project manager on a part-time basis. By 1998, The community had built a small hotel with six bungalows and restaurant, installed a water supply system with fibreglass tanks and a roof catchment system, created three walking tours, a coconut crab hunt, a garden tour and custom dance. The project had trained local guides, established a handicraft centre, created two sub-committees (eco-tourism and conservation), and appointed a full time eco-tourism manager. The Vanuatu Energy Unit contributed solar powered lighting and the Department of Geology and Mines provided a new water well. A radio-telephone link was established with the main urban area to facilitate reservations. A forest fruit project and Alley cropping garden project has been established as a further source of income for the villages.
>
> *Sources:* Nari, R. 1997 and Reti, I. 1998

provide sustainable development assistance in the subregion. The Pacific Island Forestry and the Trees Support Programme awards provides a good example of an innovative SPC scheme aimed at promoting the adoption of sustainable forestry techniques. Other activities include management advice and applied research in the coastal/national fisheries sector and research on oceanic (mainly tuna) fisheries.

- **Tourism Council of the South Pacific** The role of TCSP is to market and promote tourism to the region, and to help the tourist sector enhance the quality of its product through a variety of programmes on training, tourism awareness, and preservation of the environment.
- **University of the South Pacific** The University was established in Fiji in 1969, and includes regionally focused research under its four schools of agriculture, humanities, pure and applied social and economic development.
- **Pacific Island Development Programme** It has 22 members including Pacific island developing countries and territories, and draws academic resources both subregionally and from international organizations to plan and conduct projects across a range of development issues.

2. *Other Cooperation Programmes in the Subregion*

Fisheries departments in the Cook Islands, Tonga Vanuatu, the Solomon Islands, Papau New Guinea, Fiji, and Samoa (see Box 18.2) are at various stages of developing partnerships with local communities to sustainably manage their coastal resources, often with the support of external funding. The United Nations is supporting climate change programmes in Niue and Papau New Guinea, and integrated coastal zone management programmes are being implemented in Fiji, Marshall Islands, Samoa, and Tuvalu with support from Japan. Much attention has also be paid to the analysis of sea level fluctuations and their impacts to the subregion by the governments of USA, Australia, and Japan. Numerous projects are also underway to address the issues of potable water supply and sanitation, primarily with support through ADB, USAID, the European Union, and AusAID. Total financial injection to the sector amounts to around US$36 million. National programmes in New Zealand, Tonga, the Solomon Islands, and Vanuatu include the establishment of species and habitat conservation areas, and the promotion of eco-tourism activities.

CHAPTER EIGHTEEN

Box 18.2 The Samoa Fisheries Extension and Training Project

Coastal fisheries are an important source of protein for the people of the South Pacific islands. The advent of commercial fishing has, in some areas, resulted in over harvesting of fish and damage to coral reef ecosystems. Subsistence fisheries continue to operate, and expand, in parallel with efforts at commercializing coastal fishing efforts. Because of the diffuse, and multi-species nature of the fisheries, the small size of the fisheries departments, and the distances between islands, practical management of the subsistence fishery is beyond the capability of national fisheries officers. There is, therefore, little information available to document the state of the fisheries other than community perceptions of a decreasing catch.

If the local communities willingly assume responsibility and control of their own fisheries resources, with guidance of experienced fishery biologists, some form of management may be possible. AusAID funded the Samoa Fisheries Extension and Training Project to work out how Government/Community partnerships might be developed. The project is one of the most innovative and successful examples of community/government partnerships for coastal fisheries management in the South Pacific region.

The Samoa Fisheries Extension and Training Project began in 1995 by producing a series of information sheets (written in Samoan and English) intended for the Village Councils. Project personnel met with Village Councils and explained the need for local management and suggested the national government would be willing to assist the villages in producing local management plans. By 1998, 54 villages had joined the plan. Many had created and approved fisheries management plans. The plans include bans on use of explosives and chemicals, a reserve area, recognition of size limits and other restrictions. The plans also provide for strict enforcement of the regulations.

The extension process takes about three months. The socio-biologist on the team believes this is the minimum time required for the people to take ownership of the project. The key to a successful village fisheries management plan was having the villagers develop it because they understood the issues and wished to protect their own fishery resources. If they had other motivations (expecting foreign aid or free goods and services) the project would not succeed. In helping villages to make the process their own, the extension officers do not provide answers or give instructions. They ask questions – *What is the condition of your fisheries?* If the villagers report various problems the officer asks, *What do you think is causing the problem?* and later, after this is discussed, *What do you think might be done to solve the problem?*

The project found that most village councils knew the answers to these questions, sometimes better than the fishery agents did. By asking questions, everyone learned and the villagers gained a feeling of ownership of the programme.

Source: Chesher 1998a.

SUBREGIONAL OUTLOOK

Like the countries of Southeast Asia, South Pacific countries have a strong history of environmental cooperation, and a major challenge for the subregion is therefore the continuation and development of cooperation and coordination initiatives and to balance its economic development perspective with its long-term environmental goals and responsibilities. The sectoral and hierarchical structure of most South Pacific Islands governments (and the split between traditional governance verses colonial governance) has left its mark in respect to the existing vertical and horizontal communication capacity. Communications will need to be significantly strengthened in the process of mainstreaming environment as an integral component of the planning system. Among the major accomplishments of national governments in the subregion are steps in decentralization of environmental management responsibilities, and the development of partnerships with NGO communities. Such developments will need to be built upon, and links and partnerships will also need to be established and strengthened between all levels of society, industry, and the academic community. Finally, the growing pressures of climate change, economic reform, and access to genetic resources, are stimulating the rapid development of adaptive strategies, and these will require continued subregional cooperation and the timely implementation of concrete actions.

Chapter Nineteen

Pollution and acid rain is a major problem in Northeast Asia.

19

Northeast Asia

**INTRODUCTION
SHARED ENVIRONMENTAL PROBLEMS
CAUSES
POLICY RESPONSE
SUBREGIONAL OUTLOOK**

CHAPTER NINETEEN

INTRODUCTION

The Northeast Asian subregion comprises six countries: People's Republic of China, Japan, Russian Federation, Republic of Korea, Democratic People's Republic of Korea and Mongolia. It has the highest population of all the subregions with a total of 1.48 billion people, and a growth rate of 1.2 per cent. The population is however diversely distributed, and is dominated by People's Republic of China, the largest country in the world (with 1.27 billion people). The highest levels of GDP are found in Japan, which also leads the subregion in respect to environmental performance in many areas. For example, in terms of overall GDP per kg of energy consumption, Japan's efficiency is up to ten times that of People's Republic of China and the Russian Federation (see Figure 19.1).

The subregion still maintains a rich biodiversity in its natural forests, grasslands, mountains, deserts and wetlands, although biodiversity loss remains pervasive. The Russian Federation and People's Republic of China in particular have large areas of natural forests. This chapter highlights the major environmental issues in the subregion with emphasis on shared concerns and challenges. Table 19.1 provides a summary of the key environmental issues and their causes in Northeast Asia.

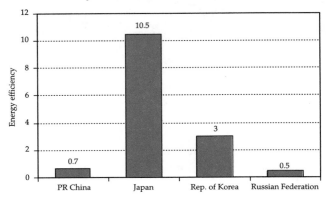

Figure 19.1   Energy Efficiency (1996 GDP $ per kg oil equivalent) in Northeast Asia

Source: World Bank 1999

Table 19.1   Key Environmental Issues and Causes in Northeast Asia

Countries	Key Issues	Key Causes
PR China	Acidification of inland waterways and acid deposition; degradation of water supply; loss of agricultural land; loss of biodiversity; vulnerability to natural disaster, especially drought and flooding.	Over reliance on low-grade coal; inadequate infrastructure for the management of municipal effluent; deforestation and soil erosion; poverty.
Japan	Excess volumes of industrial and municipal waste; pollution from dioxins, endocine disrupters and other industrial hazards; increasing greenhouse gas emissions; vehicle emissions; loss of biological diversity.	Unsustainable consumption patterns; lack of emission control in waste incineration and industrial processes (national and transboundary); increasing vehicle ownership; habitat destruction due to development projects and invasion of alien species.
Russian Federation	Air pollution in hot spots and major cities; pollution of inland and marine waters; deforestation; loss of biodiversity; soil erosion and contamination; radioactivity.	Inefficient heavy industry and reliance on coal for power generation; deficiencies in urban infrastructure – unmanaged industrial wastes and municipal effluents and waste; urban congestion and inefficient vehicles; unsustainable agricultural practices, and excessive chemical application; historical sites of nuclear weapons testing.
Rep. of Korea	Limited access to potable water; urban air pollution; environmental contamination; acidification of inland waterways and acid deposition.	High levels of uncontrolled atmospheric releases from industry; release of dioxins, endocrine disrupters, and other industrial hazards; transboundary air pollution from neighbouring countries.
Democratic People's Republic of Korea	Localized air pollution; water pollution and limited access to potable supplies; vulnerability to natural disaster, especially drought and flooding.	High levels of uncontrolled atmospheric releases from industry; deficiencies in urban infrastructure – industrial and municipal effluents; poverty.
Mongolia	Localized air pollution; soil erosion and desertification; loss of biodiversity; water pollution and limited access to potable supplies.	Overgrazing; deforestation; reliance on low-grade coal; promotion of rapid urbanization.

Source: Complied from WRI 1996

SHARED ENVIRONMENTAL PROBLEMS

Among the shared environmental problems of the subregion, the two key issues are environmental pollution and the depletion and degradation of natural resources. With the exception of Japan, air quality in major cities is still poor despite some improvements in recent years. Deterioration of the marine environment is visible, and the subregion's shared seas are receiving high levels of industrial and municipal effluent. Vast areas of People's Republic of China, Russian Federation and Mongolia also suffer from deforestation and desertification. The subregion's increasing vulnerability to natural disasters is compounded by environmental degradation and poverty. In prioritising shared environmental concerns in the Northeast Asia, the most important are:

- **Atmospheric pollution** With the exception of Japan (see Box 19.1), inefficient industry and energy generation practices have resulted in high levels of atmospheric pollution in many of the subregion's major cities, affecting the health of their inhabitants. The highest levels have been recorded for People's Republic of China and the Russian Federation (see Figure 19.2) although the latter has experienced reductions commonly attributed to industrial decline in the post Soviet era. While emissions are stabilizing for some of the key indicators, development predictions suggest these are likely to increase as a result of changes in the economic profile of the subregion, particularly with regards to People's Republic of China. Emissions associated with vehicle use (mainly NO_2) remain high and are also predicted to increase in Japan. Transboundary deposition from acid precipitation is impacting upon land and marine environments across the subregion.

Figure 19.2 CO_2 Emissions of Northeast Asian Countries for the year 1996

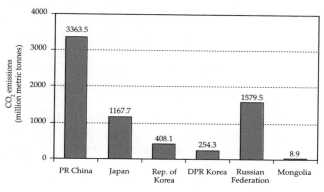

Source: World Bank 1999

Box 19.1 Air Quality Improvement in Japan

Japan experienced a period of rapid industrialization and economic growth in the 1950s and 1960s. However, little attention was paid to the accompanying air pollution, which resulted in serious health consequences by 1970. The amount of total suspended particulates (SPM) and sulphur dioxide increased enormously over this period. For example, by 1960, air pollution in the Yokkaichi City (which contains one of Japan's largest petrochemical facilities) was causing local concern, and by 1963 one-hour average sulphur dioxide levels exceeded 2 800 micrograms per m³, far in excess of WHO's suggested maximum of 350 micrograms per m³. In 1967, local residents successfully sued six companies, claiming medical costs and compensation for lost income. In total, 7 per cent of the population of the district were certified to have been medically affected by ambient air pollution.

Realizing the gravity of the situation, the Japanese Government adopted several measures which in order to introduce effective air pollution control general environmental quality standards were set for the protection of human health. Strict enforcement of the targets led to the control of smoke discharged by individual factories and other facilities, and automobile exhaust. The emission standards applied to each motor vehicle were gradually tightened and were ultimately made the most stringent in the world. Following these regulations, investments were encouraged to develop desulphurization systems, denitrification processes, and other air pollution control techniques. Finally, a Health Damage Compensation Law was introduced in 1978 to provide for seven types of compensation benefits: (a) medical care benefits and expenses; (b) disability compensation; (c) survivors' compensation; (d) lump sum survivors' compensation; (e) child compensation allowance; (f) medical care allowance; and (g) funeral expenses. The expenditure for compensation to persons suffering from pollution-related health problems was to be defrayed by those who were responsible for discharging the pollutants, as the system was to be operated on the basis of civil liability. The 1987 amendment to the law also provided for implementing a new programme intended to prevent health damage caused by air pollution (called the "Health Damage Prevention Programme"). Implementation of these measures enabled Japan to attain demonstrable success in reducing air pollution.

Source:
1. World Bank 1993
2. Environment Agency 1997
3. Environment Agency 1989

CHAPTER NINETEEN

- **Degradation of Water Quality** Issues of water scarcity and degradation vary throughout the subregion, though overall, inland water quality is observed average in a world wide context. Water quality has nevertheless been seriously degraded in the Russian Federation, whilst some countries of the region have seen an improvement in industrial pollution control (for example, heavy metal contamination in Japan), most still suffer from high levels of organic pollution, particularly from municipal sewage.
- **Degradation of the marine environment** The marine environment of Northeast Asia is suffering from increasing pollution and depletion of marine resources, for example by 1998, the fishery stocks of the northwest Pacific were fully exploited (WRI 1999). In addition, large-scale land reclamation and industrial developments have encroached into coastal wetlands. High levels of nutrients, trace metals, and organochlorines, introduced by rivers are resulting in coastal eutrophication and in many areas a high incidence of red tides, with pronounced negative effects to fisheries, recreation and health. For example, in Republic of Korea, 126 incidences of red tides were reported in 1996 alone, with losses to aquaculture estimated at US$10 million.
- **Land degradation and biodiversity loss** Land degradation severely affects areas of People's Republic of China, the Russian Federation and Mongolia. In People's Republic of China, desertification impacted 262 million hectares in 1996, accounting for just over 27 per cent of the total national territory. Of this, the area of salinized land was 44 million hectares of which nearly 6 million hectares was originally arable. The problem of salinization is also acute in the Russian Federation. With regards to biodiversity, a wide variety of habitats have been degraded in the subregion. For example, the disturbance of grassland habitats through extensive agriculture, industrial and transport-related development has played a particularly major role in the depletion of biodiversity. Natural disasters, especially floods, are also degrading habitats and depleting resources within the subregion. These impacts particularly affect the poor in People's Republic of China and the Democratic People's Republic of Korea.

CAUSES

With a few exceptions inefficiencies of resource use in industry and over-reliance on coal for power generation in a number of countries, is contributing to high levels of pollution. In addition, excessive agricultural development and forestry exploitation are also compounded by large-scale practices of overgrazing and unsustainable land-use policies. The incidence of poverty, which characterizes parts of People's Republic of China and the Democratic People's Republic of Korea, is also increasing peoples vulnerability to natural disaster. The population residing in urban regions is low, at 39 per cent, but urbanization is growing at two per cent per annum, which is outstripping the assimilative capacity of urban environmental infrastructure in many areas. Major socio-economic indicators of Northeast Asia are provided in Table 19.2.

Table 19.2 Major Socio-Economic Indicators for Northeast Asian Countries

Countries	Population (2000) (thousands)	Annual growth rate in population (%)	Population density (person per km²)	Urban population as % of total	Annual growth rate of urban population (%)	Total GNP (US$ millions) 1998	Per capita GNP (US$) 1998
PR China	1.3	0.8	133	32	2.4	923 560	750
Democratic People's Rep. of Korea	24	1.3	192	60	1.8	–	–
Hong Kong, China	7	2.3	7 555	100	2.3	158 238	23 660
Japan	126.9	0.2	336	79	0.3	4 089 139	32 350
Macao, China	473	1.3	60	99	1.3	–	–
Mongolia	2 662	1.6	2	64	2.4	995	380
Republic of Korea	47 275	0.9	470	82	1.6	398 825	8 600
Russian Federation	146.9	-0.2	9	78	0.3	331 776	2 260

Source: 1. World Bank 1999 and 2000
2. ESCAP 2000
3. United Nations 1998

NORTHEAST ASIA

POLICY RESPONSE

A. National Initiatives

Transboundary environmental problems, especially acid rain and marine pollution, have pushed governments in Northeast Asia to take action and adopt policies and programmes of subregional environmental cooperation, which are being pursued through inter-governmental and non-governmental channels. However, progress in this regard has been slow. This may in part be a reflection that the concept of Northeast Asia as a regional unit, either politically, economically, or environmentally, is a relatively new one following the Cold War. Despite recent improvements, networking among scientific institutions and staff also remains at a low level. Nevertheless, as regional economies become more inter-dependent, closer environmental cooperation in Northeast Asia is likely. However, a number of common limitations are observed in relation to achieving sustainable development in the subregion:

- **Policy** Environment was initially accorded a low priority among governments of the subregion. For example, in the Russian Federation, less than one per cent of the national budget was allocated to environment in 1994-1995. In particular, many countries of the subregion have found it difficult, if not impossible, to raise the necessary investment needed to improve environmental infrastructure through fiscal means alone. However, many countries of the region are now exploring alternative ways of generating investment, for example through the introduction of private sector financing (see Box 19.2).

- **Industry** The development of environmentally sound technologies is fairly advanced within Japanese industry, which is leading the way within the subregion in terms of research and development. Technologies which are being developed are also being exchanged with countries such as People's Republic of China and the Republic of Korea, for example through the establishment of industrial technology transfer centres such as, the ADB-funded CEST (Centre for Environmentally Sustainable Technology) in People's Republic of China. Nevertheless, large scale inefficient industry remains a significant problems in much of the subregion, with industrial hot spots of polluting

---

Box 19.2 Improvement of Urban Environment in Northeast Asia through Public-Private Partnerships: Delivering Urban Environmental Services in People's Republic of China

Since the late 1970s, the pace of urbanization has kept accelerating in People's Republic of China. The number of cities soared from 223 in 1980 to 668 in 1997, while the urban population increased from 191 million (9.14 per cent of the whole population) in 1980 to 370 million (29.94 per cent of the whole population) in 1997. A further 80 million people originating from rural areas live and work in the cities all year around.

With the rapidly increasing urbanization and continuous reform, development and investment, China's urban infrastructure has also been rapidly expanding. Total fixed-asset investment (FAI) in urban infrastructure projects climbed from 2.5 billion yuan in 1990 to 12.83 billion yuan in 1997. Based upon 1990 constant prices, the average annual growth rate from 1990 to 1997 was 17.23 per cent, nearly 10 per cent higher than the GDP growth rate during the same period. In spite of the fast pace of urban infrastructure development, however, the gap between supply and demand remains substantial. Until recently, fiscal expenditure has been the primary source of investment. However, there is a growing realization that government investment alone cannot satisfy the investment needs. Therefore, market-based approaches are now being introduced for the construction and operation of urban environmental infrastructure, including Public-Private Partnerships.

Shenyang City, located in the northeast of People's Republic of China, for example, has been experimenting in reforming the construction and operation of its municipal wastewater treatment facilities. Shenyang built the Northern Sewage Treatment Plant with a handling capacity of 400 000 m^3 per day. The capital investment in the plant amounted to 600 million yuan while its recurrent annual operation/maintenance (O/M) cost exceeds 40 million yuan. In order to recover the cost, the Shenyang Municipal Government collects over 100 million yuan annually by charging the sewage fee at a rate of 0.2 yuan per m^3. The task for revenue collection was entrusted to the Shenyang Special Environmental Protection Equipment Manufacturing Co., Ltd. (SSEPEC), China's only publicly listed firm in the environmental industry, to manage and operate the Northern Wastewater Treatment Plant. A total of 40 million yuan from the revenue is used to cover the O/M costs, and the remaining 60 million yuan is used to procure SSEPEC's shares. It is projected that all the shares acquired by the Shenyang Municipal Government will be sold for reinvestment in building two new municipal wastewater treatment plants that are currently planned for the city. Another interesting example of privatization is that of the Guangdong province, which opened the handling of infrastructure such as highways, water supply, wastewater treatment and municipal solid waste to private investors in January 2000.

Source: Centre for Environmentally Sound Technology Transfer 2000

CHAPTER NINETEEN

industries presenting serious threats to human health and the environment.

- **Legal** Although recent efforts by governments in the subregion to implement environmental laws and regulations have culminated in comprehensive framework legislation, in many cases the old legislation has remained in place, and often contradicts and/or counteracts the new legislation. Further problems stem from the large geographic scale of the subregion and the significant resources necessary for the effective deployment of monitoring and enforcement functions. Nevertheless, dramatic progress is being made with fast-track response initiatives, particularly in Japan, People's Republic of China, and Republic of Korea.
- **NGO** While there are numerous NGOs operating in Northeast Asia, and public concern for the environment is growing, the efforts of the NGO community remain somewhat broad-based, disorganized and fragmentary. To operate at their full capacity, sophisticated and innovative budgetary and profile building mechanisms need to be developed for project planning and implementation. Some success is already being witnessed by subregional groups, including, for example, the Atmosphere Action Network (focusing on issues of acid rain), and the Asia-Pacific Environmental Information Network (which is assisting in the enhancement of environmental awareness among children).

B. Subregional Cooperation

Environmental cooperation in Northeast Asia is being pursued through three different channels:

- Inter-governmental fora;
- Inter-agency initiatives; and
- Inter-national and bi-lateral initiatives.

1. Inter-governmental Fora

(a) Northeast Asian Subregional Programme on Environmental Cooperation (NEASPEC)

Initiated by ESCAP in 1993 through the integration of various informal regional talks, NEASPEC has held numerous consecutive meetings throughout the subregion: Seoul, February 1993; Beijing, November 1994; Ulaanbataar, September 1996; Moscow, January 1998; and Kobe, February 1999. The focus of NEASPEC activities rest primarily on capacity building in the management and mitigation of air pollution, deforestation and desertification, and environmental monitoring. During the 1996 Ulaanbataar meeting, a framework was adopted addressing geographical coverage, objectives, participation, coordination and management, collaborating agencies, financial mechanisms, and criteria for project/activity selection. The evolution of NEASPEC activities into legally binding agreements for action, however, are still awaited.

(b) Northwest Pacific Action Plan (NOWPAP)

NOWAP plays a key role in inter-governmental cooperation for protection of the marine environment, and has been promoted under the UNEP's Regional Seas Programme. Its objectives are the control of marine pollution and the effective management of marine areas and resources. Following its inception in 1989, NOWPAP was eventually launched in 1994. Four subregional meetings have taken place in the subsequent period: Seoul, September 1994; Tokyo, November 1996; Vladivostok, April 1998; and Beijing, April 1999. NOWPAP has five operational objectives: (i) assessing the state of the regional marine environment by coordinating and integrating monitoring and information collecting systems; (ii) establishing a comprehensive database and information management system on the regional marine environment; (iii) adopting integrated region-wide approaches in establishing coastal and marine environmental programmes; (iv) developing an integrated management system for environment and resources of marine and coastal areas; and (v) preparing legal and institutional mechanisms for subregional marine environmental conservation.

2. Inter-agency Initiatives

(a) Northeast Asian Conference on Environmental Cooperation (NEAEC)

NEAEC is facilitated through a series of seminars at which information, experiences and views among officials of central environment ministries and regional administrations, experts, and the public are exchanged. The first meeting was held in October 1992, in Niigata, Japan, and there have been annual meetings since, with the latest in 1999 in Kyoto. Subsidiary initiatives have also been born through the NEAEC framework. Specifically this includes the East Asia Acid Rain Monitoring Network, or EANET, which was proposed by the Japanese Environment Agency during the 1992 meeting. Over the period 1993-1995, three consecutive EANET meetings were organized by Japan, resulting in the development of "Guidelines for Acid Rain Monitoring" and "Conceptual Design for the Establishment of Acid Rain Monitoring Network in Northeast Asia."

(b) Tripartite Environment Ministers Meeting (TEMM).

TEMM was initiated in Seoul in 1999 under the initiative of the Republic of Korea, and has held a subsequent meeting in February 2000 in Beijing. Environment Ministers from People's Republic of China, Japan and Republic of Korea participate in the forum which provides a mechanism for fostering subregional environmental cooperation for sustainable development. Projects identified for cooperation include the development of a website; holding a roundtable for subregional environmental cooperation in industry; development of a joint education curriculum on environmental issues in Northeast Asia; and development of water renovation systems and the prevention of land based marine pollution. The significance of the 2nd TEMM is that it has established a basis for substantial cooperation between the three countries and the agreement to implement concrete actions.

3. *International and Bi-lateral Initiatives*

At the bilateral level, there are several cooperation mechanisms between the economies in Northeast Asia aimed at environmental protection. These include agreements between the Republic of Korea and People's Republic of China; the Republic of Korea and Japan; the Republic of Korea and the Russian Federation; Japan and People's Republic of China; Japan and the Russian Federation; and People's Republic of China and Mongolia.

A good example of international cooperation is supplied by the Tumen Coastal and Marine Biodiversity Conservation Programme (TCMBCP) supported by the Russian Federation, People's Republic of China and Democratic People's Republic of Korea. The programme is sponsored by UNDP, and is responding to proposals to develop an international free economic zone where the borders of the three countries intersect. It is hoped that the area might in the future become comparable to Hong Kong, China, Singapore, or Rotterdam, however, the region is also environmentally sensitive and contains significant biodiversity value. The TCMBCP proposal is also supported by other countries of Northeast Asia.

Moreover, additional financing has been committed by GEF. Specific issues addressed are, inter alia, resource allocation, transboundary environmental issues, environmental assessment of development proposals, prevention of pollution of wetlands and marine waters, and development and coordination of environmental standards. The Australian Great Barrier Reef Marine Park is being used as an example of proven viability and sustainability of multiple-use development and conservation land-use approaches.

SUBREGIONAL OUTLOOK

Transboundary environmental problems especially acid rain and marine pollution in Northeast Asia have pushed governments to take action and adopt initiatives for subregional environmental cooperation. Nevertheless, cooperation in the subregion is still at the stage of formulation, and discussions on more concrete cooperative actions are ongoing. In the longer term, it is expected that the ever-increasing regional economic inter-dependency will make closer cooperation in the region inevitable. In terms of national environmental management, following its rapid development and environmental degradation, Japan is now leading the countries of the subregion. The Republic of Korea has also encountered severe environmental degradation though has responded with comprehensive legislation and environment action. Recent efforts of the Chinese Government to implement environmental laws and regulations have also culminated in significant positive action. Common environmental problems including those of transboundary nature will, however, demand stronger cooperative approaches. Cooperation therefore needs to be furthered by focusing on seeking mutual benefits through the establishment of effective and efficient mechanisms. While the question concerns which type of institutional and financial arrangement should be chosen, the problem of contributing to institutional costs has also been one of the main hindrances to date, and it will therefore be important to seek ways to reduce these costs wherever possible.

Chapter Twenty

Desertification is a cause for serious concern in Central Asia.

20

Central Asia

**INTRODUCTION
SHARED ENVIRONMENTAL PROBLEMS
CAUSES
POLICY RESPONSE
SUBREGIONAL OUTLOOK**

CHAPTER TWENTY

INTRODUCTION

Central Asia comprises the countries of Azerbaijan, Kazakhstan, Kyrgyzstan, Tajikistan, Turkmenistan, and Uzbekistan. At over four million square kilometres, the subregion covers an area larger in size than India, Pakistan, and Bangladesh combined, but has a total population of just under 64 million people, and an annual population growth rate of just under one per cent. Central Asia is an area of the world as yet little known beyond its own borders, yet the region's contribution to the resolution of many global environmental problems will be critical in coming years.

The subregion occupies the intersection of Europe and Asia. With northern taiga forests and large southern deserts, as well as the largest mountains in the former Soviet Union, Central Asia exhibits a great diversity of ecosystems. Species include more than 7 000 flowering plants, and in some areas up to twenty per cent of plant species are endemic. Similarly, the region has nearly a thousand species of vertebrates, including over one hundred species of reptile (Krever 1998). Notable species of fauna are the snow leopard, Caspian seal, bearded vulture, sturgeon, groundhog, and hyena. This chapter highlights the major environmental issues in Central Asia with particular emphasis on shared concerns and challenges. Table 20.1 provides a summary of the region's key environmental issues and their causes.

SHARED ENVIRONMENTAL PROBLEMS

Environmental hazards and disasters are a significant concern in the subregion, which includes some of the USSR's principal nuclear, chemical, and biological weapons production and testing sites, and the desiccated Aral Sea, once the planet's fourth largest lake with a thriving fishery. Most Central

Table 20.1 Key Environmental Issues and Causes in Central Asia

Country	Key Issues	Key Causes
Azerbaijan	Degradation of inland waters and marine environment; natural habitat and biodiversity loss; urban air pollution.	Unmanaged exploitation of oil and gas resources; inefficient industrial operations and deficient industrial and municipal infrastructure; illegal trade and poaching; military conflict; reliance on old and inefficient private transport; poverty and increasing pressure on resources.
Kazakhstan	Scarcity of potable water sources; degradation of in-land waters, particularly Aral sea; loss of arable land and reduction in land productivity; desertification; salinization; land contamination and radioactivity; industrial pollution 'hot spots', particularly mercury and petroleum; contamination of groundwater.	Large scale weapons testing; inappropriate management of nuclear power plant; over exploitive and extensive mining activities; relics of large-scale inefficient industry; poverty and increasing pressure on resources.
Kyrgyzstan	High incidence of water-borne disease; land degradation; salinization and loss of productivity.	Deficiencies in municipal infrastructure; inappropriate agricultural practices; poverty and increasing pressure on resources.
Tajikistan	Scarcity of potable water sources; land degradation and reduction in productivity; salinization.	Deficiencies in municipal infrastructure; inappropriate agricultural practices, chemical application and extensive irrigation systems; mono-culture; civil war; poverty and increasing pressure on resources.
Turkmenistan	Land degradation; salinization; desertification; contamination of surface and groundwater, particularly Caspian and Aral Sea basins; scarcity of potable water sources.	Inappropriate agricultural practices, chemical application and extensive irrigation systems; mono-culture; poverty and increasing pressure on resources.
Uzbekistan	Scarcity of potable water sources; degradation of in-land waters, particularly Aral sea basin; desertification; salinization; habitat loss; land contamination.	Inappropriate agricultural practices, chemical application and extensive irrigation systems; mono-culture; inefficient industrial operations and deficient industrial and municipal infrastructure; poverty and increasing pressure on resources.

Source: WRI 1996

Asian states attained their independence with plans for environmental reforms high on their agendas, and with their citizenry highly supportive of such reforms. Yet both society and the political establishment did not anticipate the subsequent sharp economic decline which accompanied reform and which resulted in a decline in investment in urban and rural infrastructure, welfare systems, public education, and medical services. As a result, despite the wave of environmental interest and activism that initially emerged, environmental management reforms and initiatives have struggled to win support in recent years. Undoubtedly, one of the regions' major environment concerns in the Aral Sea (see Box 1.1). In prioritizing the region's shared environmental concerns, the most important are:

- **Water Degradation** Central Asia has a substantial amount of water, however, scarcity problems stem from poor management distribution and pollution (particularly during the Soviet era). For example, the widespread adoption of unlined irrigation canals, the introduction of unsustainable large irrigation schemes, and the perception of water as a free resource have all contributed to degrading water resources. Water supply and sanitation systems are also in poor repair in many areas. Virtually all major urban/industrial centres throughout the subregion suffer from significant water pollution with tests indicating that approximately 12.5 per cent of household water contains biological contaminates and over 3.5 per cent is chemically polluted beyond WHO standards.
- **Soil Erosion and Land Degradation** Several factors cause land degradation in the region including: loss of vegetative cover (i.e. from over-grazing, expanding human populations, and pollution); erosion (both wind and water); depletion of soil resources (i.e. from non-rotation of crops); and salinization (from poor irrigation practices). Despite their sensitivity and the expense of reclamation, desert and semi-deserts are extensively used for agriculture and animal husbandry in all the states of the region. Arable land in the subregion is also heavily degraded (see Box 20.1).
- **Loss of Biodiversity** Central Asia contains a well developed network of nature reserves inherited from the USSR; roughly three per cent of the region was designated under some form of conservation regime in the late Soviet era, equivalent to about 100 000 km^2 (Sievers et al 1995). While much is being done to protect these areas, the last decade or so has witnessed a significant decline in the quality and health of many of the reserves due to a combination of impoverishment of park rangers; widespread appropriation or lease of reserve lands for fishing and agriculture; military activities; and trophy hunting of endangered species.
- **Pollution of the Caspian Sea** The Caspian Sea, covering more than 370 000 km^2, is the planet's largest inland body of water. Its littoral states are Azerbaijan, the Islamic Republic of Iran, Kazakhstan, the Russian Federation, and Turkmenistan. Today, the sea is severely polluted from a concentration of the 100 or so rivers which enter it, and the uncontrolled oil and gas extraction from it (Zonn 1999). While the Caspian still yields 90 per cent of the world's Sturgeon, the annual yield of sturgeon has fallen from tens of thousands of tonnes per year to under ten thousand tonnes. Primary reasons for sturgeon decline are cited as overfishing, dams, poaching, and pollution of spawning grounds in rivers like the Volga, which provides 80 per cent of the Caspian's annual inflow. In 1988, almost no sturgeon spawned successfully in the Volga River due to heavy metal poisoning (Zonn 1997).

Box 20.1 Human Induced Land Degradation in Turkmenistan

A high level of salinization is a major feature of soil degradation in Turkmenistan. It varies by types and form, although anthropogenic factors have contributed a great deal to land degradation. An example is the Khauzkhan area, the zone covered by the second phase of the Karakum Canal. Since the groundwater was not too deep, and the initial salinity was low, it was erroneously thought possible to apply lowland farming without construction of a drainage network. This, however, led to the intensification of secondary salinization and reduction in productivity of the land. Similar negative consequences of the irrigation on the virgin lands are observed in the piedmont plain of the Kopetday in the Akhal Velayat, in the Shakhsenem area (Dashoguz Velayat), and in the Yulangyz area (Lebap Velayat).

In summary, an overall analysis of soil degradation process in Turkmenistan shows that human induced factor such as over-irrigation, heavy use of machinery, cultivation of unsuitable slopes, non-rotation of crops, as well as excessive use of nitrogenous fertiliser have all had a major contribution towards soil degradation.

Source: Government of Turkmenistan

CHAPTER TWENTY

> ### Box 20.2 Radiation and Human Health
>
> During the mid nineties, a study on several indicators of health in the town of Atbasar (Akmola region in central Kazakhstan) revealed a number of startling facts. Radiation and heavy metal pollution have been documented in the town's soil, air, flora, and water. The town is located in a valley which is not connected to national electricity infrastructure, and coal is consequently used as fuel for domestic heating. For the past forty years, uranium ore has been transported through the town without regard to human safety. Results of the study identified high levels of radionucleides in the coal, together with saturation of the local river with radionucleides as the result of dumpings at the uranium mines 100 kilometres upstream. In addition, atomic tests were carried out at a rocket test site not far from the town. In 1992, the average life expectancy was 62 years; in the first half of 1993 it was 59 years. Infant mortality is 37 in 1000.
>
> After researching several population groups with an average age of 36, it was revealed that 100 per cent had dental pathologies and a quarter had lost more than four teeth. Cardiograms of 90 per cent of the group evidenced abnormalities. Psychological disorders among adults are increasing by almost five per cent a year, and by almost seven per cent a year among children. Over the past ten years, the incidence of ectopic pregnancies has increased by over ten times. 70 per cent of tenth grade boys have evidenced serious morphological abnormalities in their sperm. Morbidity rates for malignant tumours are increasing by three per cent each year. Observations of human leukocytes revealed that 16 per cent evidenced abnormalities. This snapshot of Atbasar is a living example of serious health problems that can emerge as a result of exposure to radiation.
>
> *Source:* Rubezhansky 1994

- **Air pollution** While hydropower meets an appreciable amount of the region's energy needs, reliance on coal, inefficient power plants, industrial practices, and private transport have all led to high levels of both local and transboundary air pollution. In addition, the subregion also receives acid precipitation from the wider region, including the Russian Federation, and although aggregate emission and concentrations of most major pollutants have dropped significantly since independence (due to industrial decline), air quality is still a major threat to human health.
- **Radiation** Hundreds of nuclear, chemical, and biological weapons tests were conducted at the Semipalatinsk (Kazakhstan) nuclear test area, the Naryn (Kazakhstan) testing area, and on Resurrection Island (Kazakhstan/Uzebekistan) in the Aral Sea. Moreover, much of the uranium mined for weapons came from open mines near Atbasar (Kazakhstan), Chkalovsk (Tajikistan), and Maili Su (Kyrgyzstan), among other places. An installation to recover uranium from the waters of Lake Issyk-kul also operated in Kyrgyzstan for nearly 40 years (Charsky and Tishkova 1998). The incidence of ill-health and death which has been caused by radioactive contamination at many or all of these sites is significant (see Box 20.2), and remains a major environmental and health concern for the subregion.

CAUSES

The inheritance of environmentally inefficient policies, including over exploitation of natural resources and lack of appropriate industrial controls, have left a severe development burden on the subregion. The sharp economic decline, poverty, and the lack of investment in infrastructure following independence also pose some serious issues, both currently and in terms of the pent-up potential in the expansion of Central Asia's industrial base. The region suffered a severe economic shock in the wake of the 1998 Russian economic collapse, and in 1999 real income and GNP fell by more than 50 per cent for a large proportion of the subregion's residents. With the exception of Kazakhstan (the population of which has fallen by 10 per cent), the region shows high birth rates together with increasing death rates; in a decade the demographic structure of the subregion has skewed remarkably in the direction of having a majority of the population under 18 years old. Major socio-economic indicators of Central Asia are provided in Table 20.2.

Table 20.2 *Major Socio-Economic Indicators for Central Asian Countries*

Countries	Population (2000) (million)	Annual growth rate in population (%)	Population density (person per km²)	Urban population as % of total	Annual growth rate of urban population (%)	Total GNP (US$ millions) 1998	Per capita GNP (US$) 1998
Azerbaijan	7.8	0.6	91	57	1.2	3 821	480
Kazakhstan	16.2	-0.03	6	56	-0.1	20 856	1 340
Kyrgyzstan	4.7	0.7	24	33	0.3	1 771	380
Tajikistan	6.2	1.4	43	28	1.4	2 256	370
Turkmenistan	4.5	1.7	10	45	2.0	–	–
Uzbekistan	24.3	1.6	58	37	1.1	22 900	950

Source: 1. World Bank 1999 and 2000
2. ESCAP 2000
3. United Nations 1998

POLICY RESPONSE

A. National Initiatives

A major challenge facing the subregion is the inclusion of environment as a central component within its economic transition and recovery planning framework, linking issues such as public health and productivity, risks of irreversible damage to natural resources, and the diversification of its industrial base. A number of common limitations are observed in relation to their achievement of sustainable development in the subregion.

- **Policy** A variety of command and control mechanisms embraced during the Soviet period remain in-place in many of the states of the subregion, however the capacity for monitoring and enforcement, the duplication of responsibility, and subsequent economic decline, have all placed real pressure on the ability of many central and regional administrations to effectively deliver on policy objectives.
- **Legal** Legislation applicable to sustainable development does exist within the subregion, and includes approaches to establish clear liability, jurisdiction over polluters, incentives against pollution, public oversight, and transparency guarantees. However, in practice, the enforcement of compliance is severely lacking throughout Central Asia.
- **Industry** Factories and agricultural systems in the subregion are based on technologies and techniques that generally still rely on massive resource throughput, and operate at low efficiency often causing high levels of pollution. Policies aimed at increased output and the development of large-scale industry, together with the constraint of sectoral diversity, have hampered the development of a small or medium sized industrial base and presented the countries of the subregion with some unique transitional problems.
- **NGO** Environmental NGOs have been numerous in Central Asia since 1989, although despite gaining experience in the last decade, they have lost much public resonance and have failed to expand their membership base. Nevertheless the continuation of subregional communication, conferences and meetings, particularly among the scientific community, has led to the proliferation of strong personal ties and several subregional NGO coalitions are present and vocal.

B. Subregional Cooperation

Strong national and regional identities and affinities for the environment are evident in Central Asia, together with a common technical language. Furthermore the levels of harmonization in systems and legislation for environmental management are significantly advanced in comparison to other subregions. Numerous international conventions have been ratified for environmental protection in many of the states (see Chapter 21), for example, Central Asian states were among the very first countries to join the United Nations Convention to Combat Desertification (UNCCD 1994). Issues of economic and military security, together with the attraction of international financing have also played a major role in recent subregional cooperation initiatives (see Box 20.3). Major environmental cooperation projects in the subregion include:

- the Caspian Environment Programme (CEP)
- the Tien-Shan Biodiversity Project
- Cooperation Initiatives for the Aral Sea

CHAPTER TWENTY

> **Box 20.3 Joint Declaration of the Environmental Protection Ministers of Central Asia**
>
> 22 April 1998
>
> We, the Ministers of Environmental Protection of the 5 countries of the Central Asian region (Kazakhstan, Kyrgyzstan, Tajikistan, Turkmenistan, and Uzbekistan), with participation of representatives of international organizations (UNDP, UN ECE, World Bank, EU, TACIS, OESR, and USAID) meeting in Almaty, Kazakhstan on 22 April 1998 within the framework of the Environment for Europe process:
>
> – noting the high vulnerability of Central Asia's natural ecosystems to anthropogenic influence and noting that the past irrational uses of natural resources has led to substantial degradation of natural ecosystems, including the Aral Sea catastrophe, which creates serious barriers to the sustainable development of the region,
>
> – affirming once again the will of our countries to cooperate in the field of environmental protection and rational use of Central Asia's natural resources according to the principles of the 1995 Nukus and 1997 Almaty declarations.
>
> – proceeding from the fact that the heads of state of Central Asia declared 1998 the Year of Environmental Protection in Central Asia under the auspices of the United Nations, commit themselves to the principles of Agenda 21, and support the aspiration of developing and implementing a unified strategy for the sustainable development of the countries of Central Asia,
>
> – stressing the immediate necessity of integrating environmental and economic policies,
>
> – noting the importance of donor assistance in resolving the environmental problems of the region and the necessity of considerable increases in such assistance in accordance with the Rio Declaration,
>
> – expressing aspiration for more complete integration into the Environment for Europe process and noting endorsement of the Conference of European Ministers to take place in Denmark in June 1998,
>
> – committing ourselves to contribute decisively to resolution of the environmental problems of our region, and guided by the 17 March 1998 intergovernmental Agreement between the Government of the Republic of Kazakhstan, the Government of the Kyrgyz Republic, and the Government of the Republic of Uzbekistan on Cooperation in the Field of Environmental Protection and Rational Use of Central Asia's Natural Resources, declare it necessary to:
>
> 1) With consideration to the particularities and interests of each country, develop unified approaches to the creation and realization of national environmental policies, which includes:
>   a) harmonizing nature protection legislation and institutional structures;
>   b) completing development and implementation of harmonized economic mechanisms of nature use and effective realization of national environmental action plans;
>   c) carrying out joint nature protection projects in response to high priority environmental problems;
>   d) forming a regional information network on environmental protection and rational use of natural resources;
>   e) developing and implementing harmonized economic mechanisms in the process of carrying out environmental policies and encouraging environmentally clean production; and
>   f) facilitating local initiatives to develop sustainable structures of consumption and attracting NGOs to participate in the resolution of these issues.
>
> 2) Continue the process of acceding to international nature protection conventions and UN Economic Commission for Europe programmes, as well as other global conventions and programmes.
>
> 3) Develop a regional environmental action plan for the states of Central Asia.
>
> 4) Encourage international organizations, donor states, and other interested parties to support the efforts of the states of Central Asia to resolve regional and global environmental problems with the maximum use thereby of local specialists.
>
> Source: Ministry of Ecology and Natural Resources of the Republic of Kazakhstan 1998

1. *Caspian Environment Programme (CEP)*

CEP originated in 1995 and is channeled through the Global Environment Facility (GEF) while being supported mainly by funding from the European Commission's TACIS programme, and a number of bilateral donors.

Legal entitlements of the Caspian were the subject of treaties in 1921 and 1940 between the (now) Russian Federation and Islamic Republic of Iran. In 1994, at a meeting of deputy foreign ministers in Moscow, Islamic Republic of Iran proposed establishment of a Caspian Organization for Cooperation. In recent years, a variety of meetings between the littoral states have continued to yield gradual progress in resolution of the sea's legal status and regimes for hydrocarbon rights, shipping, fishing, and environmental protection. CEP is now the largest water management programme in the world, and is being coordinated out of Baku in Azerbaijan. The programme is conceived as a framework for coordinated actions by the Caspian littoral governments and international partners to achieve sustainable environmental management in the region. Key priorities for year four of the programme include the close involvement of both the public and industry, and the reinforcement of the programmes regional institutions and management structures.

2. *Tien-Shan Biodiversity Project*

The Tien-Shan Biodiversity Project was developed by GEF to support the protection of vulnerable and unique biological communities within the West Tien Shan Range and to assist in strengthening and coordinating national polices, regulations and institutional arrangements for biodiversity protection. The project is supported by a number of donors and is currently observed as the best example of transboundary biodiversity protection in the subregion, in that it is actively embracing the goal of creating habitat corridors between four roughly adjacent zapovedniki (reserves): Aksu-Dzhabagly (Kazakhstan), Besh-Aral and Sary-Chelek (Kyrgyzstan), and Chatkal (Uzebekistan).

3. *Cooperation Initiatives for the Aral Sea*

Combating the Aral Sea disaster in Central Asia is seen as a symbolic display of subregional cooperation, and was initiated prior to independence. Since independence, an Interstate Commission for Water Coordination (ICWC) to determine annual water allocations has been created, although this did not receive international funding until 1993. With active donor participation, especially the World Bank, an Interstate Council on the Aral Sea (ICAS) was established and subsequently assumed the activities of ICWC. However, failure to secure the massive capital injections anticipated for the recovery programme (both nationally and internationally), contributed to the shift of the now Aral Sea Basin project to the regional office of the United Nations Development Programme. In 1997, ICAS was itself absorbed into a new Interstate Federation for the Aral Sea (IFAS) which remains in operation and is governed through an Executive Board composed of the deputy prime ministers and environment ministers of the five Aral Sea Basin states. IFAS is presently assisted by a GEF mission to address the root causes of the overuse and pollution of international waters in the Aral Sea basin by contributing to the formulation and implementation of the first stage of a regional Strategic Action Programme.

SUBREGIONAL OUTLOOK

The economic down-turn in Central Asia in this decade has continued to mitigate against many of the environmental burdens of industrial production, with industry operating at significantly reduced levels of capacity. However, the potential effects of future capacity expansion present real concerns for the environment and are stimulating the necessity for cooperation in strategic subregional development, particularly in relation to the exploration of oil and gas reserves. In this context, Central Asia's educational and scientific strengths present extremely valuable tools in meeting the goals of long-term resource efficiency and pollution prevention objectives. To date the economic declines over the transition period have contributed to cooperation for environmental management occurring through the facilitation of western donor investments. While the states themselves donate a great deal to this process, it has been proposed that without such third party action, the level of environmental cooperation in the region would be minimal. Furthermore, competition for funding may have worked against cooperation in the subregion with the struggle to maximize national allocations of funds earmarked for "Central Asian" projects. For the transition period to be completed in an economically and environmentally sustainable manner, such trends will need correction through the development of open policies, plans, and programmes that are practical, accountable, and implemented. The challenge for Central Asia is to apply emerging concepts and champion new insights into sustainable development, as the subregion may be one of the most in need of, and most capable of, pioneering innovative action.

PART FIVE

CHAPTER 21-22

PART I

CHAPTER	1	LAND
CHAPTER	2	FORESTS
CHAPTER	3	BIODIVERSITY
CHAPTER	4	INLAND WATER
CHAPTER	5	COASTAL AND MARINE
CHAPTER	6	ATMOSPHERE AND CLIMATE

PART II

CHAPTER	7	URBAN ENVIRONMENT
CHAPTER	8	WASTE
CHAPTER	9	POVERTY AND ENVIRONMENT
CHAPTER	10	FOOD SECURITY

PART III

CHAPTER	11	INSTITUTIONS AND LEGISLATION
CHAPTER	12	MECHANISMS AND METHODS
CHAPTER	13	PRIVATE SECTOR
CHAPTER	14	MAJOR GROUPS
CHAPTER	15	EDUCATION, INFORMATION AND AWARENESS

PART IV

CHAPTER	16	SOUTH ASIA
CHAPTER	17	SOUTHEAST ASIA
CHAPTER	18	SOUTH PACIFIC
CHAPTER	19	NORTHEAST ASIA
CHAPTER	20	CENTRAL ASIA

PART V   FUTURE OUTLOOK

| CHAPTER | 21 | GLOBAL AND REGIONAL ISSUES |
| CHAPTER | 22 | ASIA AND THE PACIFIC INTO THE 21st CENTURY |

Chapter Twenty-One

ESCAP uses various mechanisms to promote regional cooperation for solving global and transboundary problems.

21

Global and Regional Issues

INTRODUCTION
FOLLOW-UP TO UNCED AND AGENDA 21
REVIEW OF THE IMPLEMENTATION OF
KEY INTERNATIONAL CONVENTIONS
GLOBAL PARTNERSHIPS: PROBLEMS AND PROSPECTS
ECONOMY AND THE ENVIRONMENT
REGIONAL AND SUBREGIONAL COOPERATION
CONCLUSION

CHAPTER TWENTY-ONE

INTRODUCTION

The close of the 20th Century marked the realization of the serious impacts of unsustainable economic growth, poverty and population increase. These are manifested in the destruction of the region's forests, massive soil erosion, accumulation of greenhouse gases in the atmosphere, degradation of marine life, a decline in the ocean's ability to absorb humanity's waste and loss of biodiversity. It is now also being understood that local infractions on the environment can have serious international and global repercussions.

At the regional scale, issues such as transboundary pollution, desertification, migratory species, species trade, the management of large eco-regional landscapes, and the equitable use of shared natural resources require regional and international cooperation. Over the past few decades, the international community has been formulating legal instruments as a framework for cooperation in meeting the numerous challenges presented by such problems. However, despite this, and the proliferation of projects at the regional and subregional level, progress towards reversing unsustainable trends has been slow. Moreover, beyond the usual limitations of financial and technological resources, there is a fundamental need to transform the attitudes of decision-makers at the highest level throughout the region, from one of 'business-as-usual', to one of sustainable development. Ultimately, such a transformation is absolutely necessary for the achievement of global and regional environmental goals.

This chapter provides a review and assessment of the major international conferences and key legal instruments, such as international conventions and agreements, protocols, treaties, and action programmes, that are of significance to the region. It discusses the problems and prospects of global partnerships and the relevance of the concept of environmental security to Asia and the Pacific within the context of the region's prevailing social and economic conditions. Finally, it covers the issue of regional and subregional cooperation which, besides offering excellent synergy for capacity building, also provides excellent opportunity for coordinated response to global initiatives, such as Agenda 21, and the various regional and international conventions and agreements which have been established.

FOLLOW-UP TO UNCED AND AGENDA 21

The many international and regional environment conferences which have taken place over the years have led humanity to recapitulate on the environmental crisis facing the world, and have provided many avenues for international and regional cooperation. Of these conferences, perhaps the most significant was the United Nations Conference on Environment and Development (UNCED), held in June 1992, involving nearly a hundred world leaders, making it the world's largest meeting of its kind in the history of diplomacy (Sitarz 1994). The principal outcome of UNCED was Agenda 21, a comprehensive blueprint for the sustainable development of the world. The principles of UNCED are eloquently expressed in the six core themes of Agenda 21: enhancement of the quality of life on Earth; efficient use of the earth's resources; protection of global commons; management of human settlements; management of chemicals and wastes; and promotion of sustainable economic growth.

Unlike many of international conventions, or Multilateral Environment Agreements (MEAs), it has inspired, Agenda 21 is not a legally binding instrument. Therefore, the assessment of its impact cannot be undertaken in terms of compliance, but rather how sustainable development – the development ideology it espouses – has captured the mainstream themes of global development and influenced the fundamental principles of the various MEAs. In this regard, In June 1997, the Rio+5, a Special Session of the United Nations General Assembly to assess implementation of Agenda 21 found that despite progress in many areas, the global environment continues to deteriorate. Five years after Rio, the world is still engaged in the same kind of economics and industry, employing the same technologies, viewing security from the same narrow perspectives, and engaged in the same consumption and production patterns that have brought about the current state of environmental decline. The general consensus of participants echoed that the Rio+5 was a sobering reminder that little has been accomplished in moving toward sustainable development, and that in the five years leading up to the meeting (1992-1997), there was little progress in implementing the key components of Agenda 21.

A. Implementation at the Global Level

The assessment made by the Special Session may be considered as an appraisal of progress in Agenda 21 at the global level. The meeting correctly noted the five years after UNCED was a period of accelerated globalization and intensive interaction between nations in the areas of trade, investment, and expansion of capital markets. As such, globalization presents new opportunities for the economic and social development of nations; unfortunately, few countries have been able to take advantage of these opportunities, and incremental differences in economic development between North and South have continued to increase in most cases.

In particular, the Special Session noted that, commitments made at UNCED-both before and after-remain largely unfulfilled. There is a need to strengthen the provisions of the various MEAs, as well as putting in place practical implementing mechanisms. One major achievement, however, has been the establishment, funding, restructuring, and replenishment of the Global Environmental Facility (GEF). However, much more remains to be done in the areas of finance, technology transfer, and capacity building. Although foreign direct investment has benefited some countries, official development assistance has fallen (Figure 21.1). The debt situation in some developing countries is a big constraint in efforts to their attaining sustainable development, and the technology gap between the industrial and developing countries continues to widen. Moreover, the benefits from information technologies have sharpened the relative competitiveness of the OECD countries as a block, placing further pressure on the world's developing economies.

Finally, the Special Session recognized that efforts have been made by governments and international organizations to integrate environmental, economic, and social objectives into decision-making by formulating new policies. Many lessons have been learned about the effectiveness of measures that take into account the consensus of all stakeholders, especially grassroots' concerns. In addition, many United Nations conferences have been held to strengthen commitment to achieving the goals of Agenda 21, and have played a crucial role in achieving progress in this regard.

Figure 21.1 Trends in ODA and Private Direct Investments

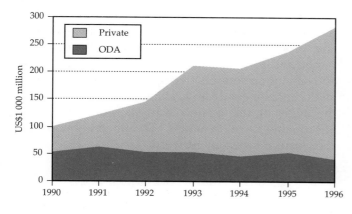

Source: UNEP 2000

1. *Assessment of Implication By Issue*

The Special Session of the General Assembly concentrated mainly on the reiteration and amplification of some of the main themes of Agenda 21, and a list of urgent actions was developed under two broad categories: cross-sectoral and sectoral issues.

(a) Cross-Sectoral Issues

The Special Session noted that economic development, social development, and environmental protection were inter-dependent and mutually reinforcing components of sustainable development, which also requires adherence to democracy, transparency, accountable governance, fundamental human rights, in addition to effective participation by civil society and equitable growth between countries. Integrating economic, social, and environmental objectives requires a broad package of policy instruments, including regulation, market-based instruments, and the internalization of environmental costs. In this light, it was recommended that environmental and social impact analysis be applied in country-specific situations to ensure that integrated approaches are effective and cost-efficient, and that sufficient impetus be given by countries to develop their national strategies for sustainable development by 2002.

Another key issue which requires action is the creation of an enabling international economic framework; globalization and the emergence of new trade regimes have made external factors a critical component in determining the success or failure of developing countries in their national efforts at sustainable development. Such issues could be approached effectively through constructive dialogue and genuine partnership, which takes into account the relative contributions to global environmental degradation between developed and developing countries, and their differentiated responsibilities.

(i) Eradication of Poverty

The eradication of poverty is an overarching theme of sustainable development. However, five years after UNCED, the number of people living in absolute poverty, particularly in developing countries, was still increasing. The enormity and complexity of the poverty issue could endanger the political stability of some developing countries, thereby undermining economic development and further degrading the environment. Priority actions should include the provision of, and access to sustainable livelihoods; access to basic social services; safety nets to those who cannot support themselves; cushions to the disproportionate impact of poverty on women. In addition, interested donors and recipients should work together to allocate bigger shares of official development assistance (ODA) to poverty eradication.

(ii) Changing Consumption and Production Patterns

Actions identified to address changing consumption and production patterns included, the identification of best practices that promote environmental effectiveness and efficiency as well as social equity; consideration of the linkages between urbanization and the environmental effects of consumption and production; improvement of the quality of information regarding the environmental impact of products and services; encouraging business and industry to develop and apply environmentally sound technology; promoting the role of business in shaping consumption patterns; and developing core indicators. The Special Session also agreed that in the long term, industrialized countries should achieve a 10-fold improvement in the energy and material efficiency of production, with a 4-fold improvement in the next three to four decades. Furthermore, there was a need for industrialized countries to pay special attention to the negative impacts of export and access to markets of developing countries.

(iii) Making Trade and Environment Mutually Supportive

The Special Session noted the need to establish favourable macroeconomic conditions to enable all countries to benefit from globalization. It further recommended that the Bretton Woods institutions, WTO, and the United Nations should enhance their responsiveness to sustainable development objectives. Decisions on further trade liberalization should take into account the impact of sustainable development. It called for a timely and full implementation of the results of the Uruguay Round, promotion of the universality of WTO, analysis of the environmental effects of the transport of international goods, and institutional cooperation between UNCTAD, WTO, and UNEP towards identifying ways to make trade and sustainable development mutually supportive.

(iv) Population

The Special Session noted that there was a need to recognize the critical linkages between demographic trends and sustainable development. The decline in population growth rates must be underpinned through national and international policies that promote, economic development; social development; environmental protection; poverty eradication; further expansion of basic education with full and equal access for girls and women; and health-care, including both family planning and sexual health, consistent with the report of the International Conference on Population and Development.

(v) Health

The overriding goal for the future was identified to be the implementation of the "Health for All" strategy, to enable populations to achieve a higher level of health and well-being; recognizing also that the achievement of good health would also improve their economic productivity and social welfare. To underpin these efforts, the Special Session supported actions such as the provision of safe drinking water and accelerated research in vaccine development. It also noted the importance of accelerating the process of eliminating the unsafe uses of lead, including its use in gasoline, and called for strategies to make parents, families and communities aware of the dangers of tobacco-use.

(vi) Sustainable Human Settlements

The Special Session called for urgent action to implement the commitments made in the United Nations Conference on Human Settlements, and in Agenda 21, and in particular, to identify additional financial resources to attain the goal of shelter-for-all and sustainable human settlements. It was also noted that the CSD should establish global targets to promote Local Agenda 21 campaigns.

(b) Significant Sectoral Issues

The Special Session called for international cooperation in support of national efforts in a number of key sectors, as follows.

(i) Freshwater

It was noted that the growing demand for water would become a major limiting factor in socio-economic development unless early action was taken. In particular, urgent responses were required in the formulation and implementation of policies and programmes for integrated watershed management; strengthening of regional and international cooperation for technology transfer; financing of integrated water resources programmes and projects; encouragement of investments from public and private sources to improve water supply and sanitation services (recognizing water as a social and economic good); and calling for a dialogue under the aegis of the United Nations Commission on Sustainable Development (CSD) aimed at building a consensus on the way forward.

(ii) Oceans and Seas

Urgent requirements for action in the following areas: the ratification or accession to the relevant MEAs; strengthening of inter-governmental links for the implementation of integrated coastal zone management; identification of priorities at the global level; prevention or elimination by governments of overfishing; consideration by governments of the positive and negative impacts of subsidies; and improvements in the quality and quantity of scientific data used by governments as the basis for effective decisions.

(iii) Forests

The Special Session noted that the Intergovernmental Panel on Forests' (IPF) proposals for action represented significant progress and consensus on a wide range of forest issues. To maintain momentum, urgent actions needed were, the development of national forestry programmes; clarification of all issues arising from the IPF process, in particular, international cooperation in financial assistance and technology transfer, and trade and environment in relation to forest products and services; and, further collaboration of international organizations and the high-level Inter-agency Task Force on Forests.

A significant agreement in this sector called for the establishment of the Intergovernmental Forum on Forests under the aegis of the CSD, whose task was to facilitate the implementation of the IPF recommendations and to review and monitor progress on sustainable forest management. The Forum would also identify possible elements of, and work towards, an international forestry agreement (see later in Chapter).

(iv) Biodiversity

The Special Session noted the urgent need for decisive action to conserve and maintain genes, species and ecosystems, including: ratifying the CBD and implementing it fully; undertaking concrete actions for the fair and equitable sharing of the benefits from the use of genetic resources; respecting, preserving, and maintaining knowledge, innovations and practice of indigenous and local communities; and rapidly completing the biosafety protocol following the UNEP International Technical Guidelines for Safety in Biotechnology.

(v) Atmosphere

There was a consensus on the importance of political will and effort as a requirement to protect the global climate, however, it was noted that there had been insufficient progress by many industrialized countries in meeting their commitment to reduce greenhouse gas emissions to 1990 levels by 2000. Very little progress was achieved in this sector during the Special Session; the agreed text only noted that the international community confirmed its recognition of the problem of climate change as one of the biggest challenges facing the world in the next century.

(vi) Small Island Developing States

The Special Session noted the international community's reaffirmation of its commitment to the implementation of the Programme of Action for the Sustainable Development of Small Island Developing States (SIDS). It was also noted that efforts to implement the Programme of Action needed to be supplemented by, effective financial support from the international community; and the SIDS information network and technical assistance programme.

2. *Assessment by Means of Implementation*

In view of the disappointing record of the implementation of Agenda 21, the Special Session of the General Assembly focused on the improvement of the means of implementation. The most controversial topic was in regard to the financial mechanisms designed to support the implementation of Agenda 21, whereby calls were made for the urgent fulfilment of all financial commitments which had been made, including the replenishment by the donor community of the International Development Association (IDA) and the GEF. No agreement was reached on this subject. There were, however, suggestions for new and innovative financial mechanisms to support Agenda 21 (for example a tax on airline travel), but recommendations were only made for further research in this regard.

The transfer of environmentally sound technologies (ESTs) is crucial to the implementation of Agenda 21. There were agreements calling for the urgent fulfilment of all UNCED commitments concerning the transfer of ESTs to developing countries, with a regular review as part of the CSD programme, and also stating the importance of identifying barriers and restrictions to the transfer of publicly and privately owned ESTs. Other agreements included affirming the role of governments in providing research and development to promote and contribute to the development of institutional and human resources, and calling for the creation of an enabling environment to help stimulate private sector investment and public private partnerships (PPPs).

In the area of capacity building, the Special Session noted the need for renewed support by the international community for national efforts for capacity building in developing countries and economies in transition. It also mentioned the UNDP Capacity 21 programme, and recommended that it should give priority to building capacity for the elaboration of sustainable development based on participatory approaches.

CHAPTER TWENTY-ONE

In summary, the progress of implementing Agenda 21 had been disappointing, compared to the goals of the 1992 Earth Summit. Moreover, the downward trend in official development assistance (ODA) (Figure 21.1), and the lack of agreement on replenishment of the GEF, suggested grave problems in implementation. These problems mirrored, in many ways, some of the very real barriers which seem to exist for North-South partnerships and cooperation in achieving the goals of Agenda 21. Nevertheless, the Special Session raised the profile of the concept of sustainable development and the work of the CSD. Governments and civil society could capitalize on this heightened public awareness to explore new initiatives at the international, regional, national and local levels.

B. Implementation at the Regional and Subregional Levels

Earlier chapters of this report (in particular, Part One) have shown the significant complexity and scope of the environmental problems facing Asia and the Pacific. Such complexity makes the region one of most suitable for assessing the performance of Agenda 21 to date; success or failure being indicative of global performance. As early as February 1993, the region responded to the call of Agenda 21, when ESCAP, in collaboration with other United Nations agencies, organized a high-level meeting to translate Agenda 21 into a coherent regional programme. The product of the meeting was the Framework of Regional Action for Sustainable Development. Subsequent key milestones representing progress since this date have been as follows.

1. *The Third Ministerial Conference on Environment and Development*

The Third Ministerial Conference on Environment and Development (November 1995, Bangkok) raised the level of awareness on Agenda 21 by giving it prominence in the media, enhancing support by governments, and actually demonstrating alternative clean technologies. An NGO/Media Symposium was also held in conjunction with the Conference. This symposium elaborated on the concept of sustainable development by identifying areas where NGOs and media could be most effective. The main issues that emerged were increasing access to government information; promotion of public participation in environmental projects; recognition and facilitation of the role NGOs and the media in providing critiques; government funding support to NGOs; and the development of strategies for cooperation between NGOs, the media, and communities. A private sector symposium of senior business and industry leaders was also held for the occasion, and a display of viable clean technologies was complemented by a seminar on environmental business management.

One important output of the Ministerial Conference was the Ministerial Declaration on Environmentally Sound and Sustainable Development in Asia and the Pacific. This declaration affirmed the support of the leaders of the region to the goals of Agenda 21 and highlighted the special concerns of the region, with emphasis on the need for regional cooperation.

On a more pragmatic level, the most significant achievement of the Ministerial Conference was the adoption of the Regional Action Programme for Environmentally Sound and Sustainable Development 1996-2000. This Action Programme was the blueprint for concrete action, and identified 141 tangible results for achievement in 24 programmatic areas, which in turn supported the six core themes of Agenda 21 (Table 21.1).

2. *The Global Conference on the Sustainable Development of Small Island Developing States (SIDS)*

The key instrument that emerged from this Conference was the Programme of Action (POA), which was adopted in Barbados on May 1994. The Conference has its roots in Chapter 17 of Agenda 21. The priority areas identified by the POA included, climate change and sea level rise; natural and environmental disasters; management of wastes; coastal and marine resources; freshwater resources; land resources; energy resources; tourism resources; biodiversity resources; national institutions and administrative capacity; regional institutions and technical cooperation; transport and communication; science and technology; human resources development; and implementation, monitoring and review. The POA also recognized several cross-sectoral areas that require attention, such as capacity building; human resource development; institutional development at all levels; cooperation in the transfer of environmentally sound technologies; trade and economic diversification; and finance.

The Conference also adopted the Barbados Declaration, which embodies the political will and the unanimity of the SIDS in pursuing the POA. In its 22nd Special Session (September 1999), the United Nations General Assembly undertook a review and appraisal of the implementation of the POA. In general, most delegates to the Special Session were of the view that SIDS have lived up to their commitments made in Barbados. However, they lamented the failure of the international community to do the same. The SIDS are facing decreasing ODA when there is an urgent need for additional financial

GLOBAL AND REGIONAL ISSUES

Table 21.1 Programme Areas of the Regional Action Programme and the Core Themes of Agenda 21

Programme areas of RAP	Core themes of Agenda 21*
1. Air quality	1, 3
2. Water quality	1, 3
3. Toxic chemicals and hazardous wastes	5
4. Urban environmental issues	4
5. Energy	2
6. Forests	2
7. Biodiversity	2
8. Coastal and marine environment	2, 3
9. Desertification and land degradation	2
10. Wetlands and lakes	2
11. Integrated mountain development	1, 2, 4
12. Implementation of the international conventions and appropriate regional conventions	3, 5, 6
13. Institutions and legislation	1, 2
14. Environmental standards	1, 2
15. Environmetal impact, audit, and risk assessment	1, 5, 6
16. Use of economic instruments	6
17. Mutually supportive trade and environment policies	6
18. Natural resource accounting	2
19. Combating poverty to achieve sustainable development	1, 6
20. National strategies and action plans	1, 2, 4, 5, 6
21. Environmental education, public awareness and training	1
22. Sustainable development indicators	6
23. Environment and natural resource monitoring and assessment	2
24. Reporting on sustainable development	2, 6

* themes are summarized as:
1. enhancement of the quality of life on Earth;
2. efficient use of the earth's resources;
3. protection of our common global resources;
4. management of human settlements;
5. management of chemicals and wastes; and
6. promotion of sustainable economic growth.

resources from the international community, as they are increasingly becoming vulnerable to globalization and trade liberalization, in addition to natural and man-made disasters such as sea level rise and hurricanes.

The findings of the Special Session are contained in the "Declaration on the State of Progress and Initiatives for the Future Implementation of the POA for SIDS" (IISD 1999). It indicated the necessity for a sharply focused implementation of the POA, particularly in relation to the following areas.

- **Climate change** SIDS and the international community will have to work together to build capacity to respond and adapt to climate change, for example by improving climate prediction capabilities. They should also improve collaboration with United Nations DESA and the UNFCCC Secretariat, with the goal of incorporating information on SIDS into the long-term planning of these organizations.
- **Natural and environmental disasters and climate variability** SIDS and the international community should aim for the improved scientific understanding of severe weather events, particularly El Nino. They should also work for the better strategies for prediction and reduction of damages of natural disaster, and should forge stronger partnerships with the private sector.
- **Freshwater, Coastal and marine resources** Because of the urban expansion in SIDS and their limited watersheds, activities should focus on the integrated water resources management approach. In regard to coastal and marine resources, SIDS and the international community should seek stronger regional partnerships with the Global Programme for Action for the Protection of the Marine Environment from Land-based Activities, and the regional seas programmes. They should also work together to develop, and implement, guidelines for practices and techniques to achieve sustainable management of coastal and marine resources (including the maintenance of health coral reefs, and implementation of coral reef actions plans), and strengthen skills in scientific research relevant to the management and conservation of fish stocks. The document also urges greater regional cooperation in the management, monitoring, control and surveillance of marine areas, and in the assessment of land-based sources of pollution, and reaffirms the assertion of SIDS in the POA concerning the ban on importation of hazardous wastes and the prohibition of their transboundary movement.
- **Energy** Noting the heavy dependency of SIDS on conventional energy sources, the document focuses on the need to mobilize technical, financial, and technological assistance to encourage energy efficiency and to accelerate the shift to new and renewable energy resources.
- **Tourism** The Special Session calls on SIDS and the international community to establish regional and national environmental assessment programmes to address the social,

economic, and cultural implications of tourism development; strengthen institutional capacity in the tourism sector (e.g. through public-private partnerships); promote environmental protection and the preservation of cultural heritage through local community awareness and participation; encourage the use of modern data collection technologies and communication systems, and improve the collection and use of data as a means to facilitate the development of sustainable tourism. The document also urges SIDS to enact new legislation in pursuit of sustainable tourism, and to implement regulations of the IMO and the International Civil Aviation Organization. Overall, resource mobilization and finance is recognized as being an issue of particular significance. The difficulty of SIDS in accessing concessional financing was traced to their relatively high GDP per capita. This singular consideration does not take into account their levels of development, real standard of living, and their vulnerability. There is a need to enhance the responsiveness of GEF to SIDS.

3. *Implementation of the Regional Action Programme*

As discussed earlier, the adoption of the Regional Action Programme at the aforementioned Ministerial Conference, strengthened regionally-based activities in strategy formulation, legislation, institution building, and research in relation to Agenda 21. As planned, national governments of the region bear the primary responsibility for implementation of the programme, and as such, there are over 600 projects distributed across some 16 countries in the region (notably including India, Bangladesh, Indonesia and People's Republic of China). These projects are primarily focused on the following key issues.

- **Air quality** This is a major problem in most countries, particularly in relation to industrial and vehicular emissions. Projects and activities are concentrated on, capacity building for air quality monitoring and management; transport infrastructure improvements; monitoring networks for acid deposition; and phasing out of ozone depleting substances (Chapter 6).
- **Water quality** The most significant problems relate to the limitation and contamination of supplies. Projects are focused on management, conservation, and enhancement of water quality including the rehabilitation of rivers.
- **Toxic chemicals and hazardous wastes** Countries of the region have focused activities on the containment, treatment and safe disposal of hazardous and chemical wastes. Projects include capacity building in legislation, institutional mechanisms, database development, training, and the adoption of environmentally acceptable methods of recycling wastes.
- **Urban environment** Projects are increasingly focusing on improving urban environmental quality in critical areas, through a combination of, information exchange through strengthening of networks such as CITYNET (see Chapter 7); integrated urban environmental action plans; private sector involvement in urban environmental improvement; and public participation through awareness programmes and citizen action groups.
- **Energy** Projects are focused on increasing the efficiency of energy production, transmission, distribution, and consumption. Research on environmentally-sound energy sources and new and renewable energies is also a focal area.
- **Forests** In response to the widespread problem of deforestation, projects are focusing on, the assessment of forest cover; enforcement of legal mechanisms to prevent deforestation; strengthening capacity to promote sustainable forestry; and the promotion of social forestry (Chapter 2).
- **Biodiversity** In an effort to fulfil national obligations under CBD (see Chapter 3), projects are concentrating on, the survey and assessment of existing biodiversity; conservation of endangered species; establishment of protected areas, and protected area plans; promotion of regional and subregional conservation programmes; research and development; and community participation.
- **Coastal and marine environment** Projects are focusing on, integrated coastal zone management; adherence to the regional seas programmes; combating oil spills; prevention of pollution from land and offshore activities; conservation of marine species and their habitats; and implementation of actions arising from relevant MEAs (Chapter 6).
- **Desertification and land degradation** Projects include survey and assessment of land degradation; drought mitigation; reclamation of desertified land; soil and water conservation (Chapter 1), including the improved application of agro-chemicals; and integrated land and water management conservation and management strategies.

GLOBAL AND REGIONAL ISSUES

- **Integrated mountain development** Mountain environments are known for their fragility. Projects are focusing on the concept of integrated mountain development, and include land use planning; watershed management; rehabilitation of degraded hilly areas; and alternative sustainable livelihood for mountain inhabitants.
- **Wetlands** In an effort to fulfil national obligations under RAMSAR (see Chapter 4), projects are focusing on, developing inventories of wetland resources; development of conservation strategies for critical areas (including legal provisions); prevention of pollution; and the promotion of scientific research.
- **Policy support** In support of the sustainable development principle, projects include the development of environmental standards and the introduction/promotion of environmental management techniques, such as EIA, risk assessment, and environmental audits, market-based instruments, natural resources accounting, and environmentally supportive trade policies (Chapter 12). Other national projects are those relating to the preparation of national strategies and action plans, and environmental education, and public awareness programmes.

REVIEW OF THE IMPLEMENTATION OF KEY INTERNATIONAL CONVENTIONS

The last three decades have seen remarkable progress in the globalization of the environment and the greening of international law, driven by humankind's growing awareness of the environment, and the momentum created by initiatives such as Agenda 21. As a consequence, international agreements have proliferated, embodying some of the landmark concepts and principles expressed in the Rio Declaration, most notably the following.

- **The precautionary principle** In simple terms, this principle asserts that the implementation of environmental protection measures should consider the best available knowledge of the time. Lack of full scientific certainty, which previously guided social action, should not delay these measures.
- **The legal notion of common responsibility** In practical terms, the principle of commonality means that states are required to take into account the needs of all the parties to a convention, i.e. no country could exploit a common global resource that will result in adverse effects to the others.
- **The principle of common but differentiated responsibility** Contained in Principle 7 of the Rio Declaration, this principle includes the notion of taking into account the differing situations of various countries and the need to consider a state's contribution to the environmental problem, as well as its ability to contribute to its solution. In practical terms, international law now considers the needs and capacities of developing countries *vis-à-vis* the industrialized countries.
- **Integration of pollution control; cradle-to-grave approach; and total life-cycle approach** As opposed to the traditional approach of regulating and/or controlling specific pollutants to individual media, the concept of integrated pollution control takes into account the effects of pollutants on the environment as a whole, the regulation of entire production processes and the entire life-cycles of substances and products. This holistic approach to pollution management is now a common feature of international agreements.

Although at first there was some resistance to these concepts, they have gradually taken root as countries worldwide, including within Asia and the Pacific, have signed up to the MEAs which embody them (see Figure 21.2). In this light, some of the key MEAs may be viewed amongst the great achievements of humankind, as they have become

Figure 21.2 Percentage of Countries in Asia and the Pacific that are Parties to Conventions

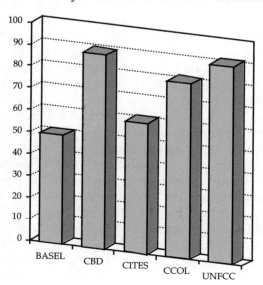

Source: GEO 2000

399

CHAPTER TWENTY-ONE

the major channels for international and regional cooperation on a number of issues which affect all of our lives.

This section assesses the implementation of the key conventions, looking at factors such as, the status of their ratification; the promulgation of supporting laws and regulatory measures; the existence of coordinating and funding mechanisms; and the effectiveness of enforcement at the regional and subregional level. The enactment of legislation is often taken as an indicator of implementation. However, this alone is insufficient to guarantee success, since it must be accompanied by programmes and activities on the ground, all of which require funding, both to implement, and to monitor.

Basic information about the most significant MEAs is presented in Table 21.2, each of which is discussed in more detail below (and also in their respective Chapters in Part One).

A. Atmosphere-related Conventions

1. The Vienna Convention and the Montreal Protocol

The Coordinating Committee on the Ozone Layer (CCOL) was created to periodically assess the condition of the ozone layer. In 1981, intergovernmental negotiations began to phase out chlorofluorocarbons (CFCs), which led to the adoption of the Vienna Convention for the Protection of the Ozone Layer (CPOL) in March 1985. This convention encourages Parties to undertake general measures to protect human health by protecting the ozone layer from human activities that may deplete it.

The discovery of the Antarctic ozone hole in 1985 prompted the international community to take stronger measures against ozone-depleting substances (ODS). Recognizing the need to reduce the production and consumption of several types of CFCs and halons, the Montreal Protocol on Substances that Deplete the Ozone Layer was adopted in 1987. The Protocol provided for the phase-out schedules of CFCs and halons and the potential revisions of the schedules based on the periodic scientific and technological assessments. Several amendments to the Protocol further accelerated the phase-out of these substances and new substances were added to the list of ODS, for example, the 1992 Copenhagen Amendment added methyl bromide to the list. The historical record on the pace of ratification of the Vienna Convention, the Montreal Protocol and its amendments in Asia and the Pacific is shown in Figure 21.3, and Table 21.2.

The Montreal Protocol is exemplary in the sense that it truly embodies the principle of common but differentiated responsibility between the industrial and developing countries. It recognizes the fact that industrialized countries are responsible for the bulk of emissions of ODS. Moreover, they have the financial and technological resources to find proper replacements for these substances. Meanwhile, the developing countries are given a grace period before they must start out their phase-out schedules.

The other strong feature of the Montreal Protocol and its various amendments is the financial mechanism for implementation. The Protocol established the Multilateral Fund in 1990. The Fund is used to pay the incremental costs incurred by developing countries in phasing out their consumption and production of ODS. Moreover, an equal number (7) of industrialized and developing countries administer the Fund. Some US$903 million has been disbursed by the Fund for projects and capacity building. An additional financing mechanism is provided by the Global Environmental Facility (GEF) to help developing countries, as well as countries with economies in transition (CEIT). The CEITs are not eligible for assistance under the Multilateral Fund. To date, some US$115 million has been approved for some CEIT. The impact of the Multilateral Fund on the phasing out of CFC is clearly seen in Figure 21.4.

Effective sanctions are provided in the Protocol to promote participation and honest compliance, for example, trade sanctions. The objective is to

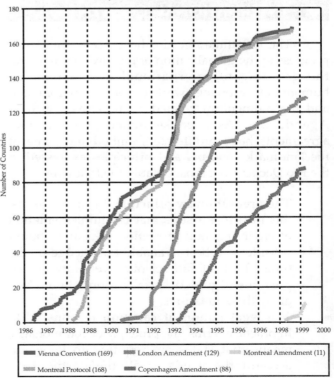

Figure 21.3 Ozone Protocol and Amendments Ratification Status

Source: United Nations Office of Legal Affairs 1999

Table 21.2 Basic Information on the Different Conventions on the Environment

Item	CPOL	UNCLOS	CBD	CITES	BASEL	MARPOL	RAMSAR
Legal scope	Open to all States	Open to all States, certain self-governing associated States and territories, and international organizations to which their member States have transferred competence over matters governed by the Convention	Open to all States and regional integration organizations	Open to all States recognized by the UN. Not yet open to regional integration organizations	Open to all States and political and/or economic regional organizations, regional integration organizations	Open to all States. Not open to regional integration organizations. NGOs participate with observer status at IMO meetings	Membership open to all member States of the UN or members of the specialized agencies and the IAEA. Open to regional integration organizations
Geographical scope	Global	Global	Global	Global	Global	Global seas. The Convention designates the Antarctic, Mediterranean, Baltic, Red, and Black Seas, the Gulf of Aden, and the Persian Gulf area as special areas in which oil discharge is virtually prohibited and the wider Caribbean and the North Seas as special areas subject to more stringent requirements governing the disposal into the sea of ship-generated garbage.	Global
Time and place of adoption	March 22, 1985, Vienna	December 10, 1982, Montego Bay, Jamaica	The agreed text of the Convention was adopted in May 22, 1992 at Nairobi. It was opened for signature in Rio de Janeiro on June 5, 1992	March 3, 1973, Washington, D.C.	March 22, 1989, Basel	November 2, 1973 and February 17, 1978 (protocol), London	February 2, 1971, Ramsar, Islamic Republic of Iran
Entry into force	September 22, 1988	November 16, 1994	December 29, 1993	July 1, 1975	May 5, 1992	October 2, 1983	December 21, 1975
Status of participation	166 Parties, including EU by April, 1998 No signatory without ratification	Signed by 158 Parties, ratified by 132 Parties as of October 29, 1999	175 Parties, including the EU, January 15, 1999 (http://biodiv.org.ca); 12 signatories without ratification, acceptance or approval	145 Parties as of March 1, 199 (GEO 2000)	118 Parties, including EU by April 15, 1998. Three signatories without ratification	104 Parties (94 per cent of world tonnage by January 31, 1998); 38 States have made exceptions for Annexes III (19) and IV (17). No signatories without ratification, acceptance or approval.	106 Parties by May 1, 1998. Three signatories without ratification, acceptance or approval.
Publications	Montreal Protocol Handbook; Action on Ozone (quarterly newsletter); Reports of the 23 meetings of the Executive Committee of the Multilateral Fund	Report of the Secretary General on Oceans and the Law of the Sea; The Law of the Sea Bulletin; Law of the Sea Information Circular	Global Biodiversity Outlook, a periodical report on the state of biodiversity worldwide and the implementation of the Convention	Notifications to the Parties; Identification Manual; The Evolution CITES, A Reference Book	Annual Reports; Technical Guidelines (Series); Managing Hazardous Wastes	IMO News and reports of the MEPC and MSC of IMO	Ramsar Newsletter (six times a year); Directory of Wetlands of International Importance (triennial); Ramsar Manual; Towards Wise Use of Wetlands, a collection of guidelines and case studies; The Legal Development of the Ramsar Convention; The Economic Valuation of Wetlands 1997; Wetlands, Biodiversity and the Ramsar Convention (1977)
Websites	http://www.unep.org/unep/secretar/ozone/home.htm http://www.unep.ch/ozone...	http://www.un.org/depts/losgopher://gopher.un.org:70/11/LOS for ISBA: http://www.isa.org.jm	http://www.biodiv.org E-conference biodiv.conv <APC>		http://www.unep.ch/basel	http://www.imo.org	http://www.iucn.org/themes/ramsar

Source: Bergesen, et al 1999

Figure 21.4 Multilateral Ozone Cumulative Funds Approved and CFC Tonnes Phase Out in the World

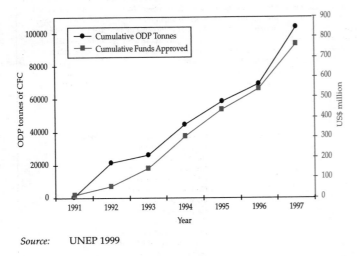

Source: UNEP 1999

Figure 21.5 Impact of the Montreal Protocol on Ozone Depletion

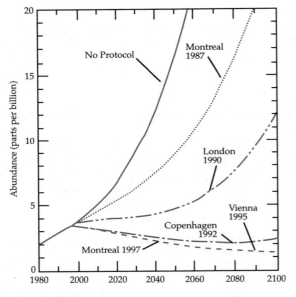

Source: UNEP 1999

encourage countries to participate in the Protocol by preventing non-participating countries from gaining competitive advantage. The sanctions discourage the movement of ODS manufacturing facilities. Each Party is prohibited from importing controlled substances from states not party to the Protocol. Similarly, since 1993, no Party may export ODS to any State not party to the Protocol nor import specific products to any non-Party.

The relative impacts of the Montreal Protocol and its amendments are shown in Figure 21.5. Without the Protocol, it is estimated that ozone depletion would have risen by 50 per cent by 2050 in the Northern Hemisphere's mid-latitudes, and 70 per cent in the Southern Hemisphere's mid-latitudes. The implications would have been tragic, as there would most likely have been 19 million more cases of non-melanoma cancer, 1.5 million more cases of melanoma cancer, and 130 million more cases of eye cataracts

Despite the phenomenal success of the Montreal Protocol compared to other MEAs, it still faces some challenges. The most important of these are: the continuous illegal increase in the flow of CFC products to developing countries, from countries who have adopted ozone-safe products; the illegal flow of CFCs to industrialized countries; the increased consumption of ODS in some developing countries, as allowed by the Protocol prior to their phase-out by July 1, 1999; and the increase in ozone depletion due to global warming. However, these outstanding issues are not thought to be too significant, and the success of the Montreal Protocol should pave the way towards a benchmark convention, from which the effectiveness of the other MEAs may be assessed.

2. *The Framework Convention on Climate Change and the Kyoto Protocol*

The objectives of the Convention (UNFCCC) are: to stabilize greenhouse-gas (GHG) concentrations in the atmosphere at a level that would prevent dangerous anthropogenic impacts on the climate, within a timeframe sufficient to allow ecosystems to adapt naturally to climate change; to ensure that food production is not threatened; and, to enable economic development to proceed in a sustainable manner.

An important feature of UNFCCC was the establishment of a system whereby governments report on their GHG emissions and their strategies for climate change. These reports enable the Secretariat to closely monitor the progress of the Convention. The developed countries also committed to take measures in returning to their 1990 levels of emissions by 2000. Moreover, they agreed to transfer financial resources and technology to the developing countries to help them achieve their goals, in particular through a financial mechanism which is operated by the GEF, under the guidance of the Conference of Parties (COP). The projects that GEF supports are those that have global environmental benefits, particularly those that reduce emissions of GHGs, improve energy efficiency and promote the use of new and renewable energy resources. It also supports capacity building in developing countries to implement the Convention. A Trust Fund for Participation in the UNFCCC Process was also established (Bergesen et al 1999) to support developing countries/CEIT participation in the COPs. Similarly, a Trust Fund For Supplementary Activities has been established to support the Secretariat and its subsidiary bodies. Both of these Funds are from voluntary contributions.

(a) The Kyoto Protocol

The Kyoto Protocol, which was the product of the Conference of Parties (COP) 3, is central to the implementation and issues surrounding the UNFCCC. Before it comes into force, no less than 55 of the Parties accounted for at least 55 per cent of the total CO_2 emissions in 1990 by Annex I countries. Unfortunately, as of 1999, only seven have ratified the Protocol, while 84 countries have signed.

The central and controversial issue of the Protocol was the emission targets for the 39 industrialized countries (Annex B of the Protocol). The targets are eight per cent for the European Union (collectively), seven per cent for the United States, and six per cent for Japan, all below the 1990 levels, and for the period 2008-12. Some countries are allowed a 10 per cent increase. The overall result is only a 5.2 per cent reduction in emissions of the six GHGs from the 1990 levels of the industrialized countries. Another controversial provision is the treatment of carbon sinks. There were no provisions in the Protocol requiring non-Annex B countries to reduce their GHG emissions. However, there is recognition that these countries would eventually need to take meaningful climate protection measures, without spelling out what those measures might be.

Upon ratification of the Protocol, those parties which are legally bound to reduce emissions, will be able to participate in the initial rounds of an emissions trading programme (to be developed). The so-called "emissions trading" allows an Annex I country with an excess of emission units, presumably from below commitment levels, to sell the credits to another Annex I country. This is the most contentious of all the mechanisms within the Protocol. Some of the other important provisions of the Protocol include, that each party in Annex I should make demonstrable progress in its commitments by 2005; and that calculation of net removals of GHGs by sinks resulting from direct human-induced activities be limited to afforestation, reforestation, and deforestation since 1990.

Another important conceptual innovation of the Protocol is the Clean Development Mechanism (CDM), provided for in Article 12.2. The CDM allows governments or private entities in industrialized countries to comply with their emission reduction targets by implementing projects in developing countries. The industrialized country will receive credits in the form of Certified Emission Reductions (CERs). This is in contrast to the AIJ, where Annex I countries help one another in meeting their reduction targets.

(b) COP4 and COP5

COP4 yielded the two-year Buenos Aires Plan of Action in November 1998. The plan established deadlines for finalizing the outstanding details of the Kyoto Protocol so that the agreement will become operational when the Protocol enters into force sometime in 2000. The Plan of Action deals mainly with the Protocol's flexible mechanisms, compliance issues, policies and measures. It also addresses the issue of transferring climate-friendly technologies to developing countries and examines the special needs and concerns of countries adversely affected by the impact of global warming, as well as the economic implications of response measures.

The agenda of COP5, held in Bonn in October 1999, was based on the Buenos Aires Plan of Action. The main item was the finalization of the rulebook of the Protocol to be presented for approval in COP5 in 2000 at the Hague. The important decision had to do with improving the rigor of national reports from industrialized countries and the strengthening of those from the developing countries. There was also an agreement addressing the bottlenecks in the delivery of communications from developing countries. Over the year, the various bodies of the Convention will finalize regimes for non-compliance, capacity building, emissions trading, joint implementation, and CDM. They will also point a way forward to determining the adverse effects of reduction measures on developing countries. Moreover, they will formulate measures on how to account for the role of carbon sinks such as forests.

(c) Current Status

The UNFCCC is a framework convention, and therefore does not enjoy the status of international law. However, its enunciated principles could be considered as part of international customary laws, in particular the reaffirmation of the precautionary principle and the principle of common but differentiated responsibility. Remarkably, the UNFCCC has the highest number of ratifications amongst the key MEAs. Some 91 per cent of the countries of the world have ratified it, with the highest percentage of ratification in Asian and Pacific region (Table 21.3). This means that the principles of the Convention have almost gained universal acceptance. However, the Kyoto Protocol, the legally binding document, is expected to encounter difficulties before it enters into force.

The earliest expected date for ratification of the Protocol is 2002, ten years after the signing of the UNFCCC in the Earth Summit, assuming that it is indeed ratified in that year. On the other hand, the time between the signing of the Vienna Convention (1985) and the entry into force of the Montreal

Table 21.3   Percentage of Parties to UNFCC in Asia and the Pacific (As of 1 March 1999)

Region	Number of countries	UNFCCC Parties	Ratification (%)
South Asia	9	8	89
Southeast Asia	5	4	80
Greater Mekong	5	5	100
Northeast Asia	5	5	100
Pacific	16	15	9
Asia and the Pacific	40	37	93
World	193	176	91

Source: UNEP 2000

Protocol (1989) was only four years. The reasons for this difference in speed of ratification could be the fact that the scientific foundation of the Montreal Protocol is less controversial than that of the Kyoto Protocol. Also, the Kyoto Protocol is more complex, involving many stakeholders (Annex I countries, CEIT, G7/People's Republic of China, JACANNZ, NGOs, energy intensive industries, etc.) operating in an environment of a widening gulf between industrial and developing countries, and with a commonly held perception that climate change is an industrialized country problem.

The debate in the United States on the Kyoto Protocol is significant to its coming into force. Since the US is credited with 25 per cent of the required 55 per cent of 1990 levels of emissions, its ratification of the Protocol carries a lot of weight. In the US, some scientists and industry-supported organizations claim that the global climate models (GCMs) used by the IPCC as the basis for the Protocol may have overestimated the projected impacts of global warming by as much as 100 per cent. They dispute the findings of the IPCC that there is "discernible human influence on and dangerous anthropogenic interference with" the Earth's climate system. They argue that the IPCC data do not justify the urgency of the emissions cut in the Protocol, or even the application of the precautionary principle. However, the second assessment report of the IPCC has been reviewed positively by some 2000 scientists and experts in the world and also confirmed the availability of the so-called "no-regrets options" and other cost effective strategies for combating climate change (United Nations 1999a).

B.   Ocean and Marine-related Conventions

1.   *The UN Convention on Law of the Sea (UNCLOS)*

The United Nations Law of the Sea governs all aspects of the world's oceans, such as the delimitation of national jurisdiction, environmental control, marine scientific research, economic and commercial activities, transfer of technology and the settlements of disputes relating to ocean matters. UNCLOS deals with all sources of the marine environment, from the coastal areas to the seabed and subsoil. Flag and port states are urged to act on violations of pollution regulations by vessels. Coastal states must promote compliance with the international environmental standards of the Convention, especially under the IMO.

The Convention also covers marine scientific research. It grants jurisdiction over marine scientific research to coastal states in its Economic Exclusion Zone (EEZ) and continental shelf. Other States must obtain consent for such research. However, some claim that this would limit research because some of the strongest capabilities are in countries (such as the United States) that are not Parties to the Convention. The other distinctive feature of UNCLOS is the coupling of environmental obligations with compulsory arbitration and adjudication. The Convention provides for dispute settlement over conflicting interpretations of its provisions. The goal of the convention is to promote compliance with its provisions and insure that disputes are settled by peaceful means. Thus, two agencies were created and empowered: the International Seabed Authority and the International Tribunal on the Law of the Sea.

A separate agreement relating to the implementation of Part XI of the Convention (on deep seabed mineral resources) was passed and this was entered into force on July 28, 1996. It had 79 signatories and 94 ratifications and/or accessions (United Nations 1999c). From the perspective of conservation, the most important and concrete outcome of UNCLOS is the 1999 Fish Stock Agreement. The underlying objective of the Agreement is the conservation and management of straddling fish stocks. It elaborates on the fundamental principle established in the Convention, that states should cooperate to ensure conservation and promote the optimum utilization of fisheries resources within and beyond the EEZ. It introduces a number of innovative measures such as the obligation of states to adopt a precautionary approach to fisheries exploitation. It gives expanded powers to port states to enforce certain obligations in safeguarding the proper management of fishery resources. To date this agreement has not yet entered into force, having been signed by 59 States, with 24 ratifications (UN, 1999c).

2.   *The IMO Convention (MARPOL 73/78)*

Oil pollution of the oceans and coast of the Asian and Pacific Region is a major problem (see

Chapter 5). The Indian Ocean, the Malacca Straits and Lombok-Makasiar Straits, the South China Sea, and the Sea of Japan are some of the world's major highways for the transport of petroleum. As such, the IMO Convention (or MARPOL 73/78) is of vital importance for the region.

The main objectives of the Convention are, to eliminate pollution of the sea by oil, chemicals, and other harmful substances which might be discharged in the course of operations; minimize the amount of oil which could be released accidentally in collisions or strandings by ships, including also fixed or floating platforms; and improve further the prevention and control of marine pollution from ships, particularly oil-tankers.

MARPOL 73/78 is comprehensive in the sense that it covers not only oil, but also other forms of pollution from ships. The Convention now refers to a number of conventions, protocols, amendments, annexes and other agreements dealing with shipping and the oceans, which could be classified into three major groups, such as maritime safety; prevention of marine pollution; and liability and compensation, especially in relation to damage caused by pollution. Outside these major groupings are a number of other conventions dealing with facilitation, tonnage measurement, and unlawful acts against shipping and salvage. The Convention contains special provisions for the control of pollution from more than 400 liquid noxious substances, as well as for sewage and garbage disposal.

An important feature of MARPOL is the concept of "special areas". These are sea areas where conditions such as oceanography, ecology, and traffic require special or stricter provisions. Two special areas are the Great Barrier Reef of Australia, and the Black Sea. There are efforts to make the Marine World Heritage Sites in the region as special areas under the convention.

The history of the IMO Convention illustrates the tediousness of the international legal process. It took five years for the Convention to come into force. The adoption and enforcement of protocols and amendments takes even longer. The circulation of reports made by the IMO to the contracting parties constitutes the basic process of monitoring compliance. After inspections, port states make reports on the compliance of flag states to the IMO. The reports required are annual enforcement reports by port and flag states; annual summary report by Party State's administration of incidents involving spillages of oil of more than 50 tonnes; and annual assessments.

MARPOL 73/78 has been successful in bringing the world's merchant fleet under the IMO management regime. Since its entry into force in 1983, about 94 per cent of the world's tonnage is now under MARPOL 73/78, with 104 contracting parties (as of 1998). As its reach expanded, there has been a notable decline in marine pollution. However, in the past few years, only a small percentage of parties in the developing countries of Asia and the Pacific have complied with the reporting requirements, primarily due to a lack of resources. One shortcoming of the reporting system in the region is the lack of provision for an independent verification of data or information in the reports. Furthermore, the lack of effectiveness is hampered by lack of trained personnel in flag state administration; the inability to retain skilled personnel; inappropriate delegation of inspection authority; and lack of sufficient financial resources.

C. Biological Diversity-related Conventions

1. The Convention on Biological Diversity

The CBD affirms that biodiversity conservation is a common concern of humankind and reaffirms the principle that states have sovereign rights over their own biological resources. The key commitments made by the Parties are: to develop national strategies and plans for the conservation and sustainable use of biodiversity (see Chapter 3); integrate biodiversity into relevant sectoral plans; monitor biodiversity and associated impacts; protect biodiverse habitats and ecosystems, and the environs; regulate, manage or control risks associated with use and release of organisms produced from biotechnology; prevent the introduction of exotic species; preserve knowledge and practices of indigenous and local communities; develop necessary legislation and other regulatory provisions for the protection of threatened species; establish programmes and promote research that contributes to knowledge of biodiversity; and promote programmes for public education and awareness building.

The provisions concerning genetic resources are, that national governments have the authority to determine access to genetic resources as provided for in national legislation; each Party shall endeavour to facilitate access to genetic resources for environmentally sound uses by other Parties; and access is on mutually agreed terms and is subject to mutually informed consent.

The acceptance of the CBD is remarkably high in the Asian and Pacific Region: 90 per cent of countries have already ratified the Convention (GEO 2000) (Table 21.4). This is understandable in view of the fact that most of the biodiversity of the world is in the region. In fact, the CBD has the second highest number of ratifications in the world, second only to UNFCCC. However, in addition to the ongoing

Table 21.4 Percentage of Parties in CBD in Asia and the Pacific (As of 1 March 1999)

Region	Number of countries	CBD Parties	Ratification (%)
South Asia	9	8	89
Southeast Asia	5	4	80
Greater Mekong	5	4	80
Northeast Asia	5	5	100
Pacific	16	15	94
Asia and the Pacific	40	36	90
World	193	174	90

Source: UNEP 2000

Table 21.5 Percentage of Parties in CITES in Asia and the Pacific (As of 1 March 1999)

Region	Number of countries	CITES Parties	Ratification (%)
South Asia	9	7	78
Southeast Asia	5	5	100
Greater Mekong	5	4	80
Northeast Asia	5	4	80
Pacific	16	5	31
Asia and the Pacific	40	25	63
World	193	145	75

Source: UNEP 2000

problem of financial and technical resources, it is thought that two significant issues will hound the future COPs: biosafety and intellectual property rights.

2. *Convention on International Trade in Endangered Species of Wild Fauna and Flora (CITES)*

CITES aims to ensure, through international cooperation, that trade in species of wild flora and fauna does not threaten the survival of the species in the wild, and, moreover, to protect certain endangered species from over-exploitation by a system of import/export permits issued by the management authority, under the control of a scientific authority (see Chapter 3).

Some of the requirements of Parties include the need for: permits for species listed in Appendices I and II of the Convention, stating that export/import will not be detrimental to the species; export permits whether trade in species listed in Appendix III of the Convention is from the State that listed the species, or otherwise; entry in a reservation if a Party does not accept the placing of a species in a certain Appendix; and detailed records of trade in species to be maintained by each Party. The COP to the Convention meets about every two years to examine progress in the restoration and conservation of protected areas and to revise the Appendices as appropriate. If a two-thirds majority endorses them, the amendments enter into force with formal ratifications.

Although there are prescribed penalties for non-compliance, or even trade sanctions, the Parties in general avoid being cited in alleged infractions. In some very serious cases, the Secretariat recommends to Parties not to accept permits from a particular country, pending the correction of problems of implementation.

Surprisingly, despite the rather long history of and important stake in the Convention, its percentage of ratification in the Asian and Pacific Region is low. Only 25 countries of the region have ratified it, which is lower than the world average (Table 21.5). However, the percentage ratification is high in the megadiversity subregions.

3. *The RAMSAR Convention*

The main objective of the Convention is the conservation and wise use of wetlands, through actions at both the national and international level. The Parties to the Convention are primarily required to: designate at least one national wetland for inclusion in the List of Wetlands of International Importance; promote the conservation of the wetlands included in the List; establish wetland nature reserves; cooperate in the exchange of information, and train personnel for wetlands management; and cooperate with other countries concerning shared wetland species. The Convention also promotes regional cooperation through various regional-level dialogues and provides linkages to other MEAs.

The rate of ratification is low in the Asian and Pacific region (Table 21.6), particularly in the South Pacific subregion. As of May 1999, the Contracting Parties in the region are Bangladesh, People's Republic of China, India, Indonesia, Islamic Republic of Iran, Japan, Malaysia, Mongolia, Nepal, Pakistan, Philippines, Republic of Korea, Sri Lanka, Thailand, Viet Nam, Australia, and New Zealand. Many of the Small Island Developing States (SIDS) of the Pacific are not Parties to the Convention due to a common misunderstanding that SIDS do not have may wetlands, or that the Convention is mainly focused on migratory birds (the broad definition of wetlands in the Convention could actually cover rivers, lakes, seagrass habitats, and coral reefs).

Significant headway has been made with implementing the Convention in the region in the last few years. Major achievements include: the completion or planning of National Wetland Policy/

Table 21.6 Percentage of Parties to RAMSAR in Asia and the Pacific (As of 1 March 1999)

Region	Number of countries	RAMSAR Parties	Ratification (%)
South Asia	9	6	67
Southeast Asia	5	3	60
Greater Mekong	5	2	40
Northeast Asia	5	4	80
Pacific	16	3	19
Asia and the Pacific	40	18	45
World	193	114	59

Source: UNEP 2000

Table 21.7 Parties to the Basel Convention and Ratification (As of 1 March 1999)

Region	Number of countries	BASEL Parties	Ratification (%)
South Asia	9	7	78
Southeast Asia	5	4	80
Greater Mekong	5	2	40
Northeast Asia	5	4	80
Pacific	16	4	25
Asia and the Pacific	40	21	53
World	193	121	63

Source: UNEP 2000

Strategy/Action Plans in 14 countries of the region; establishment of National Wetland Bureaus in nine countries; implementation of significant number of actions to enhance education and public awareness at all levels, with the active involvement and contributions of NGOs; greater consideration given to wetland concerns in national environmental conservation planning initiatives; enhancement of efforts to foster greater cooperation among institutions in wetland management and to coordinate the implementation of various conventions and agreements; and an increase in the participation of local stakeholders in site management (RAMSAR COP 7 Doc. 12).

D. Hazardous Waste (the Basel Convention)

The main objectives of the Basel Convention (Convention on the Control of Transboundary Movements of Hazardous Wastes and Their Disposal) are: to control and reduce transboundary movements of hazardous wastes; minimize the generation of hazardous wastes; and assist developing countries and CEIT in practicing environmentally sound management of the hazardous and other wastes which they generate. The major obligations of the Parties are to prohibit and/or control the transport, disposal or export of hazardous wastes in designated areas, and to establish one or more competent authorities as a focal point for implementation of the Convention at national level.

The percentage of ratification in the Asian and Pacific Region is low compared to the world average, however it is quite high in South, Southeast and East Asia (Table 21.7). It is not surprising that the percentage of ratification in the South Pacific is only 25 per cent, since the Waigani Convention already covers the subregion. However, there is still a distinct advantage for SIDS to accede to the Basel Convention. For one thing, it is global in scope, and most of the industrial countries are Parties to the Convention, which will introduce benefits from the financing and technology transfer provisions of the Convention.

The most significant development under the Convention is Decision II/12 agreed by consensus on 25 March 1994 to ban all exports of hazardous wastes from OECD countries to non-OECD countries. The great environmental significance of this (now famous) Basel Ban (Decision II/12) is that it recognized and closed the recycling loophole through which it is thought that more than 90 per cent of hazardous waste was flowing. Since 1998, OECD countries have been forced to minimize their wastes because hazardous waste traders will no longer be able to justify their claims that they export these wastes for recycling in developing countries.

Those who have vested interests in continuing trade in hazardous wastes challenged the quick implementation of the Basel Ban. The argument was that the ban was not part of the Convention, and therefore not legally binding. Furthermore, the Decision does not use the distinction between the OECD and non-OECD countries. Rather, it bans exports from what is called Annex VII countries (EU, OECD, Liechtenstein to all Parties of the Convention). For this amendment to the Convention to come into effect, 62 countries need to ratify it. As of 11 November 1999, only 17 had done so.

E. Desertification (Convention to Combat Desertification)

The main objectives of the Convention to Combat Desertification (CCD) are: to combat desertification and mitigate the effects of drought, by establishing strategies and priorities within the framework of sustainable development plans; to address the underlying causes of desertification; to promote the awareness and facilitate participation of local populations; and to prove an enabling environment by strengthening relevant existing legislation, or by enacting new laws and establishing long-term policies. It is also stipulated that developed countries should actively support the efforts of affected developing-countries, for example, by providing funding, promoting the mobilization of new and additional funding (e.g. through the GEF), mobilizing funding from the private sector and other non-government sources, and promoting and facilitating access by affected developing country Parties to appropriate technology and knowledge.

Key decisions and developments during the various COPs which have taken place since the Convention was established have included:

- COP1, held in Rome in 1997 established a Committee on Science and Technology (CST), which advises and meets simultaneously with the COP. An innovative feature of COP1 was the active participation of the NGOs in the plenary sessions, although the Conference was haunted by inter-regional issues and issues surrounding the choice of memberships to the Bureau.
- COP2 held in Dakar in 1998 established an *ad hoc* panel to consider the links between traditional knowledge and modern technology. The other dominant theme of this meeting was fostering partnerships between and among a variety of actors at all levels including NGOs and people in communities.
- The main output of COP3, held in Recife, Brazil in 1999, was the Recife Initiative, which calls for enhanced implementation of the CCD. It also reiterates the need to mobilize resources and the transfer of technology, and reaffirms the need to integrate CCD implementation into mainstream national development efforts.

The percentage ratification in the Asian and Pacific Region is significantly lower than the world average, particularly in the South Pacific and Southeast Asia (Table 21.8). However, considering the scepticism expressed in Rio regarding the viability of a global convention on desertification, the recent progress of the CCD indicates a good outlook for the future. As of November 1999, 159 countries had ratified the convention (UN 1999b). The convention has also broken new ground in terms of NGO participation.

F. Prospects for Future International Conventions

There are at least two good prospects for future conventions: persistent organic pollutants (POPs) and forests. These are discussed below.

1. Towards a Global POPs Treaty

POPs are long-lived organic compounds that become more concentrated as they move up the food chain. They can travel thousands of kilometres from

Table 21.8 Percentage of Parties to CCD in the Subregion of Asia and the Pacific (As of 1 March 1999)

Region	Number of countries	CCD Parties	Ratification (%)
South Asia	9	7	78
Southeast Asia	5	2	40
Greater Mekong	5	4	80
Northeast Asia	5	3	60
Pacific	16	6	38
Asia and the Pacific	40	22	55
World	193	144	75

Source: UNEP 2000

their point of release. Although POPs cover a broad range of chemical classes, the concern and much of the research has focused on about a dozen chemicals. These include the industrial PCBs, polychlorinated dioxins and furans, pesticides such as DDT, chlordane, and heptachlor. Most of these chemicals are now banned in industrialized countries, but remain in use in many developing countries. The majority of these chemicals, however, are still being produced in the United States.

The Earth Summit created the Inter-governmental Forum on Chemical Safety (IFCS) to deal with the proliferation and spread of hazardous chemicals into the environment. In 1996, the IFCS concluded that there was sufficient evidence for international action. They recommended that a global legally binding instrument was necessary to address the serious risks posed by POPs to human health, and to the earth's ecosystems.

The UNEP Governing Council answered the pressing concerns about POPs by authorizing the creation of an Inter-governmental Negotiating Committee (INC) to prepare a global POPs treaty. The first round of negotiations started in Montreal in June of 1998, followed by a second round in Nairobi in January 1999. A total of five negotiating sessions are planned before the end of 2000. This will be followed by a diplomatic conference for the signing of the treaty.

The raging controversy during the Montreal and Nairobi meetings revolved around the issues of whether to manage POPs, or reduce and ultimately eliminate POPs. As expected, the industry wants to manage their use, and the NGOs are on the side of reduction and eventual elimination. However, there was a positive spirit in all camps during the two sessions, representing the commonly held view amongst all camps that urgent action needs to be taken.

2. A Forest Treaty

The prevailing North-South polarization on forestry issues at the time of the Earth Summit did not permit the adoption of an international agreement, although there was universal acceptance of the importance of forests to the ecological well-being of the Earth. The previous outcome of Rio for forests was the Non-legally Binding Authoritative Statement of Principles for a Global Consensus on the Management, Conservation and Sustainable Development of All Types of Forests, now known as the Forest Principles.

In 1995, the UNCSD was able to create consensus to establish the Inter-governmental Panel on Forests (IPF) with a limited life of two years. The two years of work resulted in over 100 proposals covering a wide range of topics in sustainable forest management. As was expected, consensus could not be reached on issues relating to financial mechanisms, transfer of technology, trade and environment, and legal instruments.

The recommendations of the IPF were endorsed by the CSD in 1997 and by the UN General Assembly in the same year. However, in view of the many unresolved issues, the Economic and Social Council of the United Nations (ECOSOC) decided on the continuation of the inter-governmental policy dialogue by creating the Inter-governmental Forum on Forests (IFF), which is *ad hoc* and open-ended. Its mandate covers promoting and facilitating the implementation of the proposals for action of the IPF, and reviewing, monitoring and reporting on progress in the management, conservation and sustainable development of all types of forest; taking into consideration those matters left pending and other issues arising from the programme elements of the IPF process; and international arrangements and mechanisms to promote the management, conservation and sustainable development of all types of forests.

The IFF reported its work to the CSD in 1999. The report will be used by the CSD in making the decision on whether the Forum will engage in further action on establishing an inter-governmental negotiation process on new arrangements and mechanisms, or a legally-binding instrument on all types of forests. However, this plan is already delayed. The Third Meeting of IFF last May 1999 was not very encouraging and there were very few substantive agreements. The report of meeting contains a heavy load of bracketed texts that became a burden during the Fourth Meeting in January 2000. The prospect of a Forest Convention is still some way off.

GLOBAL PARTNERSHIPS: PROBLEMS AND PROSPECTS

The most outstanding outcome of the Earth Summit was the historic commitment to partnership between the developing and industrialized countries, Agenda 21 and the signing of some of the most important MEAs. Almost all of these call for the provision of financial and technological resources to the developing countries. These pleas are constantly echoed in Conferences of the Parties and other international conferences that followed Rio. However, progress has been slow. As a prelude to finding new ways of international and regional collaboration, it is therefore essential to look at the financial and technological resource flows from developed to developing countries.

A. Transfer of Financial Resources

The promise at the Earth Summit was that richer countries will help the poorer nations achieve sustainable development by stepping up aid spending. Ironically, as shown in Figure 21.1, the actual magnitude of official development assistance (ODA) has been decreasing since this time. Between 1990 and 1996, ODA decreased by about 12 per cent, while the debt burden of developing countries expanded by almost 500 per cent (Table 21.9). The surging debt flow, composed mainly of loans to commercial banks and purchase of bonds by the industrialized countries, is a cause for alarm. During the period, private flows also had a remarkable jump, from US$44.4 billion to US$243.8 billion, due to large increases in FDI and portfolio equity flows (Figure 21.1 and Table 21.10). The private flow's share as fraction of aggregate net resource flow rose from 44 per cent in 1990 to 86 per cent in 1996. FDI represents more than 45 per cent of total private flows in 1996. Grants and other concessional finance, which represent the bulk of aid flows, fell by US$1 billion below the figure in 1990 in nominal terms, and more in real terms.

The ratio between ODA and GNP in OECD countries fell from 0.33 per cent in 1992, which was maintained in the first three years of the decade, to 0.22 per cent in 1997. This is the lowest ratio recorded since 1970, when the United Nations adopted the 0.7 per cent target. In this same year the US (0.08 per cent) and Japan (0.11 per cent) record the lowest ODA to GNP ratios. However, Japan is the biggest donor in dollar terms. Furthermore, Japan, the US., France, and Germany accounted for some 58 per cent of total ODA in 1997. Among the G-7 countries, the United Kingdom has reaffirmed its commitment to meet the

Table 21.9 Aggregate Net Long-term Resource Flows to Developing Countries, 1990-96 (US$ billion)

Type of flow	1990	1991	1992	1993	1994	1995	1996[a]
Aggregate net resource flows	100.6	122.5	146.0	212.0	207.0	237.2	284.6
Official development Finance	56.3	65.6	55.4	55.0	45.7	53.0	40.8
Grants	29.2	37.3	31.6	29.3	32.4	32.6	31.3
Loans	27.1	28.3	23.9	25.7	13.2	20.4	9.5
Bilateral	11.6	13.3	11.3	10.3	2.9	9.4	-5.6
Multilateral	15.5	15.0	12.5	15.4	10.3	11.1	15.0
Total private flows	44.4	56.9	90.6	157.1	161.3	184.2	243.8
Debt flows	16.6	15.2	35.9	44.9	44.9	56.6	88.6
Commercial banks	3.0	2.8	12.5	-0.3	11.0	26.5	34.2
Bonds	2.3	10.1	9.8	35.9	29.3	28.5	46.1
Others	11.3	3.3	13.5	9.2	4.6	1.7	8.3
Foreign direct investment	24.5	33.5	43.6	67.2	53.7	95.5	109.5
Portfolio equity flows	3.2	7.2	11.0	45.0	32.7	32.1	45.7

Source: World Bank. Debtor Reporting System.
Note: Developing countries are deferred as low and middle income countries 1990 per capita incomes of less than $65 (low) and $9 385 (middle).
[a] Preliminary.

Table 21.10 Aggregate Net Private Capital Flows to Developing Countries, 1990-96 (US$ billion)

Type of flow	1990	1991	1992	1993	1994	1995	1996[a]
Total private flows	44.4	56.9	90.6	157.1	161.3	184.2	243.8
Portfolio flows	5.5	17.3	20.9	80.9	62.0	60.6	91.8
Bonds	2.3	10.1	9.9	35.9	29.3	28.5	46.1
Equity	3.2	7.2	11.0	45.0	32.7	32.1	45.7
Foreign direct investment	24.5	33.5	43.6	67.2	83.7	95.5	109.5
Commercial banks	3.0	2.8	12.5	-0.3	11.0	26.5	34.2
Others	11.3	3.3	13.5	9.2	4.6	1.7	8.3
Memo items							
Aggregate net resource flows	100.6	122.5	146.0	212.0	207.0	237.2	284.6
Private flows' share (per cent)	44.1	46.4	62.1	74.1	77.9	77.7	85.7

Source: World Bank Debtor Reporting System.
[a] Preliminary.

0.7 per cent target. Table 21.11 shows that, of the total US$243.8 billion private flows, some US$119.4 billion or 49 per cent went to the Asian and Pacific Region and US$52 billion or 43 per cent of the region's share of this went to People's Republic of China. The other top destinations in the region were Malaysia, Indonesia, Thailand, and India. It is interesting to note that the top 12 recipient countries captured 72.5 per cent of the net capital flows to developing countries in 1996. In addition, three countries of Eastern Europe were not traditional recipients of development finance.

The distribution of private capital flows in 1995 looks different when seen as percentage of GDP. The top ten recipients are also shown in Table 21.11. Although the region captured almost half of the flows, only Malaysia and Papua New Guinea were in the first ten in terms of private capital flows per GDP. People's Republic of China, India, and Indonesia were not in the list partly because of the relatively large GDPs of these countries.

Concessional finance, which includes grants, bilateral and multilateral finance, has generally been declining (Table 21.12). The level for 1996 in real terms shrunk back to 1990 level. It declined by US$0.8 billion between 1995 and 1996. The biggest component of this decline was the decrease in grants of US$1.3 billion. The outlook is not promising with decreasing commitments from donor countries.

It is interesting to examine the regional distribution of official concessional finance (Figure 21.6). In 1989, the main destination was Sub-Saharan Africa, which captured 40 per cent of the flow. The Asian and Pacific Region (East and

GLOBAL AND REGIONAL ISSUES

Table 21.11   Net Private Capital Flows to Developing Countries by Country Group, 1990-96

(billions of U.S. dollars)

Country group or country	1990	1991	1992	1993	1994	1995	1996[a]
All developing countries	44.4	56.9	90.6	157.1	161.3	184.2	243.8
Sub-Saharan Africa	0.3	0.8	0.3	0.5	5.2	9.1	11.8
East Asia and the Pacific	19.3	20.8	36.9	62.4	71.0	84.1	108.7
South Asia	2.2	1.9	2.9	6.0	8.5	5.2	10.7
Europe and Central Asia	9.5	7.9	21.8	25.6	17.2	30.1	31.2
Latin America and the Caribbean	12.5	22.9	28.7	59.8	53.6	54.3	74.3
Middle East and North Africa	0.6	2.2	0.5	3.9	5.8	1.4	6.9
Income group							
Low-income countries	11.4	12.1	25.4	50.0	57.1	53.4	67.1
Middle-income countries	32.0	44.0	64.8	107.1	104.2	130.7	176.7
Top country destinations[b]							
PR China	8.1	7.5	21.3	39.6	44.4	44.3	52.0
Mexico	8.2	12.0	9.2	21.2	20.7	13.1	28.1
Brazil	0.5	3.6	9.8	16.1	12.2	19.1	14.7
Malaysia	1.8	4.2	6.0	11.3	8.9	11.9	16.0
Indonesia	3.2	3.4	4.6	1.1	7.7	11.6	17.9
Thailand	4.5	5.0	4.3	6.8	4.8	9.1	13.3
Argentina	-0.2	2.9	4.2	13.8	7.6	7.2	11.3
India	1.9	1.6	1.7	4.6	6.4	3.6	8.0
Russia Federation	5.6	0.2	10.8	3.1	0.3	1.1	3.6
Turkey	1.7	1.1	4.5	7.6	1.6	2.0	4.7
Chile	2.1	1.2	1.6	2.2	4.3	4.2	4.6
Hungary	0.3	1.0	1.2	4.7	2.8	7.8	2.5
Percentage share of top twelve countries	83.6	76.8	87.4	84.1	75.4	73.3	72.5

Source: World Bank Debtor Reporting System and staff estimates.
Note: Private flows include commercial bank lending guaranteed by export credit agencies.
[a] Preliminary.
[b] Country ranking is based on cumulative 1990-95 private capital flows received. Private flows include commercial bank loans guaranteed by export credit agencies.

Table 21.12   Official Net flows of Development Finance, 1990-96

Category	1990	1991	1992	1993	1994	1995	1996[a]
Official development finance	56.3	65.6	55.4	55.0	45.7	53.0	40.8
Concessional finance	43.9	53.3	46.1	43.4	40.7	45.2	44.4
Grants	29.2	37.3	31.6	29.3	32.4	32.6	31.3
Loans	14.8	15.9	14.6	14.2	14.3	12.6	13.1
Bilateral	8.7	9.2	7.4	7.3	6.0	4.8	4.8
Multilateral	6.1	6.7	7.2	6.9	8.3	7.8	8.3
Nonconcessional finance	12.3	12.1	9.3	11.5	1.1	7.8	3.6
Bilateral	2.9	4.1	3.9	3.1	-3.1	4.6	-10.4
Multilateral	9.4	8.3	5.4	8.4	2.0	3.2	6.8
Memo items							
Use of IMF credit	0.1	3.2	1.2	1.6	1.6	16.8	0.6
Vectortral cooperation grants	14.2	15.7	17.9	18.5	17.4	20.7	20.0

Source: World Bank. Debtor Reporting System.
Note: Official concessional finance inflows of capital development assistance (excluding isdinical cooperation grants) and Eastern Europe and the formes Soviet Union Memo items are not included in preceding aggregates.
[a] Preliminary.

Figure 21.6 Regional Distribution of Development Finance, 1989

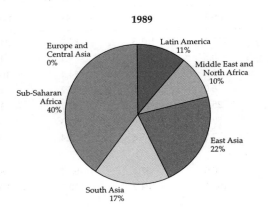

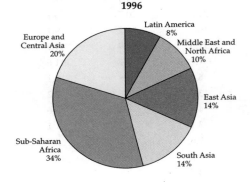

Source: OECD Development Assistance Committee and World Bank Debtor Reporting System.

Note: Includes official development assistance and official aid to middle- and low-income countries; excludes official development assistance to high-income countries.

South Asia) got 39 per cent, while Europe and Central Asia got nothing. There was an enormous share for Europe and Central Asia in 1996, which increased by as much as 20 per cent of the total. The region's share declined to 28 per cent. The trend is a decreasing share of the shrinking concessional finance for Asia and the Pacific.

The prospect for the reversal of the decreasing trend in ODA is very uncertain and there is little room for optimism. The financial crisis in East Asia could uncover some debt arrangements that must be rescued. This could lead to increased claims on global resources and to further reduction on traditional aid programmes. There were other discouraging developments in 1997, for instance, when the Asian Development Fund did not meet its targets. Moreover, there was also a decline in IBRD's income.

One of the most important reasons for the major shifts in financial flows to the region is that many countries dramatically broke from the development tradition of the past by liberalizing foreign investment rules. Policies that discouraged investments were repealed. Others, such as foreign ownership of real estate, the guarantee of the repatriation of profits and capital, and the protection of property rights were promulgated. These liberalization measures were further reinforced by privatization, which swept the countries of the region. Governments were selling utilities, power, telecommunications and other state enterprises to private bidders-domestic and international or partnerships of these two groups. On the other hand, there were complementary developments from capital-rich countries. Interest rates have been low in many countries and investors were in search of overseas opportunities where they could find higher returns. Other industrial countries have also relaxed the control on outward bound capital. Moreover, these very favourable conditions were enhanced by the increasing power of international communications and computer links that allowed investors to move money from one place to another with considerable ease. The result of all these developments was a boom in for most countries of the region, which grew by about three times the growth rates in the industrialized counties.

The Secretariat of the Earth Summit estimated the financial needs of implementing Agenda 21 in developing countries to be over US$600 billion annually for the period 1993-2000 (Sitarz 1994). About US$125 billion in grant or concessional terms will also be required from the international community. The total concessional finance (grants and loans) may not increase in the foreseeable future. The US$44.4 billion in 1996 for concessional finance did not all go to sustainable development activities. It is safe to say that the financial requirements of Agenda 21 expected from concessional financing is quite far from the target set at the Earth Summit. The estimated annual requirement of US$600 billion is just about a third of the approximately US$2 trillion total GNP of bigger developing member countries of the ADB (ADB, 1999). As a percentage of the total GNP of all the developing countries, it is much smaller.

At the Earth Summit, there was a reiteration of UN plea to the industrial countries to provide ODA of 0.7 per cent of the GNP to the countries. Based on the actual records of contributions to ODA, the biggest givers are not the same as those meeting the target of 0.7 per cent of GNP. Seven years after Rio, only four of the industrial countries have achieved the target and in fact, ODA has been declining. Moreover, ODA is necessarily earmarked for sustainable development or Agenda 21 activities.

B. Transfer of Technology

Most of the MEAs provide for the transfer of environmentally sound technologies to the developing countries. This provision has been one of the most controversial of many international agreement, and in many cases is expressed in very general terms that are not easy to translate into more operational language. Moreover, most MEAs do not have compliance mechanisms.

The coordination, as well as the principal initiative, in technology transfer resides in the mandate of the United Nations Commission on Sustainable Development (CSD). Its work programme in this sector is focused on Cleaner Production/Eco-efficiency; Environmental Accounting & Environmental Management Systems; Environmentally Sustainable Technologies (EST) Information Systems; and National Strategies and Policies.

To assist and complement the work of the CSD, the United Nations Asian and Pacific Centre for Technology Transfer (APCTT) was also established for the purpose of diffusing environmentally sustainable technologies (ESTs) in the region. The CSD through the Department of Social Affairs and Development (DESA) prepared a review of the implementation of Chapter 34 of Agenda 21 which, covers the transfer of ESTs, cooperation, and capacity building. Key areas in relation to technology transfer are discussed below.

1. *Information Networks*

One of the problems for the diffusion of ESTs is the lack of information for decision-making. There is a whole universe of choices and choosing the right one for a given situation requires not only good information, but also sound expert advice. In response to this need, particularly in developing countries, the United Nations system is providing assistance through the development of information networks and clearing houses.

There have been significant accomplishments to date. In coordination with FAO, the Agricultural Information System (AGRIS) and the Current Agricultural Information System (CARIS) were established. Both networks operate multi-lingual and offer the possibility of integrating and supporting national and regional systems. UNIDO's Industrial and Technological Information Section has been providing assistance to developing countries in terms of building techno-intelligence capabilities at the sub-sectoral level. This will enable them to monitor technological developments, assess global market trends, and analyse key competitors. UNIDO also conducts a business forum (Techmart) where seekers and suppliers of technology meet. Information in terms of samples, process flow diagrams, photographs, and product catalogues are distributed. A compendium of technologies offered and requested worldwide are provided in advance of the meeting.

UNEP and UNCHS have developed a number of databases and clearing-houses such as the ITPCT and the International Cleaner Production Information Clearinghouse under the Montreal Protocol. In the specific area of fertilizer production, the Fertilizer Advisory, Development and Information for Asia and the Pacific (FADINAP) have information for the transfer of ESTs. Two international and regional technology centres have also been established: the African Regional Centre for Technology Information System (ARCTIS) and the Asian and Pacific Centre for the Transfer of Technology (APCTT). Both have mechanisms for promoting purpose oriented information networking. APCTT focuses on small and medium scale industries.

2. *Access to Technology Transfer*

The various agencies of the United Nations have a pivotal role to play in promoting the international transfer of technology which is essential to sustainable development of the developing countries. UNEP's International Environmental Technology Centre (IETC) is seeking the barriers to technology transfer and promoting consensus to remove such barriers. UNIDO and UNCTAD have undertaken policy studies and provided policy advice on various aspects of technology transfer. FAO is formulating frameworks for technology assessment and also designing guidelines for this purpose. UNIDO and UNEP promote cleaner production through process optimization and the re-engineering of technologies to suit local conditions. WMO's Hydrological Operational Multipurpose System (HOMS) is a networked technology transfer system in science and engineering with a formal structure for the transfer of discrete and distinct items of technology. UNDP's Urban Environment Technology Initiative provides some appropriate technological alternatives to urban policy makers in developing countries. UNCTAD's programme on technology transfer places emphasis on enhancement of technological and scientific innovative capacities *vis-à-vis* the challenges and opportunities posed by technological change. World Bank's GEF-supported project foster the transfer of ESTs while strengthening institutional capacity and improving the scientific and technological base. FAO focuses on indigenous technologies while promoting technological packages that could increase productivity. WHO is promoting the transfer of vaccination technology to developing countries which includes the establishment of advisory panels and the provision of databases on international and national regulations. The International Civil Aeronautics Organization (ICAO) is working on a coordinated global plan for communications, navigation, surveillance, and air traffic management and transferring these technologies in a progressive manner.

To understand the demand side of technology transfer, the technology needs of developing countries must be assessed.

3. Capacity Building for Managing ESTs

Whether imported or self-generated, it is essential for developing countries to have the capacity for technology management. Capacity-building policies for technology management should be basically country- or demand-driven. Traditional supply-driven approaches in technical assistance in this area have failed to produce the desired results in capacity building. It is important to involve the users and beneficiaries in programmes of this sort.

Some progress have already been achieved. The UN System has continued to provide technical assistance aimed at capacity building in terms of training, scholarships, etc. Recent efforts have emphasized participatory approaches and involved the private sector. In industry, UNIDO promoted the concept of EST and is gaining ground over the traditional end-of-pipe solution. The emphasis is increasingly on process improvement towards waste minimization. ICAO is assisting civil aviation authorities of developing countries in maintaining academic standards and self-sufficiency through its global resource-sharing network. ILO is addressing the issue of the employment effects of technological change and the issue of flexible specialization and organization of production. The World Bank has continued to expand its support for projects aimed at rehabilitating research institutes and strengthening university-level educational capacities. UNESCO has approached capacity building by encouraging partnerships among governments, industry, research and educational institutions, and non-government organizations at national and regional levels. These partnerships involve the pooling of resources.

4. Network of Research Centres

It is imperative for countries to have their own research and development systems in order to enhance the generation of indigenous ESTs, and to successfully engage the international community in cooperative activities.

Some achievements have been recorded in this aspect. The UN System has fostered collaborative networks through South-South and North-South institutional cooperation and partnership. UNESCO's Man and the Biosphere Programme and IOC's International Hydrological Programme have established regional and subregional networks for research, training, and knowledge sharing. These networks now constitute a global network. The Global Change System for Analysis, Research, and Training (START) Regional Research Networks has been promoting research on the regional origins and impacts of environmental changes. In the START concept, distinct bio-geographical areas have been identified for the development of regional networks. These networks would then gradually evolve into a global network. The Consultative Group on International Agricultural Research (CGIAR) has been helping in the strengthening of national research systems. The National Cleaner Production Centres (NCPCs) have promoted cooperation with other countries by establishing national units doing local demonstration projects and research in industrialized and developing countries. The International Atomic Energy Authority (IAEA) has coordinated research programmes that link national institutes in developing countries and industrialized countries.

5. Support of Cooperation and Assistance

Agenda 21 indicates mechanisms whereby capacity-building in assessing, adapting, managing ESTs in developing countries can be supported. Significant advancements in this area have already been accomplished. The United Nations system considers the training of local scientists and engineers to be the crucial aspect of the technology transfer of ESTs. Activities include the organization of short-term courses, follow-up programmes to provide continuous information and technical assistance to former participants, and refresher regional courses. UNEP, ILO, and WHO have jointly undertaken "train the trainer" programmes on environmental management in industry, reflecting the recognition that utilizing local trained personnel is more effective than technology transfer by foreign experts. UNIDO has focused on institutional capacity building and institutional support through the provision of information on ESTs to environmental management agencies.

6. Technology Assessment

Assessment of the technology needs enable countries to make intelligent choices. Activities in this area include capacity building in technology assessment with environmental impact and risk assessment and the strengthening of the networks for technology assessments at various levels. In the United Nations system, most of the activities are part of the capacity building process in developing countries that is related to the promotion of better access to technology transfer. In general, there are two types of activities: those that are aimed at strengthening capabilities in technology assessment capabilities in developing countries; or assessments undertaken by the United Nations itself which would benefit the developing countries. UNDP and FAO also undertook a project called Farmer-Cantered Resource Management involving eight Asian countries. This project has strong elements of technology assessment through participatory approaches. UNEP initiated the preparation of an Environmental Technology Assessment Newsletter for the purpose of sharing information and experience in this field.

Some progress has been made in preparing guidelines and manuals for conducting technology assessment in the region. UNEP in consultation with UNCTAD initiated studies on the guidelines to avoid the transfer of hazardous technologies. ITU approached technology assessment as an integral component of technology transfer in the process of adaptation to local conditions. The science and technology programme of UNCTAD includes a series of activities in technology assessment, including workshops in technology assessment and forecasting. The Regional Commissions are also promoting this type of activity.

7. *Collaborative Arrangements and Partnerships*

Collaboration and partnerships between enterprises in developing and developed countries are requisites to the promotion of technology transfer. Joint ventures and foreign direct investments play a big role in the transfer of technologies that benefit both the supplier and recipients of technologies. However, there are not nearly enough of these North-South or South-South arrangements between enterprises. The potential contributions of trans-national corporations in this area have also not been fully explored.

Several achievements have been reported in this area. FAO is making use of these arrangements to promote technology transfer among developing countries through the TCDC workshops. UNEP jointly with Tuffs University and the Prince of Wales Business Leaders Forum (PWBLF) published a handbook of successful partnerships among various entities. In a few Asian countries, there has been collaboration between enterprises in research and development, which could lead to commercialization in the recipient country. Compared to traditional technology transfer, the resulting technology is adapted to local conditions and has a lower cost.

There is a multitude of activities, especially in the United Nations system, aimed at the promotion of technology transfer. Obviously, they are not quite enough to close the ever-widening gap in technology between the developing countries of Asia and the Pacific and the OECD countries. Overall, the region, with the exception of Japan, is a net importer of technology. While it may be true that technology is being transferred, it is not at a rate that enhances rapid industrial progress, nor is it enough to promote technological self-reliance. There are still many barriers to successful and equitable transfer of ESTs in the region.

In 1996, the Asian and Pacific Centre for the Transfer of Technology (APCTT) organized a workshop to study the mechanisms for and barriers to the technology transfer of ESTs. Accordingly, the most important impediments include weak or distorted demand for EST due to lack of awareness on the part of the small and medium enterprises (SME) and little or no enforcement capability; low technical capability-lack of knowledge, resources and linkages to access ESTs; very little accessible information about technological alternatives; misallocation of financial resources in favour of larger projects or end-of-pipe solutions; missing connections between potential partners; and technology transfer mindset that technology is equated with hardware, seen as a single, point-in-time transaction.

The facilitating activities reported by CSD above and the barriers reported by APCTT meet head on, resulting in the present technological reality in the developing countries of the region.

ECONOMY AND THE ENVIRONMENT

Economy is the most important driver of environmental change, whether in a global, regional or national context. The production, distribution and consumption of goods, and the pursuit of leisure constitute the principal components of the human footprint on the environment. Population and the state of the economy are therefore crucial to meeting the challenge of sustainable development.

A. **Economic Landscape**

In recent years, the world economy has been moving in diametrically opposite directions. Many developing economies of Asia contracted because of capital flight, the turmoil in financial institutions and the sharp decline of the values of currencies. The so-called "tiger economies" or the "newly industrializing economies" contracted, reversing the trends of the last decade. The previously fastest growing economies of the world, went into recession. Even Japan, the world's second largest economy was dragged into recession because of the economic decline of most Asian countries, in addition to some of its own domestic problems. On the other hand, North America and Europe were experiencing high economic growth. The United States is experiencing unprecedented long-term growth of its equity

CHAPTER TWENTY-ONE

markets. Its stock exchange is on the longest bull run in history. Similarly, European stock exchanges were gaining values because of the high expectations on the stronger economic integration of the European Union and the widespread adoption of a single currency.

The recent economic contraction in East Asia had caused social and political dislocation (Box 21.1). In a region that has grown accustomed to increasing prosperity, the doubling or tripling of unemployment was a serious social shock, with political ramifications. Displaced urban workers returned to rural areas of their origins, and added pressure to the already stressed ecosystems there. In Indonesia, the social impacts were particularly severe. The number of people below the poverty line was 11 per cent in 1996, but went up to 48 per cent during the financial crisis. This is expected to increase even more in the early part of this century, to over 66 per cent, if there is no improvement in the economy (DESA 1999).

The economic development landscape is being transformed, thus presenting new challenges to the sustainable development of the region. After more

Box 21.1 The Social Impact of the Asian Economic Crisis

The Asian financial crisis is causing turmoil throughout the affected economies in such areas as employment, prices, human development, poverty and social capital. The social consequences of the crisis vary across countries according to the extent of the downturn and dislocation. While the poor and vulnerable groups (such as women, children, and migrant workers) are invariably at risk, the impact of the crisis is pervasive, hurting all social classes, particularly middle- and lower-middle-income ones. The damage is proportionately more severe in urban centres than in rural areas. Unemployment rates are on the rise. In Indonesia, the rate rose to 5.5 per cent in 1998, up from 4.7 the previous year. In Thailand, it reached 5.3 per cent, compared with just above 1 per cent in 1997. The unemployment rate in Korea more than doubled to 6.8 per cent, while in the Philippines, it climbed to 9.6 per cent. However, these figures probably underestimate the true extent of the problem, as underemployment is also likely to have increased markedly. Standards of living are falling as inflation outstrips any increase in nominal incomes. For example, in Indonesia the increase in food prices contributed a large portion of the 58 per cent rise in the consumer price index in 1998.

The crisis has reduced household expenditure in a number of important social areas, including health and nutrition, education, and family planning. Although social services are subsidized, households still have to incur costs, either directly or indirectly, when they use these services. With reduced incomes and higher prices for such items as medicines and school supplies, households tend to cut down far too much on their consumption of these important items. Investment in human resources takes time, which becomes scarcer as household members work longer hours to cope with falling incomes. The quality and quantity of government-provided services are declining because of budget reductions and massive shifts of clients from private to public providers. Thus, observers cite notable enrolment decreases or drop-out increases, particularly at the secondary level. Estimates indicate that in Indonesia more than 6 million students have dropped out of school and in Thailand about 250 000 students have dropped out. In Malaysia private hospitals and clinics have prepared a fall of up to 50 per cent in the number of patients seeking treatment. In Indonesia, many people are shifting from modern medical care to traditional healers and to self-treatment, while the higher cost of contraceptives hampers the government's family planning programme. Consequently, a large number of women are dropping out of the programme, which is likely to lead to a rise in the birth rate and a substantial increase in the number of abortions. In Thailand, higher drug prices are adversely affecting the treatment of AIDS patients.

The crisis is setting back the spectacular strides made in the fight against poverty during the last two decades. Estimates made in the late 1990s indicated that GDP shrinkages in countries like Indonesia would result in a rise in the incidence of poverty of as much as 4 per cent in the last few years of the decade, with a larger relative increase in urban centres than in rural areas. The increases in the incidence in poverty correspond to about 6 million Indonesians falling below official poverty lines, a huge number that require urgent policy attention, but also far less than claimed by earlier reports. If these later estimates are correct, there are some grounds for encouragement, as reductions in poverty brought about by years of broadly based economic growth may not be quickly reversed all that quickly.

The environment is also suffering from serious degradation because of the crisis. Household attempts to obtain additional income often lead to increased environmental destruction, such as deforestation, erosion and overfishing. In Thailand, the devaluation of the baht has provided a strong stimulus to agricultural exports, resulting in expansion and intensification of shrimp farming and hence, destruction of wetlands and increased salinity of paddy fields. An increase in illegal logging has also occurred in Thailand and in neighbouring countries such as Cambodia and Myanmar.

Finally, in many countries the informal norms and social relationships that enable people to cooperate in pursuit of a common benefit are breaking down as a result of the crisis. This is seen in an increase in crime and violence, and in weakening of community cooperation. The adverse social consequences of the crisis are clearly severe and require urgent action on the part of governments to provide adequate social safety nets. However, the only long-term solution is to ensure a return to broadly based and sustained growth.

Source: ADB 1999

than half a century of experience in economic development, the World Bank (1999) claims four important critical lessons: macroeconomic stability is essential for economic growth; growth does not trickle down, and so basic human needs must be addressed directly; a comprehensive approach is required, there is no single policy that will trigger economic development; and sustainable development should be rooted in processes that are socially inclusive and responsive to changing circumstances.

While lessons are being learned, populations in Asia and the Pacific are constantly growing, natural resources are degrading, and income gaps are increasing. The challenge to sustainable development is to continue to tap the pool of human inventiveness and creativity to find ways of dealing with continuously changing economic and social landscape, and to utilize environmentally sound and sustainable technologies. However, even when technologies are available, as in industrialized countries, the resistance to the adoption of cleaner and more efficient production is heavily influenced by the economic climate. In developing countries, economic decline heightens the imperative of exporting commodities even at the price of uncontrolled expansion into wilderness areas and the logging of already over-logged forests.

B. Global Economic Perspectives

The Asian and Pacific Region's economy is now growing at its slowest rate since the early years of the decade. The growth of world output declined dramatically too, from 3.3 per cent in 1997 to 1.7 per cent in 1998 (DESA 1999). The slowdown had many causes, such as the contraction of many Asian economies, the recession in Japan and the Russian economic crisis. The world economy in the last few years of the 20th century was not uniformly bleak. There were some countries enjoying prosperity such as the United States and Europe. Furthermore, the rate of economic output rose in some 100 countries, including some transition economies and in India and People's Republic of China. Together, the latter two Asian countries comprise about half of the world's population. The economic contagion in East Asia is not a universal phenomenon. However, there were fears that the recession in this part of the world could affect the rest of the world. The recession in East Asia is expected to recover slowly.

Japan has the greatest influence in the economies of Asia and the Pacific. Like other countries in the region, it too slipped into recession in the fourth quarter of 1997. Output contracted further in 1999. The source of economic doldrums in Japan is, however, mostly domestic. One such problem is the weakness of the banking sector and the failure of the government to provide stimulus in terms of robust spending programmes. Also, there is the problem of weak exports due in part too weak demand from the crisis-affected countries.

The developments in East Asia had also profound consequences for other developing countries, particularly those most integrated into the global financial markets and those that are dependent on oil exports. The long-term regional economic outlook is good. A mild economic recovery of the region's economy occurred in 1999, and expected to improve further in the new millennium. The reasons for this optimism are the first signs of recovery in Southeast Asia; strong non-inflationary growth in the US; improving economic prospects in the Russian Federation; some positive indications of recovery in Japan; and a lack of contagion effects from the Brazilian crisis. However, the region also faces some risks. First, the US economy could falter and the Japanese economy might fail to recover. The fate of sustainable development in the Asian and Pacific Region is sadly very dependent on the economic fortunes of the world's two largest economies.

The picture for economies of the developing countries of Asia and the Pacific is dominated by the dramatic events of 1997. The first sign was the deceleration of exports in 1996. This situation worsened as a result of the currency crisis in 1997, which evolved into a widespread economic decline in the region in 1998. Economic growth rates in Southeast Asia declined sharply by -7.4 per cent. The corresponding decrease in the newly industrialized economies of the region was -1.6 per cent. However, the economies of the People's Republic of China and Mongolia maintained their hefty growth rates of about 7.8 per cent in 1998 and declined only slightly in 1999. On the average, the island states of the South Pacific have also been affected by the crisis of 1997. In contrast, the average growth rate in South Asia even registered growth of 5.7 per cent and is predicted to maintain the momentum from 2000 onwards.

The major factor that led to the crisis in East Asia was the generally weak and over regulated banking sector. This drove away the private investors resulting in reversed capital flow. Since the crisis, some of the worst hit countries like Thailand, Korea, and even to some extent, Indonesia, have instituted banking reforms but the ratio of non-performing loans to total lending continued to rise in 1998. The immediate impact of capital outflow was currency devaluations, and these in turn pushed inflation rates upwards. The average inflation rate in Southeast Asia rose from 5.6 per cent in 1997 to 21 per cent in 1998. The average across Asia rose from 4.6 per cent in 1997 to 6.5 per cent in 1998. As countries continued to defend their currencies, the average inflation rate in Asia was expected to decrease to 3.7 per cent in 1999 and 4.1 per cent in 2000 (ADB 1999).

Private capital flows to Asia, which amounted to US$105 billion in 1996, changed direction in 1997 as investments were withdrawn. The central banks of the region were forced to adopt policies for monetary contractions, which reduced local demand worsening the already bad debt problem. Exports did not increase appreciably despite the favourable exchange rates. As expected, the current account balance as percentage of GDP increased as capital flowed out. It expanded from -1.3 per cent in 1997 to 3.6 per cent in 1998. Most of this increase came from import reductions and shortfalls in export. The current account balance is predicted to be 2.5 per cent in 1999 and 1.4 per cent in 2000 under the assumption that the capital inflow is negligible and that imports recover only slowly. The debt-service ratio for Asia as a whole has also increased from 1996 to 1997. The value for 1998 was 19.2 per cent. However, this is expected to decrease to 17.7 per cent in 1999 and 13.5 per cent in 2000.

In the last quarter of 1998, strong deflationary polices and fiscal stimulus from the crisis-affected countries prevented the further deterioration of currency exchange rates. Stability was restored in the Asian currencies (ADB 1999b) and its financial markets. The Japanese economy seemed to be finally responding to fiscal stimulus in the latter part of 1998. But even more impressive was the recovery of the Korean economy, which appeared sooner than expected. All indications point to full recovery of the Asian economy. It is expected that the growth rate for Asia will be 5.5 per cent in 2000 (Figures 21.7 and 21.8), and even higher in some developing countries of the region. Fortunately, People's Republic of China and South Asian countries with a large share of poverty in the region were not greatly affected by the financial crisis. In addition, Southeast Asia is showing early signs of economic recovery. There is growing optimism that that the outlook for sustainable development in the region will become better than it was a few years ago.

The first essential step towards sustainable development is the elimination of mass poverty. This is not a problem of individual countries alone. Nor are the impacts of poverty confined within national borders. Extreme poverty will push people to migrate to more prosperous countries using illegal means. Unsustainable development that leads to the depletion of natural resources may cause serious scarcity of vital resources such as water, fuels, or arable land that might provoke some countries to take extreme measures endangering the security of others. Sustainable development not only calls for

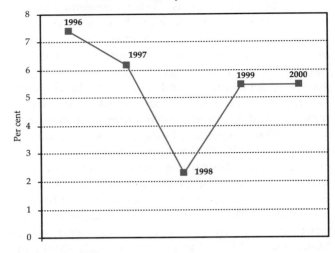

Figure 21.7 GDP Growth Rates in Developing Asian and Pacific Economies

Source: ADB 1999

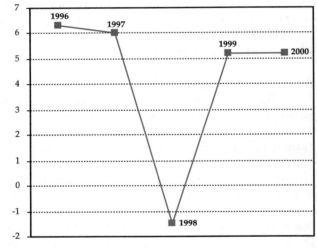

Figure 21.8 GDP Growth Rates in Newly Industrializing Economies of Asia and the Pacific

Source: ADB 1999

international partnership but also a reconsideration of the traditional concept of security. Scholars are now adopting a broader view of security that includes the concept of environmental security.

C. Environmental Security

The Rio Principles recognized that war is destructive to the environment and sustainable development. The opposite is also true-unsustainable development that depletes natural resources resulting in scarcity could induce social effects that cause violent conflicts. In fact, it is possible to think of sustainable development as a long-term negation of conflict. This is the core idea behind the concept of environmental security. This concept has become popular over the last few years in the context of the post-cold war world, which made the re-engineering of the international system imperative. The issues surrounding environmental degradation have triggered a number of initiatives in foreign policy and international law. The number of meetings related to the MEAs has increased and has inundated foreign offices with new responsibilities. Complementing these initiatives are efforts in the security field that study environmental degradation as one of the unconventional threats to the security of the state. The nature of foreign policy is evolving in the direction of influencing the internal affairs of countries so as to prevent self-destructive behaviour that could result in human suffering and political instability as an outcome of severe environmental stress (e.g., use of trade sanctions on products of unsustainable processes). The research and studies on the relationships of environmental pressure and violent conflict is one of the proactive approaches to foreign policy.

Global environmental issues such as population growth in developing countries, ozone depletion, climate change, the loss of biodiversity, and the phenomenon of environmental refugees have security implications. These may turn out to be the greatest threats to global security. The scientific uncertainties and the apocalyptic scale of these problems make them difficult to analyse. The implications of global environmental issues could have also evolved from two causal steps: first, environmental scarcity causes social effects; and second, social effects could lead to violent conflicts (Figure 21.9; Homer-Dixon 1999). Moreover, there are three sources of scarcity: supply-induced, demand-induced, and structural scarcities.

Supply-induced scarcity is due to diminishing supply because of depletion in quantity or degraded in quality. For instance, diminished timber stand because of deforestation or decline of fertility of soils because of improper irrigation. Demand-induced scarcity is due to increasing utilization of a fixed supply due to increased population, which decreases the supply per capita. Lastly, structural scarcity is due inequitable access to the supply when some groups are able to obtain disproportionate share of the supply when the supply to others are too small. Five main social effects could increase the probability of violent conflict. These are constrained agricultural activity, often in ecologically marginal regions; constrained economic productivity, mainly affecting people who are highly dependent on environmental resources and who are ecologically and economically marginal; migration of these affected people in search of better lives; greater segmentation of society, usually along existing ethnic cleavages; and disruption of institutions, especially the state.

Within these perspectives, international environmental insecurity is enhancing in the wake of such social trends as an increase in inequality, both among and within nations; a continuation of hunger and poverty, despite the fact that globally, enough

Figure 21.9 Causal Links between Environmental Scarcity and Violence

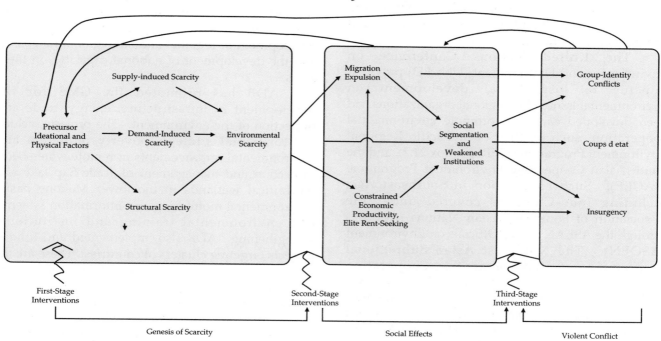

Source: Homer-Dixon, T. 1999

CHAPTER TWENTY-ONE

food is available; and greater human health risks resulting from continuing resource degradation and chemical pollution.

REGIONAL AND SUBREGIONAL COOPERATION

A. General Initiatives

Regional and subregional cooperation provides opportunities for promoting environmental security and adopting a coordinated response to global initiatives, such as Agenda 21, the action plan for SIDS and the various international environmental conventions. In the wake of dwindling international resources, as a methodology, cooperation also offers an efficient means for training, capacity building, implementing guidelines and sharing the findings of studies on special topics, experiences and good practices in the region.

It is for the same reason that regional and subregional cooperation figures prominently in the outcome of the 1995 Ministerial Conference on Environment and Development in Asia and the Pacific. The Regional Action Programme for Environmentally Sound and Sustainable Development, 1996-2000, adopted by the Conference emphasized regional and subregional cooperation as an important mechanism for the implementation of the Programme. Agenda 21 had already emphasized regional and subregional implementation. The nineteenth special session of the General Assembly held in June 1997 to review the progress in the implementation of Agenda 21 recognized the need for the exchange of information on best practices and national experiences and the need for regional and subregional cooperation.

The United Nations Conference on Environment and Development (UNCED) provided impetus to institutional development on environmental issues. Subregional organizations had been developed earlier to promote environmental cooperation, such as the South Pacific Regional Environment Programme (SPREP) (Box 21.2), and the South Asian Cooperative Environment Programme (SACEP). Such cooperation in Southeast Asia including Indo-China, is covered under the Association of Southeast Asian Nations (ASEAN) through the ASEAN Senior Officials on Environment (ASOEN). The Northeast Asian Subregional Programme of Environmental Cooperation (NEASPEC) is presently evolving with the support of the six participating countries of Northeast Asia. In Central Asia, the riparian States of the Aral Sea have formed the Interstate Council for the Aral Sea (ICAS) and the International Fund for the Aral Sea (IFAS).

There are also subregional cooperation programmes/organizations sharing common resources such as the Regional seas programmes of the United Nations Environment Programme (UNEP), the International Centre for Integrating Mountain Development (ICIMOD), the Mekong River Commission (MRC), the Greater Mekong Subregion (GMS) promoted by the ADB, and the Forum Fisheries Agency. There is also cooperation through networking, such as the Regional Network of Research and Training Centres on Desertification Control in Asia and the Pacific promoted by ESCAP in 1985. Project based subregional cooperation is also promoted through inter-country participation.

The Regional Seas Programmes of UNEP have been promoted to protect common resources. Among them, the South Pacific Region, East Asian Seas Region, South-Asian Seas Region, North-West Pacific Region, and part of the Kuwait Action Plan Region are in Asia and the Pacific. Each of these programmes has an action plan adopted by the participating countries, with the exception of the plan for the Northwest Pacific, which is under discussion. ESCAP assisted UNEP and participated in the development of the action plans and institutional mechanisms for the East Asian Seas, South Asian Seas and the South Pacific. The secretariats for the South Pacific Seas regional action plan and the South Asian Seas regional action plan were entrusted to SPREP and SACEP, respectively. The support for the East Asian Seas Regional Coordinating Unit is provided by UNEP through the secretariat in Bangkok. UNEP also maintains the Environment Assessment Programme for Asia and the Pacific which is providing valuable support to the subregional organizations in their preparation of state of the environment reports, development of regional and subregional databases, and the development of national capacities in these areas.

ADB has promoted the GMS for the development of infrastructure, energy, trade and protection of the environment. The projects include environmental strategy, poverty reduction and environmental improvements in remote watersheds, protection and management of Tonle Sap lake and the critical wetlands in the lower Mekong basin, environmental monitoring and information systems, and environmental training and institutional strengthening. ADB also implemented and funded projects targeting clusters of countries. These are the Asia Least-cost Greenhouse Gas Abatement Strategy (1996-98), and Coastal and Marine Environmental Management in the South China Sea (1996).

UNDP support included building national and subregional capacities for the implementation of Agenda 21 through their Capacity 21 programme. It also supported joint resource management for cross-border cooperation for the Mekong River Commission. Another example is the Tumen River Area Development Programme in Northeast Asia, and the Himalayan Eco-regional Initiative of Bhutan, People's Republic of China, India, Myanmar, Nepal and Pakistan for biodiversity conservation and management. The sixth cycle programme of UNDP emphasizes implementation through subregional organizations.

ESCAP has established mechanisms for consultation with the subregional organizations for which meetings are organized annually with the heads of participating organizations. The discussions in these meetings include cooperation in the social and economic field as well as on sustainable development issues. The third Consultative Meeting among Executive Heads of Subregional Organizations and ESCAP that was held in Tehran in May 1997 formulated a set of recommendations for inter-subregional cooperation. ESCAP has also signed memoranda of understanding with a number of subregional organizations. Subregional organizations are an integral part in the implementation of the ESCAP work programme and whenever feasible joint activities are organized, such as the activities during the preparatory phase of the 1995 Ministerial Conference on Environment and Development in Asia and the Pacific.

An Inter-agency Sub-committee on Environment and Sustainable Development under the Regional Inter-agency Committee for Asia and the Pacific (RICAP) established by ESCAP provides a framework for cooperation between the United Nations, regional and subregional inter-governmental and other organizations active in Asia and the Pacific. The Sub-committee generally meets at least twice a year to exchange information on the activities for its members and to discuss areas of cooperation. This forum provides an excellent opportunity for assessing regional and subregional activities in the implementation of Agenda 21 the Regional Action Programme and international environmental conventions. It is also promoting a joint project on the testing of indicators of sustainable development in Asia and the Pacific.

B. **Regional and Subregional Conventions and Agreements**

Regional and subregional MEAs are concerned mainly with shared facilities and the protection and proper management of the region's abundant but severely threatened resources. Several MEAs have resulted from negotiations on problems such as sharing river basins between different countries. The regional MEAs are shown in Table 21.14. Most attention is being paid to the atmosphere, and natural resources such as water, wildlife etc. (Box 21.2) and natural disasters to which the region is very prone and which appear to be increasing in frequency and severity. Many countries have developed their own strategies, independent of MEAs, for example, to reduce air pollution (Box 21.3) and protect wildlife, but MEAs can help to reinforce the actions of planners and decision-makers. Because of the lack of resources, expertise, and sometimes political will, regional MEAs are seldom fully translated into the national legislation that would be needed to ensure implementation but these have resulted in relatively high levels of public information and awareness.

CONCLUSION

The last two decades of the 20th century witnessed the outpouring of tremendous energy in the creation of international partnerships and development of an international legal regime to protect the global environment. The Earth Summit, and the conventions it brought forth, are outstanding examples of this burst of international creativity in meeting the environmental challenges of our common future. The success of these efforts is laudable. Now, there are multilateral environmental agreements (MEAs) that cover most of the important environmental issues. The environment is also now mainstream – born of many initiatives at all levels. Ironically, this proliferation of activity in itself presents new challenges of policy coordination, avoidance of overlaps, systems integration and minimization of bureaucracies, and above all, effective implementation.

Beyond ratification, compliance and implementation of the MEAs means that the required national legislation have been enacted and complemented by specific action programmes. The word "compliance" is often used as an indicator of implementation (UNEP 1999), to mean the undertaking of activities to realize the spirit of the agreements, as well as their specific requirements. "Effectiveness" is the ultimate measure of the success of any treaty. It reflects the extent to which the quality of the environment or the behaviour of actors have been influenced by the implementation of the agreement. In hindsight, and in the light of the experience of the last several years, it is extremely difficult to assess the implementation of any treaty, not only because of the limited availability of data, but also because of the variety of national contexts on which compliance largely depends.

Eight years after Rio, the Asian and Pacific Region has learned not only about the complexity of implementing sustainable development, but also that

CHAPTER TWENTY-ONE

new challenges have made the translation of the concept into practice extremely difficult. Globalization has strengthened the economic linkages of the world. Capital and financial flows could change directions overnight and some countries could find themselves short of valuable financing. Economic development could abruptly decline and collapse, resulting in further environmental destruction. East Asia in 1997 portrayed this new reality. Rapid economic development crashed without warning and this fueled pessimism as sustainable development retreated down the list of national priorities.

Further, the developing countries of the region are also realizing that globalization has extended and strengthened the leverage of Western countries in the developing world. Trade sanctions are becoming noble means of economic control over the production of economically weak countries. These sanctions could be provoked by internal political acts of importing countries or by their violations, perceived and real, of environmental standards. On the positive side, this could serve to enhance compliance with the MEAs. On the other hand, and at worst, sanctions could provoke nationalistic fervour, where environmentalism is associated with foreign interests and therefore against national interests.

As the gap between the industrial and developing countries widens, the emerging digital economy is creating a new gap- the digital divide. Information-rich countries are becoming stronger and are beginning to use this power to enhance their competitive edge, further widening the gap. Poorer countries could, and may, postpone further implementation and invoke the principle of common but differentiated responsibility.

The problem of coordination, integration, and the extraction of synergy from the recommendations and action plans of international conferences and many MEAs are now in the international radar screens. What would be done with the information, in terms of navigating among the rocks of the contentious issues of free trade, globalization and localization, financing, technology transfers, and the historical culpability of nations, will determine the future directions of the global environmental regime.

Most of the heatedly debated issues in almost all of the meetings are those involving financial and technological resources transfer and capacity building. These problems could be condensed to the problem of finance. Pursuing the synergistic financing of the MEAs should perhaps be the core strategy of the above initiatives. The first step should be the review of financial mechanisms of the conventions with the intention of harmonizing them and expanding the magnitudes to more meaningful levels. However, the increases should be based on actual programmatic needs at the national and local levels in consultation with Parties to the conventions that provide the largest shares of the funding. It is also imperative to coordinate the provisions of mandatory budgets to the conventions with development assistance and its focus on poverty alleviation. After all, the elimination of poverty is relevant to all the targets of the conventions.

The sustainable financing of the conventions could have a better chance of becoming a reality if innovative financing modalities are explored. One possibility is the cultivation of partnerships with the private sector, particularly those that stand to benefit from the implementation of the conventions such as, for instance, the suppliers of non-fossil energy sources *vis-à-vis* the UNFCCC. An international tax on air travel is another possibility. The Green Development Mechanism and emissions trading are also promising approaches that may provide additional funds for implementation of the MEAs.

In the coming Rio+10 in 2002, the greatest unexpressed fear of the diplomatic community is that it could be a reprise of Rio+5. And yet fear can inspire people to heroic boldness. The promotion of effective regional and subregional cooperation is one outcome of such inspiration. However, in the final analysis, there is still an urgent need to make a courageous leap over the great divide between the developing and industrial countries to save our common global and regional resources.

The principle of differentiated responsibility agreed to in Rio also demands that the commitments made by the developed countries be put into action, much more so because of the time lag and continuously increasing pressure of globalization. Realization of the objective of Agenda 21 depends to a large degree on national and local governments, since they are in the frontier of development actions. They create and enforce legislation on the use of natural resources. They formulate and implement development plans. They maintain institutions that embody and monitor the implementation of environmental and development policies. Because of their crucial roles, international organizations work with national governments and their local units in designing and executing projects relevant to Agenda 21. Though regional and subregional cooperation is vital for providing opportunities for a coordinated response to global initiatives, the dwindling availability of international financial resources, lack of technology of transfer and unfavourable trade regimes demand enhanced regional units to respond proactively.

Chapter Twenty-Two

Environmental degradation for short term gains poses a major environmental challenge for the region in the 21st Century.
- a. Clean felled mangrove forests for shrimp farming;
- b. Destruction of coral reefs by the use of dynamite for fishing.

22

Asia and the Pacific into the 21st Century

INTRODUCTION
PREVAILING CONDITIONS AND TRENDS
PROJECTED TRENDS AND FUTURE SCENARIOS
CHALLENGES AND OPPORTUNITIES
CONCLUSION

CHAPTER TWENTY-TWO

INTRODUCTION

Long before the end of the last Century, declining environmental quality and increasing public concern over the environment, both locally and globally, had begun to create a demand for the strengthening of environmental protection within Asia and the Pacific. Policy-makers therefore began the decade of the 1990s with a mandate to improve the state of the environment in region. Whether or not they have succeeded is perhaps a subjective judgement, biased to some extent by an individual's social and economic welfare-improvements in some dimensions of environmental quality have been, and are continuing to be achieved, especially among the more developed economies of the Northeast Asian (and parts of South Pacific) subregion. However, what is beyond doubt is that the 1990s have seen progress in the establishment of the institutions and policy tools needed to address the region's urgent environmental concerns from hereon.

Two core findings have shaped the approach taken by policy makers to the environment in Asia and the Pacific during the last decade. Firstly, a growing body of information became available documenting the generally poor state of the environment. The UN ESCAP State of the Environment Reports, of which this is the third, were instrumental in assembling data that confirmed the everyday experience of many citizens of the region: from air to water pollution, from land degradation to desertification, the rapidly industrializing and urbanizing countries of the region experienced poor and, in many cases, declining environmental quality (United Nations 1990 and 1995). A number of topical reports by the ADB (1997), the World Bank (1994 and 1997) and other organizations also predicted that continuation of these trends would lead to a further decline in many dimensions of environmental quality during the 1990s in many parts of the region (Brandon and Ramankutty 1993; Hettige et al 1997; O'Connor 1994). Secondly, in general, the poor state of the environment has, in general, been attributed to the policy failures and institutional weakness in environmental management. Where environmental protection systems have been strengthened, this has generally led to progress in reducing pollution, land degradation and other environmentally damaging processes, and, in turn, measurable improvements in environmental quality. Moreover, contrary to the fears of some policy-makers, there is little evidence that stronger environmental regulation has undermined the competitiveness of the Asian and Pacific Region economies in world markets (ADB 1997). Progress in policy reform and institutional capacity building, however, has been highly uneven; faster and more effective in some places than in others.

This Chapter concludes the report, and summarizes the discussion therein concerning the prevailing conditions and trends in the physical environment across the region, their impacts on the health and well-being of the region's population, and the management and policy responses which have been adopted to address them. It then goes on to look at projected trends and future scenarios for the region's environment, and concludes with a discussion of future prospects.

PREVAILING CONDITIONS AND TRENDS

A snapshot of prevailing and likely trends for environmental and socio-economic conditions in Asia and the Pacific is provided in Figure 22.1, which shows an improvement in economic and some quality of life indicators, but portrays a picture of the overall degradation of the environment. These trends are discussed further below.

A. Socio-Economic Trends

Asia and the Pacific is the most populous region in the world, with a population that has more than doubled in the latter half of the last Century, from 1.7 billion in 1960 to 3.7 billion in 2000, and is still rising. The rate of growth of population, however, fell slightly in the last decade, from 1.6 per cent in the period 1990-95, to 1.4 per cent in the period 1995-2000. Much of this population increase has taken place in urban areas; the urban population of the region has doubled in the last 20 years alone. The number of megacities (with populations greater than one million) has also increased significantly, from 3 in 1980 to 12 in 2000, which alone presently accommodate about 12 per cent of the region's urban population.

1. Economic Growth

Economic growth, measured in terms of gross domestic product (GDP), has contributed to improving social conditions across much of the region. Between 1965 and 1990, GDP in Asia and the Pacific as a whole grew by an annual average of 3.8 per cent per capita (ADB 1997). Over the 1990-1997 period, GDP soared at an annual rate of 7.9 per cent in Northeast Asia and Southeast Asia, and 5.7 per cent in South Asia (World Bank 2000). Despite the financial crisis that hit many countries of the region in 1997-98, economic growth managed to revive in 1999; the GDP of countries in developing parts of the region posted a robust growth rate of 6.2 per cent in 1999, much higher than the 2.6 per

Figure 22.1 *Environment and Development Trends in Asia and the Pacific 1995-2005*

	South Asia		North-East Asia		South-East Asia		Pacific		Central Asia	
	1995-2000	2000-2005	1995-2000	2000-2005	1995-2000	2000-2005	1995-2000	2000-2005	1995-2000	2000-2005
Socio-economic trends										
GDP growth	↑	↑	↑	↑	↑	↑	⇧	⇧	⇩	⇧
Population growth rate	⇩	⇩	↓	↓	⇩	⇩	⇩	⇩	⇩	⇩
Incidence of poverty	⇩	⇩	↓	↓	↕	↓	⇩	⇩	⇩	⇩
Urban growth	↑	↑	↑	↑	↑	↑	↑	↑	↑	↑
Slums and squatters	↑	↑	⇧	⇧	↑	↑	⇧	⇧	⇧	⇧
Life expectancy	⇧	⇧	↑	↑	↑	↑	⇧	⇧	⇧	⇧
Infant mortality	⇩	⇩	↓	↓	↓	↓	↓	↓	⇩	⇩
Traditional diseases	⇩	⇩	⇩	⇩	⇩	⇩	⇩	⇩	↕	⇩
Modern diseases	↑	↑	⇧	⇧	⇧	⇧	↑	↑	⇧	⇧
Child under nourishment	⇩	⇩	↓	↓	↓	↓	⇩	⇩	↕	↕
Nutrition	⇧	⇧	⇧	⇧	⇧	⇧	⇧	⇧	⇧	⇧
Natural disaster losses	↑	↑	↑	↑	⇧	⇧	↑	↑	⇧	⇧
Environmental trends										
Resources										
Arable land per capita	⇩	⇩	⇩	⇩	⇩	⇩	⇩	⇩	⇩	⇩
Land degradation	↑	↑	↑	↑	↑	↑	⇧	⇧	↑	↑
Desertification	↑	↑	↑	↑	↑	↑			↑	↑
Deforestation	↑	↑	↑	⇧	↑	↑	↑	↑	↑	↑
Tree plantation	⇧	↑	↑	↑	↑	↑	⇧	⇧	⇧	⇧
Loss of habitat and species	↑	↑	↑	↑	↑	↑	↑	↑	↑	↑
Water consumption	↑	↑	↑	↑	↑	↑	⇧	⇧	↑	↑
Marine resources loss	↑	↑	↑	↑	↑	↑	↑	↑		
Commercial use of energy	↑	↑	↑	↑	↑	↑	↑	↑	↑	↑
Food security	↓	↓	⇩	⇩	⇩	⇩	⇩	⇩	⇩	↓
Resource use by industry	↑	↑	↑	↑	↑	↑	↑	↑	↑	↑
Environmental degradation by tourism	↑	↑	⇧	⇧	↑	↑	↑	↑	⇧	⇧
Pollution										
Freshwater pollution	↑	↑	↑	↑	↑	↑	↑	↑	↑	↑
Coastal pollution	↑	↑	↑	↑	↑	↑	↑	↑		
Air pollution	↑	↑	↑	⇧	↑	↑	⇧	⇧	⇧	↑
Greenhouse gases	↑	↑	↑	↑	↑	↑	⇧	⇧	↑	↑
Solid waste generation	↑	↑	↑	↑	↑	↑	↑	↑	↑	↑
Agro-chemical use	↑	↑	↑	↑	↑	↑	↑	↑	↑	↑
Pollution by energy generation	↑	↑	↑	↑	↑	↑	↑	↑	↑	↑
Vehicular pollution	↑	↑	↑	↑	↑	↑	↑	↑	↑	↑
Industrial pollution	↑	↑	↑	↑	↑	↑	↑	↑	↑	↑
Environmental policies/actions										
Public authorities action	↑	↑	↑	↑	↑	↑	↑	↑	↕	⇧
Business sector's response	↑	↑	↑	↑	↑	↑	↑	↑	↕	⇧
Env. monitoring & research	↑	↑	↑	↑	↑	↑	↑	↑	↕	⇧
Env. education & awareness	↑	↑	↑	↑	↑	↑	↑	↑	↑	↑
Activities of major groups	↑	↑	↑	↑	↑	↑	↑	↑	⇧	⇧
Int'l Conventions (participation)	↑	↑	↑	↑	↑	↑	↑	↑	↑	↑
Subregional cooperation	↑	↑	↑	↑	↑	↑	↑	↑	⇧	⇧

Note: ↑ Increase   ⇧ Slight increase   ↓ Decrease   ⇩ Slight decrease   ↕ No change

Red color shows deteriorating trend
Green color shows improving trend
GDP and urban growth have not been indicated by red or green color because their impact could be good or bad

cent rate recorded in 1998. This rate is expected to remain virtually unchanged through 2000 (ADB 2000).

2. *Income and Quality of Life Trends*

In 1975, over half of the population of Asia and the Pacific was classed as poor, based on the World Bank "dollar a day" threshold for poverty. Moreover, using a more generous poverty threshold of 2 dollars a day, almost 2 billion people, or the majority of the region's population could be classed as poor. Today, using the "dollar a day" criteria, around a quarter of the population is classed as poor. However, due to the rapid population growth, the absolute number of poor has remained extremely high, at 900 million in 1998 (which is about twice as many people as in the rest of the developing world combined). For example, in South Asia, where the overwhelming majority of the poor are found to live (some 522 million), numbers have increased by more than 30 million since 1980, although the ratio of poor to total population has declined considerably over this period (World Bank 1999). Likewise, People's Republic of China in 1998 still had 213 million poor (World Bank 1998), some 17 per cent of its population, and it has also been estimated that 10 million people may have been added to the ranks of the poor between 1996 and 1998 due to the financial and economic crisis that hit the countries in Northeast and Southeast Asia.

Standards of health and nutrition have improved substantially over the past few decades. Average life expectancy across the region rose from 58 years in 1975, to 67 years at end of the Century. In particular, these rates of increase were exceeded in the subregions of Northeast, Southeast and South Asia. The biggest influence on raising life expectancy has been reduction in infant mortality. The lowest infant mortality rate, at around 6 per thousand live births in Central Asia, while the highest is in South Asia, at 77 per thousand live births; these are compared to a worldwide developing country average of 58 per thousand live births (ADB 2000). These variations to a large extent reflect differences in factors such as nutritional intake, access to health-care, safe water and sanitation. The incidence of child malnutrition in Northeast and Southeast Asia is almost the lowest in the developing world (only Latin America is lower). By contrast, incidence is the highest in South Asia, where in 1990, six out of ten children had stunted growth from malnutrition, compared with four out of ten in Sub-Saharan Africa. In addition, one in three babies in South Asia was born underweight in the same year, compared to 1 in 6 in Sub-Saharan Africa (ADB 1997).

B. **Resource Trends**

1. *Land, Forest and Biodiversity*

The assessment of the condition of natural resources in Asia and the Pacific shows an overall deteriorating trend (Figure 22.1). Land scarcity, along with its degradation, is putting serious stress on food production and security in the region. The arable land per capita of agricultural population has already declined by about 20 per cent since 1970. Currently, per capita arable land availability is 0.18 hectare, which is much below the world average of 0.24 ha (UNEP 1999). Land degradation is also contributing to a rising number of landless, who are moving to environmentally fragile areas such as steep slopes and forests, further contributing to land degradation. Of the world's 1 900 million hectares of land affected by soil degradation since 1945, the largest portion of this area (over 850 million hectares) was in Asia and the Pacific. Arid and semi-arid areas of the region are particularly vulnerable; it is estimated that 1.3 billion (over a third of the region's population) live in areas prone to drought and desertification.

Agricultural intensification, including the expansion of irrigation and the increased application of agrochemicals, has allowed a substantial expansion in crop yields, and also permitted the cultivation of vast expanses of arid lands. Irrigated areas alone increased by about 6 per cent between 1995 and 1998. However, the chances of a their further expansion at such a rate are low, due to increasing pressures on finite (and in some cases dwindling) water resources. Fertilizer and pesticide consumption in developing countries of the region has also grown steadily over the years (Figure 22.2). The region's share in world's total fertilizer consumption swelled from 32 per cent in 1988, to 50 per cent in 1998. Pesticide use increased from an annual average rate of 3 per cent in 1983/93, to 4.4 per cent in 1993/98. However, the growth in production attributable to these factors is unlikely to be sustainable, since poor irrigation practices and over-application of agro-chemicals have taken their toll on both land and water resources, including an increase in pollution, Stalinization and desertification.

Forests and biodiversity are also under extreme pressure. The region has the highest rate of deforestation, at 1.2 per cent per annum (FAO 1998), the highest rate of commercial logging and biggest volume of fuelwood removal in the world. Already, over half of the region's forest base has disappeared and another three quarters of a million hectares of forests are being lost annually. Habitat modification, fragmentation and loss, along with over-exploitation of resources and introduction of exotic species, have placed the rich biodiversity of the region under serious threat. About two thirds of the region's

Figure 22.2 Fertilizer Consumption Trends in Asia and the Pacific 1980-1997

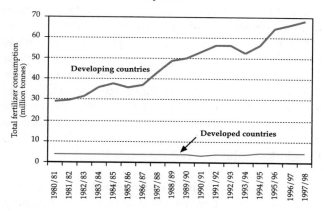

Source: FAO 1999

wildlife habitats have already been destroyed, thousands of plant and animal species are threatened and genetic diversity is declining at a pace (Tuxill 1999). However, a positive trend is the growing rate of forest plantation, which is one of the highest in the world.

2. *Aquatic Resources*

Massive withdrawals from rivers, lakes and underground reservoirs have contributed to a growing scarcity of freshwater in parts of the region. Taking the standard benchmarks of 1 600 cubic metres per capita per year as water stress, 1 000 cubic metres per capita as water scarce, the Republic of Korea (at around 1 500 cubic metres), Singapore (at around 1 700 cubic metres), and Maldives (at around 1 000 cubic metres) are already water stressed or scarce. In addition, several other areas in the region including the Aral Sea in Central Asia, and the North China plains are also experiencing acute shortages of water. Sectoral competition and conflicts in water-use (particularly between agriculture and industry) have also become critical.

The marine and coastal environment is also under pressure from species over-exploitation and habitat degradation, and is under the looming threat of potential climate change and sea level rise. Several of the region's most important fishing areas and almost two thirds of the major fish species are either fully or over-exploited. About 80 per cent of the coral reefs in Southeast Asia, 54 per cent in the Indian Ocean, and 41 per cent in the Pacific Ocean, are at medium to high risk.

3. *Food Security*

There has been a decline in the growth rate of crop production over the past few years. Growth in cereal production, for example, declined from 3.2 per cent in 1969-71 to 1.9 per cent in 1994-96, and decreased by a further per cent or so 1996/97 (FAO SOFA). The region has been a net food importer of late, with cereal imports growing at an average annual rate of 9.5 per cent in the period 1986/96. With regards to food security, the average daily food availability in calories falls below nutritional requirements in Bangladesh, Cambodia and Mongolia, and close to them in Nepal, Sri Lanka, Lao People's Democratic Republic, Papua New Guinea and the Solomon Islands. Aquaculture is becoming increasingly important to augment food production, as the productivity of land declines and many marine fish stocks are over-exploited. The region now contributes almost 90 per cent to the total world production of marine aquaculture.

4. *Energy Consumption*

Energy consumption has escalated with the growth in industrialization and urbanization in the region. Between 1975 and 1995, commercial energy use in South Asia had an average annual growth rate of 6.6 per cent; and in Northeast Asia and the South Pacific, 5.3 per cent, between 1975 and 1995 (World Bank 1998). Per capita commercial energy use more than doubled during the same period (UNEP 1999). Overall, the region accounted for 26.8 per cent of the world's commercial energy consumption in 1995.

C. **Pollution Trends**

The urban areas are by far the worst affected by air pollution. Of the major cities in the world with the highest levels of total suspended particulates (TSP) in the air, 9 are located in Asia. The levels of TSP in many cities are two to three times those recommended by WHO. Six of the major cities with high levels of sulphur dioxide are also located in the region (UNEP 1999). The impacts of haze, acid rain and transboundary air pollution have also increased substantially.

Inland and coastal water pollution has also increased in severity as a result of factors such as unplanned and unmanaged urbanization, industrialization and agrochemical use. Four rivers in the region – the Yellow River in People's Republic of China, the Ganges River in India, and the Amu Darya and Syr Darya in Central Asia – top the list of the worlds most polluted rivers, according to a report of the World Commission on Water. In terms of levels of suspended solids, the region's rivers typically contain some four times the world average and 20 times OECD country average levels (ADB 1997). Biological oxygen demand (BOD) in the Asian and Pacific Region's rivers is also 1.4 times the world average, and they contain almost three times as much bacteria from sewage. The reported median fecal coliform count in the region's rivers is reportedly 50 times higher than the WHO guidelines (UNEP 1999). Pollution of coastal and marine environment has also intensified and is evident in the increased frequency of algal blooms in the region's seas, with major outbreaks in recent years in Australia, People's Republic of China, Republic of Korea, Japan and New Zealand.

The burden of waste has also increased markedly, with rapid population growth, urbanization and increasing affluence. Conservative estimates of waste arisings in the region give that 1.5 million tonnes of municipal solid wastes are produced per day, and 5.2 million tonnes of industrial solid wastes; figures which are expected to more than double in the next 25 years (World Bank 1999). The generation of hazardous wastes from manufacturing, hospital and health-care facilities and nuclear power and fuel processing plants has also increased tremendously. People's Republic of China alone produces more than 50 million tonnes of hazardous wastes per annum.

D. Cost of Environmental Degradation

Overall, it is estimated that the economic cost to the region from environmental degradation ranges from 1 to 9 per cent of a country's annual GNP (ADB 1997), or an average of about 5 per cent of GDP; in People's Republic of China, this may be as high as 10 per cent (World Bank 1999). Non-economic costs, that affect welfare but not GDP, are even larger, but are often difficult to quantify (ADB 1997). The economic cost of air pollution health damages is estimated at US$1 billion a year in cities such as Bangkok and Jakarta (World Bank 1999).

E. Trends in Policy Environment

While most environmental trends are negative, several positive changes can be discerned in the state of policy responses across the region. Among these are the improvement in governance by public authorities through strengthening of institutions; enhancements in the formulation and implementation of policies; growing environmental awareness and public participation (Box 22.1); increasing

---

Box 22.1 Community Action for the Environment: Bhaonta-Kolyala Village in Rajasthan

It's an unusual ritual the villages of Bhaonta-Kolyala follow. Every year, they pour water into a johad – a crescent-shaped earthen check-dam. This started in 1986, when villagers of Bhaonta-Kolyala noticed a remarkable development in Gopalpura, a nearby village. Gopalpura had water in its wells round the year. The reason – villagers had revived johads with the help of Tarun Bharat Sangh, an NGO.

Led by two local farmers, the beleaguered villagers approached the TBS. They were offered help, but on one condition – that the villagers should be ready to take upon themselves the task of regeneration.

After organizing themselves and the neighbouring villages, on March 6, 1987, the villagers started protecting forests and repairing old johads. They mapped the natural drainage system and chose tentative sites to construct new johads. The aim was to catch each and every drop of rainwater that fell on the village, say the villagers.

During the course of their search, they discovered an old johad, buried in silt, on the slope of the barren hills. In 1988, repair work on the johad started. When the monsoons arrived, the johad was filled with water. Overwhelmed by the results from a single johad, the villagers started building more such structures. Today, the village has a total of 15 water harvesting structures, including a 244 metres long, 7 metres tall concrete dam in the upper catchment of the Aravalli, the construction for which was started in 1990.

The dam was a turning point. Even those who had migrated were called back. By 1995, a year after the completion of the dam, water level in the wells downstream rose by two to three feet. "The percolation of water from this dam is three feet an hour. Its impact is felt in villages 20 km downstream. All the wells are now filled with water," says a villager. Today, all the agricultural land is under cultivation. Milk production has risen up to 10 times. Every rupee invested in a johad has increased the village's annual income by 2.5-3 times.

The most important lesson from Bhaonta-Kolyala is that when villages work with each other to regenerate the environment, there are unexpected blessings. Sometimes, they are as big as a river. In the case of Bhaonta Kolyala, it was Arvari River. In 1990, when the villagers started constructing the big dam, no one knew that the site was the origin of the river. And by catching and percolating water, they were injecting life into the river. Moreover, building water-harvesting structures was not enough for the villagers. To control soil erosion, they then went on to demarcate 12 square kilometres of the adjoining forest area for regeneration, and in 1995 they declared it as a public wildlife sanctuary, the "Bhaironath Public Wildlife Sanctuary", claimed to be the first of its kind in the country, and now home to three tigers, many bluebulls and deer.

Source: Down to Earth, April 30, 2000

> **Box 22.2 Private Sector Initiatives for Environmental Rehabilitation – the Use of Indigenous Techniques**
>
> Rehabilitation of arid lands subjected to mineral exploration has been accelerated and improved by the private sector in Australia, by the application of Aboriginal fire management techniques. Such techniques assist seed release in regenerating spinifex grasses, which in turn accelerates the rehabilitation of smaller disturbed areas, such as drill pads, hard standing areas and borrow pits. It also avoids the dominance by annual species previously experienced using other methods.
>
> The initiative by an Australian company was first started by discussing the idea with local Aboriginal people. It was noted that a striking decline in the number of native mammals had taken place in the desert, and fewer species of flora were appearing. Studies indicated that this decline had occurred recently, in the last 30 to 50 years, i.e. since Aboriginal groups had abandoned their traditional life in the Western Desert and settled in missions and other European communities. The last Aborigines moved out in the early 1960s.
>
> The source of this decline was soon revealed: when Aborigines moved out, with them went the practice of frequent burning. The great skills with which these fires were set by Aborigines to take advantage of wind conditions, humidity and topography, emerged in their conversations with scientists. Their pattern of burning meant that there were always many areas adjacent to each other at different stages of growth and decline, and there was rarely enough fuel over a big area to feed the vast bush fires which now occur.
>
> When the bushes re-burnt, the seeds stored in the sandy soil were heated. This made them receptive to the next rain, soon after which 20 or 30 species of plants, most of them herbs, blossomed and provided food and medicine for local people and food and shelter for native fauna. Even after the heat of a fire had split the seed coat, the seeds had to wait several years for water, in a region that had a sparse 150 mm to 250 mm rainfall a year.
>
> Having realized the value of the technique, the company employed local Aboriginal communities to burn the desert spinifex in a traditional way, which allowed the revival of scores of plant species and encouraged the return of many animals. The project also provided guidelines for using fire to rehabilitate areas throughout the remote arid interior of Western Australia, and the company is now conducting further trials to develop fire behaviour models that will help in planning safe burning programmes to enhance and protect the remarkable desert ecosystem.
>
> *Source:*   Government of Australia 1999

environmental consciousness in business and industry (Box 22.2); and enhanced interest and participation of NGOs and civil society in environmental management. Various efforts have been also made to explore innovative environmental regulatory policies that are sensitive to country and local context, and socio-economic situations (Aden et al 1999; Afsah and Vincent 1997; World Bank 1999), traditional command and control regulation is in many cases being supplemented by a variety of second and third generation policy tools.

However, in terms of government allocations, some countries of the region, particularly in Southeast Asia, have reduced the budget for the environment, although there are clear differences in the precise nature of these cuts. In Thailand, for instance, budgets have been reduced by about 20 per cent, especially for pollution and energy conservation; the Philippines suffered a 25 per cent mandatory reserve on many types of expenditures. In Northeast Asia, the Republic of Korea's environmental budget shrunk from 2.8 per cent in 1997 to only 0.3 per cent in 1998 (ADB 1998). Moreover, highly visible instances of policy failure, such as the forest fires and haze that have plagued Southeast Asia, only serve to highlight the opportunity for policy reform and more effective policy implementation, all on the basis of stronger systems of governance within the region.

PROJECTED TRENDS AND FUTURE SCENARIOS

A. Socio-economic

Over the next two decades, population growth rates in the region are expected to decline significantly. In Northeast Asia, for example, population growth is expected fall from 0.9 per cent per year, to 0.4 per cent, in the next 20 years. In absolute terms, however, the total population will increase substantially with the addition of 700 million people in the next 15 years in Asia alone (UNDP 1999). This increase in population could be one of the most important factors for environmental stress, including scarcity of resources such as land, forest, water and biodiversity, and may contribute further to water and air pollution (Kainuma et al 1998). The burgeoning population will be accompanied by increasing number of urban dwellers. In 20 years' time, over half of the region's population will live in cities. Migration and urban reclassification will continue to contribute between 55 to 60 per cent of urban increase, and the number of cities with a population of more than 10 million is expected to rise from 11 at present, to 18 by 2015, and will continued to increase in future (Kainuma et al 1998).

1.   Economic Growth

Asia and the Pacific is projected to continue with its high rates of output growth. Such growth is critical in creating employment, alleviating poverty and making available resources for infrastructure and human resource development, and for increasing access to basic amenities. The projections show a tendency towards convergence of growth rates between different subregions, as well as between countries within a subregion. Long-term projections indicate that People's Republic of China will consolidate its position as the second largest economy, while the region will emerge as the largest economic zone in the world (Noland 1995).

Economic growth implies a shift to higher productivity, thus the agricultural sector's share in output and employment is expected to decline, as there is a shift towards a predominance of industrial productivity. This, in turn, implies an increasing use of energy, especially coal – a major source of pollution. Moreover, growth of the industrial sector invariably occurs in clusters in and around major cities, thereby invoking the migration of the labour force into urban areas, and pressurizing urban environments.

2.   Income and Quality of Life

Based on the analysis of future growth in Asia and the Pacific, it has been projected that three out of five of the region's absolute poor could be lifted from poverty by 2025. South Asian countries are likely to see a substantial reduction in poverty, where it has been predicted to fall below 20 per cent by about 2025. However, despite these potential improvements, the region's core poverty will still be concentrated in South Asia. In People's Republic of China, if the relationship between growth and poverty continues to hold, "a dollar a day" poverty in People's Republic of China could fall to less than 10 per cent over the same period.

With the continuation of positive economic trends, the region's health and nutrition profile will also continue to improve. Child malnutrition in People's Republic of China and Southeast Asia may well decline to 15 per cent in 2020. However, South Asia will still have a large share of the undernourished children in the world; even under the optimistic scenarios, 1 in 3 South Asian children under 5 will remain undernourished. The life expectancy at birth will improve all over the region, and is expected to reach an average of 70 years in the region by 2030.

B.   Natural Resources

1.   Land, Forest and Biodiversity

Without major interventions, the rate of land degradation is likely to continue (UNEP 1999). Maintaining, let alone improving, the situation for per capita land availability will be difficult, as populations continue to increase and agricultural land is lost to urban, industrial and transport infrastructure. The challenge is to optimize land use for competing needs. Given the limited scope for expanding croplands, future food production will rely heavily on the intensification of agriculture and use of fertilizer and pesticides. As past trends in freshwater supply indicates, it may not be possible to expand irrigated lands any further, and they may even contract due to depletion of aquifers and enhanced competition with other users. Moreover, high rates of fertilization and pesticide-use may not translate to a corresponding increase in yield due to erratic and sometimes non-optional applications, and concomitant negative environmental impacts of agrochemical applications.

The region's dominance of world trade in tropical hardwoods is likely to decline in future because of the depletion of timber reserves and increased domestic wood consumption (UNEP 1998). Tropical forest destruction has gone too far to prevent some irreversible damage, and it would take many generations to replace the lost forests with plantations. If current rates of harvesting continue, the Asian and Pacific Region's timber supplies may vanish in less than 40 years. If continued, deforestation will also further aggravate the widespread incidence of desertification, soil erosion, siltation, flooding and biodiversity loss, and will be one of the major contributors to drought and potential threats of climate change. Forest plantation efforts are likely to intensify. The sustainable forest and agricultural management policies that were introduced in the 1990s will continue to be implemented and may show more promising results. Shifting cultivation may continue to pose a threat to watersheds, for instance throughout the Mekong subregion. In Northeast Asia, particularly in Japan, Republic of Korea and Democratic People's Republic of Korea, the trade-off between forest protection and development will become a critical issue where there is limited flat land for urban and industrial expansion.

Increasing population pressure and land use changes will continue to threaten the region's biodiversity. Some scientists project that a mass extinction of species may take place in the tropics within the next 20-25 years, based on the fact that only about 10 to 30 per cent of natural habitats are now left in many countries of the region (UNEP 1998).

Genetic erosion may also intensify as the reliance on high yielding varieties increase. In India, for example, 75 per cent of rice production may come only from 10 varieties by 2005 compared with the over 30 000 varieties traditionally cultivated.

Among resources, the most dramatic rise in demand is for freshwater, for example, demand for safe drinking water is anticipated to increase five-fold in the next 40 years (Kainuma, M. et al 1998; Figure 22.3). Based on an acceptable threshold of 1 600 cubic metres per capita per year, arid areas such as Afghanistan and Iran (Islamic Republic of) will become water stressed by 2025 if the population growth under the United Nations low and medium projections continues. Parts of India (e.g. Rajasthan) and People's Republic of China will narrowly avoid this situation. Increasing freshwater demand among different sectors will also intensify current sectoral conflicts. Although agriculture will continue to be the largest consumer of water, the fastest increase in water demand will occur in urban and industrial sectors, where demands are projected to climb 135 per cent over the next 40 years (UNEP 1998). Water scarcity may also be exacerbated by potential climate change.

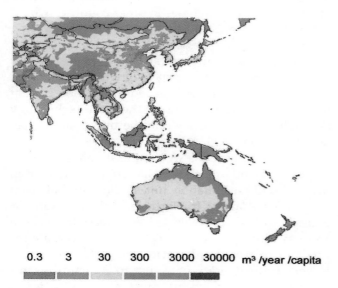

Figure 22.3 Projected Water Resources per Capita in 2050

Source: Kainuma, M. et al (1998)

Disclaimer:
1. Any boundaries and names shown and designations used on this map do not imply official endorsement or acceptance by the United Nations.
2. The dotted line represents approximately the Line of Control in Jammu and Kashmir agreed upon by India and Pakistan. The final status of Jammu and Kashmir has not yet been agreed upon by the parties.

2. *Aquatic Resources*

As most fishing areas reach their maximum potential and the production from capture fisheries dwindles, aquaculture production will become an increasingly important industry in the region. Predictions show that there may no longer be an increase in average capture fisheries output. The rate of mangrove forest depletion and coastal and marine resource degradation may not halt, but will slow down in many countries of the region due to introduction of protective measures.

3. *Food Security*

The region is projected to remain a net food importer in the near future (Alexandratos 1995). The trend in food availability in terms of calorie per capita is expected to improve by 2010, with projected calorie per capita of 2 450 for South Asia, 3 040 for Northeast Asia and 2 730 for all developing countries (Leisinger 1996). Nevertheless, a United States Department of Agriculture study indicates that the situation may deteriorate in future in some countries of the region, which may face a decline in per capita consumption of food during the next decade. The limiting factors in meeting the challenge of producing more food will be availability of productive land and supplies of fresh water, especially in populous and arid areas. The region is thus expected to continue to rely heavily on imported foods. It has been predicted that the region's share of world cereal imports may rise to about 42 per cent from its current level of 33 per cent by 2010 (ADB 1997). Even People's Republic of China, which is currently self-sufficient in grain production may begin to import around 175 billion tonnes by 2025, an amount almost equal to total current world exports (Brown et al 1999).

4. *Energy Consumption*

Demand for energy in Asia and the Pacific is expected to double every twelve years, compared to the world estimate of every 28 years (UNEP 1999). The total primary energy supply in the region (excluding Islamic Republic of Iran, Central Asia and some South Pacific countries) is projected to rise from 2 791 million tonnes of oil equivalent in 1997, to 4,392 million tonnes of oil equivalent in 2010 (IEA 1998; Figure 22.4). The region's primary energy supply is expected to soar by 78 per cent, including People's Republic of China by 59 per cent and Japan, Australia and New Zealand combined by 19 per cent. Coal will remain the major fuel choice in future, especially in India and People's Republic of China, and demand is projected to increase by 6.5 per cent a year (World Bank 1997). Despite shrinkage in their uses, traditional sources of fuel such as firewood, charcoal, and crop and animal residues will continue to be

Figure 22.4 Current and Projected Total Primary Energy Supply in Asia and the Pacific 1997 and 2010

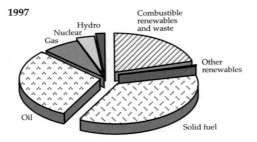

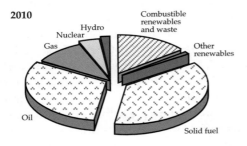

Source: IEA 1998 and 1999

Notes: Excludes Central Asia, Islamic Republic of Iran and some South Pacific countries; based on business as usual scenario; one million tonnes of oil equivalent is equal to 41.868 x 1 015 Joules

important sources of energy for rural populations in developing countries in coming decades. The use of nuclear and hydropower is expected to grow, particularly in those countries without substantial fossil fuel resources (United Nations 1995). If current trends in economic development continue, the current rate of energy consumption growth will more than triple within the next 30 years.

C. Pollution

As material inputs expand, the pollution load in terms of discharges in all of the natural spheres will inevitably increase, with a corresponding high cost to human health. With unchanged policies and technologies, emissions from electricity generation and transportation in developing countries will grow exponentially. Projections indicate that by 2030, they will be between five and ten times higher than during the 1990s. Improved policies, however, could cut the rate of emissions by about 20 per cent from their projected 2030 levels. Policy reforms in general, together with investments in low-polluting technologies, can more or less stabilize emissions at their 1990 levels.

The continuous growth in energy consumption and reliance on energy sources with relatively high carbon content, such as coal and oil, will enhance the production of greenhouse gases such as CO_2 in the region. Under the high economic growth scenario, emissions will peak by 2050 at a level 2.7 times that of 1990 and then start to fall two times the current level by 2100 at around 13 Giga tonnes of Carbon (GtC) (Kainuma et al 1998). Since transport and power generation, the fastest growing sectors, are likely to remain as the prime consumers of energy, improvement of energy efficiency and implementation of initiatives for the development of mass transport system will help reduce the growing emission trends. The movement toward accelerated decarbonization made in Kyoto, Japan in 1997 will also enable a reduction in overall greenhouse gas emissions by at least 5 per cent below the 1990 levels in between 2008 and 2012. It is estimated that suspended particulate matter (SPM) concentration in People's Republic of China's atmosphere will increase 1.4 times in 2020 and 3.5 times in 2050 larger than that in 1990 under business as usual scenario. Figure 22.5 shows regional distribution of particulate matter with diameter less than 10 μm in 1990 (PM_{10}) (Kainuma et al 1998).

Inventories of anthropogenic sulphur emissions are subject to uncertainty, particularly at the regional level. Current growth paths suggest that sulphur dioxide emissions in Asia and the Pacific will rise from their 1990 level of about 40 million tonnes by 2020. Introducing basic emission control technologies in People's Republic of China, India, and Pakistan and advanced technologies in other countries would cut emissions in 2020 by 40 per cent. Introducing advanced technologies in all countries would cut sulphur dioxide emissions by 50 per cent, but will still leave them 30 per cent above their 1990 level. Emissions could only be reduced below their current levels if all countries used the best technology available, but the costs would be huge: more than US$590 billion by 2020.

Estimates show a worsening situation in terms of national and transboundary air pollution, from SO_2 and acid precipitation, unless effective abatement measures are taken. In Northeast Asia, for example, emissions of SO_2 are expected to increase from about 15 to 41 tonnes in 2020, which could exacerbate the problem of cross-border acid rain. The areas with lowest critical loads (i.e. up to 320 milligrams per sq m) and which are most sensitive to acidic deposition are located in South China and in Southeast Thailand, Cambodia and South Viet Nam.

Water pollution may also worsen in many countries of the region. The Republic of Korea has initiated a set of ambitious water quality targets for 2001 and 2005, but considering the serious eutrophication problems, especially in lakes, it will be difficult to meet these standards; more so under a scenario of increasing fertilizer use.

Coastal and marine pollution in the region is likely to increase in future. Untreated urban and industrial wastes that find their way into the sea will continue to constitute a major threat to the coastal and marine environment. In addition, the mercury associated with long term nuclear wastes dumping and oil spills caused by tanker accidents will continue to pose major threats to the overall quality of coastal waters and damage to marine ecosystem and fishery resources.

Another important problem will be caused by the increase in wastes. Figure 22.6 shows the increase in wastes in the high and low economic growth scenarios. Municipal solid wastes would increase more than seven times in Asia and the Pacific in the high growth scenario and three times in the low growth scenario in 2030 (Kainuma et al 1998). In the short term, according to the World Bank, municipal solid wastes may more than double in the next 15 years (World Bank 1999). Furthermore, hazardous wastes generated by manufacturing, hospital and health-care facilities, and nuclear power and fuel processing plants are also estimated to more than double within the next 10 to 15 years.

D. Policy Environment

There is a growing trend in the continued advancement of policy and programmes which aim to integrate environmental consideration into the region's development framework. However, the resources and access to environmental technology will continue to present constraints in the implementation of policies for sustainable development. A lot will hinge upon the inputs and assistance of the developed countries towards provision of new and additional financial resources and the transfer of technology on concessional terms. The role of major public interest groups, particularly the private sector and NGOs, will continue to enhance.

Figure 22.5 Suspended Particulate Matters in Asia and the Pacific

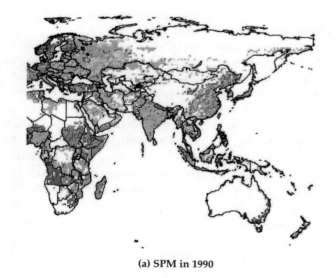

(a) SPM in 1990   (b) SPM in 2050 (under business as usual)

Source: Kainuma, M. et al, 1998

Disclaimers:
1. Any boundaries and designations used on this map do not imply official endorsement or acceptance by the United Nations.
2. The dotted line represents approximately the Line of Control in Jammu and Kashmir agreed upon by India and Pakistan. The final status of Jammu and Kashmir has not yet been agreed upon by the parties.

CHAPTER TWENTY-TWO

Figure 22.6 Projected Municipal Solid Wastes Generation in Asia and the Pacific 2030

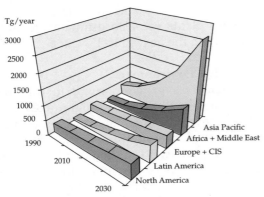

(a) Projection of municipal solid wastes under the low growth scenario

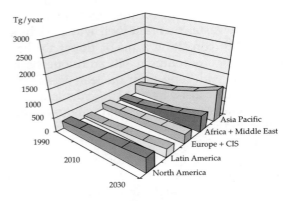

(b) Projection of municipal solid wastes under the high growth scenario

Source: Kainuma, M. et al, 1998

CHALLENGES AND OPPORTUNITIES

Whilst population and unsustainable economic growth, poverty, consumption and globalization have all been identified as key elements of change in the conditions of environment and natural resources, the failure to avoid environmental degradation in the region can to a large extent be traced to the weakness of institutions, adoption of inappropriate policy tools and the lack of effective policy enforcement. The principal environmental challenge in the 21st Century is therefore policy reform and its effective implementation in both individual and cross-sectoral areas. Meeting the sustainability challenge requires new policy approaches that integrate environmental concerns into the very core of investment, planning and technology decision-making in the region's economies.

A.   Policy Challenges

The implementation of appropriate policies and programmes, enactment of laws, rules and standards for enforcement of policies, and creation of delivery mechanisms for the implementation of plans and enforcement by institutions all provide stability for the environment. Challenges, in this respect, lie in how governments can provide clear signals and incentives to the agents tasked to carry out the underlying goals and objectives of development. Effective implementation of environmental legislation remains one of the biggest challenges in many countries of the region. It needs to be resolved by bridging the gaps between intent and action.

At the national level, the most daunting challenge is not only to guide the overall development process on a sustainable path, but also to promote vertical coordination between various tiers of government, at national, provincial and local levels, as well as horizontal coordination between the key sectors of the economy. The process of plan formulation should involve intensive deliberations on those aspects of sustainable development most relevant to national priorities and extensive participation of the public in giving their views and suggestions on adoption of appropriate means to accomplish sustainability.

An essential aspect of such plans, and one often omitted in practice, is the examination of such cross-sectoral issues as budgetary priorities, trade and investment policies, specific technology needs, research and development, and roles of trans-national corporations and international capital flows. Comprehensive analytical procedures for prior and simultaneous assessment of the impacts of decisions on social and environmental aspects of sustainability need to be applied, not only at the project and sectoral levels (Box 22.3), but also in the analysis of programmes and policies including macro-economic policies.

The 1992 Earth Summit moved the concept of sustainable development to the forefront of public discourse in Asia and the Pacific and around the world. If sustainable development remains somewhat elusive as an organizing principle for the world economy as a whole, the events of the 1990s helped to bring the concept into sharper focus in a regional context. From the Asian economic crisis, to forest burning, haze and continued destruction of coastal ecosystems, as well as many other environmental and development challenges, two key dimensions of sustainable development in the region emerge. The first is the urgent imperative to dramatically reduce severe poverty and improve socio-economic welfare for much of the population. The second, is that sustainable development in Asia and the Pacific must

> **Box 22.3 Integrated Environment and Business Planning – The Case of Two Shrimp Farms in Indonesia**
>
> The intricate relationships of environmentally and economic sustainability have formed the basis of much research at the global, national and regional levels. Recent case studies of two comparable shrimp farms in Indonesia, however, offer an interesting example of how financial injections and poor business management, in the absence of local by-in, can lead to unsustainable development at the project level.
>
> At one time, Lampung in southern Sumatra, with an area of 80 000 hectares, was reputed to be the largest shrimp farm in the world, and produced around 50 000 tonnes of shrimp per annum. At the beginning of the 1990s, a combination of bi-lateral and international financing was channelled through a financier to provide loans to 7 750 farmers taking part in the scheme, amounting to about US$65 000 each. They used this to purchase shares in a plot, the majority of which remained under the ownership of the financier. Loans were undertaken on the expectation of full repayments and personal profits within 8 years. The scale of the establishment provided work for an additional 30 000 labourers. Feed, fry, power and other necessities were all provided by the financier, and paid for by the farmers out of their returns. However, farmers were forced to take out further loans, or to default on payments of their original loan, when harvests fell short, and could not recover these losses when harvest improved, since repayments also increased during these periods. This led to growing tensions between farmers, farm management and the banks, the tripling of farmer's debts, and the recall of many of the farms. Consequently, some 18 000 ponds have fallen into disrepair or disuse to the extent that many are now potential hazards to the environment.
>
> Just north of the farm, in Wachyuni Mandira, a similar establishment was also developed and began its operations around five years later. The deal offered to the Wachyuni Mandira shrimp farmers, however, facilitated 'total ownership'. As a result of careful financial and business planning, and the use of many years of experience of local fishermen and elders, conditions on the farm have remained stable, careful observation has been taken of environmental sensitivities, and increasingly higher yields have been supported. Whilst rising profit margins have also led to some management-worker tensions at times, local buy-in has again been advanced and the operation continues to expand, demonstrating the "win win" benefits to the business and to the environment when careful and cautious business planning is observed.
>
> *Source:* Far Eastern Economic Review 30 May 2000 and Jakarta Post 30 March 2000

resolve the escalating demands for energy, materials, water and other resource inputs to production and consumption that accompany economic growth.

Much of Asia and the Pacific is in the midst of an urban-industrial transition unprecedented in scale and intensity. The development growth trends for the region outlined earlier in this Chapter make the challenge of achieving sustainable development particularly significant, not just for the Asian and Pacific Region, but for the global economy as a whole. Fortunately, those same trends also provide significant opportunities for addressing the challenge of clean (i.e. green) and shared (i.e. pro-poor) growth in the region over the next two decades.

B. Sectors and Mechanisms for Action

In response to the sustainable development challenges facing the region, a number of key sectors (or areas) for action, and mechanisms for implementing these actions, can be identified, as discussed below.

1. Environmental Quality and Human Health

Among the most significant environmental challenges for the region are those facing the region's cities. Despite their potential to offer a better quality of life, the cities are often beset by growing problems of environmental deterioration relating to the loss of natural resources, the lack of adequate shelter and dwelling, deteriorating ambient air and water quality conditions, inadequate infrastructure, and an increasing deterioration in urban services provision, including water supply and sanitation facilities, and municipal, industrial and hazardous waste management. These conditions impact directly upon the health and welfare of urban residents, and in particular the poor, leading to the widespread incidence of infectious diseases, including acute respiratory infections.

2. Globalization and Policy Integration

Economic development in Asia and the Pacific over the next two decades will occur in an international context. From flows of capital, technology, ideas and information, to the increasing global reach of corporations, non-governmental organizations, and regulatory organizations, development is now a global process. Linkages among businesses, consumers, citizens, and organizations are becoming both more extensive and more intensive in their character. For some, these processes of globalization will usher in a world with more inequality and more volatility, where the costs and benefits do not fall equally among the rich and the poor (Greider 1997), and where deepening environmental problems are the inevitable result. For others, globalization offers vast new opportunities for improvements in social-economic welfare. In many ways, the Asian economic crisis of 1997/98 was a test case for these viewpoints. But it is also important to place the events of 1997/98 in the broader context of the economic achievements of the past three decades among the newly industrializing economies of the region. The capital flows that were a source of volatility in 1997/98 fuelled much of the growth that occurred in past decades. The events of 1997/98 were more a crisis of institutions and of effective regulation than an indictment of economic integration and international investment.

Globalization is perhaps best conceptualized as an accelerator of change, rather than as a process that necessarily entails positive or negative outcomes. The challenge and the opportunity for the Asian and Pacific region are those of harnessing the processes of globalization to the goals of sustainable development, including reductions in poverty and improvements in the environmental performance of economic activity. Indeed, it is difficult to imagine that the challenges of sustainable development in the region can be met without effective deployment of the tremendous resources and know-how that are contained within the global economy as a whole. At the core of this process is the private market economy. But public policy also has a crucial role to play in shaping processes of private investment and in creating the enabling environment in which private investment takes place.

3. *Energy Efficiency and the Promotion of Clean Technology*

A major concern for the region is that the current environmental policy does not effectively address the escalating demand for energy. From agriculture to energy supply, transportation to tourism, the sustainability transition requires a substantial acceleration in current trends towards efficiency improvements in economic processes. The place to begin such an initiative is with the calculation of current levels of environmental performance and the identification of aggressive but attainable goals for reducing energy, materials, pollution and waste intensity. The policy tools for assessing these trends, such as life-cycle analysis and materials flow analysis, are now available (Allenby 1999; Graedel and Allenby 1995; Socolow et al 1994). In industry this entails assessing and reducing energy and materials use in a dynamic of continuous improvement, as well as reducing pollution and waste discharges throughout the entire process of production or manufacture. In agriculture it entails tracking the use of chemicals, energy, water and other materials in the production of food. Importantly, goals for reducing energy, materials, pollution and waste intensity need to be examined within broad industrial sectors, such as the manufacturing industry, and need to be adopted by the industrial ministries responsible for the development of these economic sectors.

Perhaps the key opportunity related to environmental intensity lies in shaping the pattern of new urban-industrial investment. Attracting and promoting clean technology investment in industry and in urban development will be critical to reducing the energy, materials, pollution and waste intensity of economic activity as a whole. To some degree, this goal will be met through the attraction of new investment and capital turnover. Evidence suggests that within many sectors of economic activity new capital tends to be cleaner than old. Nevertheless, the sustainability transition requires improvements that go beyond the modest improvements in environmental performance secured through capital turnover *per se*. Investments in research and development, in accelerated technology transfer, and in the enhanced modification and extension of clean technologies will also be required.

4. *Poverty Reduction Strategies*

Efforts to reduce poverty must be comprehensive enough to address all of its many causes, and their success is fundamentally linked to sound economic management and good governance, at all levels. To be effective, poverty reduction strategies must simultaneously address the combined goals of ensuring that all poor households are provided with the opportunity to earn a sustainable livelihood, while ecologically-vulnerable areas are handled in an integrated manner encompassing resource management, poverty alleviation and employment generation. They also need to be complemented by social development policies that permit access by the poor to education, health, social protection and other basic services.

A range of activities need to be undertaken by individual countries (including governments, NGOs and other citizen's groups), for instance, to investigate the environmental implications of investment in poverty reduction strategies, and to examine the effectiveness, practicality and appropriate forms of policy aimed at achieving poverty reduction and sustainable environmental management in different agro-ecological zones and urban environments. The concept of zoning enables different strategies to be developed to suit different zones. For example, in areas of low agricultural potential, in which the thresholds of ecological sensitivity and resilience have not yet been crossed, social welfare transfers to the poor could be channelled through public works programmes geared at supplementing natural resilience, such as tree-planting by means of rainwater harvesting in drylands.

From an international perspective, the world has, in recent decades, become increasingly interdependent with the emergence of a global economy. The formation of this economy has been accompanied by policies and practices whereby the developed countries now heavily influence the terms under which the developing countries participate in the international system (United Nations 1995). This could potentially have significant negative impacts for poverty and the environment. There is, therefore, an urgent need to examine the effects of developed countries' macroeconomic and trade policies on environmental management behaviour also remains, in particular the consequences for the poor.

5. *Strategic Environmental Management*

(a) Integrating Economic and Environmental Policy

To secure the progress in environmental performance needed to offset future growth in economic activity, policy-makers will need to focus on the fundamentals of investment, technology and production, or what have been labelled elsewhere as the "denominators" of economic activity. Traditionally environmental policy-makers have focused on pollution and other negative outputs of economic activity, and has not taken a strategic approach which recognizes core economic processes as part of the domain of responsibility. Investment, technology choices, land use planning and other key elements of economic activity have been influenced only indirectly, by means of environmental regulation, taxes, and subsidies of one form or another. Focusing directly upon core economic processes opens up the range of policy instruments that can be brought to be bear on the sustainability challenge, from education to industrial policy, from trade to technology policy. It also allows for the strategic coordination of policies across a variety of areas, such as agriculture, urban, trade and environment policy. Careful coordination of economic and environmental policy will be crucial if countries are to balance competing priorities, such as the desire to reduce pollution while also improving industrial output. The goal of reducing poverty while improving environmental performance will require careful coordinated economic planning at the national level around issues of investment and infrastructure development within urban and rural areas. The economies of the region will also need to integrate economic and environmental policy at the highest levels of government, and internalize environmental performance into the core of investment and technology planning.

(b) Setting Clearly Defined Goals

The experience of the past decade has confirmed the importance of clearly articulated and consistently implemented goals for sustainable development, as well as effective monitoring and reporting of the progress that is made toward achieving these goals. It is for this reason that organizations such as the ADB, UNDP and ESCAP have placed a high priority upon the specification of measurable goals at the national level for such issues as reducing pollution, eliminating poverty, and improving social welfare. These goals must be supported by a low cost, transparent, and scaleable programme of performance measurement that reaches up from the individual organization to the industrial sector and to the economy as a whole. Clear and consistent goals are crucial to the creation of a stable enabling environment for research, investment, and strategic planning on the part of communities, firms, industries, and financiers. A key priority for countries of the region must be the further development and specification of short and medium term goals for sustainable development, and funding of the necessary research and monitoring capability to identify indicators and to track progress toward these goals.

(c) Influencing New Investment and Technology Choices

The very processes of urban and industrial growth, which potentially constitute a shadow on the environmental future of the region, also create significant opportunities to influence processes of new investment and technological change. Much of the capital stock and infrastructure that will be in place in the region twenty years from now is not on the ground today. The opportunity exists for many developing economies in Asia and the Pacific to turn the trajectory of urban-industrial investment toward a pathway that is dramatically less energy, materials, and waste intensive, i.e. to shape the process of urban-industrial development as it happens. When the OECD countries moved to strengthen environmental regulation during the 1960s and 1970s, the challenge was one of improving the environmental performance of mature industrial economies, typically through the retrofit application of end of pipe pollution control. In Asia and the Pacific today the challenge however is one of influencing the technology and planning choices of new industrial and urban investment. Already across a wide variety of industrial sectors, newer cleaner technologies are economically viable alternatives to older dirty technologies. If the process of new industrial and urban investment in the region is based on the development, deployment and use of ever-cleaner technology and infrastructure, this will go a long way toward moving the region toward a pathway of sustainable development.

(d) Promotion of Non-Regulatory Mechanisms

One of the important corollaries of globalization is the increasing importance of non-regulatory drivers of environmental performance in economic activity. Tighter links between consumers and producers, and the growing availability of environmental performance information, are enhancing the significance of eco-labelling and similar programmes. Within large production networks, supply chain management is now an important tool for environmental management, and increasingly, issues of environmental performance are beginning to influence investment decisions. The full impact of non-regulatory drivers on environmental performance is hard to assess accurately at this time. However, it is understood that the effectiveness of non-regulatory drivers is likely to depend on the quality of information on environmental performance that is available within and among organizations. Therefore, public policy can most effectively promote non-regulatory drivers by supporting a programme of low cost, standardized and transparent environmental performance information disclosure for firms and industries.

6. *Governance, Institutions and Capacity Building*

Whether or not the challenges will be met and whether the opportunities for sustainable development will be taken depends to a significant extent upon the evolution of structures of governance within the Asian and Pacific Region and the global economy as a whole. The successful implementation of a more transformative policy framework in support of sustainable development will require broad engagement of multiple stakeholders and a strong commitment to collaborative and participative governance (see Box 22.4). The bedrock of such collaborative governance is access to information and a commitment to democratic decision-making.

Box 22.4 ECO ASIA and the Promotion of Regional Cooperation for the Environment

In order to promote regional cooperation on environmental issues, and thus to contribute to sustainable development in Asia and the Pacific, the Environment Agency of Japan has been working hard to promote regional dialogue on environmental policy. To this end, in 1991, Japan initiated the Environment Congress for Asia and the Pacific, or ECO ASIA in short, which has convened eight times since then.

ECO ASIA is regarded as one of the leading forums for environmental policy dialogue in Asia and the Pacific region. Key activities to date which have been implemented under ECO ASIA include the following.

- ECO ASIA '99 (the eight session), held in Sapporo, Japan in September 1999, was attended by 111 participants from 17 countries in the region, including 8 Environment Ministers and 12 representatives from 11 international organizations. In addition to allowing individual countries to present their own environmental status reports, some of the key outcomes of the sessions were the common points of agreement which emerged between members, which, for example, included: that links should be strengthened between ECO ASIA and parliamentarians or parliamentarian groups around the region, to promote common policy responses and raise environmental awareness in the political arena; the importance of promoting "win-win" approaches to combating climate change, and the wider and more effective participation of both developing and developed countries in this process; and the need for enhanced technology transfer to developing countries by the private sector, inter-governmental and scientific networks, for example, the Acid Deposition Monitoring Network in East Asia (EANET).

- The ECO ASIA Long-term Perspective Project was initiated at ECO ASIA '93. It aimed to provide decision-makers in the region with a scientific basis for policy formulation in order to support the process of sustainable development through to 2025. To achieve this, it forecasts, on a continuing basis, the probable state of the environment in Asia and the Pacific in 2025, and identifies policy options for the realization of sustainable development. In Phase I (1993-1997) it evaluated current and future environmental and socio-economic issues in the region and offered 'Asia-Pacific Eco-Consciousness' as a conceptual tool for building partnerships. Objectives of Phase II (1998-2001) include capacity-building for personnel and institutions of the participating countries.

- Establishment of the Environmental Information Network for Asia and the Pacific (ECO ASIA NET) was proposed by the Environment Agency of Japan at ECO ASIA '96. ECO ASIA NET aims at supporting sustainable development in the region through information sharing and dissemination by using Internet and other available communications technology. It provides a vital forum for information sharing on the environment among stakeholders, including policy-makers, businesses and NGOs across the region.

Source: http://www.ecoasia.org

Collaborative governance will need to link effectively with multiple scales of decision-making, from international organizations such as the World Trade Organization and the ADB, to decentralized systems of local and regional representative government. Only in this way will the goals and aspirations of the people of Asia and the Pacific be met.

Many of the decisions and initiatives required to promote sustainable development will take place on a local, regional and urban scale. From land use controls to approaches to water and energy supply, effective decision making requires mobilization of knowledge regarding local context, challenge and opportunity. Over the next two decades, decentralization and devolution of decision-making below the national scale is likely to accelerate within many countries of the region. As decentralization takes place, it is critically important that this process be matched by the investment of resources and building of local and regional institutional capacity. Many advances in improving environmental performance during the 1990s have derived from policies that are implemented at the local level. Much remains to be done in decentralizing and enhancing capacity, especially in the areas of sustainable forest management, rural development financing, and integrated coastal resource management.

Finally, the aforementioned integration of environmental concerns into economic policy should be accompanied by continued efforts to strengthen environmental regulation and resource management systems within the region. As indicated, evidence in this report suggests that where environmental policies are effectively designed and consistently enforced, improvements in environmental performance result. The state of environmental regulatory capability and practice is quite variable within the region. Some countries already have widely developed environmental regulatory systems and enforcement capacity, and the emphasis in these cases should be on the pushing beyond compliance, and on the use of 2nd and 3rd generation policy instruments, such as information disclosure and market-based instruments. In many of the region's developing countries, the key priority is on strengthening policy implementation and enforcement of basic systems of environmental protection. Throughout the region, environmental protection systems will be called upon to protect public health through the establishment of clear and consistently enforced ambient environmental standards.

CONCLUSION

This analysis of the state of environment in Asia and the Pacific in the year 2000 demonstrates that overall environmental conditions continue to deteriorate. Deterioration can be seen in over a third of the region's croplands, it's falling water tables, and it's diminishing forests. Recent and unprecedented forest fires not only destroyed thousands of hectares of forests, but also perpetuated the polluting haze from Indonesia to India. Perhaps, the most significant single indicator of the region's deteriorating environmental health is the declining biodiversity from wilderness, oceans, and cultivated landscape.

In this gloomy scenario, hopeful trends are evident in the decline in birth rates and fertility levels, increased life expectancy, somewhat reduced poverty levels, improved levels of nutrition, growing public awareness and participation, an improved policy environment and an enhanced role of the private sector in environmental protection, and the promotion of sustainable development. For example, some developing countries have dramatically lowered birth rates and moved forward towards population stability. However, with its heavy base even before stabilization, the population growth momentum is likely to continue in sheer scale and number at least in the early part of the 21st Century.

Implementation of appropriate policies and programmes, enactment of laws, rules and standards for enforcement of policies, and creation of appropriate delivery mechanisms are the key policy challenges for the region. Moreover, in the wake of development needs, the resources for investment in environment are low, which poses a daunting challenge in generating resources by innovative means. A new policy model based on a mix of command and control and market-based mechanisms with a strong but limited government steer is gaining ground in countries like Malaysia and the Philippines. This is characterized by the role of government as a facilitator rather than provider; by a prominent private sector and civil society; and by a pricing reform on environmental goods and services and improved management. The model appears to have great potential for developing countries of Asia and the Pacific as it can be achieved by a government that is deficit of financial resources.

If a strategy to promote sustainable development in the region is to succeed, it must also focus on poverty eradication. Although there is a repeated acknowledgment of both the vicious "cycle of poverty" and its intrinsic linkages with the environment and the urgency to address poverty alleviation, little evidence has emerged to show that effective and concerted actions that have been taken in that direction. Empowerment of communities and provision of access to resources could be most powerful mechanisms towards alleviation of poverty while improving the environment.

CHAPTER TWENTY-TWO

The entry into the new millennium has also brought forward the key limits to economic growth in terms of natural resources, such as freshwaters, forests, rangelands, fisheries and biological diversity. There is a need to recognize these natural limits and adjust national economies accordingly, promote efficiency in the use of water, energy and materials, curb growing profligacy in the use of resources, and reflect the cost of loss of natural resources and enhancing pollution in the estimation of national incomes. It is extremely important not to let accelerating change outstrip the assimilative capacity of the natural environment, and overwhelm political institutions through an extensive breakdown of the ecological system on which the economy depends.

These issues are enshrined in Agenda 21, which advocates utilization of natural resources in harmony with nature. The ultimate realization of the objectives of Agenda 21 depends to a large degree on national and local governments because they are in the frontier of development actions. They create and enforce legislation on the use of natural resources. They formulate and implement development plans. They maintain institutions that embody and monitor the implementation of environmental and development policies. However, the supportive role of the developed countries in the provision of financial resources and technology transfer remains equally important.

Globalization, while opening the awareness for development has also made developing countries of the region vulnerable to trade sanctions and international pressures. These sanctions could be provoked by countries' perceptions of a regional violation in environmental standards. Moreover, new and emerging issues and challenges have now surfaced as a result of bioprospecting and recent developments in biotechnology (particularly those related to genetically modified organisms). Multinational companies are exercising property rights and patent regimes presenting dangers of biopiracy, and monopolization of improved varieties of seeds which are threatening the cultural and traditional rights of indigenous communities and oriental farmers.

Whilst Agenda 21 and Rio+5 clearly recognized the principle of "common but differentiated responsibilities", translation of this principle into reality has not risen beyond the narrow self-interest of states. The commitments made by developed nations regarding financing and technology transfer are still far from their realization.

Under these troublesome scenarios, the need for acquiring information and promoting regional and subregional cooperation was never so pressing in Asia and the Pacific as it is today. Regional cooperation is vital for providing opportunities for ensuring a coordinated response to continuously increasing pressures of globalization. Moreover, the dwindling availability of international financial resources, technology transfers, and unfavourable trade regimes, demand enhanced regional unity in a proactive response.

REFERENCES

CHAPTER ONE

ADB/SPREP, 1992. *Salinization by Sea Level Rise*, fr. p-15, No. 25, appeared in p-12 of the 2nd draft.

AFFA Joint Release, 1999. March 19, http://www.affa.gov.au/tuckey/releases/99/99_32tuj.html

Ali, M. and Byerlee, D., 1998. *Productivity growth and resource degradation in Pakistan's Punjab: A decomposition analysis*, Draft, World Bank, Washington, DC.

Anonymous, 1996. Kangarooburgers? It's enough to make Aussies hopping mad, *Asia-Pacific Agribusiness Report*, vol. 114, No. 4.

Baitulin, I. and Beiturova, G., 1997. National strategy to combat desertification in Republic of Kazakstan, *Desertification Control Bulletin*, No. 33, pp. 19-27.

CIESIN, 1999. CIESIN Thematic Guides, Draft, http://www.ciesin.org/TG/LU/policy.html

Costanza, R., d'Arge, R., de Groot, R., Faeber, S., Grasso, M., Hannon, B., Limburg, K., Naeem, S., O'Neill, R.V., Paruelo, J., Raskin, R.G., Sutton, P. and van den Belt, M., 1997. The value of the world's ecosystem services and natural capital, *Nature*, 387, May 15.

Crosson, P.R., 1995b. *Soil Erosion and its On-farm Productivity Consequences: What Do We Know?* Resources for the Future Discussion Paper 95-29, Resources for the Future, Washington, DC.

Darkoh, M.B.K., 1996. Land degradation and sustainable development in Papua New Guinea, *Desertification Control Bulletin*, No. 29.

De, G.J. and Wiersum, K.F., 1992. Rethinking erosion on Java: A reaction, *Netherlands Journal of Agricultural Science*, No. 40, pp. 373-379.

Dent, F.J., 1990. *Approaches of Land Vulnerability Assessment for Food Security in Asia*, FAO, Bangkok.

Dent, F.J., 1990. Land resources of Asia and the Pacific, in Problem Soils of Asia and the Pacific, *RAPA Report 1990/6*, Bangkok.

Diemont, W.H., Smiet, A.C. and Nurdin, 1991. Rethinking erosion on Java, *Netherlands Journal of Agricultural Science*, No. 39, pp. 213-224.

Dregne, H.E., 1998. Land degradation: assessment and monitoring, *Land Degradation*, Newsletter of the International Task Force on Land Degradation.

Dregne, H.E. and Chou, N.T., 1992. Global desertification dimensions and costs, in Dregne, H.E., ed., *Degradation and Restoration of Avid Lands*, Texas Tech. University.

Eswaran, H., 1999. Meeting the challenges of land degradation in the 21st century, *Proceedings International Conference on Land Degradation*, January 25-29, Khon Kaen, Thailand.

FAO/RAP, 1999. *Poverty Alleviation and Food Security in Asia: Land Resources*, Bangkok.

FAO, 1995. *Drylands Development and Combating Desertification*, A Bibliographic Study of Experiences in Countries of the CIS, FAO Environment and Energy Paper 14, Rome.

FAO, 1997. *Dryland Development and Combating Desertification*, A bibliographic study of experiences in China.

FAO, 1997. *FAO Production Yearbook*, Rome.

Forman, R.T.T., 1995. *Land Mosaics: The Ecology of Landscapes and Regions* (Cambridge, Cambridge University Press).

Glazovsky, N.F. and Shestakov, A., 1995. Environmental migration caused by desertification, in *Desertification and Migration*, Geoforma Ediciones, Lograno, Espana, pp. 147-158.

Glazovsky, N., 1997. An integrated approach to inter-regional co-operation and major activities within the Inter-Regional Programme of Action to Combat Desertification and Drought, *Desertification Control Bulletin*, No. 31.

Government of Turkmenistan, 1996. *National Action Programme to Combat Desertification, in Turkmenistan*, Desert Research Institute/UNEP, Ashkhabad, 106 pp.

Hillel, D.J., 1991. *Out of the Earth: Civilisation and the Life of the Soil* (New York, Free Press).

Huang, J. and Rozelle, S., 1994. Environmental stress and grain yields in China, *American Journal of Agricultural Economics*, No. 77, pp. 246-256.

Huang, J., and Rozelle, S., 1996. Technological change: Rediscovering the engine of productivity growth in China's real economy, *Journal of Development Economics*, No. 49, pp. 337-369.

Huang, J., Rosegrant, M. and Rozelle, S., 1996. *Public investment, technological change, and reform: A comprehensive accounting of Chinese agricultural growth*. Working paper, International Food Policy Research Institute, Washington, DC.

Hyams, E., 1952. *Soil and Civilisation* (New York, Haper Colophon).

Kaul, R.N., et. al., 1999. India's approach to combat drought and desertification, *Desertification Control Bulletin*, No. 34.

Kharin, N.G., 1997. Strategy to combat desertification in Central Asia, *Desertification Control Bulletin*, No. 29, pp. 29-34.

Kharin, N.G., Kalenov, G.S., and Kurochin, V.A., 1993. Map of human-induced land degradation in Aral Sea Basin, *Desertification Control Bulletin*, No. 23, pp. 24-28.

Kharin, N., 1994. Ecological castrophe in Central Asia. *Our Planet*, UNEP, vol. 6, No.5, pp. 27-28.

Kharin, N., 1996. Strategy to combat desertification in Central Asia, *Desertification Control Bulletin*, No. 29.

Kharin, N., 1998. Estimation of desertification damage and the cost of land rehabilitation: A methodological approach, *Desertification Control Bulletin*, No. 33.

Kowsar, A.S., 1998. Aquifer management: A key to food security in the deserts of the Islamic Republic of Iran, *Desertification Control Bulletin*, No. 33, pp. 24-28.

Kruzhilin, I.P., 1995. The state and social and economic significance of relations in the countries of Commonwealth of Independent States, in *Protection and Utilization of Agricultural Land Subject to Salinization*, Moscow, pp. 75-82 (in Russian).

Lopez, R., 1996. *Policy Instruments and Financing Mechanisms for the Sustainable Use of Forests in Latin America,* Working paper, Environment Division, Inter-American Development Bank, Washington, DC.

Magrath, W.B., and Arens, P., 1989. *The Cost of Soil Erosion on Java: A Natural Resource Accounting Approach*, Environment Department, Working Paper 18, The World Bank, Washington, DC.

Mainguet, M. and Rene, L., 1998. Human-made desertification in the Aral Sea Basin: Planning and Management Failures, *Desertification Control Bulletin*, No. 33.

Ministry of Nature and Environment, 1997. Mongolia.

NEPA, 1993. *China Environment News*, December 8.

Noor, M., 1993. *Desertification Control Using Shelter Belts in Thal, Pakistan*, Rangeland research literature, Pakistan Agricultural Research Council, Islamabad.

Oldeman, L.R., 1998. *Soil Degradation: A Threat to Food Security?*, Report 98/01. The Netherlands: International Soil Reference and Information Centre, Wageningen.

Oldeman, L.R., Hakkeling, R.T.A. and Sombroek, W.G., 1991. *World Map of the Status of Hunam-induced Soil Degradation: An Exploratory Note,* The Netherlands and Nairibi. Kenya: International Soil Reference and Information Centre and United Nations Environment Programme, Wageninigen.

Oldeman, L.R., 1994. The global extent of soil degradation, in Greenland, D.J. and Szaboles, T., eds., *Soil Resilience and Sustainable Land Use*, Commonwealth Agricultural Bureau International, Wallingford.

Olk, D.C., Cassman, K.G., Randall, E.W., Kinchesh, P., Sanger, L.J. and Anderson, J.M., 1996. Change in chemical properties of organic matter with intensified rice cropping in tropical lowland soil, *European Journal of Soil Science*, vol. 45, pp. 293-303.

Plit, et. al., 1995. *Dryland Development and Combating Desertification: Bibliographic Study of Experiences in Countries of the CIS*, FAO Environment and Energy Paper 14, Rome.

Repetto, R., Magrath, W., Welk, M., Beer, C. and Rossini, F., 1989. *Wasting Assets,* World Resources Institute, Washington, DC.

Rozelle, S., Huang, J. and Zhang, J.L., 1997. Poverty, population, and environmental degradation in China, *Food Policy*, vol. 22, No. 3, pp. 229-251.

Scherr, S.J. and Yadav, S., 1996. *Land Degradation in the Developing World: Implications for Food, Agriculture, and the Environment to 2020*, Food, Agriculture, and Environment discussion paper No. 14 (Washington, DC, I.F.P.R.I.).

Scherr, S.J., 1999. *Soil Degradation: A Threat to Developing Country Food Security by 2020*, Food, Agriculture, and the Environment Discussion Paper 27, International Food Policy Research Institute, Washington, DC.

Scherr, S.J., 1997a. *Is Soil Degradation a Threat to Developing Country Food Security?*, Draft,

Agricultural and Resource Economics Department, University of Maryland, Collage Park, Maryland.

Sharma, P.N., and Wagley, M.P., 1996. *Case Studies People's Participation in Watershed Management in Asia Part I: Nepal, China, and India*, PWMTA-WHTUH-FARM Field Document 4.

Sheehy, D.P., 1992. A perspective of desertification of grazing land ecosystem in North China, *Ambio*, vol. 21, No. 4, pp. 303-307.

SOE-China. 1997. *State of the Environment in China*, http://svrl-pek.unep.net/soechina.htm

UNCCD, 1998. *The Social and Economic Impact of Desertification in Several Asian Countries: Inventory Study for the Interim Secretariat of the Convention to Combat Desertification*, Geneva.

UNCCD, 1999. *The United Nations Convention to Combat Desertification, Fact Sheet*, http://www.unccd.de/publicinfo.htm

UNCED, 1992. *Status of Desertification and Implementation of the United Nations Plan of Action to Combat Desertification*, Report of the Extension Director, UNEP, Nairobi.

UNSO, 1999. UNDP Office to Combat Desertification and Drought (UNSO), http://www.undp.org/seed/unso.htm

Van, L.G. and Oldeman, L., 1997. *Soil Degradation in South and Southeast Asia*, The Netherlands: International Soil Reference and Information Centre for the United Nations Environment Programme, Wageningen.

WRI, 1999. Intensification of agriculture: land conversion, in *World Resources 1998-99*.

Young, A., 1993. *Land Degradation in South Asia: Its Severity, Causes, and Effects upon the People*, Final report prepared for submission to the Economic and Social Council of the United Nations (ECOSOC), Food and Agriculture Organisation of the United Nations, United Nations Development Programme, and United Nations Environment Programme, Rome.

Asia Week, 1997. Dark Cloud of Death, October 10.

Asian Development Bank, 1996. *Annual Report.*

Evans, B., 1996. *Technical and Scientific Elements of Forest Management Certification Programmes,* Paper for the Conference on Economic, Social and Political Issues in Certification of Forest Management, May 12-16, 1998, Malaysia.

FAO, 1997. *State of the World's Forests,* Rome.

FAO, 1998. *Asia-Pacific Forestry Towards 2010,* Report of the Asia-Pacific Forestry Sector Outlook Study, Rome.

FAO, 1998. *Selected Indicators of Food and Agriculture Development in Asia-Pacific Region, 1987-1997* (Rome, RAP Pub.).

FAO, 1998. *State of the World's Forests,* Rome.

Government of Papua New Guinea, 1996. *Country Report.*

Hildyard, N., et. al., 1998. *Same Platform, Different Train: The Politics of Participation,* UWASYLVA, 1994, vol. 49.

ITTO, 1998. *Tropical Forest Update,* Newsletter from the International Tropical Timber Organization, vol. 8, No. 4.

Johnson, N. and Ditz, D., 1997. *Sustainable Challenges to Forests in the United States,* WRI, Island Press.

Montalembertde, M.R., 1997. *Cross-Sectoral Linkages and the Influence of External Policies on Forest Development,* UNASYLVA, No. 182.

Ostrom, E., 1999. *Self-Governance and Forest Resources,* Center for International Forestry Research, Occasional Paper No. 20, ISSN 0854-9818, pp. 1-11.

Repetto, R. and Gillis, 1988. *Public Policies and the Misuse of Forest Resources,* New York.

Ruitenbeek, J. and Cartier, C., 1998. *Rational Exploitations: Economic Criteria and Indicators for Sustainable Management of Tropical Forests.* Center for International Forestry Research, Occasional Paper No. 17, pp. 1-12.

United Nations, 1995. *State of the Environment in Asia and the Pacific Report,* New York.

World Bank, 1999. *The World Bank Annual Report 998,* Washington, DC.

WRI, 1997. New York.

WRI, 1998. *Frontier Forest Index,* Washington, DC.

WRI, 1999. *World Resources 1998-1999.* New York.

WWF, 1999. *Forest for Life: Global Annual Forest Report.*

Abas, A., 1999. *Sharp Decline in Turtle Population*, http://www.cyberct.com.my/arbec/new_tur.htm

ADB and IUCN, 1995. *Biodiversity Conservation in the Asian and Pacific Region: Constraints and Opportunities*, Asian Development Bank, Manila.

Aerni, P., et. al., 1999. An indication of public acceptance of transgenic rice in the Philippines, *Biotechnology and Development Monitor*, No. 38, pp. 18-21.

Amstilavskii, A., 1991. Sturgeon and salmon on the verge of extinction, *Environmental Policy Review*, vol. 5, No. 1.

Anonymous, 1995. *Background Paper on Biodiversity and Tourism*, Paper presented during the Second Meeting of the Conference of the Parties to the Convention on Biological Diversity, Jakarta.

Anonymous, 1997. *Wetlands and Integrated River Basin Management: Experiences in Asia and the Pacific*, UNEP/Wetlands International, Asia-Pacific, Kuala Lumpur.

Daintree, B.C., 1996. Threats to tortoises and freshwater turtles, *Oryx*, April.

Baille, J. and Groombridge, B., eds., 1996. *The 1996 IUCN Red List of Threatened Animals*, World Conservation Union, Gland, Switzerland.

Baxter, J., 1995. *Chromolaena odorata*: weed for the killing or shrub for the tilling?, *Agroforestry Today*, April-June.

Birstein, V., 1993. Sturgeons and paddlefishes: threatened fishes in need of conservation, *Conservation Biology*, December.

Botanic Gardens Conservation International (no date) BGCI leaflet.

Braatz, S., 1992. *Conserving Biological Diversity: A Strategy for Protected Areas in the Asian and Pacific Region*, World Bank Technical Paper No. 193.

Bright, C., 1996. *State of the World 1996*. Worldwatch Institute and W.W. Norton and Co. Inc., New York.

Butman, C., Carlton, J. and Pulumbi, S., 1995. Whaling effects on deep-sea biodiversity. *Conservation Biology*, April.

Campbell, T. and Schlarbaum, S.E., 1994. *Fading Forests: North American Trees and the Threat of Exotic Pests*, Natural Resources Defense Council, New York.

Carlton, J., 1992a. Marine species introductions by ships' ballast water: an overview, in Richard De Voe, M., ed., *Introductions and Transfers of Marine Species: Achieving a Balance Between Economic Development and Resource Protection*, Proceedings of a conference and workshop, October 20-November 2, 1991, Hilton Head, S.C.

Carlton, J., 1992b. Dispersal of living organisms into aquatic ecosystems as mediated by aquaculture and fisheries activities, in Rosenfield, A. and Mann, R., eds., *Dispersal of Living Organisms Into Aquatic Ecosystems*, College Park, Md.: Maryland Sea Grant.

CITES, 1999a. *Botanic Gardens Conservation News*, vol. 3, No. 2, June.

CITES, 1999b. *Protected Species*, http://www.cites.org/CITES/english/species.shtml

CITES, 1999c. *Making CITES Work: Partnerships and Cooperation – National Coordination Committee (NCC) of India*, http://www.traffic.org/making-CITES-work/mcw_ncc.html

CITES, 1999d. *Making CITES Work: Public Awareness – Protect Endangered Species Campaign*, Hong Kong, http://www.traffic.org/making-CITES-work/mcw_hk-protect.html

CITES, 1999e. *Making CITES Work: Training Initiatives – Orchid Identification Guide*, Thailand, http://www.traffic.org/making-CITES-work/mcw_tl-orchid.html

CMS, 1999a. *Parties to the Convention on the Conservation of Migratory Species of Wild Animals*, http://www.wcmc.org.uk/cms/part_lst.htm

CMS, 1999b. *African-Eurasian Migratory Waterbird Agreement (AEWA)*, http://www.wcmc.org.uk/cms/aew_bkrd.htm

CMS, 1999c. *Conservation Measures for the Slender-billed Curlew*, http://www.wcmc.org.uk/cms/sbc_bkrd.html

CMS, 1999d. *Memorandum of Understanding Concerning Conservation Measures for the Siberian Crane*, http://www.wcmc.org.uk/cms/sib_bkrd.html

Cogger, H., 1992. *Reptiles and Amphibians of Australia* (New York, Cornell University Press).

Commonwealth of Australia, 1998. *Australia's national report to the fourth conference of the parties to the Convention on Biological Diversity,* http://www.biodiversity.environment.gov.au/biocon/natrep/index.htm

Conservation International, 1998a, *Biodiversity of Papua New Guinea.* Press Release, July 16. http://www.conservation.org/web/news/pressrel/98-0716.htm

Conservation International, 1998b. *Megadiversity Country Tables,* http://www.conservation.org/web/fieldact/megadiv/maps.htm

Conservation International, 1998c. *Global Diversity Hotspots: Indonesia,* http://www.conservation.o.../fieldact/REGIONS/ASPAREG/Indonesi.htm

Conservation International, 1998d. *Global Diversity Hotspots: Irian Jaya,* http://www.conservation.o.../fieldact/REGIONS/ASPARGE/Irianjay.htm

Conservation International, 1998e. *Global Diversity Hotspots: Togian Island,* http://www.conservation.o.../fieldact/REGIONS/ASPAREG/Togian.htm

Conservation International, 1998f. *Global Diversity Hotspots: Philippines,* http://www.conservation.o.../fieldact/HOTSPOTS/philippi.htm

Conservation International, 1998g. *The Asian and Pacific Region,* http://www.conservation.o...b/fieldact/REGIONS/aspareg/aspareg.htm

Courtenay, W., 1993. Biological pollution through fish introductions, in McKnight, B., ed., *Biological Pollution: The Control and Impact of Invasive Exotic Species* (Indianapolis, Indiana Academy of Science).

Craig, G.B., 1993. The diasphora of the Asian tiger mosquito, in McKnight, B., ed., *Biological Pollution: The Control and Impact of Invasive Exotic Species* (Indianapolis, Indiana Academy of Science).

Cromarty, 1996. Cited by Moser, M., Prentice, C. and Frazier, S., 1996. *A Global Overview of Wetland Loss and Degradation,* Wetlands International, http://www.ramsar.org/about_wetland_loss.htm

Culotta, E., 1991. Biological immigrants under fire, *Science,* December 6.

Department of Environment, Ministry of Local Government, Housing and Environment, 1997. *Convention on Biological Diversity 1997 National Report to the Conference of the Parties by the Republic of Fiji,* http://www.biodiv.org/natrep/Fiji/Fiji.pdf

Dobson, A., 1995. The ecology and epidemiology of rinderpest virus in Serengeti and Ngorongoro conservation area, in Sinclair, A.R.E. and Arcese, P., eds., *Serengeti II: Research, Management and Conservation of an Ecosystem* (Chicago, University of Chicago Press).

ESCAP, 1996. *Implementation of Projects: Biodiversity Management and the Development of a Research and Information Base on Forests and Grasslands,* Third meeting of senior officials on environmental cooperation in North-East Asia, September 17-20, Ulaanbaatar.

FAO, 1998. *Asia-Pacific Forestry Towards 2010,* Bangkok.

Frazier, S., 1999. *Ramsar Sites Overview: A Synopsis of the World's Wetlands of International Importance,* Wetlands International.

Frazier, S., 1999. *Wetlands International Newsletter,* No. 7, May.

Fritts, T.H. and Rodda, G.H., 1995. Invasions of the brown tree snake, in La Roe, E.T., et. al., eds., *Our Living Resources: A Report to the Nation on the Distribution, Abundance and Health of U.S. Plants, Animals, and Ecosystems,* National Biological Service, U.S. Department of the Interior, Washington, DC.

Goss, H., 1995. The mysterious case of the wobbly possum, *New Scientist,* August 5.

Government of Australia, 1999. Media Release by the Minister for Forestry and Conservation, Canberra.

Government of Bhutan, 1997. *Biodiversity Action Plan for Bhutan.* http://www.biodiv.org/natrep/Bhutan/Bhutan.pdf

Government of China, 1997. *China's National Report on the Implementation of the Convention on Biological Diversity,* http://www.biodiv.org/natrep/China/China.pdf

Government of Indonesia, 1997. *National Report on the Implementation of Convention on Biological Diversity,* http://www.biodiv.org/natrep/Indonesia/Indonesia.pdf

Government of Japan, 1997. *The First National Report Under the Convention on Biological Diversity,* http://www.biodiv.org/natrep/Japan/Japan.pdf

Government of Korea, 1998. *National Biodiversity Strategy of the Republic of Korea,* http://www.biodiv.org/natrep/Korea%20(Republic%20of)/Korea%20(Republic%20of).pdf

Government of Marshall Islands, 1997. *Convention on Biological Diversity 1997: Preliminary Report to the Conference of the Parties,* http://www.biodiv.org/natrep/Marshall%20Islands/Marshall%20Islands.pdf

Government of New Zealand, 1997. *National Report: Convention on Biological Diversity,* http://www.biodiv.org/natrep/New%20Zealand/New%20Zealand.pdf

Government of Sri Lanka, 1997. *First National Report on the Implementation of the Convention on Biological Diversity,* http://www.biodiv.org/natrep/Sri%20Lanka/Sri%20Lanka.pdf

Government of Turkey, 1997. *National Biodiversity Strategy and Action Plan,* http://www.biodiv.org/natrep/Turkey/Turkey.pdf

Government of Uzbekistan, 1997. *Biodiversity Conservation: National Strategy and Action Plan,* http://www.biodiv.org/natrep/Uzbekistan/Uzbekistan.pdf

Government of Viet Nam, 1997. *Report of the Vietnamese Delegation at the Fourth Conference of the Parties to the Convention on Biological Diversity,* http://www.biodiv.org/natrep/Vietnam/Vietnam.pdf

Green, M. and Paine, J., 1997. *State of the World's Protected Areas at the End of the Twentieth Century,* Paper presented at IUCN World Commission on Protected Areas Symposium on Protected Areas in the 21st Century: From Islands to Networks, November 24-29, Albany.

Groombridge, B. and Jenkins, M., 1998. *Freshwater biodiversity: a preliminary global assessment,* in WCMC (World Conservation Monitoring Centre), World Conservation Press.

Gubler, D.J. 1993. Emergent and resurgent arboviral diseases as public health problems, in Mahy, B.W.J. and Lvov, D.K., eds., *Concepts in Virology, From Ivanovsky to the Present* (Chur, Switzerland, Harwood Academic Publishers).

Hallegraeff, G. and Bolch, C.J., 1991. Transport of toxic dinoflagellate cysts via ships' ballast water, *Marine Pollution Bulletin,* January.

Hancock, J., Kushlan, J. and Philip K.M., 1992. *Storks, Ibises and Spoonbills of the World* (London, Academic Press).

Hanna L., Carr, J.L. and Lankerni, A., 1995. Human disturbance and natural habitat: a biome level analysis of a global data set, *Biodiversity and Conservation,* No. 4.

Hannah, L., et. al., 1994. A preliminary inventory of human disturbance of world ecosystems, *Ambio,* July.

Hayward, D., 1991. Poisonous jam prescribed for possum power, *Financial Times,* July 31.

Heywood, V.H., 1989. Patterns, extents and modes of invasions by terrestrial plants, in Drake, A., et. al., eds., *Biological Invasions: A Global Perspective* (Chichester, UK, John Wiley and Sons, Inc.).

International Network for Bamboo and Rattan, 1997. Bamboo/rattan worldwide, *INBAR Newsletter,* Nos. 4 and 5.

Iqbal, M., 1995. *Trade Restrictions Affecting International Trade in Non-wood Forest Products,* Rome.

IUCN and the International Academy of the Environment, 1994. *Widening Perspectives on Biodiversity,* Switzerland and UK.

Jackson, P.W., 1999. *Botanic Gardens Conservation News,* vol. 3, No. 2, June.

James, D., 1999. The threat of exotic grasses to biodiversity of semi-arid ecosystems, *Aridlands Newsletter,* No. 37, Tuscon, http://www.ag.arizona.edu/OALS/ALN/aln37/james.html

Jenkins, R., 1999. Bt crops are unsustainable, *Biotechnology and Development Monitor,* No. 38, pp. 24.

Kyodo News Service, 1999. *Global Warming: Threat to Coral Reef,* July 6, http://...com/pages/updatecontent/Reef.html

Lever, C., 1985. *Naturalized Mammals of the World* (London, Longman).

Liu, J. and Hills, P., 1997. Environmental planning, biodiversity and the development process: the case of Hong Kong's Chinese white dolphins, *Journal of Environmental Management,* No. 50, pp. 351-367.

MacDonald, I., et. al., 1989. Wildlife conservation and the invasion of nature reserves by introduced species: a global perspective, in

Drake, J.A., et. al., eds., *Biological Invasions: A Global Perspective* (Chichester, UK, John Wiley and Sons, Inc.).

Mainguet, M. and Letolle, R., 1998. Human-made desertification in the Aral Sea basin: planning and management failures, *Desertification Control Bulletin*, No. 33.

Matthiesen, P., 1997. The last wild tigers, *Audubon*, March-April.

Ministry for Nature and the Environment of Mongolia-United Nations Development Programme-Global Environment Facility (GEF), 1998. *Biological Diversity in Mongolia (First National Report)*, http://www.biodiv.org/natrep/Mongolia/Mongolia.pdf

Ministry of Forests and Soil Conservation, 1997. *National Report on the Implementation of the Convention on Biological Diversity in Nepal*, http://www.biodiv.org/natrep/Nepal/Nepal.pdf

Ministry of Planning, Human Resources and the Environment, 1997. *Biodiversity Conservation in Maldives: Interim Report to the Convention on Biological Diversity*, http://www.biodiv.org/natrep/Maldives/Maldives.pdf

Ministry of Science, Technology and the Environment, 1998. *First National Report to the Conference of Parties to the Convention on Biological Diversity, Malaysia*, http://www.biodiv.org/natrep/Malaysia/Malaysia.pdf

Moser, M., 1999. Wetlands International and the Ramsar Convention-an intimate partnership, *Wetlands International Newsletter*, No. 7, May.

Moser, M., Prentice, C. and Frazier, S., 1996. *A Global Overview of Wetland Loss and Degradation*, Wetlands International, http://www.ramsar.org/about_wetland_loss.htm

NRC, 1993. *Managing Global Genetic Resources: Agricultural Crop Issues and Policies* (Washington, DC, National Academy Press).

Oud, E. and Muir, T.C., 1997. Engineering and economic aspects of planning, design, construction and operation of large dam projects, in IUCN-The World Conservation Union and The World Bank Group, *Large Dams: Learning from the Past, Looking at the Future*, Workshop Proceedings, IUCN, Gland, Switzerland and Cambridge, UK and The World Bank Group, July, Washington, DC.

Oxfam, 1998. *Biotechnology in Crops: Issues for the Developing World*. Oxfam Policy Papers-Research paper for Oxfam, GB, May.

Phillips, O. and Meilleur, B., 1998. Usefulness and economic potential of the rare plants of the United States: a statistical survey, *Economic Botany*, vol. 52, No. 1.

Plucknett, D., et. al., 1987. *Gene Banks and the World's Food* (Princeton, New Jersey, Princeton University Press).

Protected Areas and Wildlife Bureau-Department of Natural Resources and Environment, 1998. *The First Philippine National Report to the Convention on Biological Diversity*, http://www.biodiv.org/natrep/Philippines/Philippines.pdf

Pryor, L.D., 1991. Forest plantations and invasions in he Mediterranean zones of Australia and South Africa, in Groves, R.H. and di Castri, F., eds., *Biogeography of Mediterranean Invasions* (Cambridge, Cambridge University Press).

Putterman, D.M., 1999. *Genetic resources utilization: critical issues in conservation and community development*, http://bcnet.org/whatsnew/biopros.html

RAFI, 1999. *Bioprospecting/Biopiracy and Indigenous Peoples*, http://www.Latinsynergy.org/bioprospecting.htm

Ramsar Convention Bureau, 1999. *Botanic Gardens Conservation News*, vol. 3, No. 2, June.

Risler, J. and Mellon, M., 1995. *Perils Amidst the Promise: Ecological Risks of Transgenic Crops in a Global Market* (Cambridge, Mass., Union of Concerned Scientists).

Savidge, J., 1987. Extinction of an island forest avifauna by an introduced snake, *Ecology*, June.

Schaik, C.V., 1999. *Situation at Orangutan Station Suaq Balimbang Critical: Orangutan Viability in the Wild Increasingly Questionable*, http://www.noord.bart.nl/~edcolijn/anarchy.html

Scott, 1989. And Scott and Poole, 1989., cited by Anonymous, 1997. *Wetlands and Integrated River Basin Management: Experiences in Asia and the Pacific*, UNEP/Wetlands International, Asia-Pacific, Kuala Lumpur.

Shelton, J.W., Balick, M.J. and Laird, S.A., 1997. *Medicinal Plants: Can Utilization and Conservation Coexist?* (New York, New York Botanical Garden).

Shengji, P., 1998. Biodiversity in the Hindu Kush Himalayas, *ICIMOD Newsletter*, No. 31, Autumn.

SPREP, 1997. *SPREP: South Pacific Regional Environment Programme*, http://www.sidsnet.org/pacific/sprep/whatsprep_.htm

State Committee of Russian Federation for Environmental Protection, 1997. *Biodiversity Conservation in Russia: The First National Report of Russian Federation*, http://www.biodiv.org/natrep/Russian%20Federation/Russian%20Federation_eng.pdf

Sunderland, T., 1998. *The Rattan Palms of Central Africa and Their Economic Importance*, Society for Economic Botany Annual Meetings, July, Aarhus.

Swanson, T., Pearce, D. and Cervigni, R., 1994. *The Appropriation of the Benefits of Plant Genetic Resources for Agriculture: An Economic Analysis of the Alternative Mechanisms for Biodiversity Conservation*, FAO Commission on Plant Genetic Resources, Rome.

The National Parks Board and the Report Drafting Committee, 1997. *First National Report Under the Convention on Biological Diversity, Singapore*, http://www.biodiv.org/natrep/Singapore/Singapore.pdf

The Nature Conservancy, 1999. *Coastal and Marine Conservation*, http://www.tnc.org/infield/intprograms/asiapacific/coastal.html

The Office of the Environmental Policy and Planning, 1997. *National Report on the Implementation of the Convention on Biological Diversity, Thailand*, http://www.biodiv.org/natrep/Thailand/Thailand.pdf

Tripp, R. and van der Heide, W., 1996. The erosion of crop genetic diversity: challenges, strategies and uncertainties, *Natural Resource Perspectives*, No. 7, March.

Tuxill, J. and Bright, C., 1998. Losing strands in the web of life, in L.R. Brown, et. al., 1998. *State of the World*, New York, USA, Worldwatch Institute and W.W. Norton and Co. Inc., New York and London.

Tuxill, J., 1999. Appreciating the benefits of plant biodiversity, in L.R. Brown, et. al., 1999. *State of the World*, New York, USA, Worldwatch Institute and W.W. Norton and Co. Inc., New York and London.

UNDP, 1999. *Biodiversity Conservation and Sustainable Livelihood Options in the Grasslands of Eastern Mongolia*, Eastern Steppe Biodiversity Project, August 13.

UNEP and Wetlands International-Asia-Pacific, 1997. *Wetlands and Integrated River Basin Management*, pp. 70.

UNEP, 1995. *Global Biodiversity Assessment*.

UNEP, 1997. *Global State of the Environment Report: Global Environment Outlook-1*, http://www.grida.no/prog/global/geo1/ch/ch3_14.htm

UNEP, 1997d. *Global State of the Environment Report: Global Environment Outlook-1 – Asia and the Pacific Regional Initiatives*, http://www.grida.no/prog/global/geo1/ch/ch3_14.htm

UNEP, 1999a. *The Malaysia Environmental Quality Report 1996: Executive Summary*, Department of Environment, Jalan Raja Laut, Kuala Lumpur, http://www.jas.sains.my/doe/eqr95/eqr96sum.html

UNEP, 1999b. *East Asian Seas*, http://www.unep.org/water/regseas/easian.htm

UNEP, 1999c. *Regional seas programme: International Coral Reef Initiative (ICRI)*, http://www.unep.org/water/icri.htm

UNEP, 1999e. *Regional seas programme*, http://www.unep.org/water/regseas/regseas.htm

United Nations, 1995. *State of the Environment in Asia and the Pacific Report*, New York.

Usher, M.B., 1989. Biological invasions into tropical nature reserves, in P.S. Ramakrishnan, ed., *Ecology of Biological Invasions in the Tropics*, Proceedings of an International Workshop held at Nainital, India, (New Delhi, International Scientific Publications).

Vincent, A., 1996. *The International Trade in Seahorses* (Cambridge, TRAFFIC International).

Visser, B., 1998. Effects of biotechnology on agro-biodiversity, *Biotechnology and Development Monitor*, No. 35, June.

Watkins, P., 1999. *Wetlands International Newsletter*, No. 7, May.

Watson, T.R., et. al., eds., 1996. *Global Biodiversity Assessment* (London, Cambridge University Press).

WCMC, 1992. *Global Biodiversity: Status of the Earth's Living Resources*, Cambridge.

WCMC, 1998. Freshwater biodiversity: a preliminary global assessment, *WCMC Biodiversity Series*, No. 8 (WCMC-World Conservation Press).

Whitmore, T.C., 1989. Invasive woody plants in perhumid tropical climates, in Ramakrishnan, P.S., ed., *Ecology of Biological Invasions in the Tropics*, Proceedings of an International Workshop held at Nainital, India (New Delhi, International Scientific Publications).

Wijk, J.V., 1994. Hybrids: bred for superior yields or for control?, *Biotechnology and Development Monitor*, No. 19, pp. 3-5.

World Bank, 1999. *Conserving Biodiversity through Ecodevelopment*, Washington, DC.

WRI, 1999a. *Biodiversity in Freshwater Ecosystems*, http://www.wri.org/wri/biodiv/b03-gbs.html

WRI, 1999b. http://www.wri.org/wri/wr-98-99/fragment.html

WRI, 1999c. *The Last Frontier Forests: Regional Overview-Asia*, http://www.igc.apc.org/wri/ffi/lff-eng/asia2.htm

WRI, 1999d. *Tropical Forest Species Richness*, http://www.wri.org/biodiv/b01-koa.html

WRI, 1999e. *Marine Species Richness and Conservation*, http://www.wri.org/wr-96-97/bi_b1.html

WRI, 1999f. *Marine Biodiversity*, http://www.wri.org/wr-96-97/bi_txt4.html

WRI, 1999g. *Coral Reef Ecosystem*, http://www.wri.org/wri/biodiv/bo2-koa.html

WRI, 1999h. *Great Barrier Reef*, http://www.wri.org/wri/indictrs/australi.html

WRI, 1999i. *East Asia*, http://www.wri.org/wri/indictrs/reefasia.html

WRI, 1999j. *Coral Reefs: Assessing the Threat*, http://www.wri.org/wri/wr-98-99/coral.html

WRI, 1999k. *Status of the World's Coral Reefs*, http://www.wri.org/wri/indictrs/rrstatus.html

WRI, 1999l. *Pacific Ocean*, http://www.wri.org/wri/indictrs/rrstatus.html

WRI, 1999m. *Threats to Reefs*, http://www.wri.org/wri/indictrs/threatrr.htm

WRI, 1999n. *Coral Reef Bleaching*, http://www.wri.org/wri/indictrs/rrbleach.htm

WRI, 1999n. *Pressures on Marine Biodiversity*, http://www.wri.org/wri/wr-96-97/bi_txt5.html

WRI, 1999o. *Tools for Protecting Marine Biodiversity*, http://www.wri.org/wr-96-97/bi_txt6.html

Wright, B.D., 1996. *Crop Genetic Resource Policy: Toward a Research Agenda*, EPTD Discussion Paper 19, International Food Policy Research Institute, Washington, DC.

Youth, H., 1994. Flying into trouble, *World Watch*, January/February.

ADB, 1997. *Emerging Asia: Changes and Challenges*, Manila.

ADB, 1998. *The Bank's Policy on Water*, Working Paper, Manila.

Afzal, M. and Talib H., 1996. *Environmentally Sustainable Management of Irrigation System in Pakistan*, Paper presented in International Conference on Management and Business, LUMS, June 3-5, Lahore.

Anonymous, 1997. *Wetlands and Integrated River Basin Management: Experiences in Asia and the Pacific*, UNEP/Wetlands International-Asia-Pacific, Kuala Lumpur.

AMICC, 1997. *Water: Asia's Environmental Imperative*, Nanyang Technological University, Singapore.

Avakyan, A.B. and Iakovleva, V.B., 1998. Status of global reservoirs: The position in the late twentieth century, *Lakes and Reservoirs: Research and Management*, vol. 3, No. 7, March (International Lake Environment Committee, Blackwell Science).

Baltazar, M.O., 1996. *Water and Sustainable Development in the Philippines*, Country Paper presented to ESCAP Ad-Hoc Expert Group Meeting on Sustainable Development of Water Resources, July 10-12, Bangkok.

BGS, ODA, UNEP and WHO, 1996. *Characterization and Assessment of Groundwater Quality Concerns in Asia-Pacific Region*, UNEP/DEIA/AR. 96-1, Nairobi.

Brown, L.R., Flavin, C. and French, H.F., 1999. *State of the World*, Worldwatch Institute and W:W. Norton & Company, New York.

Chakrapani, G.J. and Subramanian, V., 1996. Fractionation of heavy metals and phosphorus in suspended sediments of the Yamuna River, India, *Environmental Monitoring and Assessment*, vol. 43 (Netherlands, Kluwer Academic Publishers).

China State Environmental Protection Administration, 1997. *Report on the State of the Environment in China*.

Commission on Sustainable Development, 1997. Fifth Session on Comprehensive Assessment of the Freshwater Resources of the World, Report of the Secretary-General, Economic and Social Council, United Nations, New York.

Commonwealth of Australia, 1996. *Australia State of Environment 1996*, State of the Environment Advisory Council and Department of the Environment, Sport and Territories (Collingwood, Australia, CSIRO Publishing).

Das Gupta, A., 1996. *Groundwater and the Environment*, Inaugural Lecture, Asian Institute of Technology, Bangkok.

David, L.J., Goluveb, G. and Nakayama, M., 1988. *The Environmental Management of Large International Basins*, The EWINMA Programme of UNEP, Water Resources Development, vol. 4, No. 2, pp. 103-107.

DOE (Department of Environment) Malaysia, 2000. Written Communication.

ESCAP, 1990. *Waterlogging and Salinity Control in Asia and the Pacific*, ESCAP/UNDP Project on Strengthening of the Regional Network of Research and Training Centres on Desertification Control in Asia and the Pacific (DESCONAP).

ESCAP, 1995. Integrated water resources management in Asia and the Pacific, *Water Resources*, Series No. 75, United Nations, New York.

ESCAP, 1996. *Protection of Water Resources, Water Quality and Aquatic Ecosystems in Asia and the Pacific*, United Nations, New York.

ESCAP, 1996a. *Report of the Ad-Hoc Group Meeting on Sustainable Development of Water Resources*, July 10-12, ENR/rep. July 12, pp. 3/Annex 2, Bangkok.

ESCAP, 1997. *Implementation of Sustainable Development Programmes for Agenda 21, Chapter 18-Freshwater Resources*, Report of ESCAP on the Implementation of Freshwater-Related Recommendations of Agenda 21, Bangkok.

ESCAP, 1997a. *Sustainable Development of Water Resources in Asia and the Pacific: An Overview*, United Nations, New York.

ESCAP, 1998. *Sources and Nature of Water Quality Problems in Asia and the Pacific*, ESCAP, United Nations, New York.

ESCAP, 1998a. *Towards Efficient Water Use in Urban Areas in Asia and the Pacific*, United Nations, New York.

ESCAP, 1998b. *National Experience in the Integration of Water Resources Management into Economic* and Social Development Plans, Survey Report of the National Experience in the Asian and Pacific Region, Bangkok.

ESCAP, 1999. *ESCAP Population Data Sheet*, Population and Development Indicators for Asia and the Pacific 1999, Population and Rural Development Division, ESCAP, Bangkok.

ESCAP, 1999a. *Water Quality of Selected Rivers in Asia: Protection and Rehabilitation*, ESCAP, United Nations, New York.

FAO, 1999. *Irrigation in Asia in Figures*, Water Reports No. 18, Rome.

FAO, 1990. *Problem soils of Asia and the Pacific*, FAO RAPA Publication No. 1990/6. Bangkok.

Goluveb, G.N., 1993. Sustainable Water Development: Implications for the Future, *Water Resources Development*, vol. 9, No. 2.

Hiscock, K., 1997. Groundwater Pollution and Protection, Chapter 13, in O'Riordan, T., ed., *Environmental Science for Environmental Management* (England, Addison Wesley Longman Ltd.).

Hodgson, G., 1999. *Using Reef Check to Monitor Coral Reefs, Scientific Aspects of Coral Reef Assessment*, Monitoring and Restoration, National Coral Reef Institute (NCRI), April 14-16, Ft. Lauderdale, Fl., http://www.nova.edu/ocean/ncri/conf99.html

Hughes, R., Adnan, S. and Dalal-Clayton, B., 1994. *Floodplains or Flood Plans? A Review of Approaches to Water Management in Bangladesh*, IIED, London and RAS, Dhaka.

IFDCR, 1996. *IFDCR Newsletter*, vol. 4, Issue 2, July-December.

Independent Newspaper, 1999. *Headline: Half of the World's Rivers Polluted or Running Dry*, Byline by Mary Dejevsky in Washington, DC, November 30, London.

Jacobs, J.W., 1996. Planning for change and sustainability in water development in Lao PDR and the Mekong River Basin, *Applied Geography*, vol. 20, No. 1, Pergamon, Elsevier Science Ltd., UK.

Kharin, N., 1996. Strategy to combat desertification in Central Asia, *Desertification Control Bulletin*, No. 29.

Kwun, S., 1999. *Water for Food and Rural Development*, Regional Consultation Meeting for ICID Vision for Sub-Sector, "Water for Food and Rural Development", Country Paper of Republic of Korea, May 17-19, Kuala Lumpur.

MacIntosh, A.C. and Yniguez, C.E., 1997a. *Second Water Utilities Data Book*, ADB, Manila.

Mainguet, M. and Letolle, R., 1998. Human.made desertification in the Aral Sea Basin: planning and management failures, *Desertification Control Bulletin*, No. 33, 1998.

Mantell, L., 1998. *Millions in Bangladesh Face Slow Poisoning from Arsenic-Contaminated Water*, World Specialist Website News and Analysis: Asia (Indian Sub-continent), http://wsws.org/news/1998/dec1998/bang-d02.shtml

NEPA, 1992. *Report on the State of the Environment in China*, National Environmental Protection Administration, China.

Newson, M., 1992. *Land, Water and Development: River Basin Systems and their Sustainable Management*, Routledge, London.

Pinlac, E.M., 1992. *Groundwater Contamination Assessment in the Philippines*, Technical Workshop, Groundwater Contamination in Sub-humid and Humid Tropical Asia. UNESCO/ROSTSEA Thailand National Committee for IHP, Department of Mineral Resources, Bangkok.

Postel, S., 1996. Forging a Sustainable Water Strategy, in Brown, L.R., ed. al., 1996. *State of the World*, Worldwatch Institute, New York.

Ramnarong, 1991. Thailand: groundwater quality monitoring and management, groundwater quality and monitoring in Asia and the Pacific, *Water Resources*, Series No. 70, ESCAP, Bangkok.

Ramnarong, V. and Buapeng, S., 1991. Mitigation of Groundwater Crisis and Land Subsidence in Bangkok, in *Proceedings of the 4th International Symposium on Land Subsidence*, Houston, May.

Rao, R.V., 1996. *Water and Sustainable Development in the Philippines*, Country Paper presented to ESCAP Ad-Hoc Expert Group Meeting on Sustainable Development of Water Resources, July 10-12, Bangkok.

Raskin, P., Gleick, P., Kirshen, P., Pontius, G. and Strzepek, K., 1997. *Water Futures: Assessment of Long-Range Patterns and Problems*, Stockholm Environment Institute, Stockholm.

Republic of Korea's Ministry of Environment, 1997. *Environmental Protection in Korea*, Seoul.

SCARM, 1998. *Sustainable Agriculture: Assessing Australia's Recent Performance*, A Report to the Standing Committee on Agriculture and Resource Management of the National Collaboration Project on Indicators for Sustainable Agriculture, SCARM Technical Report 70 (Collingwood, Australia, CSIRO Publishing).

Seckler, D., Amarasinghe, U., Molden, D., de Silva, R. and Barker, R., 1998. *World Water Demand and Supply, 1990 to 2025: Scenarios and Issues*, International Water Management Institute (IWMI), Colombo.

Seckler, D., Barker, R. and Amarasinghe, U., 1999. Water scarcity in the 21st century, *Water Resource Development*, vol. 15, Nos. 1 & 2, Taylor and Francis, Ltd.

SEPA, 1997. *National Report on Sustainable Development*, State Environmental Protection Administration, China (Beijing, China Environmental Science Press).

SEPA, 1997. *State of the Environment in China*.

Sham, S., 1997. *Encyclopedia of Malaysia*, vol. 1, (Kuala Lumpur, Archipelago Press).

Sharma, M., 1986. *Role of Groundwater in Urban Water Supplies in Bangkok, Thailand and Jakarta, Indonesia*, Working Paper of the East-West Centre, Honolulu.

Shiklomanov, I.A., 1997. *Assessment of Water Resources and Water Availability of the World*, World Metoerological Association, Geneva.

Smith, C.M., Wilrock, R.J., Vant, W.N., Smith, D.G. and Cooper, A.B., 1993. *Towards sustainable agriculture in New Zealand: Freshwater quality in New Zealand and the influence of agriculture*, MAF Policy Technical paper 93/10. Ministry of Agriculture and Fisheries, Wellington.

Soetrismo S. and Hehanussa, P., 1992. *Groundwater Contamination in Indonesia*, Technical Workshop, Groundwater Contamination in Sub-humid and Humid Tropical Asia. UNESCO/ROSTSEA Thailand National Committee for IHP, Department of Mineral Resources, Bangkok.

Somasundaram M.V., Ravindarn, G. and Tellam, J.H., 1993. Groundwater Pollution for the Madras Urban Aquifer, *Groundwater*, vol. 31, No. 1, pp. 4-11.

Soussan, J., 1998. *Water/Irrigation and Sustainable Rural Livelihoods: What Contributions Can We Make?*, edited by Carney, D., Paper presented at the Department for International Development's Natural Resources Advisers Conference, July 1998.

South Asia Technical Advisory Committee, 1999. *South Asia Regional Water Vision 2025*, Global Water Partnership.

Speers, A., 1999. Australia's Urban Water Program: An Integrated, Nationwide Initiative, *Aridlands Newsletter*, No. 45, Spring/Summer, http://ag.arizona.edu/OALS/ALN/aln45/australia.html

Sweden's Ministry of Foreign Affairs, 1999. *Our Future with Asia: Proposal for a Swedish Asia Strategy*, The Asian Strategy Project, Ministry of Foreign Affairs, Sweden.

Tolba, M.K., 1988. EMINWA and sustainable water development, *Water Resources Development*, vol. 4, No. 2, pp. 76-79.

UNEP, 1999. *Global Environment Outlook 2000*. (London, Earthscan).

UNEP Freshwater Programme, 1995. *Environmentally Sound Management and Sustainable Use of Freshwater Resources: Support for Decision Making*, Draft, Nairobi.

UNEP, 1991. *Freshwater Pollution*, UNEP/GEMS Environment Library No. 6. Nairobi.

UNEP, 1994. *The Pollution of Lakes and Reservoirs*, UNEP Environment Library No. 12, Nairobi.

UNEP, 1997. *Worlds Atlas of Desertification*, second Edition, Arnold London.

UNEP, 1996. *Groundwater: A Threatened Resource*, UNEP Environment Library No. 15, Nairobi.

UNESCO, 1992. *Small Tropical Islands: Water Resources of Paradise's Lost*, Water-related Issues and Problems of the Humid Tropics and Other Warm Humid Regions, IHP, Humid Tropics Programme, Series No. 2, Paris.

UNIDO, 1996. *Global Assessment of the Use of Freshwater Resources for Industrial and Commercial Purposes*, Industry, Sustainable Development, and Water Programme Formulation, Technical Report, New York.

United Nations, 1995. *State of Environment in Asia and the Pacific 1995*, New York.

WHO, 1996. *Planning and Developing Drinking Water Quality Surveillance and Control Programmes*, Report of a Regional Consultation, September, Kathmandu.

World Water Council, 1998. *Water Policy*, Official Journal of the World Water Council, edited by Priscoli J.D., Elsevier Science Ltd., USA.

Worldwatch Institute, 1996. *The Worldwatch Report: Water-borne Killers*, Los Angeles Times Syndicate, April, http://www.enn.com/enn-news-archive/1996/04/041796/features.asp

WRI, 1999. *Water Resources and Human Health: Water Pollution and Human Health in China*, http://www.wri.org/health/prchealt.html

WRI, UNEP, UNDP and World Bank, 1998. *World Resources 1998-99: A Guide to Global Environment* (and the World Resources Database diskette) (New York, Oxford University Press and Oxford, Oxford University Press).

ADB/NACA, 1999. *State of Asian Aquaculture: Regional Overview*, Preliminary Report.

Anon, 1997. The Holmenkollen Guidelines for Sustainable Aquaculture, in *Proceedings of the Second International Symposium on Sustainable Aquaculture*, Oslo, November 2-5, Norwegian Academy of Technological Sciences, Trondheim, Norway, 9 pp.

Australia's Oceans Policy, 1998. Government Printing Office, Canberra, Australia, http://www.environment.gov.au/portfolio/minister/env/98/mr23dec98.html

Bour, W., 1990. *Coastal and Coral Reefs Studies in New Caledonia, using SPOT Images, for Environment and Management Monitoring*, Seminar on Remote Sensing Applications for Oceanography and Fishery Environment Analysis, May, Beijing.

Bryant. D., et. al., 1998. *Reefs at Risk World: A Map- Based Indicator of Threats to the World's Coral Reefs*, WRI, Washington, DC.

Buddemeier, R.W., 1999. *Is It Time To Give Up? Scientific Aspects Of Coral Reef Assessment, Monitoring, and Restoration*, National Coral Reef Institute (NCRI), April 14-16, Ft. Lauderdale, Fl. http://www.nova.edu/ocean/ncri/conf99.html

Caldecott, J. and Salmon, M., 1999. *Deep Water*, Elipsis, London, 128 pp.

Cartright, I., 1996. *The South Pacific Forum Fisheries Agency: Past, Present and Possible Future Roles in Fisheries Management in the Central and Western Pacific*, FFA, Honiara, Solomon Islands.

Cartright, I., 1998. *Multilateral High Level Conferences on South Pacific Tuna Fisheries*, FFA, http://www.ffa.int

Chesher, R.H., 1969. *Acanthaster planci: Impact on Pacific Coral Reefs*, Final Report, U.S. Department of Commerce, National Technical Information Service, PB 187 631, 168 pp.

Cisin-Sain, B., Knecht, R.W. and Fisk, G.W., 1995. Growth in capacity for integrated coastal management since UNCED: an international perspective, *Ocean and Coastal Management*, vol. 29, No. 1-3, pp. 93-123.

Clucas I.J. and James, D.G., eds., 1997. *Reduction of Wastage in Fisheries*, Papers presented at the Technical Consultation, Tokyo, October 28-November 1, 1996, FAO Fisheries Report No. 547, Suppl., Rome, 338 pp.

Coastcare. 1999. *Coastcare Australia*, http://www.environment.gov.au/marine/coastcare/index.html

CSIRO, 1996. *State of the Environment of Australia*, Ministry of Environment, Sports and Territories, Canberra.

Done, T.J., 1999. *Useful Science For Coral Reef Management: The Cooperative Research Centre Model*, Scientific Aspects Of Coral Reef Assessment, Monitoring, and Restoration. National Coral Reef Institute (NCRI), April 14-16, Ft. Lauderdale, Fl, http://www.nova.edu/ocean/ncri/conf99.html

FAO, 1997a. Review of the state of world fishery resources: marine fisheries, *FAO Fisheries Circular*, No. 920 FIRM/C920.

FAO, 1997b. *Fisheries Management Frameworks of the Countries Bordering the South China Sea*, Bangkok, Asia and Pacific Fisheries Commission.

FAO, 1998 *Economic Viability and Sustainability of Marine Capture Fisheries – Findings of a Global Study and Recommendations of an Interregional Workshop*, FAO Fisheries Technical Paper No. 377.

FAO, 1999. *The State of the World Fisheries and Aquaculture 1998*, FAO code: 43 AGRIS: M11; M12 ISSN 1020-5489, Rome.

Feldman, M., 1994. *Greenpeace Sues the Minister – On What Grounds?*, New Zealand Fisherman, June, July, August.

Fortes, M.D., 1994. Philippine Seagrasses: status and perspectives, 291-310, in Wikinson, C., Sudara, S. & Chou. L.M., Proc. Third ASEAN-Australia Symposium on Living Coastal Resources, May 16-24, Thailand, and Volume 1: Status Reviews, Australian Agency for International Development, 454 pp.

Froese, R. and Pauly, D., eds., 1997. *FishBase 97: Concepts design and data sources*, ICLARM, Manila, 256 pp.

Gabrie, C., Licari, M.L. and Mertens, D., 1995. *L'etat de l'environment dans les Territoires Francais du Pacifique Sud*, Preliminary Report L'Institute Francais de L' Environment.

Gillett, R., 1997. *The Importance of Tuna to Pacific Island Countries*, Report for the Forum Fisheries Agency.

Government of China, 1997. *State of the Environment in China*, State Environmental Protection Administration, Beijing.

Government of China, 1999. *State of the Environment in China*, State Environmental Protection Administration, Beijing.

Grandperrin, R., de Forges, R.B., Auzende, J.M., eds., 1997. *Marine Resources of New Caledonia*, The ZoNeCo programme, SMAI BP 8231, Noumea, New Caledonia, Zoneco@smai.nc; http://www.smai.nc

Guch I.C., 1999. The Use Of The NOAA/NESDIS Interactive Coral Reef "Hot Spot" Web Page To Monitor Coral Bleaching During The 1997/98 El Nino And The 1998/99 La Nina Events, Scientific Aspects Of Coral Reef Assessment, Monitoring, And Restoration, National Coral Reef Institute (NCRI), April 14-16, Ft. Lauderdale, Fl. http://www.nova.edu/ocean/ncri/conf99.html

Hardy, J.T., 1997. Biological effects of chemicals in the sea-surface microlayer, in Liss, P.S. and Duce, R.A., eds., *The sea surface and global change*, Cambridge University Press, Chapter 11, pp. 339-370.

Heck, K.L. Jr. & McCoy, E.D., 1978. Biogeography of Seagrasses: evidence from associated organisms, 109-128, *Proc. International Sympsoium on Marine Biogeography and Evolution in the Southern Hemisphere*, July 1978, Auckland, N.Z. Volume 1, N.Z. DSIR Information Series 137, 355 pp.

Hinrichsen D., 1997. Requiem for Reefs?, *International Wildlife*, March/April, 8 pp.

Hodgson, G., 1999. *Using Reef Check To Monitor Coral Reefs*, Scientific Aspects Of Coral Reef Assessment, Monitoring, And Restoration, National Coral Reef Institute (NCRI), April 14-16, Ft. Lauderdale, Fl. http://www.nova.edu/ocean/ncri/conf99.html

IMO, 1996. *Technical Cooperation Committee – 42nd session: June 20.*

International Workshop on Integrated Coastal Management in Tropical Developing Countries, Lessons Learned from Successes and Failures, 1996. *Enhancing the Success of Integrated Coastal Management: Good Practices in the Formulation, Design, and Implementation of Integrated Coastal Management Initiatives*, MPP-EAS Technical Report No. 2, GEF/UNDP/IMO Regional Programme for the Prevention and Management of Marine Pollution in the East Asian Seas/Coastal Management Center, Quezon City, the Philippines.

Jameson S.C., et. al., 1995. *State of Reefs: Regional and Global Perspectives*, ICRI., U.S. Department of State, Washington, DC.

Kushairi, M.R., 1999. *The Extent Of 1998 Coral Bleaching Catastrophe In The Marginal Seas Of The Indo-Pacific*, Scientific Aspects Of Coral Reef Assessment, Monitoring, And Restoration. National Coral Reef Institute (NCRI), April 14-16, Ft. Lauderdale, Fl. http://www.nova.edu/ocean/ncri/conf99.html

Langdon et. al., 1999. Geo-chemical consequences of increased atmospheric CO_2 on coral reefs, *Science*.

Li-Jones, X., 1998. *What has been learned from the Indian Ocean Experiment?* http://www-indoex.ucsd.edu/whatlearned/whatlearned.html#11

Liss, P.S. and Duce, R.A., eds. 1997. *The Sea Surface and Global change*, Cambridge University Press.

McGinn, A.P., 1999. *Safeguarding the Health of Oceans*, Worldwatch Paper 145, Worldwatch Institute, Washington, DC, 87 pp.

Michaelis, F.B., 1998. *International Year of the Oceans-1998*, Australia's policies, programs and legislation Science, Technology, Environment and Resources Group Research Paper 6, 1998-99.

Nerem, R.S., Haines, B.J., Hendricks, J., Minister, J.F., Mitchum, G.T., and White W.B., 1997. Improved determination of global mean sea level variations using TOPEX/POSEIDON altimeter data, *Geophysical Research Letters*, vol. 24, pp. 1331-1334.

NOAA, 1999. Climate Hotspots and Coral Bleaching Events, http://psbsgi1.nesdis.noaa.gov:8080/PSB/EPS/SST/climohot.html

Norse, E.A., 1993. *Global Marine Biological Diversity*, Island Press, 383 pp.

Parfit, M. and Kendrick, R., 1995. Diminishing Returns: Exploiting the Ocean's Bounty, *National Geographic*, vol. 188, No. 5, pp. 2-37.

Pauly, D. and Christensen, V., 1993. Stratified models of large marine ecosystems: a general approach and an application to the South China Sea, 148-174, in Sherman, K., Alexander, L.M. and Gold, B.D., eds., *Large Marine Ecosystems: Stress, Mitigation and Sustainability*, AAAS Press (American Association for the Advancement of Science publishing division), Washington, DC, 376 pp.

Resource Assessment Commission, 1993. *Resource Assessment Commission Coastal Zone Inquiry,* Final Report, Government of Australia, Canberra, http://www.environment.gov.au/marine/natmis/aust_programs/coastal_zone/contents.html

Robbins, J., 1999. *Coral Reefs,* the focus at the International Tropical Marine Ecosystem Management Symposium in Townsville Australia, ITEM: 38 IOC news January 14, 1999.

Russ, G.R. and Alcala, A.C., 1988. *A Direct Test of the Effects of Protective Management on a Tropical Marine Reserve,* SPC/Inshore Fishery Resources, Res./BP. 29.

Sorensen, J., 1997. National and international efforts at integrated coastal management: definitions, achievements and lessons, *Coastal Management,* No. 25, pp. 3-41.

SPREP, 1998. *Annual Report: 1996/1997,* South Pacific Regional Environment Programme SPREP, Apia, Samoa, 72 pp.

Sudara et. al., 1994. Human uses and destruction of ASEAN sea grass beds, 110-113, in Wilkinson, C.R., ed., *Living Coastal Resources of Southeast Asia: Status and Management,* Report of the Consultative Forum Third ASEAN-Australia Symposium on Living Coastal Resources, May 1994, Thailand, Australian Agency for International Development, 188 pp.

Talaue-McManus, L., 2000. *Transboundary Diagnostic Analysis for the South China Sea,* EAS/RCU Technical Report, UNEP, Bangkok.

Uherbelau, V. and Cartwright, I., 1998. *Regional Cooperation In Fisheries – How Do We Get There?,* FFA Regional Cooperation Paper For Presentation At The 17th General Assembly Association Of Pacific Island Legislators, Tinian Commonwealth Of The Northern Mariana Islands May 25-29, South Pacific Forum Fisheries Agency.

UNDP/Government of Sri Lanka/FAO, 1991. Marine Fisheries Management Project SRL/91/022.

UNEP, 1999. *Global Environment Outlook 2000* (London, Earthscan).

United Nations, 1995, *Guidelines on Environmentally Sound Development of Coastal Tourism,* New York.

Weber, P., 1993. *Abandoned Seas: Reversing the Decline of the Oceans,* Worldwatch Paper 116, Washington, DC, 66 pp.

Wilkinson, C., 1998. *Status of Coral Reefs of the World 1998,* Australian Institute of Marine Science.

WRI, 1997. *Marine Biodiversity,* www.wri.org/wri/wr-96-97/bi_txt4.html

WRI, 1999i, *East Asia,* http://www.wri.org/wri/indictrs/reefasia.html

Wright, A. and Doulman, D., 1991. Drift-net fishing in the South Pacific: from controversy to management, *Marine Policy,* vol. 15, No. 5, pp. 303-337.

Yanagawa, H., 1997. Small pelagic resources in the South China Sea, 365-380, in Devaraj, M. and Marftusubroto, P., eds., *Small Pelagic Resources and their Fisheries in the Asia-Pacific Region,* Proc. APFIC Working Party on Marine Fisheries, First Session, May 13-16, RAP Publication 1997/31, Bangkok, 445 pp.

Zann, L., 1995. *Our Sea, Our Future,* Major findings of the State of the Marine Environment Report for Australia, Ministry for the Environment, Sports and Territories, Canberra, http://www.environment.gov.au/marine/publications/somer/chapter1.html

ADB, 1994. *National Response Strategy for Global Climate Change: People's Republic of China*, edited by Siddiqi, A.T., Streets, D.G., Zongxin, W. and Jiankun, H., Environment Division, Office of Environment and Social Development, Manila.

ADB, 1994b. *Climate Change in Asia*, edited by Qureshi, A. and Hobbie, D., Environment Division, Office of Environment and Social Development, Manila.

ADB, 1996. *Megacity Management in the Asian and Pacific Region*, Manila.

ADB, 1998. *Asia Least-Cost Greenhouse Gas Abatement Strategy*. 13 volumes, Environment Division, Office of Environment and Social Development, Manila.

ADB, 1999. *Sustainable Cities: Environmental Challenges in the 21st Century*, Environment Division, Office of Environment and Social Development, Manila.

Asia-Pacific Economic Cooperation, 1997. *Study on Atmospheric Emissions Regulations in APEC Economies and Their Compliance at Coal-Fired Plants*, Report of the Clean Fossil Energy Experts Group, Regional Energy Cooperation Working Group, Coordinated by the Department of Primary Industries and Energy, Australia, East-West Center, Program on Resources, Honolulu.

Association of Southeast Asian Nations, 1998. *News release*, ASEAN Secretariat, Jakarta.

British Petroleum, 1998. *BP Statistical Review of World Energy June 1998*, London.

Carbon Dioxide Information Analysis Center, 1997. *1995 Estimates of CO_2 Emissions from Fossil Fuel Burning and Cement Manufacturing….*, Oak Ridge, Tennessee, Environmental Sciences Division, Oak Ridge National Laboratory.

Central Pollution Control Board, 1994. *Notification on National Ambient Air Quality Standards*, CPCB, Ministry of Environment and Forests, New Delhi.

Commonwealth Scientific and Industrial Research Organization, 1995. *Collaboration in Scenario Development and Impact Projects 1990-95*, edited by Hennessy, K.J., Whetton, P.H. and Pittock, A.B., Mordialloc, Victoria.

Dinar, A., et. al., 1997. *Measuring the Impact of Climate Change on Indian Agriculture*, World Bank Technical Paper No. 402, The World Bank, Washington, DC.

Elvingson, P., 1996. Southeast Asia future pollution projected, *Acid News*, vol. 3, pp. 11-14.

Intergovernmental Panel on Climate Change, 1995. *Climate Change 1994: Radiative Forcing of Climate Change and An Evaluation of the IPCC IS92 Emission Scenarios* (Cambridge, Cambridge University Press).

International Institute for Applied Systems Analysis, 1995. Transboundary Air Pollution, *IIASA Options*, Fall/Winter 1995, IIASA, Laxenburg, Austria.

Johnson, T.M., Junfeng, L., Zhongxiao, J. and Robert, P.T., eds., 1996. *China: Issues and Options in Greenhouse Gas Emissions Control*, World Bank Discussion Paper No. 330, The World Bank, Washington, DC.

Kato, N. and Akimoto H., 1992. Anthropogenic Emissions of SO_2 and NO_x in Asia, *Atmospheric Environment*, vol. 26, pp. 2997-3017.

Ministry of Non-Conventional Energy Sources, Government of India, 1997. *Annual Report*, New Delhi.

Oberthur, S., 1999. *Status of Montreal Protocol Implementation in Countries with Economies in Transition*, A Report to GEF and UNEP.

Ostro, B., 1994. *Estimating the Health Effects of Air Pollutants: A Method with an Application to Jakarta*, World Bank Policy Research Working Paper No. 1301, The World Bank, Washington, DC.

Photovoltaic Insider's Report, 1999. PIR, Dallas, Texas.

Porter, G., et. al., 1998. *Study of GEF's Overall Performance*, Global Environment Facility, Washington, DC.

Shah, Jitendra, J. and Naqpal, T., eds., 1997. *URBAIR: Urban Air Quality Management Strategy in Asia*, World Bank Technical Paper No. 379 (Jakarta Report) and 380 (Metro Manila Report), The World Bank, Washington, DC.

Siddiqi, T.A., 1995. Asia-Wide emissions of greenhouse gases, in *Annual Review of Energy and the Environment*, vol. 20, pp. 213-232.

Soud, H. and Wu, Z., 1998. *East Asia-Pollution Control and Coal-Fired Power Generation*, IEA Coal Research, The Clean Coal Center, London.

Stares, S. and Zhi, L., eds., 1995. *China's Urban Transport Development Strategy*.

Streets, D.G., and Waldhoff, S.T., 1998. Biofuel use in Asia and acidifying emissions, *Energy*, the International Journal, vol. 23, pp. 1029-1042.

Streets, D.G., Carmichael, G.R., Amann, M. and Arndt, R.L., 1999. Energy consumption and acid deposition in North-East Asia, *Ambio*, vol. 28, pp. 135-143.

Tata Energy Research Institute, 1998. *TERI Energy Data Directory and Yearbook 1998/99*, TERI, New Delhi.

UNDP, 1996. *China Environment and Sustainable Development Resource Book II: A Compendium of Donor Activities*, Beijing.

UNDP, UNEP, and the World Bank, 1994. *Global Environment Facility: Independent Evaluation of the Pilot Phase*, Washington, DC.

UNEP, 1998a. *Ozonaction*, Industry and Environment Department, No. 28, October, Paris.

UNEP, 1998b. *Environmental Effects of Ozone Depletion: 1998 Assessment*, Nairobi.

UNEP, 1999. *State of the Environment in China 1997*, UNEP web site.

UNESCAP, 1997. *Regional Cooperation on Climate Change*, New York.

Walsh, M. and Shah, J.J., 1997. *Clean Fuels for Asia: Tactical Options for Moving Toward Unleaded Gasoline and Low-Sulfur Diesel*, World Bank Technical Paper No. 377, The World Bank, Washington, DC.

Wang, W., and Wang, T., 1995. On the origin and trend of acid precipitation in China, *Water, Air, Soil Pollution*, vol. 85, pp. 2295-2300.

WHO and UNEP, 1992. *Urban Air Pollution in Megacities of the World* (Oxford, Blackwell).

WHO, 1996. *Climate Change and Human Health*, edited by McMichael, A.J., et. al., Geneva.

WHO, 1997. *Health and Environment in Sustainable Development: Five Years after the Earth Summit*, Geneva.

WHO, 1998. *Report of the Bi-regional Workshop on Health Impacts of Haze – Related Air Pollution*, held at Kuala Lumpur, Malaysia, June 1-4, Manila.

WMO, 1998. *Scientific Assessment of Ozone Depletion: 1998*, Global Ozone Research and Monitoring Project Report No. 44, 2 volumes, Nairobi.

World Bank, 1992. *World Development Report 1992: Development and the Environment* (New York, Oxford University Press).

World Bank, 1997. *Clear Water, Blue Skies: China's Environment in the New Century*, Washington, DC.

World Bank, 1998. *World Development Indicators 1998*, Development Data Center, Washington, DC.

WRI, 1998. *World Resources 1998-99: A Guide to the Global Environment* (New York, Oxford University Press and Oxford, Oxford University Press).

Zhangfu, W. and Soud, H.N., 1998. *South Asia-Air Pollution Control and Coal-Fired Power Generation*, IEA Coal Research, London.

WWF, 1998. In the British Broadcasting Corporation, *Special Report*, April.

Zhao, D. and Wang, A., 1994. Estimation of anthropogenic methane emissions in Asia, *Atmospheric Environment*, vol. 28, pp. 689-694.

CHAPTER SEVEN

AAMA, 1993. World Motor Vehicle Data, Edition Washington, DC, quoted in WRI, *World Resources 1996-7* (Oxford, Oxford University Press and New York, Oxford University press).

Ansari, J.H., 1997. *Land Use Planning for Disaster Risk Reduction in Urban Areas.*

DFID, 1999. *Ubanization*, Department for International Development, Issue 9, November, UK.

DOE, 1990. *This Common Inheritance: Britain's Environmental Strategy*, HMSO Cm 1200, London.

Douglass, M., 1992. The political economy of urban poverty and environmental management in Asian: access, empowerment and community based alternatives, *Environment and Urbanization*, vol. 4, No. 2.

Douglass, M., 1992. *Urban Poverty and Policy Alternatives in Asia*, Prepared for the Division of Industry, Human Settlements & Environment, UNESCAP, 155 pp.

Dowall, D., 1991. *The Land Market Assessment: A New Tool for Urban Management*, Paper prepared for UNDP/World Bank/UNCHS, Urban Management Programme, Washington, DC.

Dowall, D., 1992. A second look at the Bangkok land and housing market, in *Urban Studies*, vol. 4, No. 2.

ECOTECT, 1993. *Reducing Transport Emissions through Planning*, HMSO, London.

Fernandes, E. and Varley, A., eds., 1996. *Illegal Cities Land and Urban Change in Developing Countries*, pp. 233-273.

ESCAP, 1990. *Regional Assessment of Progress on Water Management Issues in Asia and the Pacific*, United Nations, New York.

ESCAP, 1993. *The State of Urbanization in Asia and the Pacific*, New York.

Hardoy, J., Mitlin, D. and Satterwaite, D., 1992. *Environmental Problems in Third World Cities* (London, Earthscan).

Institut d'Estudis Mertropolitans de Barcelona, 1998. *Cities of the World: Statistical Administrative and Graphical Information on the Major Urban Areas of the World*, vol. 5, institut d'Estudis Metropolitons de Barcelona.

Istanbul Declaration on Human Settlements.

Kingsley T., et. al., 1994. *Managing Urban Environmental Quality in Asia*, World Bank Technical Paper No. 220.

Momin, M.A., 1993. Housing in Bangladesh: The Bangladesh Observer, February 12, 1992, quoted in UNESCAP, 1993. *State of Unbanization in Asia and the Pacific*, United Nations, New York.

Nath, K.J. and others, 1993. Urban solid waste: appropriate technology, Proceeding of the 9th Water and Waste Engineering for developing countries conference, Loughborough University of Technology, England, 1993, quoted in UNCHS, 1998. *Refused Collection Vehicle for Developing Countries*, HS/138/88E, UNCHS (Habitat), Nairobi.

Paboon, C., Kenworthy, J.R., Newman, P.W.G. and Arter, P., 1994. *Bangkok Anatomy of a Traffic Disaster*, Paper presented at the Asian studies of Australia Biennial Conference 1994, Murdoch University, Perth.

Surjadi, C., 1993. Respiratory diseases in mother and children and environmental factors among the household in Jakarta, *Environment and Urbanization*, vol. 5, No. 2, October, pp. 78-86.

Tahir, M.A., et. al., 1994. *A Non-Renewable Resource Being Wasted Unscrupulously*, Soil Survey of Pakistan, Lahore.

Ser, T.T., ed., 1998. *Megacities, Labour, Communications*, Institute of South-East Asian Studies, Stamford Press Pty. Ltd., Singapore, Malaysia, pp. 1-63.

Sharma, V.K., 1997. *Preparedness for National Disasters in Urban Areas.*

UN ESCAP, 1993. *State of Urbanization in Asia and the Pacific*, New York.

UN ESCAP, 1996. *Living in Asian Cities*, Bangkok.

UN ESCAP, 1999, *Population data sheet*, Bangkok.

UNCHS (Habitat II), 1996. *Best Practices Initiative*, Report of the Secretary General.

UNCHS (Habitat), 1996. *An Urbanizing World*, Oxford University Press.

UNCHS (Habitat), 1999 *Community Development Programme (CDP)*, Nairobi.

UNCHS (Habitat), 1999. *Cost Effective and Appropriate Sanitation Sulabh International India*, Nairobi.

UNCHS (Habitat), 1999. *Land Management Programme (LMP)*, Nairobi.

UNCHS (Habitat), 1999. *Local Leadership and Management Training Programme (LLMTP)*, Nairobi.

UNCHS (Habitat), 1999. *Localizing Agenda 21: Action Planning for Sustainable Urban Development (LA21)*, Nairobi.

UNCHS (Habitat), 1999. *Review of Current Global Trends in Economic and Social Development*, Nairobi.

UNCHS (Habitat), 1999. *Safer Cities Programme (SCP)*, Nairobi.

UNCHS (Habitat), 1999. *Sustainable Cities Programme (SCP)*, Nairobi.

UNCHS (Habitat), 1999. *The State of World Cities*, Nairobi.

UNCHS (Habitat), 1999. *Women and Habitat Programme (UMP)*, Nairobi.

UNCHS (Habitat), 1999. *Urban Management Programme (UMP)*, Nairobi.

UNCHS, 1998. *Refuse Collection of the Vehicle for the Developing Countries*, HS/138/88E, Nairobi.

UNCHS, 1999. *Best Practices: Improving Urban Infrastructure Development in Foshan City-China*, Nairobi, pp. 1-9.

UNCHS, 1999. *Best Practices: Integrated Environmental Programme, Zhangjiagang City-China*, Nairobi. pp. 1-5.

UNCHS, 1999. *Best Practices Database: The Comprehensive Improvement of the Urban Environment of Zhuhai-China*, pp. 1-10.

UNCHS, 1999. *Best Practices Database: Slum Networking – A Holistic Approach for Improvement of Urban Infrastructure and Environment*, Nairobi, pp. 1-5.

UNCHS, 1999. *Best Practices Database: Weilhai – an Ecologically Balanced Coastal City*, Nairobi, pp. 1-3.

UNDP, 1999. *Good Governance for Environmental Sustainability*.

UNESCAP, 1995. *Population Data Sheet*, Bangkok.

United Nations, 1995. *State of Environment in Asia and the Pacific*, New York.

United Nations, 1996. *Urban Agglomerations*, Population Division, DESA, New York.

United Nations, 1998. *World Urbanization Prospects (The 1996 Revision)*, DESA, New York.

World Bank, 1999. *Impact Evaluation Report on building institutions and financing Local Development challenge: Lesson from Brazil and the Philippines*.

World Bank, 1999. *Making Cities Livable Now: Priorities of the world Bank*, Washington, DC.

World Bank, 1999. *The Urban Development Challenge: Putting People First*.

ACT, 1996. *A Waste Management Strategy for Canberra: No Waste by 2010*, Department of Urban Services, Canberra.

ADB, 1997. ADB Loan to Pakistan for Wastewater Management in Karachi, *ADB News Release*, No. 77, September 18.

Agarwal, R., 1998. India: The world's final fumpyard, *Besel Action News*, vol. 1, No. 1.

Anjello, R., and Ranawana, A., 1996. Death in slow motion – India has become the dumping ground for the west's toxic waste, *Asia Week*.

Anonymous, 1994. Manila incineration planned, *ENR*, New York, vol. 233, No. 4, pp. 13.

Anonymous, 1995. The Philippines: Smoky Mountain Blues, *The Economist*, London, vol. 336, No. 7931, pp. 41.

ASEAN/UNDP, 1998. *Technology and Environment: The Case for Cleaner Technologies*, ASEAN Secretariat, Jakarta.

Asian Water, vol. 14, No. 1-12.

Asian Water, vol. 15, No. 1-8.

Aziz, M.A., 1997. *ASEAN/UNDP ASP-5 Technology Element IV: Technology and Environment*, Draft Report, ASEAN Secretariat, Jakarta.

Basel Convention, 1994. *Framework of Document on Preparation of Technical Guidelines for the Environmentally Sound Management of Wastes Subject to the Basel Convention*. Document 94/005, Geneva.

Batstone, R., Smith, J.E. and Wilson, D., eds., 1989. *The Safe Disposal of Hazardous Wastes: The Special Needs and Problems of Developing Countries*, Washington, DC, World Bank Technical Paper 93, vols. 1, 2 and 3.

Becht, M., 1999. Solid waste management, *Asian Water*, vol. 15, pp. 28-30.

Blum, D.L., 1995. Environmental legislation in Asia, *Petromin Journal*, January Issue.

Carpenter, D.L., Richardson, A.C. and Khaliq, R., 1996. Booming growth in Asia raises demand for environmental technologies, Washington, DC, *Business America*, vol. 117, No. 4, pp. 4.

Carr, E., 1998. Survey: The sea on the age. *The Economist*, vol. 347, No. 8069, pp. 54-56.

Chin, K.K., 1998. Performance of small wastewater treatment plants, *Proc. Water Environment Federation Technical Conference and Exhibition*, Singapore, March 7-11, pp. 99-104.

China, 1997. *China Environment News*, Issue No. 77.

Chua, H., Yu, P.H.F. and Sin, N.S., 1999. Hong Kong's sewage strategy, *Asian Water*, pp. 22-26.

Chynoweth, E., 1995. Asia/Pacific moves to address mountainous waste problems, *Chemical Week*, New York, vol. 152, No. 11, p. 42-46.

CITYNET/UNCRD, 1994. *Report of the CITYNET/UNCRD/City of Makati Seminar on Recycling in Asia*, November 21-23, Metro Manila.

CITYNET/UNDP, 1996. Wastewater management in Asia Pacific, *Proc. CITYNET/LIFE Regional Training Workshop on Wastewater Management*, July 23-26, Colombo.

Collin, J., 1996. Polluters will pay, *Far Eastern Economic Review*, Hong Kong, vol. 145, No. 38, pp. 6-8.

Cost-Pierce, B., 1998. Constraints to the sustainability of cage aquaculture for resettlement from hydropower dams in Asia: An Indonesian case study, *Environment and Development*, vol. 7, No. 4, pp. 335-363.

ENV, 1997. *Annual Report*, Ministry of Environment, Singapore.

Environment News: *Waste Killing One of the Asia's Largest Lakes*.

Erdin, E. and Ozdaglar, D., 1998. *Solid Wastes Reuse, Recycling and Recovery in Turkey*, http://members.xoom.com/basak/swretr.html

ESCAP, 1992. *Technology Planning for Industrial Pollution Control*, United Nations, New York.

ESCAP, 1994. *Guidelines for Development of a Legal and Institutional Framework to prevent Illegal Traffic in Toxic and Dangerous Products and Wastes*, United Nations, New York.

ESCAP, 1996. *Guidelines on Monitoring Methodologies for Water, Air and Toxic Chemicals/Hazardous Wastes*, United Nations, New York.

ESCAP, 1997. Agricultural biomass energy technologies for sustainable rural development, *Proc. Expert Group Meeting on Utilization of Agricultural Biomass as an Energy Source*, July 16-19, United Nations, New York.

ESCAP/CITYNET, 1993. *ESCAP/CITYNET Regional Seminar-cum-Study Visit on Final Disposal of Solid Waste*, September 1-4, China.

FAO/UNDP, 1997. Promotion of sustainable agricultural and rural development in China: elements for a policy framework and national Agenda 21 Action Programme, *Ministry of Agriculture Report*, China.

Fauzia, C.I. and Rosenani, A.B., 1996. Animal wastes and the environment: Malaysian perspective, *ENSEARCH*, vol. 9, pp. 33-41.

Fernandez, A., 1993. Public-Private partnerships in solid waste management, Nagoya, Japan, *Regional Development Dialogue*, vol. 14, No. 3, pp. 3-26.

Government of Malaysia, 1998, *Malaysia Environmental Quality Report 1998*, Department of Environment, Kuala Lumpur.

Government of the Republic of Korea, 1999. *Environmental White Book*, Seoul.

Greenpeace, 1997. *The Waste Invasion,* http://www.greenpeace.org/~comms/no.nukes/p970131.html

Greenpeace, 1998. *Dutch PVC Waste Still Exported to Asia – Call for an End to Delayed Dumping,* http://www.greenpeace.org/pressrelease/toxics/1998feb4.html

Hara, T., 1997. Waste recycling in Japan, *TECH MONITOR*, vol. 14, No. 2, pp. 45-50.

Henderson, J.P. and Chang, T.J., 1998. *Solid Waste Management in China,* http://www.ecowaste.com/swanabc/papers/hend01.htm

Hernandez, J., 1993. *The Hazardous Waste Manager's Manual*, Workbook and Database.

Hiebert, M., 1995. Recycling: A fortune in waste, *Far Eastern Economic Review*, December Issue, Hong Kong.

Higham, S., 1998. Saving Pasig River: Manila's central sewer, *Asia's Journal of Environmental Technology*, vol. 14, No. 8, pp. 12-15.

Hirose, T., 1997. Waste management in food processing industries in Japan, *TECH MONITOR*, vol. 14, No. 6, pp. 23-26.

Hunt, C., 1996. Child waste pickers in India: The occupation and its health risk, *Environment and Urbanization*, vol. 8, No. 2, pp. 111.

Indonesia, 1997. *The Indonesian Environmental Almanac*, Jakarta.

In-na, Y., 1998. Samut Prakarn Wastewater Management Project: Collection and Treatment Turnkey Contract, Bangkok, *ASEP Newsletter*, vol. 14, No. 1, pp. 10-15.

Keen, M., Hunt, C. and Sisto, N., 1997. Economic strategies to promote environmentally sound waste management, UNEP/SPREP, Canberra, *Proc. Workshop on Waste Management in Small Island Developing States in the Pacific*, vol. 2, pp. 69-85.

Kim, 1998. Country Focus – South Korea, *Asian Water*, vol. 14, No. 5, pp. 9-11.

Kiravanich, P. and In-na, Y., 1995. Breakthrough in the sewerage sectors in Thailand – Establishment of Wastewater Management Authority in BMA, Bangkok, *Journal of Asian Society of Environmental Protection*, vol. 11, No. 3 and 4, pp. 7, 13.

Kiser, J.V.L., 1998. Thailand: A kingdom of waste management opportunities, Atlanta, *World Wastes*, vol. 41, No. 1, pp. 15-19.

Koe, C.C. and Aziz, M.A., 1995. *Regional Programme of Action on Land-based Activities Affecting Coastal and Marine Areas in East Asian seas*, UNEP/EAS/RCU, Technical Report Series, No. 5.

Leong, L.T. and Quah, E., 1995. Management of non-hazardous solid waste, in Institute of Policy Studies, *Environment and City*, Singapore.

Malaysia, 1998. *The Encyclopedia Malaysia*, vol. 1: The Environment (Kuala Lumpur, Archipelago Press).

Hill, M., 1997. Growing waste stream in pushing China to recycle, New York, *Modern Plastics*, vol. 74, No. 9, pp. 18.

Medina, M., 1998. Scavenger cooperatives in developing countries, *Biocycle, Emanus*, vol. 39, No. 6, pp. 70-72.

Mowbray, D., 1997. The Waigani Convention, Canberra, *Proc. UNEP/SPREP Workshop on Waste Management in Small Island Developing States in the South Pacific*, pp. 177-196.

Nelson, D., 1997. *Toxic Waste: Hazardous to Asia's Health*, East-West Center, Honolulu, Hawaii.

Nishigaya, N., 1998. Recycling of waste plastics in Japan, *TECH MONITOR*, vol. 15, No. 2, pp. 18-22.

Ogawa, H., 1993. Improving management of hospital wastes, Nagoya, Japan, *Regional Development Dialogue*, vol. 14, No. 3, pp. 108-117.

Perla, M., 1997. Community composting in developing countries, *Biocycle, Emanus*, vol. 38, No. 6, pp. 48-51.

Philippines, 1997. *Essential Statistics on Environmental and Natural Resources-An Update*, Manila.

Pitot, H.A., 1996. Domestic waste management and recycling in Delhi 'Basti' in India, *Environment and Urbanization*, vol. 8, No. 2.

Ratra, O.P., 1998. Plastic waste management in India: systems and policies, *TECH MONITOR*, vol. 15, No. 4, pp. 28-36.

Rummel-Bulska, I., 1993. The Basel Convention: A global approach for the management of hazardous wastes, UNEP, Bangkok, *NETTLAP Publication*, No. 5.

Sandra, C.L., 1994. *Private Sector Participation in Municipal Solid Waste Services in Developing Countries*, UNDP/UNCHS/World Bank Urban Development Programme, Washington, DC.

Saywell, T., 1997. Fishing for trouble, *Far Eastern Economic Review*, vol. 60, No. 11, pp. 50-52.

Strait Times, 1995. September 14, Singapore.

Suyanto, W. and Yatim, 1993. Handling and transport of low level radioactive wastes in Indonesia, *RAMTRANS*, vol. 4, No. 3/4, pp. 213-218.

Tabucanon, 1998. Country Focus – Thailand, *Asian Water*, vol. 14, No. 3, pp. 10-14.

Tanaka, Y. and Takahashi, K., 1998. Waste management in the Japanese chemical industry, *TECH MONITOR*, vol. 15, No. 6, pp. 36-39.

Thom, N., 1997. *Composting and Vermicomposting of Municipal Solid Waste*, Canberra, UNEP/SPREP Report on Regional Workshop, pp. 111-115.

UNEP, 1994. Management of toxic chemicals and hazardous wastes in Asia-Pacific Region, *NETTLAP Publication*, No. 5, UNEP/ROAP, Bangkok.

UNEP, 1998. *Municipal Solid Waste Management*, Newsletter and Technical Publication, International Environmental Technology Center.

UNEP/SPREP, 1997. *Waste Management in Small Island Developing States in the South Pacific*, Report of the Regional Workshop Organized by UNEP/SPREP in Collaboration with Environment Australia, Canberra, vols. 1 and 2.

UNIDO, 1997. *Minimizing Waste by Desire*, Document Reference ID: Earthsummits.htmls.

United Nations, 1992. *Agenda 21*, New York.

United Nations, 1995. *State of the Environment in Asia and the Pacific*, New York.

Wahyono, S. and Sahwan, F.L., 1998. Solid waste composting trends and projects, *Biocycle*, vol. 39, No. 10, pp. 64-68.

WHO, 1996. *Action Plan for the Development of National Programme for Sound Management of Hospital Wastes*, New Delhi.

World Bank, 1995. *Viet Nam*, The World Bank's Agricultural and Environment Division Report, Country Department I-East Asia and Pacific Region, Washington, DC.

World Bank, 1998. *Pollution Prevention and Abatement Handbook: Towards Cleaner Production*, Washington, DC.

WQI, 1999. January/February Issue, pp. 5, 28, 32.

WRI, 1995. *World Resources 1995*.

Yan, S., 1998. Singapore's DTSS: more than just a Pipeline Project, *Asian Water*, vol. 14, pp. 14-15.

ADB, 1999. *The Asian Development Bank's Poverty Reduction Strategy*, Working Draft, July, Manila.

Anonymous, 1999. Liberal times: Globalization and poverty alleviation, a forum for liberal policy, in *South Asia*, vol. VII, No. 3.

Carew-Reid J, 1997. *Strategies for Sustainability: Asia*, IUCN.

Commonwealth Group of Experts, Commonwealth Secretariat, 1991. *Sustainable Development, An Imperative for Environmental Protection*.

DFID, 1997. *Eliminating World Poverty: A Challenge for the 21st Century*, A summary, November, London.

ESCAP, 1995. *Quality of Life in the ESCAP Region*, United Nations, New York.

ESCAP, 1996. *Guidelines on the State of the Environment Reporting in Asia and the Pacific*, United Nations, New York.

ESCAP, 1997. *Asia and the Pacific in Figures*, United Nations, New York.

ESCAP, 1998. *Reducing Poverty by Improving Accessibility: Transport Intervention in Lao Peoples Democratic Republic*, Bangkok.

FAO, 1998. *Rural Women and Food Security: Current Situation and Perspectives*, Rome.

Hasan, A., 1997. *Working with Government, Orangi Pilot Project (OPP)*, Karachi.

Himmerfarb, G., 1984. *The Idea of Poverty*, New York, Knopf.

Johanston, R.J. and Taylor, P.I., 1988. *A World in Crisis? Geographical Perspectives*.

Kleinman, D., 1995. Human adaptations and population growth: A non-malthusian perspective, quoted in ESCAP, *Quality of Life in the ESCAP Region*, United Nations, New York.

Lewis, O., 1966. The culture of poverty, *Scientific American*, No. 215.

Melissa, L. and Mearns, R., 1991. *Poverty and Environment in Developing Countries*, ESRC Society and Politics Group, Global Environmental Change Programme and the DFID.

Nordstorm, H. and Vaughan, G., 1999. *Trade Liberalization Reinforces the Need for Environmental Cooperation*, World Trade Organization.

ODI, 1999. *ODI Poverty Briefing*, February.

Quarterly Report No. 76, Orangi Pilot Project, October–December 1998.

Red Cross & Red Crescent, 1999. *World Disaster Report*.

Repetto, R., et. al., 1989. *Wasting Assets: Natural Resources in the National Income Accounts*, WRI, Washington, DC.

Stanton, B., 1994. Child health: equity in the non-industrialized countries, *Social Sciences Medicines*, vol. 38., No. 10.

UNDP, 1998. *Human Development Report*, New York.

UNDP, 1999. *Human Development Report*, New York.

UNDP, http://www.undp.org/

UNEP, 1997. *Global Environment Outlook I, Global State of the Environment Report*, Nairobi.

UNEP, http://www.unep.org/

UNESCAP, http://www.unescap.org/

UNICEF, 1997. *The Progress of Nations*, New York.

United Nations, 1995. *State of the Environment in Asia and the Pacific*, New York.

United Nations, 1997. *Global Changes and Sustainable Development: Critical Trends*, Department for Policy Coordination and Sustainable Development, New York.

United Nations, 1997. *Manila Declaration on "Accelerated Implementation of the Agenda for Action on Social Development in the ESCAP Region"*.

United Nations, http://www.un.org/

University of Sussex, *Information for Development in the 21st Century*, http://www.id21.org/

WHO, 1997. *Health and Environment in Sustainable Development: Five Years After the Earth Summit*, Geneva.

WHO, 1997. *The World Health Report 1997: Conquering Suffering, Enriching Humanity*, Geneva.

Wilkinson, R.G., 1994. The epidemiological transition: from material scarcity to social disadvantage, *Daedalus*, Fall, vol. 123, No. 4.

World Bank, 1999. *Environment Matters: Annual Review, Towards Environmentally and Socially Sustainable Development*, Washington, DC.

World Bank, 1999. *Project Appraisal Document: Poverty Alleviation Fund Project for Islamic Republic of Pakistan*, May.

World Bank, 1999. *World Development Report, Knowledge for Development 1998/99.*

World Bank, http://www.worldbank.org/

WCED, 1987. *Our Common Future.*

WRI, UNEP, UNDP and World Bank, 1996. *World Resources 1996-97: A Guide to the Global Environment.*

WTO, http://www.wto.org/

ADB, 1993. *Environmental Assessment Requirements and Environmental Review Procedures of the Asian Development Bank*, Manila.

Ahmad, Y., 1988. *Issues of sustainability in agricultural development in Asia*, Paper prepared for the International Consultation on Environment, Sustainable Development and the Role of Small Farmers, IFAD, October 11-13, Rome.

Alamgir, M. and Poonam, A., 1991. *Providing food security for all* (New York, New York University Press).

Alamgir, M., 1997. Training program on rural poverty, food security and sustainable development, Rome, unpublished.

Alamgir, M., 1998. Paths of sustainable development – regional dimensions of the challenge: the case of Asia, in Leihner, D.E. and Mitschein, T.A., eds., 1998. *A Third Millennium for Humanity?: The Search for Paths of Sustainable Development,* (Frankfurt am Main, Peter Lang), pp. 97-168.

Alexandratos, N. and Jelle, B., 1999. *Land Use and Land Potentials For Future World Food Security*, Paper for the International Conference on Sustainable Future of the Global System, Unite Nations University and Institute for Global Environmental Strategies, February 23-24, Tokyo.

Alexandratos, N., ed., 1995. *World Agriculture: Towards 2010* (New York, John Wiley & Sons).

Azimi, A.M. and Henry, T., 1998. *Environment Profile: Thailand*, ADB, Manila.

Bingsheng, K., 1998. Food problems and outlook in China, in Horiuchi, H. and Tsubota, K. eds., *Sustainable Agricultural Development: Compatible with Environmental Conservation in Asia*, Japan International Research Center for Agricultural Sciences, pp. 31-43.

Brown, L.R., 1998. Struggling to raise cropland productivity, in World Watch Institute, 1998. *State of the World 1998*.

China Daily, 1998a. 18 (5735).

China Daily, 1998b. 18 (5736).

China Daily, 1998c. 18 (5738).

China Daily, 1998d. 18 (5739) (BW No. 301).

Cohen, J.E., 1995. *How Many People can the Earth Support?* (New York, W.W. Norton & Company).

Dyson, T., 1996. *Population and Food: Global Trends and Future Prospects*. (London, Routledge).

Eldredge, N., 1998. *Life in the Balance: Humanity and the Biodiversity Crisis*, (New Jersey, Princeton).

FAO, 1990. *International Code of Conduct of the Distribution and Use of Pesticides*, Rome.

FAO, 1994. *Medium-term Prospects for Agricultural Commodities: Projections to the Year 2000*, Rome.

FAO, 1995a. *The Director General's Programme of Work and Budget for 1996-97*, Rome.

FAO, 1995b. *Dimensions of Need, An Atlas of Food and Agriculture*, Rome.

FAO, 1996a. *The Sixth World Food Survey: 1996*, Rome.

FAO, 1996b. *Food For All*, Rome.

FAO, 1997a. *Selected Indicators of Food and Agriculture Development in Asia-Pacific Region, 1986-96*, Bangkok.

FAO, 1997b. *FAO's Emergency Activities*, Rome.

FAO, 1998a. *FAO Production Yearbook*, vol. 51, Rome.

FAO, 1998b. *The State of Food and Agriculture 1998*, Rome.

FAO, 1998c. *The Right to Food in Theory and Practice*, Rome.

FAO, 1999a. *Food Outlook*, No. 2, Rome.

FAO, 1999b. *Foodcrops and Shortages*, No. 2, Rome.

FAO, 1999c. *Poverty Alleviation and Food Security in Asia: Land Resources*, Rome.

FAO, Undated. http://www.fao.org

FAO, Undated. *The Global Information and Early Warning System on Food and Agriculture*, Two documents, Rome.

FAO, Undated. *The Special Program for Food Security*, Two documents, Rome.

FAO/CWFS, 1993. *Improving Food Security through Sustainable Productivity Increases in Fragile Areas of Developing Countries*, Rome.

FAO/CWFS, 1994. *Assessing the Contribution of High Potential Areas in Developing Countries to Improving Food Security on a Sustainable Basis*, Rome.

FAO/CWFS, 1999a. *Assessment of the World Food Security Situation*, Rome.

FAO/CWFS, 1999b. *Investment in Agriculture for Food Security: Situation and Resource Requirements to Reach the World Food Summit Objectives*, Rome.

FAO/WFP, 1996a. *Special Alert No. 267 – Democratic People's Republic of Korea*, Rome.

FAO/WFP, 1996b. *Special Alert No. 270 – Democratic People's Republic of Korea*, Rome.

FAO/WFP, 1996c. *Special Alert No. 275 – FAO/WFP Crop and Food Supply Assessment Mission to the Democratic People's Republic of Korea*, Rome.

FAO/WFP, 1997a. *Special report: FAO/WFP Crop and Food Supply Assessment Mission to the Democratic People's Republic of Korea*, November 25, Rome.

FAO/WFP, 1997b. *Special alert No. 277 – FAO/WFP Crop and Food Supply Assessment Mission to the Democratic People's Republic of Korea*, Rome.

FAO/WFP, 1997c. *Commonwealth of Independent States (CIS): Developments in Food Production and Marketing and Preliminary Assessment of 1997 Food crop Production and 1997/98 Cereal Import Requirements*, Special Report, December 15, Rome.

FAO/WFP, 1998a. *FAO/WFP Crop and Food Supply Assessment Mission to Indonesia*, Special report, April 17, Rome.

FAO/WFP, 1998b. *FAO/WFP Crop and Food Supply Assessment Mission to the Democratic People's Republic of Korea*, Special report, June 25, Rome.

FAO/WFP, 1998c. *FAO/WFP Crop and Food Supply Assessment Mission to Indonesia*, Special report, October 6, Rome.

FAO/WFP, 1998d, *FAO/WFP Crop and Food Supply Assessment Mission to the Democratic People's Republic of Korea*, Special report, November 12, Rome.

FAO/WFP, 1998e. *FAO/WFP Crop and Food Supply Assessment Mission to the Democratic People's Republic of Korea*, Special Report, December 6, Rome.

FAO/WFP, 1999a. *FAO/WFP Crop and Food Supply Assessment Mission to Lao People's Democratic Republic*, Special report, March 4, Rome.

FAO/WFP, 1999b. *FAO/WFP Crop and Food Supply Assessment Mission to Indonesia*, Special report, April 8, Rome.

Freedonia Group, 1999. *World Pesticides to 1998*, Report 636, Cleveland.

Gardner, G., 1996. Preserving agricultural resources, in World Watch Institute, New York, *State of the World*, pp. 79-94.

Horiuchi, H. and Tsubota, K., eds,. 1998. *Sustainable Agricultural Development: Compatible with Environmental Conservation in Asia*, Japan International Research Center for Agricultural Sciences.

Idriss, J., Alamgir, M. and Panuccio, T., 1992. *The State of World Rural Poverty* (New York, New York University Press).

IFAD, 1988 through 1998. *Annual Report*, Rome.

IFAD, 1988a. Environment, sustainable development and the role of small farmers: issues and options, unpublished.

IFAD, 1988b. Mongolia: country strategic opportunities paper, Rome, unpublished.

IFAD, 1990. Kingdom of Nepal: Hills Leasehold Forestry and Forage Development Project, Appraisal Report, Rome, unpublished.

IFAD, 1993. Report and recommendation of the President to the Executive Board on a proposed loan to the Socialist Republic of Viet Nam for the participatory resource management project Tuyen Quang Province, Rome, unpublished.

IFAD, 1994. Report and recommendation of the President to the Executive Board on a proposed loan to the Lao People's Democratic Republic for the Bokeo food security project, Rome, unpublished.

IFAD, 1995a. Report and recommendation of the President to the Executive Board on a proposed loan to the Republic of Indonesia for the Eastern islands smallholder farming systems and livestock project, Rome, unpublished.

IFAD, 1995b. Report and recommendation of the President to the Executive Board on a proposed loan to the Socialist Republic of Viet Nam for the rural micro finance project, Rome, unpublished.

IFAD, 1996. LAO PDR: country strategic opportunities paper, Rome, unpublished.

IFAD, 1996. Viet Nam: country strategic opportunities paper, Rome, unpublished.

IFAD, 1997. Democratic People's Republic of Korea: Hill areas livestock development project, (2 Volumes), Rome, unpublished.

Karim, A., 1999. The endangered coastal biodiversity of Bangladesh, *BETS Quarterly*, Bangladesh Environment and Technical Services, Dhaka, Issue No. 31, pp. 2.

Khan, M.A., 1998. In *Proceedings of Regional Expert Group Meeting on IPM in Rural Poverty Alleviation*, November 11-13, ESCAP, Bangkok.

Leihner, Dietrich E. and Thomas A.M., eds., 1998. *A Third Millennium for Humanity: The Search for Paths of Sustainable Development*, (Frankfurt am Main, Peter Lang).

Mazur, L., A., 1994. *Beyond the Numbers* (Washington, DC, Island Press).

Patend, D., H., 1996. *Biodiversity* (New York, Clarion).

Pesticide Manufacturers Association of India, *http://www.pmfai.org*

Poapongsakorn, N., 1998. Problems and outlook of agriculture in Thailand, in Horiuchi, H. and Tsubota K., eds., *Sustainable Agricultural Development: Compatible with Environmental Conservation in Asia*, Japan International Research Center for Agricultural Sciences, pp. 45-62.

Roa, K.P., 1998. Food problems and outlook in India, in Horiuchi, H. and Tsubota K., eds., *Sustainable Agricultural Development: Compatible with Environmental Conservation in Asia*, Japan International Research Center for Agricultural Sciences, pp. 63-72.

Sasson, A., 1992. Prospects for biotechnology in selected Asian Countries, in United Nations, Department of Economic and Social Development, *Biotechnology and Development: Expanding the Capacity to Produce Food*, United Nations, New York, pp. 65-72.

Timmer, C.P., 1997. *Food Security Strategies: the Asian Experience*, FAO Agricultural Policy and Economic Development Series 3, Rome.

UNEP, 1997. *Global Environment Outlook I*, New York.

United Nations, Department of Economic and Social Development, 1992. *Biotechnology and development: Expanding the Capacity to Produce Food*, New York.

WFP, 1995. *Annual Report 1995*, Rome.

WFP, 1998. *Annual Report 1998*, Rome.

World Bank, 1999a. *Knowledge for Development- World Development Report*, Oxford.

World Bank, 1999b. *World Development Indicators: 1999*, Washington, DC.

World Food Summit, 1996. *Rome Declaration on World Food Security and World Food Summit Plan of Action*, Rome.

World Watch Institute, 1998. *State of the World 1998*, New York.

WRI, 1994. *World Resources 1994-95*, Oxford.

WRI, 1996. *World Resources 1996-97*, Oxford.

WRI, 1998. *World Resources 1998-99*, Oxford.

Zhao, Q.G., 1998. Technology for conservation of soil and water resources in China, in Horiuchi, H. and Tsubota, K., eds., *Sustainable Agricultural Development: Compatible with Environmental Conservation in Asia*, Japan International Research Center for Agricultural Sciences, pp. 99-104.

ADB, 1997. *Asian Development Outlook, 1997 and 1998*, ADB, Manila.

Barrow, C.J., 1995. Sustainable development: concept, value and practice, *Third World Planning Review*, vol. 17, No. 4, November, pp. 369-386.

Chaudhry, M.G. and Sahibzada, Shamin, A., 1997. Integrating environmental considerations into economic decision making processes: The case of Pakistan's cotton sector, ESCAP, Bangkok, unpublished.

Chesher, R., 1997. Modalities for environmental assessment: fishery resources development and management in Western Samoa, ESCAP, Bangkok, unpublished.

Chesher, R., 1997. Modalities for environmental assessment: land resources development and management in the Solomon Islands, ESCAP, Bangkok, unpublished.

Chesher, R., 1997. Modalities for environmental assessment: water resources development and management in Kiribati, ESCAP, Bangkok, unpublished.

Coase, R.H., 1992. The institutional structure of production, *American Economic Review*, 82, September, pp. 713-19.

Dasgupta, P. and Karl-Goran M., 1990. The environment and emerging development issues, in World Bank, 1990, *Proceedings of the World Bank Annual Conference on Development Economics*, supplement to the World Bank Economic Review and The World Bank Research Observer, Washington, DC.

Elahi, M., 1997. Institutional arrangements and mechanisms at the provincial level in Pakistan: NWFP Province, ESCAP, Bangkok, unpublished.

Esrom, D.K., 1997. Institutional arrangements and mechanisms at local level: the case of Port Vila in Vanuatu, ESCAP, Bangkok, unpublished.

Ganapin, Jr, Delfin, J., 1991. Effective environmental regulation: the case of the Philippines, in Denizhan, E., ed., 1991. *Environmental Management in Developing Countries*, Development Centre of the OECD, Paris.

Government of India, 1992. *National Conservation Strategy and Policy Statement on Environment and Development*, Ministry of Environment and Forests, New Delhi.

Guthman, J., 1997. Representing crisis: the theory of Himalayan environmental degradation and the project of development in Post-Rana Nepal, *Development and Change*, vol. 28, pp. 45-69.

Han, T.W. and Yoo, S.H., 1997. Synthesis paper: institutional arrangements for integrated policy making: the cases of China, Malaysia, Myanmar, The Philippines and The Republic of Korea, ESCAP, Bangkok, unpublished.

Hardin, G., 1981. The Tragedy of the commons, *Science*, 162, December 13, 1968.

His Majesty's Government of Nepal, 1992. *The Eighth Plan, 1992-97*, Planning Commission, Nepal.

Jha, A.N., 1997. Institutional mechanisms in Nepal, Philippines and Fiji, ESCAP, Bangkok, unpublished.

Kaniaru, D., Kurukulasuriya, L., Abeyegunawardene, P.D. and Martino, C., eds., 1997. *Report on the Regional Symposium on the Role of the Judiciary in Promoting the Rule of Law in the Area of Sustainable Development*, 4-6, SACEP/UNEP/NORAD, Colombo.

Kaniaru, D., Kurukulasuriya, L. and Abeyegunawardene, P.D., eds., 1997. *Compendium of Summaries of Judicial Decisions in Environment Related Cases (with Special Reference to Countries in South Asia)*, SACEP/UNEP/NORAD, Colombo.

Kere, F.T., 1997. Institutional arrangements and mechanisms at sector level: the case of Squash in Vanuatu, ESCAP, Bangkok, unpublished.

Kurukulasuriya, L., ed., 1995. *UNEP's New Way Forward: Environmental Law and Sustainable Development*, UNEP, Nairobi.

McGregor, A., 1997. Synthesis paper: institutional arrangements and mechanisms, Pacific Islands, ESCAP, Bangkok, unpublished.

McGregor, A., 1997. Institutional arrangements and mechanisms at local level: the case of Suva in Fiji, ESCAP, Bangkok, unpublished.

Mohamed, M.Z. and Jaafar, A.B., 1997. Institutional arrangements and mechanisms at the provincial level: the case of Kuala Lumpur, ESCAP, Bangkok, unpublished.

Munasinghe, M. and Wifrido, C., 1994. *Economy wide Policies and the Environment: Lessons from Experience*, World Bank, Washington, DC.

Murty, M.N., 1997. Institutional arrangements and mechanisms at national level in India, ESCAP, Bangkok, unpublished.

Nijkamp, P., Bergh, V. and Soeteman, F.J., 1990. Regional Sustainable Development and Natural Resource Use in *Proceedings of the World Bank Annual Conference on Development Economics, 1990,* Supplement to the World Bank Economic Review and World Bank Research Observer, World Bank, Washington, DC.

Nikpay, N., 1997. Modalities for environmental assessment: assessment and valuation of dry lands and desertified areas in the Islamic Republic of Iran, ESCAP, Bangkok, unpublished.

North, D.C., 1992. *Transaction Costs, Institutions and Economic Performance,* International Centre for Economic Growth Publications (San Fransisco, ICG Press).

Panayotou, T., 1995. *Internationalization of Environmental Costs,* paper for the United Nations Conference on Trade and Development, October 26.

Pearce, D., Markandya, A. and Barbier, E., 1989. *Blueprint for a Green Economy* (London, Earthscan).

Pearce, D., Markandya, A. and Barbier, E., 1990. *Sustainable Development: Economics and Environment in the Third World* (London, Earthscan).

Pradhan, B.B., 1997. Institutional arrangements and mechanisms at the national level in Nepal, ESCAP, Bangkok, unpublished.

Rahim, K.A., 1997. Institutional arrangements and mechanisms at sector level: the case of agriculture in Malaysia, ESCAP, Bangkok, unpublished.

Ranade, A. and Guha-Khasnobis, B., 1997. Synthesis paper: Institutional arrangements and mechanisms, South Asia, ESCAP, Bangkok, unpublished.

Robinson, N., ed., 1993. *Agenda 21: Earth's Action Plan* (New York, Oceania Publications).

Root, H.L., 1995. *Managing Development through Institution Building,* Occasional Papers, No. 12, ADB, Manila.

Sevele, F.V., 1997. Institutional arrangements and mechanisms at national level in Tonga, ESCAP, Bangkok, unpublished.

Shupeng, C., 1997. Institutional arrangements at the provincial/local level: the case of Shenyang (China), ESCAP, Bangkok, unpublished.

SACEP, 1997. *South Asia Handbook of Treaties and Other Legal Instruments in the Field of Environmental Law,* Colombo.

Sterner, T., 1994. *Economic Policies for Sustainable Development,* Boston.

Tietenberg, T., 1994. *Environmental Economics and Policy* (New York, Harper and Collins).

Tillekeratne, L.S.G., 1997. Institutional arrangements and mechanisms at sector level: tea sector in Sri Lanka, ESCAP, Bangkok, unpublished.

United Nations, 1995. *State of the Environment in Asia and the Pacific,* New York.

United Nations, 1996. *Indicators of Sustainable Development: Framework and Methodologies,* New York.

Van A.B., 1989. Role of institutions in development, in World Bank, 1989. *Proceedings of the World Bank Annual Conference on Development Economics,* supplement to the World Bank Economic Review and World Bank Research Observer, World Bank, Washington, DC.

Webster, D. and Muller, L., 1997. Synthesis paper: modalities for environmental assessment, East and South-East Asia, ESCAP, Bangkok, unpublished.

William G.A., Professor of Natural Products Chemistry, University of the South Pacific

Williamson, O.E., 1994. The institutions and governance of economic development and reform, in World Bank, 1994, *Proceedings of the World Bank Annual Conference on Development Economics,* supplement to the World Bank Economic Review and World Bank Research Observer, World Bank, Washington, DC.

World Bank, 1995. *Monitoring Environmental Progress: A Report on Work in Progress,* Washington, DC.

World Commission on Environment and Development, 1987. *Our Common Future (The Brundtland Report)* (New York, Oxford University Press).

ADB, 1992. Environmental legislation and administration: briefing profiles of selected developing member countries of the Asian Development Bank, *ADB Environment Paper, No. 2*, Asian Development Bank, Manila, 59 pp.

ADB, 1997. *Central Asian Environments in Transition*, ADB, Manila, pp. 18.

ADB, 1997b. *Emerging Asia: Changes and Challenges*, ADB, Manila.

Afsah, S., Laplante, B. and Wheeler, D., 1996. *Controlling Industrial Pollution: A New Paradigm, Policy Research Working Paper No. 1672*, World Bank, Policy Research Department, Environment, Infrastructure, and Agriculture Division, Washington, DC.

ENVIRONMENT, 1998. Environmental taxes, *China's Bold Initiative*, vol. 40, No. 7, pp. 11-13 and 33-38.

ESCAP, 1997. Selected issues with reference to the work of the Committee on Environment and Sustainable Development: refinement and promotion of methodologies for the integration of environment and development, E/ESCAP/ESD(4)/1, August 14, unpublished.

ESCAP, 1999. Integrating environmental considerations into economic policy making processes, background readings volume III: institutional arrangements and mechanisms at sector level, United Nations, New York, unpublished.

Gray, D., 1995. *Reforming the Energy Sector in Transition Economies: Selected Experience and Lessons*, World Bank Discussion Paper No. 296, Washington, DC.

HIID, 1996a. *Ostenka Ushcherb Zdaroviye ot Zagrazneniya Vozdukha goroda Almaty* (Health Damage Assessment of Air Pollution in Almaty, Kazakhstan), HIID, International Environment Programme, Environmental Economics and Policy Project Working Group (Dzhabasov, A., et. al.).

HIID, 1996b. *Ostenka Ushcherb Zdaroviye ot Zagrazneniya Vozdukha goroda Tashkent* (Health Damage Assessment of Air Pollution in Tashkent, Uzbekistan). HIID, International Environment Program, Environmental Economics and Policy Project Working Group (Samojlev, S.V., et. al.).

Kraemer, R., 1995. Substitution of energy-intensive household appliances in Australia, in Janicke, M. and Weidner, H., eds., *Successful Environmental Policy: A Critical Evaluation of 24 Cases*, Berlin.

Kummer, D., and Sham, C.H., 1994. The causes of tropical deforestation: A quantitative analysis and case study from the Philippines, in Brown, K. and Pearce, D.W., eds., *The Causes of Tropical Deforestation* (London, University College of London Press).

Lohani, B.N., Evans, J.W., Everitt, R., Ludwig, H., Carpenter, H., Tu, H.L., 1997. *Environmental Impact Assessment for Developing Countries in Asia*, vol. I, pp. 1-32.

Lopez, R., 1991. *The Environment as a Factor of Production: the Economic Growth and Trade Policy Linkage*, Maryland, Economics Department, University of Maryland, mimeographed, December.

Mok, S.T., 1998. Current State of Forest Certification and Role of Forest Stewardship Council (FSC), in *Proceedings of the National Seminar on Sustainable Forest Management and Forest Certification*, Ho Chi Minh City, February 10-12, (Hanoi, Agriculture Publishing House), pp. 20-23.

Panayotou, T. and Somthawin Sungsuwan, 1994. An econometric analysis of the causes of tropical deforestation: The case of Northeast Thailand, Development Discussion Paper No. 284, HIID, Cambridge, Massachusetts, also in Brown, K. and Pearce, D.W., eds., *The Causes of Tropical Deforestation* (London, University College of London Press).

Smith, D.B. and van der Wansem, M., 1995. *Strengthening EIA Capacity in Asia*, A Synthesis Report of Recent Experience with Environmental Impact Assessment in Three Countries: The Philippines, Indonesia and Sri Lanka, prepared for the WRI, 100 pp.

United Nations, 1995. *State of the Environment in Asia and the Pacific*, New York.

Wang, X., 1996. *China's Coal Sector: Moving to a Market Economy*, World Bank, China Country Department, Washington, DC.

World Bank, 1992. *Water Resources Management*, Policy Paper, Washington, DC.

Zhang H. and Richard J.F. Jr., 1998. Shaping an Environmental Protection Regime for the New Century, *Asian Journal of Environmental Management*, May, vol. 6, No. 1, pp. 45.

AMTA's Australian Battery Recycling Program, 1999. http://www.amta.org.au/recycle/backgrd.htm

ASER, 1999. Vol. 5, Issue 5, October, Issue 6, November, 2000. Issue 7, December/January and Issue 9, March.

ASIAWEEK, 2000. Special Report "Chew on It", March 10, pp. 44.

ASIAWEEK, 2000. Special Report "Etched in the Mind", May 10, pp. 46.

ASIAWEEK, 2000. Special Report "I'll Drink to That", May 10, pp. 46.

ASIAWEEK, 2000. Special Report Environment, The Chill Factor, May 10, pp. 49.

Baalu, T.R., 1999. *Central Pollution Control in India*, Central Pollution Control Boards, Ministry of Environment and Forest.

Business and the Environment, 1998, November.

Commonwealth of Australia, 1994. *Australia's National Report 1994* (Canberra, AGPS Press).

Dias, A.K. and Begg, M., 1994, Environmental policy for sustainable development of natural resources: mechanisms for implementation and enforcement, *Natural Resources Forum*, (Cambridge: MIT Press), vol. 18, No. 4, pp. 275-286.

Eco Services International, *The global Directory for Environmental Technologies*, 2000. www.eco-web.com

ENFO News, 1997, Environmental management in China: current practices, December, Beijing.

ENFO, vol. 18, No. 4, December 1996, a quarterly newsletter of environment systems information center (ENSIC), Asian Institute of Technology (AIT), Bangkok.

Environment Agency of Japan, 1996, *Quality of the Environment in Japan*, Partnerships for a Rich Environment for the Future, Tokyo.

ESCAP, 1999. *Interrelationship Between Trade and Environment in Asia and the Pacific*, (ST/ESCAP/2025), United Nations, New York, pp. 143.

Freeman, H., 1998. *Waste Minimization* (New York, McGraw Hill Book Co., Ltd.).

Government of India, 1992. *Environment and Development: Traditions, Concerns and Efforts in India*, National Report to UNCED, Ministry of Environment and Forests, New Delhi.

Government of Indonesia, 1995. *Indonesian Country Report on Implementation of Agenda 21*, The State Ministry of Environment, Jakarta.

Government of Japan, 1997. *Quality of the Environment in Japan 1997: New Approaches and Responsibilities to Arrest Global Warming*, Tokyo.

GRI, http://www.globalreporting.org/

Holme, R. and Watts, P., 2000. *Corporate Social Responsibility: Making Good Business Sense*, World Business Council for Sustainable Development, pp. 9-13.

International Institute for Sustainable Development, 1996. *A Trade and Sustainable Development Perspective*, October.

International Organization for Standardization, 1998. *The ISO Survey of ISO 14000 Certificates*, Eighth cycle.

Khew, E., 2000, Personal Communication of SAFECO.

Levine, S.C., 1994. *Private Sector Participation in Municipal Solid Waste Services in Developing Countries*, vol. I: The Formal Sector, Urban Management Programme Discussion Paper 13, UNDP/UNCHS/World Bank.

Media Release by Senator Campbell's office, 1999. *Mobile Industry Recycling Scheme*, Canberra, November 24.

Park, J., 1997. *Corporate Environmental Management Strategies in the Asian and Pacific Region*, A report prepared for Sustainable Assessment Management.

Peglau, R., 2000. Federal Environment Agency of Germany, Berlin.

Philippine Council for Sustainable Development, 1997. *Philippine Agenda 21: A National Agenda for Sustainable Development*, Manila.

Recycle 2000 user-friendly, *www.adal.com/adal/events/recycle/stats.html*

Schmidheiny, S., 1992. Changing course: a global business perspective on development and the environment, in *Technology Cooperation*, Ch. 8.

Serageldin, I., et. al., 1995. The business of sustainable cities, World Bank, *ESD Series*, No. 7, Washington, DC.

TEI, 1996. *Towards Environmental Sustainability-Annual Report 1996*, Thailand Environment Institute, Bangkok.

TEI's, http://www.tei.or.th/bep/Bep.cfm

The Sunday Telegraph Report, 2000. May 7, pp. 55.

United Nations, 1999. *Strategic Environmental Planning*, ENRD, ESCAP, No. 1, pp. 43, Bangkok.

Wastewise Construction Programme, Australia 2000. http://www.environment.gov.au/epg/wastewise

World Bank, 1997. *Infrastructure in East Asia: How to Increase Private Investment*, World Bank Press release, September 20, http://www.worldbank.org/html/extdr/extme/ampr_003.htm

World Bank, 2000. *PPI Project Database*.

World Bank, 2000. *World Development Indicators 2000*, PPT-slides on Private Sector Participation.

AFEJ-UNEP, 1996. *Reporting on Tourism and Environment: A Backgrounder*, Asia-Pacific Forum of Environmental Journalists and United Nations Environment Programme, Bangkok.

CEE, 1997. *CEE Annual Report 1997-8*, Centre for Environmental Education, Ahmedabad.

EEPSEA, 1998. *EEPSEA 1993-98*, Singapore, Environmental Economics Programme for Southeast Asia, International Development Research Center.

ESCAP, 1997. Report of the regional workshop on promotion of environmental awareness, June 30 - July 2, 1997, Bangkok, unpublished.

Khan, M.A., 1997. NACOM and SEHD, Paper, *Consultation on the Use of Multimedia Technologies for Poverty Eradication*, UNDIP/SAPNA/AMIC, Lake Badhkal, Haryana.

Laird, J., 1999. *The Role of Media in Promoting Sustainable Consumption, Cheju Island, Korea*, paper presented to International Expert Meeting on Sustainable Consumption Patterns.

Librero, F., 1998. Environmental communication in the Philippines: what the research says," *Journal of Development Communication*, Asian Institute for the Development of Communication, Kuala Lumpur, vol. 9, No. 1, pp. 46-53.

Lubis, A.R., (Comp.), 1998. *Water Watch: A Community Action Guide*, Asia-Pacific People's Environmental Network, Penang.

Oh, C., 1998. *Role of NGOs in Regional and Sub-Regional Sustainable Development*, paper presented to Regional Consultative Meeting on Sustainable Development in Asia and the Pacific, Manila.

People & the Planet, 1998. Biodiversity Hotspots Revealed, Planet 21, London, vol. 7, No. 4, pp. 10-13.

Ramanathan, S., ed., 1997. *Water: Asia's Environmental Imperative*, Asia Media Information and Communication Centre, Singapore.

Saeed, S., Goldstein, W. and Shrestha, R., eds., 1998. *Planning Environmental Communication and Education: Lessons From Asia* (Cambridge, IUCN Publication Services Unit).

Satterthwaite, D., et. al., 1996. *The Environment for Children* (London, Earthscan).

SEC, 1998. *Green Groups Directory* (3rd Ed.), Singapore Environment Council, Singapore.

Shameem, S., (Comp.), 1998. *Ideas for Action: Making Urban Areas Child-Friendly*, Malaysian Council for Child Welfare, Kuala Lumpur.

Sofjan, S.H. and Arokiasamy, E.R., (Comp.), n.d. *Our Cities, Our Homes*, Penang.

Southbound Sdn. Bhd. and Asia-Pacific 2000, UNDP.

SUSTRANS, 1996. *Streets for People*, Sustainable Transport Network for Southeast Asia, No. 3, Penang.

UNEP & AFEJ, 1995. *Media and Environment Handbook*, UNEP, Bangkok.

United Nations, 1995. *State of the Environment in Asia and the Pacific*, New York.

United Nations, 1998. *Case Studies-Major Groups in Sustainable Development Education*, Commission on Sustainable Development, Sixth Session, April 20 - May 1, New York.

United Nations, 1999. *Case Studies: The Role and Contribution of Major Groups to Promoting Sustainable Consumption and Production Patterns*, DESA, New York.

United World College of Southeast Asia, 1997. *Global Concerns*, Singapore.

Vincentian Missionaries, 1998. The Patayas Environmental Development Programme: Micro-enterprise promotion and involvement in solid waste management, in Quezon City, *Environment and Urbanization*, International Institute for Environment and Development, vol. 10, No. 2, London.

Wickramaratne, D., ed., 1998. *Environment Is Their Vision and Mission: Sri Lanka's Best Environmental Success Stories*, Nugegoda, Sri Lanka Environmental Journalists Forum, Sri Lanka.

Zhao, X. and Halbertsma, T., 1997. Taoist sacred mountains, *News From ARC*, Autumn, Alliance of Religions and Conservation, Manchester.

Websites

1. AMIC, 1999. http://www.ecanetnet
2. Asian Ecotechnology Network, 1999. http://www.mssrf.org.sg/aeis/aen.html
3. CAP, 1999. http://www.capside.org.sg/

4. CEE, 1997. *CEE Annual Report 1997-98*, Ahmedabad, India

5. Green Map, 1999. http://www.greenmap.com/grmaps/grindex.html

6. Greenpeace, 1999. greenpeace.china@dialb.greenpeace.org

7. TWN, 1999. http://www.twnside.org.sg/

8. United Nations, 1999. http://iisd1.iisd.ca/50comm/commdb/list/co7.htm

9. United Nations, 1999. http://iisd1.iisd.ca/50comm/commdb/list/co8.htm

10. United Nations, 1999. http://www.un.org/esa/earthsummit/abhiyon.htm

11. United Nations, 1999. http://www.un.org/esa/earthsummit/paki.htm

12. United Nations, 1999. http://www.un.org/esa/sustdev/success/cb5.htm

13. United Nations, 1999. http://www.un.org/esa/sustdev/success/os2.htm

14. United Nations, 1999. http://www.un.org/esa/sustdev/success/vietnam.htm

15. United Nations, 1999. http://www.un.org/esa/sustdev/success/watenfed.htm

16. United Nations, 1999. http://www.un.org/esa/sustdev/wedo.htm

17. United Nations, 1999. http://www.un.org/era/sustdev/success/os4.htm

18. United Nations, 1999. http://www.un.org/esa/sustdev/success/environm.htm

19. United Nations, 1999. http://www.un.org/esa/sustdev/success/gis_sp.htm

20. United Nations, 1999. http://www.un.org/esa/sustdev/success/pendeba.htm

21. WWF Malaysia, 1999. http://www.geocities.com/RainForest/2701/

CHAPTER FIFTEEN

ADB, 1992. *Environment and Development: A Pacific Island Perspective*, Manila.

Agenda 21 – New Zealand, http://www.un.org/esa/agenda21/natlinfo/countr/newzea/social.htm

Agenda 21 – Thailand, http://www.un.org/esa/agenda21/natlinfo/countr/thai/social.htm

AMIC, 1997. *Asian Communications Handbook 1997*, Singapore.

AMIC, 1997. *Asian Communications Handbook 1997*. Singapore.

AMIC, 1999. Web site of ECANET, http://www.ecanet.net/

ARC, 1995. *Faith and Nature: The Ohio Declaration on Religions, Land and Conservation*, adopted on May 3, 1995, UK.

Baitullin, I. and Bekturova, G., 1997. National strategy to combat desertification in the Republic of Kazakhstan, *Desertification Control Bulletin*, No. 30, 1997.

Bishop, P., et. al., 1997. Environmental issues and environmental education in the Mekong Region, in *Environmental Education in Viet Nam*, UNESCO-PROAP, Bangkok, pp. 41-48.

Calimag, P., 1996. The Dalaw Turo strategy and action plan, *The SASEANEE Circular*, vol. 4, No. 2, October, pp. 14-15.

CEE, 1994. *The SASEANEE Circular*, vol. 2, No. 1, January 1994, Ahmedabad.

CEE, 1994. *The SASEANEE Circular*, vol. 2, No. 1, January, Ahmedabad.

CEE, 1996. *The SASEANEE Circular*, vol. 4, No. 2, October, Ahmedabad.

CEE, 1997. *CEE Annual Report 1996-97*, Ahmedabad.

CSD, 1997 and 1998. *Fifth & Seventh Sessions Of The United Nations Commissions on Sustainable Development*, April 1997 and 1998.

CSD, 1999. *Implementation of The International Work Programme on Education, Public Awareness And Training*, Commission on Sustainable Development, Seven Sessions, April, E/CN-17/1999/11.

DA, 1998. *Development Alternatives Newsletter*, February issue (special issue on Delhi Environmental Action Network), Delhi.

Deo, S. and Wendt, N., 1999. Paper presented at the NETTLAP Regional Conference in Phuket, Thailand, September 30 - October 2, SPREP Secretariat, Apia.

Department of Environment, Government of Malaysia, 1996. *Malaysia Environmental Quality Report 1996*. Ministry of Science, Technology and Environment, Kuala Lumpur.

Fien, J. and Tilbury, D., 1996. *Learning for a Sustainable Environment: An Agenda for Teacher Education in Asia and the Pacific*, by UNESCO-PROAP (UNESCO Principal Regional Office for Asia and the Pacific), Bangkok.

Fien, J., 1999a. Promoting education for sustainable future: approaches to regional cooperation in Asia and the Pacific, in *IGES, International Conference on Environmental Education in the Asia-Pacific Region*, February 27-28, Yokohama.

Fien, J., 1999b. Environmental education in Australia with special reference to postgraduate education at Griffith, in *IGES, International Conference on Environmental Education in the Asia-Pacific Region*, February 27-28, Yokohama.

Government of Pakistan, 1996. *Environment in Pakistan: Challenges and Achievements*, published by the Ministry of Environment, Urban Affairs, Forestry and Wildlife, Pakistan.

Government of Thailand, 1997 and 1998. *Thailand's Report to the Commission on Sustainable Development Fifth and Sixth Sessions*, New York.

IGES, 1999. *International Conference on Environmental Education in the Asia-Pacific Region*.

Infoterra: 1st Asia-Pacific Roundtable on Clean Production http://www.pan.cedar.univie.ac.at/archives/infoterra/msgo1155.html

ITO, 1999. Environment agency's policy for the promotion of environmental education, in *GES, International Conference on Environmental Education in the Asia-Pacific Region*, February 27-28, Yokohama.

IUCN and IUCN-CEC, 1998. *Planning Environmental Communication and Education: Lessons from Asia*, Saeed, S., Goldstein, W. and Shrestha, R., eds. (Bangkok, IUCN Regional Coordinating Office).

IUCN-CEC, 1997. *Public Education and Awareness Case Studies series*, Community involvement in Marine and Coastal Management: Australia's Marine and Coastal Community Network, Gland.

Junior Eco Club, http://www.wnn.or.jp/wnn-jec.html

Kartikeya, V.S., Raghunathan, M. and Ahnihotri, J., eds., 1995. *Environmental Education in Asia: Regional Report for the UNESCO Inter-regional Workshop on Reorienting Environmental Education for Sustainable Development,* Centre for Environment Education, Ahmedabad.

King Mahendra Trust for Nature Conservation (KMNCT), 1996. *APCA: A New Approach in Protected Area Management,* Kathmandu.

LAND Cover in Australia, http://www.brs.gov.au/land&water/landcov/lc_what.html

Mihikatha Institute, 1997. *Annual Report 1995-97,* Dehiwala.

Ministry of Environment, Republic of Korea, 1995. *Environmental Protection in Korea 1995,* Seoul.

Natural Resource Aspects of Sustainable Development In Kazakhstan, http://www.un.org/esa/agenda21/natlinfo/countr/kazakh/natur.htm

NEFEJ, 1998. *Annual Report of the Nepal Forum of Environmental Journalists,* Kathmandu.

Networking Needs and Problems, http://www.grida.no/prog/cee/enrin/htmls/rusia/arf_p4.htm

Palmer, J.A., 1998. *Environmental Education in the 21st Century: Theory, Practice, Progress and Promise,* pp. 27 and 169-170.

Palmer, J., Goldstein, W. and Curnow, A., eds., 1995. *Planning Education to Care for the Earth,* IUCN Commission on Education and Communication, IUCN-CEC, Gland.

Panos Institute South Asia, 1998. Director's Report of activities carried out in 1997-98 Kathmandu, unpublished.

Panos, 1998. *The Internet and Poverty: Real Help or Real Hype?,* Panos Media Briefing No. 28, April, The Panos Institute, London.

Ravuvu, A., 1998. *Environmental Education in the Pacific: Challenges for Millennium 2000,* paper presented at the CEE/ASPBAE Environmental Education Workshop in Ahmedabad, India, January.

Royal Government of Bhutan, 1999. *Bhutan: The Path Towards Sustainable Development,* An official brochure, Thimphu.

Sony, 1999. *For the Next Generation: Sony Environmental Report 1999,* Sony Corporation, Tokyo.

SPREP, 2000. Personal Communication, May 4.

Supetran, A. and Dulce, D., 1999. *Environmental Education-Experiences and Plans of the Philippines,* paper presented at NETTLAP Regional Conference in Phuket, Thailand, September 30 - October 2.

TVE, 1999. *Harnessing Television's Power to Communicate for Sustainable Development,* Booklet produced for private circulation by Television Trust for the Environment, Asia Pacific Regional Office, Colombo.

UNEP, 1995. *The Global 500: Roll of Honour for Environmental Achievement 1995,* Nairobi.

UNEP, 1997. *The Global 500: Roll of Honour for Environmental Achievement 1997,* Nairobi.

UNEP-ROAP, 1999. http://www.unep.org

UNESCO, 1997a. *Declaration of Thessaloniksi,* prepared at the International Conference on Environment and Society: Education and Public Awareness for Sustainability, held in Thessaloniski, Greece, December 8-12, UNESCO-EPD-97/CONF.401/CLD2).

UNESCO, 1997b. *World Communication Report: The Media and the Challenge of the New Technologies* (Paris, UNESCO Publishing).

United Nations, 1993. *Agenda 21,* New York.

US-AEP, 1998. *United States-Asia Environmental Partnership: Year in Review 1998,* USAID (United States Agency for International Development) Washington, DC.

Wijayadasa, K.H.J., ed., 1997. *Harmonizing Environment and Development in South Asia,* SACEP, Colombo.

ZERI, 1999. Public Information Brochure, United Nations University, Tokyo, http://www.zeri.org

ADB, 1994. *Asian Development Outlook 1994*, Oxford University Press.

ADB, 1997. *Emerging Asia Changes and Challenges*, Manila.

Ali, A., 1996. Vulnerability of Bangladesh to climatic change and sea level rise through tropical cyclones and storm surges, *Water, Air, and Soil Pollution*, vol. 92, No. 1-20, pp. 171-179.

Ansari, J.H., 1996. *Improving Urban Land Management in India*, Journal of the Institute of Town Planners, India, vol. 14, No. 3/4, pp. 34-50.

Balakrishnan, L., 1996. Energy conservation and management: the role of women, *Renewable Energy*, an International Journal, vol. 9, No. 1-4, pp. 1165-1170.

Bangladesh, 1994. *Country Presentation of Bangladesh*, Regional Meeting on the State of the Environment in Asia and the Pacific.

Bhutan, 1994. *Conservation in Bhutan 1994, Country Presentation of Bhutan*, Regional Meeting on the State of the Environment in Asia and the Pacific.

Biswas, D., 1996. Environmental legislation and enforcement mechanism, *Asia Pacific Monitor*, vol. 13, No. 3, pp. 16-20.

Biswas, D.K., Pandey, M. and Sharma, D.C., 1996. Emerging cleaner technologies in India, *Invention Intelligence*, vol. 31, No. 6, pp. 253-255.

Chattopadhyaya, D. and Majumadar, S., 1996. Impacts of oil price shock on carbon emissions in India: an econometric analysis, *Energy Sources*, vol. 18, No. 6, pp. 711-726.

Devgupta, S.R., Koirala, B.P. and Gautum, C., 1996. Industry pollution inventory and management in Nepal, *UNEP Industry and Environment*, vol. 19, No. 1, pp. 37-42.

Dhittal, K., 1996. Resource conservation and economic development: an example of Royal Chitwan National Park, *The Economic Journal of Nepal*, 16:4:64 , pp. 179-208.

Dreze and Sen, 1995. *India: Economic Development and Social Opportunity* (Delhi, Oxford University Press).

Dutt, A.K. and Rao, J.M., 1996. Growth, distribution and environment: sustainable development in India, *World Development*, vol. 24, No. 2, pp. 287-305.

Erda, L., 1996. Agricultural vulnerability and adaptation to global warming in India, *Water, Air and Soil Pollution*, vol. 92, No. 1-2, pp. 63-73.

ESCAP, 2000, *Population Data Sheet*, Bangkok.

FAO, 1999. *State of the World's Forests*.

FAO, 1999. *Statistical Databases: Agriculture*, http://www.fao.org

Government of United States, 1999. *Guide to Country Profiles*, The World Factbook.

India, 1995. Ministry of Environment and Forests, Government of India, Tata Energy Research Institute, July, India.

Islam, M.N. and Jolley, A., 1996. Sustainable development in Asia: the current state and policy options, *Natural Resources Forum*, vol. 20, No. 4, pp. 263-279.

Jayadevarappa, R. and Chhatee, S., 1996. Carbon emission tax and its impact on a developing country economy – a case study of India, *The Journal of Energy and Development*, vol. 20, No. 2, pp. 229-247.

Jha, P.K., 1995. Pollution preventing efforts and strategies for the Kathmandu Valley, *Water, Air and Soil Pollution*, vol. 85, No. 4, pp. 2643-2648.

Karim, Z., Hussain, S.G. and Ahmad, M., 1996. Assessing impacts of climatic variation on foodgrain production in Bangladesh, *Water, air and Soil Pollution*, vol. 92, No. 1-2, pp. 53-62.

Khadka, R.B. and Tuladhar, B., 1996. Developing an environmental impact assessment system in Nepal – a model that ensures the involvement of people, *Environmental Impact Assessment*, vol. 14, No. 4, pp. 435-448.

Khan, N.A., 1996. Revisiting community forestry developments in Nepal: a selected review of performance, *Asian Journal of Environmental Management*, vol. 4, No. 2, pp. 95-101.

Khanna, P., 1996. Policy options for environmentally sound technologies in India, *Water Science Technology*, vol. 33, No. 3, pp. 131-134.

Madhu B.D., 1997. Public transport and urban environment: the challenges, *Indian Journal of Transport Management*, vol. 21, No. 2, pp. 89-96.

Madhusri, K. and Sharma, K.R., 1996. Urban development and its effect on environment – a case study (India), in *National Seminar on Environmental Conservation and Management (ECOMAN)*, March 7-8, Vishakakhapatnam.

Maldives, 1994. *State of the Environment*, Ministry of Planning, Human Resources and the Environment, Male.

Matthew, T. and Unnikrishnan, S., 1997. A comparative study of hazardous waste legislation and management, *Encology*, vol. 11, No. 9, pp. 13-21.

Morgan, M.G. and Dowlatabadi, H., 1996. Recent climatic change, greenhouse gas emissions and future climate: implications for India, *Climatic Change*, vol. 34, No. 3-4, pp. 337-368.

Nepal, 1994. *National State of the Environment*, prepared for submission to SACEP, Colombo.

Norman, O., 1993. The environment and economic policy trends of Pakistan, *Journal of International Development*, 5:2, pp. 225-235.

Pakistan, 1994. *Country Report on State of Environment in Pakistan*, Regional Meeting on the State of the Environment in Asia and Pacific, Myanmar.

Parikh, J.K. and Sharma, V.K., 1996. Economic and policy analysis of trade and environment linkages in India, *Journal of Indian Association for Environmental Management*, vol. 23, No. 2, pp. 71-78.

Pohit, S., 1997. The impact of climate change on India's agriculture: some preliminary observations, *Proceedings of the 20th Annual Conference of the International Association for Energy Economics: Energy and Economic Growth: Is sustainable Growth possible?*

Prakash, S., 1997. Poverty and environment linkages in mountains and uplands: reflections on the poverty trap thesis, in *Collaborative Research in the Economics of Environment and Development (CREED)*, Working Paper series No. 12/IIED 12.

Ramanathan, R., 1996. Indian Transport Sector: energy and environmental implications, *Energy Sources*, vol. 18, No. 7, pp. 791-805.

Rao, J., 1995. Economic reform and ecological Refurbishment: a strategy for India, *Economic and Political Weekly*, July 15, pp. 1749-1761.

Sahai, R. and Bkade, U.K., 1996. An update on environmental legislation, *The Indian Mining and Engineering Journal*, vol. 35, No. 7, pp. 31-34.

Sharma, C.K., 1996. Overview of Nepal's energy sources and environment, *Atmospheric Environment*, vol. 30, No. 15, pp. 2717-2720.

Shreshtha, R.M. and Malla, S., 1996. Air pollution from energy use in a developing country city: the case of Kathmanu valley, Nepal, *Energy, the International Journal*, vol. 21, No. 9, pp. 785-794.

Singh, R.B. and Pandey, B.W., 1996. *Environmental Monitoring of Particulate Mater and Pollution in Major Rivers of India*, Archives of Hydrobiology Special issues on Advances in Limnology, vol. 47, pp. 557-561.

Singh, R.K. and Anand, H., 1996. Water quality index of some Indian Rivers, *Indian Journal of Environmental Health*.

Sri Lanka, 1994. State of the Environment in Sri Lanka, Ministry of Environment and Parliamentary Affairs, Submitted to the SACEP, Colombo.

Srinivas, D.K., 1996, *Energy Environment Linkages of Urban Transport Sector in India*.

Tariq, A.S. and Purvis, M.R.I., 1997. Industrial use of biomass energy in Sri Lanka, *International Journal of Energy Research*, vol. 21, No. 5, pp. 447-464.

UNDP, 1996. *Human Development Report 1996*, Oxford University Press.

UNDP, 1998. *Human Development Report*, Oxford University Press.

UNEP, 1997. *Global Environment Outlook-I*, New York.

UNEP, 2000. *Global Programme of Action Regional Technical Workshops: South Asian Seas*, http://www.gpa.unep.org/seas/workshop/workshop.htm#saseas

UNEP, 2000. *Regional Seas Programme*, http://www.gsf.de/UNEP/oca.html

UNFPA, 1996. *The State of World Population 1996, Changing Places: Population, Development and the Urban Future*, United Nations Population Fund.

United Nations, 1990. *State of the Environment in Asia and the Pacific 1990*, Bangkok.

United Nations, 1998. *World Urbanization Prospects (1996 revision)*, DESA, New York.

United Nations, 1999. *World Population Prospects (1998 revision)*, DESA, New York.

Watal, J., 1996. The relationship between the TRIPS agreement and the environment, *The Asia Pacific Technology Monitor*, vol. 13, No. 5, pp. 22-25.

WCMC, 1999. Global overview of forest conservation (for protected area), in *Forest Information Services*.

Wickramsinghe, 1997. Women and minority groups in environmental management, *Sustainable Development*, vol. 5, No. 1, pp. 11-20.

Wijaratne, M.A., 1996. Vulnerability of Sri Lanka tea production to global climate change, *Water, Air and Soil Pollution*, vol. 92, No. 1-2, pp. 82-97.

World Bank, 1998. *Industrial Pollution in Economic Development*.

World Bank, 2000. *Meeting India's Future Power Needs-Planning for Sustainable Development*.

WRI, 1999. *World Resources 1997-98*, Washington, DC.

ADB, 1995. Coastal and marine environmental management, *Proceedings of a Workshop*, Manila.

ADB, 1999. *Asian Development Outlook 1999* (New York, Oxford University Press).

ADB, http://www.adb.org

Akella, A.S., 1999. *The East Asian Financial Crisis: Evolution and Environmental Implications for Indonesia*, WWF Macro-economics Program Office, Washington, DC.

ASEAN, http://www.aseansec.org

Asia Pacific Economic Cooperation, 1997. Marine resources conservation working group, *Proceedings of the Workshop on the Impacts of Destructive Fishing Practices on the Marine Environment*, December, Hong Kong.

Asiaweek, 1999. May 14, vol. 25, No. 19, Hong Kong.

BAPEDAL, http://www.bapedal.go.id

Brookfield, H. and Byron, Y., eds., 1993. *South-East Asia's Environmental Future: The Search for Sustainability* (Kuala Lumpur/Singapore, United Nations University Press and Oxford University Press).

Brown, L.R., Renner, M. and Halwell, B., 1999a. *Vital Signs 1999. The Environmental Trends that are Shaping Our Future* (New York and London, W.W. Norton & Co.).

Brown, L., Flavin, C. and French, H., 1999b. *State of the World: A Worldwatch Institute Report on Progress Toward a Sustainable Society* (New York and London, W.W. Norton & Co.).

Brunner, J., Talbott, K. and Elkin, C., 1998. *Logging Burma's Frontier Forests*, WRI, Washington, DC.

Bryant, D., Burke, L., Mcmanus, J. and Spalding, M., 1998. *Reefs at Risk, A Map-Based Indicator of Threats to the World's Coral Reefs*, WRI, Washington, DC.

Eng, C.T., Ross, A. and Yu, H., 1997. *GEF/UNDP/IMO Regional Seas Programme for the Prevention and Management of Pollution in the East Asian Seas*, Quezon City, Philippines.

DFA, 1998. *Winning the Challenges of the New Millennium*, a commemorative book on the occasion of the 31st ASEAN Ministerial Meeting and Related Meetings, Philippines, Office of ASEAN affairs, July 20-29.

U.S. Global Change Research Program Seminar Series, 1998, *Development of Asian Mega-Cities: Environmental, Economic, Social, and Health Implications*, June 11.

FAO, 1995. *Review of the State of World Fishery Resources: Aquaculture*, FAO Fish, Cir. 886, pp. 1-127.

FAO, 1997. *State of the World's Forests 1997*, http://www.fao.org

FAO, 1999. FAO Statistical Database 1990-1998, http://www.fao.org

FAO, 1999. http://www.fao.org

Garcia, M.O., 1997. *Ecologia Filipina, The Almanac*, Environmental Center of the Philippines Foundation, Quezon City, Philippines.

GCTE Impacts Centres Overall Coordinator, http://www.bogor.indo.net.id/IC-SEA

Glover, D. and Jessup, T., 1999. *Indonesia's Fire and Haze The Cost of Catastrophe*, Institute of South East Asian Studies and the International Development Research Centre, Singapore.

Goodland, R. and Daly, H., 1996. Environmental sustainability: universal and non-negotiable, *Ecological Applications*, 6, pp. 1002-1017.

Goreau, T.J. and Hayes, R.L., 1995. Monitoring and calibrating sea surface temperature anomalies using satellite and in-situ data to study the effects of weather extremes and climate changes on coral reefs, *Proceedings Of The Conference For Remote Sensing And Environmental Monitoring For The Sustainable Development Of The Americas*, March 21-22, San Juan, Puerto Rico, http://www.fas.harvard.edu/~goreau/frame.index.html

Great Barrier Reef Marine Park Authority, the World Bank and the World Conservation Union (IUCN), 1995. *A Global Representative System of Marine Protected Areas*, vol. IV, IBRD and World Bank, Washington, DC.

Dixon, H. and Thomas F., 1999. *Environment, Scarcity and Violence* (New Jersey, Princeton University Press).

Indonesian Forest Fire Management, WWF-Indonesia, http://www.iffm.or.id

Josupeit, H., 1984. A survey of external assistance to the fisheries sector in developing countries, *FAO Fish*, Cir. 755, Rev. 1, pp. 1-54.

Primavera, J.H., 1994. Shrimp farming in Asia and the Pacific: environment and trade issues and regional cooperation, in *Trade and Environment: Prospects for Regional Cooperation*, Nautilus Institute, Berkeley, California, pp. 161-186.

Rosenberry, B., 1996. World shrimp farming 1996, *Shrimp News International*, San Diego, California.

Schweithelm, J., 1999. *The Fire This Time: An Overview of Indonesia's Forest Fire in 1997/1998*. WWF-Indonesia, Jakarta.

Sharma, N.P., ed., 1992. *Managing the World's Forests. Looking for Balance Between Conservation and Development*, World Bank, Washington, DC.

Siegert, F. and Hoffmann, A.A., 1998, *Evaluation of the Forest fires 1998 in East-Kalimantan (Indonesia) using multitemporal ERS-2 SAR images and NOAA-AVHRR data*, 1998, International Conference on Data Management and Modelling Using Remote Sensing and GIS for Tropical Forest Land Inventory, October, Jakarta, pp. 26-29.

Silvestre, G. and Pauly, D., eds., 1997. *Status and Management of Tropical Coastal Fisheries in Asia*, ICLARM and ADB, Makati City and Mandaluyong City.

UNEP, 1997. *Global Environment Outlook I, Global State of the Environment Report 1997*, http://grid2.cr.usgs.gov/geo1

United Nation, 1995. *State of the Environment in Asia and the Pacific*, New York.

Vayda, A.O., 1999. *Finding Causes of the 1997-98 Indonesian Forest Fires: Problems and Possibilities*, WWF-Indonesia, Jakarta.

Wilson, E., 1992. *The Diversity of Life* (Cambridge, Mass., The Belknap Press of Harvard University Press).

World Bank, http://www.worldbank.org

World Bank, 1998. *Environment Matters*, Fall, Washington, DC.

WRI, 1994. *World Resources 1994-1995, People and the Environment* (New York and Oxford, Oxford University Press).

WRI, 1999. *World Resources 1998-1999*, The World Bank and UNEP, http://www.wri.org

Cartwright, I., 1996. *The South Pacific Forum Fisheries Agency: Past, Present and Possible Future Roles in Fisheries Management in the Central and Western Pacific*, FFA, Honiara.

Chesher, R., 1993. Giant Clam Sanctuaries in the Kingdom of Tonga, *Marine Studies*, Series Number 95/2, University of the South Pacific.

Craig, P., Green, A. and Saucerman, S., 1994. *Coral Reef Troubles in American Samoa*, Department of Marine and Wildlife Resources, American Samoa.

CSIRO, 1996. *State of the Environment of Australia*, CSIRO, Canberra.

Gabrie, C., Licari, M.L. and Mertens, D., 1995. *L'etat de l'environment dans les Territoires Francais du Pacifique Sud*, Preliminary Report L'Institute Francais de L' Environment.

IDEC, 1990. *Environmental Management Plan for the Kingdom of Tonga*, ESCAP, Bangkok.

Johnston, R., 1999. Water Crisis, *Island Business*, February.

Kaluwin, C., 1996. *Case Studies: Climate Change and Seal Level Rise Background*, Regional Heads of Planning Meeting on Sustainable Development, Apia, Western Samoa, June 12-16, SPREP, Apia.

Lavea, I., 1996. *Mainstreaming Environment*, Regional Heads of Planning Meeting on Sustainable Development, Apia, Western Samoa, June 12-16, SPREP, Apia.

Liss, P.S. and Duce, R.A., eds., 1997. *The Sea Surface and Global Change*, Cambridge University Press.

Paeniu, B., 1996. *Sustainable Development and Planning in the Pacific*, Regional Heads of Planning Meeting on Sustainable Development, Apia, Western Samoa, June 12-16, SPREP, Apia.

Postawko, S., 1997. *Schools of the Pacific Rainfall Climate Experiment (SPaRCE)*, 3rd SPRE, Meeting on Climate Change and Sea Level Rise in the Pacific, August 18-22, Noumea.

Reti, I., 1998. *South Pacific Biodiversity Conservation Programme*, Report of the Project Manager, SPREP.

Robbins, J., 1999. Coral reefs the focus at the International Tropical Marine Ecosystem Management Symposium in Townsville Australia, ITEM: 38, *IOC news*, January 14.

Rosenberg, E., 1998. *Concept Paper Proposing the Framework for a Pacific Island Regional Conservation Trust Fund*, ESCAP Pacific Operations Centre, Vanuatu.

Sloth, B., 1988. *Nature Legislation and Nature Conservation as Part of Tourism Development in the Island Pacific*. Pacific Regional Tourism Development Programme, Tourism Council of the South Pacific, Suva, 82 pp.

SPREP, 1997. *The SPREP Action Plan 1997-2000*, SPREP, Apia.

SPREP, 1998. *SPREP Annual Report: 1996/1997*, South Pacific Regional Environment Programme, SPREP, Apia, 72 pp.

Thaman, R., 1976. *The Tongan Agricultural System*, University of the South Pacific Press, 433 pp.

Watts, M., 1994. *The Poisoning of New Zealand*, Auckland Institute Press, 223 pp.

Wilkinson, C., 1998. *Status of Coral Reefs of the World: 1998*, Australian Institute of Marine Science.

WRI, *WRI Database*, http://www.worldbank.org

Carmichael, G. and Arndt, R., 1995, Long Range Transport and Deposition of Sulfur in Asia, RAINS-ASIA, March, unpublished.

Carmichael, G. and Arndt, R., 1997. *Baseline Assessment of Acid Deposition in North East Asia*, ESENA, The Nautilus Institute (from Internet).

Cooper, R.N., 1995. The Coase Theorem and international economic relations, *Japan and the World Economy*, 7.

Dua, A. and Daniel, C.E., 1997. *APEC and Sustainable Development*, mimeo.

ESCAP, 1997. *Framework for the North East Asian Subregional Programme of Environmental Cooperation: Institutional Aspects and the Feasibility of Establishing a Trust Fund*, ENR/SO/ECNA(4)/3, September.

ESCAP, 2000, *Population Data Sheet*, Bangkok.

Haas, P.M., 1998. *Prospects for Effective Marine Governance in the Northwest Pacific Region*, paper presented at ESENA Workshop Energy-Related Marine Issues in the Sea of Japan, July 11-12, Tokyo.

Chung, H. and Yoo, R.S., 1994. *A Study on Korea-China Environmental Cooperation*, Korea Institute for International Economic Policy, September, Korea.

Hayes, P. and Lyuba Z., 1995. Acid rain in a regional context, in *The Role of Science and Technology in Promoting Environmentally Sustainable Development*, Seminar Proceedings, Science and Technology Policy Institute and the United Nations University, June 13-15, Seoul.

Jin, K.H., 1998. *Marine Environmental Cooperation in North-East Asia*, paper presented at ESENA Workshop on Energy-Related Marine Issues in the Sea of Japan, July 11-12, Tokyo.

Don, L.S. and Lee, S.E., 1994. *The Proposals for Concrete Action Program for North East Asian Environmental Cooperation and the Activation of Cooperative Network*, Korea Environmental Science Research Council, July, Korea.

Don, L.S. and Han, T.W., 1997. *North East Asia's Transboundary Pollution Problems: A Pragmatic Approach*, Korea Institute for International Economic Policy, Seoul.

Min, B.S., 1996. *Environmental Issues and Policy Recommendations for the Regional Environmental Cooperation in North-East Asia*, Korea Environmental Technology Institute, December, Korea.

OECD, 1990. *The use of international financial transfers in resolving transfrontier and global pollution problems*, ENV/EC, (90) 25, November.

Park, B.K. and Chung, S.T., 1997. *A Study on the North East Asia's Environmental Problem Relating to China's Air Pollution*, Research Report 97-99, LG Economic Institute, October, Seoul.

Razavi, H., 1997. Innovative Approaches to Financing Environmentally Sustainable Energy Development in North-East Asia," draft, ESENA, The Nautilus Institute (from Internet).

Streets, D.G., 1997. *Energy and Acid Rain Projections for North-East Asia*, ESENA, The Nautilus Institute (from Internet).

Tahvonen, O., et. al., 1993. A Finnish-Soviet Acid Rain Game: noncooperative equilibria, cost efficiency, and sulfur agreements, *Journal of Environmental Economics and Management*, 24.

United Nations, 1998. *World Urbanization Prospects (1996 revision)*, Population Division, DESA, New York.

United Nations, 1999. *World Population Prospects (1998 revision)*, Population Division, DESA, New York.

Valencia, M.J., 1995. North-East Asian marine environmental quality and living resources: transnational issues for sustainable development, Seminar Proceedings on the *Role of Science and Technology in Promoting Environmentally Sustainable Development*, Science and Technology Policy Institute and the United Nations University, June 13-15, Seoul.

World Bank, 1998. *World Development Indicators 1998-99*.

Yamamoto, W., 1994. *Japanese Official Development Assistance and Industrial Environmental Management in Asia*, Workshop on Trade and Environment in Asia-Pacific: Prospects for Regional Cooperation, East-West Center, September 23-25.

Zarsky, L., 1995. APEC and the environment: any balance at all, *The Nation*, November 4, Bangkok.

Akhmadov, Kh.M., 1998. Opustynivaniye i bor'ba s nim v razlichnykh prirodo-lesokhozyaistvennykh zonakh Tajikistana, Paper presented on April 16, at National Seminar on Implementation in the Republic of Tajikistan of the United Nations Convention to Combat Desertification, Dushanbe, unpublished.

Anonymous, 1998. *Tajikistan State of the Environment*, http://www.grida.no/prog/cee/enrin/htmls/tadjik/soe/index.htm

Atamuradov, K., 1998. *Environmental Information Systems in Turkmenistan* (Arendal, UNEP GRID).

Atamuradov, K., 1999. Osobo okhranyaemye prirodnye territorii i buduschie zapovednye zony, map of Turkmenistan Ministry of Nature Use and Environmental Protection, unpublished.

Baratov, M.B., et. al., 1998. Osnovnye napravleniya zemledeliya v Respublike Tajikistan i ikh vliyaniye na strukturu i sostoyaniye zemel' v usloviyakh perekhodnogo perioda, Paper presented on April 16, at National Seminar on Implementation in the Republic of Tajikistan of the United Nations Convention to Combat Desertification, Dushanbe, unpublished.

Berkeliev, T., 1997. Radiation wastes and pollution in Turkmenistan, *Ecostan News*, vol. 5, No. 6, pp. 2-4, http://web.mit.edu/sts/leep//Ecostan/Ecostan506.html

Birnstein, V., 1998. Sturgeon in Central Asia: Is IUCN a successor to the CPSU?, *Ecostan News*, vol. 6, No. 6, pp. 6-11, http://web.mit.edu/sts/leep//Ecostan/Ecostan606.html

Bokonbaev, K.J., et. al., 1999. Barskoon, Mai, 1998. Chto bylo i chego ne bylo, unpublished.

Burkhanova, M.A., ed., 1999. *Materialy respublikanskogo seminara obschestvennykh organizatsii Tajikistana "Sotsial'no-ekonomicheskie problemy opustynivaniya*, Dushanbe, FSCI.

Caspian Environment Programme, 1998. *Proposal approved by GEF Council*, http://www.gefweb.org/wprogram/Oct98/UNDP/caspian.doc

Central Asia Transboundary Biodiversity Project, 1997. *Proposal approved by GEF Council*, http://www.gefweb.org/wprogram/nov97/capcd4.doc

Charsky, V.P. and Tishkova, N.M., 1998. Tsekh No. 7. *Adam zhana Gera* 9, May/June, pp. 13-15.

Chayes, A. and Chayes, A.H., 1994. *The New Sovereignty* (Cambridge, Harvard University Press).

Sergei, E., 1999. Kazakhstan and international greenhouse gas emissions trading, Unpublished talk presented on April 30, at Harvard Institute for International Development.

Murray, F. and Friendly, A. Jr., 1992. *Ecocide in the USSR* (New York, Basic Books).

Food and Agriculture Organization, 1997. Contaminated water devastates health across the Aral Sea region, http://www.fao.org/NEWS/1997/970104-e.htm

Vyacheslav, G., 1999. The Legal Status of the Caspian Sea, paper presented on June 22, at the Seventh Annual Conference on Central Asia and the Causasus, Tehran, unpublished.

Glantz, M.H., 1999. Global environmental problems in the Caspian region, Presented on March 15, at NATO Advanced Research Workshop on the Caspian Sea, Venice, unpublished.

Gribnev, I., ed., 1988. *Narodnoe khozyaistvo Tadzhikskoi SSR v 1987 gody* (Dushanbe, Irfon).

Marjukka, H., 1998. *Environmental Development Co-operation Opportunities: Kazakhstan, Kyrgyz Republic, Turkmenistan*, Finnish Environment Institute, Helsinki.

International Fund for the Aral Sea, 1997. *Water and environmental management in the Aral Sea basin* (GEF Project Proposal), http://www.gefweb.org/wprogram/JULY97ARALSEA.DOC

Jackson, R.H., 1993. *Quasi-States* (Cambridge, Cambridge University Press).

Kamakhina, G., 1997. An alternative approach to saving Turkmenistan's biodiversity, *Ecostan News*, vol. 5, No. 9, pp. 2-4, http://web.mit.edu/sts/leep//Ecostan/Ecostan509.html

Kazakh SSR, 1980. *Narodnoe xozhyaistvo Kazakstana za 60 let*, Alma-Ata.

Kazakh SSR, 1990. *Narodnoe xozhyaistvo Kazakstana za 70 let*, Alma-Ata.

Kazakhstan Academy of Sciences, 1993. Obraschenie k glavam gosudarstv i pravitelstv respublik Srednej Azii, Kazakstana i Azerbaidzhana,

Vestnik Akademii nauk Respubliki Kazakstana, 1, pp. 8-9.

Kazakhstan Ministry of Ecology and Natural Resources, 1998. Tsentral'no-Aziatskaya regional'naya konferentsiya ministrov okhrany okruzhayuschei sredy, *Ekologicheskii byulleten*, 2, pp. 90-92.

Kazakhstan Ministry of Natural Resources and Environmental Protection, 1999. Ekologicheskaya obstanovka v Respublike Kazakstan, *Ekologicheskii byulleten'*, 1, pp. 5-44.

Kirgiz SSR, 1990. *Narodnoe khozyaistvo Kirgizskoi SSR v 1982 godu*, Frunze.

Koltsov, A.V., 1988. *Rol' Akademii nauk v organizatsii regionalnykh nauchnykh tsentrov SSSR* (Leningrad, Nauka).

Kostina, T., ed., 1999. *Gorod XXI veka* (Almaty, Greenwomen).

Kotov, V. and Elena N., 1998. Implementation and effectiveness of the acid rain regime in Russia, in David G. Victor et. al., eds., *The Implementation and Effectiveness of International Environmental Commitments* (Cambridge, MIT Press), pp. 519-547.

Kovshar, A. and Zatoka, A., 1991. Printsipy razmeshcheniia i infrastruktura osobo okhraniaemykh prirodnykh territorii v aridnoi zone SSSR, *Problemy osvoeniia pustyn'*, (3-4), pp. 155-160.

Krever, V., et. al., eds., 1998. *Biodiversity Conservation in Central Asia* (Moscow, WWF).

Kudat, A., et. al., 1995. *Needs Assessment for the Proposed Uzbekistan Water Supply, Sanitation and Health Project*, World Bank.

Kurbanov, P., ed., 1999. Sostoyaniye okruzhayuschei sredy v Turkmenistane, Ministry of Nature Use and Environmental Protection, unpublished.

Kustareva, L.A., 1998. Dorogoi moi Issyk-kul, *Adam zhana Gera* 9, May/June, 11.

Makhsudov, D.M., 1998. Oroshaemoe zemledeliye i problemy opustynivaniya zemel' v Tajikistane, Paper presented on April 16, at National Seminar on Implementation in the Republic of Tajikistan of the United Nations Convention to Combat Desertification, Dushanbe, unpublished.

Mansimov, M.R., 1999. Kolebanie urovnya i zatoplenie pribrezhnoi zony Kaspiiskogo morya, Presented on March 15, at NATO Advanced Research Workshop on the Caspian Sea, Venice, unpublished.

Matveenko, I., 1998a. Issyk-kul. *Ecostan News*, vol. 6, No. 6, pp. 2-5, http://web.mit.edu/sts/leep//Ecostan/Ecostan606.html

Matveenko, I., 1998b. Kyrgyzstan NGOs fight paper mill, *Ecostan News*, vol. 6, No. 3, pp. 2-5, http://web.mit.edu/sts/leep//Ecostan/Ecostan603.html

Melnikova, G.K., et. al., 1997. A four state crisis: the Sarezsky Lake. *Ecostan News*, vol. 5, No. 10, pp. 2-3, http://web.mit.edu/sts/leep//Ecostan/Ecostan510.html

Miller, J., 1999. U.S. and Uzbeks agree on chemical arms plant cleanup, *New York Times*, May 25.

Mirkhashimov, I.Kh., et. al., 1997. Osobo okhranyaemye prirodnye territorii Kazakstana, in V.I. Drobzhev et. al., eds., *Novosti nauki Kazakstana* (Almary, Ministry of Science- Academy of Sciences of the Republic of Kazakhstan).

Mirkhashimov, I.K., ed., 1997. *Biologicheskoe i landshaftnoe raznoobrazie Kazakstana*, Ministry of Ecology and Natural Resources of the Republic of Kazakhstan, Almaty.

Mnatsakanian, R.A., 1992. *Environmental Legacy of the Former Soviet Republics* (Edinburgh, University of Edinburgh).

Mukhina, E., 1995. Soaproot Poaching. *Ecostan News*, vol. 3, No. 5, pp. 6-7, http://web.mit.edu/sts/leep//Ecostan/Ecostan305.html

Nifadiev, V.I., et. al., undated. *Sostoyaniye okruzhayuschei sredy Kyrgyzstana*, http://www.grida.no/prog/cee/enrin/htmls/kyrghiz/soe/indexr.htm

Peterson, D.J., 1993. *Troubled Lands* (Boulder, Westview Press).

Rubezhansky, Y., Atbasar. *Ecostan News*, vol. 2, No. 2, pp. 5-6, http://web.mit.edu/sts/leep//Ecostan/Ecostan202.html

Safarov, N., ed., 1988. *Ekologiya i okhrana prirody*, Ministry of Nature Protection of the Republic of Tajikistan, Dushanbe.

Sievers, E., et. al., 1995. National parks, snow leopards, and poppy plantations: the development and degradation of Central Asia's preserved lands, *Central Asia Monitor*, 2, pp. 23-30 and 3, pp. 17-26.

Sievers, E., 1999a. Inviting Stolz and greeting Oblomov: positive liberty and the donor assisted death of NGOs for governance, in W. Ascher and N. Mirovitskaya, eds., *The Caspian Sea: A Quest for Environmental Security*, Kluwer Academic Publishers, Amsterdam.

Sievers, E., 1999b. Caspian Environment Program: prospects for regime formation and effectiveness, in W. Ascher and N. Mirovitskaya, eds., *The Caspian Sea: A Quest for Environmental Security*, Kluwer Academic Publishers, Amsterdam.

Stone, R., 1999. Coming to grips with the Aral Sea's grim legacy, *Science* (284/5411), pp. 30-33.

Thurman, M., 1999. *A photographic tour of irrigation in Uzbekistan*, http://homepages.infoseek.com/~mirablik/atraf-muhit/photo.html

Tsaruk, O., 1996. Big water flows to the Kzylkum or is the Aral dying? *Ecostan News*, vol. 3, No. 4, pp. 2-3, http://web.mit.edu/sts/leep//Ecostan/Ecostan304.html

Turkmen SSR, 1974. *Turkmenistan za 50 let*, Ashgabad.

Turkmen SSR, 1987. *Turkmenistan v tsifrakh 1986*, Ashgabad.

UNDP, 1998. *Making Things Happen- Getting it Done*, UNDP Regional Bureau for Europe and CIS, New York.

Uzbek SSR, 1988. *Narodnoe khozhaistvo Uzbekskoi SSR 1987*, Tashkent.

World Bank, 1998. *Transition toward a healthier environment: environmental issues and challenges in the Newly Independent States*, http://www.esd.worldbank.org/ecssd/aarhus/NISmain.html

World Health Organization, 1999. *The World Health Report 1999*, Geneva, http://www.who.int/whr/1999/en/pdf/whr99.pdf

Yesekin, V.K., et. al., undated. *Sostoyanie okruzhayuschei sredy v Respublike Kazakstan*, http://www.grida.no/prog/cee/enrin/htmls/kazahst/soe/start.htm

Zhukov, Y., ed., 1991. *Narodnoe khozyaistvo Kirgizskoi SSR v 1989 godu* (Frunze, Kyrgyzstan).

Zonn, I.S., 1997. *Kaspiiskii memorandum* (Moscow, Korkis).

Zonn, I.S., 1999. Ecological consequences from oil and gas development in the Caspian Region. Presented on March 15, at NATO Advanced Research Workshop on the Caspian Sea, Venice, unpublished.

ADB, 1999a. *Key Indicators of Developing the Asian and Pacific Countries* (New York, Oxford University Press).

ADB, 1997. *Emerging Asia: Changes and Challenges,* ADB, Manila.

ADB, 1999. *Outlook 1999* (Oxford, Oxford University Press).

Bergesen, H.O., et. al, eds., 1999. *Yearbook of International Cooperation on Environment and Development 1998/99* (London, Earthscan).

Daly, H., 1997. *Beyond Growth,* UNDP, New York.

Dempsey, P.S., 1984. Compliance and enforcement in environmental law on oil pollution of the marine environment by ocean vessels, *NW Journal of International Law and Business,* vol. 6, pp. 459-460.

DESA, 1999. *World Economic Situation and Prospects for 1999,* United Nations, New York.

ESCAP, 1997. *Agenda 21 and Challenges for Asia and the Pacific,* United Nations, New York.

Etkins, P. and Jacobs, M., 1999. *The North The South, Ecological Constraints and the Global Economy* (Tokyo, United Nations University Press).

Greenpeace, 1999. http://www.greenpeace.org

Hardi, P., et. al., 1997. *Measuring Sustainable Development: Review of Current Practice,* Canada Industry Occasional Paper, No. 17, Canada.

Homer-Dixon, T., 1999. *Environmental Scarcity and Violence* (New Jersey, Princeton University Press).

Myers, N., 1993. *Ultimate Security: The Environmental Basis of Political Stability* (New York, W.W. Norton & Company).

PSD, 1996. *Sustainable America: A New Consensus* (New York, U.S. Government Printing Office).

Sanders, C., 1999. Twenty years on and five years, in Bergesen, H.O. et. al., eds., *Yearbook of International Cooperation on Environment and Development* (London, Earthscan).

Sitarz, D., 1994. *Agenda 21: The Earth Summit Strategy To Save Our Planet* (Boulder, Colorado, Earth Press).

UNEP, 1999. http://www.unep.ch/ozone

UNEP, 1992. *Caring for the Earth: A Learner's Guide to Sustainable Living,* New York.

UNEP, 1999. *Global Environment Outlook 2000* (London, Earthscan).

United Nations, 1997a., Summary of the Nineteenth United Nations General Assembly Special Session to Review Implementation of Agenda 1997, *Earth Negotiations Bulletin,* 21, June 23-27, vol. 5, No. 88.

United Nations, 1997b. Report of the Commission on Sustainable Development on Preparations for the Special Session of the General Assembly for the Purpose of an Overall Review and Appraisal of the Implementation of Agenda, (E/97/100).

United Nations, 1999a. http://www.unfccc.de

United Nations, 1999b. http://www.unccd.org.ch

United Nations, 1999c. doalos@un.org

Wackernagel, M. and Rees, W., 1996. Our *Ecological Footprint: Reducing Human Impact on Earth* (Gabriola Island, New Society Publishers).

World Bank, 1997. *Global Development Finance: Extracts,* A World Bank Book.

World Bank, 1999. *World Development Report,* Washington, DC.

ADB, 1997. *Emerging Asia Changes and Challenges*, Manila.

ADB, 1998. *Asia Least-Cost Greenhouse Gas Abatement Strategy*, Manila.

ADB, 2000. *Asian Development Outlook 2000* (New York, Oxford University Press).

Aden, J., Kyu-Hong, A. and Rock, M., 1999. What is driving the pollution abatement expenditure behaviour of manufacturing plants in Korea?, *World Development*, 27, 1203-1214.

Afsah, S. and Vincent, J., 1997. *Putting Pressure on Polluters: Indonesia's PROPER Program* (Cambridge, Mass: Harvard Institute for International Development).

Allenby, B., 1999. *Industrial Ecology* (New York, Prentice Hall).

Angel, D. and Rock M., 2000. *Asia's Clean Revolution* (Sheffield, UK, Greenleaf Publishing).

Brandon, C. and Ramankutty, R., 1993. *Toward an Environmental Strategy for Asia*, World Bank, Washington, DC.

Costanza, R. and Folke C., 1997. Valuing ecosystem services with efficiency, fairness and sustainability as goals, in Daily, G., ed., *Nature's Services: Societal Dependence on Natural Ecosystems* (Washington, DC, Island Press).

Dua, A. and Esty D., 1997. *Sustaining the Asia Pacific Miracle: Environmental Protection and Economic Integration* (Washington, DC, Institute for International Economics).

Graedel, T.E. and Allenby, B.R., 1995. *Industrial Ecology* (Upper Saddle River, NJ, Prentice Hall).

Greider, W., 1997. *One World, Ready or Not: the Manic Logic of Global Capitalism* (New York, Simon and Shuster).

Hettige, H., Mani, M. and Wheeler, D., 1997. *Industrial Pollution in Economic Development: Kuznets Revisited*, World Bank, Development Research Group, Washington, DC.

International Energy Agency (IEA), 1998. *World Energy Outlook*.

IEA 1999. *Energy Balance of OECD Countries 1996-97*.

Kainuma, M. et. al., 1998. *Asia Pacific into the 21st Century: Forward Looking Studies Based on the AIM Model*, National Institute of Environmental Studies, Tokyo.

Lee, K., 1993. *Compass and Gyroscope: Integrating Science and Politics for the Environment* (Washington, DC, Island Press).

National Research Council, 1997 *Environmentally Significant Consumption: Research Directions* (Washington, DC, National Academy Press).

National Research Council, Board on Sustainable Development, 1999. *Our Common Journey: A Transition Toward Sustainability* (Washington, DC, National Academy Press).

Noland, M., 1995. "Trade and exchange rate policies in growth-oriental adjustment programmes", in V. Corbo, M. Goldstein, and M. Khan, eds., 1995, *Growth-Oriented Adjustment Programmes*, International Monetary Fund and the World Bank, Washington, DC.

O'Connor, D., 1994. *Managing the Environment with Rapid Industrialization*, OECD, Paris.

Rock, M., Angel, D. and Feridhanusetyawan, T., 2000. Industrial ecology and clean development in Asia, *Journal of Industrial Ecology*, 3, pp. 29-42.

UNDP, 1999. *Human Development Report* (New York, Oxford University Press).

UNDP, 2000. *Overcoming Human Poverty, UNDP Poverty Report 2000* (New York, Oxford University Press).

UNEP, 1999 *Global Environment Outlook* (London, Earthscan).

United Nations, 1990. *State of the Environment in Asia and the Pacific*, Bangkok.

United Nations, 1995. *State of the Environment in Asia and the Pacific*, New York.

United States Department of Energy, 1999. *Energy Outlook 1999* (Washington, DC, Government Printing Office).

World Bank, 1994. *Indonesia: Environment and Development*, Washington, DC.

World Bank, 1997. *Can the Environment wait? Priorities for East Asia*, Washington, DC.

World Bank, 1998a. *World Development Indicators*, Washington, DC.

World Bank, 1998b. *East Asia: the Road to Recovery*, Washington, DC.

World Bank, 1999. *Greening Industry: New Roles for Communities, Markets and Governments*, Washington, DC.

World Bank, 2000. *World Development Indicators*, Washington, DC.

WRI, 1998. *World Resources 1998-99* (New York, Oxford University Press).

World Watch Institute, 1998 and 1999. *State of the World*, W.W. Nation and Company, New York.